Digital Twins in the Smart Classroom

About the Series

This series, ***Advances in Digital Twin Computing and Sensor Networks***, presents insights into digital twin and sensor network technologies as they become embedded in human activity, such as health care, city planning, manufacturing, distribution, and security. Digital twin computing allows researchers to replicate objects (humans or things) in digital form by capturing real-time data from sensor networks. The real and virtual/digital worlds are synchronized in real time, enabling the model to be used for planning, research, and a wider understanding of health care, the built environment, etc.

The series features contributions from a diverse group of authors, including experts from various geographical regions, academic backgrounds, and professional experiences. It covers topics such as smart cities, smart education, smart health care, smart agriculture, smart manufacturing, smart grids, and smart transport.

The series is edited by **Dr. Tuan Anh Nguyen**, a Senior Principal Research Scientist at the Institute for Tropical Technology, Vietnam Academy of Science and Technology, Vietnam. Dr. Nguyen received a PhD in Chemistry from the Paris Diderot University (France) in 2003. His research activities include smart sensors, smart networks, smart hospitals, smart cities, and advanced nanomaterials.

Dr. Nguyen was the volume editor for several Elsevier titles: *Sensors and Nanosensors for Smart Hospitals; Digital Twin and Blockchain in Smart Cities*; *Digital Twin and Blockchain in Smart Hospitals*; *Sensor Networks for Smart Hospitals*; Digital Twin, *Blockchain, and Sensor Networks in the Healthy and Mobile City*; *and Blockchain and Digital Twins for Smart Hospitals.*

Advances in Digital Twin Computing and Sensor Networks

Digital Twins in the Smart Classroom

Utilising Sensor Networks in the Educational Environment

Edited by

Tuan Anh Nguyen
Institute for Tropical Technology, Vietnam Academy of Science and Technology, Vietnam

ELSEVIER

Elsevier
Radarweg 29, PO Box 211, 1000 AE Amsterdam, Netherlands
125 London Wall, London EC2Y 5AS, United Kingdom
50 Hampshire Street, 5th Floor, Cambridge, MA 02139, United States

ISBN: 978-0-443-41632-3

For information on all Elsevier publications visit our website at
https://www.elsevier.com/books-and-journals

Publisher: Mara Conner
Acquisitions Editor: Craig Smith
Editorial Project Manager: Prashant Mahajan
Production Project Manager: Greetal Carolyn
Cover Designer: Arumugam Kothandan

Typeset by TNQ Tech

Dedication

To all teachers and students of the Newton Grammar School and Trung Vuong Secondary School (Hanoi, Vietnam)—thank you for inspiring this book through many years of companionship!

Contents

Contributors xix
About the editor xxiii
Foreword xxv

1. Introduction to smart classroom

Ahmet Göçen and Mehmet Fatih Döğer

1. **Introduction** 1
2. **Why smart schools?** 2
3. **Ecosystem of smart schools** 3
   - 3.1 Internet of Things (IoTs) 3
   - 3.2 Blockchain 3
   - 3.3 Smart sensors 3
   - 3.4 Smart wireless sensor networks 4
   - 3.5 Smart school applications 4
   - 3.6 Digital twin 4
4. **Introduction to digital twin technology** 4
   - 4.1 Fundamental principles of digital twin technology 5
   - 4.2 Areas of application for digital twin technology 6
   - 4.3 The usage of digital twin in education 6
5. **Major issues and challenges in smart technologies in schools** 9
   - 5.1 Inadequate infrastructure and hardware 9
   - 5.2 Tackling the complexities of data control and analysis 9
   - 5.3 Challenges of inefficiency and adaptation difficulties 9
   - 5.4 Security, curriculum, and industry collaboration issues 9
   - 5.5 Insufficient human capital resources 10
6. **Summary** 10

References 11

2. Smart classroom: Evolution, core components, benefits and challenges

Anucha Somabut and Kulthida Tuamsuk

1. **What is smart classroom?** 13
2. **Evolution of smart classrooms** 13
   - 2.1 Early technology integration in education 13
   - 2.2 Rapid growth in educational technology 14
   - 2.3 Shift to digital and smart learning environments 14
3. **Core components of a smart classroom** 15
   - 3.1 Hardware technologies 15
   - 3.2 Software and digital platforms 16
   - 3.3 IoT and sensor networks 16
4. **Pedagogical shifts in smart classrooms** 17
5. **Benefits of smart classrooms** 17
   - 5.1 Enhanced student engagement and participation 17
   - 5.2 Real-time data and performance analysis 17
   - 5.3 Personalized learning experiences 18
   - 5.4 Cost-efficiency and sustainability 18
6. **Challenges and barriers to implementation** 18
   - 6.1 Technical challenges 18
   - 6.2 Data privacy and security concerns 18
   - 6.3 Teacher readiness and training 18
7. **Future trends and innovations in smart classrooms** 18
   - 7.1 Integration of AI-driven chatbots and virtual tutors 19
   - 7.2 Metaverse and extended reality (XR) learning environments 19
   - 7.3 Smart campus integration and interconnectivity 19
   - 7.4 Classroom digital twins 19
8. **Conclusion** 19

References 19

3. Introduction to virtual smart classrooms, schools, and campuses

Eylül Akboğa, Gizem Merve Demir and Tufan Ekin

1. **Definition and scope** 23
2. **Implementation examples** 24
   - 2.1 Implementations using VR and AR 24
   - 2.2 Implementations using Digital Twins (DTs) 24
   - 2.3 Implementations using AI 25

2.4 Implementations using hardware and software virtualization 25
2.5 Implementations using The Internet of Things (IoT) 25
2.6 Implementation using face recognition technology 26
2.7 Other implementations 26
3. Benefits 26
3.1 Increased learning efficiency and safety 26
3.2 Cost-effectiveness 27
3.3 Effective school and campus management 27
3.4 Enhancing professional development of educators 28
3.5 Improved sustainability 28
4. Potential disadvantages and limitations 28
4.1 Privacy and security concerns 28
4.2 Financial cost 28
4.3 Health and social implications of prolonged device usage 29
4.4 Distraction and reduced academic performance 29
4.5 Digital divide and accessibility issues 29
4.6 Challenges in teacher training 29
4.7 Environmental and energy impact 29
4.8 Other issues 29
5. Technologies enabling virtual smart classrooms/schools/campuses 29
5.1 VR and AR 29
5.2 IoT, sensor, and blockchain 30
5.3 Digital Twins 30
5.4 Other technologies 30
6. Future innovations 30
6.1 VR, 6G and Symmetric Deep Q-Network (SDQN) combination 30
6.2 IoT-based real-time school safety and security system 30
6.3 Gamification-based personalized learning for non-academic skills 31
6.4 AI-based dynamic learning paths 31
6.5 AI-based early intervention for learning disabilities 31
6.6 AI-powered career and skill development programs 31
7. Impacts on society, economy and sustainability 32
References 32

4. Sensors for smart education

Ram Dular Singh Yadava

1. Introduction 37
1.1 Smart education learning environment 37
1.2 Role of sensors in AI-enabled IoT network 38
1.3 Present chapter 39
2. Sensors for smart classrooms 39
2.1 Air quality monitoring sensors 39
2.2 Interactive learning sensors 42
2.3 Occupancy and motion sensors 43
2.4 Biometric and behavioral sensors 45
3. Virtual and augmented reality sensors 46
3.1 Inertial Measurement Unit (IMU) sensors 46
3.2 LiDAR sensors 47
3.3 Haptic sensors 47
4. Digital twins in sensor-enabled learning 48
4.1 Digital twin for electronic nose—A learning tool for machine learning 49
4.2 Digital twin for swarm intelligence—A learning tool for cooperative phenomena 50
4.3 Digital twins in higher education and research—A glimpse of the future 50
5. Overview and conclusion 51
References 52

5. IoT sensors for smart classroom

Ruslan Saad Abdulrahman, Aymen Jaber Salman and Fawzi Mohammed Munir Al-Naima

1. Introduction 57
2. Basic configuration of IoT sensors 57
3. Characteristics of IoT sensors 58
4. IoT development history 59
5. Basics of sensors 59
5.1 The functioning of sensors 59
5.2 Key components of a sensor 60
5.3 Classification of sensors 60
5.4 Sensor characteristics and parameters 61
6. IoT sensors architecture 61
7. IoT networking technology 62
8. IoT data communication protocols 62
9. Essential IoT applications 65
10. Smart sensors and IoT technologies in academic environments 65
10.1 Varieties and roles of smart sensors in universities 65
10.2 IoT architectures applicable to educational settings 65
10.3 Data flow: Collection, transmission, and analytical processes in universities 66
10.4 Security and privacy considerations for IoT in academia 66
11. Integration of smart sensors and IoT in University classes 66
11.1 Key applications in academic settings 66

11.2 Challenges in deployment and mitigation strategies 66
11.3 Impact on pedagogy and learner outcomes 67
12. Discussion 67
12.1 Advantages of IoT-enabled smart sensors in university classrooms 67
12.2 Challenges to adoption: Financial, infrastructure, and ethical dimensions 67
12.3 Comparative review of implementations: Achievements and persistent issues 67
12.4 Strategies to foster successful integration 68
13. Challenges and limitations 68
14. Future trends in IoT sensors 70
15. Conclusion 70
References 71

6. Sensor network, blockchain, and digital twin as the backbones of a connected ecosystem of smart schools

Xavier Fernando and George Lăzăroiu

1. Introduction 73
2. Computer vision- and sensor network-based performance assessment and classroom management in Internet of Things and extended reality educational environments 74
3. Big educational data-driven personalized teaching and learning simulation interoperability and predictability in blockchain-based virtual shared-school environments 75
4. Digital twin classroom monitoring and analysis in artificial intelligence- and cloud computing-based immersive educational and training environments 76
5. Conclusions 78
References 78

7. Digital twins for smart education: Utilizing sensor networks for education

Mana Saleh Al Reshan, Darakhshan Syed and Asadullah Shaikh

1. Introduction 81
1.1 Background 81
1.2 Importance of digital twins in education 81
1.3 Objectives 82
1.4 Problem statement 82
2. Digital twins in education 83
3. Utilization of sensor networks for education 84
3.1 Types of sensors in smart education 85
3.2 Architecture of sensor networks for digital twins 85
4. Advantages of digital twins in education 87
4.1 Increase the quality of learning 87
4.2 Customized learning support 87
4.3 Enhance practical expertise 87
4.4 Encourage innovation in classroom and educational models 87
4.5 Promoting future skills 88
5. Challenges and future directions 88
5.1 Data security and privacy 88
5.2 Design and cost of infrastructure 88
5.3 Complex system integration 88
5.4 Hardware connectivity and maintenance 88
6. Analysis of the existing research 90
7. Conclusion 90
References 91

8. Smart sensor networks in the student activities

Usama Kalim, Sadia Irfan, Asha Kanwar and Ronghuai Huang

1. Introduction 93
1.1 Purpose of this chapter 94
2. An introduction to Smart Sensor Networks 94
2.1 Concept definition 94
2.2 Technologies involved 94
2.3 Valuation in learning 95
3. SSN applications in student activities 95
3.1 Classroom observation 95
3.2 Welfare and well-being 95
3.3 University safety and navigation 95
3.4 Innovative attendance systems 96
3.5 Leisure and social events 96
4. Real-world applications and case studies 96
4.1 Smart South Korea classrooms 96
4.2 IoT-enabled American campus 96
4.3 European institutions' wearable tech 96
5. SSNs' advantages for student activities 97
5.1 Enhanced learning results in adaptive learning contexts 97
5.2 Student engagement improves through behavioral cue assessment and response analysis 97
5.3 Enhanced safety via emergency systems and surveillance 97

5.4 Effective resource management with predictive maintenance and tracking of use 98
5.5 Data-driven decision making 98
6. Ethical considerations and smart sensor network challenges in higher education 98
6.1 Consent and data privacy 98
6.2 Infrastructure and cost control 99
6.3 Technical restrictions 99
6.4 Opponents of change 99
7. Future directions 100
7.1 Integration for active interventions with predictive analytics and artificial intelligence 100
7.2 Ecosystems of personalized learning inspired by sensor-driven feedback 100
7.3 Green campuses with energy-efficient sensor deployments 100
7.4 Blockchain for transparency and data security enhancement of trust 101
8. Conclusion 101
References 102

9. Blockchain for smart education

Antonella Petrillo, Fabio De Felice and Mizna Rehman

1. Definition of blockchain and smart education 105
2. Importance of blockchain for digital education 105
2.1 Connection between blockchain, digital twins, and smart classroom 109
2.2 The role of blockchain in education 109
3. Current applications of blockchain in education 110
3.1 The European Smart Campus Initiative 110
3.2 National Natural Science Foundation of China project EduRSS 110
3.3 Advanced technology-based smart and immersive classroom 112
4. Challenges and limitations of blockchain for education 114
4.1 Scalability and throughput constraints in blockchain-based educational systems 114
4.2 Privacy and data security challenges in educational blockchains 114
4.3 Interoperability and standardization challenges 115
4.4 Energy consumption and sustainability issues 115
4.5 Legal, regulatory, and governance challenges 115
5. The future of blockchain in smart education 115
6. Conclusion 116
References 117

10. Student check-in scheme based on blockchain for smart classroom

QingYang Ding, JianBin Chen and Xiao Chen Hao

1. Introduction 119
2. Related work 120
2.1 Smart classroom 120
2.2 Check-in 120
3. Blockchain-based students check-in scheme for smart class-room 121
3.1 Proposed scheme overview 121
3.2 Data block structure 122
3.3 Consensus mechanism 123
3.4 Data interaction mechanism 125
4. Conclusion 127
References 127

11. Electronic student records management through blockchain

Sohaib Latif

1. Introduction 129
2. Understanding blockchain technology 130
2.1 Foundations of blockchain 130
2.2 Key characteristics 130
2.3 Blockchain versus traditional databases 131
2.4 Types of blockchains 131
2.5 Smart contracts and automation 131
2.6 Blockchain interoperability and standards 131
2.7 Educational relevance 132
3. Benefits of blockchain in student records management 132
3.1 Enhanced data security and protection against fraud 132
3.2 Improved data integrity and auditability 132
3.3 Student empowerment and ownership of credentials 132
3.4 Streamlined administrative processes and reduced operational costs 132
3.5 Facilitation of cross-border academic mobility 133
3.6 Real-time verification and transparency 133
4. Case studies 133

4.1 University of Melbourne—Verifiable digital credentials 133
4.2 MIT Media Lab—Digital diplomas on the blockchain (MIT Media Lab, 2017) 134
4.3 Holberton School—Blockchain for lifelong learning records (Capetillo et al., 2022) 134
4.4 National government initiatives—Malta and Singapore (OpenCerts, 2019) 134
4.5 European Blockchain Services Infrastructure (EBSI) (European Commission, 2018) 134
5. Challenges and considerations 135
5.1 Data privacy and confidentiality 135
5.2 Legal and regulatory compliance 135
5.3 Technical complexity and infrastructure readiness 135
5.4 User adoption and change management 136
5.5 Financial costs and sustainability 136
5.6 Standardization and interoperability 136
5.7 Ethical and philosophical implications 136
6. Future directions 137
6.1 Toward interoperable and standardized credential ecosystems 137
6.2 Enhanced privacy through advanced cryptography 137
6.3 Integration with artificial intelligence and learning analytics 137
6.4 Cross-institutional and cross-sector collaboration 138
6.5 Decentralized learning marketplaces and smart contracts 138
6.6 Designing for equity, inclusion, and accessibility 138
6.7 Sustainable and scalable blockchain architectures 138
7. Conclusion 138
References 139

12. Digital twin application in school facilities management

Sanjeev Salunke, Swagatika Nanda, Shabnam Kumari and Amit Kumar Tyagi

1. Introduction 141
2. Literature review 142
3. Digital twin technology framework and concept 144
3.1 Physical layer 145
3.2 Data acquisition layer 145
3.3 Data integration layer 145
3.4 Digital model layer 145
3.5 Application and analytics layer 145
3.6 Presentation layer 145
4. Application of digital twin in school facilities management 146
4.1 Energy efficiency and optimization 146
4.2 Predictive and preventive maintenance 146
4.3 Space and occupancy management 146
4.4 Emergency management and safety 146
4.5 Indoor environmental quality 147
4.6 Asset and resource tracking 147
4.7 Sustainability monitoring and compliance 147
4.8 Integration with educational outcomes 147
5. Proposed digital twin framework 147
6. Result evaluation 149
6.1 Case study 1: Smart secondary school (urban setting) 150
6.2 Case study 2: Rural government school (pilot deployment) 151
6.3 Simulating emergency evacuation planning 153
7. Conclusion and future work 154
References 154

13. Cloud, fog and edge computing for school digital twin

Anita Mohanty, Ambarish G. Mohapatra, Abhro Mukherjee and Bright Keswani

1. Introduction 157
1.1 Definition of digital twin in education 157
1.2 Importance of real-time digital representation of school environments 157
1.3 Need for distributed computing: Cloud, fog, and edge 158
1.4 Chapter overview 158
2. Digital twin concept for schools 158
2.1 What constitutes a school digital twin 158
2.2 Physical layer: IoT devices and sensors 159
2.3 Cyber layer: Data processing and visualization 159
2.4 Interaction layer: Control, prediction, and optimization 159
2.5 Benefits: Predictive maintenance, smart learning, safety, resource efficiency 159
3. Role of cloud, fog, and edge computing 159
3.1 Cloud computing: Features, advantages, limitations for schools 160

3.2 Fog computing: Near-user processing, latency reduction, orchestration 161
3.3 Edge computing: Real-time analytics at the device level 162
3.4 Comparative analysis table: Cloud versus fog versus edge for School Digital Twin 163
4. Proposed architecture 163
4.1 Layered architecture: Device Layer, edge layer, fog layer, cloud layer 163
4.2 Data flow and processing mechanism 164
4.3 Security and privacy considerations 164
4.4 Advantages of the layered architecture 164
5. Key applications and use cases 164
5.1 Smart classroom management (lighting, HVAC, attendance) 165
5.2 Emergency monitoring and evacuation systems 165
5.3 Real-time resource optimization 165
6. Challenges and research opportunities 165
7. Conclusion 166
References 167

14. Blockchain and zero-knowledge proof for security and privacy preservation in student sensor networks

Vincent Omollo Nyangaresi, Ahmad A. AlRababah, Ali Hamzah Najim, Zaid A. Abduljabbar and Mohammad Shariq

1. Introduction 169
2. Related works 171
3. System model 172
3.1 System initialization phase 172
3.2 Registration phase 172
3.3 Login and authentication 174
4. Security analysis 176
4.1 Theorem 1: Our approach ensures robust authentication 176
4.2 Theorem 2: Single point of failure threats are thwarted 176
4.3 Theorem 3: User identity leakage and forgery attacks are prevented 176
4.4 Theorem 4: Our protocol mitigates eavesdropping and session hijacking threats 176
4.5 Theorem 5: Data privacy is preserved 176
4.6 Theorem 6: Message replay attacks are kept at bay 176
4.7 Theorem 7: The proposed protocol mitigates privileged insider 176
4.8 Theorem 8: Denial of service attacks are kept at bay 176
4.9 Theorem 9: Side-channeling attacks are ineffective against our protocol 177
4.10 Theorem 10: Man-in-the-middle and impersonation threats are prevented 177
4.11 Theorem 11: Scalability is achieved 177
5. Performance evaluation 177
6. Conclusion 178
References 178
Further reading 182

15. Data privacy considerations in implementing digital twins in educational settings

Assunta Di Vaio and Luisa Mastellone

1. Data privacy of Digital Twins in educational settings 183
2. Digital Twins in educational settings 184
3. GDPR compliance and its application in educational settings 187
4. Consent management in educational settings 187
5. Ethical implications for responsible management of Digital Twins in educational settings 189
6. Sustainability accounting and accountability issues from Digital Twins in educational settings 190
Acknowledgments 192
References 192

16. Digital twin technology for enhancing school operations efficiency

Wei Ni and Sai Zou

1. Introduction 195
2. Related work 196
2.1 Theoretical foundations 196
2.2 Industry applications 196
2.3 Research gaps and contributions 196
3. System architecture 196
3.1 System framework 196
3.2 Operational objectives 198
4. Data center architecture 198
4.1 System architecture 198
4.2 Data processing pipeline 199
4.3 Intelligent resource repository 199
5. Decision support platform 199
5.1 Architecture overview 199
5.2 Individual-level analytics 200

- 5.3 Regional decision optimization 201
- 5.4 Strategic governance modules 201
- 5.5 Technical implementation 201

6. Management support platform 201
- 6.1 Framework design 201
- 6.2 Evaluation metrics system 201
- 6.3 Precision profiling engine 202
- 6.4 Implementation outcomes 202

7. Instructional support platform 202
- 7.1 System architecture 202
- 7.2 Design principles 203
- 7.3 Implementation roadmap 203
- 7.4 Core functional modules 203
- 7.5 Empirical validation 203

8. Ubiquitous application platform 203
- 8.1 System architecture 203
- 8.2 Implementation strategy 204
- 8.3 Core functional modules 204
- 8.4 Empirical validation 204

9. Conclusion 204
- 9.1 Digital transformation of student learning 204
- 9.2 Intelligent decision-making in education 205
- 9.3 Personalized educational services 205

References 205

17. Digital twin-assisted adaptive video streaming optimization in the smart classroom

Xinyu Huang

1. Introduction 207

2. DT-assisted smart-classroom framework 208

3. System model 209
- 3.1 Service delay model 209
- 3.2 PQoE model 210
- 3.3 PQoE model formulation 211
- 3.4 DT-assisted PQoE model 211

4. DRL-based resource management algorithm 212
- 4.1 Problem formulation 212
- 4.2 Proposed algorithm 212
- 4.3 Markov Decision Process 212
- 4.4 DRL-based communication and computing resource allocation (DCTRA) algorithm 213

5. Performance evaluation 214
- 5.1 Simulation setup 214
- 5.2 PQoE and convergence performance evaluation 214

6. Conclusion 216

References 216

18. Digital twins of clothing: A transformative approach to educating fashion designers

Victor Kuzmichev, Aleksei Moskvin and Mariia Moskvina

1. Introduction 219

2. Concepts and tools for educational applications of digital twins 219
- 2.1 Digital model versus digital twin: Advantages for fashion design 219
- 2.2 Software programs 220
- 2.3 Opportunities and benefits for teachers and students 221
- 2.4 How digital twins help the teacher 222
- 2.5 How digital twins help the student 224

3. Body—clothing system: A multidisciplinary object 225
- 3.1 Overview of the body + clothing interaction 225
- 3.2 Multidisciplinary foundations 225
- 3.3 Digital twins in historical and contemporary contexts 226
- 3.4 Educational and scientific implications 226

4. Twins of the human body: Principles of generation and practical application 226
- 4.1 Historical context and evolving methodologies 226
- 4.2 Parametric modeling and digital avatar generation 227
- 4.3 Practical applications and case studies 227
- 4.4 Challenges, limitations, and future directions 228

5. Twins of textile materials: Principles of creation and testing 229
- 5.1 Capturing the complexity of fabrics 229
- 5.2 Methodologies and simulation techniques 229
- 5.3 Applications in design, conservation, and education 229
- 5.4 Future directions in textile simulation 230

6. Twins of historical and modern clothing: Cross-cultural, aesthetic, and functional aspects 230
- 6.1 Bridging historical reconstruction and contemporary design 230
- 6.2 Cross-cultural and aesthetic considerations 230
- 6.3 Functional aspects and wearability analysis 231
- 6.4 Educational and research implications 231

7. Methodological recommendations for the use of software for working with digital twins 232
8. Conclusions 233
Funding 233
References 233

19. Digital twins as a tool for optimizing social interaction: A case study in Indonesian higher education with STELE approach

Ahmad Al Yakin, Muthmainnah Muthmainnah, Ali Said Al Matari, Luis Cardoso and Ahmed J. Obaid

1. Introduction 239
2. Literature review 240
   2.1 A Smart Technology-Enhanced Learning Environment (STELE) approach 240
   2.2 The use of digital twins in education 7.0 241
3. Methodology 244
   3.1 Participants and setting 245
   3.2 Procedure 245
   3.3 Data analysis 246
4. Findings and discussion 246
5. Discussion 250
6. Conclusion 256
References 256

20. Ethics in the Metaverse classroom: A case study for EFL education in Indonesia

Muthmainnah Muthmainnah, Sajed S. Ingilan, Luis Cardoso, Hamed Barjesteh, Ali Said Al Matari, Ahmad Al Yakin and Ahmed J. Obaid

1. Introduction 261
2. Related work 263
   2.1 Metaverse ecosystem 263
   2.2 Metaverse as tools and technology in EFL education 265
   2.3 Digital twin-based technology in English education ethics 267
3. Methodology 268
   3.1 Participants 268
   3.2 Instruments 269
   3.3 Data analysis 269
4. Findings and discussion 270
5. Discussion 271
6. Conclusion 274
References 274

21. Essential requirements of smart learning environments: A case study in the Arab region

Ruba Fahmi Bataineh, Rommel Mahmoud AlAli, Rula Fahmi Bataineh and Ali Ahmad Al-Barakat

1. Introduction 279
2. Methodological approach 282
   2.1 Method design 282
   2.2 Participants 282
   2.3 Data collection and analysis 283
3. Findings 283
   3.1 Theme 1: Effective interaction and engagement 283
   3.2 Theme 2: Support for modern technologies 284
   3.3 Theme 3: Continuous assessment 284
   3.4 Theme 4: Technical and academic support 284
   3.5 Theme 5: Flexibility of access to content 285
   3.6 Theme 6: Adaptation to and alignment with learners' needs 285
4. Discussion 286
5. Conclusions, implications, and recommendations 287
Funding 287
Acknowledgments 287
References 287

22. IoT sensors and devices in smart schools: A case study in the Universitas Negeri Makassar (Indonesia)

Abdul Wahid

1. Introduction 291
   1.1 Background 291
   1.2 Role of Digital Twins 291
   1.3 Objectives and scope 292
2. Sensor technologies in education 292
   2.1 Sensor typologies 292
   2.2 Integration methodologies 293
3. Case studies in IoT-enabled smart schools 294

3.1 Security Village (smart fingerprint access) 294
3.2 Facial recognition with LBPH algorithm 294
3.3 Smart Blended Learning with AI object detection 295
3.4 Smart Campus Framework implementation 296
4. Systems architecture and AI integration 297
4.1 Infrastructure design 297
4.2 AI and big data integration 298
5. Ethical and governance frameworks 298
6. Conclusion and future directions 299
7. Recommendations 299
References 299

23. Recent developments in smart classrooms/smart schools in India

Shalini Kumari, Chander Prabha and Nabanita Roy

1. Introduction 301
2. Components of the school digital twin 301
2.1 Benefits of DT in schools 301
3. History of digital twin models 302
4. Digital twin technology in schools 304
4.1 The framework of DTs in school 304
5. Application of digital twin for schools 305
5.1 Campus management 305
5.2 Learning and curriculum enhancement 305
5.3 Student and teacher performance monitoring 305
6. Current developments in smart classrooms/smart schools in India 306
6.1 Technology adoption trends 307
6.2 Regional analysis 308
6.3 Usage of smart classrooms 308
6.4 Challenges in implementation 309
7. Conclusion 309
References 309

24. Digital twin in smart education: Challenges, limitations, and case studies

Mohamed S. Abdalzaher

1. Introduction 311
2. Concept of digital twins in education 312
3. Sensor networks and data collection in smart classrooms 312
3.1 Types of sensors in smart classrooms 313
3.2 Data acquisition and processing 313
3.3 Privacy and ethical considerations 313
4. Implementing digital twins in smart classrooms 313
4.1 Architecture of digital twins in educational settings 313
4.2 Role of AI and machine learning in digital twins 314
4.3 Challenges in implementation 315
4.4 Strategies for effective implementation 315
5. Benefits of digital twins in smart learning environments 315
5.1 Enhanced personalization of learning 315
5.2 Real-time monitoring and feedback 315
5.3 Improved resource management 315
5.4 Data-driven decision-making 316
5.5 Facilitating remote and hybrid learning 316
5.6 Supporting professional development for educators 316
6. Challenges and limitations of digital twins in education 316
6.1 High implementation costs 316
6.2 Technical complexity and interoperability 316
6.3 Data privacy and security concerns 317
6.4 Ethical and surveillance concerns 317
6.5 Digital divide and accessibility issues 317
6.6 Scalability challenges 317
7. Case studies of digital twins in education 318
7.1 Case study 1: Enhancing personalized learning at Nanyang Technological University, Singapore 318
7.2 Case study 2: Optimizing campus management at the University of Southern Denmark 319
7.3 Case study 3: Predictive maintenance and safety at Purdue University, USA 319
7.4 Key lessons 319
7.5 Challenges faced by early adopters 319
8. Future trends of digital twins in education 319
8.1 Integration with augmented and virtual reality 320
8.2 Real-time adaptation and autonomous classrooms 320
8.3 5G and edge computing for faster, real-time data processing 320
8.4 Blockchain for secure and transparent data management 320
8.5 Enhanced collaboration through cloud-based digital twins 320
8.6 AI-driven predictive analytics for institutional decision-making 321
9. Conclusion and future perspectives 321
References 321

25. Assistant agent for smart classroom application: Past, present and future direction

Anh Tuan Nguyen, Thi Ngoc Yen Huynh and Yonghan Ahn

1. **Introduction** 325
2. **Methodology** 326
3. **Application review of assistant agents** 326
   - 3.1 Pedagogical applications 326
   - 3.2 Administrative automation 334
   - 3.3 Collaborative learning enhancement 335
   - 3.4 Accessibility & inclusion 336
4. **Challenges and future directions** 338
   - 4.1 Privacy, bias, and fairness considerations 338
   - 4.2 Integration of assistant agents into existing educational ecosystems 338
   - 4.3 Hallucination and overreliance 339
5. **Conclusion** 339

References 340

26. Digital twin technology for student healthcare in smart schools

Zahra Rizvi and Aida Ghazizadeh

1. **Introduction** 347
2. **Digital twin technology: A technological overview** 348
   - 2.1 Definition and fundamentals 348
   - 2.2 Key technological components 348
3. **Benefits for schools** 349
4. **Smart schools and healthcare: Bridging technology and wellness** 351
   - 4.1 Concept of smart schools 351
   - 4.2 Integration of sensor technology 351
   - 4.3 Case studies of successful implementations 351
   - 4.4 Benefits of sensor integration in smart school ecosystems 352
5. **Healthcare applications of digital twin in schools** 353
6. **Technological infrastructure for digital twins in smart schools** 353
7. **Medical and healthcare perspectives** 355
8. **Challenges and solutions** 355
   - 8.1 Solutions 356
9. **Conclusion** 356
10. **Acronyms** 357

References 358

27. Transformative impact of smart classroom/school/campus on the educational environment

Semiu Olawale Makinde and Wing Cheung Tang

1. **Introduction** 361
   - 1.1 Definition and significance of smart learning environment 361
   - 1.2 Transition from traditional classrooms to digital spaces 362
   - 1.3 Key components, advantages, challenges, and future trends of smart classrooms 362
2. **Understanding smart classrooms** 362
   - 2.1 Definition and main features 362
   - 2.2 Challenges and considerations 363
3. **Expanding to smart schools** 363
   - 3.1 Concepts and infrastructure 363
   - 3.2 Key technologies of smart schools 364
   - 3.3 Advantages of smart schools 364
   - 3.4 Implementation barriers 365
4. **Scaling up: Smart campuses in higher education** 366
   - 4.1 Defining smart campuses 366
   - 4.2 Core technologies in smart campuses 367
   - 4.3 Benefits for universities and students 368
   - 4.4 Challenges in deployment 369
5. **Future trends and innovations in smart learning environments** 370
6. **Conclusion** 370

References 370

28. Role of artificial intelligence in enhancing school digital twin

Prabhakar Krishnan, Amit Kumar Tyagi and Shabnam Kumari

1. **Introduction** 373
2. **Related work** 374
3. **Architecture and design of AI-enhanced school DTs** 378
   - 3.1 Conceptual model 378
   - 3.2 Roles of sensors, data aggregators, and AI inference engines 378
   - 3.3 System components 379
   - 3.4 Data flow and feedback loops 379
   - 3.5 Security and ethical considerations 380

4. **Functional applications of AI in school DTs** 381
   4.1 Smart learning environments 381
   4.2 Student performance monitoring 381
   4.3 Infrastructure and facilities management 381
   4.4 Administrative and decision support 382
   4.5 Well being and mental health 382
5. **Case studies and practical implementations** 382
   5.1 Case study 1: AI-enhanced DT in a K–12 smart school 382
   5.2 Case study 2: Virtual campus twins for administrative simulations 383
   5.3 Pilot projects and prototypes 384
6. **Challenges and limitations** 385
   6.1 Technical challenges 385
   6.2 Pedagogical issues 386
   6.3 Ethical and privacy concerns 386
7. **Future directions and research opportunities** 386
   7.1 Emerging technologies 387
   7.2 Intelligent school ecosystems 387
   7.3 Policy and governance models 388
   7.4 Cross disciplinary research 388
8. **Conclusion** 388

References 389

29. Multiple digital student attendance check through blockchain

S. Ramamoorthy and R. Poorvadevi

1. **Introduction** 391
2. **Literature survey** 392
3. **Proposed method** 394
4. **DLedger-based student check data management** 395
5. **Digital student check-in process steps** 396
   5.1 Pre-check-in: Registration and onboarding 396
   5.2 Class schedule and QR code generation 396
   5.3 Check-in by the student (live attendance) 397
   5.4 Data validation and blockchain logging 398
   5.5 Post-check-in reporting and analysis 398
6. **Implementation work** 398
   6.1 Administration activities and roles 398
   6.2 Faculty roles and purposes 398
   6.3 Student involvement and role 399
   6.4 Integration of blockchain 399
   6.5 A summary of workflow 399
   6.6 Advantages of this method 399
7. **Results** 399
8. **Conclusion** 400

References 400

Index 401

Contributors

Mohamed S. Abdalzaher, Department of Seismology, National Research Institute of Astronomy and Geophysics, Helwan, Cairo, Egypt; Department of Electrical Engineering, American University of Sharjah, Sharjah, United Arab Emirates

Zaid A. Abduljabbar, Department of Computer Science, College of Education for Pure Sciences, University of Basrah, Basrah, Iraq; Department of Business Management, Al-Imam University College, Balad, Iraq

Ruslan Saad Abdulrahman, Department of Computer Engineering, College of Engineering, Al-Nahrain University, Baghdad, Iraq

Yonghan Ahn, Department of Architecture Engineering, Hanyang University ERICA Campus, Ansan, Gyeonggi-do, South Korea

Eylül Akboğa, Department of Information Systems and Technologies, Faculty of Computer and Information Sciences, Yeditepe University, Istanbul, Turkey

Ali Ahmad Al-Barakat, Department of Curriculum and Methods of Instruction, Yarmouk University, Irbid, Jordan; Department of Education, University of Sharjah, Sharjah, United Arab Emirates

Fawzi Mohammed Munir Al-Naima, Department of Computer Engineering, College of Engineering, Al-Nahrain University, Baghdad, Iraq; Department of Electrical Engineering Techniques, University of Kut, Wasit, Iraq

Rommel Mahmoud AlAli, The National Research Center for Giftedness and Creativity, King Faisal University, Al-Ahsa, Saudi Arabia

Ali Said Al Matari, A'Sharqiyah University, Ibra, Oman

Ahmad A. AlRababah, Faculty of Computing and Information Technology in Rabigh, Rabigh, KSA, King Abdulaziz University, Bando, Kenya

Mana Saleh Al Reshan, Department of Information System, College of Computer Science and Information Systems, Najran University, Najran, Saudi Arabia; Emerging Trends and Technologies Lab (ETRL), College of Computer Science and Information Systems, Najran University, Najran, Saudi Arabia

Ahmad Al Yakin, Universitas Al Asyariah Mandar, West Sulawesi, Indonesia

Hamed Barjesteh, Islamic Azad University, Amol, Iran

Ruba Fahmi Bataineh, Department of Curriculum and Methods of Instruction, Yarmouk University, Irbid, Jordan

Rula Fahmi Bataineh, Department of English for Applied Studies, Jordan University of Science and Technology, Irbid, Jordan

Luis Cardoso, Polytechnic Institute of Portalegre and Centre for Comparative Studies of the University of Lisbon, Lisbon, Portugal

JianBin Chen, School of Management, Beijing Union University, Beijing, China

Fabio De Felice, Department of Engineering, University of Naples "Parthenope", Napoli, Italy; Isola C4, Centro Direzionale Napoli, Napoli, Italy

Gizem Merve Demir, Department of Information Systems and Technologies, Faculty of Computer and Information Sciences, Yeditepe University, Istanbul, Turkey

QingYang Ding, School of Management, Beijing Union University, Beijing, China

Assunta Di Vaio, University of Naples Parthenope, Naples, Italy

Tufan Ekin, Department of Management Information Systems, Faculty of Computer and Information Sciences, Yeditepe University, Istanbul, Turkey

Mehmet Fatih Döğer, Ministry of National Education, Directorate General for Innovation and Educational Technologies, Ankara, Türkiye

Xavier Fernando, Toronto Metropolitan University, Toronto, ON, Canada

Aida Ghazizadeh, Old Dominion University, Norfolk, VA, United States

Ahmet Göçen, Afyon Kocatepe University, Faculty of Education, Educational Sciences, Afyonkarahisar, Türkiye

Xiao Chen Hao, College of Applied Arts and Science, Beijing Union University, Beijing, China

Ronghuai Huang, Smart Learning Institute of Beijing Normal University, Beijing, P.R. China

Xinyu Huang, University of Waterloo, Electrical and Computer Engineering, Waterloo, ON, Canada

Thi Ngoc Yen Huynh, Department of Smart City Engineering, Hanyang University ERICA Campus, Ansan, Gyeonggi-do, South Korea

Sajed S. Ingilan, University of Southeastern Philippines, Davao, Philippines

Sadia Irfan, Faculty of Education, Southwest University, Chongqing, China

Usama Kalim, Smart Learning Institute of Beijing Normal University, Beijing, P.R. China

Asha Kanwar, Smart Learning Institute of Beijing Normal University, Beijing, P.R. China

Bright Keswani, Department of Computer Science & Engineering, Poornima University, Jaipur, Rajasthan, India

Prabhakar Krishnan, Department of Management, St. Joseph's University, Bengaluru, Karnataka, India

Shabnam Kumari, SRM Institute of Science and Technology, Chennai, Tamil Nadu, India

Shalini Kumari, Chitkara University Institute of Engineering and Technology, Chitkara University, Punjab, India

Victor Kuzmichev, Department of Clothes Design, Ivanovo State Polytechnic University, Ivanovo, Russia

Sohaib Latif, Department of Computer Science and Software Engineering, Grand Asian University, Sialkot, Pakistan

George Lăzăroiu, Curtin University, Perth, WA, Australia; Cardiff Metropolitan University, Cardiff, United Kingdom; Spiru Haret University, Bucharest, Romania; Toronto Metropolitan University, Toronto, ON, Canada

Semiu Olawale Makinde, Institute of Education, Al-Hikmah University, Ilorin, Nigeria

Luisa Mastellone, Marinter Shipping Agency Srl, Naples, Italy

Anita Mohanty, Electronics Engineering, Silicon University, Bhubaneswar, Odisha, India

Ambarish G. Mohapatra, Electronics Engineering, Silicon University, Bhubaneswar, Odisha, India

Aleksei Moskvin, Department of Clothes Design and Technology, Saint Petersburg State University of Industrial Technologies and Design, Saint Petersburg, Russia

Mariia Moskvina, Department of Clothes Design and Technology, Saint Petersburg State University of Industrial Technologies and Design, Saint Petersburg, Russia

Abhro Mukherjee, Techno India University, Kolkata, West Bengal, India

Muthmainnah Muthmainnah, Universitas Al Asyariah Mandar, West Sulawesi, Indonesia

Ali Hamzah Najim, Department of Computer Technical Engineering, Imam Al-Kadhum College (IKC), Baghdad, Iraq

Swagatika Nanda, Narayana Group of Institutions, Bhubaneswar, Odisha, India

Anh Tuan Nguyen, Department of Smart City Engineering, Hanyang University ERICA Campus, Ansan, Gyeonggi-do, South Korea

Wei Ni, School of Engineering, Edith Cowan University, Perth, WA, Australia

Vincent Omollo Nyangaresi, Department of Computer Science and Software Engineering, Jaramogi Oginga Odinga University of Science and Technology, Bondo, Kenya; Department of Applied Electronics, Saveetha School of Engineering, SIMATS, Chennai, Tamil Nadu, India

Ahmed J. Obaid, Faculty of Computer Science and Mathematics, University of Kufa, Najaf, Iraq

Antonella Petrillo, Department of Engineering, University of Naples "Parthenope", Napoli, Italy; Isola C4, Centro Direzionale Napoli, Napoli, Italy

R. Poorvadevi, Department of Computational Intelligence, SRM Institute of Science and Technology, Kattankulathur, Tamil Nadu, India

Chander Prabha, Chitkara University Institute of Engineering and Technology, Chitkara University, Punjab, India

S. Ramamoorthy, Computing Technologies, SRM Institute of Science and Technology, Kattankulathur, Tamil Nadu, India

Mizna Rehman, Department of Engineering, University of Naples "Parthenope", Napoli, Italy; Isola C4, Centro Direzionale Napoli, Napoli, Italy

Zahra Rizvi, College of William & Mary, Williamsburg, VA, United States

Nabanita Roy, Chitkara University Institute of Engineering and Technology, Chitkara University, Punjab, India

Aymen Jaber Salman, Department of Computer Engineering, College of Engineering, Al-Nahrain University, Baghdad, Iraq

Sanjeev Salunke, Soundarya Group of Institutions, Bengaluru, Karnataka, India

Asadullah Shaikh, Department of Information System, College of Computer Science and Information Systems, Najran University, Najran, Saudi Arabia; Emerging Trends and Technologies Lab (ETRL), College of Computer Science and Information Systems, Najran University, Najran, Saudi Arabia

Mohammad Shariq, Department of Information and Systems Security, College of Information Technology, United Arab Emirates University, Abu Dhabi, United Arab Emirates

Anucha Somabut, Department of Educational Technology, Faculty of Education, and the Smart Learning Innovation Research Center, Khon Kaen University, Khon Kaen, Thailand

Darakhshan Syed, Computer Science Department, Bahria University, Karachi, Pakistan

Wing Cheung Tang, Jesselton University College, Kuala Lumpur, Malaysia; Institute of Vocational Education, Hong Kong Metropolitan University, China

Kulthida Tuamsuk, Department of Information Science, Faculty of Humanities and Social Sciences, and the Smart Learning Innovation Research Center, Khon Kaen University, Khon Kaen, Thailand

Amit Kumar Tyagi, National Forensic Sciences University, Gandhinagar, Gujarat, India

Abdul Wahid, Universitas Negeri Makassar, Makassar, Indonesia

Ram Dular Singh Yadava, Department of Physics, Banaras Hindu University, Varanasi, Uttar Pradesh, India; Solid State Physics Laboratory, Delhi, India; Independent Researcher (Condensed Matter Physics and Sensor Systems), Varanasi, Uttar Pradesh, India

Sai Zou, College of Big Data and Information Engineering, Guizhou University, Guizhou, China

About the editor

Tuan Anh Nguyen is a Senior Principal Research Scientist at the Institute for Tropical Technology, Vietnam Academy of Science and Technology, Hanoi, Vietnam. He received a BS in Physics from Hanoi University in 1992, a BS in Economics from Hanoi National Economics University in 1997, and a PhD in Chemistry from the Paris Diderot University, France, in 2003. He was a Visiting Scientist at Seoul National University, South Korea, in 2004, and the University of Wollongong, Australia, in 2005. He then worked as a Postdoctoral Research Associate and a Research Scientist at Montana State University, United States, from 2006 to 2009. In 2012, he was appointed as the Head of the Microanalysis Department at the Institute for Tropical Technology. His research areas of interest include smart sensors, smart networks, smart hospitals, smart cities, digital twins, complexiverse, and bio-wave. He has edited more than 75 books for Elsevier, 12 books for CRC Press, 1 book for Springer, 1 book for RSC, and 2 books for IGI Global. He is the Editor-in-Chief of *Kenkyu Journal of Nanotechnology & Nanoscience* and Editor of the Elsevier book series *Advances in Digital Twin Computing and Sensor Networks.*

Foreword

Multidisciplinary collaboration has long been acknowledged as a driving force in pushing the boundaries of scientific and technological innovation. This is especially true at the intersection of education, digital technology, and data science, where insights from computer science, cognitive psychology, pedagogy, and network engineering coalesce to shape the next generation of learning experiences. In this context, digital twins (DTs) in smart classrooms—supported by sensor networks and real-time analytics—represent a transformative approach to optimizing educational environments.

Digital twins have already demonstrated their value in fields such as manufacturing (Onaji et al., 2022), healthcare (Okegbile et al., 2023), and urban management (Jafari et al., 2023), offering predictive insights and data-driven decision-making. Their adaptation to educational settings marks a significant leap forward, enabling real-time feedback on classroom conditions, student engagement, and cognitive load. By integrating IoT-based sensors, AI algorithms, and interactive learning platforms, a digital mirror of the physical classroom can provide educators with actionable data to refine their teaching strategies.

Despite its considerable potential, the adoption of digital twins in education remains in its early stages. Recent systematic reviews highlight a growing body of research on digital twins in educational settings, demonstrating their evolving role in enhancing interactive and data-driven learning environments (Razzaq et al., 2023). In parallel, higher education institutions are increasingly integrating AI-driven learning environments, yielding notable benefits. Empirical studies indicate that AI-powered tutoring systems enhance student engagement and provide more personalized academic support. Research conducted at IU International University of Applied Sciences in Germany found that students using AI-based learning assistants were able to complete their coursework 27% faster, illustrating the efficiency gains provided by such tools (Möller et al., 2024). These findings underscore the need for a structured and well-regulated integration of digital twins and AI-driven technologies in education. Addressing technological feasibility, data privacy, and pedagogical adaptability will be essential to ensuring that these innovations enhance learning outcomes while maintaining ethical and educational integrity.

Just as significantly, ethical and privacy considerations remain central to the conversation. Sensor-rich environments risk encroaching on student autonomy and educational equity if data is not managed responsibly. Historical lessons from data-driven education initiatives remind us that technology alone cannot replace effective pedagogy, and blind reliance on automation can introduce biases into learning processes. The key, therefore, is to design digital twins that enhance, rather than override, human-driven education.

Encouragingly, the development of digital twins for smart classrooms seems to be following a responsible trajectory, much like early applications of AI in healthcare and industrial automation. The collaboration between technologists, educators, and policymakers is shaping a framework where functionality, ethics, and pedagogical value are equally prioritized. This approach may serve as a model for future educational technology integration, ensuring that digital twins contribute to, rather than disrupt, the learning experience.

The chapters in this book arrive at a pivotal moment, providing a thorough examination of digital twins in educational contexts. Topics include sensor network architecture, real-time data processing, learning analytics, and the institutional frameworks that support successful implementation. By addressing both technical design and ethical-pedagogical considerations, this volume offers valuable guidance for academics, educators, and policy professionals who seek to harness digital twins effectively.

Readers interested in smart classrooms, AI-enhanced education, and the transformative power of digital twins will find in these pages not only a deep dive into emerging research but also a reflective perspective on best practices and future directions. By combining technological ingenuity with responsible, learner-centric strategies, digital twins can help usher in an era of truly dynamic, adaptive education.

Jun Cai

PERFORM Research Chair, Network Intelligence and Innovation Lab (NI2Lab), Department of Electrical and Computer Engineering, Concordia University, Montreal, Quebec, Canada

References

Jafari, M., Kavousi-Fard, A., Chen, T., & Karimi, M. (2023). A review on digital twin technology in smart grid, transportation system and smart city: challenges and future. *IEEE Access, 11*, 17471–17484.

Möller, M., et al. (Feb. 21, 2024). *Revolutionising distance learning: A comparative study of learning progress with AI-driven tutoring. arXiv*: arXiv:2403.14642.

Okegbile, S. D., Cai, J., Yi, C., & Niyato, D. (March/April 2023). Human digital twin for personalized healthcare: vision, architecture and future directions. *IEEE Network Magazine, 37*(2), 262–269.

Onaji, I., Tiwari, D., Soulatiantork, P., Song, B., & Tiwari, A. (Aug. 2022). Digital twin in manufacturing: conceptual framework and case studies. *International Journal of Computer Integrated Manufacturing, 35*(8), 831–858.

Razzaq, S., Shah, B., Iqbal, F., Ilyas, M., Maqbool, F., & Rocha, A. (Apr. 2023). DeepClassRooms: a deep learning based digital twin framework for on-campus class rooms. *Neural Computing and Applications, 35*(11), 8017–8026.

Chapter 1

Introduction to smart classrooms

Ahmet Göçen[1] and Mehmet Fatih Döger[2]
[1]Afyon Kocatepe University, Faculty of Education, Educational Sciences, Afyonkarahisar, Türkiye; [2]Ministry of National Education, Directorate General for Innovation and Educational Technologies, Ankara, Türkiye

1. Introduction

Virtual reality is the first step in a grand adventure into the landscape of the imagination.

Frank Biocca.

Frank Biocca's observation illustrates the transformative potential of emerging technologies that we face across various domains, particularly within the field of education. Modern "smart schools" integrate a diverse array of cutting-edge innovations which include the Internet of Things (IoT), generative artificial intelligence (GAI), blockchain, and digital twin technology to merge digital and physical realms in ways that were previously unimagined (Egunjobi & Adeyeye, 2024; Zhang & Li, 2021). Moreover, these technological advancements which also include augmented reality (AR), virtual reality (VR), and extended reality (XR) are catalyzing immersive learning experiences that significantly enhance teaching, resource management, and student engagement.

Smart schools or smart classrooms are characterized by the integration of various technological systems and applications. Smart sensor networks, for instance, gather real-time data on factors such as classroom occupancy, temperature, and even student participation. This data is then utilized by artificial intelligence driven algorithms to optimize learning settings, personalize instructional strategies, and streamline administrative tasks (Zhang et al., 2024). Meanwhile, IoT devices ranging from wearable health monitors to interactive whiteboards enable seamless communication between the physical environment and digital platforms, further enriching the learning experience. In parallel, blockchain technology safeguards the integrity of academic records and supports secure credentialing processes, while digital twin technology virtualizes school infrastructure and systems, offering predictive insights into maintenance, resource allocation, and student progress.

Incorporating these diverse tools, smart classrooms transcend traditional educational models. They facilitate a holistic learning environment that supports multiple modes of instruction such as blended learning, flipped classrooms, and online or distance learning. Students benefit from interactive lessons that leverage 3D content, simulations, and virtual communities, while teachers gain robust support in monitoring academic progress and customizing lesson plans. At the institutional level, administrators and policymakers can rely on data-driven insights to address infrastructure challenges, update curricula, and ensure that digital resources are equitably distributed. Ultimately, the smart school ecosystem holds the promise of delivering education as a service, an idea that becomes increasingly relevant as more schools adopt cloud-based platforms and online resources to supplement or even replace physical classrooms (Zahedi, 2024).

The chapters that follow delve into the key components and benefits of these emerging technologies, discuss major implementation challenges (ranging from infrastructure gaps to privacy concerns), and outline best practices for seamlessly embedding digital tools in modern educational contexts. In the first section below, readers will gain an understanding of why *smart schools* are essential, focusing on the need for advanced technologies in education. In the second section, the discussion centers on *the ecosystem of smart schools*, including key technologies such as artificial intelligence, IoT, blockchain, smart sensors, networks, smart school applications, and digital twins, and their roles in transforming education. The third section provides insight into *digital twin technology*, its underlying principles, and its applications in education, including proactive maintenance, resource optimization, and student tracking. Subsequently, the fourth section addresses *the major issues and challenges associated with implementing smart technologies in schools*, such as infrastructure, data management, security, and resource limitations. Finally, the fifth

Digital Twins in the Smart Classroom. https://doi.org/10.1016/B978-0-443-41632-3.00023-0

section, under *summary*, underscores the importance of overcoming these challenges for the successful integration of smart technologies in educational environments.

2. Why smart schools?

As technology continues to evolve, the traditional concept of schools is being transformed into a connected, intelligent ecosystem, driven by the integration of smart sensor networks, IoT devices, blockchain, and digital twins. With the help of artificial intelligence, these systems become even smarter, allowing for better decision-making and streamlined processes. As we venture into the future, virtual smart classrooms and schools promise to redefine how education is delivered, making school operations more efficient, secure, and ultimately, more impactful.

The integration of smart technologies within smart classrooms has emerged as a central focus for both schools and policymakers, with the objective of optimizing the efficiency of teaching resources and improving overall management practices with digital solutions. Extensive research in the field highlights the significant benefits of IoT and smart classroom technologies for shaping the future of "smart schools," ultimately enhancing both daily teaching activities and school administration (Zhang & Li, 2021).

Smart technologies are fundamental to digital transformation and the development of smart schools. They underpin augmented and virtual reality, which facilitate the creation of immersive virtual worlds that digitally extend the physical environment. These innovations offer three-dimensional (3D) learning spaces that integrate physical, social, and normative dimensions, complemented by virtual communities in which avatars represent users interacting via application programming interfaces. Additionally, the evolution of immersive visualization techniques such as augmented reality (AR) and virtual reality (VR) results in more intuitive user interfaces that enhance users' understanding of complex systems and their interactions within these systems (Hananto et al., 2024). Furthermore, a notable advancement is the implementation of "digital twins" in education, which bridge the real and virtual worlds and contribute to the global digitalization of information (Zahedi, 2024). In the context of smart schools, digital twins can enhance learning by providing virtual replicas of laboratory equipment, supporting campus infrastructure development, evaluating course designs, and promoting teachers' professional growth (Pathak & Upadhyay, 2024).

By harnessing the power of these technological systems, schools can improve communication between teachers and students, enhance student autonomy in learning, and implement innovative approaches that facilitate deeper learning experiences. For example, IoT applications in smart classrooms can significantly enhance information sharing and data analysis, thereby transforming the learning process while simultaneously creating an engaging environment that boosts student interest and elevates teaching quality. These advanced technologies also allow for real-time tracking of student participation and mental states through tools such as thermal infrared imaging, thereby providing valuable insights that can further personalize the learning experience. Moreover, the integration of IoT in schools not only enables more efficient resource management and improves classroom organization but also helps bridge the gap between theoretical knowledge and practical application, thus offering a holistic educational experience that caters to diverse student needs. Furthermore, smart devices embedded in smart classrooms contribute to an optimized learning environment by maintaining ideal temperature and lighting conditions, ensuring both comfort and concentration for students and teachers alike. Through continuous monitoring with sensors, these systems can track and adjust temperature and lighting levels in real time, dynamically responding to external changes and guaranteeing data accuracy. To enhance the reliability of these systems, a backup mechanism and robust software infrastructure are often included, ensuring uninterrupted monitoring and accurate data collection. This ensures the seamless operation of smart classrooms, facilitating a learning environment that is both adaptive and efficient (Zhang & Li, 2021). In light of these benefits, the key features of smart schools are summarized in Fig. 1.1.

The use of these tools, whose functions and benefits are outlined above, has led to the emergence of many other gadgets in classrooms. Tools that can be used in smart classrooms include microcontroller development boards, advanced embedded systems, electronic whiteboards, mobile phones, iPads, laptops, and tablets. Additionally, automatic electric lighting, smart HVAC systems, RFID-enabled student ID cards, wireless door locks, security cameras, video conferencing, biometric attendance tracking, eBooks, augmented reality, virtual reality,

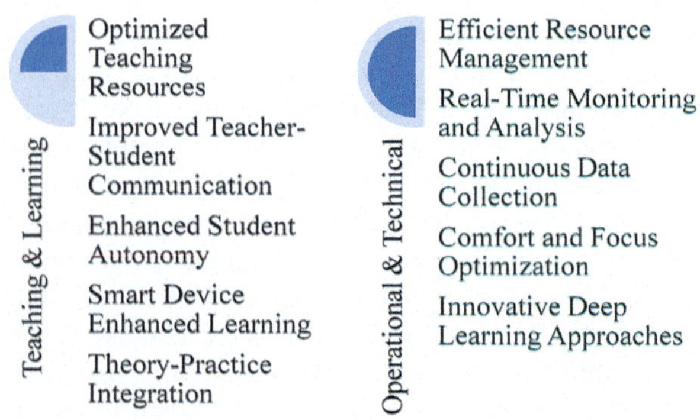

FIGURE 1.1 Key benefits and features of smart schools.

additive manufacturing devices for 3D objects, and MOOCs can enhance the learning experience (Pervez et al., 2018, p. 793).

3. Ecosystem of smart schools

The concept of smart classroom represents a significant outcome of the integration of technology and education. The advent of smart classrooms has profoundly transformed the educational system by incorporating advanced tools such as smart boards, computer technology, information networks, and emerging technologies including artificial intelligence, IoT, blockchain, smart sensors, smart school applications, and digital twin technology. These innovations have radically altered the processes of knowledge creation, sharing, and dissemination. In the following parts, we examine these technologies which are part of smart school ecosystem and assess their impact on contemporary educational practices.

Artificial Intelligence (AI): AI seeks to replicate human cognitive abilities such as learning, reasoning, and self-correction (Li et al., 2023) through algorithms that transform data into actionable insights. Among the most widely used approaches for developing intelligent systems are machine learning and deep learning (Rathore et al., 2021). AI finds broad application in data processing and analysis, supporting tasks that range from advanced information retrieval to machine learning–driven automation. Its diverse applications include virtual assistants, image recognition, search engines, and both speech and facial recognition systems, among others (Figueroa et al., 2024).

Effective AI implementation demands not only the formulation of pertinent questions and a critical assessment of responses but also the cultivation of creative problem-solving that integrates empathy, logic, and innovative thinking. Moreover, incorporating humanities into technology education is crucial. Additionally, the ability to adapt rapidly and acquire new skills is essential for effectively navigating and contributing to the fast-evolving digital ecosystem (Dhamiwal and Saini, 2024).

3.1 Internet of Things (IoTs)

We are now witnessing the emergence of the IoT era, characterized by ubiquitous computing and networks of interconnected, internet-enabled objects, with micro-controllers, sensors, and actuators serving as its fundamental building blocks (Pervez et al., 2018). This technology automates the communication and sharing of information across devices and networks (Ağca, 2023) and enhances personalized learning by integrating attendance and monitoring systems that employ tags and sensors to track student engagement and learning styles, thereby assisting teachers in tailoring curricula to individual needs (Pervez et al., 2018). Furthermore, IoT enables students to rapidly acquire new skills and knowledge through secure integration in e-campus activities, leveraging smart devices connected across university campuses, including classrooms, laboratories, and parking areas. The rapid advancements in IoT and AI in recent years underscore the imperative for today's generation to swiftly embrace these technologies in education and beyond (Dhamiwal & Saini, 2024).

3.2 Blockchain

Blockchain is a secure, tamper-proof distributed ledger technology that uses cryptography to protect transactions within a decentralized network, ensuring data integrity and transparency. Its architecture, comprising peer-to-peer networks, consensus mechanisms, cryptography, and smart contracts, collectively provides a solution for secure, transparent transactions without intermediaries. Furthermore, when integrated with digital twins, which simulate the behavior of physical components, blockchain authenticates digital asset ownership and enables real-time interactions, thereby enhancing security, trust, privacy, scalability, and interoperability (Zhang, 2024).

Within the realm of cyber security, blockchain plays a crucial role in preventing data tampering, particularly by safeguarding sensitive student data and ensuring secure electronic records. It also contributes to improving the security and flexibility of digital twin systems by supporting the creation of secure edge networks with techniques such as local model verification and asynchronous learning aggregation (Zhang, 2024). Moreover, permissioned blockchain bolsters security further by controlling access and managing participants, especially in cases involving private data or where mutual trust among end users is lacking.

In the field of education, blockchain is transforming credentialing, data security, and student record management. By offering secure storage and sharing academic records, blockchain overcomes vulnerabilities inherent in traditional centralized systems, ensuring verifiable, tamper-proof credentials. When combined with AI-driven analytics that personalize learning, blockchain helps to identify knowledge gaps and recommend tailored pathways. Additionally, blockchain simplifies the management and transfer of academic transcripts, ensuring smooth record exchange across institutions and reducing the risk of credential fraud (Egunjobi & Adeyeye, 2024).

3.3 Smart sensors

In smart classrooms, sensors are essential for collecting learning data throughout the educational process, thereby forming the foundation for intelligent systems that provide educators with valuable insights to enhance student experiences and improve teaching effectiveness (Zhang et al., 2024). These sensors, including those that measure environmental parameters such as carbon dioxide, pH, and heart rate, as well as devices that monitor physiological and behavioral data, serve to not only improve the physical classroom environment but also foster student engagement, streamline attendance management, and enable personalized learning (Silva et al., 2024). Additionally, the collected data can be systematically analyzed to support informed decision-making, further enhancing the learning process.

Beyond environmental monitoring, advanced sensor systems capture subtle indicators such as facial expressions, body movements, sounds, and brain waves, providing an objective evaluation of both teachers' and students' statuses. These systems act as teaching supervisors by generating academic prediction reports, supporting personalized learning plans, facilitating educational chats, and delivering comprehensive intelligent support (Zhang et al., 2024). When combined with AI-driven software algorithms, the capabilities of these sensors are significantly enhanced, leading to improved data processing, analysis, and overall classroom management. This fusion of sensor technology and artificial intelligence significantly expands the potential of smart classrooms, making data usage more efficient and insightful while providing real-time insights into student participation and attendance (Zhang et al., 2024).

3.4 Smart wireless sensor networks

These networks, often combined with Internet of Things (IoT) technologies, artificial intelligence (AI), and edge computing, are revolutionizing how schools operate, teach, and engage students. The integration of IoT devices and advanced sensor technologies in schools enables real-time monitoring, asset tracking, and the creation of digital twins—virtual representations of physical systems designed to optimize and maintain them (Rathore et al., 2021). Smart wireless sensor networks can help authorities collect data on various metrics, such as pulse rates and quiz scores, which can be analyzed to provide personalized learning suggestions through deep reinforcement learning (DRL) (Abdulkareem et al., 2023).

The rapid development of wireless technologies, alongside the growing demand for IoT services, has spurred a global expansion of interconnected smart devices, thereby enabling the creation of dynamic learning environments. While these advancements offer significant benefits, they also heighten the complexity of accurately modeling such systems in virtual environments (Zhang, 2024). Consequently, smart sensor networks and IoT sensors have become indispensable for actively monitoring student activities and school facilities, ensuring a responsive and efficient educational environment.

3.5 Smart school applications

Smart school applications comprise a suite of features such as quizzes, leaderboards, attendance tracking, notice boards, and fee management systems that are designed to support students, teachers, and administrative staff in enhancing learning outcomes, promoting healthy competition, and streamlining administrative tasks. The benefits of smart school applications include improved communication, enhanced administrative efficiency, more personalized learning experiences, and increased parental involvement (Shahzad et al., 2023).

3.6 Digital twin

Digital twin technology constitutes a fundamental component of the ongoing technological revolution by creating virtual replicas of physical school systems. These digital replicas enable schools to optimize operations and enhance their overall learning experience. More detailed information regarding digital twin technology will be provided in subsequent sections; therefore, this section offers only a brief overview.

4. Introduction to digital twin technology

The term "digital twin" was first introduced by Michael Grieves in 2002 during a presentation on product life cycle management (Zhang, 2024). In essence, a digital twin is a virtual representation of a physical object, system, or environment designed to simulate and mirror its real-world counterpart (Ağca, 2023). The concept of creating identical replicas, or "twins," to simulate and test real-world systems was not entirely new. For example, NASA employed a similar approach during the Apollo missions by creating duplicate spacecraft to simulate potential system failures on Earth. This strategic decision compensated for the limited communication and computational resources available at the time, enabling engineers to troubleshoot issues and make informed decisions before transmitting them to astronauts, thereby ensuring their safety without direct access to the spacecraft (Barbie & Hasselbring, 2024).

A digital twin is a virtual replica of a physical entity created by integrating data from multiple sources such as academic performance, campus behavior, and digital material consumption which can be analyzed to offer personalized insights, identify knowledge gaps, and recommend targeted educational resources (Furini et al.,

2022). By replicating real-world entities, digital twins support decision-making and risk assessment, thereby facilitating a deeper understanding of system behaviors and interactions (Chamorro-Atalaya et al., 2024).

The primary goal of digital twins is to enhance efficiency. These systems integrate the real and virtual worlds by processing real-time data to develop effective solutions, and a digital twin is deemed successful only when its virtual counterpart closely matches the functionality of the physical model (Rathore et al., 2021). Digital twins can now be generated from diverse data sources including social media platforms (e.g., Instagram, Facebook, LinkedIn, Twitter), government databases, healthcare systems, retail platforms such as Amazon, workplaces, and educational institutions that together contribute to a detailed digital representation of an individual's knowledge, skills, and interests (Saracco, 2018). In this context, digital twins are emerging as a revolutionary tool in education, enabling learners to engage with digital replicas of real-world entities through technologies such as virtual reality.

Digital twin technology has proven particularly useful in applications such as predictive maintenance, production scheduling, and performance enhancement. The development of digital twins has been facilitated by advancements in technologies such as the Internet of Things (IoT), artificial intelligence (AI), and machine learning, which together provide the necessary foundation for creating these virtual models. In this process, AI represents an advanced, integrative element that enables models to predict future actions based on historical data. Through digital simulations, processes can be visualized and optimized before actual implementation, thereby simplifying operations and contributing to environmental and climate protection efforts. For instance, in healthcare, simulating patient conditions can lead to more accurate decisions and improved treatment planning.

4.1 Fundamental principles of digital twin technology

The creation of a digital twin relies on three essential components: an information model that simplifies the physical object, a communication system that facilitates data exchange between the digital twin and its corresponding real-world object, and a data processing unit that interprets and utilizes this information to construct an accurate digital representation (Hananto et al., 2024). In this framework, the virtual twin continuously collects updated data from the physical system to predict its future states. Dynamic, real-time interactions occur between physical objects and their virtual counterparts, as illustrated in Fig. 1.2. This interaction is inherently bidirectional: physical systems transmit real-time data to the virtual space, where digital twin models are generated; in turn, the digital twins analyze the incoming data, update their models, and formulate optimization strategies to enhance the performance of the physical systems (Zhang, 2024).

As seen in Fig. 1.2, digital twins establish precise mappings between physical objects and digital spaces, synchronizing historical and real-time data to reflect the full status of physical systems with minimal errors. They operate in real time, continuously updating models via edge computing, are distributed across multiple entities and servers, and incorporate intelligent techniques—such as AI—to process data, optimize performance, and provide predictions or decisions for physical systems (Zhang, 2024). In brief, the relationship between IoT, big data, AI, and digital twins involves IoT devices harvesting data from the physical environment, which are then processed by AI models to create digital twins capable of detecting, predicting, optimizing, and making decisions—ultimately enabling process optimization across various industries (Rathore et al., 2021).

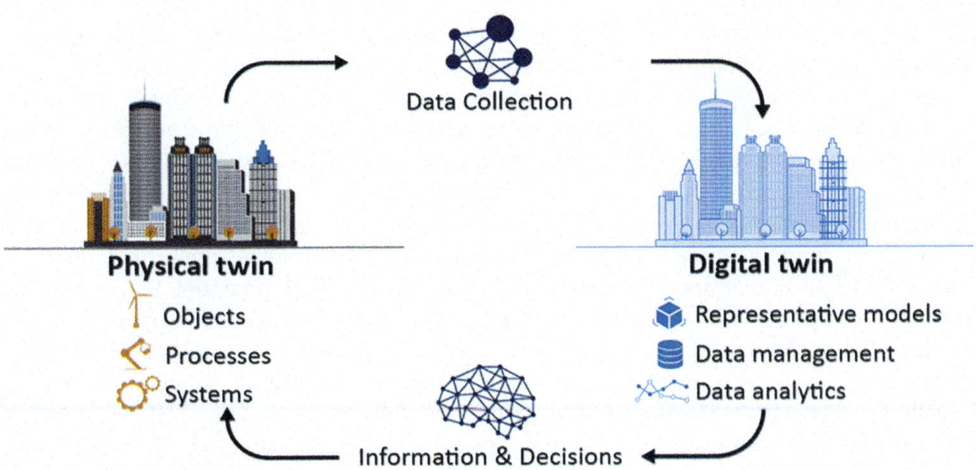

FIGURE 1.2 Illustration of the digital twin model (taken from https://www.gao.gov/products/gao-23-106453).

4.2 Areas of application for digital twin technology

These technologies have the potential to detect issues and offer solutions during natural disasters such as earthquakes and floods by providing early warnings and optimized responses. Much like the autonomous capabilities in Tesla vehicles, digital twins could enable vehicles to perform self-maintenance by utilizing a variety of data inputs and algorithms, thereby reducing the need for frequent manual interventions. As the transportation systems of cities are mapped and integrated with this technology, autonomous taxis might emerge as a viable solution for urban mobility, offering enhanced efficiency and safety. Furthermore, the concept of a digital twin could extend to individuals from birth, enabling continuous monitoring of health data that facilitates the early detection of diseases and the provision of tailored, optimal healthcare solutions, thus enhancing overall well-being throughout a person's life. Digital twins can be applied in virtual labs, libraries, training robotics, virtual classrooms, blended learning, and virtual gaming, all of which are pivotal in supporting online or distance learning initiatives and driving the digitization of education (Ağca, 2023).

The growing interest in digital twins underscores its transformative potential across various industries, unlocking new opportunities for optimization, efficiency, and innovation. Their applications now span diverse sectors such as manufacturing, aviation, intelligent transportation, urban intelligence, energy, agriculture, construction, smart cities, and emerging wireless communication technologies, all of which benefit from the enhanced decision-making and predictive capabilities offered by digital twin technology (Ağca, 2023; Hananto et al., 2024).

4.3 The usage of digital twin in education

Digital twins in education are tools that simulate physical educational infrastructures, enabling the observation, analysis, and simulation of real-world objects to enhance learning environments. Although digital twin technology has been around for more than 20 years, it has not been widely adopted in industries or education, whereas virtual and augmented reality remain more popular for creating digital models (Ağca, 2023).

As an educational innovation, digital twins can play a crucial role in the digital transformation of education, particularly by supporting online and distance learning through the integration of various technologies for virtual learning design and deployment (Ağca, 2023). The use of digital twin prototypes revolutionizes the development of embedded software systems by enabling an iterative process that begins with hardware emulation such as sensors, actuators, and communication protocols. This approach fosters collaboration between engineers and domain experts, integrates operational prototypes, and allows for the reuse of control logic, thereby reducing parallel development efforts and minimizing technical debt (Barbie & Hasselbring, 2024).

Digital twins in education can be employed in various contexts, including digital, virtual, or online classrooms, where they help create simulated environments that enable students to interact with and explore all dimensions and functionalities of real-world objects or systems. This technology enhances learning by providing immersive, interactive experiences that allow students to observe, analyze, and engage with physical entities or educational infrastructures in a virtual setting (Ağca, 2023). Digital twins can also enable students to monitor their progress and receive personalized lesson plans tailored to their abilities and advancements (Pathak & Upadhyay, 2024). Additionally, digital twins have proven beneficial for maintenance, cost reduction, optimization, simulation performance, monitoring, product life cycle understanding, assessment validation, performance evaluation, product design, and safety and risk mitigation (Weichbroth et al., 2024, p. 774). Based on these findings and competencies, the application of digital twin technology in education is summarized in the table presented in Fig. 1.3.

As illustrated in Fig. 1.3, the integration of digital twins in education yields a comprehensive array of transformative advantages. This integration not only optimizes the learning experience and streamlines resource management but also fosters a safer, more cost-efficient environment while enhancing the efficiency and effectiveness of teaching, learning, and administrative operations. When these overarching benefits and capabilities are considered, the following specific advantages can be identified in schools and other educational contexts.

4.3.1 Proactive maintenance and predictive monitoring for school facilities

One key advantage is the use of digital twins for proactive maintenance, which enables more effective monitoring of school facilities. This technology facilitates predictive maintenance by allowing school administrators to identify issues before they escalate, thus ensuring a more seamless and cost-efficient environment for both staff and students. Furthermore, predictive analytics integrated into digital twin models can assist school management in forecasting resource needs and identifying potential issues such as equipment failures or overcrowded classrooms. In areas such as system monitoring and control, digital twins can be used to monitor and manage school networks as well as heating, ventilation, and air conditioning (HVAC) systems in real time, thereby ensuring a safe and efficient learning environment (Teke et al., 2023). Ultimately, digital twins

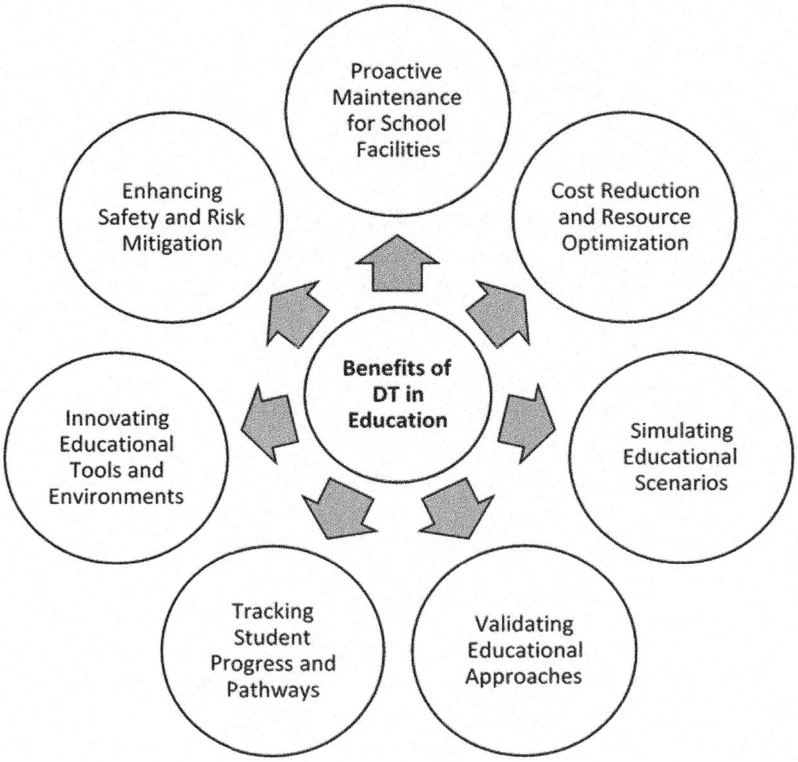

FIGURE 1.3 Benefits and functions of digital twin.

enable schools to make data-driven decisions, enhance learning experiences, and optimize overall educational operations.

4.3.2 Cost reduction and resource optimization in schools

By providing a clear understanding of the school's infrastructure and resource utilization over time, digital twins can help identify areas for material and energy savings. This results in more efficient budgeting and reduced operational costs (Mohamed et al., 2023). Additionally, digital twins offer insights into how resources—such as classroom space, technology, and teaching materials—can be better allocated to maximize educational outcomes, ensuring that schools deliver optimal value to their students.

4.3.3 Simulating educational scenarios for improved decision-making

With recent technological advancements integrated into teaching, educators can place personalized education at the center of student-centered learning, thus enabling students to customize their learning experiences. IoT technology offers an effective solution to challenges faced by students in large classes, addressing individual needs and preventing disengagement by integrating technology into education through careful planning (Pervez et al., 2018). Moreover, digital twins can support performance simulations, allowing schools to test various scenarios—such as different teaching methods or campus layouts—to determine which approaches yield the most successful educational outcomes. Continuous monitoring of both physical assets (such as buildings and equipment) and academic performance provides a dynamic, real-time view of school operations, thereby helping educators make data-driven decisions that improve the learning environment and student outcomes. For example, digital twins can create virtual replicas of school buildings or classrooms to optimize space and resource allocation before implementing physical changes, thus improving efficiency and reducing costs. In addition, teachers and students can use digital twins to simulate learning environments, making the educational experience more interactive and immersive. To illustrate, digital twins enable students to conduct virtual scientific experiments, embark on global field trips, engage with historical figures, and interact with instructors in remote learning environments, thereby overcoming limitations related to safety, resources, and physical presence (Pathak & Upadhyay, 2024).

4.3.4 Tracking student progress and optimizing educational pathways

Another critical benefit is the ability of digital twins to track student progress over time, providing insights into learning patterns, strengths, and areas for improvement. This data-driven approach enables educators to optimize lesson plans and interventions, ensuring that each student's educational journey is as effective as possible. Moreover, digital twins can monitor the degradation of knowledge and skills over time, laying the foundation for personalized, proactive education programs. Educational institutions can leverage this data to customize learning experiences, and in a future interconnected ecosystem, devices such as smartphones, cameras, and virtual assistants (e.g., Alexa, Siri, Cortana) could contribute to a symbiotic relationship by identifying gaps in learning and recommending relevant content (Saracco, 2018). Although current digital twins are still in their early stages, ongoing innovations in this field may lead to more integrated learning systems, ultimately transforming education and career development.

4.3.5 Validating educational approaches, ensuring consistency and inclusiveness

Validation and assessment are also crucial in education. Digital twins can be used to evaluate teaching methods, curriculum effectiveness, and student performance, thereby ensuring that the tools and strategies employed are reliable and consistent. This capability supports the development of alternative educational scenarios, allowing educators to experiment with different approaches before full-scale implementation. Furthermore, training and simulation with digital twins can help students understand complex systems, such as through virtual labs for science experiments or training tools for vocational education, thus enhancing learning without the risk of physical damage or accidents. Studies indicate that digital twins enhance collaboration between students and educators by providing shared learning experiences, interactive content, and virtual labs that allow students to engage with complex concepts and gain hands-on experience outside traditional classroom hours (Ezeoguine & Kasumu, 2024).

Digital twin technology offers a valuable alternative for students who lack access to physical laboratory equipment, ensuring that learning is not limited by resource constraints. By enabling educators to combine theoretical knowledge with practical applications, this technology enhances overall teaching strategies (Chamorro-Atalaya et al., 2024). To foster a more equitable educational system for individuals with limited access to education, individualized learning systems must be implemented. Personalized learning models can be developed through three layers: an infrastructure layer that tracks activities via multimedia and IoT devices; a data layer that creates a digital twin of the student using a combination of on-site, online, and offline activities; and a presentation layer that delivers personalized, accessible interfaces to enhance the learning experience (Furini et al., 2022). Additionally, the integration of technologies such as AI-powered interactive chatbots, digital twin technology, and virtual proctoring can create a dynamic, interactive learning environment that personalizes experiences for students, enhances comprehension, and provides valuable insights for educators—particularly benefiting students with speech disorders (Dhananjaya et al., 2024).

4.3.6 Innovating educational tools and environments with digital twins

Digital twins facilitate product design, such as the development of new digital learning tools or classroom layouts, thereby enabling schools to test different variations of educational technologies and environments without significant resource investments. This approach also promotes collaborative learning by allowing students to work together in virtual environments, fostering joint problem-solving, virtual experiments, and real-time communication. These experiences enhance teamwork skills and prepare students for the collaborative nature of the modern workplace (Jankovskis et al., 2024). Moreover, digital twin prototypes in education can significantly improve students' ICT skills by simulating physical systems that require precise data and substantial computational power (Acker et al., 2023). The use of digital twins for laboratory equipment allows students to experiment both inside and outside the classroom, integrating theoretical learning with practical applications, fostering situational awareness, and providing opportunities to learn from mistakes. For instance, digital twins used in airplane flight simulations have been shown to increase curiosity and engagement, while also preparing students to handle high-risk equipment and offering cost-saving benefits by reducing the need for physical, sensitive, or expensive materials (Pathak & Upadhyay, 2024).

4.3.7 Enhancing safety and risk mitigation in schools through real-time monitoring

Safety and risk mitigation are enhanced through the continuous monitoring of a school's physical environment, which enables the early identification of potential hazards or health risks. This proactive approach allows administrators to take timely steps to safeguard both students and staff.

5. Major issues and challenges in smart technologies in schools

Although the adoption of digital twin technology has surged in recent years, with numerous successful applications reported, several challenges and issues remain unresolved (Weichbroth et al., 2024).

5.1 Inadequate infrastructure and hardware

There is often a significant shortage of accessible hardware and software for implementing digital twin technology in schools and educational settings (Chamorro-Atalaya et al., 2024). To effectively monitor and analyze student behavior, it is essential to integrate existing infrastructures, such as smart building systems, Wi-Fi networks, and surveillance cameras, with advanced technologies like IoT devices. Technologies such as thermographic cameras, sensors, and Wi-Fi passive tracking provide valuable insights into classroom occupancy, student positioning, social distancing, and group study activities while ensuring compliance with privacy regulations (Furini et al., 2022). Despite their potential, the deployment of these tools is frequently hindered by high costs and the complexity of implementation.

5.2 Tackling the complexities of data control and analysis

The implementation of digital twins faces several challenges, particularly related to IoT data harvesting, integration, and sharing, which can result in high costs. In some instances, the cost of creating a digital twin may even exceed that of the physical asset itself. Moreover, the data collected is often large, heterogeneous, unstructured, and noisy, necessitating extensive cleaning, organization, and homogenization before effective utilization. Managing the flow of such vast amounts of data adds another layer of complexity. To improve the accuracy of digital twins, machine learning models require sufficient, high-quality data for training. The rapid growth of IoT technologies has led to the generation of large volumes of sensor data, which in turn demands advanced tools, architecture and algorithms for efficient processing and analysis. While edge and cloud computing offer potential solutions, aggregating data in the cloud can increase response times. Furthermore, selecting the appropriate machine learning model for digital twin applications is challenging due to varying accuracy and efficiency among models, and practical applications of AI in digital twinning remain relatively limited (Rathore et al., 2021).

5.3 Challenges of inefficiency and adaptation difficulties

Although the rapid advancement of AI technology has endowed smart machines, vehicles, and various IoT devices with enhanced capabilities in environmental perception, knowledge acquisition, decision-making, and state prediction, these AI technologies and sensors can be inefficient and difficult to adapt. Since the exploration of machine learning applications is still in its early stages, challenges such as data and model selection, system complexity, lack of user-friendliness, issues with data quality, and limited interdisciplinary collaboration constrain performance and prediction accuracy (Tahmasebinia et al., 2023). For instance, training AI for autonomous driving on actual vehicles can cause accidents, and optimizing cellular networks with AI is slowed by delays in receiving feedback from complex systems, thereby hindering learning effectiveness. While simulation software can speed up feedback, creating accurate simulations for dynamic systems is challenging due to nonlinearity and uncertainty, making it difficult to directly transfer strategies from simulations to real-world applications (Zhang, 2024).

5.4 Security, curriculum, and industry collaboration issues

Despite the promising potential of digital twins across various sectors, their implementation faces considerable challenges, including curricular limitations. Current curricular documents do not include IoT topics, either as core subjects or optional modules, necessitating revisions to incorporate IoT concepts such as device connectivity, sensors, data analysis, and automation (Bisik & Takuneno, 2024). Moreover, obstacles such as employee reluctance—driven by job security concerns—and difficulties in fostering effective collaboration among diverse industry stakeholders further complicate adoption (Weichbroth et al., 2024). Additionally, critical digital twin systems, particularly those related to human and defense data, require stringent security and privacy safeguards, with an emphasis on securing communication protocols and ensuring the safe storage of asset-related data to protect against breaches from both internal and external threats. In educational settings, the implementation of AI-based systems—especially those employing facial recognition and biometric data—raises significant concerns regarding information security and privacy (Figueroa et al., 2024). As computer systems are vulnerable to various attacks designed to steal information or disrupt services, either temporarily or permanently, data stored within these systems is at risk of theft and manipulation. Therefore, protecting sensitive student data and ensuring adherence to data protection regulations are crucial challenges that must be addressed.

5.5 Insufficient human capital resources

Initially, technology replaced workers by automating tasks that were monotonous, hazardous, or disruptive. Over time, it evolved to augment workers' capabilities, and today, emerging technologies not only supplant or support workers but also enhance their skills, enabling them to be more efficient and effective (Zahedi, 2024). However, an increasing number of teachers are not equipped with the knowledge or skills required to instruct students in these new technologies, and existing professional development programs are insufficient to bridge this gap. In other words, many teachers lack familiarity with IoT technologies and lack the formal training needed to integrate them effectively into classroom instruction, highlighting the need for specialized professional development. Consequently, targeted training programs must be implemented to equip educators with the knowledge and skills necessary to effectively teach emerging technologies in smart classrooms (Bisik & Takuneno, 2024).

To address these needs, schools should partner with technology companies and government agencies to share resources, expertise, and funding, while providing continuous training for educators to effectively use new technologies. Furthermore, technology companies must focus on data privacy and security, and policymakers should update curricula to include these technologies while working to bridge the digital divide, ensuring that all students, regardless of socio-economic background, have access to these advancements.

6. Summary

As technology continues to evolve, schools are transforming into connected, intelligent ecosystems by integrating AI, smart sensor networks, IoT devices, blockchain, and digital twins, thereby enhancing facilities management and educational experiences. Innovations such as augmented reality, virtual reality, and digital twins are redefining classroom experiences by offering immersive learning environments, improving decision-making, and streamlining administrative processes. Notably, the rapid expansion of AI applications positions it as a key enabler for future classrooms, where data can be seamlessly extracted from computers, sensors, cameras, and learning platforms (Figueroa et al., 2024). With the support of artificial intelligence, these technologies foster improved communication, personalized learning, and efficient resource management, ultimately making education more effective, secure, and impactful. Furthermore, these systems support functions such as user registration, facial recognition for attendance, online exams, teaching assessments, and real-time result tracking, while also managing classroom conditions, such as temperature through automated control. In addition, blockchain technology enhances transparency, ensures data integrity, improves cybersecurity, safeguards student records, streamlines credentialing processes, and enables secure transactions in the education sector. The classroom structure also facilitates a seamless learning experience through pre-class self-study, in-class activities, and post-class feedback, using resources such as micro videos, teaching materials, and assignments. This approach fosters a personalized, intelligent educational environment that allows students to thrive in a digitalized world (Zhang & Li, 2021).

The integration of sensor, IoT, and blockchain technologies in education is transforming the way educational environments function, enhancing both teaching and learning experiences. For instance, smart classrooms utilize advanced sensor technologies, including environmental and physiological sensors alongside AI and IoT and enable real-time data collection that augments teaching methods, student engagement, and classroom management. Similarly, IoT has profoundly impacted education by reshaping infrastructure and pedagogical approaches. By interconnecting devices and sensors, IoT facilitates personalized learning experiences, allowing for tailored curricula and more effective skill development through smart school integration. This interconnectedness also promotes a more efficient and responsive educational environment. In parallel, blockchain technology fortifies the educational landscape by ensuring security, transparency, and trust. By protecting transactions and preserving data integrity, blockchain reinforces privacy and scalability, offering tamper-proof credentials and secure management of student records, while also facilitating seamless data exchanges between institutions and preventing fraud. Together, these technologies create a dynamic, secure, and personalized learning experience for all stakeholders in the education system.

Digital twin is essentially an algorithm that replicates the behavior of a physical counterpart, generating identical outputs from given inputs. Initially employed in smart manufacturing, this technology has since been adopted across diverse fields including construction, education, business, transport, healthcare, and sports (Rathore et al., 2021). Digital twins serve as a bridge between the physical and virtual realms by creating virtual replicas of physical assets, enabling real-time monitoring and data analysis that improves equipment reliability, optimizes performance, and fine-tunes maintenance schedules. By facilitating early failure detection, digital twins reduce downtime and enhance efficiency. Moreover, by enabling advanced analytics and real-time data processing, digital twins provide critical insights for decision-making, predictive maintenance, and system optimization (Hananto et al., 2024). We also noted the growing adoption of digital twins in various sectors such as aviation for maintenance and risk prediction, healthcare for patient monitoring, intelligent transportation for traffic management, and wireless communications for network performance enhancement. However, challenges such as data integration, privacy concerns, and scalability remain obstacles to widespread adoption. Nonetheless, the transformative potential of digital twins in enhancing operational efficiency, reducing costs, and supporting informed decision-making renders them invaluable across industries, while also advancing education, training, collaboration, and decision-making processes (Weichbroth et al., 2024).

To adequately prepare today's students for the future, it is essential to embrace technological advancements that have reshaped traditional classrooms by minimizing reliance on physical books. Digital devices, such as computers and smart gadgets, have replaced traditional textbooks, offering convenient access to a vast array of resources without the burden of physical materials. The integration of online learning systems, open courses, and digital resources has substantially improved education by reducing time expenditure, enhancing efficiency, and providing immediate access to information (Pervez et al., 2018). Consequently, there exists a need to restructure the educational system's architecture around a sharing culture. In this new paradigm, physical schools and classrooms are increasingly supplemented by educational platforms that offer flexible, intelligent learning environments essentially, schools as a service (Zahedi, 2024).

Lastly, Shahzad et al. (2023) emphasize the importance of developing secure, user-friendly smart school applications that cater to the diverse needs of students, teachers, and parents. These applications must guarantee equitable access to technology and reliable internet connectivity, particularly in underserved communities. Achieving an inclusive, tech-driven educational ecosystem demands strategic investments in digital infrastructure, comprehensive teacher training, and supportive policy frameworks that ensure every stakeholder can fully benefit from these innovations.

References

Abdulkareem, S. A., Jabbar, K. A., Ahmed, I., Zearah, S. A., Diame, H. A., & Sadiq, S. N. (2023). Deep reinforcement learning-based smart education system with wireless sensor network virtualization. In *2023 Annual International Conference on Emerging Research Areas: International Conference on Intelligent Systems (AICERA/ICIS)* (pp. 1–6). IEEE.

Acker, J., Rogers, I., Guerra-Zubiaga, D., Tanveer, M. H., & Moghadam, A. A. A. (2023). Low-cost digital twin approach and tools to support industry and academia: A case study connecting high-schools with high degree education. *Machines, 11*(9), 860. https://doi.org/10.3390/machines11090860

Ağca, R. K. (2023). Using digital twins in education from an innovative perspective: Potential and application areas. *Education Mind, 2*(2), 65–74. https://doi.org/10.58583/Pedapub.EM2306

Barbie, A., & Hasselbring, W. (2024). From digital twins to digital twin prototypes: Concepts, formalization, and applications. *IEEE Access, 99*, 1. https://doi.org/10.1109/ACCESS.2024.3406510

Bisik, D., & Takuneno, K. J. (2024). Analysis of the need for internet of things curriculum in vocational high schools: A case study of SMK Negeri 1 Kupang. *Journal of Innovative Technologies in Learning and Education, 1*(1), 112–124. https://doi.org/10.37792/jitle.v1i1.1400

Chamorro-Atalaya, O., Flores-Velásquez, C. H., Flores-Cáceres, R., Arévalo-Tuesta, J. A., Zevallos-Castañeda, M., Tomás-Quispe, G., Aranda-Dextre, C. C., Cruz-Telada, Y., Caller-Luna, J., & Alarcón-Anco, R. (2024). Use of digital twin technology in the teaching-learning process in the field of university education: A bibliometric review. *International Journal of Learning, Teaching and Educational Research, 23*(2), 313–329. https://doi.org/10.26803/ijlter.23.2.15

Dhamiwal, A., Shubham, & Saini, M. (2024). Artificial intelligence and IoT trends in education. In *National Conference on Future Computing Technologies for Sustainable Development (NCFCTSD-24)*. Rohtak: DCSA, M D University, 978-93-80544-57-1.

Dhananjaya, G. M., Goudar, R. H., Kulkarni, A., Rathod, V. N., & Hukkeri, G. S. (2024). A digital recommendation system for personalized learning to enhance online education: A review. *IEEE Access, 12*, 34019–34041. https://doi.org/10.1109/access.2024.3369901

Egunjobi, D., & Adeyeye, O. J. (2024). Blockchain in education: Transforming credentialing, data security, and student records management. *International Research Journal of Modernization in Engineering Technology and Science, 6*(10). https://doi.org/10.56726/IRJMETS62066

Ezeoguine, P. E., & Kasumu, R. Y. (2024). Undergraduate students' perception of digital twins technology in education: Uses and challenges. *International Journal of Education and Evaluation (IJEE), 10* (2), 381–396. https://doi.org/10.56201/ijee.v10.no2.2024.pg381.396

Figueroa, E., Batista, E., Palau, R., Unciti, O., Ferre, M., & Martínez-Ballesté, A. (2024). The use of artificial intelligence techniques in smart classrooms is in its infancy. *IEEE Access, 12*, 125179–125193. https://doi.org/10.1109/ACCESS.2024.3454372

Furini, M., Gaggi, O., Mirri, S., Montangero, M., Pelle, E., Poggi, F., & Prandi, C. (2022). Digital twins and artificial intelligence: As pillars of personalized learning models. *Communications of the ACM, 65*(4), 98–104. https://doi.org/10.1145/3478281

Hananto, A. L., Tirta, A., Herawan, S. G., Idris, M., Soudagar, M. E. M., Djamari, D. W., & Veza, I. (2024). Digital twin and 3D digital twin: Concepts, applications, and challenges in industry 4.0 for digital twin. *Computers, 13*, 100. https://doi.org/10.3390/computers13040100

Jankovskis, D., Cirule, I., & Carbone, A. (2024). Digital twins and e-learning: Navigating challenges and opportunities. In G. Casalino, et al. (Eds.), *Higher education learning methodologies and technologies online. HELMeTO 2023. Communications in computer and information science* (Vol 2076, p. 12). Cham: Springer. https://doi.org/10.1007/978-3-031-67351-1_12

Li, X., Zeng, L., Lu, X., Chen, K., Yu, M., Wang, B., & Zhao, M. (2023). A review of artificial intelligence in the rupture risk assessment of intracranial aneurysms: Applications and challenges. *Brain Sciences, 13*(7), 1056. https://doi.org/10.3390/brainsci13071056

Mohamed, N., Lazarova-Molnar, S., & Al-Jaroodi, J. (2023). Digital twins for energy-efficient manufacturing. In *2023 IEEE International Systems Conference (SysCon)* (pp. 1–7). IEEE.

Pathak, R. K., & Upadhyay, P. (2024). Integration of digital twins technologies in education for experiential learning: Benefits and challenges. *International Research Journal on Advanced Engineering Hub (IRJAEH), 2*(3), 442–449. https://doi.org/10.47392/IRJAEH.2024.0064

Pervez, S., Rehman, S. U., & Alandjani, G. (2018). Role of internet of things (IoT) in higher education. In *Proceedings of ADVED 2018-4th international conference on advances in education and social sciences* (pp. 792–800). Turkey: Istanbul, 978-605-82433-4-7.

Rathore, M. M., Shah, S. A., Shukla, D., Bentafat, E., & Bakiras, S. (2021). The role of AI, machine learning, and big data in digital twinning: A systematic literature review, challenges, and opportunities. *IEEE Access, 9*, 32030–32052. https://doi.org/10.1109/ACCESS.2021.3060863

Saracco, R. (2018). What would education be like in 2050? Digital twins. *IEEE Future Directions*. https://cmte.ieee.org/futuredirections/2018/02/21/what-would-education-be-like-in-2050-digital-twins/.

Shahzad, I., Walke, S., Dahat, A., & Kurmi, P. (2023). Smart school application: Enhancing administrative efficiency, communication, and personalized learning in education. *International Journal of Scientific Research in Computer Science, Engineering and Information Technology, 9*(3), 524. https://doi.org/10.32628/CSEIT2390267

Silva, M. J., Rodrigues, M., & Tempera, T. (2024). Framework for a research-based and interdisciplinary use of sensors in elementary teacher education. *Sensors, 24*(17), 5482. https://doi.org/10.3390/s24175482

Tahmasebinia, F., Lin, L., Wu, S., Kang, Y., & Sepasgozar, S. (2023). Exploring the benefits and limitations of digital twin technology in building energy. *Applied Sciences, 13*(15), 8814. https://doi.org/10.3390/app13158814

Teke, İ., Teke, O., & Kılınç, M. (2023). The future of smart campuses: Combining digital twin and green metrics. *Acta Infologica, 7*(2), 384–395.

Weichbroth, P., Jandy, K., & Zurada, J. (2024). Toward sustainable development: Exploring the value and benefits of digital twins. *Telecom, 5*(3), 774–791. https://doi.org/10.3390/telecom5030039

Zahedi, M. H. (2024). Digital transformation in higher education: Education as a service. *AI and Tech in Behavioral and Social Sciences, 2* (3), 43–56. https://doi.org/10.61838/kman.aitech.2.3.6

Zhang, Y. (2024). Blockchain for digital twin. In *Digital twin. Simula SpringerBriefs on computing* (Vol 16)Cham: Springer. https://doi.org/10.1007/978-3-031-51819-5_5

Zhang, X., Ding, Y., Huang, X., Li, W., Long, L., & Ding, S. (2024). Smart classrooms: How sensors and AI are shaping educational paradigms. *Sensors, 24*(17), 5487. https://doi.org/10.3390/s24175487

Zhang, M., & Li, X. (2021). Design of smart classroom system based on internet of things technology. In *Mobile information systems* (Vol 2021, p. 5438878). https://doi.org/10.1155/2021/5438878

Chapter 2

Smart classroom: Evolution, core components, benefits and challenges

Anucha Somabut[1] **and Kulthida Tuamsuk**[2]

[1]Department of Educational Technology, Faculty of Education, and the Smart Learning Innovation Research Center, Khon Kaen University, Khon Kaen, Thailand; [2]Department of Information Science, Faculty of Humanities and Social Sciences, and the Smart Learning Innovation Research Center, Khon Kaen University, Khon Kaen, Thailand

1. What is smart classroom?

Smart classrooms are educational environments equipped with advanced technologies such as interactive whiteboards, digital projectors, and IoT devices to facilitate a more engaging and effective learning experience (Ashwin et al., 2023). They incorporate AI and sensor technologies to monitor and enhance student engagement, attendance, and learning outcomes (Zhang et al., 2024). These classrooms are designed to support personalized learning paths, allowing students to progress at their own pace and receive targeted support (Aggarwal et al., 2024).

Smart classrooms enhance student engagement and motivation through interactive and multimedia content, making learning more appealing and effective. They provide immediate feedback to students, helping them understand their mistakes and improve their learning outcomes. By offering flexible learning schedules and remote access to educational resources, smart classrooms make education more accessible, particularly for students in remote or underserved areas (Aggarwal et al., 2024; Ashwin et al., 2023).

While smart classrooms offer numerous benefits, challenges such as data privacy concerns, high implementation costs, and the need for algorithm optimization remain significant hurdles (Zhang et al., 2024). Additionally, the lack of uniform evaluation criteria for smart classroom construction and the need for better integration of online and offline resources are areas that require further research and development (Ren, 2024). Despite these challenges, the continued advancement of educational technologies promises to further enhance the capabilities and impact of smart classrooms in the future.

Smart classrooms signify a major shift from traditional educational environments, driven by technological advancements. They incorporate digital tools, sensors, artificial intelligence (AI), and Internet of things (IoT) to foster interactive, personalized, and efficient learning. This evolution transitions from teacher-centered to student-centered learning, focusing on engagement, real-time feedback, and tailored educational experiences.

2. Evolution of smart classrooms

The historical development of smart classrooms reflects a significant evolution in educational technology, transitioning from basic tools like projectors to sophisticated AI-based systems. This transformation has been driven by the need to enhance learning environments, making them more interactive, personalized, and efficient. The journey from traditional to digital and smart learning environments has been marked by several milestones including smart classrooms, each contributing to the current state of educational technology (Liu, 2024; Long, 2024) (Fig. 2.1).

2.1 Early technology integration in education

The integration of technology in education has transformed learning environments, fostering engagement and accessibility. This early adoption has paved the way for innovative teaching methods and enhanced student outcomes (Long, 2024).

- **Projectors and overhead transparencies**: Initially, classrooms integrated simple technologies such as projectors and overhead transparencies, which allowed teachers to present information visually, enhancing engagement and understanding.

Digital Twins in the Smart Classroom. https://doi.org/10.1016/B978-0-443-41632-3.00017-5

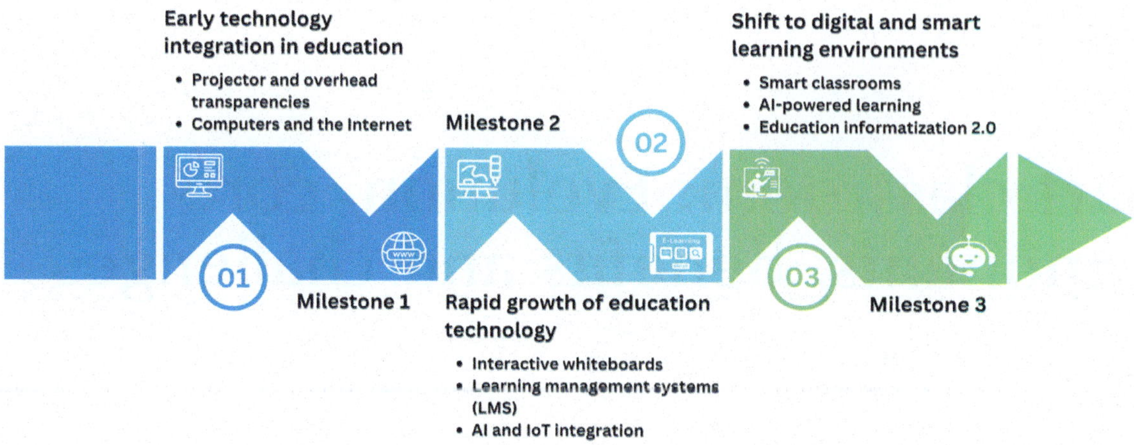

FIGURE 2.1 Evolution of educational technology—Smart classrooms.

- **Computers and the internet**: The introduction of computers and the internet in the late 20th century marked a significant shift, providing access to vast resources and enabling new forms of learning and communication.

Technology has transformed education by enhancing engagement and accessibility, evolving from projectors to computers and the internet, which improved visual learning and communication methods significantly.

2.2 Rapid growth in educational technology

The evolution of educational technology has transformed teaching and learning, marking significant milestones that enhance accessibility, engagement, and effectiveness in education. Here are some key developments in this dynamic field.

- **Interactive whiteboards**: These replaced traditional blackboards, allowing for dynamic presentations and interactive lessons, which increased student participation and engagement (Long, 2024).
- **Learning management systems (LMS)**: Platforms like Moodle and Blackboard emerged, facilitating online course management, resource sharing, and communication between students and educators (Long, 2024).
- **AI and IoT integration**: The integration of AI and IoT has been pivotal, enabling personalized learning experiences through intelligent tutoring systems and adaptive learning platforms. These technologies allow for real-time data collection and analysis, tailoring educational content to individual student needs (Akter & Asaduzzaman, 2023; Dimitriadou & Lanitis, 2023; Vashishth et al., 2024a).

Educational technology has evolved with key milestones like interactive whiteboards, Learning Management Systems, and AI integration, enhancing accessibility, engagement, and personalized learning experiences for students.

2.3 Shift to digital and smart learning environments

As education evolves, the transition from traditional methods to digital and smart learning environments is reshaping how knowledge is imparted and acquired, enhancing accessibility and engagement for learners everywhere.

- **Smart classrooms**: These environments utilize sensors and AI to monitor and enhance the physical classroom setting, manage attendance, and provide personalized learning experiences. AI technologies improve data processing and analysis, expanding the capabilities of smart classrooms (Dimitriadou & Lanitis, 2023; Zhang et al., 2024).
- **AI-powered learning**: AI has revolutionized learning environments by customizing instruction to individual needs, promoting personalized learning experiences, and preparing students for future demands. This includes the use of intelligent tutoring systems and immersive technologies like virtual and augmented reality (Kalyani, 2023, p. 8).
- **Education informatization 2.0**: This era emphasizes the integration of big data, cloud computing, and mobile learning, further advancing the capabilities of smart classrooms and enhancing educational outcomes (Long, 2024).

The shift to digital learning enhances education through smart classrooms, AI, and personalized experiences, integrating big data and cloud computing to improve accessibility, engagement, and educational outcomes.

While the integration of advanced technologies in education has brought numerous benefits, it also presents challenges. Ethical considerations, data privacy, and equitable access to technology are critical issues that need to be addressed. Moreover, the cost and complexity of implementing these technologies can be prohibitive for some institutions. Despite these challenges, the potential of AI and smart technologies to democratize education and provide inclusive learning experiences is significant, suggesting a promising future for educational innovation (Kalyani, 2023; Vashishth et al., 2024b).

3. Core components of a smart classroom

Smart classrooms represent a significant evolution in educational environments, integrating advanced technologies to enhance learning experiences. These classrooms are characterized by using hardware technologies, software platforms, and IoT systems to create interactive, efficient, and personalized learning spaces. The core components of a smart classroom can be categorized into three main areas: hardware technologies, software and digital platforms, and IoT and sensor networks. Each of these components plays a crucial role in transforming traditional classrooms into smart learning environments (Fig. 2.2).

3.1 Hardware technologies

Hardware technologies play a crucial role in shaping modern computing and electronics, driving advancements in performance, efficiency, and functionality across various industries. Key hardware technologies in a smart classroom may include:

- **Interactive whiteboards**: These instruments and devices serve as fundamental components within the context of contemporary smart classrooms, facilitating the execution of dynamic presentations that captivate audiences and promote interactive lessons that foster a participatory learning environment. Furthermore, these tools empower educators to showcase a diverse array of multimedia content while simultaneously engaging students through various touch-based interactions that significantly enhance the overall learning experience and make it more immersive and enjoyable for all participants involved (Saini & Goel, 2019).
- **Smart projectors and AR/VR devices**: Smart projectors, which are innovative devices designed to enhance the way digital content is displayed, enable users to present information in a visually appealing manner that captures the attention of the audience, while at the same time, AR/VR devices provide a remarkable opportunity for immersive and interactive learning experiences that transport users into different environments and scenarios. These advanced technologies play a crucial role in facilitating the visualization of complex concepts and ideas, thereby transforming traditional learning methods into a more engaging and effective educational experience that keeps learners motivated and eager to explore further (Alfarsi et al., 2020).

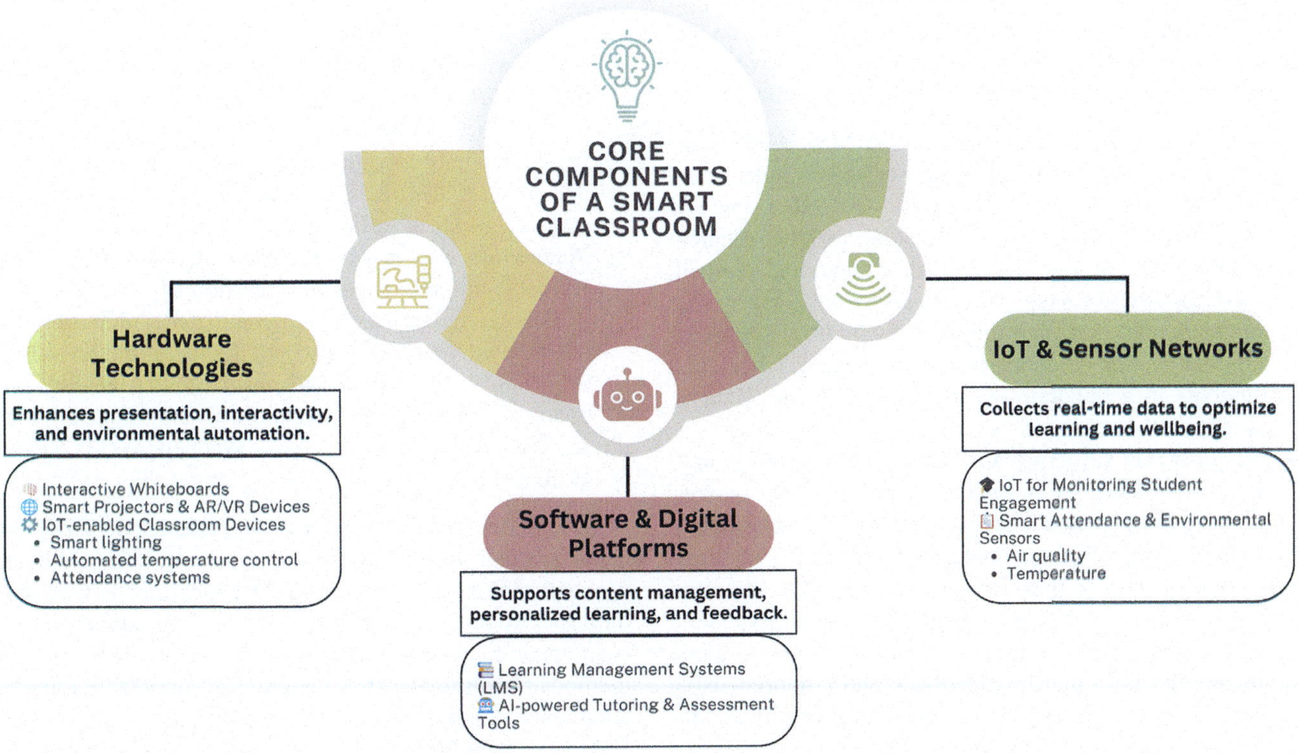

FIGURE 2.2 Core components of a smart classroom.

- **IoT-enabled classroom devices**: Devices, which include innovative technologies such as smart lighting that adjusts to the needs of the users, temperature controls that ensure optimal comfort levels, and automated attendance systems that streamline the process of tracking student presence, all play a significant role in contributing to the establishment of a conducive learning environment that fosters engagement and productivity among learners. These advanced tools assist in the optimization of classroom conditions by creating an atmosphere that is not only comfortable but also conducive to learning, while simultaneously alleviating the administrative workload that teachers face daily, allowing them to focus more on delivering quality education and less on bureaucratic tasks (Pradeep, 2023).

Hardware technologies are essential in modern computing and electronics, enhancing performance and efficiency across various industries. Key developments include interactive whiteboards, which facilitate dynamic presentations and engage students in smart classrooms; smart projectors and AR/VR devices that provide immersive learning experiences; and IoT-enabled classroom devices like smart lighting and automated attendance systems that optimize learning environments and reduce teachers' administrative tasks.

3.2 Software and digital platforms

In today's interconnected world, software and digital platforms play a pivotal role in shaping how we communicate, work, and conduct business, driving innovation and enhancing efficiency across various sectors. Software and digital platforms in a smart classroom may include:

- **Learning management systems (LMS)**: LMS platforms play a crucial role in the effective functioning of smart classrooms, serving as a comprehensive and centralized system that is essential for the efficient management of a wide array of educational content, meticulously tracking the progress and performance of students, and significantly facilitating seamless communication and interaction between teachers and students in a dynamic learning environment (Kaur & Bhatia, 2024).
- **AI-powered tutoring and assessment tools**: These innovative tools are designed to deliver highly personalized learning experiences that are tailored specifically to meet the diverse and unique needs of each individual student, ensuring that everyone receives the attention and resources that align with their learning styles and preferences. Additionally, these platforms are equipped to provide instantaneous feedback and comprehensive assessments, which play a crucial role in aiding students to enhance their understanding of the material and significantly improve their overall academic performance (Alfarsi et al., 2020).

Software and digital platforms are essential in today's interconnected world, influencing communication, work, and business operations, while fostering innovation and efficiency. LMSs are crucial for smart classrooms, enabling the management of educational content and tracking student progress. AI-powered tutoring and assessment tools enhance personalized learning by adapting to student needs and offering real-time feedback, thereby improving understanding and performance.

3.3 IoT and sensor networks

The IoT and sensor networks represent a transformative shift in technology, enabling seamless connectivity and data exchange between devices, enhancing efficiency and decision-making across various sectors.

- **IoT for monitoring student engagement**: IoT devices, which encompass a wide range of interconnected technological gadgets, have the remarkable capability to meticulously track student engagement through sophisticated sensors that continuously monitor various indicators of attention levels and participation rates during classroom activities. This invaluable data, once collected and analyzed, empowers educators to thoughtfully tailor their teaching strategies and methodologies to significantly enhance student involvement and foster a more interactive learning environment (Mohanty et al., 2024).
- **Smart attendance and environmental sensors**: Automated attendance systems, which are sophisticated technological solutions, leverage the capabilities of IoT technology to significantly streamline and enhance the process of tracking student attendance, thereby making it more efficient and less time-consuming. Meanwhile, a variety of environmental sensors are strategically implemented to continuously monitor and evaluate crucial classroom conditions, including but not limited to air quality and temperature levels, all of which work together to ensure that a healthy and conducive learning environment is maintained for all students and educators alike (Pradeep, 2023; Shayer et al., 2024).

The IoT and sensor networks facilitate connectivity and data exchange between devices, improving efficiency and decision-making across sectors. In education, IoT devices monitor student engagement by tracking attention and participation, allowing educators to adapt their teaching methods. Additionally, IoT technology supports automated attendance systems and environmental sensors that monitor classroom conditions like air quality and temperature, promoting a healthy learning environment (Arya & Verma, 2024; Okhanashvili, 2024).

While the integration of these technologies in smart classrooms offers numerous benefits, it also presents challenges. Issues such as data privacy, cost, and resistance from educators need to be addressed to fully realize the potential of smart classrooms (Zhang et al., 2024). Additionally, the variability in the implementation of smart classroom technologies can lead to inconsistent outcomes, highlighting the need for standardized approaches and further research to optimize their use (Chen & Liu, 2024).

4. Pedagogical shifts in smart classrooms

The integration of technology in education has led to significant pedagogical shifts, particularly in the context of smart classrooms. These environments leverage advanced technologies to enhance learning experiences, offering a stark contrast to traditional classroom settings. Smart classrooms facilitate active learning strategies and personalized learning models, which are increasingly being adopted to improve educational outcomes.

The traditional model of education, which has been widely accepted and practiced for many years, is predominantly defined by a method that places the teacher at the center of the learning experience, where knowledge is conveyed primarily through a series of expository lectures, while the students, unfortunately, find themselves in a position that is largely passive and somewhat disengaged from the active learning process that ideally should take place (Routray & Mohanty, 2022). This instructional framework, which relies heavily on a one-size-fits-all approach, frequently exhibits a significant lack of adaptability and personalization tailored to the unique needs of individual learners, thereby resulting in a considerable limitation on student engagement and ultimately hindering the ability of the educational system to effectively address and accommodate the diverse range of individual learning requirements that exist within any classroom setting (Peirats-Chacón, 2023).

Smart classrooms leverage the exceptional and continuously evolving technological innovations that are persistently arising within contemporary society to formulate educational settings that are not only vibrant but also profoundly adaptive to the fluctuating requirements of both students and educators. These meticulously crafted environments actively cultivate and advocate for a pedagogical approach that prioritizes the learner, wherein students engage with and utilize a multitude of technological tools and resources specifically designed to address a wide spectrum of learning styles, preferences, and individual needs that may be present among the student population. Moreover, these sophisticated educational environments are intentionally structured to enable the delivery of instantaneous and constructive feedback, while simultaneously providing analytics-driven insights that collaboratively work to personalize and enrich the educational experiences of each learner, thereby ensuring that the educational process is not only efficient but also profoundly individualized (Aggarwal et al., 2024; Kaur & Bhatia, 2024).

5. Benefits of smart classrooms

Smart classrooms, leveraging advanced technologies such as IoT, AI, and sensor systems, offer numerous benefits that enhance educational experience. These benefits include increased student engagement, real-time data analysis, personalized learning experiences, and cost-efficiency and sustainability. Each of these aspects contributes to a more effective and sustainable educational environment, transforming traditional teaching methods into dynamic, interactive learning experiences.

5.1 Enhanced student engagement and participation

In today's educational landscape, fostering student engagement and participation is crucial for effective learning. Engaged students are more likely to absorb information, collaborate with peers, and develop critical thinking skills. Mohanty et al. (2024) demonstrated a rise in student engagement metrics from 60% to 85% after the deployment of IoT technologies in smart classroom, underscoring augmented participation and interaction. Additionally, Aggarwal et al. (2024) emphasized that the use of multimedia resources, gamification, and interactive simulations in smart classrooms can create students with more learning engagement and facilitate understanding of complex concepts.

5.2 Real-time data and performance analysis

Real-time data and performance analysis are essential for organizations to make informed decisions and optimize operations. Intelligent educational environments equipped with sensory devices and AI can assess physiological and behavioral metrics, providing immediate insights into student performance and engagement (Zhang et al., 2024). This technology enables educators to identify struggling students, track progress, and adjust teaching strategies (Aggarwal et al., 2024). Additionally, IoT systems in these environments can monitor attendance and automatically regulate environmental conditions, improving the overall learning experience (Pradeep, 2023).

5.3 Personalized learning experiences

Personalized learning experiences are becoming vital in diverse educational settings, focusing on tailoring instruction to individual needs to boost engagement and success. Intelligent educational platforms offer customized learning paths, allowing students to progress at their own pace and receive targeted support in specific areas (Aggarwal et al., 2024). The use of artificial intelligence enhances these environments by providing personalized feedback and recommendations (Pradeep, 2023). Empirical case studies demonstrate that technology-enhanced classrooms promote student independence and collaborative learning, key components of individualized education (Huang, 2023).

5.4 Cost-efficiency and sustainability

Cost-efficiency and sustainability are essential for businesses to balance profitability with environmental responsibility. Intelligent educational environments leverage IoT and AI technologies to reduce the need for a large workforce and administrative staff by automating many functions (Shayer et al., 2024). The use of digital resources and cloud computing decreases reliance on physical materials, promoting economic efficiency and ecological sustainability (Ghonaim & Ghamdi, 2018). Additionally, energy-efficient features like automated lighting and climate control enhance sustainability in these advanced educational settings (Pradeep, 2023).

6. Challenges and barriers to implementation

The implementation of smart classrooms in developing regions faces several challenges and barriers, which can be broadly categorized into technical challenges, data privacy and security concerns, and teacher readiness and training. These barriers hinder the effective integration of technology in educational settings, impacting the quality and accessibility of education. The following sections describe each of these challenges, providing insights from various studies.

6.1 Technical challenges

Organizations are increasingly dependent on advanced systems and innovative solutions to meet operational needs, but this reliance introduces various technical challenges such as integration issues, cybersecurity threats, scalability concerns, and data management complexities. Developing regions often struggle with inadequate technological infrastructure, which is a significant barrier to the integration of educational technology. For instance, in Pakistan, the lack of ICT infrastructure limits access to digital resources and technical facilities, which are crucial for effective technology integration in education (Adnan et al., 2021). Similarly, in Bangladesh, the absence of robust infrastructure is a major impediment to the successful introduction of ICT in classrooms (Khan et al., 2012). In addition, the integration of new technologies often faces compatibility issues with existing legacy systems. This challenge is prevalent in regions where educational institutions rely on outdated technology, making it difficult to adopt new digital tools and platforms (Jain & Aggarwal, 2019).

6.2 Data privacy and security concerns

The adoption of IoT technologies in educational settings, such as in Spanish universities, raises significant cybersecurity concerns. The lack of clear legal frameworks and insufficient digital skills among faculty and students exacerbate these risks, making it challenging to protect sensitive student data. The collection and use of student data for educational purposes raises ethical concerns, particularly regarding consent and data protection. These concerns are heightened in regions with limited regulatory oversight, where the potential for misuse of student data is significant (Wilson, 2017).

6.3 Teacher readiness and training

A critical barrier to technology integration is the lack of digital literacy among educators. In Africa, for example, gaps in digital literacy hinder the effective use of digital tools in pre-service teacher education, impacting the quality of teaching and learning (Ngema & Ajani, 2024). Similarly, in Indonesia, comprehensive teacher training programs are necessary to enhance digital literacy and facilitate the adoption of IT innovations in education (Hadiningrat et al., 2024). Teachers' resistance to adopting new technologies is a common barrier, often stemming from low self-efficacy and increased workload associated with technology use. In the Maldives, for instance, teachers face challenges in implementing blended learning due to resistance and lack of readiness (Ali, 2024). This resistance is also observed in other developing regions, where cultural and institutional factors contribute to hesitancy in embracing technological change.

7. Future trends and innovations in smart classrooms

The future of smart classrooms is poised to be transformed by several key technological innovations, including AI-driven chatbots and virtual tutors, metaverse and extended reality (XR) learning environments, blockchain for secure student records, and smart campus integration.

These advancements promise to enhance educational experiences, improve efficiency, and ensure secure management of educational data. Each of these trends is supported by ongoing research and development, as highlighted in the papers provided.

7.1 Integration of AI-driven chatbots and virtual tutors

The integration of AI-driven chatbots and virtual tutors is revolutionizing education, enhancing personalized learning experiences, and providing immediate support, making quality education more accessible than ever before. AI-driven chatbots and virtual tutors, such as ChatGPT, are revolutionizing educational support by providing personalized learning experiences and real-time feedback. These tools enhance language acquisition, content creation, and assessment processes, while also addressing ethical considerations in their deployment (Ruksana, 2024). The integration of chatbots in metaverse platforms, as demonstrated in university prototypes, offers interactive and personalized support for students and faculty, improving course navigation and administrative efficiency (Xie et al., 2023). AI tools are also empowering educators by assisting in the creation of personalized learning platforms and predictive analytics for early intervention in student performance (Chopra & Arora, 2023).

7.2 Metaverse and extended reality (XR) learning environments

The XR learning environments represent a transformative approach to education, blending immersive technologies to enhance engagement and learning experiences in innovative ways. The metaverse, a virtual world where users interact through avatars, is being leveraged to create engaging and dynamic learning environments. AI-driven virtual avatar assistants enhance user interaction and engagement in these metaverse classrooms (Halim et al., 2024). Virtual reality (VR) technology is being used to construct virtual university campuses, offering on-demand VR lectures and automated evaluations, thereby linking multiple digital campuses for a comprehensive educational experience (Nagao, 2023). The integration of AI, VR, and high-speed internet (6G) is facilitating the development of smart and immersive classrooms, revolutionizing curricula and educational delivery methods (Sharrab et al., 2023).

7.3 Smart campus integration and interconnectivity

Smart campus integration and interconnectivity represents a transformative approach to education, enhancing collaboration, efficiency, and innovation through advanced technologies and seamless connectivity among campus resources and stakeholders. The integration of sustainable IoT technologies in smart classrooms is reshaping educational environments by enhancing learning experiences and improving resource efficiency. These innovations contribute to more adaptive, efficient, and environmentally conscious educational spaces (Çela et al., 2024). The convergence of digital tools and platforms, such as LMS and collaborative tools, is redefining the boundaries of teaching and learning, enabling connections beyond physical spaces (Hassan, 2023).

7.4 Classroom digital twins

Digital twin technology could not only reduce the disparity between real classroom and virtual classroom, but generate more varied new datasets. Classroom digital twin is the bridge between the real and virtual classrooms. Classroom digital twin also is the mirror that reflects the real classroom into the virtual classroom (Pang et al., 2023; Razzaq et al., 2023).

8. Conclusion

The concept of smart classrooms represents a significant shift in educational paradigms, driven by advancements in information technology and the need for more interactive, personalized, and efficient learning environments. Smart classrooms integrate technologies such as big data, artificial intelligence, cloud computing, and IoT to enhance educational experience. These technologies facilitate a more flexible, creative, and adaptive learning process, aligning with the needs of modern students and educators. The future of smart classrooms holds promising implications for global education, with potential for widespread adoption and further innovation. While smart classrooms offer numerous benefits, it is important to consider the potential challenges and limitations associated with their implementation. Issues such as digital equity, data privacy, and the digital divide must be addressed to ensure that all students can benefit from these advancements. Additionally, the rapid pace of technological change necessitates ongoing research and adaptation to keep educational practices relevant and effective.

References

Adnan, K., Shaikh, F. B., Afzaal, R., Ali, A. B. A., & Ponnusamy, V. (2021). Identification of facilitators and barriers of ICT usage in the education sector of Pakistan: A systematic literature review. In N. Jhanjhi, K. Rafique, & V. Ponnusamy (Eds.), *ICT solutions for improving smart communities in Asia* (pp. 27–53). IGI Global Scientific Publishing. https://doi.org/10.4018/978-1-7998-7114-9.ch002

Aggarwal, D., Sharma, D., & Saxena, A. B. (2024). Smart education: An emerging teaching pedagogy for interactive and adaptive learning methods. *Journal of Learning and Educational Policy, 4*(4), 1–9. https://doi.org/10.55529/jlep.44.1.9

Akter, S., & Asaduzzaman, A. (2023). Internet of things and artificial intelligence enabled smart classroom. In *The 6th International Conference on Electrical Information and Communication Technology, Khulna, Bangladesh* (pp. 1–6). https://doi.org/10.1109/eict61409.2023.10427809

Alfarsi, G., Tawafak, R. M., Eldow, A., Malik, S. I., Jabbar, J., & Sideiri, A. A. (2020). Smart classroom technology in artificial intelligence: A review paper. In *Proceedings of the International Conference on Culture Heritage, Education, Sustainable Tourism, and Innovation Technologies, Medan, Indonesia* (pp. 229–235). https://doi.org/10.5220/0010306502290235

Ali, R. (2024). How challenging? Barriers for teachers in institutional implementation of blended learning. *Open Learning: the Journal of Open and Distance Learning*, 1–18. https://doi.org/10.1080/02680513.2024.2342922

Arya, R., & Verma, A. (2024). Role of artificial intelligence in education. *International Journal of Advanced Research in Science, Communication and Technology, 4*(2), 589–594. https://doi.org/10.48175/ijarsct-19461

Ashwin, M., Kumar, E. S., Naidu, R. C. A., & Ramamoorthy, R. (2023). IoT based innovative teaching learning using smart classrooms. In *2023 International Conference on Sustainable Computing and Data communication Systems, Erode, India* (pp. 1143–1148). https://doi.org/10.1109/ICSCDS56580.2023.10104589

Çela, E., Fonkam, M., Eappen, P., & Vajjhala, N. R. (2024). Current trends in smart classrooms and sustainable internet of things. In S. G. Thandekkattu, & N. R. Vajjhala (Eds.), *Designing sustainable internet of things solutions for smart industries* (pp. 1–26). IGI Global Scientific Publishing. https://doi.org/10.4018/979-8-3693-5498-8.ch001

Chen, J., & Liu, H. (2024). Effects of smart classroom on students' learning outcomes: A meta-analysis. *International Journal of Web-Based Learning and Teaching Technologies, 19*(1), 1–16. https://doi.org/10.4018/ijwltt.356509

Chopra, D., & Arora, P. (2023). Smart classrooms, smarter teachers: A deep dive into AI-driven educational enhancement. *Journal of Informatics Education and Research, 3*(2). https://doi.org/10.52783/jier.v3i2.479

Dimitriadou, E., & Lanitis, A. (2023). A critical evaluation, challenges, and future perspectives of using artificial intelligence and emerging technologies in smart classrooms. *Smart Learning Environments, 10* (12). https://doi.org/10.1186/s40561-023-00231-3

Ghonaim, W., & Ghamdi, A. A.-G. (2018). A smart class hyperlearning and management suite. In *2018 International Conference on Computational Science and Computational Intelligence, Las Vegas, NV, USA* (pp. 555–561). https://doi.org/10.1109/CSCI46756.2018.00113

Hadiningrat, K. P. S. S., Silalahi, V. A. J. M., & Wardani, F. P. (2024). Opportunities and challenges in implementing information technology innovations in the Indonesian education sector. *EAJMR East Asian Journal of Multidisciplinary Research, 3*(8), 3763–3776. https://doi.org/10.55927/eajmr.v3i8.10686

Halim, N. A. A., Azlan, M. H., Ismail, A. W., Fazli, F. E., Ahmad, M. A., & Aladin, M. Y. F. (2024). Edu-metaverse classroom with AI-driven virtual avatar assistant. In *The 5th International Conference on Smart Electronics and Communication (ICOSEC), Trichy, India* (pp. 1548–1554). https://doi.org/10.1109/icosec61587.2024.10722453

Hassan, G. (2023). Technology and the transformation of educational practices: A future perspective. *International Journal of Economic, Business, Accounting, Agriculture Management and Sharia Administration, 3*(1). https://doi.org/10.54443/ijebas.v3i1.1136

Huang, C. (2023). Smart classroom empowering students' growth and development: A case study of Zhongliang Xiangyun Branch of Chengdu Longjiang Road Primary School. *Science Insights Education Frontiers, 15*(1), 14. https://doi.org/10.15354/sief.23.s1.ab014

Jain, A., & Aggarwal, R. (2019). Studying the interrelationship amongst barriers for implementation of technology based educational techniques in teaching institutions in India. *International Journal of Computer Application, 178*(19), 5–12. https://doi.org/10.5120/IJCA2019918936

Kalyani, D. L. K. (2023). Revolutionizing education: Artificial intelligence's pioneering role in shaping tomorrow's scholars. *International Journal of Multidisciplinary Research in Arts, Science and Technology, 1*(2), 1–7. https://doi.org/10.61778/ijmrast.v1i2.6

Kaur, A., & Bhatia, M. (2024). Smart classroom: A review and research agenda. *IEEE Transactions on Engineering Management, 71*, 2430–2446. https://doi.org/10.1109/TEM.2022.3176477

Khan, M. S. H., Hasan, M., & Clement, C. K. (2012). Barriers to the introduction of ICT into education in developing countries: The example of Bangladesh. *International Journal of Instruction, 5*(2), 61–80. https://www.researchgate.net/publication/281278856_Barriers_to_the_Introduction_of_ICT_into_Education_in_Developing_Countries_The_Example_of_Bangladesh#fullTextFileContent.

Liu, Y. (2024). Analysis of the path of integrating smart classroom into traditional classroom teaching. *New Explorations in Education and Teaching, 2*(5), 30–32. https://doi.org/10.18686/neet.v2i5.4621

Long, T. (2024). Smart classroom in the era of education informatization 2.0. *International Journal of Education and Humanities, 12*(2), 279–282. https://doi.org/10.54097/sm67dx71

Mohanty, D., Kumar, S., Thongam, R., Kawale, S. R., Jakkani, A. K., & Vani, V. G. (2024). Enhancing classroom engagement through IoT-enabled smart learning environments. *ShodhKosh: Journal of Visual and Performing Arts, 5*(1), 1003–1010. https://doi.org/10.29121/shodhkosh.v5.i1.2024.2591

Nagao, K. (2023). Virtual reality campuses as new educational metaverses. *IEICE - Transactions on Info and Systems, E106-D*(2), 93–100. https://doi.org/10.1587/transinf.2022eti0001

Ngema, T., & Ajani, O. A. (2024). Exploring digital transformation in preservice teacher education in Africa: Prospects, challenges, and implications for sustainable lifelong learning. *International Journal of Innovative Technologies in Economy, 3*(47). https://doi.org/10.31435/rsglobal_ijite/30092024/8236a

Okhanashvili, S. (2024). Artificial intelligence in education. *In International Scientific Conference "Education, Research, Practice" Proceedings, 5*, 71–78. https://doi.org/10.52340/erp.2024.05.06

Pang, Z., Zhao, H., Tang, Z., & Kong, F. (2023). A digital twins-based smart classroom monitoring system. In *2023 IEEE International*

Conference on Advanced Learning Technologies (ICALT) (pp. 347–349). IEEE. https://doi.org/10.1109/ICALT58122.2023.00107

Peirats-Chacón, J. (2023). The smart classroom: Learning challenges in the digital ecosystem. *Education Sciences, 13*(7), 662. https://doi.org/10.3390/educsci13070662

Pradeep, A. (2023). Developing an IoT enabled smart classroom S3TH system. In *The 5th International Conference on Inventive Research in Computing Applicationsm, Coimbatore, India* (pp. 1374–1379). https://doi.org/10.1109/icirca57980.2023.10220862

Razzaq, S., Shah, B., Iqbal, F., Ilyas, M., Maqbool, F., & Rocha, A. (2023). DeepClassRooms: A deep learning based digital twin framework for on-campus class rooms. *Neural Computing & Applications*, 1–10. https://doi.org/10.1007/s00521-021-06754-5

Ren, F. (2024). The construction and application of smart classrooms in China: A literature review based on 93 studies. *Science Insights Education Frontiers, 23*(1), 3705–3724. https://doi.org/10.15354/sief.24.re385

Routray, S. K., & Mohanty, S. (2022). Smart classroom-based innovative solution toward uninterrupted education: Perspective. *International Journal of Virtual and Augmented Reality, 6*(1), 1–14. https://doi.org/10.4018/ijvar.306689

Saini, M., & Goel, N. (2019). How smart are smart classrooms? A review of smart classroom technologies. *ACM Computing Surveys, 52*(6), 1–28. https://doi.org/10.1145/3365757

Sharrab, Y. S., Almutiri, N. T., Tarawneh, M., Alzyoud, F. Y., Al-Ghuwairi, A.-R., & Al-Fraihat, D. (2023). Toward smart and immersive classroom based on AI, VR, and 6G. *International Journal of Emerging Technologies in Learning (Ijet), 18*(2), 4–16. https://doi.org/10.3991/ijet.v18i02.35997

Shayer, K. S., Medul, M. H., Badoruzzaman, M., Shuvo, J. I., Rabbu, M., & Haque, F. M. M. (2024). An integrated framework for enhanced learning environments: IoT-driven smart classrooms with multi-layered security protocols and adaptive infrastructure. In *2024 International Conference on Advances in Computing, Communication, Electrical, and Smart Systems, Dhaka, Bangladesh* (pp. 1–6). https://doi.org/10.1109/icaccess61735.2024.10499605

Ruksana, T. P. (2024). AI-driven pedagogy: Unveiling ChatGPT's influence in education. *Journal of Applied Science, Engineering, Technology and Management, 2*(01), 09–14. https://doi.org/10.61779/jasetm.v2i1.2

Vashishth, T. K., Sharma, V., Sharma, K. K., Kumar, B., Chaudhary, S., & Panwar, R. (2024). AIoT in education transforming learning environments and educational technology. In S. Rezaei, & A. Ansary (Eds.), *Artificial Intelligence of Things (AIoT) for productivity and organizational transition* (pp. 72–107). IGI Global Scientific Publishing. https://doi.org/10.4018/979-8-3693-0993-3.ch004

Vashishth, T. K., Sharma, V., Sharma, K. K., Kumar, B., Chaudhary, S., & Panwar, R. (2024). Transforming classroom dynamics: The social impact of AI in teaching and learning. In Z. Ahmed, A. Hassan, & R. Saeed (Eds.), *AI-Enhanced teaching methods* (pp. 322–346). IGI Global Scientific Publishing. https://doi.org/10.4018/979-8-3693-2728-9.ch015

Wilson, N. (2017). Sociotechnical and pedagogical barriers to technology integration. In M. Grassetti, & S. Brookby (Eds.), *Advancing next-generation teacher education through digital tools and applications* (pp. 19–37). IGI Global Scientific Publishing. https://doi.org/10.4018/978-1-5225-0965-3.CH002

Xie, Q., Lu, W., Zhang, Q., Zhang, L., Zhu, T., & Wang, J. (2023). Chatbot integration for metaverse - A university platform prototype. In *2023 IEEE International Conference on Omni-Layer Intelligent Systems, Berlin, Germany* (pp. 1–6). https://doi.org/10.1109/coins57856.2023.10189232

Zhang, X., Ding, Y., Huang, X., Li, W. D., Long, L., & Ding, S. (2024). Smart classrooms: How sensors and AI are shaping educational paradigms. *Sensors, 24*(17), 5487. https://doi.org/10.3390/s24175487

Chapter 3

Introduction to virtual smart classrooms, schools, and campuses

Eylül Akboğa[1], Gizem Merve Demir[1] and Tufan Ekin[2]

[1]*Department of Information Systems and Technologies, Faculty of Computer and Information Sciences, Yeditepe University, Istanbul, Turkey;*
[2]*Department of Management Information Systems, Faculty of Computer and Information Sciences, Yeditepe University, Istanbul, Turkey*

The subject of virtual smart classrooms, smart schools, and smart campuses will be covered according to this plan:

1. Definition and scope
2. Implementation examples
3. Benefits
4. Potential disadvantages and limitations
5. Technologies enabling virtual smart classrooms/schools/campuses
6. Future innovations
7. Impacts on society, economy and sustainability

The terms "smart classroom", "smart school", and "smart campus" are frequently used in the literature. Adding "virtual" adjective to these terms emphasizes the role of virtual technologies. These terms with or without "virtual" are used interchangeably in this chapter.

1. Definition and scope

The developments in technology have had a great impact on education. These effects have led to the emergence and frequent use of virtual smart classrooms, schools, and campuses.

A virtual smart classroom is made up of various components that work together to provide an interactive and engaging learning environment that enhances teaching techniques, develops students' abilities, raises their academic level and allows them to engage more actively in the learning process. Different technology components—such as Virtual Reality (VR), Augmented Reality (AR), management systems, and mobile computing— contribute to a virtual smart classroom. In addition, innovations in pedagogy regarding content delivery, student engagement and assessment contribute to making classrooms smart (Kaur et al., 2022; Wang & Keat, 2024). The scope of a virtual smart classroom can be outlined in four main aspects (Saini & Goel, 2020):

- Smart content which includes especially rich and interactive multimedia,
- Smart interaction and engagement,
- Smart assessment,
- Smart physical environment such as ideal temperature and humidity control.

A smart school is an educational institution that utilizes technology to enhance teaching and learning, aiming to prepare students for the information age (Omidinia et al., 2013). Schools are no longer confined to traditional methods of teaching and administration. Instead, they are evolving into "smart schools" that leverage technology to enhance learning outcomes, streamline operations, and prepare students for a digital future. The fundamental distinction between smart schools and traditional schools lies in their approach to student development and the integration of technology into the educational process. Unlike traditional schools, which primarily rely on printed textbooks and standardized tests, smart schools move beyond a fixed curriculum by incorporating technology-driven methodologies. This allows them to equip students with the skills needed to effectively utilize technological applications, even if such instruction is not explicitly included in formal coursework.

A smart campus is a growing trend where educational institutions integrate advanced technologies with physical infrastructure to enhance learning experiences, support decision-making and promote campus sustainability. Smart campus, which is an improved version of the digital campus, creates a smart learning environment by enabling seamless access to educational resources, fostering interactive and data-driven learning experiences, and

Digital Twins in the Smart Classroom. https://doi.org/10.1016/B978-0-443-41632-3.00029-1

optimizing campus services for students and faculty. As the high-end form of a smart education system, it combines cloud computing and IoT (Internet of Things) to support university management, teaching, research, and various academic activities. Moreover, real-time data analytics and virtual technologies enable more informed decision-making, improved planning, better resource allocation, and personalized learning pathways for students (Dong et al., 2020; Min-Allah & Alrashed, 2020). The term "smart" stands for self-directed, motivated, adaptive resources and technology reflecting the core principles of a smart campus that leverages advanced technologies to create an innovative, efficient, and student-centered learning environment. Smart campuses are designed to enhance student experience, operational efficiency and teaching (Imbar et al., 2020, pp. 1–7). In addition, they support collaboration between students, faculty and administrators by creating a more interactive, innovative and virtual campus experience through the use of VR, AR, and digital twin technologies (McGrath et al., 2023). Moreover, they also help drive sustainable urban growth, improve energy efficiency, and enhance the quality of life for everyone involved. By integrating these technologies, smart campuses not only boost operational efficiency but also create a space where students are better equipped to face the challenges of a fast-changing, tech-driven world (Bellaj et al., 2024, pp. 1–7).

2. Implementation examples

There are a lot of implementation examples of virtual smart classrooms, schools and campuses. Some of them are covered below.

2.1 Implementations using VR and AR

Virtualization continues to shape the future of smart schools, offering adaptable, cost-effective, and innovative solutions for modern education. It played a critical role in maintaining academic continuity during the COVID-19 pandemic by enhancing online engagement and collaboration, ensuring that students and educators could sustain interactive learning experiences despite physical distance (Infante-Moro et al., 2022). Furthermore, virtualization supports virtual labs and remote research, enabling students to conduct experiments without physical constraints, thereby expanding opportunities for scientific exploration and practical learning in digital environments (Verbytska and Syzonenko, 2020). Some specific implementation examples are provided in the following paragraphs:

A virtual smart classroom was used in the field of digital media education in a university in China and it increased the effectiveness of course goals. An AR/VR teaching system was used successfully as a project-based learning environment in this example (Zhu et al., 2022, pp. 37–42).

A blended teaching model combining smart classrooms and virtual simulation training was implemented in the education of nursing undergraduates in China. As a result, the academic performance, individualistic learning capability as well as teaching effectiveness improved due to this new model (Meng et al., 2023).

Another research from Thailand proved that a virtual smart classroom system developed learning, innovation and creativity skills of undergraduate students from different fields such as education, agriculture, and sports science (Songkram et al., 2021).

An example of improvements in conversational skills, vocabulary and grammar retention by the use of AR and VR in language education is reported in Malaysia (Annamalai et al., 2023).

VR significantly enhances the student experience by providing interactive opportunities. Stanford Library has introduced VR workstations equipped with Oculus and HTC Vive, which are advanced wearable devices that allow users to fully experience simulated environments. By incorporating virtual reality, the library enhanced the educational experience and promoted technological advancement within the university environment. Additionally, 3D printers transformed digital design ideas into physical objects within the campus. These, along with other advanced technologies, allowed students to engage in immersive experiences that simulated campus life. This initiative reflects a proactive approach to integrating emerging technologies into academic resources, providing opportunities to students for exploration and learning (Guo et al., 2024).

Another VR application that is used to create a smart campus is the virtual map service. These services have become an essential feature in many universities, allowing students to remotely explore campus spaces. For example, Clemson University Library offers a visual GIS (Geographic Information System) service that provides a fully 3D virtual tour of the entire university. This service gives students the opportunity to virtually visit campus and gain a comprehensive understanding of the university's layout, even if they are unable to visit in person. This example demonstrates how this technology transforms the campus into a virtually enhanced smart environment (Guo et al., 2024).

2.2 Implementations using Digital Twins (DTs)

Digital Twins (DTs)—virtual replicas of physical environments—enhance school management and personalized learning by providing dynamic, data-driven educational experiences.

One key application of DTs is their use in virtual science labs, which enable safe and cost-effective experiments, allowing students to explore scientific concepts without the risks associated with physical experimentation (Schleich et al., 2017). For example, DTs are used in virtual chemistry labs equipped with IoT sensors to enable students to conduct simulated experiments with real-time feedback, bridging theoretical and practical learning (Shao & Helu, 2020).

Additionally, DTs contribute to energy-efficient management by utilizing IoT sensors to optimize energy consumption, ensuring that school facilities operate sustainably while reducing overall costs (Poyyamozhi et al., 2024).

2.3 Implementations using AI

AI is one of the leading driving factors of technological developments. Hence, we will provide examples about AI-powered learning and its use as an administrative assistance in schools.

Massachusetts Institute of Technology (MIT) DAILy Workshop aimed to improve AI literacy among middle school students. Developed by the MIT Scheller Teacher Education Program (STEP) and MIT Media Lab, this 30-hour curriculum introduced students to machine learning and neural networks through interactive tools like Google's Teachable Machine. The program not only strengthened technical skills but also highlighted ethical considerations in AI. Findings indicate that students significantly improved their understanding of AI concepts and interest in AI-related careers (Zhang et al., 2023).

Arizona State University (ASU) has integrated AI-powered voice assistants to enhance student engagement and administrative efficiency. By providing students with Amazon Alexa devices, ASU streamlines access to academic resources, schedules, and campus navigation, reducing reliance on physical guides and multiple website searches. This initiative showcases how AI-driven automation can optimize information delivery and accessibility (Nikai Salcido, 2017).

AI is increasingly integrated into medical training, particularly in diagnostics, ethical decision-making, and critical thinking development. AI-powered tools support students in various aspects of medical education, including disease diagnosis through AI-based image analysis, which enhances accuracy and efficiency in identifying medical conditions. Additionally, simulated medical scenarios refine diagnostic accuracy by allowing students to practice clinical decision-making in a controlled environment. Furthermore, ethical case studies enable students to analyze AI-assisted diagnoses, helping them understand the implications of technology-driven medical assessments (Alam et al., 2023; Katznelson & Gerke, 2021).

2.4 Implementations using hardware and software virtualization

Virtualization enhances digital learning, accessibility, and resource efficiency by enabling schools to leverage technology for more flexible and scalable educational experiences. One major application of virtualization is its role in optimizing infrastructure and learning. One research indicates that 26.3% of schools have adopted virtualization specifically for cloud-based learning and administrative processes, allowing them to streamline operations and improve digital accessibility (Klement, 2017).

2.5 Implementations using The Internet of Things (IoT)

The Internet of Things (IoT) revolutionizes smart schools by improving energy efficiency, student engagement, and Science, Technology, Engineering, and Mathematics (STEM) education. One of its key applications is sustainability and energy efficiency, demonstrated by the Green Awareness in Action (GAIA) project, which successfully reduced energy consumption by 54% in an Italian high school, highlighting IoT's role in sustainable resource management (Mylonas et al., 2023). IoT plays a role in shaping the future of smart schools, ensuring a more interactive, efficient, and technologically advanced educational environment.

Wageningen University of Netherlands implemented IoT technology to improve the use of campus space by introducing a real-time occupancy monitoring system in educational areas. By installing sensors and Wi-Fi tracking systems, the university collected data on how classrooms and study areas were being used, providing valuable insights into actual occupancy patterns. This data enabled university administrators to compare real-time occupancy rates with planned usage, helping to identify inefficiencies such as underused or overcrowded spaces. By incorporating IoT-based analytics into its decision-making processes, the university was able to adjust space allocations in real time, optimize scheduling and ensure that educational facilities were utilized more effectively across the campus (Valks et al., 2020).

Another example is an IoT-based platform to optimize indoor environmental conditions and occupant well-being. The system continuously monitored key environmental factors, including air quality, temperature, noise levels, and lighting conditions, while also incorporating user feedback. This real-time data was utilized to dynamically adjust building settings, maintaining optimal environmental conditions for students and faculty. By integrating IoT-driven insights with facility management strategies. The platform contributed to enhanced energy efficiency and an improved user experience in university spaces (Romero Herrera et al., 2020).

Arizona State University (ASU) has implemented smart campus innovations within its football stadium, transforming it into a data-driven and highly connected environment. By utilizing smart sensors and AI-powered cameras, the university enhances operational efficiency and optimizes resource management. These IoT-enabled systems collect real-time data on water consumption, sales and noise levels. This data enables administrators to make informed decisions regarding lighting, utility usage, and crowd management (Manning, 2020).

2.6 Implementation using face recognition technology

One notable area of study is the application of face recognition technology to enhance campus security, automate administrative tasks, and improve user experience. Chinese universities have rapidly adopted face recognition technology as a core component of their smart campus infrastructures. This integration has significantly enhanced security, operational efficiency, and overall convenience for both students and staff. Leading institutions such as Peking University and Tsinghua University have pioneered the integration of biometric authentication systems, enabling seamless access to campus facilities without the need for traditional ID cards or passwords. In addition to strengthening campus security, these systems facilitate the automation of attendance tracking and enable cashless transactions. They also offer personalized services, fostering a more interconnected, data-driven academic environment that enhances both the administrative experience and overall campus life (Dai, 2019).

2.7 Other implementations

Sustainable transportation has become a key focus in the development of smart campuses. İstanbul Technical University (İTÜ) of Turkey Ayazağa Campus has implemented a smart bike-sharing system as part of its sustainable smart campus initiatives. The aim was to minimize carbon emissions and promote eco-friendly transportation. The campus is developing a 6-kilometer-long bicycle and pedestrian route, integrating 180 shared bicycles, including 10 pedal-assisted electric bikes, into its transportation infrastructure. These bicycles are placed across the campus. Their locations are strategically chosen, taking pedestrian and vehicle traffic flow into account. Through the mobile application, students and staff can rent bikes from designated docking stations and return them to any bike-sharing station on campus. This initiative not only improves sustainable mobility but also contributes to efficient and environmentally conscious smart campus transportation (*İTÜ'de Paylaşımlı Bisiklet Devri (Shared Bicycle Era at Istanbul Technical University)*, 2018).

Students at one of the leading technical universities of Turkey, Gebze Technical University (GTU), developed sustainability-focused projects as part of their Smart Campus Project. These projects include a vehicle-sharing application to reduce traffic and emissions, a second-hand product exchange platform to minimize waste and a resource tracking system to monitor energy and water usage. By merging technology and sustainability, these initiatives contribute to a greener campus while encouraging students to create a more environmentally friendly smart campus (Gebze Technical University Office of Press and Public Relations, 2024).

Universities are increasingly integrating smart technologies into their campuses to create an efficient learning environment and enhance the campus experience. Yonsei University of the Republic of Korea has proactively embraced these advancements through its smart campus project. Students receive real-time notifications about cultural, artistic and academic events through a mobile calendar and can access related content via online platforms. Additionally, the project enhances smart mobility and campus services by enabling students to use mobile student ID cards for library access, check real-time shuttle bus locations and monitor cafeteria occupancy. By implementing these initiatives, Yonsei University aims to create a fully integrated, tech-driven campus that boosts accessibility, efficiency, and student involvement (Director of Academic Information Center, 2025).

3. Benefits

We will cover the benefits of using smart classrooms, schools and campuses in this part.

3.1 Increased learning efficiency and safety

Conventional classroom settings frequently encounter limitations in terms of resources, such as instructors, educational materials, and apparatus. Virtual classrooms can overcome these limitations. Digitalization and cloud computing technologies enable access to extensive educational resources and high-quality educational materials worldwide, thereby increasing learning opportunities.

Research found that integrating smart technology in classrooms transformed traditional lecture-based methods into more dynamic, engaging, and student-centered learning environments, fostering critical thinking, creativity, and collaboration among students (Madhur et al., 2024).

The use of interactive tools like smart boards facilitates collaborative learning, enables access to online resources, and supports the use of educational software, thereby enhancing student engagement and improving learning outcomes (Stealey, 2025).

AI-driven systems enable educators to tailor instruction to individual student needs, enhancing engagement and academic outcomes. For instance, AI-powered intelligent tutoring systems (ITS) utilize machine learning algorithms to adapt to each student's unique learning preferences and styles, creating individualized learning paths that cater to their specific requirements and pace (Bhushan et al., 2023).

The use of metaverse technology can provide students with the opportunity to carry out experiments, operations, and other activities in a virtual setting, thereby circumventing the constraints imposed by limited experimental equipment, materials, and resources. Experiments can also benefit from increased efficiency and safety measures. Moreover, metaverse technology enables the provision of customized learning routes and resource suggestions for students via intelligent data analysis and recommendation software, thereby enhancing students' learning productivity (Li et al., 2023, pp. 196–202).

In smart schools, both teachers and students have immediate access to a vast array of online resources, facilitating real-time project work and independent learning, thereby eliminating the need for physical library visits and enabling a more flexible educational experience. This accessibility allows for a more personalized and efficient learning environment, as digital devices and resources enable students to engage in various learning modalities. A study examining the effect of smart schools on students' learning and teaching highlighted that the integration of technology in educational settings enhances the development and speed of the teaching-learning process. Furthermore, the availability and utilization of online learning resources among undergraduate students have been shown to significantly contribute to teaching and learning, providing flexibility and immediate access to information. These findings underscore the importance of incorporating digital resources in smart schools to support diverse learning needs and promote academic success (Kofo et al., 2022).

Smart campuses can enhance traditional campus experiences by incorporating virtual technologies such as AR, VR and digital twins. This will provide students with interactive and immersive environments that exceed physical limitations. These technologies enable remote access to labs, libraries, and collaborative spaces, promoting equality among students (Finnley, 2023).

3.2 Cost-effectiveness

Implementing technology in education can lead to cost savings over time by reducing the need for physical materials like textbooks and paper, and by streamlining administrative tasks, thereby enhancing efficiency and cost-effectiveness. A study evaluating the cost-effectiveness of virtual and traditional education models found that virtual education, either alone or combined with traditional methods, was as cost-effective as or more so than traditional education. Additionally, research on the cost-benefit of online learning indicates that, despite initial infrastructure investments, online education can be more cost-effective in the long term due to reduced operational costs. These findings suggest that integrating technology into educational settings can result in significant cost efficiencies over time (Pakdaman et al., 2021).

3.3 Effective school and campus management

Effective school management plays a crucial role in optimizing both administrative processes and student outcomes. Smart school management platforms streamline operations for students and administrative staff. The integration of smart school management platforms that collect and analyze real-time data is transforming educational institutions by streamlining operations and personalizing learning experiences (Bhushan et al., 2023).

Smart campus technologies are used in increasing student engagement and fostering a sense of community. Interconnected systems allow students to easily communicate with peers, faculty and administrators through digital platforms and mobile applications. Real-time updates, event notifications and personalized alerts ensure that students are always aware of academic deadlines, campus events and opportunities to get involved. This creates a dynamic environment where students can actively engage in campus life, collaborate on projects and participate in extracurricular activities. As a result, smart campus technologies are making communication more efficient, fostering a more active and connected student experience, and contributing to a stronger campus community (The Sustainability Directory, 2025).

The use of virtualization technologies in smart campuses enables predictive insights, improvements and optimized planning. By utilizing VR, AR and digital twin technologies, universities can simulate campus environments, analyze potential problems, and optimize resource allocation before implementing physical changes. VR campus tours enable students to explore the university remotely. Moreover, VR headsets can be used for gaming and social engagement, further enhancing the campus experience. In addition, AR can be a valuable tool for students with special needs, aiding in their learning process by enhancing their comfort, confidence and overall abilities in life (Mokmin & Rassy, 2024).

Finally, when it comes to security, smart campus technologies play a major role in enhancing security and emergency response systems. AI-based surveillance, biometric authentication and automated access control ensure that only authorized individuals can enter certain areas, reducing security risks. Real-time monitoring and alert systems enable faster response to potential threats, while emergency communication platforms provide instant notification to students and staff in critical situations. In addition, smart access systems in libraries and research facilities streamline resource management by preventing unauthorized borrowing and ensuring the availability of materials for students. By integrating these advanced security measures, smart campuses foster a safer and more efficient environment where students can focus on their education without concerns about their well-being or access to academic resources (Abdullah et al., 2019).

3.4 Enhancing professional development of educators

Smart schools not only transform the student experience but also enhance the professional development of educators, as they shift toward a role of facilitator and coordinator in smart learning environments (Samsudin, 2024). In traditional settings, teachers often deliver the same subject matter across different classes each year, with only minor adjustments to the curriculum. While the core content remains largely unchanged, this repetitive nature can lead to reduced engagement for both teachers and students. However, even if the fundamental curriculum remains static, smart schools leverage technology to make teaching methodologies more dynamic and interactive. By integrating digital tools, educators can personalize the learning experience based on individual student needs, allowing them to tailor instruction to suit each student's learning pace and preferences. Moreover, adaptive learning platforms enable teachers to use data analytics to track student progress and adjust instructional methods accordingly, ensuring that the teaching approach remains flexible and responsive to student performance. Additionally, AI-assisted resources help incorporate real-world applications of theoretical knowledge, allowing students to connect classroom concepts with practical problem-solving scenarios (Contrino et al., 2024).

3.5 Improved sustainability

Sustainability and environmental benefits are key benefits behind the adoption of smart campus solutions. By using data analytics and automation, campuses can optimize resource consumption, minimize waste and implement green initiatives more effectively. Smart monitoring systems track energy and water usage patterns, enabling data-driven decisions that enhance sustainability efforts. In the same way, smart tree management systems monitor moisture levels and environmental conditions to improve watering and support the greening of the campus. Additionally, digitalization reduces the reliance on paper-based administrative processes, contributing to a more eco-friendly campus and creating a healthier and more cost-effective learning environment for students and faculty (Moraes et al., 2020, pp. 111–120). Adopting energy-efficient strategies, such as renewable energy technologies, not only promotes sustainability but also supports the effective functioning of smart schools (Zeeshan et al., 2022). Main benefits of a smart campus are enhanced resource efficiency and streamlined management. IoT based monitoring systems collect real-time data on space utilization, energy consumption and occupancy rates. This allows a campus to allocate its resources in a more efficient way. Automated lighting, heating and water management systems reduce waste and costs while contributing to sustainability goals (Cata, 2015, pp. 195–197). In addition, smart campus technologies are improving mobility and accessibility with optimized transport systems, real-time shuttle bus tracking, smart parking systems, making campus navigation easier. These innovations work together to make campus life more efficient for students (*Smart Campuses*, 2019).

4. Potential disadvantages and limitations

In addition to all of the benefits, there are some potential disadvantages and limitations of virtual smart classroom/school/campus systems which is summarized in this chapter. Implementors must address these to ensure that technology enhances learning while minimizing unintended negative consequences.

4.1 Privacy and security concerns

One major concern is privacy and security. The widespread use of IoT devices in smart campuses often results in the collection of sensitive data, such as personal information and location data. This increase in the number of smart devices is a vulnerability that can have an impact on campus security. In addition, the scale and complexity of the campus network makes network monitoring and management extremely difficult, which can further amplify potential threats (Ikrissi & Mazri, 2020).

4.2 Financial cost

Implementing a smart campus comes with substantial financial costs. The investment in IoT devices, VR and AR technologies and necessary infrastructure to support a smart campus requires a significant budget. Additional costs such as maintenance, system updates and training students and staff can increase running costs. These expenses may limit further advancements and challenge the sustainability of the existing budget and innovations (Feridun & Bayraktar, 2024).

4.3 Health and social implications of prolonged device usage

Excessive screen time has been linked to eye strain, musculoskeletal issues, and sleep disturbances. Digital fatigue among students can also affect their overall well-being and academic performance (Buabbas et al., 2020).

The increased use of virtual interactions may lead to a decline in face-to-face socialization among students, which is an increasingly concerning issue (Villanueva, 2024).

Learners might get confused at installation or configuration phases of AR and VR devices, technical limitations of hardware such as the weight of wearable devices should also be taken into account (Elkoubaiti & Mrabet, 2018, pp. 334–341).

4.4 Distraction and reduced academic performance

Unregulated use of smartphones and digital devices in classrooms can lead to increased distractions, making it difficult for students to focus on lessons. Studies suggest that excessive device usage negatively impacts learning outcomes. For instance, research has shown a significant negative correlation between in-class cell phone use and overall course grades, with frequent device usage corresponding to a drop in academic performance. Additionally, studies have found that students who engage in texting during class sessions tend to have lower test scores, indicating that mobile phone use in class impairs comprehension and performance. These findings underscore the potential detrimental effects of unregulated digital device usage on student learning (Duncan et al., 2012).

4.5 Digital divide and accessibility issues

Not all students have equal access to advanced technology, leading to disparities in education. The digital divide disproportionately affects students from low-income families, limiting their ability to benefit from smart school initiatives. Research indicates that students from low-income households and minority groups are less likely to have access to digital technologies, resulting in lower academic achievement and poorer educational outcomes. Additionally, students without home Internet access face challenges in developing necessary digital literacy skills required in the 21st century. These disparities underscore the need for targeted interventions to bridge the digital divide and promote educational equity (Afzal et al., 2023).

4.6 Challenges in teacher training

Educators often receive inadequate training on how to use advanced educational technologies effectively. This lack of professional development can hinder the effective implementation of smart school initiatives and increase teacher stress (Strom, 2021).

4.7 Environmental and energy impact

Another critical issue associated with smart campus implementations is their environmental and energy impact. The large-scale deployment of smart technologies can lead to increased energy consumption. IoT devices, sensors, cameras and automated systems require continuous power to operate, significantly increasing the overall energy demand of the campus. This can be problematic in regions where energy resources are limited, leading to higher operating costs and a larger carbon footprint (Madera Doval, 2022).

4.8 Other issues

Some of the remaining disadvantages of smart classrooms, schools and campuses can be summarized as follows (Kambala & Mathe, 2022):

i. It might be difficult to create content.
ii. Demands a robust and reliable network.
iii. Susceptible to errors and failures of technological devices.

Although there are several disadvantages and limitations associated with implementing a smart classroom/school/campus, if managed properly, the benefits will overwhelmingly outweigh these challenges.

5. Technologies enabling virtual smart classrooms/schools/campuses

Educational institutions should be racing toward making their campuses smart in response to the global digitalization trend. To achieve this, they make use of various technologies and innovations. These are summarized below.

5.1 VR and AR

VR and AR technologies, commonly referred to as immersive technologies, are positively transforming campuses. As these technologies continue to advance, they provide innovative solutions that enrich the campus experience for both students, staff and faculty. With VR, immersion is achieved through wearable devices, such as haptic gloves and VR headsets, which help make abstract concepts and knowledge more tangible, thereby enhancing learning, socialization and resource management within the campus. Unlike VR, AR is a form of experience where the real environment is enhanced with virtual information, using devices such as smartphones or tablets that capture the campus. It offers only partial immersion and increases the amount of information that the users can take from the environment. As a result, both VR and AR technologies contribute to learning by encouraging participation, enhancing interactivity and boosting students' motivation and attention, ultimately creating a virtual and dynamic campus experience (Zhang et al., 2022).

5.2 IoT, sensor, and blockchain

IoT and sensor technologies are among the most influential and effective technologies in the implementation of smart campuses. They drive the rapid growth of smart campuses while offering numerous opportunities to enhance students' learning experiences and better campus experience. IoT, which refers to interconnection between devices, enables real time data collection and analysis, resulting in a smart environment. Sensors play a crucial role in collecting data about temperature, water, motion and light levels, as they enable the detection of events or changes in some states. For example, smart lighting, temperature regulation and energy monitoring systems rely heavily on IoT and sensors to optimize campus operations and reduce resource consumption, contributing to a sustainable and virtually enhanced campus environment (Arenas et al., 2022, pp. 1–7). Additionally, the integration of IoT and blockchain technology significantly enhances data management and security within educational institutions (Agarwal et al., 2021). The use of blockchain for security and privacy in student sensor networks enhances smart campus applications that need transparency, immutability, privacy and security of student data (Fernández-Caramés & Fraga-Lamas, 2019).

5.3 Digital Twins

Digital Twins play a role in shaping educational infrastructures by creating virtual replicas of physical environments, enabling real-time monitoring, simulation, and data-driven decision-making (Pathak & Upadhyay, 2024). Digital twins also provide an advanced approach to optimizing campus energy consumption and reducing emissions (Luying et al., 2024). Moreover, the data generated by Digital Twin helps identify potential issues and predict maintenance needs through its ability to analyze and evaluate effectively. Therefore, Digital Twin technology is used to improve efficiency, incorporate virtuality and enhance the overall quality of life for students, faculty, and staff within the campus (Pexyean et al., 2024).

5.4 Other technologies

Other Technologies enabling smart classrooms, schools and campuses are: Cloud Computing (CC), Wireless Sensor Networks (WSN), Big Data Analytics (BDA), Compressed Sensing (CS), 5G and 6G networks (Memos et al., 2020, pp. 174–178).

Additionally, Artificial Intelligence (AI), face recognition and hardware/software virtualization technologies are used in implementations. AI use cases are extensively provided in "implementation examples" and "future innovations" parts of this chapter.

6. Future innovations

As education systems continue to evolve, the concept of smart classroom/school/campus is expanding beyond current technological applications. The integration of artificial intelligence (AI), the Internet of Things (IoT), and data-driven decision-making has already begun transforming educational environments. However, to fully optimize learning experiences, administrative processes, and student well-being, further advancements are necessary. The following innovations represent realistic and research-backed developments that could be implemented in smart learning environments in the coming years. These concepts aim to enhance personalized learning, security, career development, and inclusivity, ensuring that education systems remain adaptable and effective in addressing diverse student needs.

6.1 VR, 6G and Symmetric Deep Q-Network (SDQN) combination

VR, 6G (sixth generation of mobile network technology) and (SDQN) were used to create an optimized learning experience and results showed that learning effectiveness increased. This combination offered a new standard for interactive learning environments (Liu, 2024).

6.2 IoT-based real-time school safety and security system

School safety systems currently operate reactively, responding to incidents after they occur. Future smart schools should incorporate real-time monitoring and predictive security systems using IoT sensors and AI analytics. A school-wide IoT-based security system would utilize biometric scanners and AI-assisted cameras to track movement and detect unauthorized entries, ensuring that potential threats are identified in real-time. Additionally, behavioral analysis algorithms would be employed to identify signs of stress, bullying, or potential security threats, allowing for proactive intervention. Furthermore, smart ID badges with geofencing technology would notify parents and staff if a student leaves school grounds unexpectedly, adding an additional layer of security. For instance, an IoT-based child security monitoring system has been proposed to assist parents in remotely monitoring their children's safety, enabling immediate action if necessary (Heng & Kamsin, 2021).

6.3 Gamification-based personalized learning for non-academic skills

While smart schools focus on academic improvement, soft skills such as communication, teamwork, and problem-solving remain underdeveloped in many educational systems. Gamification can be used to bridge this gap by making skill-building engaging and measurable. AI-powered gamification platforms assign students skill-based missions, rewarding teamwork and leadership challenges to encourage collaboration. They also use AR/VR simulations for real-world skill-building, such as financial literacy and decision-making under pressure.

A high school in Finland has introduced AI-driven teamwork simulation games where students complete project-based challenges, earning leadership points based on collaboration and decision-making abilities. The impact of gamification on student engagement and learning outcomes has been a growing area of academic inquiry. For instance, MIT Media Lab is studying the effects of gamified learning on problem-solving skills, exploring how interactive educational strategies can enhance cognitive development (Feng et al., 2024).

6.4 AI-based dynamic learning paths

One of the main limitations of traditional schooling is the rigid curriculum structure that does not account for individual student strengths and interests. AI can personalize learning paths by adapting content and pacing based on student progress. To achieve this, AI-powered platforms continuously monitor student performance, engagement, and learning preferences to adjust lesson difficulty and content accordingly. Additionally, the system suggests additional reading, exercises, and real-world projects based on a student's evolving capabilities, ensuring a more tailored educational experience. Furthermore, AI can recommend extracurricular activities that align with a student's career aspirations, helping them explore relevant opportunities beyond the classroom.

The effectiveness of AI-driven personalized learning pathways is an ongoing area of academic study. For example, Harvard University is researching AI-driven personalized learning to optimize student engagement (Manning, 2024).

The use of AI technology in teaching materials, class management and performance evaluation in virtual smart classrooms is discussed in some research. AI is identified as one of the leading technologies to make smart classrooms even smarter (Dimitriadou & Lanitis, 2023). Current research supports the potential benefits of AI-driven scheduling. For example, The Open University in the UK is actively developing AI-driven learning schedules to enhance student performance, exploring how personalized lesson timing can improve academic outcomes (Swargiary, 2024). Additionally, studies indicate that adaptive scheduling can lead to a measurable 12% improvement in student GPA, demonstrating the effectiveness of AI in optimizing learning efficiency (Marciniak et al., 2022).

6.5 AI-based early intervention for learning disabilities

Many learning disabilities, such as dyslexia, Attention Deficit Hyperactivity Disorder (ADHD), and autism, often go undiagnosed or are detected too late, affecting student performance and well-being. AI can help identify learning difficulties early and provide tailored interventions. To achieve this, AI analyzes handwriting, reading patterns, and cognitive test results to detect learning disabilities at an early stage. Additionally, AI-powered tutors adjust teaching speed, content complexity, and sensory input based on individual learning profiles, ensuring personalized support for students. Furthermore, speech-to-text and text-to-speech AI systems assist students with reading and comprehension difficulties, making the learning more accessible (Barua et al., 2022). The potential of AI-driven early detection systems is being explored in various educational research initiatives. For example, Stanford University is testing AI-driven reading tutors for dyslexic students, aiming to enhance literacy skills through personalized AI assistance (Sacks, 2025).

6.6 AI-powered career and skill development programs

Many students struggle with career choices due to a lack of guidance and exposure to evolving job markets. AI can help students identify suitable career paths by analyzing personal strengths, interests, and industry trends. To support this, AI assesses student coursework, extracurricular activities, and skill acquisition progress to recommend potential career paths tailored to individual capabilities. Additionally, AI-driven platforms match students with mentorship programs, internships, and online certifications, ensuring they gain relevant experience and professional connections. Furthermore, predictive analytics forecast future job trends, helping students align their education with industry demands and make informed career decisions (Westman et al., 2021).

7. Impacts on society, economy and sustainability

The integration of smart technologies into campus environments impacts society, economy, and sustainability, with both positive and negative implications from different perspectives. On the positive side, smart technologies can optimize energy consumption, reduce waste and efficient use of resources, contributing to a sustainable environment. For instance, automated lighting, automated heating, smart grids, air quality monitoring and recycling systems can help reduce energy consumption, improve waste management and healthier living conditions resulting in a more sustainable campus. However, the implementation and maintenance of these technologies can also have downsides to achieving sustainability. Such as the high energy consumption required to run digital systems, increased electronic waste, greater carbon footprint and the impact of manufacturing and disposing of technology products. Therefore, while smart technologies offer positive sustainability impacts, their costs must be taken into account in order to ensure a healthier and sustainable environment.

One of the primary impact areas of a smart campus is society. A smart campus influences the way students, faculty, staff and even individuals outside the campus interact with one another. It enhances social cohesion among individuals by creating an interconnected and supportive environment. However, the digitization of processes may lead to digital inequality, as not all individuals may have equal access to technology, which can limit their opportunities in the smart society. Furthermore, the integration of advanced technologies on campus transforms society's views on innovation and enhances their overall quality of life. For instance, the digitization of processes makes resources more accessible, allowing individuals with disabilities or those from underserved communities to participate more fully in campus life. In addition, health-focused technologies such as smart healthcare systems, smart gyms and wellness programs encourage both mental and physical well-being in society. This creates a sense of community, brings people together and significantly improves public health. On the other hand, technological dependency may also arise. It may lead to reduced face-to-face interaction and social isolation, which could weaken the sense of community. As a result, the adoption of smart technologies extends beyond the campus, promoting broader acceptance of smart city concepts and contributing to more connected and sustainable urban development, though it may also have some negative effects on society (Kolotouchkina et al., 2024; Wang et al., 2024).

Moreover, a smart campus has a significant impact on the economy. The integration of smart technologies contributes to the economy in most instances through automated processes, energy optimization, resource allocation, enhanced critical thinking skills and strategic planning. For instance, Birmingham City University invested nearly $128M in its new intelligent buildings, resulting in savings of $177K during construction, with forecasted savings of $105K annually. Beyond the education sector, businesses also leverage smart technologies to enhance productivity and reduce costs. By using digital twin technologies, companies benefited from a 22% annual return on investment and 19% cost savings. However, the dependency on technology and the need for continuous upgrades may strain budgets. In contrast, the continuous improvement of smart technologies such as VR, AR, IoT and digital twin leads to the creation of new jobs in technical support, maintenance, and development roles, driving economic growth. While the implementation of smart technologies brings numerous positive impacts, the high initial investment required can be a significant financial barrier for some institutions, leading to less interaction with the economy. As a result, a smart campus environment is experienced while costs are reduced, contributing to the economy in the majority of cases (Alghamdi & Shetty, 2016; Clough, 2024).

References

Abdullah, A., Thanoon, M., & Alsulami, A. (2019). Toward a smart campus using IoT: Framework for safety and security system on a university campus. *Advances in Science, Technology and Engineering Systems Journal, 4*(5). https://doi.org/10.25046/aj040512. Article 5.

Afzal, A., Khan, S., Daud, S., Ahmad, Z., & Butt, A. (2023). Addressing the digital divide: Access and use of technology in education. *Journal of Social Sciences Review, 3*(2). https://doi.org/10.54183/jssr.v3i2.326. Article 2.

Agarwal, P., Idrees, S. M., & Obaid, A. J. (2021). Blockchain and IoT technology in transformation of education sector. *International Journal of Online and Biomedical Engineering (iJOE), 17*(12). https://doi.org/10.3991/ijoe.v17i12.25015. Article 12.

Alam, F., Lim, M. A., & Zulkipli, I. N. (2023). Integrating AI in medical education: Embracing ethical usage and critical understanding. *Frontiers of Medicine, 10*, Article 1279707. https://doi.org/10.3389/fmed.2023.1279707

Alghamdi, A., & Shetty, S. (2016). Survey toward a smart campus using the Internet of Things, 239 https://doi.org/10.1109/FiCloud.2016.41.

Annamalai, N., Uthayakumaran, A., & Zyoud, S. H. (2023). High school teachers' perception of AR and VR in English language teaching and learning activities: A developing country perspective. *Education and Information Technologies, 28*(3). https://doi.org/10.1007/s10639-022-11275-2. Article 3.

Arenas, J., Palominos, P., Martin, R., Camargo, M., Bary, R., & Kaschel, H. (2022). Smart campus: Concepts, development and future. In *2022 IEEE 28th International conference on engineering, technology and innovation (ICE/ITMC) & 31st International association for management of technology (IAMOT) joint conference*. https://doi.org/10.1109/ICE/ITMC-IAMOT55089.2022.10033214

Barua, P. D., Vicnesh, J., Gururajan, R., Oh, S. L., Palmer, E., Azizan, M. M., Kadri, N. A., & Acharya, U. R. (2022). Artificial intelligence enabled personalised assistive tools to enhance education of children with neurodevelopmental disorders—A review. *International Journal of Environmental Research and Public Health, 19*(3). https://doi.org/10.3390/ijerph19031192. Article 3.

Bellaj, M., Bendahmane, A., Younes, A., Boudra, S., & Ennakra, M. (2024). A systematic review of smart campus technologies to improve students' educational experiences. *2024 Mediterranean smart cities conference (MSCC)*. https://doi.org/10.1109/MSCC62288.2024.10697051

Bhushan, M., Shingate, R., Shah, T., Vyas, N., & Shirsath, S. (2023). Intelligent tutoring system: Personalised learning plans with AI. *The European Chemical Bulletin, 12*(5), 1964–1975. https://doi.org/10.13140/RG.2.2.36573.59369

Buabbas, A. J., Al-Mass, M. A., Al-Tawari, B. A., & Buabbas, M. A. (2020). The detrimental impacts of smart technology device overuse among school students in Kuwait: A cross-sectional survey. *BMC Pediatrics, 20*(1). https://doi.org/10.1186/s12887-020-02417-x. Article 1.

Cata, M. (2015). Smart university, a new concept in the internet of things. *2015 14th RoEduNet International conference - Networking in education and research (RoEduNet NER)*. https://doi.org/10.1109/RoEduNet.2015.7311993

Clough, H. (2024). Digital twins cut costs and emissions by nearly 20%, survey finds. *Planning, Building & Construction Today*. https://www.pbctoday.co.uk/news/digital-construction-news/bim-news/digital-twins-cut-costs-emissions-by-nearly-20-survey-finds/144970/.

Contrino, M. F., Reyes-Millán, M., Vázquez-Villegas, P., & Membrillo-Hernández, J. (2024). Using an adaptive learning tool to improve student performance and satisfaction in online and face-to-face education for a more personalized approach. *Smart Learning Environments, 11*(1). https://doi.org/10.1186/s40561-024-00292-y. Article 1.

Dai, S. (2019). China universities deploy facial recognition for student registration. *Tech in Asia*. https://www.techinasia.com/day-university-china-means-face-scan-enrol.

Dimitriadou, E., & Lanitis, A. (2023). A critical evaluation, challenges, and future perspectives of using artificial intelligence and emerging technologies in smart classrooms. *Smart Learning Environments, 10* (1), 12. https://doi.org/10.1186/s40561-023-00231-3

Director of Academic Information Center. (2025). YONSEI University, Seoul, Korea | Campus life | smart campus | S-Campus. https://www.yonsei.ac.kr/en_sc/campus/scampus.jsp.

Dong, Z. Y., Zhang, Y., Yip, C., Swift, S., & Beswick, K. (2020). Smart campus: Definition, framework, technologies, and services. *IET Smart Cities, 2*(1). https://doi.org/10.1049/iet-smc.2019.0072. Article 1.

Duncan, D. K., Hoekstra, A. R., & Wilcox, B. R. (2012). Digital devices, distraction, and student performance: Does In-Class cell phone use reduce learning? *Astronomy Education Review, 11*(1). https://doi.org/10.3847/AER2012011. Article 1.

Elkoubaiti, H., & Mrabet, R. (2018). A survey of pedagogical affordances of augmented and virtual realities technologies in IoT - Based classroom. In *2018 IEEE 5th International congress on information science and technology (CiSt)*. https://doi.org/10.1109/CIST.2018.8596654

Feng, J., Tan, W. H., & Yu, B. (2024). A systematic literature review of the impact of gamification instruction on students' problem-solving skills. *International Journal of Evaluation and Research in Education, 13*(6). https://doi.org/10.11591/ijere.v13i6.29695. Article 6.

Feridun, K. B., & Bayraktar, Ü. (2024). The future of virtual reality and education. *Turkish Online Journal of Educational Technology - TOJET, 23*(3). Article 3.

Fernández-Caramés, T. M., & Fraga-Lamas, P. (2019). Towards next generation teaching, learning, and context-aware applications for higher education: A review on blockchain, IoT, fog and edge computing enabled smart campuses and universities. *Applied Sciences, 9*(21). https://doi.org/10.3390/app9214479. Article 21.

Finnley, M. (2023). Why should higher education institutions invest in VR and metaverse education? *Axon Park*. https://axonpark.com/why-should-higher-education-institutions-invest-in-vr-and-metaverse-education/.

Gebze Technical University Office of Press and Public Relations. (2024). Sustainable projects by GTU students for the smart campus project. https://www.gtu.edu.tr/en/icerik/8/24860/display.aspx.

Guo, Y., Li, S., Zhang, X., Fu, Y., Yuan, Y., & Liu, Y. (2024). Embracing the metaverse: A survey of virtual reality and augmented reality practices at the United States' top one hundred university libraries. *College & Research Libraries, 85*(7). https://doi.org/10.5860/crl.85.7.1006. Article 7.

Heng, L. Y., & Kamsin, I. F. B. (2021). IoT-based child security monitoring system. In *3rd International conference on integrated intelligent computing communication & security (ICIIC 2021), Bangalore, India*. https://doi.org/10.2991/ahis.k.210913.058

Ikrissi, G., & Mazri, T. (2020). A study of smart campus environment and its security attacks. *ISPRS - International Archives of the Photogrammetry, Remote Sensing and Spatial Information Sciences, XLIV-4/W3–2020*, 255–261. https://doi.org/10.5194/isprs-archives-XLIV-4-W3-2020-255-2020

Imbar, R. V., Supangkat, S. H., & Langi, A. Z. R. (2020). Smart campus model: A literature review. In *2020 International conference on ICT for smart society (ICISS)*. https://doi.org/10.1109/ICISS50791.2020.9307570

Infante-Moro, A., Infante-Moro, J. C., & Gallardo-Pérez, J. (2022). Key factors in the success of virtualization of teaching in Spanish universities during the COVID-19 pandemic. *Journal of New Approaches in Educational Research, 11*(2). https://doi.org/10.7821/naer.2022.7.1002. Article 2.

İTÜ'de Paylaşımlı. (2018). *İTÜ'de Paylaşımlı Bisiklet Devri (Shared bicycle era at Istanbul*. Technical University). http://yesilkampus.mozaik-test.itu.edu.tr/haberler/2018/04/20/i-tu-de-paylasimli-bisiklet-devri.

Kambala, Y. J., & Mathe, R. (2022). Importance of smart classroom: A study. *The Journal of Multidisciplinary Research, 2*(1). https://doi.org/10.37022/tjmdr.v1i2.280

Katznelson, G., & Gerke, S. (2021). The need for health AI ethics in medical school education. *Advances in Health Sciences Education, 26*(4). https://doi.org/10.1007/s10459-021-10040-3. Article 4.

Kaur, A., Bhatia, M., & Stea, G. (2022). A survey of smart classroom literature. *Education Sciences, 12*(2), 86. https://doi.org/10.3390/educsci12020086

Klement, M. (2017). Models of integration of virtualization in education: Virtualization technology and possibilities of its use in education. *Computers & Education, 105*, 31–43. https://doi.org/10.1016/j.compedu.2016.11.006

Kofo, S. A., Ochayi, O. A., & Jimoh, B. A. (2022). Access and utilization of online learning resources among undergraduate students. *Indonesian Journal Of Educational Research and Review, 5*(1). https://doi.org/10.23887/ijerr.v5i1.45959. Article 1.

Kolotouchkina, O., Ripoll González, L., & Belabas, W. (2024). Smart cities, digital inequalities, and the challenge of inclusion. *Smart Cities, 7*(6). https://doi.org/10.3390/smartcities7060130. Article 6.

Li, Y., Wei, Y., Chen, Z., & Zhang, Y. (2023). A metaverse intelligent classroom construction method for multi-domain users. In *2023 9th International conference on big data and information analytics (BigDIA)*. https://doi.org/10.1109/BigDIA60676.2023.10429394

Liu, S. (2024). Virtual reality and 6G based smart classroom teaching using artificial intelligence. *Wireless Personal Communications*. https://doi.org/10.1007/s11277-024-11223-y

Luying, L. I., Ping, C., Xingjian, W., & Chunxiao, L. I. (2024). Research and design of smart campus for digital twins. *Tongxin Xuebao, 45*, 215–223.

Madera Doval, D. P. (2022). Una revisión sistemática sobre smart classrooms. *CITAS, 8*(2). https://doi.org/10.15332/24224529.7953

Madhur, P., Desale, G. B., Annam, V., & Sharma, C. S. (2024). Impact of smart classrooms in teaching learning effectiveness in higher education: A quantitative investigation. *Journal of Informatics Education and Research*. https://doi.org/10.52783/jier.v4i2.762

Manning, K. (2020). 8 smart campus technology examples. *ProcessMaker*. https://www.processmaker.com/blog/8-examples-of-smart-campus-technology/.

Manning, A. J. (2024). Professor tailored AI tutor to physics course. *Engagement doubled*. The Harvard Gazette. https://news.harvard.edu/gazette/story/2024/06/professor-tailored-ai-tutor-to-physics-course-engagement-doubled/.

Marciniak, J., Wójtowicz, A., Kolodziejczak, B., Szczepanski, M., & Stachowiak, A. (2022). Impact of course scheduling on student performance in remote learning. *Proceedings of the 27th ACM Conference on on Innovation and Technology in Computer Science Education, 1*, 400–406. https://doi.org/10.1145/3502718.3524788

McGrath, O., Hoffman, C., & Dark, S. (2023). Future prospects and considerations for AR and VR in higher education academic technology. *Educause Review*. https://er.educause.edu/articles/2023/4/future-prospects-and-considerations-for-ar-and-vr-in-higher-education-academic-technology.

Memos, V. A., Minopoulos, G., Stergiou, C., Psannis, K. E., & Ishibashi, Y. (2020). A Revolutionary Interactive Smart Classroom (RISC) with the use of emerging technologies. In *2020 2nd International conference on computer communication and the internet*. ICCCI). https://doi.org/10.1109/ICCCI49374.2020.9145987

Meng, Y., Song, J., Yu, X., Xu, X., & Zhang, H. (2023). Design and evaluation of blended teaching in the smart classroom combined with virtual simulation training in basic nursing courses. *BMC Medical Education, 23*(1), 752. https://doi.org/10.1186/s12909-023-04721-3

Min-Allah, N., & Alrashed, S. (2020). Smart campus—A sketch. *Sustainable Cities and Society, 59*, Article 102231. https://doi.org/10.1016/j.scs.2020.102231

Mokmin, N. A. M., & Rassy, R. P. (2024). Review of the trends in the use of augmented reality technology for students with disabilities when learning physical education. *Education and Information Technologies, 29*(2). https://doi.org/10.1007/s10639-022-11550-2. Article 2.

Moraes, P., Pisani, F., & Borin, J. (2020). Towards a simulator for green smart campus systems. *Workshop de Computação Aplicada à Gestão Do Meio Ambiente e Recursos Naturais (WCAMA)*. https://doi.org/10.5753/wcama.2020.11025

Mylonas, G., Paganelli, F., Cuffaro, G., Nesi, I., & Karantzis, D. (2023). Using gamification and IoT-based educational tools towards energy savings—Some experiences from two schools in Italy and Greece. *Journal of Ambient Intelligence and Humanized Computing, 14*(12). https://doi.org/10.1007/s12652-020-02838-7. Article 12.

Nikai Salcido. (2017). Arizona State University and Amazon bring first-of-its-kind voice technology program to university classrooms and campus.

Omidinia, S., Masrom, M., & Selamat, H. (2013). An examination of the concept of smart school: An innovation to address sustainability. In *Proceedings of the 2nd International conference on advances in computer science and engineering. International conference on advances in computer science and engineering, Los Angeles,USA*. https://doi.org/10.2991/cse.2013.73

Pakdaman, M., Nazari Moghadam, M., Dehghan, H. R., Dehghani, A., & Namayandeh, M. (2021). Evaluation of the cost-effectiveness of virtual and traditional education models in higher education: A systematic review. *Health Technology Assessment in Action*. https://doi.org/10.18502/htaa.v3i1.5715

Pathak & Upadhyay. (2024). Integration of digital twins technologies in education for experiential learning: Benefits and challenges. *International Research Journal on Advanced Engineering Hub (IRJAEH), 2*(03). https://doi.org/10.47392/IRJAEH.2024.0064. Article 03.

Pexyean, T., Saraubon, K., & Nilsook, P. (2024). AI simulation, IoT, and digital twin for smart campus. In *2024 Research, Invention, and Innovation Congress: Innovative Electricals and Electronics (RI2C)* (pp. 176–182). https://doi.org/10.1109/RI2C64012.2024.10784437

Poyyamozhi, M., Murugesan, B., Rajamanickam, N., Shorfuzzaman, M., & Aboelmagd, Y. (2024). IoT—A promising solution to energy management in smart buildings: A systematic review, applications, barriers, and future scope. *Buildings, 14*(11). https://doi.org/10.3390/buildings14113446. Article 11.

Romero Herrera, N., Doolaard, J., Guerra-Santin, O., Jaskiewicz, T., & Keyson, D. (2020). Office occupants as active actors in assessing and informing comfort: A context-embedded comfort assessment in indoor environmental quality investigations. *Advances in Building Energy Research, 14*(1). https://doi.org/10.1080/17512549.2018.1488620. Article 1.

Sacks, I. (2025). *Stanford team builds tool to keep young readers from falling through the cracks*. Stanford Graduate School of Education News. https://ed.stanford.edu/news/stanford-team-builds-tool-keep-young-readers-falling-through-cracks.

Saini, M. K., & Goel, N. (2020). How smart are smart classrooms? A review of smart classroom technologies. *ACM Computing Surveys, 52*(6), 1–28. https://doi.org/10.1145/3365757

Samsudin. (2024). The impact of teacher professional development programs incorporating educational technology on student achievement: A meta-analysis. *Academy of Education Journal, 15*(2). https://doi.org/10.47200/aoej.v15i2.2544. Article 2.

Schleich, B., Anwer, N., Mathieu, L., & Wartzack, S. (2017). Shaping the digital twin for design and production engineering. *CIRP Annals, 66* (1). https://doi.org/10.1016/j.cirp.2017.04.040. Article 1.

Shao, G., & Helu, M. (2020). Framework for a digital twin in manufacturing: Scope and requirements. *Manufacturing Letters, 24*, 105–107. https://doi.org/10.1016/j.mfglet.2020.04.004

Smart Campuses. (2019). *Smart campuses: The next-generation campus*. Deloitte United States. https://www2.deloitte.com/us/en/pages/consulting/solutions/next-generation-smart-campus.html.

Songkram, N., Songkram, N., Chootongchai, S., & Samanakupt, T. (2021). Developing students' learning and innovation skills using the virtual smart classroom. *International Journal of Emerging Technologies in Learning (iJET), 16*(04), 34. https://doi.org/10.3991/ijet.v16i04.15221

Stealey, K. (2025). What are the advantages of smart boards in the classroom? [Online post] https://insights.samsung.com/2025/02/28/what-are-the-advantages-of-smart-boards-in-the-classroom-3/.

Strom, A. (2021). The negative effects of technology for students and educators [Master's thesis, Northwestern College]. NWCommons. https://nwcommons.nwciowa.edu/education_masters/326/.

Swargiary, K. (2024). The impact of AI-Driven personalized learning and intelligent tutoring systems on student engagement and academic achievement: Ethical implications and the digital divide. *SSRN*. https://doi.org/10.2139/ssrn.4897241

The Sustainability Directory. (2025). Smart campus. https://sustainability-directory.com/term/smart-campus/.

Valks, B., Arkesteijn, M., Koutamanis, A., & den Heijer, A. (2020). Towards a smart campus: Supporting campus decisions with internet of things applications. *Building Research & Information, 49*, 1–20. https://doi.org/10.1080/09613218.2020.1784702

Verbytska, A., & Syzonenko, O. (2020). Forced virtualization for research activities at the universities: Challenges and solutions. *Revista Romaneasca Pentru Educatie Multidimensionala, 12*(2Suppl.1). https://doi.org/10.18662/rrem/12.2Sup1/294. Article 2Sup.1.

Villanueva, X. (2024). Virtual reality's threat to socialization | River campus libraries. https://www.library.rochester.edu/about/news/virtual-realitys-threat-socialization.

Wang, Y., & Keat, O. B. (2024). Smart classroom: The evolution and application in teaching and learning. *International Journal of Social Science and Human Research, 7*(08). https://doi.org/10.47191/ijsshr/v7-i08-56

Wang, K.-C., Pan, H.-W., & Wu, C.-E. (2024). Smart campus innovative learning model for social practitioners of universities' third mission: To promote good health and well-being. *Sustainability, 16*(14). https://doi.org/10.3390/su16146017. Article 14.

Westman, S., Kauttonen, J., Klemetti, A., Korhonen, N., Manninen, M., Mononen, A., Niittymäki, S., & Paananen, H. (2021). Artificial intelligence for career guidance – Current requirements and prospects for the future. *IAFOR Journal of Education, 9*(4). https://doi.org/10.22492/ije.9.4.03. Article 4.

Zeeshan, K., Hämäläinen, T., & Neittaanmäki, P. (2022). Internet of things for sustainable smart education: An overview. *Sustainability, 14*(7). https://doi.org/10.3390/su14074293. Article 7.

Zhang, H., Lee, I., Ali, S., DiPaola, D., Cheng, Y., & Breazeal, C. (2023). Integrating ethics and career futures with technical learning to promote AI literacy for middle school students: An exploratory study. *International Journal of Artificial Intelligence in Education, 33*(2). https://doi.org/10.1007/s40593-022-00293-3. Article 2.

Zhang, Y., Yip, C., Lu, E., & Dong, Z. Y. (2022). A systematic review on technologies and applications in smart campus: A human-centered case study. *IEEE Access, 10*, 16134–16149. https://doi.org/10.1109/ACCESS.2022.3148735

Zhu, M., Yang, L., & Zhang, Y. (2022). Design and application of project-based teaching of convergence media smart classroom based on VR+AR technology. In *2022 International conference on education, network and information technology (ICENIT)*. https://doi.org/10.1109/ICENIT57306.2022.00016

Chapter 4

Sensors for smart education

Ram Dular Singh Yadava[1, 2, 3]

[1]*Department of Physics, Banaras Hindu University, Varanasi, Uttar Pradesh, India;* [2]*Solid State Physics Laboratory, Delhi, India;* [3]*Independent Researcher (Condensed Matter Physics and Sensor Systems), Varanasi, Uttar Pradesh, India*

1. Introduction

1.1 Smart education learning environment

Smart education broadly refers to the integration of modern digital technologies—such as the Internet of Things (IoT), Artificial Intelligence (AI), big data, cloud computing, blockchain, virtual reality, and digital twins—to enhance and transform teaching and learning environments. Traditional face-to-face classrooms with blackboards and handwritten notes are increasingly being replaced by whiteboards and multimedia tools that make learning more interactive and efficient. A typical IoT and AI-enabled classroom is equipped with interactive whiteboard connected to computer with internet access and a projector, forming the central hub for content delivery and collaboration. Interaction between the teacher and students is facilitated through personal tablets or laptops, which are wirelessly connected to the main system, allowing real-time sharing of content, quizzes, and feedback. This setup is further enhanced by an integrated audio–visual system that includes cameras for video capture, microphones for clear audio input, and speakers for sound output. The classroom becomes dynamic through integration of various types of sensors embedded in these devices—vision sensors in cameras for capturing images and video, acoustic sensors in microphones for detecting sound, and touch sensors in interactive whiteboards, laptops, and tablets for user input.

More advanced classrooms also feature IoT-enabled environmental control systems that automatically adjust indoor lighting, temperature, humidity, and air quality. These systems rely on sensor-enabled, dedicated IoT components integrated into the heating, ventilation, and air conditioning (HVAC) infrastructure, ensuring a comfortable and energy-efficient learning environment. To further enhance this system, one can incorporate advanced technologies such as motion sensors and facial recognition cameras to capture human gestures, emotions, and identity. Wearable biometric sensors could provide personalized health monitoring, while infrared cameras enable facial thermal imaging for added insight. Additionally, chemical sensors can detect abnormal vapor emissions, contributing to a more comprehensive understanding of the environment such as early detection and alarm for hazardous gas leakage and fire ignition hotspots. By seamlessly integrating these technologies with an IoT network and AI-powered data analytics, real-time interpretation of multi-source information becomes possible, enabling context-aware, adaptive learning experiences and safety in the classroom. This would allow for continuous assessment and tailored feedback, fostering a more dynamic and responsive educational environment.

Moreover, conventional ledger-based systems for managing administrative and academic operations are being migrated to blockchain-integrated infrastructures, offering immutable record-keeping, decentralized data control, enhanced transparency, and improved operational efficiency. Other campus activities—such as controlling building environments, sharing transportation resources, and managing security services—can also be optimized through IoT and AI-enabled technologies to reduce energy consumption, improve resource utilization, and enhance campus security.

The paradigm of smart education is in a constant state of advancement, drawing from the innovations in sectors, such as smart manufacturing, healthcare, and intelligent traffic systems. Just as in these domains, the infrastructure—comprising both hardware and software—undergoes continual refinement to meet the dynamic needs of learners and educators. The thoughtful integration of IoT and AI components, adaptive networking frameworks, and interactive pedagogical strategies are all being persistently reshaped. With a strong emphasis on resource optimization and customization, smart education systems

Digital Twins in the Smart Classroom. https://doi.org/10.1016/B978-0-443-41632-3.00010-2

are progressively becoming more advanced, responsive, efficient, and tailored to diverse educational goals and contexts. Numerous case studies and reports have examined the multifaceted nature of smart education technology. These contributions originate from a broad spectrum of stakeholders, including educational researchers, product developers, solution providers, and policymakers.

A curated selection of publications, cited as references (Chen et al., 2018; Dimitriadou & Lanitis, 2023; Grech et al., 2022; Huang et al., 2019; Jones et al., 2020; Jurva et al., 2020; Kaur et al., 2022; Koper, 2014; Palau et al., 2019; Richter et al., 2019; Siemens; Sharma et al., 2022; Toivonen et al., 2018; Zhang, Ding, et al., 2024, Zhang, Zhu, et al. 2024), provides a panoramic overview of the evolution of the educational landscape, from the early days of whiteboards, computers, and projectors with audio–visual integration to the recent advancements in virtual reality and digital twin technology.

The scope of smart learning environments broadens significantly when applied beyond the sciences to fields, such as the arts, social sciences, humanities, archaeology, economics, and management. While the underlying framework of sensor-enabled IoT and AI networks remains largely similar across disciplines, the design and implementation of these systems—particularly in the form of digital twins, virtual simulations, and interactive case studies—require substantial customization. Each discipline presents unique pedagogical goals and experiential learning needs, necessitating tailored solutions to create truly immersive and meaningful educational experiences; for some insight addition, one can see references (Cebrián et al., 2020; Demir, 2021; Wang & Keat, 2024).

Several factors determine the smart learning environment—the quality of teaching and learning materials, the teaching methods and strategies, the learner characteristics such as prior knowledge, motivation, and cognitive abilities, and the physical and technological aspects, which includes indoor ecosystem and access to information. A broad overview of various components involved in this domain, and the status of associated technologies can be obtained in references (Gros, 2016; Kumar et al., 2019; Masson, 2024; Saini & Goel, 2021; Ullo & Sinha, 2020). The present chapter focuses mainly on how sensor technologies enable liveliness to smart learning environments.

1.2 Role of sensors in AI-enabled IoT network

Sensors are the front-end devices that connect real-world physical phenomena and objects—humans included—to digital electronic devices and processors, much like our natural sense organs (eyes, ears, nose, tongue, and skin) interface with the brain via nervous system. A sensor is a transducer integrated with an electrical readout system that detects a desired physical quantity and delivers electrical output as signal.

The objects and phenomena of interest in smart classroom settings are represented by specific attributes associated with them—for example, presence of students by facial images, teacher–students interaction by voice and gesture communication, classroom ambience by temperature, humidity, light intensity and air quality, IoT devices by radio frequency identification (RFID) tags, and so on. The functioning of classroom activities proceeds by sensing these attributes and transmitting the sensors output to the IoT network for learning management. There are many sensors which are embedded within IoT devices network—for example, pressure sensors on whiteboard touchscreen, RFID sensors on laptop, tablets and computers, light and voice sensors in video camera, and so on. The control and automation of support facilities requires setting up a proper network of sensors.

Sensors are crucial for generating data that feeds into the IoT network, but not all types of sensors are necessary in every educational setting. Smart learning environments are designed to address the specific needs of various educational levels. In early education, for instance, basic motion and environmental sensors can effectively enhance student engagement, while more sophisticated technologies such as wearable biometric sensors and virtual and augmented (VR/AR) reality sensors become essential in secondary and higher education. At the higher education level, particularly in disciplines such as science, engineering, and medicine, the real-time interactive technologies, including adaptive animations, simulations, digital twins, and augmented reality, have the potential to revolutionize both the teaching and the learning. The application of digital twin technology can facilitate the creation of virtual classrooms and laboratories. Digital twin technology can further be extended to create virtual replicas of entire educational institutions, such as colleges or universities. By providing broader access to these educational twins, learning can become more efficient, personalized, and affordable—removing barriers imposed by geography, physical infrastructure, or limited resources.

Zhang, Ding, et al. (2024) presented a detailed survey of the research and case studies in smart education technology from the perspective of sensors. The review highlights how environmental sensors and biometric sensors in physical classroom combined with AI analytics (local or cloud based) boost student engagement, manage attendance, and provide personalized learning feedback. Dimitriadou and Lanitis (2023) presented a review paper that covers similar ground, with an emphasis on combining AR/VR and blockchain technologies in AI-enabled IoT network to facilitate multiuser virtual interaction and distance learning. Wang et al. (2021) reviewed the status of eye-tracking multisensor systems to capture student attention and understanding engagement. The review materials in these papers provide a sound understanding of the existing sensor technologies for data collection and learning analytics for real-time interaction and feedback. The next leap forward to adaptive virtual reality through virtual laboratories, augmented simulation models, and digital twins is building up—a glimpse of these can be obtained in Jones et al. (2020), Siemens; Sharma et al. (2022), Toivonen et al. (2018), and Zhang, Zhu, et al. (2024).

1.3 Present chapter

In this chapter, we provide a comprehensive overview of both existing and emerging sensor technologies designed for smart education systems. Building on prior reviews (Dimitriadou & Lanitis, 2023; Zhang, Ding, et al., 2024), our focus expands to explore the role of sensors in enabling AI-driven analytics, the seamless integration of VR and AR technologies within learning environments, and the evolving human interaction with digital twins. Section 2 discusses sensor networks deployed in physical smart classrooms. Sections 3 and 4 address VR/AR and digital twin technologies, respectively, from the perspective of sensor applications. Section 4 also explores how developing virtual digital twins based on virtual sensors for digital-real interaction could significantly impact higher education, particularly in fields such as science, engineering, and medicine.

2. Sensors for smart classrooms

Smart classrooms are dynamic environments where teachers and students interact through sensor-enabled IoT-network powered by data-driven AI decisions and feedback mechanisms. At the core of this ecosystem is a sensor network that supports real-time feedback, adaptive learning, and environmental control.

Sensors in smart classrooms can be categorized not only by what they measure but also by how they are integrated into the system and the level of autonomy they possess. Embedded sensors are integrated directly into educational devices or infrastructures such as smart desks, interactive whiteboards, or HVAC systems—where they function as part of a larger control or feedback loops. These sensors operate continuously and seamlessly within the host systems, enabling real-time data collection, processing, and actuation without the need for external intervention. Autonomous sensors, by contrast, function as independent units—often wireless, portable, and battery-powered. They are capable of monitoring environmental or behavioral parameters and transmitting data directly to centralized systems such as learning management platforms or cloud-based analytics engines. These sensors are especially valuable in settings where full system integration is not feasible, such as retrofitted classrooms, temporary setups, or mobile learning environments.

Both embedded and autonomous sensors vary widely in form, function, and deployment, depending on the specific pedagogical and infrastructural requirements of the learning environment. To understand the roles of these sensors in smart learning environments, it is useful to consider the various ways they can be classified based on their design, function, and application. One foundational classification is based on the transduction principle—that is, the physical mechanism through which sensors convert stimuli into measurable signals. Common transduction types include gravimetric, mechanical, thermal, acoustic, optical, chemical, and biological, each reflecting the underlying physics or chemistry of the sensing process. Another widely used approach classifies sensors by the type of quantity being measured. For instance, physical sensors detect attributes such as temperature, pressure, force, velocity, vibration, torque, strain, or acceleration. Chemical sensors identify the presence or concentration of gases or compounds such as oxygen, carbon dioxide, carbon monoxide, nitric oxide, or volatile organic compounds (VOCs). Biological sensors, on the other hand, are used to monitor physiological or biochemical markers such as breath alcohol levels, hydration status, or contaminants in water.

In smart learning environments, these categories often overlap, as educational applications frequently require multifunctional sensing capabilities. For example, a classroom may simultaneously employ temperature sensors to regulate indoor climate, facial recognition systems for secure and personalized access, and wearable biometric sensors to track student engagement or stress levels. Given this context, classifying sensors by their application is often the most practical and relevant approach in educational settings. Complementary classifications can also be made based on connectivity, such as wired or wireless integration into IoT/AI networks, or user interface modalities, including touch, gesture, or voice interaction, or level of signal processing integration such as digital, analog, or intelligent. The description below is along these lines.

2.1 Air quality monitoring sensors

In the educator–learner interaction space, every component—whether physical or virtual—can be considered a 'thing.' The learning environment becomes 'smart' when these components are interconnected, enabling seamless communication and information flow throughout the system—an essential characteristic of the IoT. This section provides a brief overview of the different types of sensors used to enable IoT- and AI-powered smart classrooms.

Numerous studies have shown that maintaining comfortable indoor air quality significantly enhances students focus, cognitive performance, and overall learning efficiency (Brink et al., 2023; Romero et al., 2025; Stazi et al., 2017; Uzelac et al., 2015). The primary parameters that influence indoor comfort include temperature, humidity, oxygen (O_2) and carbon dioxide (CO_2) concentrations, ambient light intensity, and acoustic noise.

In addition, several other air quality indicators are critical for assessing health risks and ensuring safety in smart classrooms. Key among these are fine particulate matter (SPM2.5/10), which represents the concentration of aerosol or smoke particles typically produced by combustion processes or atmospheric chemical reactions. Elevated hydrogen (H_2) levels may signal overheating in batteries, electronic devices, or server equipment. Carbon monoxide (CO) and volatile organic compounds (VOCs)—such as formaldehyde, benzene, toluene, and acrolein—are often released from smoldering or overheated building materials, including paints, plastics, and synthetic textiles. Toxic gases such as hydrogen cyanide (HCN), sulfur dioxide (SO_2), nitrogen dioxide (NO_2), and nitric oxide (NO) may be emitted during the combustion of synthetic materials like plastic and foam. In addition, gases such as hydrogen sulfide (H_2S), methane (CH_4), ammonia (NH_3), and biogenic amines such as histamine, tyramine and serotine from food spoilage, and cadaverine and putrescine from dead animals can invade indoor spaces, posing further biological and olfactory hazards.

To effectively monitor and regulate the indoor physical environment, smart classrooms deploy an interconnected network of sensing nodes strategically distributed throughout the space. Each node is positioned to capture environmental changes from specific locations, ensuring comprehensive spatial coverage. A sensing node typically comprises a group of sensors, with each sensor dedicated to measuring a distinct environmental parameter. These sensors are integrated with a local signal processor and a wireless communication module, enabling seamless data transmission to IoT-connected devices or central servers. The signal processor within each node performs essential functions such as signal amplification, noise filtering, denoising, and analog-to-digital conversion, ensuring that the sensor outputs are accurate and ready for digital interpretation. The communication module facilitates the wireless transmission of sensor data, along with unique digital identifiers for both the node and its constituent sensors. The network architecture—whether hierarchical, mesh, or a hybrid topology tailored to the specific context—enables real-time data collection, centralized monitoring, and adaptive environmental control, all of which are essential for maintaining comfort, efficiency, and responsiveness in smart learning environments; one may see references (Abawajy & Hassan, 2017; Abraham & Li, 2014; Kandris et al., 2020; Manu et al., 2022) for deeper insight.

The sophistication of sensor nodes and their supporting network can vary based on factors such as the types of sensors included in each node, the required performance levels, onboard signal processing capabilities, and their placement within the environment. One can either procure professional solutions from commercial vendors, for example, Honeywell, Smarter Technologies, or OZIOM (Honeywell; Smarter Technologies; OZIOM), or customize using commercially available off-the-shelf (COTS) IoT compatible components and local engineering support. While selecting sensors, key considerations include the transducer type (mechanical, dielectric, semiconductor, electrochemical), degree of onboard signal processing (filtering, amplification, analog-to-digital conversion), communication mode (wired or wireless), sensitivity, response time, packaging, and power consumption, and compatibility for integration with IoT network. Moreover, it is also important to be aware of contextually relevant air quality parameters and their prescribed standards to set the design goals for sensor network; documents cited as references (EPA (United States Environmental Protection Agency); Seguel et al., 2016; Thoua et al., 2022; WHO guidelines for Indoor Air, 2010) are useful for this purpose.

Indoor air quality in a smart classroom for physical comfort, which is a subjective feeling and can vary across people of different geographical and social background, is mainly determined by the combination of temperature and humidity. Typically, room temperature between 20–26°C and humidity between 30%–50% RH provides a good balance. The other important parameters from the comfortability perspective are O_2 and CO_2 concentration levels in air, which affect ease of breathing and mental alertness, and indoor light intensity and background noise, which affect focus and engagement. The level of CO_2 in classroom indicates the condition of ventilation; higher CO_2 concentration indicates poor ventilation. Provided the O_2 and CO_2 levels in natural external environment of the classroom are good, regulating ventilation can achieve target levels for both. In brief, sensor-enabled IoT network for HVAC regulation is the basic requirement for smart classroom physical environment.

The information given below is a brief on typical air quality monitoring sensors having low power consumption and fast response, appropriate for real-time monitoring and control of HVAC system; Baldelli (2021), Farahani et al. (2014), and Garcia et al. (2022) provide extensive survey of developments in sensor technology for air quality monitoring.

The following sensors are basic minimal required to provide comfortable physical ambience for focused educator–learner interaction in a smart classroom.

Temperature Sensor. Common types include thermistors, Resistance Temperature Detectors (RTDs), and semiconductor-based sensors. Measures ambient temperature with accuracy within ±0.5°C and response time <1 s, and is critical for regulating HVAC system.

Humidity Sensor. Capacitive and resistive sensors based on changes in dielectric constant or resistance. Detects moisture content in the air, often expressed as relative humidity with accuracy within ±2% to ±5% RH and response time <10 s.

Oxygen (O_2) Sensor. Electrochemical or zirconia-based sensors. Monitors oxygen concentration in indoor air with ±1% of accuracy and response time <30 s. These sensors are important to maintain safe and breathable limits in densely occupied or poorly ventilated classrooms.

Carbon Dioxide (CO_2) Sensor. Nondispersive infrared (NDIR) sensors. Measures CO_2 concentration in range 400–5000 ppm with accuracy ±50 ppm and response time <30 s. It assesses ventilation adequacy in indoor air atmosphere as elevated CO_2 levels can impair cognitive performance and concentration.

Light Sensor. Photodiodes and phototransistors are typical technologies used in such sensors. Detects ambient light intensity in range 0–100,000 lux with accuracy ±5% and response time <1 s. It allows automation of artificial lighting in classroom for energy efficiency and optimal visibility.

Acoustic Sensor. MEMS microphones or piezoelectric sensors are common types. Captures sound levels in the environment 30–120 dB range with accuracy ±2 dB and response time <1 s. Useful for monitoring indoor noise pollution and optimizing room acoustics through automated volume control of microphones and speakers.

The following sensors are useful to monitor and alarm health hazards and safety risks in science laboratories, cafeteria zones, or waste-handling areas of educational buildings.

Suspended Particulate Matter (SPM) Sensor. These sensors use laser scattering or optical methods to measure particulate density in real time in range 0–5000 μg/m^3 with accuracy ±5 μg/m^3 (for PM2.5) and response time <10 s. Monitor risks to respiratory problems due to allergens in air and regulate room air purifiers.

Hydrogen Sensor. Common types include electrochemical, metal-oxide semiconductor (MOS), and palladium-based catalytic sensors. These detect hydrogen through changes in conductivity, catalytic reaction, or gas diffusion rates. Typical detection range is 0–1000 ppm (for indoor safety applications), and response time <30 s. Monitor hydrogen leaks from batteries, electronics, or server equipment preventing fire and explosion risks.

Toxic Gases Sensor. Electrochemical sensors for CO, NO_2, and SO_2, and Photoionization detectors (PID) and metal-oxide semiconductors for VOCs. Typical detection ranges: 0–1000 ppm for CO, 0–20 ppm for CO, NO_2/SO_2, and 0–500 ppm for VOCs with accuracies ±3%–5%, and response times 10–60 s depending on compound and sensor type. Detect emissions from building materials, cleaning agents, or combustion byproducts enhancing air quality and occupant safety, especially in older buildings or enclosed learning spaces.

Hydrogen Cyanide (HCN) Sensor. Primarily electrochemical sensors. Typical detection range: 0–50 ppm with limit of detection below 0.1 ppm, and response time <30 s. Monitors HCN emissions in chemistry laboratories or from burning synthetic materials (e.g., polyurethane, plastics) during accidents or malfunctions.

Nitric Oxide (NO) Sensor. Typically, electrochemical or metal-oxide semiconductor (MOS) sensors; NO is a reactive gas, so detection requires rapid response and high specificity. Typical detection range: 0–100 ppm with limit of detection <1 ppm, and response time 10–60 s. NO sensors have cross-sensitivity for NO_2 and ozone and require filtering. Monitors air quality in spaces affected by vehicle emissions, fume hoods, or nearby industrial zones, and useful in environmental science laboratories or smart ventilation systems.

Methane (CH_4) Sensor. Catalytic bead sensors for combustible gas detection, NDIR sensors for improved accuracy and selectivity, and MOS sensors for cost-effective indoor use. Typically, detection range: 0–10,000 ppm, limit of detection 100–1000 ppm, and response time <30 s. Detects gas leaks in laboratories or maintenance areas for enhancing fire prevention and energy safety systems.

Ammonia (NH_3) Sensor. Electrochemical and MOS sensors are most common. Detection Range: 0–100 ppm with detection limit ~1 ppm, and response time 10–60 s. Sensitive to humidity and temperature, and robust packaging. Monitors ammonia release from cleaning agents, laboratory chemicals, or decaying organic materials important in biology laboratories, animal science laboratories, or waste storage areas.

Biogenic Amines Sensor. Primarily chemical resistance–based or chemiresistive sensors using materials such as conductive polymers, carbon nanotubes, or bio-functionalized nanomaterials. Typical detection range: ppm to ppb (very low concentrations), and response time: <60 s. Selectivity enhanced with surface modification or enzyme-functionalized layers. Used for detecting putrefactive odors or decomposing organic materials, such as laboratory wastes and food spills.

Electronic Nose. An electronic nose (e-Nose) is a system consisting of an array of chemical sensors, each sensitive to different types of gases or volatile organic compounds (VOCs). These sensors are coupled with a pattern recognition system to detect and identify specific odors or volatile compounds in the air. The e-Nose functions similarly to a mammalian olfactory system (Fang et al., 2022; Rabehi et al., 2024). Additionally, sensors for ambient air temperature, humidity, light, and sound can be integrated with the gas sensor array to create an all-in-one air quality monitoring system for smart classrooms, such as the Bosch Sensortec BME688 (Bosch). The pattern recognition software processes the sensor data, enabling the system to provide real-time insights and trigger corrective actions, such as regulating HVAC, air purification, ventilation, lighting, and acoustics.

2.2 Interactive learning sensors

In a smart classroom, interactive learning involves dynamic and continuous interactions between teachers, students, and IoT/AI-driven digital platforms designed to enhance the learning experience. This encompasses several key components, including the selection, delivery, and personalization of educational content, the promotion of student engagement, and the management of their cognitive load and emotional well-being. Learning analytics plays a pivotal role in this process by capturing, measuring, and analyzing student interactions in real time. By providing actionable insights, this software helps optimize the learning environment, support adaptive teaching strategies, and facilitate personalized learning experiences tailored to each student's needs (Wise et al., 2019).

The data that fuel AI-powered analytics come from a wide array of sensors and technologies, including thermal sensors, biometric sensors, audio–visual sensors, computer vision, motion sensors, noise sensors, and touch sensors which monitor interactions with devices such as whiteboards, tablets, and augmented reality tools. A wide range of sensors, utilizing various transduction principles and technologies, and compatible with IoT operating systems, are commercially available to support these interactions. The selection of sensors and the design of the sensor network depend on the desired sophistication level and the specific goals of the data analytics, which often involve multisensory data fusion and cloud-based deep learning capabilities.

The following presents a representative overview of various image sensors and their functional roles in the learning environment.

2.2.1 *Optical imaging sensors and computer vision*

In smart classrooms, optical image sensors play a crucial role in enabling computer vision applications such as facial expression analysis for emotional assessment, gesture recognition for interactive learning, student presence detection, and monitoring attention and engagement levels. These sensors, embedded in cameras and vision systems, capture visual information and serve as the foundation for computer vision, which uses machine learning algorithms—often supported by cloud computing—to interpret and analyze visual data in real time.

An optical image sensor works by detecting and converting photons (light particles) reflected from objects into electrical signals. This transduction process occurs when photons are absorbed by a silicon (Si) semiconductor, generating charge carriers (electron-hole pairs) in proportion to the intensity of the incident light. The resulting charge is read by either Complementary Metal-Oxide Semiconductor (CMOS) or Charge-Coupled Device (CCD) technology. In CMOS sensors, the change in voltage is measured directly at each pixel, while in CCD sensors, charge packets are transferred across the chip and read at one corner. Both sensor types use arrays of pixels to form an image frame, with each pixel representing a small portion of the scene. CMOS sensors are more widely used today due to their advantages in power efficiency, cost, and readout speed, making them ideal for real-time applications in educational environments. These sensors are inherently more sensitive in the blue region of the spectrum, due to silicon's wavelength-dependent absorption properties.

Most cameras used in classrooms capture images in the standard RGB (red-green-blue) color model and in 2D format. However, stereo cameras can capture both color and depth information, enabling 3D scene reconstruction. These additional spatial data enhance the accuracy of learning analytics by providing richer context about student posture, movement, and interaction with the environment. For more information, one can see references (CMOS SENSOR INC; Hu & Assaad, 2023).

2.2.2 *Thermal imaging sensors and thermography*

All living and nonliving objects emit thermal radiation in the form of infrared (IR) waves, which fall within 780 nm to 1 mm range of the electromagnetic spectrum. The intensity of emitted infrared radiation is directly proportional to temperature—higher temperature results in greater IR emission. Thermal imaging, or infrared thermography, is a technique that captures surface temperature distribution of an object without physical contact using IR sensor array—analogous to a photograph produced by a visible-light camera. Each IR sensor corresponds to a thermal pixel, and the sensor array—typically arranged in CMOS or CCD readout configuration—forms a thermal image frame. When integrated with front-end IR optics and back-end image processing, these sensors generate a visual representation of the surface temperature variation across the objects in their field of view—known as a thermogram or thermal image or thermal print. A thermogram is just like a photograph but based on emitted heat rather than reflected light. Unlike optical cameras, which rely on light reflection to create images, IR cameras detect thermal emissions, revealing heat-based features of an object. To understand fundamentals, one can refer to the book by Rogalski (2011).

A variety of thermal imaging technologies exist, ranging from low-cost microbolometer arrays—which use temperature-sensitive resistive materials—to high-end semiconductor photon detector arrays, capable of high-resolution, long-range thermal imaging with greater sensitivity and faster response times. IR imaging technology is highly advanced mainly because of the early emphasis being on military applications for security and surveillance. It has now extended to other important domains such as industrial process/plant monitoring for security and safety, medical disease diagnostics, and agricultural field monitoring (Bagavathiappan et al., 2013; Kesztyü et al., 2023; Ring & Ammer, 2012; Vadivambal & Jayas, 2011).

In educational and psychophysiological research, thermal imaging is used to monitor subtle skin temperature changes, particularly on the face, including areas such as the forehead, nose, and periorbital regions. These regions are thermally responsive to emotional and cognitive states due to autonomic nervous system regulation. For instance, emotional arousal, whether sympathetic (stress, anxiety, excitement) or parasympathetic (sleep, relaxation, happy)—can produce distinct facial thermal patterns, such as rapid warming around the eyes or cooling at the nasal tip (Cardone & Merla, 2017; Fardian et al., 2022; Ioannou et al., 2014; Kreibig, 2010; Merla & Romani; Pavlidis et al., 2002; Uribe et al., 2025).

The connection between thermal patterns and emotional states has been well-established in psychophysiology. When combined with other physiological signals—such as respiration rate, galvanic skin response, heart rate, blood volume pulse, and cutaneous thermal variation—and analyzed for their onset speed, intensity, and duration, thermal imaging can contribute to the identification of emotion-specific patterns.

In smart classroom settings, thermal cameras offer a noninvasive, real-time tool for assessing students' psychophysical states, including stress, fatigue, cognitive load, and engagement. Integrated with AI and other biometric or environmental sensors, thermal imaging enhances learning analytics by providing deeper insights into students' physiological responses to instructional content, teacher's delivery skill, and classroom conditions—enabling timely and personalized interventions to improve learning outcomes and well-being.

2.3 Occupancy and motion sensors

Occupancy and motion detection play a vital role in enabling smart classroom functionalities such as automated attendance tracking, space utilization analysis, adaptive lighting and climate control, and personalized learning experiences. These features contribute to creating a more responsive, energy-efficient, and student-centered learning environment.

A variety of sensor technologies are available to monitor student's presence, movement, and behavioral patterns, either as alternatives or complementary systems to one another. As discussed earlier, optical and thermal imaging sensors, when combined with customized image analytics, can effectively detect the presence and spatial position of occupants within the classroom. While these advanced imaging solutions offer high accuracy and rich contextual data, they may be more complex or costly to implement. Therefore, simpler and more cost-effective alternatives—such as passive infrared (PIR) sensors, ultrasonic sensors, or pressure-sensitive floor mats—are also commonly deployed for occupancy detection in smart classroom environments.

A brief overview of these commonly used sensor types is presented below.

2.3.1 Passive Infrared sensors

PIR sensors are thermal detectors that convert infrared (IR) radiation—naturally emitted by warm objects—into electrical signals. This energy-to-signal conversion is achieved using various transduction mechanisms, including thermoelectric (thermopile), electrothermal (bolometer), and pyroelectric principles. Among these, pyroelectric sensors, particularly those based on triglycine sulfate (TGS) or similar ferroelectric materials, are the most commonly used in PIR systems for motion and presence detection in indoor environments.

Pyroelectric sensors operate similarly to capacitors, where a ferroelectric material acts as the dielectric between two electrodes. When this material absorbs IR radiation, thermal fluctuations cause a change in dielectric polarization, which induces a small electric current. This current is then converted into a voltage signal using an external resistor circuit. Pyroelectric PIR sensors are widely adopted due to their low power consumption, ruggedness, cost-efficiency, and adequate sensitivity for short-range detection. They also offer excellent compatibility with microelectronic systems, making them ideal for embedded applications such as smart classroom infrastructure and building automation.

A PIR sensor only responds to changes in IR radiation within its field of view, which is defined by a Fresnel lens at the sensor's front end. When the thermal scene remains constant, the output remains stable (DC). However, when a person moves through the field of view, the changing IR radiation levels generate an alternating (AC) signal, as the sensor detects differences between the moving object and the static background. This dynamic response makes PIR sensors highly effective for motion detection, although they are not suitable for detecting stationary occupants.

To enhance performance, PIR sensors are commonly configured in dual or quad-element arrays. In a dual-element design, two symmetrically positioned sensing elements with split fields of view detect movement across the sensor's coverage area, producing two successive current pulses of opposite polarity. In a quad-element configuration, four sensing elements are arranged to improve background cancellation and noise reduction, offering greater accuracy in cluttered thermal environments.

For motion tracking, PIR sensors can be mounted on a servo-controlled platform, allowing the sensor to follow motion within its field of view. For more advanced occupancy detection, ceiling-mounted 2D thermal sensor arrays—integrated with CMOS or CCD readout circuits—are deployed as thermal imagers, with resolution tailored to the specific spatial monitoring needs of the classroom. The reference materials (AUTODESK Instructables; Caniou, 1999; Chidurala & Li, 2021; Chen et al. 2022; Doan et al., 2024; STMicroelectronics, 2013) provide a helpful overview of these sensors, covering both fundamental principles and advanced state-of-the-art developments.

2.3.2 Ultrasonic sensors

Ultrasonic sensors operate similarly to miniature sonar systems, using high-frequency sound waves (typically above 20 kHz, beyond the range of human hearing) to detect the presence, position, and movement of objects or people through time-of-flight and Doppler shift analysis. The sensor emits sound pulses and measures the time it takes for the reflected echoes to return, allowing it to estimate the distance and location of an object. If the object is moving, the change in frequency of the returning wave—the Doppler shift—can also indicate its velocity and direction of motion.

Ultrasonic sensors are active devices, meaning they require an external power source to operate. Typically, the same transducer element functions as both the transmitter and receiver, emitting and detecting ultrasonic sound waves. The transducers used are generally based on either capacitive or piezoelectric principles. In capacitive transducers, one plate of a capacitor is designed as a flexible membrane, while the other remains rigid. When an alternating current (AC) signal is applied, the membrane vibrates, generating ultrasonic sound waves. The same membrane can also detect returning echoes by converting the pressure variations (caused by reflected sound waves) back into electrical signals. In piezoelectric transducers, the ultrasonic waves are produced and detected through mechanical deformation of a piezoelectric crystal oscillator, which generates an electric charge in response to mechanical stress and vice versa. Microelectromechanical systems (MEMS)-based, gate-controlled CMOS technology enables the on-chip integration of ultrasonic sensors, making them particularly well-suited for embedding in IoT devices.

This dual-function vibration mechanism enables the sensor to both emit ultrasonic pulses and detect their reflections, enabling accurate rangefinding and motion detection. Unlike PIR sensors, which detect motion based on heat signatures, ultrasonic sensors can detect both moving and stationary occupants, making them especially useful in scenarios where individuals may be seated or relatively still. They perform reliably in enclosed spaces and are generally insensitive to ambient light and temperature variations, offering broad coverage when properly positioned.

In the context of smart classrooms, ultrasonic sensors are employed to support automated lighting control, HVAC system optimization, and space utilization analytics by continuously monitoring room occupancy and student movement patterns. Their real-time data help facilitate adaptive environmental responses and support energy-efficient operations. When integrated with PIR and computer vision systems, ultrasonic sensors contribute to more robust and engagement-aware sensing frameworks, enhancing the responsiveness and intelligence of smart learning environments.

Additionally, IoT-compliant ultrasonic sensor modules are readily available from a variety of commercial and open-source vendors, making them accessible for integration into modern classroom infrastructures. For further technical details, refer to Chapter 21 of Kleeman and Kuc (2008) and Monti et al. (2022).

2.3.3 Microwave radar sensors in smart classroom

Microwave radar sensors are active sensing devices that detect occupancy, motion, and micromovements by emitting microwave signals—typically in the 5–24 GHz frequency range—and analyzing the characteristics of the reflected waves. These sensors measure both time delay and Doppler shift of the returned signal to determine not only the presence and position of individuals but also subtle movements, such as breathing, body posture changes, and gestures.

Unlike passive sensors that rely on thermal or optical cues, microwave radar can detect both dynamic and stationary human presence, including seated or resting individuals, making it especially valuable in smart classroom settings. Applications include automated occupancy tracking, gesture-based interaction monitoring, real-time activity recognition, and adaptive control of lighting, HVAC, and learning systems.

Microwave radar technology, though more commonly associated with long-range applications in communications, navigation, surveillance, and medical imaging, has been adapted into compact, short-range systems for indoor human sensing. These specialized radar systems can detect not only inanimate objects but also fine-scale human motions such as gait, posture shifts, and biometric signals like breathing rate and heartbeats. Such data enhance learning analytics, enabling the system to infer student engagement, stress, or alertness levels in real time.

A comprehensive overview of short-range microwave sensing principles and applications is provided in the work by Song et al. (2024). Buyukakkaslar et al. (2024) and Costa et al. (2021) further explore advanced research and developments in radar-based human sensing. For practical implementation insights and sensor deployment guidelines, technical documents from commercial solution providers (INFINEON; OCTIOT) are recommended.

2.3.4 Radio frequency identification (RFID) sensors

An RFID sensing system comprises a wireless communication link between an RFID tag—which may be integrated with additional sensors—and an RFID reader (also known as an interrogator). Both components are equipped with antennas that enable the transmission and reception of radio frequency (RF) signals. The RFID tag includes a microchip (either passive or active) connected to an antenna and stores a unique digital identifier in its memory. Acting as a transponder, the tag responds when it receives a query signal from the reader.

The RFID reader functions as a transceiver, typically connected to a host computer or IoT platform. It emits an RF signal to activate nearby tags. The tag then modulates this incoming signal with its identifier and transmits the response back to the reader for decoding and data processing.

RFID tags can be classified as:

- Passive tags, which derive their power from the RF field emitted by the reader. They are inexpensive and compact, but limited to short-range communication.
- Active tags, which are battery-powered. These offer longer range, support on-chip processing, and can be used in remote sensing applications, such as monitoring environmental or physiological conditions.

In smart classrooms, RFID tags are often embedded into student ID cards, wearable devices, or educational tools. These systems enable automated attendance tracking, access control, and location-based monitoring without manual input. Tags attached to objects or individuals transmit data with unique IDs, allowing for the real-time monitoring of presence, movement, or participation in specific classroom zones.

Advanced RFID systems may also use sensor-enabled tags that can detect parameters such as temperature, motion, or biometric signals, enhancing the classroom's context-awareness. The nonintrusive nature, low cost, and scalability of this technology make RFID a practical component in the infrastructure of smart and adaptive learning environments. These data inputs also contribute to learning analytics, informing decisions on student engagement, performance trends, and personalized instruction. More about recent developments and emerging trends can be found in Meng and Li (2016), RF Page, Suresh and Chakaravarthi (2022), and Zhang et al. (2017).

2.4 Biometric and behavioral sensors

Biometric data refer to unique personal information obtained from precise measurements of an individual's physical, physiological, or behavioral traits. Common examples include fingerprints, facial and iris scans, voice patterns, and behavioral signatures such as gait or keystroke dynamics, which are widely used for access control, security authentication, and attendance tracking in modern workplaces and smart devices. As discussed earlier, visible and thermal facial imaging can capture distinct emotional states—such as anger, boredom, or anxiety—based on facial temperature patterns and expressions. Advanced sensing technologies, including ultrasound and microwave radar, further expand this capability by detecting body posture, gait, position, and subtle physiological signals such as respiration rate and heartbeats.

Wearable biometric sensors—such as smartwatches, fitness bands, and EEG headsets—add another valuable layer to this data ecosystem. These devices monitor vital physiological metrics, including heart rate, electrodermal activity (EDA), skin temperature, and brainwave patterns. When combined, facial imaging and wearable sensor data form a robust multimodal framework for analyzing student behavior and engagement in smart learning environments. Leveraging machine learning and big data analytics, this integrated approach yields deeper and more accurate insights than any single modality alone. For example, a spike in EDA detected by a wearable, alongside thermal imaging cues of stress and a disengaged facial expression, may signal cognitive overload—prompting an intelligent learning system to adjust content complexity, offer a pause, or modify the pacing of instruction. A comprehensive review of foundational work in this field can be found in the work by Ba and Hu (2023), Menendez et al. (2021), and Mustieles et al., 2024.

3. Virtual and augmented reality sensors

Virtual Reality (VR) involves creating a fully immersive 3D simulated environment in which users feel as though they are physically present and able to interact with the virtual world. In contrast, augmented reality (AR) overlays digital elements onto the real world, enabling users to interact with both physical and virtual objects, often using natural body movements and sensory systems (vision, hearing, touch, and smell). The sense of presence and level of immersion in VR and AR systems are achieved through a network of sensors, each serving a specific purpose. For example, a combination of stereo cameras and motion sensors tracks users' hand, eye, and body movements to generate input for control and navigation within a VR environment. In an AR environment, touchscreen interfaces and sensors that detect the position and movement of users or objects in space allow users to directly interact with virtual elements superimposed on the real world.

Sensor technologies used in smart educational setups with VR and AR functionalities can vary widely—from basic systems such as interactive whiteboards in schools to advanced configurations employed in medical surgery or engineering construction. The type and number of sensors, along with their level of integration with microelectronics, depends on the complexity of the learning content and the performance requirements. Elkoubaiti and Mrabet (2018), Górski et al. (2017), Kim et al. (2020), and Pant (1993) offer valuable insights into the foundational concepts, developmental approaches, and practical applications of VR and AR technologies within smart classroom settings.

Many of the sensors previously discussed in the context of smart classrooms—such as image, voice, environmental, and motion sensors—are also used in VR/AR environments. However, certain specialized sensors are essential for effective VR/AR-based learning. The following section will briefly highlight these key sensor types.

3.1 Inertial Measurement Unit (IMU) sensors

Inertial measurements refer to the detection of changes in position, velocity, angular, orientation, and directional sense of movement of an object in space, enabling systems to track the movement of a user or a device in real time. These measurements typically involve using three core sensors: accelerometer, gyroscope, and magnetometer for the detection of linear acceleration, angular velocity, and Earth's magnetic field, respectively. These sensors are often integrated into a single platform, along with the associated microelectronics, collectively forming the Inertial Measurement Unit (IMU). The output from the IMU is transmitted wirelessly to a host computer, where it is used to interact with VR/AR objects. The latter are the simulation graphics of the subject of interest to the user generated by computer software and projected on a screen for interaction.

Microelectromechanical systems (MEMS) technology is now preferred for fabricating these sensors because it leverages silicon planar–integrated circuit technology, enabling on-chip integration with CMOS readout and signal processing. This offers clear advantages such as compact size, low power consumption, fast response, and low noise, making it ideal for high-performance, real-time applications. The reviews cited (Fitzgerald, 2020; Shaeffer, 2013) are very useful for knowing more about this topic.

3.1.1 MEMS accelerometers

MEMS accelerometers commonly operate on the principle of capacitive sensing, which detects acceleration by measuring changes in capacitance. The core sensing element consists of a proof mass connected to a movable electrode in a parallel-plate capacitor configuration, where one plate remains fixed and the other is suspended to allow movement. Typically, an interdigitated comb-like structure is used, in which the proof mass is attached to a series of movable fingers interleaved with fixed ones. When acceleration is applied—particularly along the axis perpendicular to the gap between the plates—the inertial force causes the proof mass to displace. This displacement alters the overlap or spacing between the comb fingers, leading to a measurable change in capacitance. To enable three-dimensional motion sensing, three such capacitive structures are fabricated on a single chip, each aligned along one of the orthogonal X, Y, and Z axes. This configuration allows for precise detection of linear acceleration in all spatial directions. The linear velocity and displacement are calculated by single and double integrations of measured acceleration data, respectively.

3.1.2 MEMS gyroscopes

MEMS gyroscopes measure angular velocity based on the Coriolis effect. If a body moves linearly within a rotating frame of reference, it experiences an apparent force perpendicular to both its direction of motion and the axis of rotation, known as Coriolis force. At the core of a MEMS gyroscope is a vibrating proof mass, typically suspended by microfabricated springs. When the device undergoes rotational motion, the Coriolis force acts on the moving mass, causing it to deflect perpendicularly to both the vibration direction and the axis of rotation. This deflection is typically sensed through capacitive, piezoresistive, or piezoelectric methods. The change in capacitance resulting from the Coriolis-induced displacement is proportional to the angular rate of rotation. To enable detection of rotation about all three orthogonal axes (X, Y, Z), multiple vibrating structures are integrated into a single chip, each sensitive to rotation along a specific axis.

3.1.3 MEMS magnetometers

MEMS magnetometers are miniaturized sensors used to measure the strength and direction of magnetic fields, typically Earth's geomagnetic field, making them vital components for digital compassing and orientation. These sensors often rely on magneto-resistive, Hall-effect, or Lorentz-force-based sensing principles. In Lorentz-force-based MEMS magnetometers, a current-carrying microstructure experiences a force when exposed to a magnetic field, causing a measurable displacement of a mechanical structure. This displacement is then detected using capacitive or piezoresistive sensing. Alternatively, Hall-effect magnetometers generate a voltage perpendicular to both the magnetic field and the direction of current flow in a conductor. MEMS magnetometers can be fabricated in triaxial configurations to sense magnetic field components along the X, Y, and Z axes. In some IMUs, magnetometers help with the orientation by providing a compass-like reference, allowing the systems to correct drift in orientation over time.

The integration of MEMS accelerometers, gyroscopes, and magnetometers with signal processing CMOS electronics forms a 9-axis IMU, enabling robust motion tracking, heading detection, and spatial orientation in applications such as smartphones, autonomous navigation, VR/AR headsets and controllers, drones, wearable technologies, and smart classroom environments.

3.1.4 Role of IMUs in VR/AR systems

IMUs, comprising MEMS accelerometers, gyroscopes, and often magnetometers continuously, track a user's head, hand, and body movements by measuring linear acceleration, angular velocity, and orientation. In VR headsets and AR glasses, IMUs detect head tilts, nods, and turns, allowing virtual environments to react in real time with minimal latency (delay). Similarly, motion controllers and smartphones equipped with IMUs interpret hand gestures and device orientation to manipulate virtual objects or navigate digital spaces. This real-time motion tracking is critical for achieving natural interaction and maintaining the illusion of presence in immersive environments.

Beyond simple motion tracking, IMUs contribute to accurate position estimation, latency reduction, and motion sickness mitigation. In AR applications, inertial data are often fused with inputs from cameras or depth sensors to anchor digital content precisely within the user's physical surroundings. This sensor fusion approach reduces drift and enhances spatial awareness, ensuring that virtual overlays remain stable and contextually accurate. Moreover, the responsiveness of inertial sensors helps prevent disorientation by aligning system feedback with user movement almost instantaneously. As a result, IMUs serve as the core enablers of fluid, intuitive, and reliable interactions in both VR and AR systems, from gaming and education to industrial training and beyond.

3.2 LiDAR sensors

Light Detection and Ranging (LiDAR) is a sensing technology that uses laser light to measure distances with high precision. It works by sending out laser pulses and measuring how long it takes for the light to bounce back after hitting nearby objects—similar to how radar uses radio waves or sonar uses sound. By scanning an area with thousands of these pulses, LiDAR builds a detailed 3D map of the surroundings, known as a "point cloud," capturing the shape and depth of surfaces.

In smart education environments utilizing virtual and augmented (VR/AR) reality applications for learning, LiDAR enables room-scale tracking and spatial awareness. This allows students to move safely within a real space while interacting with immersive virtual content, reducing the risk of collisions with furniture or walls. In AR, where digital content is blended with the real world, LiDAR plays a key role by helping devices understand depth and surface structure. This leads to more accurate placement of virtual objects—such as 3D models, interactive labels, or simulations—making AR-based learning more stable, realistic, and engaging.

Although LiDAR offers precise depth-sensing, similar results can also be achieved using other technologies such as infrared sensors, camera-based simultaneous localization and mapping (SLAM), or IMUs, which track motion and orientation. Each of these has different advantages and may be used alone or together, depending on the learning environment and the hardware available.

This is a highly developed and continuously evolving field. A wealth of academic literature, along with commercial technical reports and products, is available on the subject. The review papers and research articles cited (Azuma, 1997; Stefano et al., 2021; Venkatesan et al., 2021; Xiong et al., 2021) are rich in technical content, covering developments from the field's foundational days to the most recent advancements.

3.3 Haptic sensors

Humans naturally interact with the outside real world through five senses: vision, hearing, touch, smell, and taste. The virtual world is a digital replica of the real-world objects and phenomena simulated by computer software. Therefore, for fully immersive experiences in VR/AR environment, human users must be able to engage with the computer software through all five senses in a reciprocal manner. Currently, visual, auditory, and haptic interactions are well-developed and commonly used, while smell and taste remain largely unexplored—particularly in educational settings, where they may be less critical. Visual and auditory interactions are essential minimum requirements for any VR/AR experience. Haptic interactions significantly enhance scope and intensity of virtual world experience. The following is a brief overview of haptic sensors deployed for human–machine interactions in VR/AR environments.

Haptic sensation—our perception of touch—arises from the complex interaction of mechanical stimuli detected by mechanoreceptors located in the skin, muscles, tendons, and joints. The sense of touch originates from receptors in skin, and is called tactile sensation. It allows us to perceive surface properties such as texture, pressure, temperature, and vibration. The sense of body position and movement is generated by receptors in muscles, tendons, and joints, and is called kinesthetic sensation (or proprioception). This enables us to perceive the position, direction, and force of our limb movements.

To facilitate haptic communication between humans and computers, VR/AR systems incorporate haptic interfaces capable of replicating both tactile and kinesthetic experiences. Achieving this requires a distributed network of diverse haptic sensors and actuators (devices that generate physical responses), ideally with densities approaching those of human skin, which contains hundreds of receptors per square centimeter. These systems are typically implemented in the form of haptic gloves, suits, or other wearable devices.

3.3.1 Tactile sensors

Tactile sensors are essential components of haptic interfaces. These sensors measure parameters such as pressure, texture, vibration, and contact location, mimicking the sensory capabilities of human skin. Tactile sensors are broadly categorized based on their transduction principles, including resistive, capacitive, piezoelectric, and optical types. Resistive sensors detect changes in electrical resistance due to applied pressure, while capacitive sensors measure variations in capacitance caused by deformation. Piezoelectric sensors generate voltage in response to mechanical stress, making them suitable for dynamic touch and vibration detection. Optical tactile sensors use light modulation to detect surface contact and deformation. In VR and AR applications, tactile sensors are integrated into haptic gloves, suits, or handheld devices to provide users with real-time touch feedback, enhancing immersion by simulating interactions such as grasping objects, feeling surface textures, or experiencing impact forces in virtual environments.

3.3.2 Kinesthetic sensors

Kinesthetic sensors play a crucial role in haptic systems by enabling the detection of limb position, movement, force, and torque, which together support realistic physical interaction with virtual environments. These sensors help replicate the body's sense of motion and effort, providing essential feedback for motor control and spatial awareness in VR and AR applications. Kinesthetic sensing devices rely on various transduction principles, including strain gauges, piezoelectric elements, optical encoders, and IMUs. Strain gauges measure deformation in mechanical structures to detect applied forces, while piezoelectric components convert mechanical stress into electrical signals for dynamic force sensing. Optical encoders track angular displacement and joint movement with high precision, and IMUs combine accelerometers and gyroscopes to monitor motion and orientation. In VR/AR haptic interfaces, kinesthetic sensors are integrated into exoskeletons, force-feedback devices, and robotic manipulators to simulate weight, resistance, and mechanical constraints. This allows users to feel physical properties, such as object mass, joint stiffness, or tool interaction, significantly enhancing the realism and effectiveness of virtual experiences.

The integration of haptic sensors with haptic feedback systems—such as mechanical actuators, electrical stimulators, and elastic actuators—forms a comprehensive interface that simulates touch sensations. This synergy is pivotal for effective interaction and learning. Detailed information about this topic can be found in some excellent recently published review articles (Hayward et al., 2004; Meribout et al., 2024; Shi & Shen, 2024; Xu et al., 2023; Yang et al., 2021). Together, these systems enable real-time bidirectional communication between the user and the virtual environment: sensors capture user inputs (such as touch, movement, or force), while actuators deliver corresponding physical feedback. This closed-loop interaction is essential for creating immersive, responsive, and realistic experiences in VR and AR. By replicating both tactile and kinesthetic cues, such systems support complex tasks, such as virtual object manipulation, remote collaboration, surgical training, and skill acquisition, closely mimicking real-world touch-based interactions.

4. Digital twins in sensor-enabled learning

A digital twin is a dynamic, data-driven virtual representation of a real-world object, system, or process, continuously updated through live sensor inputs and computational models. Unlike traditional simulations or augmented reality models, a digital twin is directly linked to a physical counterpart through real-time data streams.

Sensors embedded in the physical system feed key operational parameters into the virtual model, allowing it to evolve in tandem with the real-world entity. This continuous feedback loop enables real-time monitoring, analysis, optimization, and predictive insights. Enabled by advances in IoT connectivity (such as 4G and 5G), big data infrastructure, and AI-driven analytics, digital twins can be accessed via conventional displays or immersive technologies such as virtual and augmented reality, providing actionable intelligence throughout a system's lifecycle.

In smart education environments, virtual reality, augmented reality, and digital twins are increasingly used to create immersive, interactive, and data-driven learning experiences. Although all three rely on digital simulations to enhance understanding, they differ in how they connect with the real world, the types of sensors input they use, and the user experiences they deliver.

Virtual reality offers a fully immersive digital environment, often based on predesigned models. For example, in a school-level virtual biology laboratory, students can explore the anatomy of the human body by virtually "walking through" organ systems, or in physics, they can simulate experiments like measuring gravitational acceleration using virtual pendulums—without needing physical equipment.

Augmented reality overlays digital information onto the real-world using devices such as tablets or AR glasses. In a classroom, a student can point a tablet at a real beaker and instantly see animations of chemical reactions or molecular structures appearing on the screen, aligned with the beaker's position. In a physics context, pointing a device at a ramp could display live speed and angle measurements as a ball roll down, helping students visualize motion and forces.

Digital Twins go further by creating a live, virtual replica of a real-world system or experiment that is continuously updated through sensor data. For instance, in a biology experiment tracking plant growth, sensors embedded in soil can transmit real-time moisture and nutrient levels to a digital twin, allowing students to monitor environmental effects on growth remotely. In physics, a digital twin of a real electrical circuit can show voltage and current changes as students adjust resistors or power sources, with the virtual model instantly reflecting those changes.

By integrating simulation with real-time data, digital twin technology—empowered by sensors—transforms traditional laboratories into dynamic, adaptive learning environments that foster deeper understanding and safer experimentation. The resources (Jones et al., 2020; Sharma et al., 2022; Siemens; Toivonen et al., 2018; Zhang, Zhu, et al., 2024) and (Adibi et al., 2024; MathWorks: Digital Twin; Klinger, 2024; Vohra and Vohra, 2023; Zhang et al., 2022) provide valuable insights, ranging from foundational concepts to some advanced applications through illustrative examples. In the following, a few cutting-edge prospective applications of digital twins in higher educational environments are discussed.

4.1 Digital twin for electronic nose—A learning tool for machine learning

A powerful example of a digital twin can be found in the electronic nose (e-Nose)—a system designed to mimic the human sense of smell using an array of gas sensors and advanced computational models for pattern recognition. e-Nose systems employ a variety of sensor technologies based on principles such as gravimetric, electrical, acoustic, electromechanical, electrochemical, and fiber-optic transduction. Among these, surface acoustic wave (SAW) and MEMS resonator sensors are often preferred for compact, chip-level sensor systems due to their compatibility with CMOS circuitry.

Each individual sensor in the array is engineered to have specific selectivity for different chemical analytes. However, due to the inherent nature of chemical sensors, they often exhibit some degree of cross-selectivity. When exposed to a gas sample, the sensor array generates a response pattern, but the identities of the chemical analytes remain obscured. Multivariate statistical analysis and neural network methods are then used to extract and decode these patterns, identifying the chemical analytes—much like the way the mammalian nose processes scent information.

When used in a learning environment, particularly in teaching multivariate statistical analysis and machine learning, the e-Nose becomes more than just a technological tool: it functions as an intelligent digital twin, translating complex chemical interactions into measurable, interpretable data.

This digital twin does not just simulate odor detection—it evolves. Built on a network of virtual sensors that replicate the response patterns of physical–chemical sensors, the e-Nose collects real-time or simulated data about volatile compounds. These multivariate inputs are processed through trained machine learning algorithms, such as principal component analysis (PCA), support vector machines (SVM), and neural networks, to classify, identify, or even predict odor profiles.

What makes it a compelling educational digital twin is its dual nature:

Virtual representation of a physical sensing system—Students can experiment with modeled sensor arrays without needing expensive laboratory setups.

Real-time adaptivity and learning—The system can simulate aging sensors, noise, cross-sensitivity, or environmental drift, helping students understand model calibration, data preprocessing, and performance evaluation in dynamic conditions.

Moreover, when visualized through interactive dashboards or immersive environments (for example, AR interfaces showing sensor responses or decision boundaries in PCA space), students experience not just data, but insight—watching how a machine "learns to sense smell."

In short, the e-Nose as a digital twin embodies the very essence of modern, sensor-rich, intelligent learning systems: a living model, shaped by data, responsive to change, and deeply tied to real-world phenomena. It offers educators a versatile, tangible platform to teach statistical modeling, clustering techniques, pattern recognition, feature engineering, and beyond—all within a dynamic, exploratory framework.

e-Nose technology has advanced significantly, with a vast body of literature and a wide range of commercially available products. Gupta et al. (2018), Singh et al. (2020), (2021), Somvanshi and Yadava (2010), Verma and Yadava (2015), Yadava and Chaudhary (2006), and Yadava (2012) offer both fundamental insights and a solid foundation for developing e-Nose digital twins based on SAW and MEMS technology, making them highly suitable for advanced engineering education and training.

4.2 Digital twin for swarm intelligence—A learning tool for cooperative phenomena

Digital models of swarm intelligence—systems in which collective behavior emerges from the interactions of simple agents, such as insects, cells, or robots, without centralized control—present compelling platforms for understanding many cooperative phenomena occurring in nature, which inspire solution optimization strategies for many engineering, management, and healthcare issues. These models offer deep insights into complex biological, ecological, and therapeutic phenomena. Virtual swarms are represented as digital twins, where each individual agent is modeled with its specific interaction dynamics, dependent on control parameters generated by virtual sensors. These digital twins simulate and analyze swarm behavior in real time, capturing emergent patterns and guiding systems toward convergence. This paradigm has wide-ranging applications, from disease control and cooperative robotics to smart drug delivery systems.

In epidemiology, digital twins of mosquito populations are used to simulate the spread of vector-borne diseases, such as malaria or dengue. These models integrate real-world sensor data—temperature, humidity, breeding site density, and human movement patterns—to predict how virtual mosquito swarms interact with their environment and affect transmission dynamics. By tuning swarm behaviors in the digital twin (for example, mating, feeding, or habitat preferences), public health researchers can simulate interventions such as insecticide distribution, gene drive strategies, or urban sanitation policies before deploying them in the field.

In the study of biological synchronization, digital twins of firefly swarms are used to model how individual agents (fireflies) coordinate their flashing through simple coupling rules. These simulations, grounded in nonlinear dynamics and complex systems theory, are valuable tools in physics and robotics education for demonstrating how local interactions can lead to emergent global behavior. Beyond education, they also support research into decentralized synchronization protocols—principles that are being applied in real-world technologies such as sensor networks, drone coordination, and autonomous systems used in search-and-rescue operations and environmental monitoring.

In biomedical engineering, digital twins are being developed to model nanoparticle swarms for targeted drug delivery in cancer treatment. These virtual swarms simulate the collective movement of drug-loaded nanoparticles navigating through the bloodstream, responding to chemical gradients or tumor microenvironment signals. Real-time physiological data (for example, blood flow, tissue permeability) can be fed into these twins to optimize dosing strategies, delivery timing, and targeting precision. This helps researchers and learners visualize how cooperative microscale agents can outperform single-agent therapies, reducing side effects and improving efficacy.

These examples illustrate the emerging power of digital twins not just as static replicas, but as adaptive, multiagent systems that mirror complex, decentralized behaviors. In higher education and interdisciplinary research, they offer an engaging and sophisticated framework for teaching system thinking, predictive modeling, and data-informed intervention strategies (Lei et al., 2021; Li et al., 2024).

4.3 Digital twins in higher education and research—A glimpse of the future

Digital twins are increasingly shaping the landscape of higher education and academic research, offering immersive, data-driven environments for learning, experimentation, and innovation. Their ability to integrate real-time sensor data with advanced simulation models enables educators and researchers to replicate complex systems and make them accessible in virtual form—transforming traditional pedagogy and expanding research capabilities (Banu et al., 2020; Kwet & Prinsloo, 2020; Valks et al., 2020).

4.3.1 Virtual laboratories and instrumentation

In science and engineering education, digital twins enhance laboratory experiences by allowing students to engage with live sensor data, control instruments remotely, and observe real-time experimental outputs. For example, digital twins of spectroscopic equipment such as NMR and FTIR, or imaging tools such as SEM, TEM, and AFM, offer highly effective alternatives to physical laboratories—particularly valuable for remote or hybrid learning environments. Similarly, in mechanical and electrical engineering, digital twins of turbines, engines, or robotic arms enable students to run diagnostics, simulate stress conditions, and predict failures using real sensor–fed data—eliminating the risks and costs associated with wear-sensitive machinery.

4.3.2 Campus systems and infrastructure

Universities are also leveraging digital twins to manage and model their physical infrastructure. Smart building twins integrate sensor data (temperature, occupancy, and energy consumption) to simulate HVAC performance and energy flow—used as a learning platform in environmental science and architecture courses. In urban studies and civil engineering, real-time GPS and foot traffic data feed mobility digital twins that allow students to analyze pedestrian safety, design green transport systems, and address congestion through simulation-based problem-solving.

4.3.3 Biomedical and health sciences applications

In medicine and health sciences, digital twins are used for personalized, patient-specific learning. Virtual models of human organs, such as the heart and lungs, built using individual clinical data, enable students to explore the progression of diseases, and simulate the effects of various treatments. Physiological twins, created from real-time biometrics (heart rate, EEG, muscle activity) collected via wearables, help students study biofeedback, rehabilitation techniques, and brain–body interactions—supporting advanced learning in sports science, neuroscience, and medical training.

4.3.4 Research-driven system modeling

Digital twins also serve as platforms for complex system modeling in research. Smart grid simulations allow institutions to model energy systems using real-time data from solar panels, batteries, and microgrids to study load balancing and fault detection. In environmental science, ecosystem and water body twins, fed by satellite and ground sensors, help researchers and students analyze climate change impacts, water quality, and land-use patterns.

4.3.5 Personalized learning and student-centered education

In educational technology and data science, digital twins are used to model student learning behavior by combining learning analytics with cognitive and behavioral data. These models support personalized education by enabling adaptive learning paths, predictive interventions, and real-time feedback. They also play a central role in student-led projects, such as capstone work, involving smart home systems or autonomous drones—where learners build and deploy their own digital twins using real sensors and machine learning algorithms.

4.3.6 Emerging trends and future directions

Looking ahead, emerging trends such as human digital twins (for cognitive modeling), Digital Twin-as-a-Service (DTaaS), and integration with VR/AR platforms are expanding the reach of this technology. These innovations make digital twins more scalable, immersive, and accessible—poised to support the next generation of simulation-based learning and interdisciplinary research.

Collectively, digital twins offer a transformative framework for experiential education, collaborative exploration, and cutting-edge research. As technology continues to evolve, their role in smart education will deepen—bridging virtual and physical worlds to equip learners and researchers for the complex systems of the future.

5. Overview and conclusion

The evolution of smart education technology, combined with innovations in pedagogical strategies, is transforming the educational landscape, ensuring that it keeps pace with the ongoing digital revolution in modern society. At the heart of this transformation are sensors and the IoT, which enable the seamless integration of physical and digital learning environments. By embedding sensors in classrooms, learning tools, and even wearables, educators can collect real-time data on students' engagement, behaviors, and learning preferences. These data feed into AI-powered data analytics platforms that process vast amounts of information to create personalized learning experiences, predict academic outcomes, and even offer timely interventions when students are struggling. AI itself plays a pivotal role in automating tasks such as grading, curating tailored learning materials, and guiding students through adaptive learning paths, all while facilitating real-time feedback. The synergy between sensors, IoT, AI, and data analytics fosters a dynamic, interconnected educational ecosystem where learning is increasingly learner-centric, data-driven, and responsive to individual needs. As a result, this paradigm not only empowers students to take control of their learning but also equips educators with insights that help them optimize teaching strategies, making education more effective, accessible, and aligned with the demands of the digital age.

The development of sensor technologies is primarily driven by cutting-edge demands across specialized sectors such as avionics, space, military, and robotics, on one hand, and industrial automation, healthcare, automotive, and environmental monitoring, on the other. The performance requirements for sensors—such as sensitivity, specificity, detection limits, stability, and reliability—vary significantly, with high demands in military and space applications contrasting with more moderate needs in environmental monitoring. This disparity results in sensor systems that are often expensive and not accessible to many societies, particularly in the humanitarian-focused education sector. In smart education, sensors and sensor networks are typically adapted from commercially available products, but high-tech solutions remain out of reach for many due to budget constraints. To bridge this gap, there is a pressing need for affordable, rugged sensor technologies specifically tailored for education and home applications, with performance levels just sufficient to meet the sector's needs without excessive costs.

As we look toward the future of smart education, the development of advanced sensor and digital technologies will play a critical role in shaping more effective, accessible, and scalable learning environments. One promising direction is the creation of dedicated wireless IoT-compatible sensor chips that are embedded with machine learning algorithms. These sensors will not only enable real-time data collection but also allow for adaptive, intelligent systems that can tailor learning experiences based on individual student needs. Furthermore, the integration of virtual sensors, powered by simulation models and supported by sensor fusion algorithms, presents an exciting opportunity to reduce costs while enhancing the versatility and affordability of sensor systems. By enabling digital twins—virtual replicas of real-world educational environments—this approach can offer rich, dynamic simulations for experimentation and learning. However, it is important to recognize that such technologies may not be universally applicable, as some complex fields, such as physiology or cell biology, lack reliable theoretical models for accurate sensor simulation. Therefore, a balanced approach, combining both virtual and real digital twins, must be optimized for specific educational contexts to ensure the best outcomes, making future educational technologies more personalized, cost-effective, and adaptable.

In conclusion, the integration of sensors in smart education systems is driving significant improvements in how learning environments are designed and experienced. From monitoring air quality to enabling advanced virtual and augmented reality experiences, sensors empower both physical and digital classrooms. The rise of technologies such as digital twins and swarm intelligence further enhances the potential for real-time, data-driven decision-making and personalized educational pathways. As these sensor technologies continue to evolve, they promise to redefine the future of education, offering more engaging, dynamic, and efficient learning environments tailored to individual needs and preferences. The future of smart education is bright, with sensors at its core, creating a more connected and responsive world for students and educators.

References

Abawajy, J. H., & Hassan, M. M. (2017). Federated internet of things and cloud computing pervasive patient health monitoring system. *IEEE Communications Magazine*, 48–53. https://doi.org/10.1109/MCOM.2017.1600374CM

Abraham, S., & Li, X. (2014). A cost-effective wireless sensor network system for indoor air quality monitoring applications. *Procedia Computer Science, 34*, 165–171. https://doi.org/10.1016/j.procs.2014.07.090

Adibi, S., Rajabifard, A., Shojaei, D., & Wickramasinghe, N. (2024). Enhancing healthcare through sensor-enabled digital twins in smart environments: A comprehensive analysis. *Sensors, 24*(9). Article 2793, 30 pp. https://doi.org/10.3390/s24092793

AUTODESK Instructables. PIR motion sensor tutorial. https://www.instructables.com/PIR-Motion-Sensor-Tutorial.

Azuma, R. T. (1997). A survey of augmented reality. *Presence: Teleoperators and Virtual Environments, 6*(4), 355–385. https://doi.org/10.1162/pres.1997.6.4.355

Ba, S., & Hu, X. (2023). Measuring emotions in education using wearable devices: A systematic review. *Computer Education, 200*(3). Article 104797, 20 pp. https://doi.org/10.1016/j.compedu.2023.104797

Bagavathiappan, S., Lahiri, B. B., Saravanan, T., Philip, J., & Jayakumar, T. (2013). Infrared thermography for condition monitoring – A review. *Infrared Physics & Technology, 60*, 35–55. https://doi.org/10.1016/j.infrared.2013.03.006

Baldelli, A. (2021). Evaluation of a low-cost multi-channel monitor for indoor air quality through a novel, low-cost, and reproducible platform. *Measurement: Sensors, 17*. Article 100059, 11 pp. https://doi.org/10.1016/j.measen.2021.100059

Banu, J. F., Revathi, R., Suganya, M., & Merlin, N. R. G. (2020). IoT based cloud integrated smart classroom for smart and sustainable campus, 9th world engineering education forum, WEEF 2019. *Procedia Computer Science, 172*, 77–81. https://doi.org/10.1016/j.procs.2020.05.012

Bosch S. Digital nose for air quality. https://www.bosch-sensortec.com/news/digital-nose-for-air-quality.html.

Brink, H. W., Krijnen, W. P., Loomans, M. G. L. C., Mobach, M. P., & Kort, H. S. M. (2023). Positive effects of indoor environmental conditions on students and their performance in higher education classrooms: A between-groups experiment. *The Science of the Total Environment, 869*(15). Article 161813, 13 pp. https://doi.org/10.1016/j.scitotenv161813

Buyukakkaslar, M. T., Erturk, M. A., & Aydin, M. A. (2024). A review on radar-based human detection techniques. *Sensors, 24*(17). Article 5709, 21 pp. https://doi.org/10.3390/s24175709

Caniou, J. (1999). Thermal detectors. *Passive infrared detection: Theory and applications* (pp. 351–377). Boston: Kluwer Academic.

Cardone, D., & Merla, A. (2017). New frontiers for applications of thermal infrared imaging devices: Computational psychophysiology in neurosciences. *Sensors, 17*(5). Article 1042, 21 pp. https://doi.org/10.3390/s17051042

Cebrián, G., Palau, R., & Mogas, J. (2020). The smart classroom as a means to the development of ESD methodologies. *Sustainability, 12*. Article 3010, 18 pp. https://doi.org/10.3390/su12073010

Chen, X. Y., Wen, C. Y., & Sethares, W. A. (2022). Multi-target PIR indoor localization and tracking system with artificial intelligence. *Sensors, 22*(23). Article 9450, 16 pp. https://doi.org/10.3390/s22239450

Chen, G., Xu, B., Lu, M., & Chen, N. S. (2018). Exploring blockchain technology and its potential applications for education. *Smart Learning Environments, 5*. Article 1, 10 pp. https://doi.org/10.1186/s40561-018-0072-8

Chidurala, V., & Li, X. (2021). Occupancy estimation using thermal imaging sensors and machine learning algorithms. *IEEE Sensors Journal, 21*(6), 8627–8638. https://doi.org/10.1109/JSEN.2021.3049311

CMOS SENSOR INC. CCD vs CMOS: A review of sensor technology. https://www.csensor.com/ccd-vs-cmos.

Costa, F., Genovesi, S., Borgese, M., Michel, A., Dicandia, F. A., & Manara, G. (2021). A review of RFID sensors, the new frontier of internet of things. *Sensors, 21*(9). Article 3138, 34 pp. https://doi.org/10.3390/s21093138

Demir, K. A. (2021). Smart education framework. *Smart Learning Environments, 8*. Article 29, 36 pp. https://doi.org/10.1186/s40561-021-00170-x

Dimitriadou, E., & Lanitis, A. (2023). A critical evaluation, challenges, and future perspectives of using artificial intelligence and emerging technologies in smart classrooms. *Smart Learning Environments, 10*. Article 12, 26 pp. https://doi.org/10.1186/s40561-023-00231-3

Doan, T. K., Nguyen, T. N. L., Hoang, N. C., Le, M. T., & Nguyen, Q. C. (2024). Intelligent passive infrared sensor based on learning vector quantization for human detection. *Engineering Research Express, 6*. Article 015326, 13 pp. https://doi.org/10.1088/2631-8695/ad351e

Elkoubaiti, H., & Mrabet, R. (2018). How are augmented and virtual reality used in smart classrooms?. In *ICSDE'18: Proceedings of the 2nd international conference on smart digital environment* (pp. 189–196). New York: Machinery.

EPA (United States Environmental Protection Agency). Air sensor technology and indoor air quality. https://www.epa.gov/indoor-air-quality-iaq/air-sensor-technology-and-indoor-air-quality.

Fang, C., Li, H. Y., Li, L., Su, H. Y., Tang, J., Bai, X., & Liu, H. (2022). Smart electronic nose enabled by an all-feature olfactory algorithm. *Advanced Intelligent Systems, 4*. Article 2200074, 15 pp. https://doi.org/10.1002/aisy.202200074

Farahani, H., Wagiran, R., & Hamidon, M. N. (2014). Humidity sensors principle, mechanism, and fabrication technologies: A comprehensive review. *Sensors, 14*(5), 7881–7939. https://doi.org/10.3390/s140507881

Fardian, F., Mawarpury, M., Munadi, K., & Arnia, F. (2022). Thermography for emotion recognition using deep learning in academic settings: A review. *IEEE Access, 10*, 96476–96491. https://doi.org/10.1109/ACCESS.2022.3199736

Fitzgerald, A. M. (2020). MEMS inertial sensors, chapter 45. In Y. T. J. Morton, B. F. van Diggelen, G. J. J. Spilker, & B. W. Parkinson (Eds.), *Position, navigation, and timing technologies in the 21st century—Integrated satellite navigation, sensor systems, and civil applications* (pp. 1435–1446). IEEE. doi:10.1002/9781119458555.ch45.

Górski, F., Buń, P., Wichniarek, R., Zawadzki, P., & Hamrol, A. (2017). Effective design of educational virtual reality applications for medicine using knowledge-engineering techniques. *EURASIA Journal of Mathematics, Science and Technology Education, 13*(2), 395–416. https://doi.org/10.12973/eurasia.2017.00623a

Garcia, M. R., et al. (2022). Review of low-cost sensors for indoor air quality: Features and applications. *Applied Spectroscopy Reviews, 57* (9–10), 747–779. https://doi.org/10.1080/05704928.2022.2085734

Grech, A., Balaji, V., & Miao, F. (2022). *Education and blockchain*. Paris, France: UNESCO and COL. https://dx.doi.org/10.56059/11599/4131.

Gros, B. (2016). The design of smart educational environments. *Smart Learning Environments, 3*, 15, 11 pp. https://doi.org/10.1186/s40561-016-0039-x

Gupta, A., Singh, T. S., & Yadava, R. D. S. (2018). MEMS sensor array-based electronic nose for breath analysis—A simulation study. *Journal of Breath Research, 13*(1). Article 016003, 17 pp. https://doi.org/10.1088/1752-7163/aad5f1

Hayward, V., Astley, O. R., Hernandez, M. C., Grant, D., & Torre, G. R. (2004). Haptic interfaces and devices. *Sensor Review, 24*(1), 16–29. https://doi.org/10.1108/02602280410515770

Honeywell. IAQ Sensors. https://hbtmkto.honeywell.com/IAQ-Sensors.html.

Hu, X., & Assaad, R. H. (2023). The use of unmanned ground vehicles (mobile robots) and unmanned aerial vehicles (drones) in the civil infrastructure asset management sector: Applications, robotic platforms, sensors, and algorithms. *Expert Systems with Applications, 232*, Article 120897. https://doi.org/10.1016/j.eswa.2023.120897

Huang, L., Su, J., & Pao, T. (2019). A context aware smart classroom architecture for smart campuses. *Applied Sciences, 9*(9). Article 1837, 34 pp. https://doi.org/10.3390/app9091837

INFINEON. Radar sensors. https://www.infineon.com/cms/en/product/sensor/radar-sensors.

Ioannou, S., Gallese, V., & Merla, A. O. (2014). Thermal infrared imaging in psychophysiology: Potentialities and limits. *Psychophysiology, 51*, 951–963. https://doi.org/10.1111/psyp.12243

Jones, D., Snider, C., Nassehi, A., Yon, J., & Hicks, B. (2020). Characterizing the digital twin: A systematic literature review. *CIRP Journal of Manufacturing Science and Technology, 29*, 36–52. https://doi.org/10.1016/j.cirpj.2020.02.002

Jurva, R., Blue, M. M., Niemelä, V., Nenonen, S., & Chae, Y. (2020). Architecture and operational model for smart campus. *Wireless Personal Communications, 113*, 1437–1454. https://doi.org/10.1007/s11277-020-07221-5

Kandris, D., Nakas, C., Vomvas, D., & Koulouras, G. (2020). Applications of wireless sensor networks: An up-to-date survey. *Applied*

System Innovation, 3(1). Article 14, 24 pp. https://doi.org/10.3390/asi3010014

Kaur, A., Bhatia, M., & Stea, G. (2022). A survey of smart classroom literature. *Education in Science, 12*(2). Article 86, 30 pp. https://doi.org/10.3390/educsci12020086

Kesztyüs, D., Brucher, S., Wilson, C., & Kesztyüs, T. (2023). Use of infrared thermography in medical diagnosis, screening, and disease monitoring: A scoping review. *Medicina, 59*. Article 2139, 31 pp. https://doi.org/10.3390/medicina59122139

Kim, H., Kwon, Y. T., Lim, H. R., Kim, J. H., Kim, Y. S., & Yeo, W. H. (2020). Recent advances in wearable sensors and integrated functional devices for virtual and augmented reality applications. *Advanced Functional Materials, 31*(39). Article 2005692, 12 pp. https://doi.org/10.1002/adfm.202005692

Kleeman, L., & Kuc, R. (2008). Sonar sensing. In B. Siciliano, & O. Khatib (Eds.), *Springer Handbook of Robotics* (pp. 491–519). Berlin, Heidelberg: Springer. https://doi.org/10.1007/978-3-540-30301-5_22.

Klinger, N. (2024). Computer vision in AR and VR - The complete guide. https://viso.ai/computer-vision/augmented-reality-virtual-reality.

Koper, R. (2014). Conditions for effective smart learning environments. *Smart Learning Environments, 1*, 5, 17 pp. https://doi.org/10.1186/s40561-014-0005-4

Kreibig, S. D. (2010). Autonomic nervous activity in emotion: A review. *Biological Psychology, 84*, 394–421. https://doi.org/10.1016/j.biopsycho.2010.03.010

Kumar, S., Tiwari, P., & Zymbler, M. (2019). Internet of things is a revolutionary approach for future technology enhancement: A review. *Journal of Big Data, 6*. Article 111, 23 pp. https://doi.org/10.1186/s40537-019-0268-2

Kwet, M., & Prinsloo, P. (2020). The 'smart' classroom: A new frontier in the age of the smart university. *Teaching in Higher Education, 25*(4), 510–526. https://doi.org/10.1080/13562517.2020.1734922

Lei, L., Shen, G., Zhang, L., & Li, Z. (2021). Toward intelligent cooperation of UAV swarms: When machine learning meets digital twin. *IEEE Network, 35*(1), 386–392. https://doi.org/10.1109/MNET.011.2000388

Li, Q., Zhao, F., Zhao, L., Qin, X., Wang, Y., & Zhu, Y. (2024). Digital twin system based on swarm intelligence scheduling. In *6th International Conference on Internet of Things, Automation and Artificial Intelligence (IoTAAI)* (pp. 319–325). https://doi.org/10.1109/IoTAAI62601.2024.10692590

Manu, K. S. A., Abdulai, J. D., Engmann, F., Akazue, M., Appati, J. K., Baiden, G. E., & Kantanka, G. S. (2022). WSN architectures for environmental monitoring applications. *Journal of Sensors, 2022*. Article 7823481, 18 pp. https://doi.org/10.1155/2022/7823481

Masson, J. F. (2024). Roadmap for the use of machine learning and artificial intelligence in sensing. *ACS Sensors, 9*(8), 3805–3807. https://doi.org/10.1021/acssensors.4c02069

MathWorks: Digital twin. https://in.mathworks.com/?s_tid=gn_logo.

Menendez, M. H., Menendez, R. M., Escobar, C. A., & Arinez, J. (2021). Biometric applications in education. *International Journal on Interactive Design and Manufacturing, 15*, 365–380. https://doi.org/10.1007/s12008-021-00760-6

Meng, Z., & Li, Z. (2016). RFID tag as a sensor - A review on innovative designs and applications. *Measurement Science Review, 16*(6), 305–315. https://doi.org/10.3390/s21093138

Meribout, M., Takele, N. A., Derege, O., Rifiki, N., Khalil, M. E., Tiwari, V., & Zhong, J. (2024). Tactile sensors: A review. *Measurement, 238*, Article 115332. https://doi.org/10.1016/j.measurement.2024.115332

Merla, A., & Romani, G. L. (August 22–26, 2007). Thermal signatures of emotional arousal: A functional infrared imaging study. In *29th Annual International Conference of the IEEE Engineering in Medicine and Biology Society* (pp. 247–249). Lyon, France: IEEE. https://doi.org/10.1109/IEMBS.2007.4352270.

Monti, L., Tse, R., Tang, S. K., Mirri, S., Delnevo, G., Maniezzo, V., & Salomoni, P. (2022). Edge-based transfer learning for classroom occupancy detection in a smart campus context. *Sensors, 10*. Article 3692, 16 pp. https://doi.org/10.3390/s22103692

Mustieles, M. A. H., Carmona, Y. E. L., Ramírez, M. A. P., Armenta, A. A. M., Gómez, J. E. R., Gómez, C. F. C., ... Santos, J. J. L. (2024). Wearable biosensor technology in education: A systematic review. *Sensors, 24*(8). Article 2437, 43 pp. https://doi.org/10.3390/s24082437

OCTIOT. What is a microwave sensor and how does it work? https://octiot.com/what-is-a-microwave-sensor-and-how-does-it-work; Microwave Sensors Working Principle. https://octiot.com/microwave-sensors-working-principle.

OZIOM. Air quality monitoring system. https://oizom.com/product/polludrone-air-pollution-monitoring.

Palau, R., & Mogas, J. (2019). Systematic literature review for a characterization of smart learning environments. In A. M. Cruz, & A. I. Aguilar (Eds.), *Propuestas Multidisciplinares de Innovación e Intervención Educativa* (pp. 55–71). Valencia, Spain: Universidad Internacional de Valencia. https://www.universidadviu.es/modelo-de-investigacion.

Pant, V. S. (1993). Virtual reality in the classroom. *Educational Technology Magazine, 33*(4), 23–27. http://www.jstor.org/stable/44428033.

Pavlidis, I., Eberhardt, N., & Levine, J. (2002). Seeing through the face of deception. *Nature, 415*, 35. https://doi.org/10.1038/415035a

Rabehi, A., Helal, H., Zappa, D., & Comini, E. (2024). Advancements and prospects of electronic nose in various applications: A comprehensive review. *Applied Sciences, 14*. Article 4506, 46 pp. https://doi.org/10.3390/app14114506

RF Page: Components of RFID technology and applications. https://www.rfpage.com/components-of-rfid-technology-and-applications.

Richter, O. Z., Marín, V. I., Bond, M., & Gouverneur, F. (2019). Systematic review of research on artificial intelligence applications in higher education – Where are the educators? *International Journal of Educational Technology in Higher Education, 16*, 39, 27 pp. https://doi.org/10.1186/s41239-019-0171-0

Ring, E. F. J., & Ammer, K. (2012). Infrared thermal imaging in medicine. *Physiological Measurement, 33*, R33–R46. https://doi.org/10.1088/0967-3334/33/3/R33

Rogalski, A. (2011). *Infrared detectors*. Boca Raton: CRC Press.

Romero, P., Amaro, V. V., Arranz, J. I., Sepúlveda, F. J., & Miranda, M. T. (2025). Indoor air quality and thermal comfort in university classrooms in Southwestern Spain: A longitudinal analysis from aandemic to post-pandemic. *Buildings, 15*(5), 829, 34 pp. https://doi.org/10.3390/buildings15050829

Saini, M. K., & Goel, N. (2021). How smart are smart classrooms? A review of smart classroom technologies. *ACM Computing Surveys, 1* (1), 1–29. https://doi.org/10.1145/nnnnnnn.nnnnnnn

Seguel, J. M., Merrill, R., Seguel, D., & Campagna, A. C. (2016). Indoor Air Quality. *American Journal of Lifestyle Medicine, 11*(4), 284–295. https://doi.org/10.1177/1559827616653343

Shaeffer, D. K. (2013). MEMS inertial sensors: A tutorial overview. *IEEE Communications Magazine, April*(4), 100–109. https://doi.org/10.1109/MCOM.2013.6495768

Sharma, A., Kosasih, E., Zhang, J., Brintrup, A., & Calinescu, A. (2022). Digital twins: State of the art theory and practice, challenges, and open research questions. *Journal of Industrial Information Integration, 30*. Article 100383, 15 pp. https://doi.org/10.1016/j.jii.2022.100383

Shi, Y., & Shen, G. (2024). Haptic sensing and feedback techniques toward virtual reality. *Research: Ideas for Today's Investors, 7*. Article 0333, 13 pp. https://doi.org/10.34133/research.0333

Siemens. (n.d.). Digital twin – Driving the industrialization of the future. Siemens AG. https://assets.new.siemens.com/siemens/assets/api/uuid:cda2b14c-23bf-4685-8fdf-adb7fd25e3a7/Whitepaper-Digital-Twin-EN.pdf.

Singh, P., & Yadava, R. D. S. (2020). Nanosensors for healthcare, chapter 26. In B. Han, V. K. Tomer, T. A. Nguyen, A. Farmani, & P. K. Singh (Eds.), *Nanosensors for smart cities* (pp. 433–450). Elsevier. https://doi.org/10.1016/B978-0-12-819870-4.00025-6.

Smarter Technologies, Smart air quality monitoring devices. https://smartertechnologies.com/iot-solutions/smart-building-management/air-quality-monitoring.

Singh, T. S., Singh, P., & Yadava, R. D. S. (2021). Perspectives for electronic nose technology in green analytical chemistry, chapter 10. In N. K. Rawat, I. Stoica, & A. K. Haghi (Eds.), *Green polymer chemistry and composites — Pollution prevention and waste reduction* (pp. 167–201). New York: Apple Academic. https://doi.org/10.1201/9781003083917.

Somvanshi, D., & Yadava, R. D. S. (2010). Boosting principal component analysis by genetic algorithm. *Defence Science Journal, 60*(4), 392–398. https://doi.org/10.14429/dsj.60.495

Song, C., Li, Z., & Xu, W. (2024). Short-distance radar sensing application, chapter 11. In C. Gu, & J. Lien (Eds.), *Short-range micro-motion ssensing with radar technology* (pp. 291–310). The IET. https://doi.org/10.1049/PBCE125E_ch11.

Stazi, F., Naspi, F., Ulpiani, G., & Perna, C. D. (2017). Indoor air quality and thermal comfort optimization in classrooms developing an automatic system for windows opening and closing. *Energy and Buildings, 139*(15), 732–746. https://doi.org/10.1016/j.enbuild.2017.01.017

Stefano, F. D., Chiappini, S., Gorreja, A., Balestra, M., & Pierdic, R. (2021). Mobile 3D scan LiDAR: A literature review. *Geomatics, Natural Hazards and Risk, 12*(1), 2387–2429. https://doi.org/10.1080/19475705.2021.1964617

STMicroelectronics. Signal conditioning for pyroelectric passive infrared (PIR). STMicroelectronics Application Note AN4368. https://www.st.com/resource/en/application_note/an4368-signal-conditioning-for-pyroelectric-passive-infrared-pir-sensors-stmicroelectronics.pdf.

Suresh, S., & Chakaravarthi, G. (May 31, 2022). RFID technology and its diverse applications: A brief exposition with a proposed machine learning approach. *Measurement, 195*, Article 111197. https://doi.org/10.1016/j.measurement.2022.111197

Thoua, C., Cooper, E., Stamp, S., Mavrogianni, A., & Manovich, D. (2022). Indoor air quality in schools, chapter 64. In Y. Zhang, P. K. Hopke, & C. Mandin (Eds.), *Handbook of indoor air quality* (pp. 1891–1934). Singapore: Springer Nature. https://doi.org/10.1007/978-981-16-7680-2

Toivonen, V., Lanz, M., Nylund, H., & Nieminen, H. (2018). The FMS training center - A versatile learning environment for engineering education. *Procedia Manufacturing, 23*, 135–140. https://doi.org/10.1016/j.promfg.2018.04.006

Ullo, S. L., & Sinha, G. R. (2020). Advances in smart environment monitoring systems using IoT and sensors. *Sensors, 20*(11). Article 3113, 18 pp. https://doi.org/10.3390/s20113113

Uribe, S. C., Hernandez, L. A. M., Sandoval, V. M. G., Trejo, B. D., & Albarran, I. A. C. (2025). Emotion detection based on infrared thermography: A review of machine learning and deep learning algorithms. *Infrared Physics & Technology, 145*(3), 105669. https://doi.org/10.1016/j.infrared.2024.105669

Uzelac, A., Gligoric, N., & Krco, S. (2015). A comprehensive study of parameters in physical environment that impact students focus during lecture using internet of things. *Computers in Human Behavior, 53*, 427–434. https://doi.org/10.1016/j.chb.2015.07.023

Vadivambal, R., & Jayas, D. S. (2011). Applications of thermal imaging in agriculture and food industry—A review. *Food and Bioprocess Technology, 4*, 186–199. https://doi.org/10.1007/s11947-010-0333-5

Valks, B., Arkesteijn, M. H., Koutamanis, A., & den Heijer, A. C. (2020). Towards a smart campus: Supporting campus decisions with internet of things applications. *Building Research & Information, 49*(1), 1–20. https://doi.org/10.1080/09613218.2020.1784702

Venkatesan, M., Mohan, H., Ryan, J. R., Schurch, C. M., Nolan, G. P., Frakes, D. H., & Coskun, A. F. (2021). Virtual and augmented reality for biomedical applications. *Cell Reports Medicine, 2*. Article 100348, 13 pp. https://doi.org/10.1016/j.xcrm.2021.100348

Verma, P., & Yadava, R. D. S. (2015). Polymer selection for SAW sensor array based electronic noses by fuzzy c-means clustering of partition coefficients: Model studies on detection of freshness and spoilage of milk and fish. *Sensors and Actuators B: Chemical, 209*, 751–769. https://doi.org/10.1016/j.snb.2014.11.149

Vohra, M. (2023). An overview of digital twin, chapter 1. In M. Vohra (Ed.), *Digital twin technology: Fundamentals and applications* (pp. 1–16). Scrivener Publishing. https://doi.org/10.1002/9781119842316.ch1.

Wang, Y. J., & Keat, O. B. (2024). Smart classroom: The evolution and application in teaching and learning. *International Journal of Social Science and Human Research, 7*(8), 6271–6284. https://doi.org/10.47191/ijsshr/v7-i08-56

Wang, Y., Lu, S., & Harter, D. (2021). Multi-sensor eye-tracking systems and tools for capturing student attention and understanding engagement in learning: A review. *IEEE Sensors Journal, 21*(20), 22402–22413. https://doi.org/10.1109/JSEN.2021.3105706

WHO guidelines for indoor air quality: Selected pollutants. (2010). World Health Organization.

Wise, A. F. (2019). Learning analytics: Using data-informed decision-making to improve teaching and learning. In O. O. Adesope, & A. G. Rud (Eds.), *Contemporary Technologies in Education* (pp. 119–143). Cham: Palgrave Macmillan. https://doi.org/10.1007/978-3-319-89680-9_7.

Xiong, J., Hsiang, E. L., He, Z., Zhan, T., & Wu, S. T. (2021). Augmented reality and virtual reality displays: Emerging technologies and future perspectives. *Light: Science & Applications, 10*. Article 216, 30 pp. https://doi.org/10.1038/s41377-021-00658-8

Xu, J., Pan, J., Cui, T., Zhang, S., Yang, Y., & Ren, T. L. (2023). Recent progress of tactile and force sensors for human–machine interaction. *Sensors, 23*(4). Article 1868, 27 pp. https://doi.org/10.3390/s23041868

Yadava, R. D. S. (2012). Modeling, simulation and information processing for development of a polymeric electronic nose system, chapter 10. In G. Korotcenkov (Ed.), *Chemical Sensors Simulation and Modeling, Volume 3: Solid-State Devices* (pp. 411–502). New York: Momentum Press. https://www.researchgate.net/publication/285826202.

Yadava, R. D. S., & Chaudhary, R. (2006). Solvation, transduction and independent component analysis for pattern recognition in SAW electronic nose. *Sensors and Actuators B: Chemical, 113*(1), 1–21. https://doi.org/10.1016/j.snb.2005.02.031

Yang, T., Kim, J. R., Jin, H., Gil, H., Koo, J. H., & Kim, H. J. (2021). Recent advances and opportunities of active materials for haptic technologies in virtual and augmented reality. *Advanced Functional Materials, 31*(39). Article 2008831, 30 pp. https://doi.org/10.1002/adfm.202008831

Zhang, J., Zhu, J., Tu, W., Wang, M., Yang, Y., Qian, F., & Xu, Y. (2024). The effectiveness of a digital twin learning system in assisting engineering education courses: A case of landscape architecture. *Applied Sciences, 14*(15). Article 6484, 30 pp. https://doi.org/10.3390/app14156484

Zhang, X., Ding, Y., Huang, X., Li, W., Long, L., & Ding, S. (2024). Smart classrooms: How sensors and AI are shaping educational paradigms. *Sensors, 24*(17). Article 5487, 33 pp. https://doi.org/10.3390/s24175487

Zhang, J., Tian, G. Y., Marindra, A. M. J., Sunny, A. I., & Zhao, A. B. (2017). A review of passive RFID tag antenna-based sensors and systems for structural health monitoring applications. *Sensors, 17*(2). Article 265, 33 pp. https://doi.org/10.3390/s17020265

Zhang, Z., Wen, F., Sun, Z., Guo, X., He, T., & Lee, C. (2022). Artificial intelligence-enabled sensing technologies in the 5G/internet of things era: From virtual reality/augmented reality to the digital twin. *Advanced Intelligent Systems, 4*(7), Article 2100228. https://doi.org/10.1002/aisy.202100228

Chapter 5

IoT sensors for smart classroom

Ruslan Saad Abdulrahman[1], Aymen Jaber Salman[1] and Fawzi Mohammed Munir Al-Naima[1,2]
[1]Department of Computer Engineering, College of Engineering, Al-Nahrain University, Baghdad, Iraq; [2]Department of Electrical Engineering Techniques, University of Kut, Wasit, Iraq

1. Introduction

The IoT is a transformative framework connecting various devices and the Internet. This connectivity allows devices to collect, transmit, and analyze data. Sensor devices are essential to the IoT ecosystem as they gather data from the surrounding environment. This chapter explores the types and functionalities of sensor devices, discusses the different architectures and networking technologies proposed for IoT sensor networks, and outlines the communication protocols utilized. Additionally, it addresses the limitations and challenges IoT sensor networks face and highlights future trends and areas for open research in this field.

The convergence of smart sensors and IoT within university settings addresses the increasing demand for smarter, more interactive educational experiences. Integrating these technologies in classrooms and campus facilities supports not only efficient energy utilization and safety but also enables novel pedagogical approaches, facilitating active student participation and personalized instruction (Majeed et al., 2024; Shahbaz et al., 2023). Embracing this intersection creates a dynamic environment tailored to modern learning challenges, fostering innovation on both organizational and instructional levels. This chapter explores the acceptance and practical implications of IoT-enabled smart sensors in university education contexts. It examines key technologies, implementation challenges, and their impact on teaching and campus management.

2. Basic configuration of IoT sensors

Numerous technologies are used to connect various physical and virtual objects within the IoT. The architecture of the IoT should enable the incorporation of various systems and technologies. The IoT is viewed as a revolutionary paradigm that facilitates seamless data exchange via interconnected smart devices. IoT sensors are devices that identify and quantify physical attributes or environmental states, transforming this data into signals suitable for processing and transmission via a network. These sensors are essential to the IoT ecosystem, as they allow for the gathering of real-time data from multiple sources (Brida et al., 2021; Collett, 2020).

The first step in planning the IoT system is to select the required configuration, for instance, sensors, the protocol that is used for communication, and how the data will be processed and stored; all these requirements are dependent on the particular application being used. The main parts of IoT are illustrated in Fig. 5.1. As shown in the figure, the IoT devices (sensors, actuators, IoT gateways, or any device) represent the second part of this platform (Hossein Motlagh et al., 2020; Jia et al., 2019; Kumar et al., 2019).

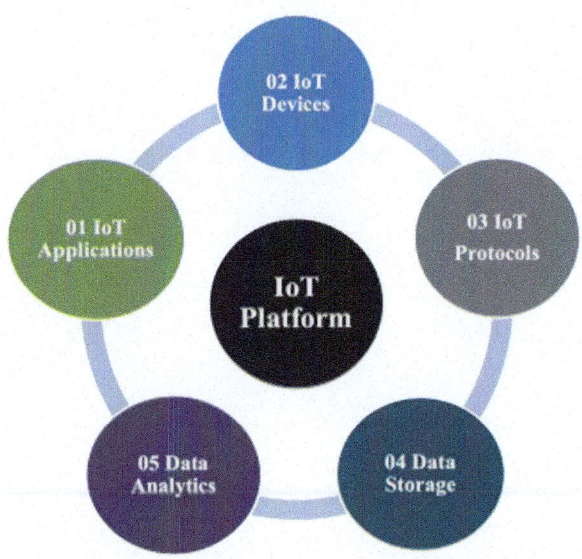

FIGURE 5.1 IoT main components (Motlagh et al., 2020).

Digital Twins in the Smart Classroom. https://doi.org/10.1016/B978-0-443-41632-3.00028-X

3. Characteristics of IoT sensors

The fundamental concept of the IoT is to connect devices to the internet, allowing them to collect, exchange, and communicate information to create a self-decision-making system. Consequently, IoT possesses several characteristics that define its primary functions, as illustrated in Fig. 5.2 (Mansour et al., 2023).

✓ Interconnectivity: The IoT utilizes a worldwide information and communication framework that links different devices and systems together.
✓ Things-related Services: The IoT can provide various services associated with these "things," as long as privacy protection is upheld and semantic consistency between physical objects and their digital representations is maintained. Improvements in both physical technology and information systems will depend on the provision of these services.
✓ Heterogeneity: The diversity of IoT devices means they are not uniform; they are diverse and constructed on different hardware platforms and networks. These devices can connect and engage with each other via various networks, making it easier to integrate with additional devices and service platforms.
✓ Dynamic Changes: The status of IoT devices can fluctuate dynamically, encompassing states like sleeping or waking, connecting or disconnecting, as well as variations in context, such as location and speed. Moreover, the quantity of linked devices may vary over time.

As they collect real-time data from the physical environment and transmit it for analysis and decision-making, IoT sensors are crucial for the IoT. The efficiency and functionality of these sensors are determined by specific characteristics, which are outlined below as illustrated in Fig. 5.3 (Brida et al., 2021).

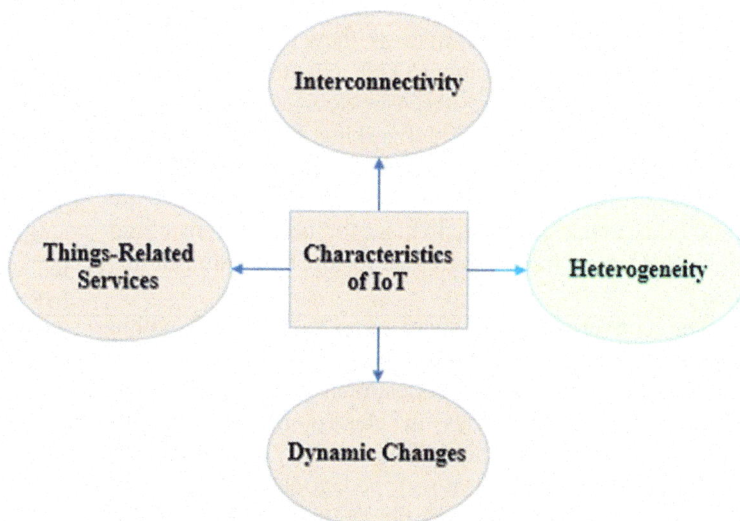

FIGURE 5.2 IoT characteristics.

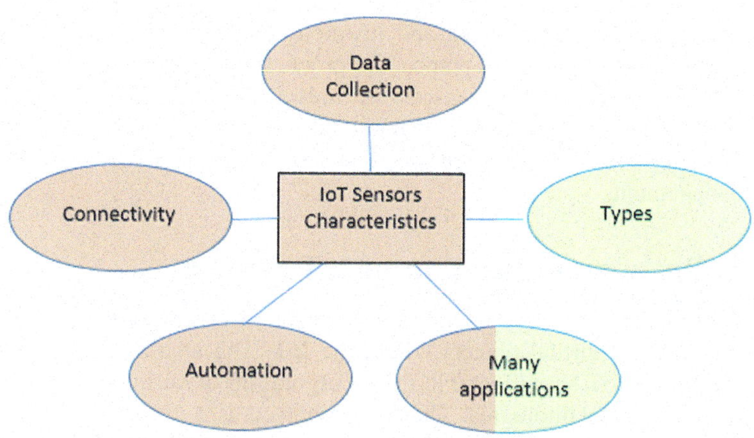

FIGURE 5.3 IoT sensors characteristics.

➢ Data Collection: They collect information on variables like motion, pressure, temperature, humidity, light, and chemical composition.

➢ Connectivity: One of the most crucial features of IoT sensors is connectivity. Sensors should support many wireless protocols, including Wi-Fi, Bluetooth, Zigbee, LoRa, and NB-IoT, for data transfer, and be able to interface with a variety of devices and platforms.

➢ Automation: They enable automated reactions to environmental changes, improving productivity across a range of applications, including healthcare, smart homes, and industrial automation.

➢ Many Applications: Internet of Things sensors are used in many different industries, including agriculture (soil condition monitoring), healthcare (patient vitals tracking), smart cities (traffic monitoring), and environmental monitoring (air quality evaluation).

➢ Types: Depending on their application and design, sensors can be classified as either active or passive, analog or digital, scalar or vector, and each has a distinct purpose.

4. IoT development history

The spread of Internet access, shifting consumer demands, and technology breakthroughs have all contributed to the considerable evolution of the IoT over time. Table 5.1 represents a timeline of the significant turning points in the development of IoT technology (Collett, 2020; Shehu Yalli et al., 2024).

5. Basics of sensors

To fully understand sensors and their operations, this section will highlight the key concepts related to these fields.

5.1 The functioning of sensors

Sensors are instruments that measure and identify physical characteristics in the surroundings and then translate this information into signals that can be deciphered. They are essential to many applications, particularly to the IoT, which makes it possible to link and gather data from the real world (Raghava Deepthi et al., 2023).

Sensors respond to specific stimuli, including light, temperature, motion, humidity, or pressure. When a sensor detects a change in its environment, it generates an output signal, which can be either analog (continuous) or digital (discrete). This output can then be transmitted to a monitoring system or processed locally.

Additionally, various processes occur to enhance the sensed data flow. These processes include pre-processing techniques such as filtering and noise reduction, as well as cloud computing for further analysis. After processing, decision-making takes place, which may involve automated algorithms or optional human input. Finally, commands and notifications are sent to actuators, such as turning on a fan or adjusting lighting.

All of these steps are part of the feedback loop. The overall flow of sensor data can be summarized in a visual representation, as shown in Fig. 5.4.

TABLE 5.1 IoT history.

Era	Key developments
Early concepts (1960s–1980s)	ARPANET introduced connected devices. The first IoT device—a modified toaster at Carnegie Mellon university—could transmit its toasting status over the internet.
IoT emergence (1990s)	The phrase "internet of things" was first used at MIT by Kevin Ashton. Automatic identification and data collection were made possible by RFID technology.
Growth of connectivity (2000)	The ITU released a framework after seeing the potential of IoT. An important turning point in IoT growth was reached when the number of linked devices skyrocketed.
Standardization and development (2010s)	With support for billions of devices, IPv6 increased the number of IP addresses available. Big businesses like IBM, Cisco, and GE made significant investments in IoT technologies.
Mainstream adoption (late 2010s)	The market for linked devices was strengthened by the rise in popularity of consumer IoT goods like Fitbit and nest.
AI and edge integration (2020s)	Real-time decision-making and more intelligent data processing are made possible by integration with AI. By processing data closer to the source, edge computing boosted processing speed and decreased latency.

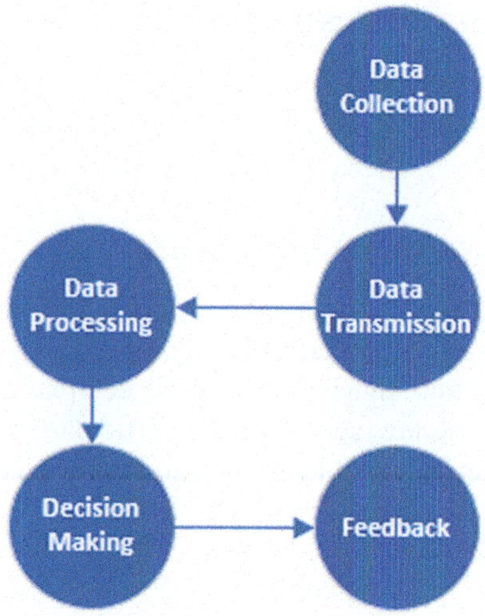

FIGURE 5.4 Sensor data flow.

5.2 Key components of a sensor

As explained previously, sensors are devices for event detection to provide the user with the required data type based on the specific application. Fig. 5.5 represents the basic elements of the sensors, such as the sensing section, transduction section, and signal processing section. The details of these sections are as follows (Domínguez-Bolaño et al., 2022; Meng, 2020):

➢ Sensing section: The component of the sensor that reacts to changes in physical variables and interacts with the surroundings.
➢ Transduction section: Produces a readable signal from the physical change that the sensing element detects.
➢ Processing section: This can be digital signals, such as binary data, or analog signals, such as voltage or current, that are received by this section for preparation using different processes, such as amplification and filtering, to be transmitted to other systems or devices.

5.3 Classification of sensors

Sensors are grouped into various categories, as summarized in Fig. 5.6. These categories depend on different properties such as operation power requirements (active and passive sensors), environment contact (contact and non-contact sensors), Signal Types (analog and digital sensors), and type of signal detection. Details of each classification are as follows (Manhas et al., 2020; Naresh & Lee, 2021):

➢ Active and passive sensors: Active sensors require an external power source to operate, such as microphones and thermistors. In contrast, passive sensors do not need any external power, for example, thermocouples and photodiodes.

FIGURE 5.5 Sensor's main components (Meng, 2020).

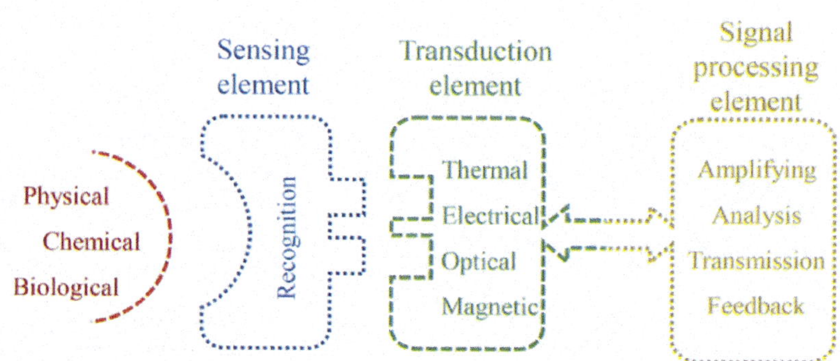

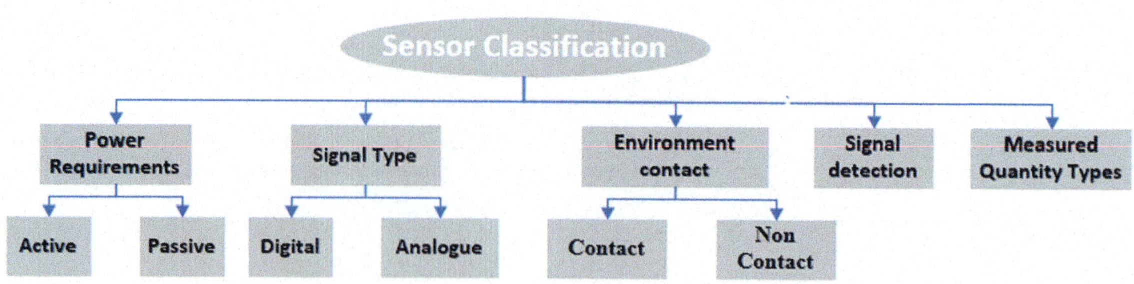

FIGURE 5.6 Sensor classifications.

- Contact and non-contact sensors: Contact sensors require physical contact with their environment, for example, temperature sensors. On the other hand, non-contact sensors do not require any physical contact, such as optical and magnetic sensors.
- Analog and digital sensors: Analog sensors convert physical quantities into analog signals, while digital sensors produce output as pulses.
- Signal detection: The sensors can be categorized based on the type of signal detection, including physical, chemical, and thermal.
- Measured Quantity Types: In this category, sensors are classified according to the data sensing type such as Temperature Sensors (Thermocouples, Thermistors), Pressure Sensors (Barometers), Proximity Sensors (Inductive, Capacitive), Motion Sensors (Accelerometers, Gyroscopes), Light Sensor (Photodiodes, Phototransistors), Humidity Sensors (Hygrometers, Capacitive humidity sensors) and Gas and Chemical Sensors(Gas detectors, pH sensors).

5.4 Sensor characteristics and parameters

The performance and appropriateness of sensors for certain applications are determined by several important features and criteria. Selecting the appropriate sensor for an IoT system or any other application requires an understanding of these criteria. The following are the primary traits and factors to take into account, as illustrated in Table 5.2 (Amalraj et al., 2019; Collett, 2020; Jangra, 2010; Krishnamurthi et al., 2020).

6. IoT sensors architecture

To accomplish the basic tasks of an IoT system effectively, it should be developed based on a reliable and well-defined architecture. Layered architecture is the most used model in IoT systems, especially the 3-tier architectures, which encompass the perception, network, and application layers. This layered architecture approach ensures scalability, interoperability, and functionality.

The 3-tier architecture is widely adopted due to its simplicity in developing effective IoT systems by addressing the application requirements, starting from smart cities and healthcare to industrial automation and smart agriculture. According to Mansour et al. (2023) the 3-tier architecture provides a robust foundation to design a scalable and efficient IoT system, tackling most of the current difficulties, such as data security, interoperability, and handling a massive amount of sensed data that is generated from the sensor devices. Stakeholders may ensure optimal performance and functionality by developing IoT systems tailored to their specific cases by understanding these designs. Table 5.3 summarizes the concept and function of the basic IoT layers.

Besides the 3-tier architecture, the research community has proposed a variety of alternative architectures to address the evolving needs of IoT systems, such as interoperability, security, and domain-specific requirements. The work by Kanellopoulos et al. classified IoT architecture based on the application fields and the domain-related requirements for each one (Kanellopoulos et al., 2023). For example, smart city systems focus on real-time data

TABLE 5.2 Sensor characteristics.

Characteristic	Description
Accuracy	The degree to which the true value and the measured value agree. Crucial in industries like medical gadgets and environmental monitoring.
Precision	Consistency across several measurements made in the same circumstances. Pertinent to guaranteeing consistent, repeatable readings.
Sensitivity	The ability to detect even the slightest changes in the variable being observed. It is essential for applications that require precise measurement.
Range	The range of values that the sensor can precisely measure. Must be in line with the application's anticipated values.
Linearity	The degree to which the input and output of the sensor are proportionate. Makes data interpretation easier and guarantees correct correlations.
Response time	After the measured variable changes, the amount of time it takes for the sensor to stabilize. Crucial in dynamic settings.
Power consumption	The energy consumption of the sensor. Vital for energy-sensitive applications or systems that run on batteries.
Operating conditions	Environmental elements that affect the sensor's operational efficiency include temperature, humidity, and pressure. Crucial for selecting the appropriate sensor for the task.

TABLE 5.3 Summary of the IoT 3-tier architecture.

Layer	Description	Key components	Functions	Challenges	Advancements
Perception layer (sensing/ Physical layer)	The foundational layer captures real-time data from the physical environment.	Sensors (e.g., temperature, motion, gas), actuators	• Data acquisition and control • Interaction with the environment	• Limited processing capability • Security vulnerabilities	• Advanced sensors • Improved communication protocols
Network layer	Ensures seamless data transfer between perception and application layers.	• Wireless: Wi-Fi, bluetooth, Zigbee • Wired: Ethernet, Fiber optics	• Reliable data transmission • Device-to-device communication	• Scalability issues • Privacy concerns • Network reliability	• 5G and IoT • Specific protocols • Standardization efforts
Application layer	Delivers data analytics and machine learning to end-users.	Data analytics, machine learning, cloud computing	• Data interpretation and decision making • Enhancing user experience	• Data privacy • Complexity in data management • Security threats	• AI-driven analytics • Edge computing capabilities

processing and integration with city infrastructure, while healthcare systems may prioritize patient data security. These domain-specific architectures enable the IoT systems to meet the unique requirements of each application field, ensuring optimal performance and functionality.

Kassab and Darabkh highlighted the importance and effectiveness of the layered architecture and showed the limitations of the classic 3-tier architecture (Kassab & Darabkh, 2020). Therefore, many researchers suggested the 4- and 5-tier architectures. Regarding the 4-tier architecture, a middleware layer has been added for data processing, protocol translation, and interoperability, while, in a 5-tier architecture, two extra layers were added: the data link and transport layers. They extend beyond the original architecture. Mrabet et al. (2020) introduced a 5-layer architecture: perception, data link, network, transport, and application layer, which augmented the traditional 3-tier reference architecture. The proposed architectures offer different and more granular ways through which scalability and performance can be improved in complex IoT systems.

On the other hand, Lombardi et al. explained how the Service-Oriented Architecture (SOA) and middleware-based architectures can be a good replacement for the traditional layered architecture model (Lombardi et al., 2021). Where SOA architecture focuses on the modularity and reusability that enable the IoT system to integrate with other heterogeneous systems seamlessly, this type of architecture addresses the interoperability challenges in heterogeneous environments.

In recent years, the research community has focused on integrating IoT systems with modern solutions such as edge and cloud computing because of their important roles in enhancing and leveraging the IoT system capabilities. Edge computing reduces latency and bandwidth usage by bringing the data processing closer to the IoT system, while cloud computing provides scalable storage and advanced data analysis capabilities. Hybrid architectures that include both edge and cloud computing are increasingly becoming popular.

7. IoT networking technology

As mentioned previously, the basic function of IoT is to connect sensors to different devices. According to this, the communication protocols need to be understood. To achieve better data transmission, these protocols guarantee the efficient collection, processing, and analysis of sensor data. Some of the most utilized wireless communication protocols in IoT applications are listed in Table 5.4 (Malathy et al., 2023; Mansour et al., 2023; Shehu Yalli et al., 2024; Čolaković et al., 2021).

Additionally, selecting the appropriate communication protocol for IoT sensors is crucial for the success of any IoT project. This choice is influenced by several aspects, as shown in Table 5.5, all of which affect the system's overall efficacy, performance, and dependability.

8. IoT data communication protocols

In the realm of the IoT, communication protocols play a vital role in facilitating the exchange of data between devices, ensuring interoperability, and optimizing performance. Many data communication protocols have been found, such as MQTT and CoAP, to offer flexibility and

TABLE 5.4 IoT networking technology.

Technology	Range	Benefits	Drawbacks
Wi-Fi	Up to 100 m (indoor)	Fast data transfer speeds, well-established for residential/commercial use.	High power consumption and limited battery life for sensors.
Bluetooth/BLE	Up to 100 m (BLE shore)	Low power usage (especially BLE), easy to implement.	Limited data transfer speeds and shorter range compared to Wi-Fi.
Zigbee	Up to 100 m	Low power, ideal for battery-powered devices, supports mesh networking for larger networks.	A more complex setup. Requires a dedicated gateway and lower data speeds.
LoRaWAN	Up to 15 km (rural)	Long-range capabilities, mesh networking for devices, work through obstructions.	Requires a dedicated gateway and lower data speeds.
NB-IoT	Similar to cellular coverage	Long battery life, can be used in cities/buildings.	Requires cellular service and may incur service charges.
Sigfox	Long-range	Server for low-cost M2M and IoT applications that need to connect across long distances.	Reduced downlink capability, increased latency, and decreased data rates
Thread	15 and 50 m, depending on the individual environmental conditions	Strong security, self-healing mesh, suitable for wide applications.	Not compatible with every IoT device, not suitable for fast applications, such as streaming videos, and is costly to implement.

TABLE 5.5 Communication protocol selection criteria.

Factor	Description
Range and Coverage	The distance between the sensors and the hub. Some protocols (e.g., bluetooth) are short-range, while others (e.g., LoRaWAN) are long-range.
Environment	Assesses environmental factors like outdoor versus indoor settings, interference, and urban density, which can affect signal propagation and coverage.
Required bandwidth	Determines the data rate needed. Low data rates are acceptable for simple tasks, while applications that require high data rates may need more bandwidth.
Payload size	Optimizing payload size in an IoT network means choosing the right protocol, minimizing data, and applying techniques like aggregation and compression to ensure reliable and efficient communication.
Energy efficiency	Evaluate how power-efficient the technology is. Low-energy protocols like BLE and LoRaWAN are ideal for battery-operated devices.
Duty Cycle	Considers whether the protocol supports sleep modes or low-power states to conserve energy.
Device Count	Estimates how many devices will be connected. Protocols like Zigbee and thread support many devices.
Connection type	Decides whether point-to-point, mesh, or star topology is best suited for the system, depending on the communication needs and device density.
Future development	Assess how easily the system can scale to accommodate more devices or sensors, and how it can evolve.
Real-time communication	Determines whether your application can tolerate low latency in essential tasks like industrial automation, health monitoring, etc.

adaptability in addressing unique application and environment requirements. Table 5.6 summarizes some of the available data communication protocols used in the IoT in terms of features, roles, use cases, and the corresponding layer they deal with in the context of IoT's 3-tier architecture.

TABLE 5.6 Summary of the data communication protocols.

Protocol	Features	Corresponding layer	Role in IoT	Use case
MQTT	Lightweight, publish/subscribe model, low bandwidth consumption	Application layer	Facilitates efficient messaging between devices	Home automation, industrial monitoring
CoAP	Request/response model, optimized for constrained devices	Application layer	Enables resource-constrained devices to communicate over the internet	Smart lighting, building automation
HTTP/HTTPS	Widely used, supports RESTful APIs, and secure via HTTPS	Application layer	Enables data exchange between clients and servers	Web services integration, device management
DDS	Real-time data exchange, publish/subscribe model	Application layer	Supports high-performance communication for critical applications	Autonomous vehicles, industrial automation
XMPP	Extensible, real-time communication, decentralized	Application layer	Facilitates instant messaging and presence information	Smart home systems, real-time monitoring
OPC UA	Platform-independent, secure, and reliable	Application layer	Enables data exchange in industrial automation	Factory automation, process control
LwM2M	Device management, lightweight	Application layer	Manages IoT devices, including firmware updates	Smart metering, remote monitoring
WebSocket	Full-duplex communication, low latency	Application layer	Provides real-time data exchange	Live updates in smart home applications
gRPC	High-performance, supports streaming	Application layer	Enables efficient microservices communication	Real-time data processing in IoT
SNMP	Network management, monitoring	Network layer	Facilitates the management of networked devices	Performance monitoring, fault detection
Z-wave	Wireless, low-power, mesh networking	Network layer	Controls home automation devices	Smart lighting and security systems
BACnet	Building automation, interoperable	Application layer	Manages HVAC, lighting, and security	Smart building management
Modbus	Simple, robust for industrial applications	Network layer	Connects supervisory computers with remote devices	SCADA systems in industrial settings
REST	Stateless uses HTTP requests	Application layer	Facilitates easy integration with web technologies	IoT APIs, web services
QUIC	Reduces latency, secures connections	Network layer	Provides fast and secure data transmission	Real-time IoT applications requiring speed
RPL	Routing for low-power networks	Network layer	Supports efficient routing in IoT networks	Smart metering, environmental monitoring

9. Essential IoT applications

The IoT has many applications that affect every part of our lives, including the automation of buildings and homes. Many businesses offer a wide range of advanced IoT solutions for controlling and monitoring smart homes and intelligent buildings. The most popular applications of IoT are summarized in Table 5.7 (Gazis et al., 2025; Hadi Masmali et al., 2023).

10. Smart sensors and IoT technologies in academic environments

10.1 Varieties and roles of smart sensors in universities

Smart sensors deployed in academic settings encompass diverse types tailored to distinct functions critical for campus management and educational enhancement. Environmental sensors measure parameters such as temperature, humidity, lighting, and air quality, supporting the creation of comfortable, energy-efficient learning spaces (Muangpoo & Pullteap, 2023). Occupancy sensors monitor classroom usage and facilitate automated control of lighting and HVAC systems, optimizing resource consumption (Muangpoo & Pullteap, 2023). Additionally, biometric and RFID sensors serve in access control and attendance monitoring, streamlining administrative processes while enhancing security (Blazevic & Riehle, 2023).

Specialized sensors address safety concerns, such as gas detectors installed in chemical laboratories to identify toxic emissions, mitigating risk for students and staff in sensitive environments (Majeed et al., 2024). Motion and proximity sensors enable real-time tracking of asset and personnel movement, contributing to spatial analytics that inform campus facility planning and user behavior studies (Koszewski et al., 2025). Collectively, these smart sensor categories enable universities to realize responsive, data-driven ecosystems.

10.2 IoT architectures applicable to educational settings

University IoT systems typically adopt multilayered architectures designed to support scalability, interoperability, and real-time data processing. A common paradigm involves the perception layer, consisting of sensor nodes and actuators that gather raw data from physical environments (Blazevic & Riehle, 2023; Walczak et al., 2023). This layer connects to the network layer, which ensures reliable, secure transmission of data using wireless protocols such as Wi-Fi, Zigbee, and cellular networks.

The processing layer, often leveraging edge computing capabilities, performs local data analysis and preliminary decision-making to reduce latency and bandwidth usage, a crucial factor in dynamic academic scenarios like occupancy detection and environmental adjustments (Walczak et al., 2023). Cloud platforms complement this arrangement by providing centralized storage, advanced analytics, and integration with institutional IT systems (Blazevic & Riehle, 2023). This layered approach balances immediate responsiveness in classrooms with

TABLE 5.7 IoT applications.

Application area	Description
Smart cities and transport	IoT enhances urban infrastructure by improving fleet management for performance and geolocation, lowering traffic congestion with smart cars, and more.
Healthcare	Smart hospital beds and wearable sensors allow for remote patient health monitoring, enhancing treatment and facilitating the early identification of health problems.
Wearable devices	IoT devices that provide real-time location and health data, such as GPS belts, smart glasses, and fitness trackers, improve user experience.
Maintenance management	Predictive maintenance is made easier by IoT in conjunction with computerized maintenance management systems (CMMSs) and enterprise asset management (EAM).
Smart grids and energy saving	Energy grid sensors and meters assist in tracking and managing power consumption, giving consumers and energy suppliers feedback.
Water supply	IoT optimizes water distribution by tracking water use, identifying issues, and sending real-time data to customers and utilities.
Agriculture	IoT-enabled agriculture increases agricultural output by automating environmental control, increasing crop yields, and conserving energy.
Smart education	School/University IoT systems typically adopt multilayered architectures designed to support scalability, interoperability, and real-time data processing.

comprehensive oversight at the campus level, enabling tailored interventions and scalable management.

10.3 Data flow: Collection, transmission, and analytical processes in universities

Smart sensor data flows through integral steps essential to unlocking actionable insights within educational contexts. Initially, sensors continuously or intermittently collect environmental or contextual data relevant to classroom conditions, safety, or attendance (Koszewski et al., 2025; Shahbaz et al., 2023). This data transits from perception devices to edge gateways, facilitating preprocessing tasks such as filtering, aggregation, or encoding to enhance transmission efficiency and data quality (Walczak et al., 2023).

Subsequent cloud-based platforms aggregate multiple data streams, enabling sophisticated analytics including pattern detection, usage forecasting, and behavior modeling (Koszewski et al., 2025). These insights empower facility managers to optimize energy consumption, security staff to respond proactively to hazards, and educators to customize learning environments (Muangpoo & Pullteap, 2023; Shahbaz et al., 2023). Furthermore, integration with learning management systems can enable personalized feedback loops supporting student engagement.

10.4 Security and privacy considerations for IoT in academia

The proliferation of IoT devices in universities raises significant concerns related to security and privacy, demanding rigorous safeguards. Data collected from smart sensors often includes sensitive personal information, such as biometric identifiers or location data, necessitating compliance with privacy regulations and ethical standards (Blazevic & Riehle, 2023). Network vulnerabilities inherent in wireless IoT communications must be mitigated through encryption, authentication protocols, and intrusion detection mechanisms to prevent unauthorized access or tampering (Majeed et al., 2024).

Edge computing offers privacy benefits by limiting raw data exposure beyond the local environment, but it also requires secure device management and firmware updates to address evolving threats (Walczak et al., 2023). In addition, universities must establish clear policies governing data ownership, sharing, and retention, coupled with user awareness programs to foster trust and transparency (Blazevic & Riehle, 2023). Addressing these security and privacy dimensions is fundamental to sustaining effective and responsible deployment of IoT-enabled smart sensors in academic environments.

11. Integration of smart sensors and IoT in University classes

Recent IEEE literature reveals diverse implementations of smart sensors and IoT within university classrooms, showcasing tangible improvements in operational efficiency and educational quality. For example, Blazevic and Riehle (Blazevic & Riehle, 2023) report on a smart campus infrastructure leveraging IoT sensors to automate attendance tracking and facility management with minimal human intervention. Shahbaz et al. demonstrate how IoT integration can enrich e-learning environments by enabling real-time interaction and adaptive content delivery, fostering more immersive educational experiences (Shahbaz et al., 2023).

11.1 Key applications in academic settings

Attendance Monitoring: Traditional attendance systems often suffer from manual errors and inefficiencies. RFID and biometric sensors embedded in IoT networks enable automated, precise attendance tracking, reducing administrative overhead and ensuring data accuracy (Blazevic & Riehle. 2023). These systems can integrate seamlessly with academic information systems, providing instantaneous attendance reports and analytics.

Energy Management: IoT-driven environmental sensors and occupancy detectors allow dynamic control of lighting, heating, ventilation, and air conditioning within classrooms. Muangpoo and Pullteap illustrate how integrating AI with sensor networks optimizes energy consumption without compromising comfort, leading to significant cost savings (Muangpoo & Pullteap, 2023). Similarly, smart green campus initiatives employ sensor data to monitor resource usage and adjust operations sustainably (Mootien & Nagowah, 2023).

Interactive Learning Experiences: IoT devices facilitate interactive pedagogical modes, such as sensor-enabled student response systems and real-time feedback tools. Shahbaz et al. describe how sensor integration supports personalized learning paths, immediate performance tracking, and virtual collaboration, thereby enhancing student engagement and motivation (Shahbaz et al., 2023).

11.2 Challenges in deployment and mitigation strategies

Integrating smart sensors and IoT in educational environments encounters several obstacles. Hardware installation complexity and sensor interoperability can hinder cohesive system functioning (Blazevic & Riehle, 2023). Diverse communication protocols and legacy IT infrastructure present compatibility issues, demanding careful selection of standard-compliant technologies and modular system design. Privacy remains paramount, particularly when handling biometric and location data; thus, stringent encryption, access controls, and anonymization techniques are essential (Majeed et al., 2024).

Scalability is another concern, as expanding sensor networks across entire campuses requires robust edge computing and network capacity planning. Incorporating edge analytics reduces latency and bandwidth demand, as evidenced by Walczak et al.'s approach to universal design-focused sensor acceptance (Walczak et al., 2023). Additionally, stakeholder training and clear policy frameworks contribute to smoother adoption and sustained operation.

11.3 Impact on pedagogy and learner outcomes

The infusion of IoT and smart sensors into classrooms reshapes teaching practices by enabling data-driven instruction and adaptable environments. Educators gain granular insights into student attendance patterns, engagement levels, and environmental factors influencing learning, allowing timely interventions (Shahbaz et al., 2023). Active learning strategies are enhanced through sensor-assisted interaction, fostering collaboration and responsiveness within class sessions.

Student outcomes improve as personalized feedback mechanisms encourage continuous progress monitoring and tailored support. The ability to automate routine tasks liberates instructors to focus on pedagogical innovation and student interaction (Blazevic & Riehle, 2023; Shahbaz et al., 2023). Moreover, smart classrooms contribute to inclusivity by adjusting physical conditions to diverse learner needs, aligning with principles of universal design (Walczak et al., 2023).

In sum, the integration of smart sensors and IoT creates a synergistic platform that advances classroom management, resource efficiency, and educational effectiveness, setting a precedent for future academic innovations.

12. Discussion

12.1 Advantages of IoT-enabled smart sensors in university classrooms

The deployment of smart sensors integrated with IoT technologies significantly enhances student engagement, operational efficiency, and the generation of data-driven insights within university settings. These systems support interactive learning environments where real-time feedback and adaptive pedagogies actively involve students, fostering participation beyond traditional lectures (Shahbaz et al., 2023). Automation of administrative tasks such as attendance tracking eliminates manual errors, reducing workload and freeing educators to focus on instructional quality (Blazevic & Riehle, 2023).

Efficiency gains are evident in resource management, with sensor-based energy control optimizing lighting, heating, and ventilation to match occupancy patterns. This dynamic adjustment not only lowers operational costs but also aligns with sustainability goals central to many academic institutions (Mootien & Nagowah, 2023; Muangpoo & Pullteap, 2023). Furthermore, continuous data collection enables granular analysis of classroom usage, environmental conditions, and learner behavior, offering actionable insights for refining educational practices and campus infrastructure (Koszewski et al., 2025). Such intelligence supports proactive strategies in facility management and targeted pedagogical interventions, driving measurable improvements in learning outcomes.

12.2 Challenges to adoption: Financial, infrastructure, and ethical dimensions

Despite clear benefits, several obstacles hinder the broad adoption of smart sensors and IoT integration in university classrooms. Initial investment costs for sensor procurement, network expansion, and system maintenance remain substantial, often deterring institutions with limited budgets (Blazevic & Riehle, 2023). Infrastructure challenges arise from the need to retrofit existing buildings with IoT-compatible hardware and reliable connectivity, compounded by diverse sensor standards that complicate seamless interoperability (Walczak et al., 2023).

Privacy concerns present critical barriers, especially where biometric or location data are collected. The potential for unauthorized data access, surveillance, or misuse requires rigorous safeguards and transparent governance frameworks to protect individual rights (Majeed et al., 2024). Ethical considerations extend to informed consent, data ownership, and responsible use of sensitive information in research and academic administration (Blazevic & Riehle, 2023; Majeed et al., 2024). Furthermore, addressing these issues demands continuous training and policy development to ensure compliance and foster trust among stakeholders.

12.3 Comparative review of implementations: Achievements and persistent issues

Recent deployments, documented in IEEE sources, reveal both encouraging successes and unresolved complexities. Implementations such as automated attendance and energy-efficient smart classrooms demonstrate measurable improvements in operational efficiency and user satisfaction (Blazevic & Riehle, 2023; Muangpoo & Pullteap, 2023). IoT-enhanced e-learning environments also exhibit increased student motivation and personalized instruction capabilities (Shahbaz et al., 2023). These case studies validate integrated smart sensor systems' technical feasibility and pedagogical promise.

However, unresolved issues remain prominent. Integration challenges persist due to heterogeneous sensor ecosystems and legacy IT infrastructure constraints, leading to fragmented system performance (Blazevic & Riehle, 2023). Scalability difficulties appear in managing and securing expanding sensor networks, with edge computing proposed as a partial solution yet requiring further refinement (Walczak et al., 2023). Privacy protections, although addressed through encryption and anonymization strategies, face ongoing risks amid evolving cyber threats and policy gaps (Majeed et al., 2024). These factors underscore the necessity for holistic frameworks that balance technological innovation, security, and user-centric values.

12.4 Strategies to foster successful integration

Effective adoption of IoT-based smart sensors in university classrooms hinges on multi-dimensional strategies. Prioritizing modular, interoperable sensor designs facilitates smoother integration with existing systems and accommodates future expansion (Walczak et al., 2023). Emphasizing edge computing capabilities reduces latency and bandwidth demands while enhancing data privacy by limiting transmission of sensitive raw data (Walczak et al., 2023). Financially, phased implementation coupled with demonstrable pilot projects can secure institutional buy-in and justify budget allocations (Blazevic & Riehle, 2023).

Establishing clear privacy policies and ethical standards, aligned with legal requirements and stakeholder expectations, builds trust and ensures responsible data stewardship (Majeed et al., 2024). Comprehensive training programs for technical staff and end-users enhance system usability and security awareness (Blazevic & Riehle, 2023). Finally, collaborative efforts involving educators, administrators, IT specialists, and students promote alignment of technological solutions with educational objectives, ensuring sustained impact and continuous improvement (Shahbaz et al., 2023). These recommendations collectively form a foundation for harnessing the full potential of IoT-driven smart classrooms while mitigating associated risks.

13. Challenges and limitations

In terms of challenges and limitations, IoT challenges and limitations are increasing more and more as the IoT evolves and engages more in our daily lives. In this section, we will classify the IoT limitations and challenges per layer concerning the IoT 3-tier default architecture. The IoT perception layer experiences many challenges and limitations. Sensor reliability and accuracy, energy consumption, data overload and management, environmental adaptability, and security risks were considered the main limitations and challenges that affected the perception layer performance. Table 5.8 discusses these challenges and limitations in detail.

TABLE 5.8 The IoT perception layer challenges and limitations.

Challenge	Description	Implications
Sensor reliability and accuracy	Ensuring sensors provide accurate data despite influences like temperature and humidity.	Inaccurate data readings require regular checks and maintenance, which can be resource-intensive (Mansour et al., 2023).
Energy consumption	Battery-operated devices must use energy efficiently to extend battery life, which can restrict their capabilities and data collection rates.	Necessitates the development of energy-harvesting solutions and low-power designs.
Data overload and management	The increase in sensors can lead to an overwhelming amount of data, making management and analysis challenging.	Requires effective strategies for filtering, aggregating, and storing data; may need more computational resources.
Environmental adaptability	Sensors must function well in various conditions (e.g., dust, moisture, extreme temperatures).	The development of durable sensors with protective features is essential for reliable performance (Kassab & Darabkh, 2020).
Security risks	The perception layer faces risks like physical tampering and unauthorized access, necessitating strong security measures.	Increases costs and complexity; requires secure measures such as secure booting and hardware protections (Bashir et al., 2022).

The IoT network layer is crucial, as it facilitates the data communication between IoT system components. However, it encounters many challenges and limitations that affect its performance, security, and scalability. Table 5.9 exposes some of these challenges in detail.

Finally, the IoT application layer utilizes the collected data from sensors and analyzes it to make the appropriate decision. Some of the key challenges and limitations that affected this layer are discussed in Table 5.10.

TABLE 5.9 The IoT Network layer challenges and limitations.

Challenge	Description	Implications
Scalability	Managing the increased volume of data from numerous IoT devices is complex, requiring efficient transmission to minimize latency and bandwidth use.	A scalable architecture is necessary for future growth and seamless data flow (Kassab & Darabkh, 2020).
Interoperability	The lack of standardized protocols creates communication barriers among devices from different manufacturers.	Efforts like IPv6 aim to enhance interoperability; however, security during data transmission remains critical (Mansour et al., 2023).
Security and privacy	Sensitive data transmission raises security concerns, making IoT networks vulnerable to attacks.	Robust encryption and authentication are essential, but they can complicate network performance (Bashir et al., 2022).
Load balancing and irregular traffic	Effective load balancing is needed to manage irregular data traffic and prevent congestion.	Strategies must ensure even distribution of traffic to optimize network resource utilization (Kanellopoulos et al., 2023).
Cloud/Fog/Edge engagement	Integration of cloud, fog, and edge computing enhances capabilities but complicates data management and security.	Seamless interaction between layers is crucial for maximizing IoT benefits (Hamdan et al., 2020).
Interference problems	Operating alongside other technologies leads to interference, affecting communication reliability.	Effective mitigation strategies are necessary to ensure stable network performance (Mansour et al., 2023).
Lack of standardization solutions	The absence of standardized communication protocols complicates device integration.	Establishing universal standards is critical for interoperability (Bashir et al., 2022).
Device-to-device (D2D) communications	Increasing D2D communication demands can overwhelm conventional gateways, leading to inefficiencies.	Addressing these demands is vital for enhancing network performance (Kassab & Darabkh, 2020).
Short communicating Messages	IoT devices often send short messages, posing challenges for traditional protocols optimized for longer packets.	Adapting communication strategies is essential to reduce congestion (Ghayvat et al., 2015).
Bandwidth and power consumption	Varying bandwidth requirements must be managed to prevent congestion; power consumption is critical for battery-operated devices.	Efficient strategies are needed to balance performance and energy usage (Ali et al., 2025).
Delay tolerance and reliability	Many applications require low latency; any delay can negatively impact performance.	Ensuring reliable data transmission while meeting delay constraints is a challenge (Mansour et al., 2023).
Heterogeneity of network protocols	Diversity in protocols complicates seamless communication among devices.	Protocol designers must ensure compatibility and performance (Bashir et al., 2022).
Mobility	Moving devices present challenges in maintaining stable connections and efficient data transfer.	Sophisticated routing strategies are required in dynamic environments (Kanellopoulos et al., 2023).
MTU size Adaptation	Determining the optimal Maximum transmission unit size is challenging; incorrect sizes can lead to fragmentation.	Proper MTU sizing is crucial for maintaining performance and data integrity (Burhan et al., 2018).
Wireless communication vulnerabilities	Wireless networks face vulnerabilities due to channel errors and interference, leading to packet loss.	These issues complicate traditional TCP/IP operations and may reduce throughput (Ali et al., 2025).

TABLE 5.10 The IoT application layer challenges and limitations.

Challenge	Description	Implications
Data management and processing	Managing the extensive data generated by IoT devices requires effective strategies for storage, retrieval, and real-time analysis.	Robust database systems and advanced analytics capabilities are necessary (Ali et al., 2025).
User interface design	Creating intuitive user interfaces is essential for user adoption, requiring a balance between functionality and usability.	Enhances user experience while meeting performance requirements (Kassab & Darabkh, 2020).
Integration with existing systems	Seamless integration with existing systems and infrastructure is complex due to compatibility issues with various platforms and protocols.	Extensive testing and validation are vital for creating a unified IoT ecosystem (Mansour et al., 2023).
Security and privacy concerns	Handling sensitive data raises significant security and privacy issues, necessitating stringent measures like encryption and secure authentication.	Compliance with regulations such as GDPR complicates data handling practices (Bashir et al., 2022).
Scalability and performance	Applications must scale to accommodate growing numbers of devices and data traffic while maintaining high performance under varying loads.	Cloud-based solutions or edge computing may be needed to distribute processing workloads effectively (Ghayvat et al., 2015).
Interoperability issues	Diversity among IoT devices can lead to interoperability challenges, requiring effective communication across different protocols and standards.	Standardization efforts are essential for seamless interactions between devices (Kassab & Darabkh, 2020).
Real-time data processing	Many applications require immediate data processing, which can be challenging due to complexities in data transmission and processing.	Implementing low-latency architectures and utilizing edge computing can help address these challenges (Ali et al., 2025).

14. Future trends in IoT sensors

The growth of the IoT is ongoing, and its integration into our daily lives is increasing. This expansion and incorporation are driven by technological advancements. The following points examine future trends that may influence IoT sensor networks:

- Enhanced Connectivity: The 5G and 6G technologies revolutionized the IoT sensor networks with a significantly higher throughput and reduced latency. It enables a massive number of devices to connect to the IoT sensor network simultaneously. This advancement will leverage the current IoT systems to enhance their responsiveness and make them intelligent IoT systems
- Edge and Fog Computing: Integrating edge and cloud computing technologies with the IoT sensor networks is a significant trend, enabling real-time analysis and decision-making.
- AI and Machine Learning Integration: The shift to integrate edge and cloud computing with IoT sensor networks opens the door for using AI and machine learning to enhance IoT sensor networks' decision-making and identify patterns, predict failures, and optimize resource utilization.
- Security and Privacy: As the IoT sensor networks expanded and the number of connected devices increased, this led to an increase in the possibility of security threats and privacy violations. Using future technologies such as blockchain, differential privacy, and federated learning becomes crucial to protect user data.
- Standardization efforts: The lack of IoT protocol standardization and architecture becomes a significant barrier to interoperability and scalability. Developing universal standards that facilitate seamless communication between devices from different manufacturers is becoming crucial. Organizations and alliances are likely to focus on creating common frameworks that support interoperability.
- Sustainable and Energy-Efficient Solutions: With the increasing number of IoT devices, sustainability and energy efficiency will become critical. Low-power sensors and energy-harvesting technologies to reduce the environmental impact of IoT deployments are needed. Besides energy-efficient communication protocols.

15. Conclusion

This study highlights the transformative role of IoT-enabled smart sensors in university environments, revealing significant advancements in educational and operational domains. Core findings demonstrate that these technologies facilitate automated attendance tracking, efficient energy management, and enriched interactive learning experiences, each contributing to enhanced classroom dynamics and campus sustainability. Smart sensors' real-time data collection and edge computing capabilities underpin responsive adjustments in physical conditions and pedagogical practices, confirming their capacity to support universal design principles and inclusive education. Furthermore, the deployment of such systems generates granular data analytics that inform strategic decisions in facility management and instructional design, improving learner outcomes and campus resource utilization.

Implications for universities focus on the strategic integration of smart sensor networks as essential components of future-ready academic infrastructure. Institutions stand to benefit from investments in interoperable, scalable sensor solutions combined with robust cybersecurity and privacy frameworks, which are critical for safeguarding sensitive information and sustaining stakeholder trust. Establishing interdisciplinary collaboration among educators, IT professionals, and administrators will optimize adoption processes and enhance the relevance of technological interventions to pedagogical goals. Emphasizing ethical data stewardship and continuous training will further fortify systemic resilience against security threats and privacy concerns, a prerequisite for broad acceptance.

Future research should explore adaptive sensor architectures that dynamically respond to evolving educational needs, incorporating advances in artificial intelligence and machine learning for predictive analytics and personalized learning enhancements. Longitudinal studies assessing the impact of IoT sensor integration on academic performance and campus sustainability metrics are necessary to validate initial successes and guide refinements. Investigations into student and faculty perceptions will shed light on social and cultural factors influencing technology acceptance, enabling more nuanced deployment strategies. Finally, expanding smart sensor applications to encompass broader aspects of campus life, including accessibility, health monitoring, and community engagement, promises to further deepen the contribution of IoT technologies to the university experience.

In sum, IoT-driven smart sensor systems present a compelling pathway for universities to embrace innovation, transforming both educational environments and administrative processes while addressing critical challenges of efficiency, inclusivity, and sustainability.

References

Ali, T. E., Ali, F. I., Dakić, P., & Zoltan, A. D. (2025). Trends, prospects, challenges, and security in the healthcare internet of things. *Computing, 107*(1), 28. https://doi.org/10.1007/s00607-024-01352-4

Amalraj, J. J., Banumathi, S., & John, J. J. (2019). IOT sensors and applications: A survey. *International Journal of Scientific& Technology Research, 8*(8), 998–1003.

Bashir, M. R., Gill, A. Q., & Beydoun, G. (2022). A reference architecture for IoT-Enabled smart buildings. *SN Computer Science, 3*(6). https://doi.org/10.1007/s42979-022-01401-9

Blazevic, M., & Riehle, D. M. (2023). University of things: Opportunities and challenges for a smart campus environment based on IoT sensors and business processes. In *International Conference Internet Things Big Data Security* (pp. 105–114). https://doi.org/10.5220/0011761900003482

Brida, P., Krejcar, O., Selamat, A., & Kertesz, A. (2021). Smart sensor technologies for IoT. *Sensors, 21*(17), 5890. https://doi.org/10.3390/s21175890

Burhan, M., Rehman, R. A., Khan, B., & Kim, B. S. (2018). IoT elements, layered architectures and security issues: A comprehensive survey. *Sensors, 18*(9). https://doi.org/10.3390/s18092796

Čolaković, A., Hasković Džubur, A., & Karahodža, B. (2021). Wireless communication technologies for the internet of things. *Science, Engineering and Technology, 1*(1), 1–14. https://doi.org/10.54327/set2021/v1.i1.3

Collett, A. (2020). *An introduction to IoT: Getting started with the internet of things*. GfA UK Ltd.

Domínguez-Bolaño, T., Campos, O., Barral, V., Escudero, C. J., & García-Naya, J. A. (2022). An overview of IoT architectures, technologies, and existing open-source projects. *Internet of Things, 20*, 100626. https://doi.org/10.1016/j.iot.2022.100626

Gazis, A., Papadongonas, I., Andriopoulos, A., Zioudas, C., & Vavouras, T. (2025). Comprehensive review: Sensor technologies, instrumentation, and signal processing in low-power IoT. *Academia Engineering, 2*(1). https://doi.org/10.20935/AcadEng7541

Ghayvat, H., Mukhopadhyay, S., Gui, X., & Suryadevara, N. (2015). WSN- and IOT-Based smart homes and their extension to smart buildings. *Sensors, 15*(5), 10350–10379. https://doi.org/10.3390/s150510350

Hadi Masmali, F., Miah, S. J., & Noman, N. (2023). Different applications and technologies of internet of things (IoT). *Lecture Notes in Networks and Systems, 464*(January), 41–54. https://doi.org/10.1007/978-981-19-2394-4_5

Hamdan, S., Ayyash, M., & Almajali, S. (2020). Edge-computing architectures for internet of things applications: A survey. *Sensors, 20*(22), 1–52. https://doi.org/10.3390/s20226441

Hossein Motlagh, N., Mohammadrezaei, M., Hunt, J., & Zakeri, B. (2020). Internet of things (IoT) and the energy sector. *Energies, 13*(2), 494. https://doi.org/10.3390/en13020494

Jangra, A. (2010). Wireless sensor network (WSN): Architectural design issues and challenges. *IJCSE) International Journal on Computer Science and Engineering, 02*(09), 3089–3094.

Jia, M., Komeily, A., Wang, Y., & Srinivasan, R. S. (2019). Adopting internet of things for the development of smart buildings: A review of enabling technologies and applications. *Automation in Construction, 101*, 111–126. https://doi.org/10.1016/j.autcon.2019.01.023

Kanellopoulos, D., Sharma, V. K., Panagiotakopoulos, T., & Kameas, A. (2023). Networking architectures and protocols for IoT applications in smart cities: Recent developments and perspectives. *Electronics, 12*(11). https://doi.org/10.3390/electronics12112490

Kassab, W., & Darabkh, K. A. (2020). A–Z survey of internet of things: Architectures, protocols, applications, recent advances, future directions and recommendations. *Journal of Network and Computer Applications, 163*. https://doi.org/10.1016/j.jnca.2020.102663

Koszewski, K., et al. (2025). Utilizing IoT sensors and spatial data mining for analysis of urban space actors' behavior in university campus space design. *Sensors, 25*(5), 1393. https://doi.org/10.3390/s25051393

Krishnamurthi, R., Kumar, A., Gopinathan, D., Nayyar, A., & Qureshi, B. (2020). An overview of IoT sensor data processing, fusion, and analysis techniques. *Sensors, 20*(21), 6076. https://doi.org/10.3390/s20216076

Kumar, S., Tiwari, P., & Zymbler, M. (2019). Internet of things is a revolutionary approach for future technology enhancement: A review. *Journal of Big Data, 6*(1), 111. https://doi.org/10.1186/s40537-019-0268-2

Lombardi, M., Pascale, F., & Santaniello, D. (2021). Internet of things: A general overview between architectures, protocols and applications. *Information, 12*(2), 1–21. https://doi.org/10.3390/info12020087

Majeed, A. S., et al. (2024). Design and implementation of IoT smart system for sensing toxic gas emissions and security system of chemical laboratories or stores of universities. In *10th International Conference on Automation, Robotics and Applications* (pp. 557–564). https://doi.org/10.1109/ICARA60736.2024.10552993

Malathy, V., Saritha, S., Latha, M., Hasmukhlal, D. J., Rawat, S., & Tiwari, M. (2023). Implementation and enabling internet of things (IoT) in wireless communication networks. In *2023 Global Conference on Information Technologies and Communications (GCITC), April, 1–5*. https://doi.org/10.1109/GCITC60406.2023.10425914

Manhas, P., Thakral, S., & Verma, J. (2020). IoT sensors: Perspectives & appliance. *International Journal of Recent Technology and Engineering, 8*(6), 3306–3310. https://doi.org/10.35940/ijrte.F8594.038620

Mansour, M., Gamal, A., Ahmed, A. I., Said, L. A., Elbaz, A., Herencsar, N., & Soltan, A. (2023). Internet of things: A comprehensive overview on protocols, architectures, technologies, simulation tools, and future directions. *Energies, 16*(8). https://doi.org/10.3390/en16083465

Meng, L. (2020). *Tailoring conducting polymer interface for sensing and biosensing* (Vol 2094). Linköping University Electronic Press. https://doi.org/10.3384/diss.diva-169499

Mootien, K., & Nagowah, S. (2023). IoT-Based smart green campus leveraging machine learning. *Internatinal Conferences on Information and Communication Technology*, 627–632. https://doi.org/10.1109/ICoICT58202.2023.10262448

Mrabet, H., Belguith, S., Alhomoud, A., & Jemai, A. (2020). A survey of IoT security based on a layered architecture of sensing and data analysis. *Sensors, 20*(13), 1–20. https://doi.org/10.3390/s20133625

Muangpoo, T., & Pullteap, S. (2023). Enhancing energy efficiency in smart classrooms through IoT and AI. *International Journal of Emerging Technology and Advanced Engineering, 13*(3), 74–80. https://doi.org/10.46338/ijetae0323_07

Naresh, V., & Lee, N. (2021). A review on biosensors and recent development of nanostructured materials-enabled biosensors. *Sensors, 21*(4), 1109. https://doi.org/10.3390/s21041109

Raghava Deepthi, J. V. N., Khan, A. K., & Acharjee, T. (2023). Energy efficient routing algorithm for WSN-IoT network. *Ingénierie des Systèmes d'Information, 28*(1), 231–238. https://doi.org/10.18280/isi.280127

Shahbaz, M., et al. (2023). Smart and advanced E-learning methodology with IoT device integration. In *6th International Conference of Women in Data Science* (pp. 217–222). Prince Sultan Univ.. https://doi.org/10.1109/WiDS-PSU57071.2023.00052

Shehu Yalli, J., Hilmi Hasan, M., & Abubakar Badawi, A. (2024). Internet of things (IoT): Origins, embedded technologies, smart applications, and its growth in the last decade. *IEEE Access, 12*, 91357–91382. https://doi.org/10.1109/ACCESS.2024.3418995

Walczak, R., et al. (2023). Acceptance of IoT edge-computing-based sensors in smart cities for universal design purposes. *Energies, 16*(3). https://doi.org/10.3390/en16031024

Chapter 6

Sensor network, blockchain, and digital twin as the backbones of a connected ecosystem of smart schools

Xavier Fernando[1] and George Lăzăroiu[1, 2, 3, 4]

[1]Toronto Metropolitan University, Toronto, ON, Canada; [2]Curtin University, Perth, WA, Australia; [3]Cardiff Metropolitan University, Cardiff, United Kingdom; [4]Spiru Haret University, Bucharest, Romania

1. Introduction

Digital twin education and training technologies configure smart classroom performance and adaptive capabilities, 3D simulation- and Internet of Things sensor-based digital pedagogy, and digital twin classroom monitoring and analysis in artificial intelligence- and cloud computing-based immersive educational and training environments. Educational big data acquisition and mapping algorithms optimize immersive and interactive student engagement and training in smart collaborative learning spaces and educational settings. Artificial intelligence and digital twin-based educational practices developed through virtual reality teaching technologies in Internet of Things-based smart classrooms are very promising.

Real-time big educational data analytics articulates interactive and integrative educational management scalability and adaptability in Internet of Things and extended reality educational environments. Pedagogical digital twin technologies enhance big data-driven personalized teaching and learning simulation interoperability and predictability in virtual shared-school environments. Educational metaverse practices integrate automated performance assessment and smart classroom technologies in immersive interactive training environments. Virtual educational systems and big educational data mining and analysis techniques shape remote personalized and adaptive learning management and classroom teaching system quality analysis in artificial intelligence-based educational spaces and for sensor network-, blockchain-, and digital twin-based smart schools.

Internet of Things educational tools and technologies assist immersive learning experiences and engagement in virtual educational and learning environments. Educational performance management and analysis tools improve digital twin-based instructional design in metaverse educational settings. Digital twin-based training and education visualization, Internet of Things-based virtualized scene teaching, and computer vision-based performance assessment and classroom management require big educational data management systems, educational virtual reality and metaverse technologies, and interactive teaching tools. Internet of Things-based educational, teaching, and learning systems further the teaching quality and learning efficiency monitoring in deep learning-based interactive context-aware smart classrooms.

Smart blockchain-based educational performance prediction technologies are pivotal in deep learning-based computational education, artificial intelligence-based smart class management and performance assessment, and big educational data mining and analysis in networked virtual interactive learning environments. Educational data mining algorithms configure big educational data collection, processing, and analysis across Internet of Things-based multiple interactive collaborative learning spaces. Multifunctional educational technologies and virtual simulation teaching systems assist artificial intelligence-based immersive and engaging pedagogical experiences and Internet of Things- and artificial intelligence-based wireless classroom knowledge.

Digital Twins in the Smart Classroom. https://doi.org/10.1016/B978-0-443-41632-3.00022-9

2. Computer vision- and sensor network-based performance assessment and classroom management in Internet of Things and extended reality educational environments

3D visualization and virtual reality teaching technologies (Xiang et al., 2025) assist in the real-time remote training process, fault diagnosis, environmental condition, equipment maintenance, and operational skill monitoring in immersive virtual reality training environments. Real-time data acquisition and knowledge transfer algorithms enable unmanned predictive maintenance management, condition assessment, and fault simulation, real-time equipment health status and immersive training experience optimization, and unplanned downtime and technical fault decrease in virtual training environments. The remote management of operational training and virtual simulation teaching systems further real-time educational data and task synchronization, equipment interaction and malfunction simulation for preventive maintenance, real-time equipment status monitoring and management, and operational training virtualization and performance analysis. Digital twin and Internet of Things technologies can reshape equipment condition monitoring and maintenance efficiency optimization, virtual reality-based operational training, fault handling, and equipment failure simulation, and maintenance training efficiency, equipment reliability and operational safety enhancement for virtual scene diagnostics and operations in immersive virtual educational environments.

Internet of Things sensor real-time data and devices, machine learning and anomaly detection algorithms, and real-time big educational data analytics further extended reality- and digital twin-based smart building operation and maintenance scalability (Salzano et al., 2025) for system performance reliability and efficiency optimization in educational and occupant-centered built environments. Extended reality and smart building technologies, digital twin real-time sensor data mapping, and Internet of Things sensor-based interoperable building management and automation systems shape environmental condition monitoring and control, smart building and urban infrastructure management practices, and system failure pattern identification. Anomaly detection real-time sensor data monitoring algorithms are pivotal in cloud computing- and in sensor data-based system failure prediction, neural network-based operational data complex pattern analysis for system degradation and failure detection. The machine learning-based real-time system feedback and maintenance prediction accuracy optimization through the use of building system and facility management practices will be immensely useful in virtual educational institutions. Cloud-based digital twin and virtual reality technologies articulate immersive manufacturing educational process automation and system integration complexity for skilled workforce training (Yang et al., 2025) in interactive personalized experiential learning outcome environments.

Big educational data management systems, k-means clustering, and pedagogical digital twin and cloud computing technologies can be deployed for Internet of Things device, machine, and system interaction control and monitoring, real-time educational process planning and evaluation, and individualized and adaptive virtual educational and teaching processes (Selim et al., 2024) across virtual educational, didactic, instructional, teaching, and learning landscapes. Decision tree and educational data mining algorithms, virtual and augmented reality technologies, and interactive teaching tools can be leveraged for sensor-based vision systems for operating condition, decision-making process assessment, and digital twin-based instructional design. Active personalized learning experiences, and behaviors for educational content accessibility and personalization for academic performance will be enhanced in virtual educational and learning environments developed through neural network analysis, near field communication technologies, and cluster analysis based k-means algorithms.

Big data-driven personalized teaching and learning simulation interoperability and predictability, digital twin-based human teacher augmentation and replacement automation, and big educational and schooling data governance structures integrate pedagogical, training, and educational practices and technologies (Arantes, 2023) in smart collaborative learning spaces and educational settings. Educational metaverse practices and decision-making processes and sociotechnological quality teaching and learning assessment require virtual educational systems and artificial intelligence-based educational data mining in digital self-construction across datafied landscapes. Artificial intelligence- and digital twin-based educational practices and personalized learning and teaching and learning automated decision-making modeling and simulation necessitate technologically personalized experiential learning and big data predictive analytics in metaverse educational settings.

Virtual reality devices and visual data tools are pivotal in multimodal digital twin environments (Sun et al., 2024) for curriculum and residual reinforcement learning. Also, smart campus digital twin for building occupancy (Roda-Sanchez et al., 2024) necessitates sensor network data, fog–edge–cloud computing technologies, and knowledge graph. On the other hand, artificial intelligence decision-making is instrumental in digital twin-based smart learning spaces (She et al., 2023) for virtual reality-based education and educational institutions that are sustainable with smart teaching infrastructure management (Muñoz Pavón et al., 2025).

Remote communication system monitoring and control for digital twin-based training and education visualization (Lei et al., 2024) effectively deploy collaborative filtering algorithms for curriculum resource simulation and prediction in teaching and learning in a digital twin campus (Tong et al., 2022). Also, the following aspects are highly recommended: digital twin technology for teaching and education practice and interactive learning experience simulation and prediction (Su et al., 2025), blockchain-, deep learning- and digital twin-based big educational data analytic performance in critical and resource-constrained operations (Dewangan & Chandrakar, 2024), and deep learning-based computational education, pedagogy, engagement, and assessment in Internet of Things-based smart classrooms (Badshah et al., 2023), Internet of Things, augmented reality, and mobile educational technologies in smart sensor- and actuator-based classrooms (Stojanović et al., 2020), image processing-based augmented reality and sensor technologies and wearable devices in smart learning environments (Yao & Liang, 2023), and interdisciplinary-integrated smart classroom and networked teaching management systems. Internet of Things-based virtualized scene teaching by image processing and deep learning-based smart classroom teaching quality data gathering analysis done by Hu et al. (Hu, 2023) articulate sensor network, blockchain, and digital twin-based smart schools.

Automated performance assessment and smart classroom technologies shape in-class and remote operations (Dimitriadou & Lanitis, 2023) for creative classes in smart classrooms and artificial intelligence-based educational spaces, artificial intelligence-based techniques for smart class management, smart teaching and educational processes, and collective social knowledge. Artificial intelligence-powered virtual pedagogical agents, smart class and classroom management technologies, automated smart adaptive educational and class systems, and attendance and action recognition task automation optimize instruction and learning processes and computer vision-based performance assessment and classroom management. Educational artificial intelligence-based systems, school education and mobile computing technologies, automated mobile social learning tools, and convolutional neural network-based facial identification and recognition software enhance automated performance assessment, integrating multisensory social interactions, motion control, and scene labeling for sensor network-, blockchain-, and digital twin-based smart schools.

Virtual and augmented reality technologies are pivotal in autonomous artificial agent collective behavior and action interpretation, complex adaptive social interaction navigation and effective cooperative cohesion, and complex social behavior and interaction dynamics (Zhinin-Vera et al., 2025) in relation to the environmental factors, contextual influences, social behaviors, cognitive processes, cultural norms, and interpersonal dynamics for the emotional and psychological well-being. Educational big data analytics is instrumental in artificial multiagent reasoning and interaction complexities, autonomous computational agent decision-making system social intelligence, socially aware artificial intelligence-based autonomous agent abilities, and autonomous system social behavior dynamics in simulated school environments and educational settings. Artificial neural network architectures configure artificial intelligence algorithmic and automated behavior prediction, context-aware socially intelligent behaviors, adaptive learning mechanisms, and simulated environment social dynamics and intelligence for socially intelligent autonomous and simulated environment multiple agents in social and educational environments.

3. Big educational data-driven personalized teaching and learning simulation interoperability and predictability in blockchain-based virtual shared-school environments

Deep and machine learning-based feedback and assessment automation, smart blockchain-based educational performance prediction technologies, and artificial intelligent tutoring systems (Dimitriadou & Lanitis, 2023) improve immersive interactive learning experiences. Machine learning algorithms, smart agent and augmented reality technologies, and fog–cloud computing and augmented display devices can be harnessed in virtual teaching processes across virtual collaborative educational and learning environments. Educational virtual reality and metaverse technologies can be deployed for artificial intelligence-based immersive and engaging pedagogical experiences, Internet of Things- and artificial neural network-based interactive and engaging educational content, and artificial intelligence-based smart class management and performance assessment for sensor network-, blockchain-, and digital twin-based smart schools.

Artificial intelligence-based collaborative multifaceted pedagogical knowledge and education achievements, school curriculum planning and practice, and active learning classroom schooling (Dai, Liu, et al., 2023) require deep reinforcement learning and convolutional neural network-based pedagogical design and conceptual teaching systems. Immersive agentic engagement across pedagogical design, educational planning, teaching and learning processes and practices, and classroom interactions necessitate artificial intelligence-based image recognition technologies. Big educational data collection, analysis, and interpretation are pivotal in socially constructed and bounded processes, virtual agentic engagement, performance, and assessment, pedagogical knowledge creation, cognitive processes, social networks, and situational logics, and artificial intelligence-based educational planning, pedagogical learning machines, devices, sensors, and actuators, and curriculum content design and agentic construction.

Multifunctional educational technologies, digital display interfaces, and near-field communication systems can be leveraged in teaching practice optimization in smart learning environments and deep learning-based interactive context-aware smart classrooms (Dai, Xiong, et al., 2023) for teaching interaction enhancement and personalized teaching and cooperative learning experiences. Learning norms and modes, big educational data mining and analysis, knowledge construction and sharing, and smart teaching practices and decisions developed through education management software for teaching and sensing equipments in networked virtual interactive learning environments can be vital in such environments. Social interactions for academic achievement, 3D extended reality integrated communication, and classroom teaching system quality analysis integrate cloud architecture technology and classroom environment data across Internet of Things-based interactive multiple collaborative learning spaces and smart normative and adaptive learning environments.

Cloud computing and fuzzy control technologies and smart wearable and acquisition devices (Alfoudari et al., 2023) configure Internet of Things sensor device- and deep learning-based education curriculum and classroom teaching and smart classroom teaching and training modes. Smart classroom and edge intelligent computing technologies articulate teaching quality and learning efficiency monitoring and artificial neural network- and big educational data-based precision teaching in Internet of Things-based smart classroom environments. Internet of Things wireless network sensors assist in teaching and learning interest assessment and smart classroom and education quality in terms of strategy-oriented planning, social interactions, educational intelligence, technology accessibility, system interconnectivity and adaptability, and management procedures for sensor network-, blockchain-, and digital twin-based smart schools.

Computer vision algorithms can be harnessed in Internet of Things-based educational institutions and settings and smart learning and collaborative virtual environments (Omrany et al., 2025) for complex learning outcomes and decision-making skills; big educational data acquisition, processing, and connectivity; and digital literacy and competence. Cyber–physical and learning management systems can be deployed in building information modeling environments for big educational data mining, fusion, integration, and visualization; learning process, procedure, and operation automation; and anomaly detection and fault finding. Predictive analytics tools, interconnected Internet of Things sensors and wearables, and digital twin technologies can be leveraged in virtual learning environments and training settings for interactive and integrative educational management scalability and adaptability, immersive and interactive student engagement and training, and simulation-based training and skill development. Digital literacy and education and learning skills and knowledge in digital twin-based learning environments and educational settings developed through immersive and collaborative virtual reality learning tools, Internet of Things-enabled building management system devices, and machine learning and cloud computing technologies for digital twin education and training technology integration can be widely beneficial.

Internet of Things-based smart education systems and deep collaborative neural networks enable remote personalized and adaptive learning management and educational engagement experiences (Amin et al., 2023) in smart learning environments. Machine learning algorithms, Internet of Things sensing devices, and big educational content-based filtering data mining further explainable artificial intelligence-based smart education, contextual knowledge sharing and educational quality optimization, and educational material network exchange digitization for sensor network, blockchain, and digital twin-based smart schools are equally important. Cyber–physical- and technological experience-based personalized smart education systems and big educational data acquisition and cloud computing technologies (Mukul & Büyüközkan, 2023) are instrumental for artificial intelligence-based smart educational agents.

Artificial intelligence-based education equipment and systems, web-based learning tools, and Internet of Things and Web 3.0 sensors, actuators, devices, and machines (Haque et al., 2023) shape technological communication and collaboration, interactive smart agents for networked gadget-based classrooms, and remote learning object configuration. Internet of Things-based educational, teaching, and learning systems and cloud computing-based educational and teaching technologies enhance big educational data sharing and synchronization and attendance and cognitive activity tracking.

4. Digital twin classroom monitoring and analysis in artificial intelligence- and cloud computing-based immersive educational and training environments

Attendance records, educational standard and performance, and course contents and delivery optimization by digital twin classroom monitoring and analysis (Razzaq et al., 2023) can improve faculty efficiency throughout the on-campus and online classes. Smart classroom boards, high-edge computing devices, audio–visual aids, convolution neural network algorithms, and multimedia articulate Internet of Things educational environments in terms of accumulated learning outcomes and teaching quality.

Virtual machine and cloud computing technologies configure technological innovation management and infrastructure digitalization processes, smart building information modeling and digital twin data interoperability, and user management techniques and multidirectional workflows (Muñoz Pavón et al., 2025) for smart laboratories, projectors, blackboards, and classrooms with regard to the academic schedule-based people flow, occupancy rate, and Internet of Things space management data. Smart facility and infrastructure management techniques articulate the sustainable development of infrastructure monitoring operations, sensor synchronization management and visualization functionalities by the use of artificial intelligence, natural language processing, and predictive maintenance algorithms across digital twin environments.

Digital twin technology and Internet of Things sensors assist machine learning-based predictive maintenance automation in extended reality educational environments (Salzano et al., 2025) with regard to the building information modeling practices, real-time building system dynamic representations and inefficiency proactive rectifications, and optimal maintenance timing. Smart building and predictive maintenance technologies enable real-time building system reliability and efficiency optimization in relation to the operational disruptions and unexpected failures across Internet of Things sensor-based smart buildings and sustainable construction landscapes. Real-time big educational data analytics streamlines maintenance processes and procedures, digital urban planning and infrastructure operation automation, and sustainable building practices and design strategies in virtual educational settings and for sensor network-, blockchain-, and digital twin-based smart schools.

3D scanning and image recognition technologies optimize big data computing for equipment failure prediction and operational monitoring in immersive interactive training environments (Xiang et al., 2025) for failure detection and prediction, fault simulation and diagnosis, condition analysis, and operational efficiency improvement by big educational data accumulation and analysis. Digital twin operational skill monitoring and virtual simulation teaching systems, educational big data acquisition and mapping algorithms, 3D visualization tools, and augmented reality technologies enhance real-time virtual modeling-based equipment status and data monitoring for tiered educational content and teaching tasks across virtual settings. Educational performance management and analysis tools, virtual reality scene construction and fault simulation algorithms, and real-time equipment operational data can be harnessed in 3D simulation environments for an interactive teaching process, pedagogical content scalability, and immersive learning experience improvement.

Internet of Things educational tools and technologies optimize extended reality- and digital twin-based structured education, learning, and training processes for theoretical knowledge and practical skill integration (Omrany et al., 2025) in artificial intelligence- and cloud computing-based immersive educational and training environments. Machine and deep learning algorithms, extended reality training and education tools, and Internet of Things sensor networks enhance immersive learning experiences and engagement, the remote operation control and monitoring capabilities, and real-time big educational data synchronization in virtual training and immersive learning environments. Environmental and physiological sensors, cyber–physical systems, and virtual reality and digital twin technologies improve 3D simulation- and Internet of Things sensor-based digital pedagogy in terms of practical knowledge and skill acquisition, interactive and integrative educational and pedagogical management processes, and simulation-based lesson planning and teaching, integrated tailored learning and training, digital literacy, and instructional design analysis for sensor network-, blockchain-, and digital twin-based smart schools.

Wireless network sensors and big educational data mining and analysis techniques improve Internet of Things- and artificial intelligence-based wireless classroom knowledge, visual educational management mechanisms, and multidimensional teaching and learning capabilities (Zhang, 2023) in smart educational and classroom environments. Educational artificial intelligence and effective computing technologies and wireless classroom teaching networks can be harnessed for interactive self-adaptive classroom teaching quality optimization, seamless personalized learning experiences, classroom teaching behavior prediction and performance, and class session emotion and behavior capturing and analysis. Smart classroom teaching behavior and monitoring systems and facial emotion and image recognition algorithms can be deployed in real-time smart classroom asset operation monitoring, academic performance and situational awareness optimization, class session image data processing and analysis, and wireless network classroom teaching processes and practices across visual communication and wireless classroom infrastructures and interactive classroom and learning environments.

Intuitive realistic autonomous artificial agent interaction symbiotic fusion, deep learning multiagent system decision-making processes, and dynamic operational changes and future state forecasting (Zhinin-Vera et al., 2025) integrate decentralized multiagent social learning and real-time Internet of Things sensory data assimilation mechanisms for prosocial skill strengthening, social dynamics and reasoning, and smart classroom performance and adaptive capabilities in virtual shared-school environments. Autonomous system social intelligence for adaptive behavior, decentralized multiagent social learning and collaborative decision-making, and socially intelligent autonomous system adaptability and effectiveness for contextual awareness require deep multiagent reinforcement learning algorithms and human digital twins in the social dynamics of autonomous agent environments. The dynamic bidirectional sensory data flow connection for social awareness, behavioral pattern and social cognitive process intricacies, real-world collaborative setting and social interaction intricacies, and accurate artificial system simulation contextualization necessitate artificial intelligence deep reinforcement learning algorithms and virtual human replicas in social contexts and educational environments.

Smart real-time context-aware teaching, tutoring, and learning techniques assist machine learning analysis (Veeramanickam et al., 2023) for knowledge sharing, educational curriculum and instructional material designing, training experience, and skill adaptive and development planning in virtual classrooms. Machine and reinforcement learning algorithms enable artificial intelligence- and Internet of Things-based smart tutoring and personalized learning systems for big educational data collection, processing, and analysis; embedded visual sensor data and knowledge transfer mapping; and Internet of Things-based motion sensor monitoring and mapping operations. Teaching and learning tools and Internet of Things-based sensor data tracking further interactive teaching and multiple learning content assessment and computation analysis, enriched multimedia content-based personalized learning integration, and knowledge acquisition and sharing for sensor network-, blockchain-, and digital twin-based smart schools.

5. Conclusions

Real-time data acquisition and knowledge transfer algorithms enable interactive and integrative educational and pedagogical management processes in digital twin-based learning environments and educational settings. Operational training and virtual simulation teaching systems can be harnessed for big educational data accumulation and analysis in immersive virtual reality training environments. Deep learning-based smart classroom teaching quality data gathering analysis, multimedia content-based personalized learning integration, and Internet of Things- and artificial neural network-based interactive and engaging educational content necessitate networked teaching management systems.

Educational big data analytics furthers individualized and adaptive virtual educational and teaching processes in Internet of Things-based educational institutions and settings. Smart real-time context-aware teaching, tutoring, and learning techniques can be deployed for artificial intelligence-based educational data mining in digital twin-based smart learning spaces. Internet of Things-based smart education systems and big educational content-based filtering data mining articulate artificial intelligence-based collaborative multifaceted pedagogical knowledge and educational achievements in smart normative and adaptive learning environments. Automated smart adaptive educational and class systems and cloud computing-based educational and teaching technologies enable Internet of Things sensor device- and deep learning-based education curriculum and classroom teaching across virtual collaborative educational and learning environments.

Extended reality training and education tools shape real-time educational process planning and evaluation in simulated school environments and educational settings. Artificial intelligence- and Internet of Things-based smart tutoring and personalized learning systems can be leveraged for personalized learning and teaching and learning automated decision-making modeling and simulation in immersive virtual educational environments. Deep reinforcement learning- and convolutional neural network-based pedagogical design and conceptual teaching systems are instrumental in interactive teaching and multiple learning content assessment and computation analysis in Internet of Things-based smart classroom environments and for sensor network-, blockchain-, and digital twin-based smart schools.

References

Alfoudari, A. M., Durugbo, C. M., & Aldhmour, F. M. (2023). Exploring quality attributes of smart classrooms from the perspectives of academics. *Education and Information Technologies, 28*, 13109–13151. https://doi.org/10.1007/s10639-022-11452-3

Amin, S., Uddin, M. I., Mashwani, W. K., Alarood, A. A., Alzahrani, A., & Alzahrani, A. O. (2023). Developing a personalized E-Learning and MOOC recommender system in IoT-Enabled smart education. *IEEE Access, 11*, 136437–136455. https://doi.org/10.1109/ACCESS.2023.3336676

Arantes, J. (2023). Digital twins and the terminology of "personalization" or "personalized learning" in educational policy: A discussion paper. *Policy Futures in Education, 22*(4), 524–543. https://doi.org/10.1177/14782103231176357

Badshah, A., Ghani, A., Daud, A., Jalal, A., Bilal, M., & Crowcroft, J. (2023). Towards smart education through internet of things: A survey. *ACM Computing Surveys, 56*(2), 26. https://doi.org/10.1145/3610401

Dai, Y., Liu, A., Qin, J., Guo, Y., Jong, M. S.-Y., Chai, C.-S., et al. (2023). Collaborative construction of artificial intelligence curriculum in primary schools. *Journal of Engineering Education, 112*(1), 23–42. https://doi.org/10.1002/jee.20503

Dai, Z., Xiong, J., Zhao, L., & Zhu, X. (2023). Smart classroom learning environment preferences of higher education teachers and students in China: An ecological perspective. *Heliyon, 9*(6), Article e16769. https://doi.org/10.1016/j.heliyon.2023.e16769

Dewangan, N. K., & Chandrakar, P. (2024). Implementing blockchain and deep learning in the development of an educational digital twin. *Soft Computing, 28*, 6619–6636. https://doi.org/10.1007/s00500-023-09501-1

Dimitriadou, E., & Lanitis, A. (2023). A critical evaluation, challenges, and future perspectives of using artificial intelligence and emerging technologies in smart classrooms. *Smart Learning Environments, 10*, 12. https://doi.org/10.1186/s40561-023-00231-3

Haque, S., Haque, M. A., Kumar, D., Mishra, K., Islam, F., Ahmad, S., et al. (2023). Assessing the impact of IoT enabled E-Learning system for higher education. *SN Computer Science, 4*, 459. https://doi.org/10.1007/s42979-023-01860-8

Hu, X. (2023). The role of deep learning in the innovation of smart classroom teaching mode under the background of internet of things and fuzzy control. *Heliyon, 9*(8), Article e18594. https://doi.org/10.1016/j.heliyon.2023.e18594

Lei, Z., Zhou, H., Hu, W., Liu, G.-P., & Guan, S. (2024). Web-based digital twin communication system of power systems for training and education. *IEEE Transactions on Power Systems, 39*(2), 3592–3602. https://doi.org/10.1109/TPWRS.2023.3308510

Muñoz Pavón, R., Alberti, M. G., Arcos Álvarez, A. A., & Jerez Cepa, J. (2025). Bim-based digital twin development for university campus management. Case study ETSICCP. *Expert Systems with Applications, 262*, Article 125696. https://doi.org/10.1016/j.eswa.2024.125696

Mukul, E., & Büyüközkan, G. (2023). Digital transformation in education: A systematic review of education 4.0. *Technological Forecasting and Social Change, 194*, Article 122664. https://doi.org/10.1016/j.techfore.2023.122664

Omrany, H., Al-Obaidi, K. M., Ghaffarianhoseini, A., Chang, R.-D., Park, C., & Rahimian, F. (2025). Digital twin technology for education, training and learning in construction industry: Implications for research and practice. *Engineering Construction and Architectural Management*. https://doi.org/10.1108/ECAM-10-2024-1376

Razzaq, S., Shah, B., Iqbal, F., Ilyas, M., Maqbool, F., & Rocha, A. (2023). DeepClassRooms: A deep learning based digital twin framework for on-campus class rooms. *Neural Computing & Applications, 35*, 8017–8026. https://doi.org/10.1007/s00521-021-06754-5

Roda-Sanchez, L., Cirillo, F., Solmaz, G., Jacobs, T., Garrido-Hidalgo, C., & Olivares, T. (2024). Building a smart campus digital twin: System, analytics, and lessons learned from a real-world project. *IEEE Internet of Things Journal, 11*(3), 4614–4627. https://doi.org/10.1109/JIOT.2023.3300447

Salzano, A., Cascone, S., Zitiello, E. P., & Nicolella, M. (2025). HVAC system performance in educational facilities: A case study on the integration of digital twin technology and IoT sensors for predictive maintenance. *Journal of Architectural Engineering, 31*(1), 1855. https://doi.org/10.1061/JAEIED.AEENG-1855

Selim, A., Ali, I., Saracevic, M., & Ristevski, B. (2024). Application of the digital twin model in higher education. *Multimedia Tools and Applications*. https://doi.org/10.1007/s11042-024-20014-3

She, M., Xiao, M., & Zhao, Y. (2023). Technological implication of the digital twin approach on the intelligent education system. *International Journal of Humanoid Robotics, 20*(2/3), Article 2250005. https://doi.org/10.1142/S0219843622500050

Stojanović, D., Bogdanović, Z., Petrović, L., Mitrović, S., & Labus, A. (2020). Empowering learning process in secondary education using pervasive technologies. *Interactive Learning Environments, 31*(2), 779–792. https://doi.org/10.1080/10494820.2020.1806886

Su, C., Huang, H., Wei, Z., & Xu, N. (2025). Application of digital twin technology in mining machinery education. *International Journal of Mechanical Engineering Education*. https://doi.org/10.1177/03064190241308472

Sun, F., Liu, N., Wang, X., Sun, R., Miao, S., & Kang, Z. (2024). Digital-twin-assisted skill learning for 3C assembly tasks. *IEEE Transactions on Cybernetics, 54*(7), 3852–3863. https://doi.org/10.1109/TCYB.2024.3368148

Tong, W., Wang, Y., Su, Q., & Hu, Z. (2022). Digital twin campus with a novel double-layer collaborative filtering recommendation algorithm framework. *Education and Information Technologies, 27*, 11901–11917. https://doi.org/10.1007/s10639-022-11077-6

Veeramanickam, M. R. M., Dabade, M. S., Murty, S. R., Borhade, R. R., Barekar, S. S., Navarro, C., et al. (2023). Smart education system to improve the learning system with CBR based recommendation system using IoT. *Heliyon, 9*(7), Article e17863. https://doi.org/10.1016/j.heliyon.2023.e17863

Xiang, Z., Tian, C., Dai, Y., Xiang, H., Huang, L., & Liu, X. (2025). Research and application of digital twin and simulation teaching methodology in DVOR. *IEEE Access, 13*, 27487–27506. https://doi.org/10.1109/ACCESS.2025.3539709

Yang, S., Mirahmadi, S. A., Zhu, E., & Solanki, B. (2025). Live digital twin with virtual reality for accessible and immersive manufacturing education. *The International Journal of Advanced Manufacturing Technology, 136*, 3577–3590. https://doi.org/10.1007/s00170-025-15078-w

Yao, S.-N., & Liang, C. (2023). Audio augmented reality in smart learning environments. *IEEE Transactions on Learning Technologies, 16*(2), 178–190. https://doi.org/10.1109/TLT.2023.3250268

Zhang, J. (2023). School wireless network classroom teaching system based on artificial intelligence. *Applied Artificial Intelligence, 37*(1), Article 2219563. https://doi.org/10.1080/08839514.2023.2219563

Zhinin-Vera, L., Pretel, E., López-Jaquero, V., Navarro, E., & González, P. (2025). Mindful human digital twins: Integrating theory of mind with multi-agent reinforcement learning. *Applied Soft Computing, 174*, Article 112939. https://doi.org/10.1016/j.asoc.2025.112939

Chapter 7

Digital twins for smart education: Utilizing sensor networks for education

Mana Saleh Al Reshan[1,3], Darakhshan Syed[2] and Asadullah Shaikh[1,3]

[1]*Department of Information System, College of Computer Science and Information Systems, Najran University, Najran, Saudi Arabia;* [2]*Computer Science Department, Bahria University, Karachi, Pakistan;* [3]*Emerging Trends and Technologies Lab (ETRL), College of Computer Science and Information Systems, Najran University, Najran, Saudi Arabia*

1. Introduction

1.1 Background

The Covid-19 pandemic accelerated the adoption of technological breakthroughs by forcing educators and educational administrators to quickly adapt technological advances in order to get beyond the constraints of isolation from the outside world. The epidemic accelerated the adoption, but it also brought attention to the already-existing digital absence. Previously, this adoption had encountered reluctance and improved slowly (Kartashova et al., 2020). However, the current state of education appears more open to implementing new technology considering these recent developments.

The emerging field of developing new services on the Internet of Things (IoT) is advancing rapidly, inspired by the dynamic breakthroughs in information technology. The landscape of IoT service development has been drastically altered by these technological advances, which have transformed service design more complicated, which required more rigorous management of devices and increased monitoring. This evolution is driven by advances in management theories, intelligent communication techniques, and the expanding range of technologies linked to computers. A move towards a more digital, intelligent, and information-rich IoT environment is indicated by the addition of advanced monitoring systems, contemporary instruments, and smart sensors (Zaballos et al., 2020). This change is essential for improving the management and visualization of the IoT's new service development workflows, which will result in the growth of creative control network designs that are more responsive and reliable (Kumar et al., 2024).

At the core of this advancement is digital twin technology, which transforms information transmission and subsequent analysis inside the Internet of Things framework. This technology, which covers a broad range of development features like real-time process detection, execution, and monitoring, has become essential to the development and production layers of IoT services. By means of wireless networks, the technology enables the seamless transmission of vital data streams, including crucial metrics like development pace, progress, and other monitoring elements. These data streams, which offer a thorough and dynamic perspective of the full development lifecycle, are essential in growing the new IoT service development practices (Minerva et al., 2020).

1.2 Importance of digital twins in education

Digital twins are virtualized versions of real-world systems, processes, or assets that are adjusted to match their real-time equivalents. Training data to recent digital twins enable previously unachievable immersive and customized learning experiences by generating dynamic and interactive digital copies of classrooms, labs, and entire institutions. Immersion and interaction in learning settings are made possible by digital twins. Without the limitations of actual resources, virtual laboratories help students to understand difficult ideas through real-world application (Maksimović & Davidović, 2022).

Additionally, by using analytics and data from sensors to modify syllabus according to each learner's requirements, digital twins enable individualized learning, increasing student engagement and boosting learning results. Beyond the classroom, digital twins increase

Digital Twins in the Smart Classroom. https://doi.org/10.1016/B978-0-443-41632-3.00019-9

operational efficiency by assisting organizations in real-time monitoring and optimization of energy use, facility utilization, and maintenance plans, resulting in smarter and more sustainable campuses. By exposing students to state-of-the-art technology utilized in professional settings, they also act as a channel between academia and industry, improving their readiness for the workforce of the future (Selim et al., 2024). All things considered, digital twins have the potential to revolutionize education by enabling individualized instruction, effective management, and the development of practical skills (Omrany et al., 2025). Fig. 7.1 presents the basic architecture of digital twins where the physical and digital environments are communicating.

1.3 Objectives

This book chapter is structured to achieve specific research objectives that facilitate a comprehensive understanding of the roles and importance of digital twins in the realm of education. The key objectives of the research are as follows:

- To provide a detailed overview of the concept of digital twins and their role in the field of education
- To explore the utilization and architectural details of sensor networks for digital twins and their applications in education.
- To address the benefits and challenges of digital twins in education.
- To provide future trends that facilitates the stakeholder engagement and present key insights from the existing research.

Achieving these goals increases the understanding, guides future investigations, and promotes the appropriate use of digital twins and the usage of sensor networks for education to improve both education and society.

1.4 Problem statement

Organizations and fields appear to be adopting digital twins too slowly, even though the idea is approximately 20 years old. Even in the real-world models and simulation, static designs are still common. The development of digital reproductions of real-world objects may be more advanced

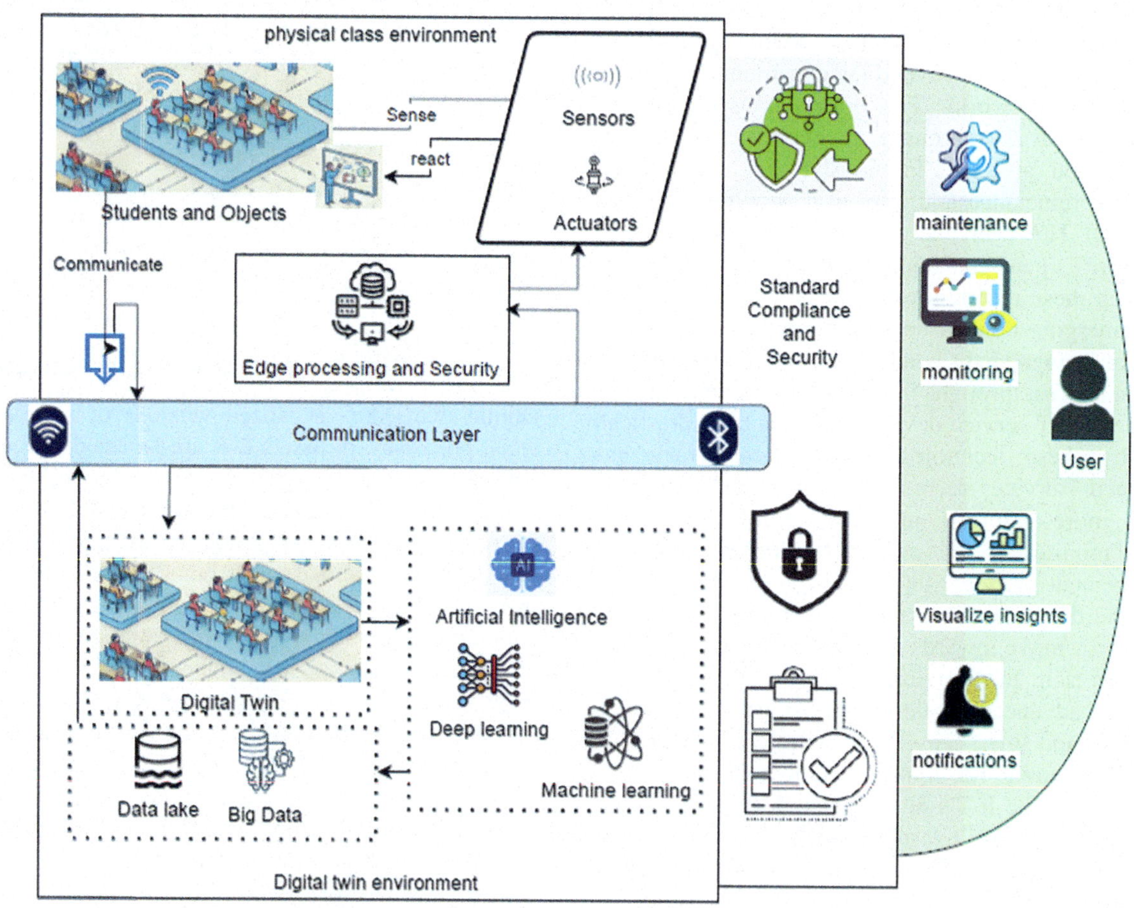

FIGURE 7.1 Architecture of digital twin for smart education.

and appear to be a better fit for virtual and augmented reality. Digital twins, however, don't seem to be as positive about anything, even in the educational field. Furthermore, in 2020 COVID-19's emergence accelerated digital change in every industry (Tao et al., 2018).

According to researchers (Fuller et al., 2020; Kartashova et al., 2020; Sepasgozar, 2020), a lot of the digital changes are expected to be long-lasting and continue long after the pandemic is over. As virtual education becomes more common, schools and other institutions won't have to worry about school environment safety like they did during the global epidemic. However, the emphasis must change to enhancing virtual learning environments in order to improve the academic achievement of students.

The primary problem is that educational institutions do not appear to properly understand the educational revolution potential of digital twins. Nowadays, a lot of research oversimplify the way digital twins are used in the classroom (Mayer, 2022). The studies fall short of realizing the full potential of digital twins in revolutionizing education by only emphasizing learner engagement or virtual educational environments. Participants in the education sector need to comprehend what digital means and the way it fits into the larger framework of educational development from the point of view of innovation. Furthermore, there are still a lot of unexplored applications thus evaluating the possible application areas is crucial (Shaikh).

This chapter fills this need and specifies its objectives; to give real-world examples, to convey a comprehensive understanding of digital twins in the context of smart education, to highlight involved the opportunities and potential challenges, and to present the role of sensor networks in smart education.

2. Digital twins in education

The digital twin paradigm is still unconventional; therefore, its universal definition remains a challenge. However, many experts agree that a digital twin involves a real-time digital representation of a physical system (Berisha-Gawlowski et al., 2021). Nonetheless, definitions frequently vary based on the opinions employed in the research, specifically, when it comes to the digital representation of the characteristics and operations of the real world. Additionally, a digital twin is an electronic replica of an actual product. Consequently, a digital twin's primary characteristic is its automatic communication of data with the real-life physical entity. Berisha-Gawlowski et al. (2021) propose that automatic data sharing develops a digital twin apart from a digital model because a model and the actual object that it represents only exchange data manually. According to computing terminology, a digital twin is an online environment or digital cyberspace that interacts, collaborates, and communicates with the real world.

In the industry and in academics, some definitions are more widely used. Nonetheless, the fundamental idea is that digital twins are digital representation created to communicate with real-world objects across their lifecycles in order to offer intelligence for improvement, evaluations, and forecasts. There are two types of digital twins: three-dimensional and five-dimensional. Digital world, real-world space, and connections are the three main parts of the three-dimensional digital twin. On the other hand, application and data are two more elements of the five-dimensional digital twins (Eriksson et al., 2022). The uses, applications, and primary technological advancements of digital twin breakthroughs are the main topics of recent literature. Many academics offer a broader perspective of the idea, concentrating on sectors that use digital twins frequently. Manufacturing, sales, services, aerospace, automotive, crop cultivation, real estate, and healthcare are among the examples. There are various applications for this technology in every industry. Digital twin technologies, for instance, are used in the manufacturing sector to improve operations, lower total engineering costs, simulate and tailor models. Digital twin technologies contribute to supply chain optimization, workforce and route management, and operational design in the retail industry (Attaran & Celik, 2023).

Regardless of the enterprise, the use of digital twins is characterized by the creation of virtual duplicates to assess the dependability, performance, or safety of the real product. Furthermore, unit and process level are when digital twin technologies are implemented (Yao et al., 2023). Unit-level applications focused on individual components and products, whereas process-level applications focused on several components or services. Similarly, the usage of digital twins is also becoming more and more popular in the education sector. The use of digital twins in classrooms and other learning settings by the education sector is the subject of an expanding database of literature. Nonetheless, the majority of academic studies examine how digital twins potentially revolutionize learning. E-learning or online education have recently emerged and proliferated in the education field. According to these notions, digital twins can improve learning practices and techniques. Digital twins, for instance, can be used to improve reinforcements, experiences in virtual reality, availability, personalized learning, and interactive education in e-learning (Hawkinson, 2022). Education institutions are also forced to manage massive amounts of data because of digitization.

Digital twins are essential to big data administration methods in education, according to recent studies (Siyan et al., 2021). As a result, how successfully universities implement and use digital twins can affect the future of e-learning. Table 7.1 presents the summary of digital twin-based applications for education along with a description. Table 7.2 presents the summary timeline-based evolution of the contributions of technology used for education based on the field used for digitization.

3. Utilization of sensor networks for education

In the context of smart education, sensor networks as a foundational technology in the "Internet of Things (IoT)" have the potential to transform how the idea of digital twins can be brought to life. These networks are made up of intelligent sensors that are networked and provide ongoing monitoring, collection, and transmission of data relating to different dimensions of learning environments. Within the educational domain, sensor networks are the primary data collection layer, which allow for instantaneous coordination of the physical and digital domains of a learning ecosystem (Hazrat et al., 2023).

By integrating these, sensor networks enable real-time digital "clones" of class in session, labs, campuses or in situ learner engagement. Recorded data includes data from

TABLE 7.1 Summary of applications of digital twins for education.

Year(s) [references]	Area application	Usage
2018 (Nikolaev et al., 2018)	Engineering education	• Introduce digital transformation in engineering fields
2020 (Sepasgozar, 2020; Zacher, 2020)	Simulations and web labs	• Simulate industry environments • Online learning through web based apps
2022 (Erdei et al., 2022; Pirker et al., 2022)	Training and education scenario	• Enable virtual education settings • Develop training models using DTs
2022 (Erdei et al., 2022; Eriksson et al., 2022; Pirker et al., 2022)	Digital labs	• Use of digital replicas • Facilitate to develop models
2022 (Han et al., 2022)	Intelligent/Smart campus	• Smart campus management • Enable disaster response
2022 (Iakovides et al., 2022)	Virtual libraries	• Introduce virtual libraries through GUI apps.
2022 (Kartashova et al., 2020)	Blended learning	• Improved blended learning • Design prototypes
2022, 2023 (Kaarlela et al., 2022; Kovbasiuk et al., 2023; Orsolits et al., 2022)	Remote education (robotics and distance)	• Enable AI based platforms • Educational robots are used

TABLE 7.2 Evolution of contributions along with the technology used for education.

Year	Technology used	Educational objective(s)
2018 (Nikolaev et al., 2018)	• GUI • virtual information spaces	• To overcome the limitations in education systems
2020 (Sepasgozar, 2020; Zacher, 2020)	• simulations • web labs • gaming platforms	• High user engagement • Interactive learning
2022 (Han et al., 2022)	• IoT • sensor networks	• Replace traditional learning with smart online systems
2023 (Razzaq et al., 2023)	• deep learning • artificial intelligence	• Enable campus monitoring
2024 (Tarng et al., 2024)	• virtual/Augmented reality	• Facilitate virtual labs for trainings • Procedure modeling
2025 (Swain et al., 2025)	• machine learning • emotion recognition systems	• Detect student emotions for adaptive education system

environment (temperature, humidity, brightness and air quality among others), performance from learners (participation, movement, attendance) and functioning of the infrastructure (use of resources, availability of devices) (Hakiri et al., 2024). This real-time data is the basis for building, analyzing, and optimizing of digital twins, which improve educational experience.

3.1 Types of sensors in smart education

In education, sensor networks exist in various sensor forms that are tailored to capture different information (Badshah et al., 2023; Gan et al., 2022; Shaikh et al., 2025; Zlatev et al., 2023). Some examples of the categories are given below.

3.1.1 Environmental sensors

These types of sensors are used to monitor classroom temperature, air quality (CO_2 concentration), humidity, and noise level in creating healthy and optimal learning environments.

3.1.2 Presence and motion sensors

These types of sensors are used to identify occupancy, watch student attendance, and track physical movements in classrooms or lecture rooms.

3.1.3 Wearable sensors

These types of sensors are embedded in smart watches or fitness bands to supervise physiological signals (like heart rate, stress levels, movement) in learning.

3.1.4 Smart cameras and expression detection sensors

These types of sensors use computer vision algorithms, to sense the facial expressions, sentiments, emotional engagement, for how long students are paying attention.

3.1.5 Interactive sensors

These types of sensors are installed on tablets, e-books, smartboards to log learning interactions, touchscreen activities, and device utilization.

3.1.6 Acoustic pattern detection sensors

These types of sensors monitor acoustic patterns to gauge the level of cooperation and identify disruptions in learning exercises.

3.2 Architecture of sensor networks for digital twins

The design architecture of the sensor networks is hierarchical, with the capability of facilitating data capture, data transmission, data analysis, and data visualization in exploration of digital twins suitable for education. It enables the lifelike connection of the physical space (classrooms, labs, and campuses) with their virtual world twin (Liu & Jiang, 2022; Razzaq et al., 2023).

According to the layered architecture of sensor networks for digital twins as shown in Fig. 7.2, the exchange of information between sensors, digital twins and educational places can be divided into the following main layers.

3.2.1 Layer 1: Real world environment

This layer comprises of physical classroom i.e., the lecture hall, the library, and the laboratory, the college, and the campus. Under this environment, students, teacher, as well as infrastructure systems including smartboards, lighting systems, and the HVAC system operate. The motivation here is that these areas represent physical domain(s) in the

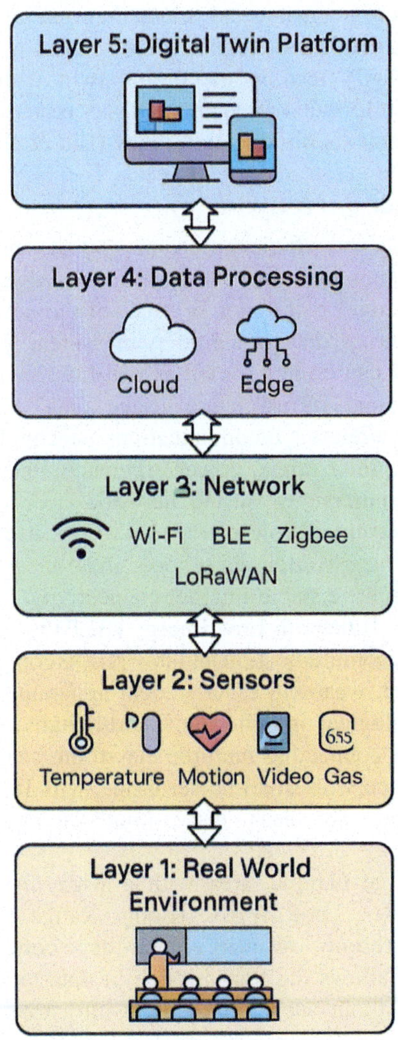

FIGURE 7.2 Layered architecture of sensor networks for smart education.

real-world, which need to obtain and mirrored with ongoing sensor data (Kurni & Srinivasa, 2024).

3.2.2 Layer 2: Sensors

The sensor layer acts as the dynamic barrier between the physical learning space and its digital counterpart. It leverages a network of smart sensors that can track dynamic changes in different aspects of the learning environment. Classroom optimal environment is regulated, includes a temperature, humidity, air quality, illumination and noise. At the same time, motion and occupancy sensors track student presence and movement, providing visibility into the interactions and use of space (Hegazy Abdelghany Mohamed Ammar, 2021). Furthermore, physiological data is collected from wearable and smart devices, like heart rate and stress indicators, that the system uses to estimate cognitive and emotional states. These are accompanied by intelligent video cameras and acoustic sensors, capturing not merely visual and audio inputs but also behavior, collaboration intensity, and interaction quality. Instead of serving as individual gadgets, these sensors work in concert, constantly feeding the digital twin with real-time, contextual information i.e., anomalies or issues that capture the fluid contexts of education today (Liu & Jiang, 2022).

3.2.3 Layer 3: Network

The reliable and efficient transmission of sensor data to the processing systems, nearby (on the edge) or at the cloud (far edge) is a major key role of the network layer. The amount of data a digital twin system that provide educational can be large, such as real-time environmental data, user and infrastructure usage data. Furthermore, multiple wireless communications technologies are selected for their range, power efficiency, and data transmission requirements by the network layer to facilitate such uninterrupted flow of data. Wi-Fi is used for high-bandwidth applications in schools and college campuses, which often have stable Internet connectivity. Short-range, low energy Bluetooth Low Energy (BLE) is preferred for short range communication and low power consumption in, for example, wearable EOG student apparatus to be used for physiological monitoring. Additionally, Zigbee is suitable for connecting multiple classroom sensors at medium distance with small power usage, with its low power consumption and mesh networking capability (Hakiri et al., 2024).

In a large campus area with a wide distribution of sensor nodes, LoRaWAN (Long Range Wide Area Network) presents low energy, long-range communication, and hence, allows the transmission of data over many kilometers without such an elaborate infrastructure (Chen, 2024). One of the primary design considerations at the network layer is to enable low latency, security, as well as energy efficient data communication so that the digital twin model is updated as soon as possible in sync with the real-world environment, such that it can be analyzed, visualized and used to derive decisions without causing an undue load on the network.

3.2.4 Layer 4: Data processing

This layer of the sensor network architecture is important for converting raw, unorganized sensor data into meaningful and actionable insights that support the digital twin of an educational environment. After the data is collected from different classroom conditions sensors, student behaviors detectors and statewide infrastructure usage, it is pre-processed in several stages. These operations are being done on premise, at the edge of the network or in the cloud. Furthermore, edge computing is commonly used for time-critical function, which need instant reaction, like turning cold/warm lighting, or air conditioning in classroom according to real-time occupation and temperature. Edge devices are also able to decrease network latency and enable faster decisions to be made, while reducing the amount of network bandwidth being used through more local data processing. The raw data is filtered to remove noise and irrelevant data, keeping only relevant, and reliable data that carries to the later stages (Koulamas & Lazarescu, 2020).

This layer also does real-time analytics, including anomaly detection e.g., sudden spike/disruption or engagement analytics i.e., when do the students attention levels drop. If deep computational analysis is required such as the long-term trends of student performance, predictive maintenance of need of classroom, aggregated reporting across many classrooms etc., the cloud processing infrastructure is used. Additionally, cloud infrastructures offer the necessary scalability and computing power to enable multi-source data integration (through sensor fusion), machine learning model application, and long-term historical data storage for insights. By smart load-balancing between edge and cloud, the data-processing layer guarantees that the digital twin is accurate, timely, and capable of quickly adapting to the dynamics of the educational environment.

3.2.5 Layer 5: Digital twin platform

It is the last, most important layer of the sensor network where all pre-processed and analyzed data is gathered to create a virtual replica of the physical learning space that is dynamic and ever-changing. This is where the core part of the system resides, converting real-time analytics into actionable knowledge. It includes visualization capabilities with detailed graphical dashboards, 2D blueprints and even immersive 3D models of the classroom that reflect the situation in the classroom, student positions, environmental factors and resource usage. This visualization could then allow both teachers, administrators, and even students themselves to intuitively grasp and get fast feedback on the state of the learning world (Yao et al., 2023).

The platform is not only used to visualize but also to monitor data, to collect live performance data such as levels of student engagement, air quality indices, usage of equipment etc. These indicators are visible on personalized dashboards, tailored to various stakeholders (teachers, campus managers), who can now supervise and intervene during implementation without manual data monitoring. Moreover, the Digital Twin Platform is an environment for decision-making that may task automated actions not just passively presenting the data. If for instance, sensors sense high carbon dioxide concentrations in a classroom, the system will work to ventilate accordingly. Similarly, if it traces patterns of student disengagement from behavior data, the platform can suggest adaptation suggestions within the educational process for example, tailor teaching materials, adapt pacing and suggest personalized learning routes.

Crucially, the digital twin is not a fixed model, but one that adapts in real time as the real-world environment changes. Any kind of interaction, movement or change in the physical environment updates the digital model instantaneously, resulting in a forward feedback loop that ultimately makes learning spaces smarter, more responsive and more optimized long term. Additionally, it enables the evolutionary process and is at the heart of the intelligent educational ecosystems that make data-informed decision-making support even more effective for learning, operationally more efficient and institutionally agile (Tagliabue et al., 2021).

4. Advantages of digital twins in education

The application of the Digital Twin (DT) in the field of education brings in numerous benefits, making the conventional learning systems enhance to smart, adaptive and learner centric environments. By generating living digital copies of classrooms, campuses, even students themselves, educational organizations can take instruction from static models to personalized and data-driven educational systems. The use of digital twins provides real-time monitoring of both physical and virtual spaces, aid predictive insights to inform decision-making, and to create engaging, interactive and immersive spaces. In addition, they contribute to better operational efficiency, by maximizing resource use and reducing energy consumption and maintenance needs, thus enabling smarter and more sustainable education infrastructures. But beyond the hardware, DTs enable personalized learning paths by dynamically adjusting content and methods according to real-time data extracted from students' interactions and behavior. In a world where agility, inclusiveness, and technology savviness are essential for education, digital twins emerge as a large driving force for the future-relevant learning environment (Chen, 2024; Hakiri et al., 2024; Koulamas & Lazarescu, 2020). Smart education is empowered by digital twin technology in a number of ways, including the following:

4.1 Increase the quality of learning

Digital twin technology can offer a more dynamic and practical learning environment, assisting students in comprehending abstract ideas and intricate theories and applying them to real-world issues. Teachers can increase teaching effectiveness by using digital twin technology to create broad and three-dimensional educational experiences that learners and increase their excitement for learning.

4.2 Customized learning support

Digital twin technology can offer customized education, engagement and assistance based on students' learning requirements and personality traits to accommodate different methods of learning (Arantes, 2024). This will assist teachers in recognizing the unique characteristics of each student, customizing learning materials and pathways, increasing learning effectiveness, and recognizing possibilities.

4.3 Enhance practical expertise

Students can investigate and enhance their skills in a controlled and secure setting by using digital twin technology to create a computer-simulated training environment. This not only eliminates security hazards and resource wastage brought on by constrained practice environments, but it also makes it easier for students to practice in a more realistic configuration.

4.4 Encourage innovation in classroom and educational models

Digital twin technology has introduced new teaching approaches and methods to smart education, including interactive learning across locations and times, and scenario-based development in learning. This encourages the variety and customization of teaching approaches and material, breaks down the constraints and restrictions placed on traditional teaching offerings, and encourages students' innovative and imaginative reasoning ideas to solve various problems.

4.5 Promoting future skills

Digital twin technology enables higher education to develop potential students with creative problem-solving, and effective skills. In addition to being one of the main objectives of smart education, these skills are crucial for adjusting to social development as well as technological and scientific advancements in the information era.

In conclusion, the importance of digital twin technology in enhancing education is to increase the effectiveness of learning, achieve personalized learning assistance, develop practical skills, encourage innovative teaching methods, and create an effective foundation for developing future talent. This supports the growth of talent, academic achievement, and better addressing societal demands.

5. Challenges and future directions

The integration of sensor networks and digital twins into educational settings brings huge benefits, but there are several important implications that need to be able to embed these technologies effectively and securely (Arantes, 2024; Çela et al., 2025).

5.1 Data security and privacy

Sensor networks in education continuously acquire data-such as biometric data, behavior, academic performance and environmental data. Securing this information is extremely necessary. Schools and universities need to have strong privacy policies, utilize encryption techniques, and practice proper data management policies. Additionally, well-defined consent procedures require particular attention, especially with data pertaining to minors. The threats of unauthenticated access, data leaks, and malicious exploitation of student data pose significant ethical and legal challenges. Therefore, such concerns require the adoption of strong cybersecurity measures alongside compliance with applicable international data protection laws, including the GDPR and FERPA.

5.2 Design and cost of infrastructure

The deployment of such large-scale sensor networks is expensive and includes extensive investment in hardware, software, the network infrastructure, and its maintenance. The financial investment in the purchase of smart sensors, installing secure communication networks, constructing data storage and analytic infrastructure significantly increased the cost. Furthermore, upgradation of existing campus infrastructure is also potentially high. In addition, the expense of electricity, bandwidth, calibration, and reparation contributes to high cost. Additionally, the educational institutions in low and middle-income countries may find it hard to fund and/or rationalize investment without defined and trackable results (Hazrat et al., 2023).

5.3 Complex system integration

A further significant obstacle is the difficulty of integrating systems of various sorts. Sensor networks are usually a mixture of sensors provided by different vendors with various communication protocols, data formats, and interoperability standards. Aggregating all these disparate systems into a single, coherent platform that can consistently deliver accurate, real-time data into a digital twin model is going to require some powerful middleware, standardized APIs and some seriously technical orchestration. However, to realize interoperability of IoT devices with cloud platforms, edge processors and educational applications is still technically challenging, needing expertise and further system development.

5.4 Hardware connectivity and maintenance

In the sensor network as the sensor nodes are hardware, therefore they have various issues related to connectivity and maintenance. These physical devices that can wear out, degrade, eventually run out of battery and suffer from software bugs and from connectivity problems. Therefore, the input to the data streams and stability of the network must be achieved by active maintenance, monitoring and occasional upgrades. When sensors fail, then incorrect data are fed into models or, worse yet, missing data, and that undermines the authenticity of digital twin representations, and the decisions based on them. In addition, one also need to come up with maintenance solutions which will reduce downtime and prevent system failures from compromising school activities, this is true especially for mission-critical applications, such as campus security and health care monitoring (Boettcher et al., 2023).

As digital twin technologies develop, the sensor network is anticipated as groundbreakingly complex and deeply entrenched in intelligent educational ecosystems. There are several positive trends that will likely continue to shape the future landscape (Çela et al., 2025). In the near future, 5G technology will revolutionize sensor networks in education. With ultra-low latency, massive number of devices connected simultaneously and high data throughput speeds, 5G networks will make live updates and interaction between physical classroom and its virtual twin possible (Ramirez et al., 2022). To learn with a second self should also be easier, more scalable and more affordable, with real-time VR simulations, augmented labs and live remote collaboration. Additionally, there's also a compelling push towards 'smart sensors' with embedded Artificial Intelligence (Edge AI) (Furini et al.; Jin et al., 2022). Instead of sending raw data to be processed in systems on central servers, Edge AI sensors can undertake local data analysis, filtering and decision-making at the source. Personalized learning with new varieties offered by emotion recognition technology (Swain et al., 2025). To achieve this some advanced sensors interpreting facial expressions, frustration and rhythm of the voice, physiological signals (such as heart rate variability or skin conductivity), which were used for real-time analysis of emotions. Furthermore, all the maintenance and connectivity related challenges of sensors networks will be addressed by self-healing networks in the future. Therefore, the emergence of self-healing sensor networks (Moustafa, 2024) will aid future educational environments since the systems automatically self-diagnose problems, skip communication pathways around faults, and repair issues without the aid of people. These networks can anticipate breakdowns and take proactive measures, like adjusting the layout of the network to withstand disruptions, using machine learning algorithms. In critical teaching moments like controlling security systems,

TABLE 7.3 Comparative analysis of digital twin applications in education based on quality parameters.

Application area	Technologies used	Impact on education	Scalability	Cost	Privacy risk
Digital labs (Erdei et al., 2022; Eriksson et al., 2022; Pirker et al., 2022)	IoT, 3D modeling, simulations	High	High	Moderate	Medium
Simulations and web labs (Sepasgozar, 2020; Zacher, 2020)	Web apps, VR/AR	High	High	High	Low
Training and education (Erdei et al., 2022; Pirker et al., 2022)	AI, cloud computing	Moderate–high	Moderate	Moderate	Medium
Remote education (Kaarlela et al., 2022; Kovbasiuk et al., 2023; Orsolits et al., 2022)	AI, robotics, IoT	High	High	High	High
Smart campus (Han et al., 2022)	IoT, sensor networks, big data	Moderate	Very high	High	High
Virtual libraries (Iakovides et al., 2022)	Cloud, AI search	Moderate	High	Low	Low
Engineering education (Nikolaev et al., 2018)	CAD, simulation software	High	Moderate	Moderate	Low
Blended learning (Kartashova et al., 2020)	LMS, IoT, AI	Moderate–high	High	Moderate	Medium

TABLE 7.4 Evolution of smart sensor networks in education based on adoption speed and future potential.

Year	Technology used	Advantage(s)	Adoption speed	Future potential
2018 (Nikolaev et al., 2018)	GUI, virtual information spaces	• Digital interface. • Addresses traditional limitations.	Slow	Low
2020 (Sepasgozar, 2020; Zacher, 2020)	Simulations, web labs, gaming	• Improved engagement. • Increased interactivity	Moderate	Medium
2022 (Han et al., 2022)	IoT, sensor networks	• Smart learning. • Data-driven learning environments.	Fast	High
2023 (Razzaq et al., 2023)	Deep learning, AI	• Online campus monitoring. • Predictive analytics.	Fast	High
2024 (Furini et al.; Boettcher et al., 2023)	VR/AR, procedural modelling	• Immersive virtual labs.	Moderate	High
2025 (Swain et al., 2025)	ML, emotion recognition systems	• Tracking student emotions. • Adaptive learning	Fast (under-progress)	Very high

monitoring health, managing VR-based assessments, and many more, self-healing features will enable uninterrupted service and maximized system life.

6. Analysis of the existing research

The application of digital twins (DTs) in integration with sensor networks in teaching and learning processes is an incorporation that shifts smart learning environments to profound levels which is fully data-driven. However, to optimize everything, it is crucial to determine the gaps in scholarly research, technological development, and practical issues of implementing them. This section describes the structural analysis of major findings about case studies of several applications, assessing advantages and disadvantages, and forecasting prospects for coming years. It demonstrates which innovations impact the most, the obstacles of scalability, costs, and AI-powered emotion detection technology (Swain et al., 2025), 5G-enabled immersive learning, and the new technologies that will transform education. Therefore, the purpose of merging these findings in Tables 7.3–7.6 is meant to prepare educators, decision makers, and technology specialists to utilize digital twins in strategizing for advanced learning environments.

7. Conclusion

An interesting topic that highlights new innovations in education is the use of digital twins. Additionally, notable gaps appear in the literature concerning this topic. One explanation for these discrepancies, though, would be that digital twins are a component of a larger trend in the education system towards digital transformation. The evolution of online and distance learning may also be reflected in the use of digital twins in education. As a result, many people may think of online learning as a form of augmented reality in education, where virtual classrooms are created with the use of AI and other interactive technologies. Digital twins are virtual copies of actual items, spheres, or surroundings, as can be determined from their definitions. Additionally, the discussions about digital twins of educational institutions are interchangeable with those about online, virtual, or digital classrooms in this situation. Digital twin technologies, as an educational innovation, are a collection of enabling technologies that make it easier to simulate the physical infrastructure of educational institutions to facilitate or support online, virtual, or remote learning.

The observations reveal that there is a widespread need for more thorough explanations of the educational

TABLE 7.5 Categorization of advantages and challenges of digital twins in education based on the critical concern.

Category	Advantage(s)	Challenge(s)	Critical concern
Learning quality (Hazrat et al., 2023)	Improves engagement through collaboration	• Infrastructure design/maintenance. • Training costs.	Cost barrier
Personalization (Arantes, 2024)	Customize education through real-time analytics.	• Privacy risks (student monitoring).	Ethical issues
Practical skills (Sepasgozar, 2020; Zacher, 2020)	Simulated hands-on training.	• Hardware and software requirements.	Resource-intensive
Innovation (Boettcher et al., 2023)	Utilization of VR classrooms and AI tutors.	• Complex integration with legacy systems.	Technological compatibility
Operational efficiency (Ezeoguine & Kasumu, 2024)	Improves campus management.	• Maintenance and scalability challenges.	Long-term viability

TABLE 7.6 Summary of future trends of digital twins in education.

Trend	Potential impact	Implementation level	Expected time duration
5G connectivity (Ramirez et al., 2022)	Revolutionary	Commercial level (2024+)	2–5 years
Edge AI sensors (Furini et al.; Jin et al., 2022)	High	Experimental level (2025)	1–3 years
Emotion recognition (Swain et al., 2025)	Very high	Pilot level (2026+)	5+ years
Self-healing networks (Moustafa, 2024)	Moderate–high	Experimental level (2025)	3–5 years

applications of digital twins. Thus, the first piece of advice is for academics to use engaging conversation while discussing digital twins in the classroom. For example, the terms "digital," "distance," and "online" have gained popularity in recent years. All industries, including education, had to undergo a rapid digital change because of the epidemic. Many academics contend that online learning is the way of the future, but they don't provide enough information on how to get there. Consequently, a more technical approach to the topic is preferred since it aids in comprehending the workings and mechanisms of these kinds of systems. In addition to all, to better comprehend how such systems operate, the architectures and overall design of the system components require more elaboration. The lack of comprehension among different stakeholders may account for the slack in the application of new technologies, even while empirical evidence has previously confirmed their effectiveness. For educational officials, the term "digital twin" may appear like a fancy concept with little real-world use. However, by describing the technology used in the new systems and how digital twins work, policymakers will be attracted and inclined to explore digital twins for easier to access, customized, and broad-ranging educational models.

References

Arantes, J. (2024). Digital twins and the terminology of "personalization" or "personalized learning" in educational policy: A discussion paper. *Policy Futures in Education, 22*(4), 524–543.

Attaran, M., & Celik, B. G. (2023). Digital twin: Benefits, use cases, challenges, and opportunities. *Decision Analytics Journal, 6*, Article 100165.

Badshah, A., et al. (2023). Towards smart education through internet of things: A survey. *ACM Computing Surveys, 56*(2), 1–33.

Berisha-Gawlowski, A., Caruso, C., & Harteis, C. (2021). The concept of a digital twin and its potential for learning organizations. *Digital transformation of learning organizations*, 95–114.

Boettcher, K., et al. (2023). Developing a real-world scenario to foster learning and working 4.0–on using a digital twin of a jet pump experiment in process engineering laboratory education. *European Journal of Engineering Education, 48*(5), 949–971.

Çela, E., et al. (2025). Current trends in smart classrooms and sustainable internet of things. *Designing Sustainable Internet of Things Solutions for Smart Industries*, 1–26.

Chen, Y. (2024). *Managing hybrid industrial IoT enterprise wireless networks*. University of Glasgow.

Erdei, T. I., Krakó, R., & Husi, G. (2022). Design of a digital twin training centre for an industrial robot arm. *Applied Sciences, 12*(17), 8862.

Eriksson, K., et al. (2022). Applying digital twin technology in higher education: An automation line case study. In *SPS2022* (pp. 461–472). IOS Press.

Ezeoguine, P. E., & Kasumu, R. Y. (2024). Undergraduate students' perception of digital twins technology in education: Uses and challenges. *International Journal of Educational Assessment and Evaluation, 10*(2), 381–396.

Fuller, A., et al. (2020). Digital twin: Enabling technologies, challenges and open research. *IEEE Access, 8*, 108952–108971.

Furini, M., et al., Digital twins and artificial intelligence as pillars of personalized learning models personalized learning models can cut student dropout rates, boost student success, improve the integration of online and on-site students, better support teachers in mixed-teaching modalities, enhance accessibility, and more.

Gan, W., et al. (2022). Iot-based multimodal analysis for smart education: Current status, challenges and opportunities. In *Proceedings of the 3rd ACM workshop on intelligent cross-data analysis and retrieval.*

Hakiri, A., et al. (2024). A comprehensive survey on digital twin for future networks and emerging internet of things industry. *Computer Networks*, Article 110350.

Han, X., et al. (2022). Intelligent campus system design based on digital twin. *Electronics, 11*(21), 3437.

Hawkinson, E. (2022). Automation in education with digital twins: Trends and issues. *International Journal on Open and Distance e-Learning, 8*(2).

Hazrat, M., et al. (2023). Developing a skilled workforce for future industry demand: The potential of digital twin-based teaching and learning practices in engineering education. *Sustainability, 15*(23), Article 16433.

Hegazy Abdelghany Mohamed Ammar, M. (2021). *Advanced digital twins for conditions monitoring, examinations, diagnosis and predictive remaining lifecycles based artificial intelligence.* London: Brunel University.

Iakovides, N., et al. (2022). Virtual library in the concept of digital twin. In *2022 International Conference on Interactive Media, Smart Systems and Emerging Technologies (IMET)*. IEEE.

Jin, A. S., et al. (2022). Resilience of cyber-physical systems: Role of AI, digital twins, and edge computing. *IEEE Engineering Management Review, 50*(2), 195–203.

Kaarlela, T., et al. (2022). Common educational teleoperation platform for robotics utilizing digital twins. *Machines, 10*(7), 577.

Kartashova, L. A., et al. (2020). Digital twin of an educational institution: An innovative concept of blended learning. In *Proceedings of the symposium on advances in educational technology, aet.*

Koulamas, C., & Lazarescu, M. T. (2020). *Real-time sensor networks and systems for the industrial IoT*. MDPI.

Kovbasiuk, K., et al. (2023). A digital twin for remote learning: A case study. In *Design, simulation, manufacturing: The innovation exchange* (pp. 379–389). Springer.

Kumar, A., Saudagar, A. K. J., & Khan, M. B. (2024). Enhanced medical education for physically disabled people through integration of IoT and digital twin technologies. *Systems, 12*(9), 325.

Kurni, M., & Srinivasa, K. (2024). *The internet of educational things.* Springer.

Liu, X., & Jiang, J. (2022). Digital twins by physical education teaching practice in visual sensing training system. *Advances in Civil Engineering, 2022*(1), Article 3683216.

Maksimović, M., & Davidović, N. (2022). The role of digital twin technology in transforming engineering education. In *9th International scientific conference technics and informatics in education.*

Mayer, I. (2022). *Digital twins for the real world.* Tilburg University.

Minerva, R., Lee, G. M., & Crespi, N. (2020). Digital twin in the IoT context: A survey on technical features, scenarios, and architectural models. *Proceedings of the IEEE, 108*(10), 1785–1824.

Moustafa, N. (2024). GH-Twin: Graph learning empowered hierarchical digital twin for optimizing self-healing networks. *Sustainable Machine Intelligence Journal, 8*(3), 35–45.

Nikolaev, S., et al. (2018). Implementation of "digital twin" concept for modern project-based engineering education. In *IFIP international conference on product lifecycle management*. Springer.

Omrany, H., et al. (2025). Digital twin technology for education, training and learning in construction industry: Implications for research and practice. *Engineering Construction and Architectural Management*. https://doi.org/10.1108/ECAM-10-2024-1376

Orsolits, H., Rauh, S. F., & Estrada, J. G. (2022). Using mixed reality based digital twins for robotics education. In *2022 IEEE International symposium on mixed and augmented reality adjunct (ISMAR-Adjunct)*. IEEE.

Pirker, J., et al. (2022). Immersive virtual reality for virtual and digital twins: A literature review to identify state of the art and perspectives. In *2022 IEEE Conference on virtual reality and 3D user interfaces abstracts and workshops (VRW)*. IEEE.

Ramirez, R., Huang, C.-Y., & Liang, S.-H. (2022). 5G digital twin: A study of enabling technologies. *Applied Sciences, 12*(15), 7794.

Razzaq, S., et al. (2023). DeepClassRooms: A deep learning based digital twin framework for on-campus class rooms. *Neural Computing & Applications*, 1–10.

Selim, A., et al. (2024). Application of the digital twin model in higher education. *Multimedia Tools and Applications*, 1–18.

Sepasgozar, S. M. (2020). Digital twin and web-based virtual gaming technologies for online education: A case of construction management and engineering. *Applied Sciences, 10*(13), 4678.

Shaikh, M.K. Digital twins and engineering education: Current status.

Shaikh, A., et al. (2025). Intelligent sensors for smart healthcare. In *Advanced sensors for smart healthcare* (pp. 347–361). Elsevier.

Siyan, C., et al. (2021). Research on the improvement of teachers' teaching ability based on machine learning and digital twin technology. *Journal of Intelligent and Fuzzy Systems, 40*(4), 7323–7334.

Swain, A., Dalei, A., & Sahoo, S. (2025). Emotional well-being of a human: A case study of proposing a digital twin solution integrating IoT sensors and cloud computing. In *2025 International conference on ambient intelligence in health care (ICAIHC)*. IEEE.

Tagliabue, L. C., et al. (2021). Leveraging digital twin for sustainability assessment of an educational building. *Sustainability, 13*(2), 480.

Tao, F., et al. (2018). Digital twin in industry: State-of-the-art. *IEEE Transactions on Industrial Informatics, 15*(4), 2405–2415.

Tarng, W., et al. (2024). Application of virtual reality in developing the digital twin for an integrated robot learning system. *Electronics, 13* (14), 2848.

Yao, J.-F., et al. (2023). Systematic review of digital twin technology and applications. *Visual computing for industry, biomedicine, and art, 6* (1), 10.

Zaballos, A., et al. (2020). A smart campus' digital twin for sustainable comfort monitoring. *Sustainability, 12*(21), 9196.

Zacher, S. (2020). Digital twins for education and study of engineering sciences. *International Journal of Engineering Science and Technology, 2*(2), 61–69.

Zlatev, Z., et al. (2023). Design and research of a Sound-to-RGB smart acoustic device. *Multimodal Technologies and Interaction, 7*(8), 79.

Chapter 8

Smart sensor networks in the student activities

Usama Kalim[1], Sadia Irfan[2], Asha Kanwar[1] and Ronghuai Huang[1]
[1]*Smart Learning Institute of Beijing Normal University, Beijing, P.R. China;* [2]*Faculty of Education, Southwest University, Chongqing, China*

1. Introduction

Higher education institutions (HEIs) increasingly adopt digital transformation to enhance the learning environment. Smart Sensor Networks (SSNs) are the recent additions to that transformation. SSNs include sensors that collect data during processing, which is an essential building block for creating information for intelligent university systems. The interlinked systems enable educational institutions to boost learning environments, security measures, and administrative protocols when they process student-related data in real-time with automated continuous monitoring (Ani, 2024; Cavus et al., 2022).

SSNs facilitate the development of smart interactive classrooms, which provide individualized learning environments in academic institutions. IOT-based smart classrooms use multiple sensors to track student engagement patterns, progress, and inside–outside classroom activities (Asad et al., 2024; Kapur et al., 2018). Real-time indicators of students' progress enhance educational delivery methods and learning capabilities to build better learning outcomes. Such monitoring of student performance allows systems to provide insights that support educational system planning across extended periods (Cavus et al., 2022).

Smart educational facilities use sensors that identify and collect data while performing facial recognition scans using various multimedia tools (HashStudioz, 2025). Implementing SSNs allows students to maximize their campus social activities and recreational entertainment. Through IoT-enabled leisure facility monitoring systems, colleges can make informed decisions about facility renovations while deciding when to schedule maintenance and construction projects (Cavus et al., 2022; Kraft-Terry & Brown, 2023). The monitoring of usage frequency by smart gym equipment functions to advise maintenance teams about necessary servicing through automatic alerts. Real-time student location information allows academic institutions to generate individual event recommendations that enhance social student interactions within the campus (Farrokhi et al., 2021; IoT World, 2023).

Wearable devices and environmental sensors included in university infrastructure help to promote student health and well-being more and more. By tracking physiological markers such as heart rate, sleep patterns, and physical activity levels, these products enable students to maintain good lifestyles (TheThings.io, 2023). Furthermore, air quality sensors in classrooms and dorms guarantee that the surroundings are fit for focus and physical comfort. If situations like excessive carbon dioxide levels or poor air quality are found, alerts can be given, and quick corrective action can be called for improvement (Khudor et al., 2023; Taheri & Salimi Beni, 2025).

Predictive analytics finds great prospects in the data produced by SSNs, especially for at-risk student identification. Analytics systems, for instance, can monitor behavioral indicators such as falling engagement in campus events, lower class attendance, or irregular study patterns and highlight them as early warning signs of academic or personal challenges (Cheung & Tai, 2023; Stratagon, 2025). By then, institutions can use focused interventions, including peer support programs, academic counseling, or counseling, to increase student retention and achievement.

Moreover, the data from SSNs allows one to design customized learning paths. Analyzing learning behaviors and preferences helps institutions customize their teaching materials and approaches to meet certain learning styles of their students (Ha et al., 2020; Kem, 2022). In addition to improving academic achievement, this raises students' motivation and contentment (HashStudioz, 2025).

Digital Twins in the Smart Classroom. https://doi.org/10.1016/B978-0-443-41632-3.00012-6

Administrative SSNs help to maximize resources. Smart attendance systems can automatically track student attendance using facial recognition software or Radio Frequency ID (RFID) cards, saving time and reducing mistakes (Technology Innovators, 2025). Energy management systems rely on occupancy sensors and environmental monitors to lower power consumption by varying temperature and lighting depending on room usage. These reductions in cost help support environmental projects that employ efficiency. Despite many benefits, using SSNs in educational environments generates serious ethical and privacy issues (Lee et al., 2024; Yılmaz & Yılmaz, 2022). Particularly when it involves sensitive information like location tracking or biometric data, constant observation of student activities can be seen as invasive (Stokel-Walker, 2020). Maintaining student confidence and following data protection rules depends on ensuring data protection by encryption, anonymization, and rigorous access control. Institutions must also get informed permission from students and clearly state how their data will be utilized (Pedagogue, 2024).

1.1 Purpose of this chapter

SSNs are essential for developing smart campuses since they provide several advantages in academic, recreational, social, and health spheres (Ha et al., 2020). Higher education institutions can guarantee student well-being, boost learning opportunities, and increase operational efficiency by using the data produced through various networks (Jiang et al., 2022; Soundari & Jyothi, 2020). Still, it is essential to solve the related privacy and data security issues to fully realize SSNs' ability to shape the direction of higher education. The chapter examines how SSNs support student involvement in several academic, recreational, social, and health spheres by monitoring and tracking. It also describes how administrative efficiency, personalized learning, and predictive analytics can benefit from data produced by these networks.

2. An introduction to Smart Sensor Networks

2.1 Concept definition

The integrated system, known as SSNs, combines separated sensor nodes that perform data collection and processing alongside communication functions. The combined nodes monitor environmental conditions through data documentation before sending the collected information to central locations for analysis and decision-making (Zhang et al., 2020). SSNs are mainly composed of:

2.1.1 Sensor nodes

Each sensor node includes hardware components for physical event detection and temperature, humidity, motion, or light measurement. Each node contains multiple features that enable data sensing and recording through its sensor device, microprocessor communication module, and power source (Hu & Hao, 2012; Sahoo et al., 2022).

2.1.2 Communication infrastructure

The communication infrastructure exists from central systems to sensor nodes through their data exchange mechanisms. Karthick Raghunath et al. (2024) state that real-time data sharing depends on wireless communication technologies.

2.1.3 Data processing units

Data processing units in centralized or distributed systems receive sensor node data for evaluation and interpretation to create valuable insights through compilation processes. The network management system determines whether SSNs operate independently or under central supervision (Pagliari & Poncino, 2020).

2.2 Technologies involved

The efficiency and usefulness of SSNs benefit from various underlying technologies that contribute to their operation.

2.2.1 Wireless Sensor Networks (WSNs)

The subset of SSNs, Wireless Sensor Networks (WSNs), functions without wires since their sensor nodes transmit data through wireless channels. WSNs deliver essential assistance to applications that need wireless data transmission since wire connections are impractical (Ghadi et al., 2024). Data streams proceed as raw bitstreams through physical media in the physical layer before the data link layer ensures safe node-to-node communication, and the application layer oversees network processing and user interface operations. The multiple applications of WSNs include environmental monitoring, healthcare services, and industrial automation (GeeksforGeeks, 2024).

2.2.2 IoT platforms: Internet of Things

IoT platforms function as intermediaries between hardware components and application layers, providing developers with essential tools and services that facilitate IoT application development, implementation, and monitoring functions. Such systems combine data acquisition components with device control, storage and processing functions, and protection measures. These systems make Real-ime decision-making possible because they provide instant communication between devices and programs. According to Hossain et al. (2024) and Rath et al. (2024), scaling SSNs within general digital ecosystems depends on IoT platforms.

2.2.3 *Cloud and edge computing*

Cloud computing enables flexible management and large-scale operations of SSNs through its centralized data storage facilities and processing capabilities. SSNs benefit from combining and analyzing an extensive quantity of collected data on sensor nodes. Edge computing operates through data processing on edge servers and local devices to bring computation near the data collection points (Asim et al., 2020). Real-time decisions become faster and bandwidth stays optimized through this processing approach. According to SSNs (Business Insider, 2024), integrating cloud and edge computing systems ensures efficient data processing and quick responses (Jihong & Xiang, 2024).

2.2.4 *AI algorithms and machine learning*

SSNs process their gathered data through artificial intelligence (AI) and machine learning (ML) techniques. ML systems detect patterns and irregularities within the data, which leads to predictive analytics functions and automated decision capabilities. AI functions across diverse technologies that mimic human cognitive abilities to handle complex operations from vision processing to natural language understanding. SSNs gain operational efficiency and adaptation capacity to changing situations by integrating ML and AI technologies (Financial Times, 2023).

2.3 Valuation in learning

SSNs transform educational settings through their ability to deliver real-time feedback, automate routine tasks, and improve decision-making methods. These networks comprise interconnected sensor devices that collect data, enabling administrators and educators to gain valuable insights (Geertsema & Lu, 2023).

SSNs function as a primary tool for automatic classroom attendance tracking. Traditional approaches to IoT-enabled smart classrooms employ technology with RFID tags, biometric scanners, and facial recognition systems to record attendance automatically. These methods are time-consuming and error-prone, yet they ensure accuracy and protect valuable teaching time (Preprints.org, 2025).

Behavioral analysis of teaching environments depends entirely on SSNs. By implementing AI-based solutions, teachers can instantly monitor student engagement levels. The system alerts teachers to change their instruction when many students appear uninterested, thus enhancing educational success (Li et al., 2023).

SSNs make the improvement of facilities management operations possible. Tracking environmental indicators through sensors allows automatic adjustments to maximize energy efficiency and maintain comfortable educational settings (Matellio, 2024). The cost reduction of running expenses enables support for environmental projects.

SSNs represent a flexible approach that improves educational spaces when implemented throughout learning environments. Schools benefit from SSNs by enabling teachers and managers to make better decisions through automated tasks and real-time data tracking, which results in improved efficiency and effectiveness of learning spaces.

3. SSN applications in student activities

Implementing SSNs has significantly transformed student life through real-time data provision, enhanced security measures, increased academic engagement, and improved student well-being. The tools enhance university operations and student customization, creating adaptable learning environments (Gao, 2022).

3.1 Classroom observation

Modern educational environments utilize SSNs to track student engagement levels and physical attendance through continuous monitoring, which controls student learning activities. The combination of RFID tags and facial recognition systems ensures dependable student identification, while motion sensors detect behavioral patterns to indicate student involvement or distraction, according to Koay et al. (2022) and Papatheocharous et al. (2024). The classroom environment is monitored by environmental sensors, which measure temperature, humidity and light levels, and air quality standards. Schools can adjust the learning environment by modifying conditions that directly impact students' comfort and cognitive performance.

3.2 Welfare and well-being

Students can track their physical state through smartwatches and fitness bands because these wearable sensors monitor their heart rate, sleep patterns, stress levels, and daily physical activity (Xiao et al., 2024). These data points produce indicators that help identify burnout symptoms and health deterioration. Early warning systems are implemented by quick mental health treatments and individualized wellness programs provided to students. Implementing proactive measures benefits students' welfare while building an improved learning environment (Shihao et al., 2025).

3.3 University safety and navigation

Implementing SSNs enhances security measures and directional assistance throughout college facilities. Students can locate their destinations instantly with the combination of GPS and Bluetooth beacons and indoor positioning systems. The systems serve as valuable tools for students new to the institution or with disability needs. Safety monitoring through SSNs detects two anomalies: unauthorized access attempts and suspicious behavioral patterns (Fährmann et al., 2024). Data integration between sensors enables automated emergency protocols that release secure areas and send alerts to emergency responders, which decreases response duration and boosts campus-wide security measures.

3.4 Innovative attendance systems

Standard attendance systems often produce errors and take up considerable time. RFID cards, Near Field Communication (NFC), and biometric systems that include fingerprint and facial recognition help simplify tracking attendance according to SSN standards. The automated attendance tracking systems reduce faculty workload and generate instant student presence reports (Hussien et al., 2024; Zhao et al., 2022). These technologies use behavioral analytics and academic performance data to identify student absenteeism and participation patterns.

3.5 Leisure and social events

The Student Surveillance System enables university students to participate in social activities alongside their academic pursuits on university grounds. The combination of sensors in university premises, including cafeterias, common areas, and sports complexes, collects data regarding user choices, areas of maximum use, and usage patterns. The collected data helps university officials optimize resource management and facility maintenance and plan university events (Teoh et al., 2021). The analysis of interactive kiosks and mobile app user feedback enables universities to boost student satisfaction, organize community events, and promote overall university involvement for all students.

4. Real-world applications and case studies

SSNs have moved from theoretical to practical deployment throughout various worldwide educational institutions. Universities from across the globe have integrated SSN technology to achieve better academic outcomes, efficient resource utilization, and enhanced student wellness (Ullah & Gul, 2023). According to the case studies, multiple regions have applied SSNs through unique and creative approaches.

4.1 Smart South Korea classrooms

With its reputation for tech-driven learning, South Korea has established modern smart classrooms based on SSNs to deliver individualized learning experiences. Real-time monitoring of student involvement, emotional responses, and classroom situations occurs through motion sensors, facial recognition, and environmental sensors installed in these classrooms (Lee & Lee, 2024). The collected data points enable teachers to adjust their teaching methods according to their students' emotional states and attention levels, thus creating a personalized learning atmosphere. Predictive analytics generated from sensor data through learning management systems help teachers locate students who need academic support before they fall behind. The South Korean Ministry of Education backs these initiatives through its national push for digital education reform, including AI applications.

4.2 IoT-enabled American campus

The implementation of IoT and sensor technology enables American colleges to establish intelligent campuses that are secure and energy-efficient. An educational institution installed motion detectors and thermal and sound sensors throughout common spaces and residential and classroom areas. The system automatically adjusts heat ventilation and lighting systems through sensors that track student movement patterns to minimize power consumption. Student safety numbers within residential buildings detect unusual sounds and movements that suggest distress or intrusion, automatically triggering security alerts for university security (Sears & Cunningham, 2024). The connection to mobile apps enables students to view real-time environmental data while requesting services and receiving safety notifications, improving comfort and campus safety.

4.3 European institutions' wearable tech

The adoption of wearable technology by European higher education institutions serves dual purposes of student health support and academic research on behavioral patterns (Hernández-Mustieles et al., 2024). Students from Finland, the Netherlands, and the UK participate in university programs that provide them with digital fitness trackers and smartwatches. The devices measure physical activity levels, sleep patterns, and heart rate variability. The researchers analyze the merged information to establish connections between physical wellness and academic success for creating evidence-based therapeutic and supportive frameworks. Public health researchers receive anonymized data from time to time to study general patterns regarding young people's choices about health and lifestyle (Joymangul et al., 2024). Wearable technology serves multiple purposes in education through health challenges and mindfulness programs and by helping schools develop their physical wellness environment.

5. SSNs' advantages for student activities

Higher education institutions became more efficient through SSNs, which enabled new opportunities for institutional management, safety procedures, and student-focused academic practices. The changing educational landscape of twenty-first-century education benefits from SSNs, which transform into responsive data-focused learning environments that demonstrate intelligence (Asad et al., 2024; Khosravi et al., 2022). The main benefits provided by SSNs within student activity frameworks are as follows.

5.1 Enhanced learning results in adaptive learning contexts

SSN applications in education represent the most essential advantage through their capability to generate learning environments that adapt to individual student needs. Real-time assessment of student cognitive state and engagement is possible through SSN systems that track facial expressions, posture, eye movement, and interaction degree (Zhang et al., 2024). AI systems utilize analyzed data to deliver personalized medical solutions or create individualized content distribution systems. The system generates warning signals and supplementary materials to notify educators when students lose interest or become unclear during educational sessions.

Environmental sensors maintain optimal cognitive function conditions in the physical classroom by monitoring air quality, temperature, and lighting levels (Deva et al., 2025; Kapur et al., 2018). According to research, high-temperature conditions and insufficient ventilation can negatively impact academic results and student concentration. SSNs prepare an environment for focused learning since these settings adjust automatically. Integrating adaptive systems with Learning Management Systems (LMS) enables student categorization based on learning styles or performance criteria through dynamic assessment while delivering personalized courses and assessments. Personalized approaches promote memory retention and cognitive load reduction, optimizing learning quality and satisfaction (VASyLyk et al., 2024).

5.2 Student engagement improves through behavioral cue assessment and response analysis

The majority of academic achievement depends on student engagement levels. SSNs serve as an advanced system that enables both evaluation and enhancement of student learning. Traditional methods of measuring engagement through classroom observations and self-reported questionnaires remain limited in their scope and arbitrary nature (Gao et al., 2021, pp. 237–242). The continuous collection of subtle behavioral inputs helps SSNs overcome these constraints. Wearable devices and furniture sensors enable tracking student involvement, body movements, and time spent on educational content.

SSNs use recorded data points to detect behavioral patterns indicating when students become disengaged and lose motivation (Lu et al., 2022). When students display a noticeable focus decline during specific times or subjects, the information allows teachers to alter their teaching plans or classroom activities. Advanced analytics systems show general pedagogical tendencies through engagement-level data from multiple classes or cohorts.

Some academic institutions have adopted gamified feedback methods that award immediate notifications to students based on their engagement performance (Kurni & K. G., 2025). The feedback mechanism decreases passive behavior while increasing active student participation and interactive peer-to-peer interactions. SSNs deliver important engagement data to educational institutions that traditional learning methods would not generate, thus providing value in virtual or hybrid educational settings.

5.3 Enhanced safety via emergency systems and surveillance

Educational institutions prioritize student protection, so SSNs have created a more secure campus environment (Hussien et al., 2024). Students can obtain real-time monitoring of university environments by integrating security cameras with motion detectors, thermal sensors, and emergency beacons. The monitoring systems detect abnormal conduct, unauthorized building entry, and environmental threats such as gas leaks and fires (Tabuenca et al., 2023). Sensor networks activate alarms while starting security protocols and send precise location data to authorities and responders through automatic alerts in dangerous situations.

Personal safety heavily depends on the implementation of smart SSNs. Wearable devices incorporate panic buttons that enable students to send emergency alerts and their exact location while present in dangerous situations. Implementing geofencing technologies in colleges during late hours aims to restrict students from entering prohibited or risky zones. SSNs enable people to monitor through proximity tracking, which supports contact tracing and social distancing enforcement during pandemics and public health emergencies (Park et al., 2021).

Through the implementation of these systems, users benefit from elevated psychological security. Students and staff members show increased safety feelings because their university actively searches for possible hazards, which leads to better mental health and academic results.

Organizations protect their legal standing and reputation by demonstrating technological efforts to reduce risks.

5.4 Effective resource management with predictive maintenance and tracking of use

The ability of SSNs to enhance operational effectiveness and resource management is a crucial advantage. Academic institutions with many buildings, such as lecture halls, labs, libraries, dorms, and recreational centers, face logistical challenges in maintaining buildings on time. Real-time monitoring of usage patterns, occupancy levels, and equipment performance by IoT-based sensors enables SSNs to simplify this process (Ong et al., 2021).

Heating ventilation and air conditioning systems employ sensors to detect equipment failure risk, enabling automatic maintenance scheduling before equipment breakdowns occur. Smart lighting solutions controlled by motion detectors ensure that no unnecessary power usage occurs in unoccupied areas, which results in significant reductions of both emissions and energy usage at the institution (Wang et al., 2024). Managers can optimize space allocation through occupancy sensors, which also help improve room booking systems for better effectiveness.

Tracking tool and region usage through SSNs enables institutions to make more accurate budgetary and resource replenishment decisions. Analyzing real-time foot traffic enables cafeterias to modify their cooking procedures, reducing food waste. Some colleges combine SSNs with smartphone apps to enable students to find accessible study rooms and parking places (Mustafa et al., 2021). Managers obtain detailed visibility from SSNs, which helps them base their decisions on student requirements and usage patterns.

5.5 Data-driven decision making

The ability of SSNs to generate real-time data-driven institutional decisions is its most substantial transformative feature. The abundance of data found in SSNs provides schools with robust evidence to construct strategic plans that enhance academic outcomes, student attendance rates, and engagement levels, as well as track health patterns, environmental conditions, and infrastructure performance (Lazaroiu et al., 2022).

Teacher identification of at-risk students becomes possible through this data, which enables them to create personalized educational solutions and optimize their instructional methods. Analysis of student participation levels through analytics can lead teachers to modify content delivery methods and introduce interactive components in the targeted modules. Administrators use campus information to create policies about safety measures, health-oriented programs, and resource management. Multiple data integration enables organizations to gain comprehensive knowledge about institutional performance alongside student experiences (Siddiqui & Shaukat, 2021).

SSNs allow organizations to collect long-term data, which supports multi-month and multi-year trend analysis. This method is vital for understanding technology interventions and educational programs. By combining SSNs with decision-support systems and dashboards, users can easily track important statistics and performance indicators while linking institutional projects to their objectives.

When organizations use this form of decision-making, an always-improving culture emerges. Organizations with SSNs' implementation speed becomes faster because they use real-time input rather than relying on anecdotal evidence or periodic polls. The agile approach enhances operational effectiveness while making institutions more aware of external problems and student requirements.

The academic and safety aspects, administrative functions, health requirements, and security measures benefit from SSNs. Predictive maintenance achieves maximum resource utilization through behavioral analytics, real-time surveillance, and emergency systems, while SSNs create personalized learning environments with adaptive features and empower stakeholders with data-driven decision-making (Fayez et al., 2021). The development of higher education in the digital age depends on SSNs to build smart learning environments that respond to and include all students. Implementing SSNs provides operations with future protection and enables institutions to establish themselves as leaders in educational innovation.

6. Ethical considerations and smart sensor network challenges in higher education

The growing use of SSNs in higher education produces numerous ethical and pragmatic problems that need resolution to ensure responsible and sustainable implementation (Predel et al., 2022). Issues affecting SSNs include privacy challenges, infrastructure limitations, technical obstacles, and cultural resistance.

6.1 Consent and data privacy

Data privacy protection and informed consent represent essential elements for an ethical application of SSNs. SSNs collect sensitive personal data through facial recognition, fingerprint scans, and behavioral and health-related metrics, including movement, attention span, and heart rate (Venticinque & Amato, 2018). The vast amount of information about student behaviors and well-being gained through SSNs becomes dangerous when used improperly or not adequately protected.

Universities must establish firm data governance principles to determine what information they will collect, its intended use, and its retention period. The rules must contain details about data ownership, access permissions, and procedures for reporting infractions. Students need a full understanding of what data collection involves, its utilization purposes, and who will have access to the gathered information. The information needs active consent from knowledgeable individuals who should receive it through straightforward language that avoids technical jargon.

Implementing privacy-by-design models requires institutions to embed data security elements into SSN system architecture during its initial development phase. Differential privacy, data anonymization, and encryption work together as tools to minimize data breaches and unauthorized monitoring incidents. Regular audits must happen to check data practices while ensuring compliance with national and international laws, including the Personal Data Protection Act (PDPA) in different nations or the General Data Protection Regulation (GDPR) in the European Union (Mohammad Amini et al., 2023).

The issue of consent remains unclear in the face of existing ethical challenges. Students often find themselves pressured into accepting monitoring as a condition of receiving teaching materials. Institutions must investigate opt-in consent methods combined with accessible alternatives and an educational effort to enhance digital rights and personal autonomy.

6.2 Infrastructure and cost control

The high costs and infrastructure requirements act as major barriers to SSN system deployment. A functional SSN ecosystem needs the installation of several hardware and software elements that integrate sensors and data servers with communication networks and analytics platforms. Institutions must allocate funds to acquire cybersecurity systems that protect against potential security threats and breaches (Raju et al., 2022).

The cost of implementing such technologies remains out of reach for colleges in underdeveloped regions. The initial expenses include premium sensors and IoT equipment and safe data center construction and installation costs. The operating expenses of IT support and data management teams, alongside system maintenance and software upgrades, personnel payments, and energy consumption, fall under continuous expenses (Teoh et al., 2021).

To overcome implementation challenges, institutions should start by launching pilot programs in specific departments and facilities as part of their staged approach. Public entities and technology-based institutions offer financial backing through grants and public-private partnerships (Tabuenca et al., 2023). Edge computing technologies and cloud-based solutions reduce expenses for expensive physical infrastructure and enable efficient data processing and analysis.

The financial blueprint for SSNs must be extended over a very lengthy period to account for technology infrastructure, staff education, ethical oversight systems, and funds to handle unexpected events.

6.3 Technical restrictions

SSNs demonstrate significant potential, but their deployment faces important technological obstacles that must be addressed. System interoperability problems, energy economy, data quality, and sensor accuracy represent the primary issues in SSN applications. The accuracy of sensors plays a vital role in data reliability because minor measurement errors can lead to significant wrong interpretations, according to Diraco et al. (2023). The incorrect classification of students as unengaged by attention–detection sensors occurs due to variations in facial position and lighting conditions, leading to unnecessary intervention and rating bias. A sensor system requires periodic calibration and environmental testing to maintain reliable accuracy.

The major problem with energy consumption primarily affects battery-operated and wearable electronic devices. Gadgets that need scheduled maintenance or battery charging can disrupt ongoing data collection, negatively affecting user experience. Research into low-power sensor design and energy-collecting technologies aims to resolve these problems. Interoperability represents the third key technical issue in this area. The functionality of SSNs depends on multiple devices that utilize different formats and connection protocols for their software programs. Uniting data from multiple sources becomes problematic and error-prone when no standardized system is in place. The expansion of SSN infrastructure requires costly system updates or component replacements because of this factor (Gao, 2022). Institutions need to focus on building flexible systems with non-proprietary solutions and join standardization efforts at both domestic and international levels to address these issues. SSN systems need ongoing system tests and updates to maintain their technical reliability.

6.4 Opponents of change

The failure of sophisticated technologies depends on how well their intended users utilize them. Staff members, teachers, and students frequently demonstrate resistance to change, which represents one of the non-technical yet vital barriers to SSN deployment (Sofi et al., 2022). People oppose new systems because they have privacy concerns, question technological benefits, fear increased monitoring, and lack digital literacy skills. Student freedom suffers from ongoing surveillance because they worry about legal infringements and the disappearance of open communication and genuine trust between them and their school authorities. SSNs meet resistance when faculty members consider them as tools for managerial authority or when they fear assessment from data they do not comprehend or trust. New systems with additional processes might create an information overload that burdens staff administrators. These problems need clear dialogue alongside comprehensive policy development supported by extensive training sessions. Organizations must involve faculty members and students in planning to ensure their input gets heard throughout implementation (Ha et al., 2020). Deploying pilot programs alongside feedback systems and town hall events helps create trust while explaining the new technology to staff members.

The development of digital literacy programs is an essential factor. Understood SSN operation requires users to decode generated data, which they can use to enhance operational or academic activities (Ruipérez-Valiente et al., 2022). Institutions that provide seminars, tutorials, and individual support sessions help stakeholders develop ownership and empowerment during their transition process. The successful implementation of SSNs demands that organizations embrace and respect different technological perspectives between generations and cultures. Sensitivity to these variations helps create more subtle and successful execution plans.

The potential benefits of implementing a Smart Sensor network in higher education face several practical and ethical barriers. Data privacy and informed consent must remain fundamental values requiring established regulations alongside advanced technological safeguards. Combining strategic planning with creative solutions solves technological limitations and infrastructure costs. Openness alongside sympathy and education form the necessary response to cultural opposition. By handling only a single issue, institutions cannot achieve a complete SSN-driven transformation of student activities, academic achievement, and institutional effectiveness.

7. Future directions

The evolution of SSNs depends on future advances and multiple domain combinations. Predictive, sustainable, customized concepts developed through these advancements offer transformational benefits for SSN capabilities within educational environments. The following portion discusses revolutionary potential approaches to implementing SSN systems within educational settings.

7.1 Integration for active interventions with predictive analytics and artificial intelligence

Combining SSNs with AI and predictive analytics systems brings one of the most promising future possibilities. Information processed by AI systems uses SSN data regarding student actions and engagement and environmental factors to develop predictive capabilities for proactive actions. The predictive analytics system will warn teachers or student advisors about students demonstrating declining academic engagement when motion sensors alongside facial recognition detection and keyboard input data suggest unfavorable indicators (Taheri & Salimi Beni, 2025).

The implementation of AI-powered SSNs enables campuses to transform into responsive environments that utilize data for the vigilant identification of distress symptoms among students. Live data inputs enable these systems to alter service delivery for students and provide material recommendations and instruction re-direction that result in better learning outcomes while maintaining student success.

7.2 Ecosystems of personalized learning inspired by sensor-driven feedback

SSNs will establish many of the upcoming personalized learning ecosystems. Monitoring factors that continue during learning provide sensors with specific profiles they use to modify educational content and approaches. Combining SSNs with Learning Management Systems allows LMS to provide specific module recommendations, pacing techniques, and interactive learning tools that satisfy individual student needs (Adibi et al., 2024). Their combination permits tracking student attention patterns, leading to better scheduling choices and workspace designs. The system uses feedback loops that enable outputs to enhance upcoming inputs in order to achieve personalized learning. Through obtaining real-time data, instructors can assign students to flexible groups based on their educational needs, academic performance, and individualized educational requirements (Agrawal & Goel, 2016; Saaida, 2023). The high level of customization helps increase academic participation while promoting student independence in learning and strengthens both inclusive educational practices and operational efficiency.

7.3 Green campuses with energy-efficient sensor deployments

Higher educational institutions have started placing sustainability at the forefront of their organizational strategy. SSNs serve as practical tools for the energy economy, enhancing the expansion of environmentally conscious campuses. Monitoring electricity together with water and the operation of Heating, Ventilation, and Air Conditioning systems (HVAC) depends on both environmental factors and occupancy patterns, which sensors detect. Empty classrooms automatically trigger motion sensors to activate energy-saving mode through temperature and light adjustments. Smart meters provide exact energy data that helps facilities managers create evidence-based energy-saving programs. SSN systems function alongside solar panels and other renewable power sources to operate sensors independent of non-renewable energy resources (Pandya et al., 2022). Green SSNs serve to decrease institutional carbon emissions and serve as educational tools by encouraging environmentally conscious conduct from students. Students can access sensor dashboards that support responsible conduct and spread information about university-wide environmental programs.

7.4 Blockchain for transparency and data security enhancement of trust

Stakeholder confidence and data security emerge as fundamental requirements because SSNs handle increasingly valuable and sensitive information. The distributed tamper-proof ledger that blockchain offers provides a convincing solution to manage SSN data. The secure data blocks linked through cryptographic hash functions enable blockchain to maintain highly accessible and intact sensor data storage. Every data entry can be timestamped along with network node verification processes, reducing the probability of unauthorized access or data modification (Ahmed, 2025). Blockchain distribution eliminates system vulnerability to cyberattacks and safeguards operations from single failure points.

Through blockchain, students can automate secure data handling processes for their learning information through controlled access management. The combination of smart contracts and self-executing codes stored on blockchain gives students the power to determine access control and usage permissions for their data. These deployment systems provide consumers with enhanced trust levels while minimizing administrative costs while adhering to data security rules (Alzoubi, 2024). The application of SSNs in higher education shows excellent potential for future development. SSNs transform universities into enlightened educational spaces through AI-based innovation, personalized design practices, and blockchain security features for data protection. The achievement of this ambition depends on exact planning while sustaining innovative concepts alongside absolute student-focused principles and ethical data management.

8. Conclusion

Adopting SSNs represents a transformative opportunity for education as an entity. These systems' ability to use real-time data for evaluation allows them to become vital components that improve student education while supporting well-being and safety operations across campuses. The learning environment created by SSNs adapts through a personalized approach that helps students interact while enabling institutional administrators to work more efficiently during the digital transformation of education.

The implementation of SSNs constitutes fundamental progress because they support flexible learning environments. Teachers understand each student better by tracking classroom actions, participation levels, and physical attendance, which helps them customize instruction. This customized approach improves academic performance and student satisfaction at high levels. The connection between SSNs allows teachers to implement faster teaching methods and deliver timely academic assistance through ongoing feedback systems between students and educators.

The implementation of SSNs enhances student welfare at both academic and non-academic levels. Wearable devices combined with environmental monitoring systems allow tracking of vital signs, stress indicators, and air quality data, which results in improved classroom environment quality. School institutions obtain these insights and use them to launch mental health prevention initiatives and wellness initiatives that reinforce student holistic education. Sensor-driven treatments reduce health hazards and enhance campus conditions, thus enabling better academic success for students over the long term.

Deploying automated alarm systems with real-time monitoring becomes more efficient because of SSNs in campus security applications. True situation awareness functions with location sensors, security devices, and emergency response technology to ensure immediate assistance in crisis situations. These security tools produce substantial advancements in educational institutions' security systems, which include maintaining safe evacuations while tracking unusual behavior.

The use of SSNs enables organizations to achieve greater control of their resources. By analyzing their energy consumption, occupancy records, and machine maintenance obligations, organizations can reduce waste, lower operating expenses, and further sustainable practices. Smart devices used for instant resource management play a crucial role in establishing sustainable educational campuses.

The implementation of SSN systems requires thorough, responsible management. The priority concerns ethical standards for protecting data privacy, gaining proper consent from users, and implementing digital surveillance systems. Weak data governance policies require institutional development for policies that promote openness through user control and responsible management. The preservation of trust and students' digital rights depend on design models for privacy protection in combination with protected data storage and dialog with stakeholder groups.

SSNs require purposeful design, which needs an examination of systems and budgetary provisions and a cultural readiness assessment. Learning institutions must allocate financial resources that support change management practices and stakeholder inclusion programs combined with employee training to achieve successful implementation. Every student, including those from varied backgrounds and those without technology access, needs consideration to benefit from these advancements.

These technologies will achieve more possibilities through their connection with artificial intelligence, blockchain systems, predictive analytics, and SSNs. These technologies enable better data security alongside enhanced decision systems and customized solutions. SSNs allow universities to revolutionize higher education practices through enhanced, more student-oriented approaches that are more effective throughout the entire educational experience.

SSNs have excellent potential to improve student life activities and educational performance. Still, their success hinges on an inclusive, ethical, forward-looking application. Institutions should optimize the benefits of SSNs by matching human values with technological innovations to deliver safe and fair educational opportunities that are enriching.

References

Adibi, S., Rajabifard, A., Shojaei, D., & Wickramasinghe, N. (2024). Enhancing healthcare through sensor-enabled digital twins in smart environments: A comprehensive analysis. *Sensors, 24*(9), 2793.

Agrawal, A., & Goel, R. (2016). Examination phobia: A common problem of students. *Review Journal of Philosophy and Social Science, 41* (2), 1–6.

Ahmed, S. (2025). Enhancing data security and transparency: The role of blockchain in decentralized systems. *International Journal of Advanced Engineering, Management and Science, 11*(1), 593258.

Alzoubi, M. M. (2024). Investigating the synergy of blockchain and AI: Enhancing security, efficiency, and transparency. *Journal of Cyber Security Technology*, 1–29. https://doi.org/10.1080/23742917.2024.2374594

Ani, K. (2024). *Advancing online student experience in the digital era: Case degree programme in aviation business*. Bachelors thesis. Haaga-Helia University of Applied Sciences. https://urn.fi.URN:NBN:fi:amk:2024052113868.

Asad, M. M., Naz, A., Shaikh, A., Alrizq, M., Akram, M., & Alghamdi, A. (2024). Investigating the impact of IoT-Based smart laboratories on students' academic performance in higher education. *Universal Access in the Information Society, 23*(3), 1135–1149. https://doi.org/10.1007/s10209-022-00944-1

Asim, M., Wang, Y., Wang, K., & Huang, P.-Q. (2020). A review on computational intelligence techniques in cloud and edge computing. *IEEE Transactions on Emerging Topics in Computational Intelligence, 4*(6), 742–763. https://ieeexplore.ieee.org/abstract/document/9171883/.

Business Insider. (2024). *New AI and 5G advancements will usher in the era of edge computing on smartphones, autonomous cars, and more*. https://www.businessinsider.com/ai-edge-computing-5g-cloud-artificial-intelligence-2024-3.

Cavus, N., Mrwebi, S. E., Ibrahim, I., Modupeola, T., & Reeves, A. Y. (2022). Internet of things and its applications to smart campus: A systematic literature review. *International Journal of Interactive Mobile Technologies (iJIM), 16*(23), 17–35. https://doi.org/10.3991/ijim.v16i23.36215

Cheung, K. K. C., & Tai, K. W. H. (2023). The use of intercoder reliability in qualitative interview data analysis in science education. *Research in Science & Technological Education, 41*(3), 1155–1175. https://doi.org/10.1080/02635143.2021.1993179

Deva, M., Wright, S. T., & Beck Dallaghan, G. L. (2025). Exploring the role of cognitive capacity and adaptive learning in professional education: A scoping review. *Medical Science Educator*. https://doi.org/10.1007/s40670-024-02271-x

Diraco, G., Rescio, G., Siciliano, P., & Leone, A. (2023). Review on human action recognition in smart living: Sensing technology, multimodality, real-time processing, interoperability, and resource-constrained processing. *Sensors, 23*(11), 5281.

Fährmann, D., Martín, L., Sánchez, L., & Damer, N. (2024). Anomaly detection in smart environments: A comprehensive survey. *IEEE Access*.

Farrokhi, A., Farahbakhsh, R., Rezazadeh, J., & Minerva, R. (2021). Application of internet of things and artificial intelligence for smart fitness: A survey. *Computer Networks, 189*, 107859.

Fayez, A. N., Ghabban, F. M., & Ameerbakhsh, O. (2021). Advantages and challenges of smart learning in higher education institutions in Saudi Arabia. *Creative Education, 12*(5), 974–982. https://www.scirp.org/journal/paperinformation?paperid=109248.

Financial Times. (2023). *The FT AI glossary*. https://www.ft.com/content/c0c8d205-e158-409e-963c-a3ac821cd7ba;ft.com.

Gao, M. (2022). Smart campus teaching system based on ZigBee wireless sensor network. *Alexandria Engineering Journal, 61*(4), 2625–2635.

Gao, N., Saiedur Rahaman, M., Shao, W., & Salim, F. D. (2021). Investigating the reliability of self-report data in the wild: The quest for ground truth. In *Adjunct proceedings of the 2021 ACM International joint conference on pervasive and ubiquitous computing and proceedings of the 2021 ACM international symposium on wearable computers* (pp. 237–242). https://doi.org/10.1145/3460418.3479338

GeeksforGeeks. (2024). *Wireless sensor network (WSN)*. https://www.geeksforgeeks.org/wireless-sensor-network-wsn/;GeeksforGeeks.

Geertsema, P., & Lu, H. (2023). Relative valuation with machine learning. *Journal of Accounting Research, 61*(1), 329–376. https://doi.org/10.1111/1475-679X.12464

Ghadi, Y. Y., Mazhar, T., Al Shloul, T., Shahzad, T., Salaria, U. A., Ahmed, A., & Hamam, H. (2024). Machine learning solutions for the security of wireless sensor networks: A review. *IEEE Access, 12*, 12699–12719. https://ieeexplore.ieee.org/abstract/document/10401918/.

Ha, N., Xu, K., Ren, G., Mitchell, A., & Ou, J. Z. (2020). Machine learning-enabled smart sensor systems. *Advanced Intelligent Systems, 2*(9), 2000063. https://doi.org/10.1002/aisy.202000063

HashStudioz. (n.d.). IoT in universities: Building smart campuses for smarter learning. Retrieved April 23, 2025, from https://www.hashstudioz.com/blog/iot-in-universities-building-smart-campuses-for-smarter-learning/.

Hernández-Mustieles, M. A., Lima-Carmona, Y. E., Pacheco-Ramírez, M. A., Mendoza-Armenta, A. A., Romero-Gómez, J. E., Cruz-Gómez, C. F., Rodríguez-Alvarado, D. C., Arceo, A., Cruz-Garza, J. G., & Ramírez-Moreno, M. A. (2024). Wearable biosensor technology in education: A systematic review. *Sensors, 24* (8), 2437.

Hossain, M., Kayas, G., Hasan, R., Skjellum, A., Noor, S., & Islam, S. R. (2024). A holistic analysis of internet of things (IoT) security: Principles, practices, and new perspectives. *Future Internet, 16*(2), 40.

Hu, F., & Hao, Q. (2012). *Intelligent sensor networks: The integration of sensor networks, signal processing and machine learning*. Taylor & Francis.

Hussien, S. H. T., Farhana, H., Vinukumar, L., Alexander, C. H. C., & Sivakumar, S. (2024). Smart campus attendance and security systems: Leveraging IoT for enhanced efficiency. *AIP Conference Proceedings, 3161*(1). https://pubs.aip.org/aip/acp/article-abstract/3161/1/020153/3310585.

IoT World. (2023). *The dawn of smart campuses – How IoT and AI can play a role*. https://iotworld.co/2023/06/the-dawn-of-smart-campuses-how-iot-can-play-a-role/.

Jiang, F., Wang, L., Li, J.-X., & Liu, J. (2022). How smart technology affects the well-being and supportive learning performance of logistics employees? *Frontiers in Psychology, 12*, 768440.

Jihong, X. I. E., & Xiang, Z. (2024). Edge computing for real-time decision making in autonomous driving: Review of challenges, solutions, and future trends. *International Journal of Advanced Computer Science and Applications, 15*(7).

Joymangul, J. S., Ciobanu, I., Angiolini, F., Lampe, J., Pedrini, C., Pinto, A., Franceschini, B., Nicolas, D., Tamburini, E., & Cecchi, F. (2024). Empowering active and healthy ageing: Integrating IoT and wearable technologies for personalized interventions. *Applied Sciences, 14*(11), 4789.

Kapur, R., Byfield, V., Del Frate, F., Higgins, M., & Jagannathan, S. (2018). The digital transformation of education. *Earth Observation Open Science and Innovation*, 25–41.

Karthick Raghunath, K. M., Koti, M. S., Sivakami, R., Vinoth Kumar, V., NagaJyothi, G., & Muthukumaran, V. (2024). Utilization of IoT-assisted computational strategies in wireless sensor networks for smart infrastructure management. *International Journal of System Assurance Engineering and Management, 15*(1), 28–34. https://doi.org/10.1007/s13198-021-01585-y

Kem, D. (2022). Personalised and adaptive learning: Emerging learning platforms in the era of digital and smart learning. *International Journal of Social Science and Human Research, 5*(2), 385–391.

Khosravi, S., Bailey, S. G., Parvizi, H., & Ghannam, R. (2022). Wearable sensors for learning enhancement in higher education. *Sensors, 22* (19), 7633. https://www.mdpi.com/1424-8220/22/19/7633.

Khudor, B. A. Q., Kheerallah, Y. A., & Alkenani, J. (2023). A new method based on machine learning to increase efficiency in wireless sensor networks. *Informatica, 46*(9).

Koay, H. V., Chuah, J. H., Chow, C.-O., & Chang, Y.-L. (2022). Detecting and recognizing driver distraction through various data modalities using machine learning: A review, recent advances, simplified framework, and open challenges (2014–2021). *Engineering Applications of Artificial Intelligence, 115*, 105309.

Kraft-Terry, S. D., & Brown, J. (2023). Data-informed planning and decision-making. In *Academic advising administration* (pp. 181–194). Routledge.

Kurni, M., & Srinivasa, K. G. (2025). IoT-Enabled gamification for education. In M. Kurni, & K. G. Srinivasa (Eds.), *The internet of educational things* (pp. 151–168). Switzerland: Springer Nature. https://doi.org/10.1007/978-3-031-67387-0_10

Lazaroiu, G., Androniceanu, A., Grecu, I., Grecu, G., & Neguriță, O. (2022). Artificial intelligence-based decision-making algorithms, internet of things sensing networks, and sustainable cyber-physical management systems in big data-driven cognitive manufacturing. *Oeconomia Copernicana, 13*(4), 1047–1080.

Lee, S. U., Paik, H., & Kanhere, S. S. (2024). Human-sensing platforms and ethical considerations throughout their data life-cycles. *IEEE Internet of Things Magazine, 7*(1), 66–73. https://ieeexplore.ieee.org/abstract/document/10397569/.

Lee, S., & Lee, K. (2024). Smart teachers in smart schools in a smart city: Teachers as adaptive agents of educational technology reforms. *Learning, Media and Technology, 49*(3), 456–477. https://doi.org/10.1080/17439884.2023.2207143

Li, L., Chen, C. P., Wang, L., Liang, K., & Bao, W. (2023). Exploring artificial intelligence in smart education: Real-time classroom behavior analysis with embedded devices. *Sustainability, 15*(10), 7940. https://doi.org/10.3390/su15107940

Lu, G., Xie, K., & Liu, Q. (2022). What influences student situational engagement in smart classrooms: Perception of the learning environment and students' motivation. *British Journal of Educational Technology, 53*(6), 1665–1687. https://doi.org/10.1111/bjet.13204

Matellio. (2024). *IoT applications in education: Transforming learning and administration*. https://www.matellio.com/blog/iot-applications-in-education/;Matellio.

Mohammad Amini, M., Jesus, M., Fanaei Sheikholeslami, D., Alves, P., Hassanzadeh Benam, A., & Hariri, F. (2023). Artificial intelligence ethics and challenges in healthcare applications: A comprehensive review in the context of the European GDPR mandate. *Machine Learning and Knowledge Extraction, 5*(3), 1023–1035.

Mustafa, M. F., Isa, M. R. M., Rauf, U. F. A., Ismail, M. N., Shukran, M. A. M., Khairuddin, M. A., Wahab, N., & Safar, N. Z. M. (2021). Student perception study on smart campus: A case study on higher education institution. *Malaysian Journal of Computer Science*, 1–20.

Ong, K. S. H., Wang, W., Niyato, D., & Friedrichs, T. (2021). Deep-reinforcement-learning-based predictive maintenance model for effective resource management in industrial IoT. *IEEE Internet of Things Journal, 9*(7), 5173–5188.

Pagliari, D. J., & Poncino, M. (2020). On the impact of smart sensor approximations on the accuracy of machine learning tasks. *Heliyon, 6*(12).

Pandya, C., Prajapati, S., & Gupta, R. (2022). Sustainable energy efficient green campuses: A systematic literature review and bibliometric analysis. *IOP Conference Series: Earth and Environmental Science, 1084*(1), 012016. https://iopscience.iop.org/article/10.1088/1755-1315/1084/1/012016/meta.

Papatheocharous, E., Buffoni, D., Maurer, M., Wallberg, A., & Ezquerro, G. (2024). Driver distraction detection using artificial intelligence and smart devices. In *Intelligent secure trustable things* (pp. 285–308). Switzerland Cham: Springer Nature.

Park, S., Michelow, I. C., & Choe, Y. J. (2021). Shifting patterns of respiratory virus activity following social distancing measures for coronavirus disease 2019 in South Korea. *The Journal of Infectious Diseases, 224*(11), 1900–1906. https://academic.oup.com/jid/article-abstract/224/11/1900/6261431.

Pedagogue. (2024). *Internet of Things (IoT) in smart campuses*. https://pedagogue.app/internet-of-things-iot-in-smart-campuses/.

Predel, C., Timmermann, C., Ursin, F., Orzechowski, M., Ropinski, T., & Steger, F. (2022). Conflicting aims and values in the application of smart sensors in geriatric rehabilitation: Ethical analysis. *JMIR mHealth and uHealth, 10*(6), e32910. https://mhealth.jmir.org/2022/6/e32910.

Preprints.org. (2025). *Data-driven education: The role of IOT and ar in enhancing learning outcomes*. https://www.preprints.org/manuscript/202502.2039/v1;. Preprints.

Raju, R., Abd Rahman, N. H., & Ahmad, A. (2022). Cyber security awareness in using digital platforms among students in a higher learning institution. *Asian Journal of University Education, 18*(3), 756–766.

Rath, K. C., Khang, A., & Roy, D. (2024). The role of internet of things (IoT) technology in industry 4.0 economy. In *Advanced IoT technologies and applications in the industry 4.0 digital economy* (pp. 1–28). CRC Press.

Ruipérez-Valiente, J. A., Martínez-Maldonado, R., Di Mitri, D., & Schneider, J. (2022). From sensor data to educational insights. *Sensors, 22*(21), 8556.

Saaida, M. B. (2023). AI-Driven transformations in higher education: Opportunities and challenges. *International Journal of Educational Research and Studies, 5*(1), 29–36. Driven_transformations_in_higher_education_Opportunities_and_challenges.pdf.

Sahoo, S. K., Mudligiriyappa, N., Algethami, A. A., Manoharan, P., Hamdi, M., & Raahemifar, K. (2022). Intelligent trust-based utility and reusability model: Enhanced security using unmanned aerial vehicles on sensor nodes. *Applied Sciences, 12*(3), 1317.

Sears, C. R., & Cunningham, D. R. (2024). Individual differences in psychological stress associated with data breach experiences. *Journal of Cybersecurity and Privacy, 4*(3), 594–614.

Shihao, L., Dahnil, D. P., & Saad, S. (2025). A survey of smart campus resource information management in internet of things. *IEEE Access*.

Siddiqui, N., & Shaukat, S. (2021). Teacher mobility in Punjab, Pakistan: Stayers and movers within the public and private schools. *Education Sciences, 11*(7), 358.

Sofi, A., Regita, J. J., Rane, B., & Lau, H. H. (2022). Structural health monitoring using wireless smart sensor network–An overview. *Mechanical Systems and Signal Processing, 163*, 108113.

Soundari, A. G., & Jyothi, V. L. (2020). Energy efficient machine learning technique for smart data collection in wireless sensor networks. *Circuits, Systems, and Signal Processing, 39*(2), 1089–1122. https://doi.org/10.1007/s00034-019-01181-3

Stokel-Walker, C. (2020 October 15). Universities are using surveillance software to spy on students. Wired. https://www.wired.com/story/university-covid-learning-student-monitoring/.

Stratagon. (2025). *Boosting student engagement with predictive analytics*. https://edu.stratagon.com/blog/boosting-student-engagement-with-predictive-analytics.

Tabuenca, B., Moreno-Sancho, J.-L., Arquero-Gallego, J., Greller, W., & Hernández-Leo, D. (2023). Generating an environmental awareness system for learning using IoT technology. *Internet of Things, 22*, 100756.

Taheri, H., & Salimi Beni, A. (2025). Artificial intelligence, machine learning, and smart technologies for nondestructive evaluation. In N. Meyendorf, N. Ida, R. Singh, & J. Vrana (Eds.), *Handbook of nondestructive evaluation 4.0* (pp. 1–29). Switzerland: Springer Nature. https://doi.org/10.1007/978-3-030-48200-8_70-1

Technology Innovators. (n.d.). IoT in education: Smart classrooms and campus management. Retrieved April 23, 2025, from https://www.technology-innovators.com/internet-of-things-iot-in-education-smart-classrooms-and-campus-management/.

Teoh, Y. K., Gill, S. S., & Parlikad, A. K. (2021). IoT and fog-computing-based predictive maintenance model for effective asset management in industry 4.0 using machine learning. *IEEE Internet of Things Journal, 10*(3), 2087–2094. https://ieeexplore.ieee.org/abstract/document/9319212/.

TheThings.io. (2023). *IoT in education: Transforming learning and campus management*. https://thethings.io/iot-solutions/iot-in-education-transforming-learning-and-campus-management/.

Ullah, A., & Gul, R. (2023). Assessing the impact of poor access to health facilities on social exclusion in children: A case study of Peshawar, Pakistan. *Dialogue, 18*(3), 75–100.

VasyLyk, M., Melnyk, N., Prytuliak, O., & ChErVINSkA, I. (2024). Adaptive learning as a factor in optimizing the education process in institutions of higher education. *Mountain School of Ukrainian Carpaty, 30*, 5–10.

Venticinque, S., & Amato, A. (2018). Smart sensor and big data security and resilience. In *Security and resilience in intelligent data-centric systems and communication networks* (pp. 123–141). Elsevier. https://www.sciencedirect.com/science/article/pii/B9780128113738000069.

Wang, E., Padhan, H., Pruseth, S. K., & Ma, J. (2024). Government efficiency, green technology, and ecological footprint: Strategic framework for natural resource management efficiency targets. *Resources Policy, 91*, 104826.

Xiao, L., Luo, K., Liu, J., & Foroughi, A. (2024). A hybrid deep approach to recognizing student activity and monitoring health physique based on accelerometer data from smartphones. *Scientific Reports, 14*(1), 14006.

Yılmaz, R., & Yılmaz, F. G. K. (2022). *Investigation of student views on data privacy and ethical use of data in smart learning environments*. https://acikerisim.bartin.edu.tr/handle/11772/6931.

Zhang, X., Ding, Y., Huang, X., Li, W., Long, L., & Ding, S. (2024). Smart classrooms: How sensors and ai are shaping educational paradigms. *Sensors, 24*(17), 5487.

Zhang, W., Su, C., & He, C. (2020). Rehabilitation exercise recognition and evaluation based on smart sensors with deep learning framework. *IEEE Access, 8*, 77561–77571.

Zhao, M., Zhao, G., & Qu, M. (2022). College smart classroom attendance management system based on internet of things. *Computational Intelligence and Neuroscience*, 1–9. https://doi.org/10.1155/2022/4953721

Chapter 9

Blockchain for smart education

Antonella Petrillo[1, 2], Fabio De Felice[1, 2] and Mizna Rehman[1, 2]
[1]*Department of Engineering, University of Naples "Parthenope", Napoli, Italy;* [2]*Isola C4, Centro Direzionale Napoli, Napoli, Italy*

1. Definition of blockchain and smart education

Blockchain is a decentralized, distributed ledger technology (DLT) that ensures secure, transparent, and immutable transaction records across multiple computers. Described by Nakamoto (2008) as a peer-to-peer network using proof-of-work, blockchain provides an unalterable history of transactions, as supported by Feld et al. (2016) and Ciaian et al. (2015). Beyond cryptocurrencies, blockchain serves as a decentralized trust system enabling secure digital transactions without intermediaries (Trivedi et al., 2021). Tripathi et al. (2023) defined blockchain's cryptographic security, such as SHA-256 hashing, ensures integrity, transparency, and resistance to tampering, making it highly secure. Smart education integrates digital technologies such as AI, IoT, and blockchain to enhance teaching and learning. It promotes adaptive, personalized learning environments, as noted by Veeramanickam et al. (2023) and Zhu et al. (2016). With blockchain, education becomes more decentralized, with transparent, tamper-proof credential verification and record management, overcoming inefficiencies in traditional systems such as manual credential verification, security vulnerabilities, and rigid learning pathways (Parmentola et al., 2022). Table 9.1 illustrates the fundamental distinctions between traditional and blockchain-enabled smart education, showing how blockchain technology overcomes key inefficiencies in the existing system.

Blockchain technology is revolutionizing smart education by improving credential verification, student identity management, academic integrity, and automated learning. Researchers have explored its potential, offering various definitions and perspectives on its role in enhancing learning environments. Table 9.2 highlights real-world blockchain applications in education, detailing use cases, underlying technologies, scholarly definitions, and academic sources.

Fig. 9.1 depicts a blockchain-based smart education ecosystem where students, educators, and authorities interact securely. The interface allows record management, while the backend automates credential issuance via smart contracts. Employers and universities verify credentials through the blockchain, ensuring authenticity. Essential data is stored on a permissioned Hyperledger Fabric ledger, enabling private channels, peer validation, and automated processes, while nonessential records use traditional storage.

2. Importance of blockchain for digital education

Blockchain technology is transforming digital education by enhancing credential verification, data security, accessibility, and administrative efficiency. One of the major challenges faced by educational institutions is ensuring the authenticity of academic credentials. With the rise of online courses and fake degrees, blockchain offers a

TABLE 9.1 Traditional versus Blockchain-Enabled Smart Education.

Aspect	Traditional education	Blockchain enabled smart education
Credential verification	Manual, time-consuming	Instant, tamper-proof (e.g., MIT digital diplomas)
Data security	Centralized, vulnerable to breaches	Decentralized, encrypted via SHA-256
Learning pathways	Fixed curriculum	Personalized via smart contracts and AI

Digital Twins in the Smart Classroom. https://doi.org/10.1016/B978-0-443-41632-3.00001-1

TABLE 9.2 Blockchain applications and definitions in smart education.

Scenario in smart education	Blockchain technology	Definitions	Sources
Tamper-proof academic certificates	Blockchain-based credential verification	Blockchain ensures the authenticity of educational credentials by creating immutable records that prevent fraud and forgery.	Raghav et al. (2019)
Student identity and data privacy	Self-sovereign identity (SSI) for students	Blockchain provides students with full control over their educational records, ensuring privacy and security without reliance on centralized institutions.	Chan et al. (2025)
Transparent research and plagiarism prevention	Timestamping and immutable publishing	Blockchain allows timestamped research publication, preventing plagiarism and ensuring intellectual property rights.	Anirudh et al. (2022)
Automated academic processes	Smart contracts for grading and certifications	Smart contracts can automate grading, certification issuance, and course completion validation, reducing administrative burdens.	Dubey and Goyal (2022)
Microcredentialing and lifelong learning	Blockchain-based digital badges and microcertificates	Blockchain enables lifelong learning records, allowing individuals to accumulate and share microcredentials from various institutions.	Chukowry et al. (2021)

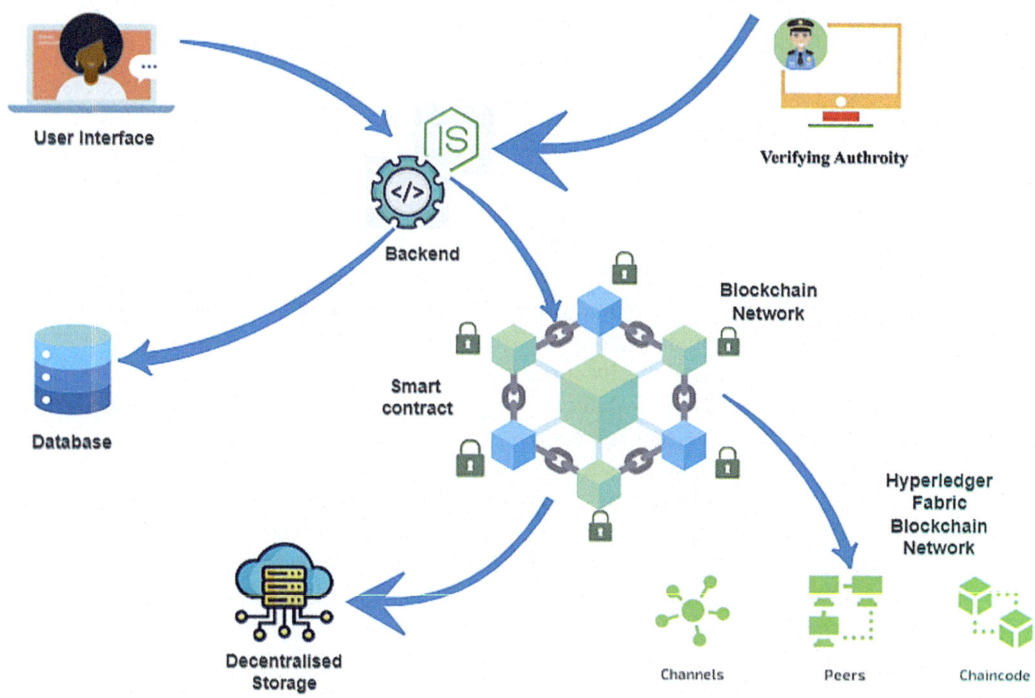

FIGURE 9.1 Blockchain-based decentralized framework for smart education ecosystem.

decentralized and immutable record of academic achievements, preventing fraud and enabling verification by authorized parties, as shown in Fig. 9.2. This provides students with greater control over their academic records, allowing them to independently manage and share their credentials without relying on institutions. Unlike traditional paper-based diplomas that can be lost or require costly retrieval processes, blockchain-stored diplomas remain secure, easily accessible, and verifiable by employers.

Beyond credential verification, the increasing digitization of education raises concerns about data security and

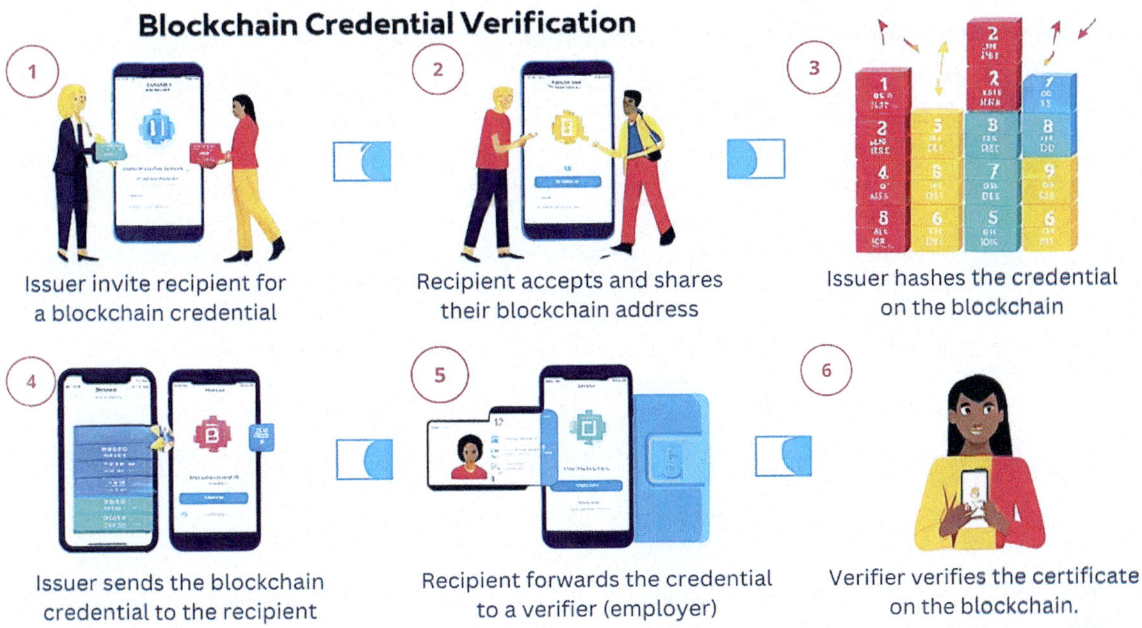

FIGURE 9.2 A blockchain-based credentialing process ensuring secure issuance, storage, and verification of digital credentials.

privacy. Blockchain addresses these issues through cryptographic encryption and decentralization, ensuring that records remain resistant to hacking. This is particularly beneficial for educational institutions, as decentralized record-keeping reduces administrative costs associated with credential verification, eliminating the need for labor-intensive manual processes. Research from the University of Rome indicates that the traditional credential verification process can cost an institution over 19,000 euros (or upwards of $20,000) annually, equivalent to nearly 36 weeks of work (Wang et al., 2023). By adopting blockchain-issued diplomas, institutions can drastically reduce both the time and money spent on manual verifications while also simplifying the process for graduate schools reviewing a student's academic history. Also, students gain greater control over their personal data, deciding what information to share while maintaining transparency through secure audit trails.

Enhancing security naturally extends to improving accessibility. Blockchain enables global reach to educational resources, certifications, and courses, breaking financial and geographical barriers. Tokenized education allows students to earn tokens for completing courses, which can be redeemed for certifications and financial incentives. Furthermore, distributed learning platforms ensure equitable access to learning materials, making education more inclusive. Employers also benefit from this transformation, as blockchain-based digital credentials allow for instant and tamper-proof verification of a candidate's qualifications, streamlining recruitment and minimizing the risk of credential fraud.

With increased accessibility comes the need for efficient management of educational processes. Smart contracts streamline these processes by automating course enrollment, certification issuance, and financial transactions. By verifying prerequisites and managing registrations based on preset conditions, smart contracts improve efficiency and reduce administrative overhead. This automation eliminates delays and ensures fairness in the distribution of scholarships and grants. Educators benefit from smart contracts that automate grading and assignment progression, freeing up time for curriculum development and enhancing overall teaching efficiency (Brown, 2022, pp. 173–194).

Educational institutions, in turn, benefit from blockchain's role in reducing administrative costs and increasing operational efficiency. Decentralized record-keeping, faster data processing, and automation of repetitive tasks minimize paperwork and simplify student records management, allowing institutions to focus more on academic quality and innovation (Sawant, 2023).

Building on administrative efficiency, blockchain fosters trust and collaboration among educational institutions. Credentialing networks powered by blockchain facilitate seamless student transitions between institutions and employers, eliminating reliance on centralized accreditation bodies. This decentralized system enhances transparency and efficiency in academic and professional credentialing (Nargis et al., 2023).

Finally, blockchain supports lifelong learning by validating informal learning experiences and enabling portable skills verification. Microcredentials and digital badges help

TABLE 9.3 Comparative analysis of blockchain enabled diplomas versus traditional credentials.

Feature	Blockchain enabled certificates	Conventional certificates
Control over records	Learners retain direct authority over their academic certificates.	Educational institutions manage and control the records.
Ease of verification	Credentials can be authenticated swiftly by employers and third parties.	Verification requires additional steps and incurs extra costs.
Verification efficiency	Institutions benefit from a rapid, cost-effective validation process.	Traditional methods are often slow and resource-intensive.
Forgery resistance	Enhances the credibility and reliability of academic certificates.	More vulnerable to forgery and unauthorized alterations.
Data protection and privacy	Employs robust encryption techniques to safeguard privacy.	Presents a higher risk of data breaches and privacy issues.
Record integrity	Ensures records remain unaltered and consistently verifiable.	Digital records can be lost or modified without stringent safeguards.

track and authenticate smaller, skill-based achievements, such as online courses and workshops, allowing individuals to build a digital portfolio of competencies (Pradeep et al., 2023). This approach acknowledges continuous skill development as a viable alternative to traditional degree-based qualifications further benefiting both learners and employers by making skills recognition more flexible and transparent.

By addressing key challenges in digital education, blockchain enhances security, efficiency, and accessibility, fostering a more connected and trustworthy global learning ecosystem. As the technology evolves, its integration into education promises a more equitable and streamlined learning experience for all. Table 9.3 details side-by-side analysis that highlights how blockchain-based academic credentials differ from and improve upon conventional methods.

Fig. 9.3 illustrates how decentralized ledger technology transforms modern education by streamlining course administration, securing creative assets, reducing costs,

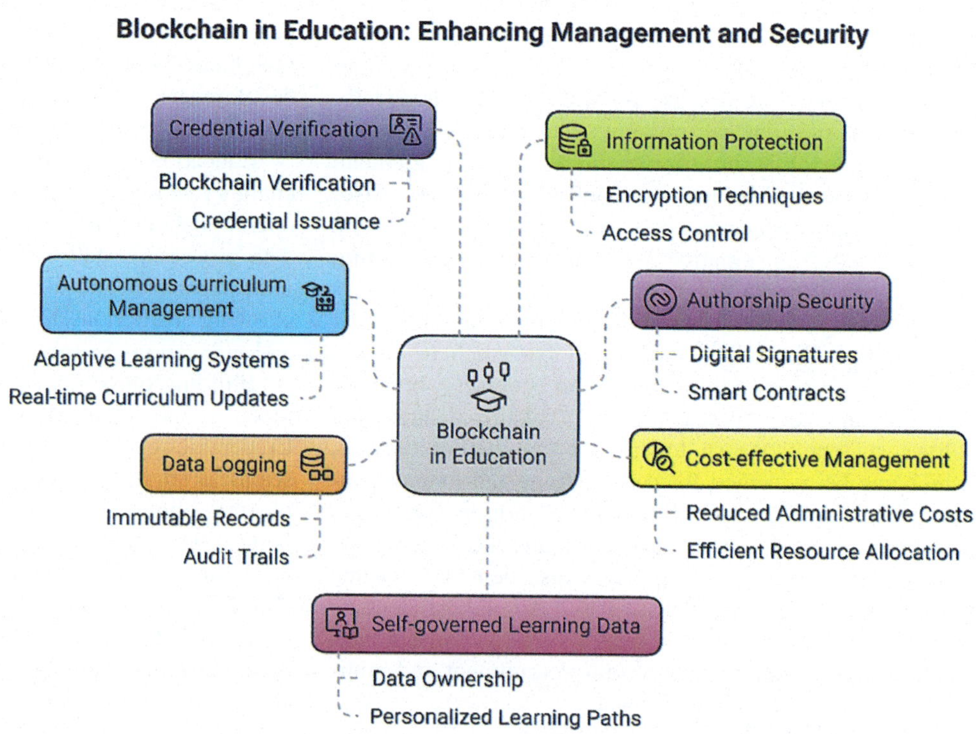

FIGURE 9.3 Blockchain as a catalyst for educational change.

maintaining academic records, preventing diploma fraud, enhancing data protection, and granting learners control over their educational information.

2.1 Connection between blockchain, digital twins, and smart classroom

Emerging technologies such as blockchain and digital twins are transforming smart classrooms by creating secure, intelligent, and data-driven learning environments that enhance academic administration and student engagement.

Digital twins provide real-time virtual replicas of classrooms, capturing student interactions and environmental conditions using IoT-enabled devices, including biometric attendance systems, interactive whiteboards, and environmental sensors. These dynamic simulations optimize learning conditions and operational efficiency.

Blockchain serves as the trust layer in this ecosystem, securing educational records, student performance metrics, and transactional data in an immutable, decentralized ledger. Smart contracts automate tasks such as credential verification, attendance tracking, and resource allocation.

Smart classroom utilizes AI, IoT sensors, and data analytics for personalized, adaptive, and interactive learning experiences (Tomczyk et al., 2020). They enable real-time feedback on engagement, cognitive performance, and environmental factors to enhance educational outcomes.

Fig. 9.4 visualizes the interconnected components of this ecosystem, including biometric attendance systems, digital annotation devices for content projection, audiovisual tools, high-resolution imaging systems for document visualization, interactive display panels and cloud-based learning resources, collectively fostering an adaptive and immersive educational experience.

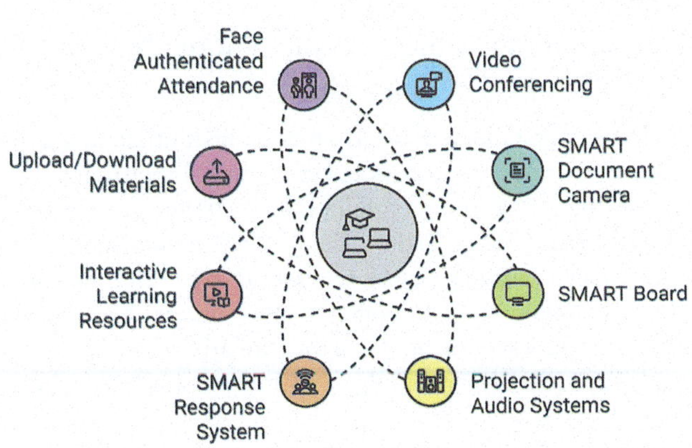

FIGURE 9.4 Components of digital classroom.

2.2 The role of blockchain in education

The blockchain–IoT framework creates a decentralized, secure infrastructure that integrates smart classrooms, cloud computing, and cybersecurity to enhance learning and administration. IoT devices such as smartboards, biometric systems, environmental sensors and AI-driven assessment tools continuously collect real-time data, which is securely recorded via blockchain's distributed ledger. This ensures data immutability, transparency, and automated credential verification while preventing fraud. Smart contracts streamline student assessments, tuition transactions, and resource allocation, reducing administrative overhead. Cloud infrastructure enables scalable, on-demand storage and interoperability through cloud-based learning management systems (LMS). Advanced cryptographic techniques, including zero-knowledge proofs (ZKPs) and the interplanetary file system (IPFS), ensure data privacy, while cybersecurity measures protect sensitive student information. Blockchain also enables tokenized credentials through nonfungible tokens (NFTs) and decentralized identity management, providing verifiable academic achievements. Moreover, smart payments using cryptocurrencies facilitate tuition fees, scholarships, and funding transactions. By integrating AI, predictive analytics, and decentralized data governance, this framework enhances efficiency, engagement, and security in digital education. Fig. 9.5 illustrates the blockchain–IoT educational system, connecting stakeholders through secure data exchange. This framework supports the following:

- *Decentralized data flow:* The diagram represents a peer-to-peer (P2P) decentralized education infrastructure, where each entity (students, teachers, devices) is securely connected through blockchain's distributed ledger. Every transaction (e.g., student performance metrics, teacher evaluations, digital identity verification) is cryptographically recorded on blockchain, ensuring data authenticity, integrity, and traceability. The hashing mechanism in blockchain ensures that any educational credential issued is immutable, preventing fraud or academic dishonesty.
- *IoT-enabled smart classrooms:* Smart sensors and AI tools personalize learning, track academic performance, and optimize classroom conditions (e.g., adjusting lighting and noise levels for better concentration).
- *Cloud-based scalability:* The cloud component in the diagram supports real-time storage, access, and synchronization of educational content and records. This ensures interoperability across multiple institutions, allowing access to academic credentials, research materials, and collaborative tools.
- *Secure student–teacher interactions:* Blockchain-based authentication restricts access to authorized users, while smart contracts automate grading and feedback.

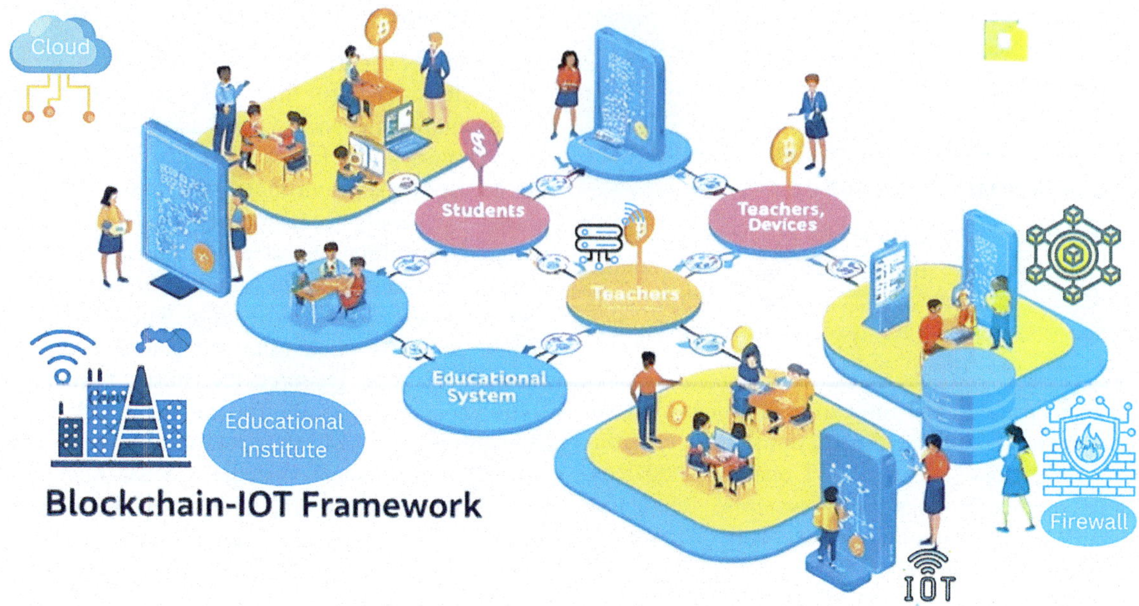

FIGURE 9.5 A IoT–blockchain integrated educational system.

- *Fraud prevention and credential verification:* The firewall component in the diagram represents multilayered security protection, ensuring that only authenticated users interact within the system preventing diploma fraud and enabling instant verification
- *Financial transactions and tokenization:* Blockchain-based payments (e.g., Bitcoin) streamline tuition fees, grants, and student loans via smart contracts.

Future advancements in smart classrooms should include quantum-resistant blockchain protocols for ultra-secure academic transactions, federated learning models for decentralized AI-driven student analysis, and edge computing with blockchain to enhance real-time analytics.

The combination of blockchain and digital twins enhances administrative efficiency, learning personalization, and fraud prevention. Table 9.4 summarizes the measurable impacts of such a system.

By merging blockchain's decentralized trust architecture with the analytical power of digital twins, smart classrooms create a dynamic, secure, and student-centric learning ecosystem that adapts to evolving educational needs.

3. Current applications of blockchain in education

The integration of blockchain technology in education is reshaping academic systems, driving efficiency, transparency, and security across various educational processes. The following are some notable implementations and their measurable impacts on educational institutions globally.

3.1 The European Smart Campus Initiative

The European Smart Campus Initiative has integrated blockchain with a digital twin model at a leading university, transforming academic administration (Lenz et al., 2021). This system enables real-time classroom simulations, automated grading, and instantaneous credential validation. The implementation has led to a 40% improvement in verification efficiency, significantly reducing credential validation time and administrative processes. Automated grading and student performance tracking have reduced faculty workload by eliminating manual assessment bottlenecks. Blockchain ensures tamper-proof credentialing and streamlines interinstitutional credit transfers.

3.2 National Natural Science Foundation of China project EduRSS

EduRSS, implemented by a Chinese institution under the Natural Science Foundation, integrates blockchain with digital twin classroom simulations to enhance academic integrity (Li & Han, 2019). This blockchain-based system has reduced administrative workload by 35%, as smart contracts autonomously authenticate student records and issue verifiable credentials. This initiative has improved employer trust by providing real-time, blockchain-secured

TABLE 9.4 The impact of blockchain and AI on educational efficiency, security, and innovation.

Category	Key insights	Quantitative results	Technology used
Cost efficiency and operational optimization (El Koshiry et al., 2023)	Automation in credential verification, record management, and financial transactions reduces administrative overhead	30%–50% cost reduction	Blockchain, smart contracts, decentralized databases
	Eliminating intermediaries in degree verification lowers authentication costs	Significant savings	Decentralized verification
	Automation of grading, attendance tracking, and improves productivity	Reduced faculty workload	Smart contracts, AI-driven automation
Fraud prevention and academic integrity (Ayub Khan et al., 2021)	High diploma fraud rates require secure academic records	58% of employers report fraud cases	Blockchain-based credentialing, immutable ledgers
	Tamper-proof digital certificates, prevent forgery	Instant verifiability	Decentralized identity (DID), cryptographic signatures
	Self-sovereign identity (SSI) ensures credentials ownership	Secure access to records	SSI frameworks
Lifelong learning and microcredentials (Peters et al., 2025)	Digital platforms facilitate microcredentialing	250,000+ badges issued	Digital badges, blockchain
	AI-driven digital twins personalize learning experiences	Improved students' outcomes	AI, digital twins, and blockchain
	Decentralized learning records validate informal education	Increased credential acceptance	Blockchain
Enhanced data security and privacy (Molina et al., 2023)	Encryption ensure data security (GDPR, FERPA compliance)	Compliance with regulations	ZKPs, encrypted storage
	DID reduces data breach risks	Fewer security incidents	DID, SSI
	Smart contracts restrict access to authorized parties	Lower cyber threats	Smart contracts
Real-time analytics and predictive learning (Martins et al., 2019)	Digital twins simulate real-time student engagement	Data-driven pedagogy	AI simulations, blockchain
	AI models predict at risk students' early intervention	Higher retention rates	Machine learning, blockchain records
	Curriculum optimized based on data analytical	Enhanced educational resources	AIs, blockchain for audit trails
Decentralized learning and interoperability (Ocheja et al., 2020)	Credential exchange between institutions and employers	Faster verification	Blockchain registries
	Cross-institutional credit transfers allow multiinstitutional learning	Reduced bureaucratic delays	Decentralized credit tracking
	Tokenized educational resources create fair content marketplaces	Increased accessibility	NFTs, smart contracts
Smart contracts for financial transactions (John et al., 2023)	Automated payments ensure transparency	Faster fund disbursement	Smart contracts
	NFTs for academic achievements incentivize learning	Increased student motivation and participation	NFTs, gamification
	Crowdfunding expands access to education financing	More educational funding	P2P blockchain financing

student records. The decentralized ledger prevents fraudulent degree claims by enhancing student identity verification.

3.3 Advanced technology-based smart and immersive classroom

The smart and immersive classroom (SIC) initiative leverages AI and blockchain to deliver dynamic, personalized learning experiences (None Awari Tejas et al., 2023). Real-time feedback and AI-driven content recommendations enhance understanding, leading to a 27% improvement in learning outcomes. Blockchain ensures that student engagement data remains secure, immutable, and resistant to manipulation. IoT-synced biometric authentication facilitates transparent attendance tracking, preventing falsification. Furthermore, advancements in AI, VR, and communication technologies, including next-generation 6G networks, are enhancing virtual learning experiences and reshaping educational curricula (Sharrab et al., 2023). Fig. 9.6 demonstrates how blockchain technology has been deployed across various domains, demonstrating its transformative potential in education and beyond.

3.3.1 Blockchain-powered educational platforms and projects

The integration of distributed ledger technology into academic credentialing systems has introduced cryptographic immutability, decentralized verification, and automated issuance, fundamentally transforming traditional certification frameworks, as depicted in Fig. 9.7. The implementation of Merkle tree structures ensures that each issued credential remains tamper-proof, allowing for on-chain verification without exposing sensitive personal data. Through elliptic curve cryptography (ECC) and asymmetric key infrastructure, institutions provide verifiable digital signatures that enable instantaneous authentication by third parties, eliminating the risk of credential forgery. Decentralized identity (DID) protocols and verifiable credentials (VC) frameworks further enhance security by allowing students to control access permissions, reducing dependence on intermediary verification entities. Moreover, the integration of blockchain oracles bridges off-chain academic records with on-chain smart contracts, enabling real-time credential validation and revocation without requiring manual intervention. Gas-optimized smart contract execution on Ethereum and other permissioned blockchain networks ensures cost-efficient operations, significantly lowering verification expenses while preserving scalability.

Several pioneering initiatives have embraced these advancements to enhance the credibility and efficiency of digital academic certification. The Blockcerts framework, developed by the MIT Media Lab in 2020, has been widely implemented across more than 600 educational institutions, enabling a dramatic reduction in credential issuance time from nearly a month to under an hour by leveraging open-source blockchain protocols for digital diploma generation. The ODEM (On-Demand Education Marketplace) initiative, formalized in its 2022 whitepaper, has optimized the cost-effectiveness of blockchain-driven credentialing, achieving a 70% reduction in verification expenditures for over 10,000 learners through automated microcredential transactions. This platform further eliminates reliance on

FIGURE 9.6 Global implementation of blockchain technology across key sectors.

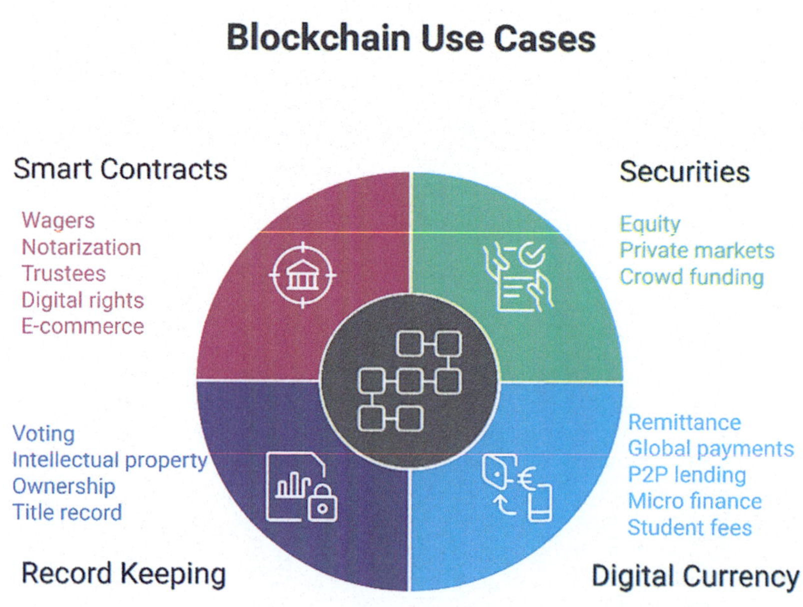

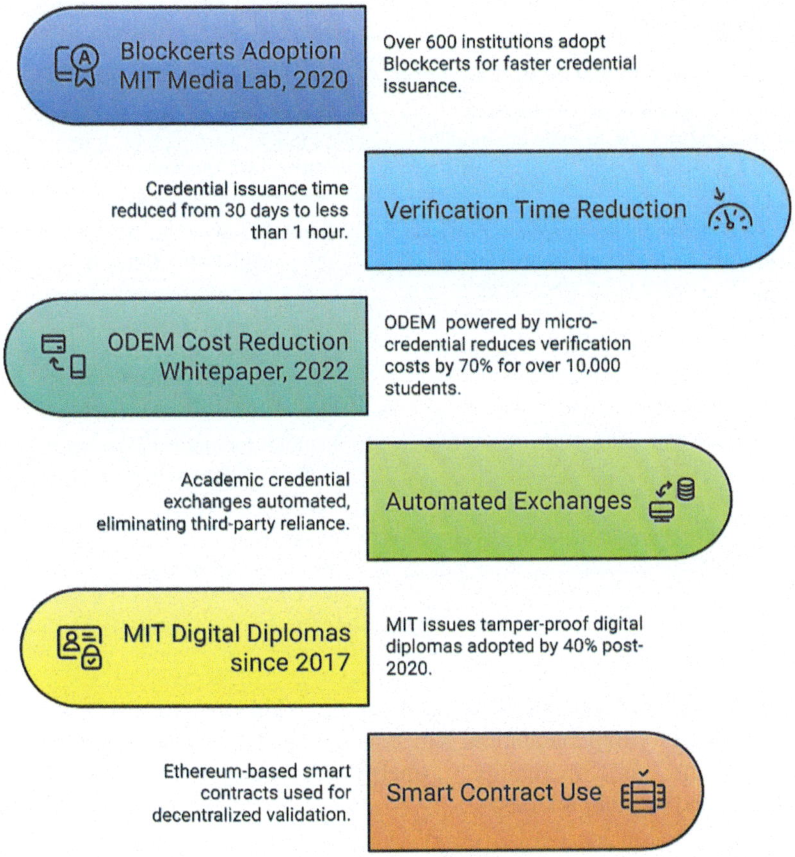

FIGURE 9.7 Institutional blockchain credentialing initiatives.

centralized authentication services by facilitating direct academic record exchanges via decentralized smart contracts. Another prominent effort, the MIT Digital Diplomas Initiative, has successfully deployed a tamper-proof, Ethereum-backed credentialing system since 2017, leading to a 40% increase in institutional adoption rates post 2020. By utilizing self-executing blockchain contracts, this initiative has eradicated diploma counterfeiting risks, reinforcing the authenticity and security of academic records. The widespread adoption of these decentralized trust models has streamlined issuance processes and reinforced institutional credibility by establishing a fraud-resistant, cryptographically secured academic record keeping model.

3.3.2 *Microcredentials and lifelong learning*

The rise of competency-based education has led to blockchain-powered microcredentialing, allowing students to earn verifiable, stackable certifications for specific skill sets. This approach enhances lifelong learning, employment matching, and credential interoperability across institutions.

- 34% of universities now issue blockchain-based microcredentials (Alsobhi et al., 2023).
- Microcredentials are increasingly recognized by employers, facilitating automated verification and seamless professional upskilling (Orman et al., 2023).

Various global initiatives have implemented blockchain-powered digital credentials to enhance verification, interoperability, and modular learning pathways. Table 9.5 highlights key projects, their credentialing approaches, adoption rates, and underlying technological frameworks.

3.3.3 *Gamification and incentives through smart contracts*

The integration of gamification strategies with blockchain-powered smart contracts has revolutionized digital learning models by significantly enhancing student engagement, improving course completion rates, and providing transparent, automated reward mechanisms through tokenized incentives Zubkov (2023). According to Major and Da Silva (2023), the introduction of gamification into Massive Open Online Courses (MOOCs) has led to a 60% increase in course completion rates, highlighting its effectiveness in sustaining learner motivation. A notable case study from the University of Nicosia demonstrated the efficacy of

TABLE 9.5 Top blockchain-based microcredential platforms.

Platforms	Credential characteristics	Implementation and adoption	Key technological features
Edubadges project, Netherlands McGreal and Olcott (2022)	Digital badges with criteria, embedded metadata, blockchain for endorsements	Piloted in 16 institutions (2017–20) to integrate modular badges	Open-source (badge), single sign-on (SSO), learning tools interoperability (LTI), blockchain functionality
NZ micro-credential system (Fisher & Leder, 2022)	Compact credentials (5–40 credits), industry-driven, annual reviews	Integrated within New Zealand Qualifications Framework (NZQF), limited university uptake	Standardized credit, quality assurance, regulatory compliance
European initiatives (Varadarajan et al., 2023)	Stackable, ECTS-aligned micro-credentials, flexible learning outcomes	Piloted by consortia (FutureLearn, digitary), supported by national grants	Digital verification, online platforms, European Qualifications Framework (EQF) alignment
Credly (Ikeda et al., 2023)	Digital badges, verified skills, criteria, and achievement evidence	Widely adopted by institutions and organizations globally	Platform for credential creation, management, and integration
Open badges (Gish-Lieberman et al., 2021)	Standardized badges for discrete skills	Adopted globally in education and professionally	Open standard, interoperability, decentralized storage.

blockchain-powered incentives, where students were rewarded with digital tokens upon reaching academic milestones, leading to a 45% improvement in engagement metrics as learners actively participated in skill-based blockchain ecosystems (Themistocleous et al., 2020). The incorporation of smart contracts within these educational frameworks has introduced several transformative advantages:

- Automated reward mechanisms: Blockchain-based incentive models autonomously issue digital tokens or academic credits upon course completion, ensuring a transparent and fair distribution of rewards.
- Instant certificate issuance: Smart contracts enable the immediate generation and verification of digital certificates upon the fulfillment of predefined academic criteria, eliminating the need for manual processing and reducing administrative overhead.
- Decentralized credential validation: Academic achievements are stored on the blockchain, allowing employers and educational institutions to verify student credentials instantly without reliance on centralized authorities, thereby enhancing trust and reducing credential fraud.

Blockchain effectively bridges gaps in learner engagement while ensuring the credibility, accessibility, and security of educational achievements in an increasingly digital and globalized academic landscape.

4. Challenges and limitations of blockchain for education

The integration of blockchain technology in education promises transformative benefits but also faces significant challenges that need to be addressed for its effective implementation. Key concerns include scalability, privacy, interoperability, energy consumption, and legal complexities.

4.1 Scalability and throughput constraints in blockchain-based educational systems

Blockchain's scalability is a critical challenge, particularly for high-volume academic transactions such as credential issuance and grading. Public blockchains, such as Bitcoin (1 MB block size, ~7 TPS) and Ethereum (~30 TPS pre-ETH 2.0 upgrade), suffer from low transaction throughput, making them inefficient for educational applications. The cost of blockchain adoption, including network maintenance, data storage, and regulatory compliance, imposes a financial burden, especially for resource-constrained institutions in developing regions. Solutions such as Layer-2 scaling (e.g., state channels and sharding), proof-of-stake (PoS) consensus, and alternative ledger structures such as directed acyclic graphs (DAGs) offer potential improvements in transaction speed and energy efficiency. Moreover, proof-of-authority (PoA) and delegated PoS (DPoS) can provide faster and cost-efficient consensus mechanisms for academic applications.

4.2 Privacy and data security challenges in educational blockchains

Data privacy is another concern, as education data is highly sensitive and subject to regulations such as GDPR, FERPA, and PDPA. Blockchain's immutable nature can conflict with the "right to be forgotten," posing challenges for erasing student data. Solutions such as zero-knowledge proofs (ZKPs), decentralized identity systems, and off-chain storage with on-chain hashing can mitigate these issues, ensuring data privacy while maintaining blockchain's security. However, blockchain's decentralized nature raises concerns about student data exposure and confidentiality, necessitating the use of privacy-enhancing technologies such as homomorphic encryption and multiparty computation.

Blockchain networks, despite their cryptographic security, face specific risks such as Sybil, 51%, and Eclipse attacks. Sybil attacks occur when fake nodes impersonate legitimate entities, compromising credential validation, mitigated by using proof-of-authority (PoA) in permissioned academic blockchains. In 51% attacks, malicious control of over half the network's computational power allows rewriting educational records, countered by combining proof-of-stake (PoS) and Byzantine fault tolerance (BFT) consensus models. Eclipse attacks isolate verification nodes, leading to fraudulent record manipulation, which can be prevented by implementing multihop overlay networks. Furthermore, blockchain's decentralized structure, while offering transparency, raises concerns over student data confidentiality due to exposure risks and metadata leakage. Solutions such as homomorphic encryption, multiparty computation (MPC), and zk-SNARKs can secure credentials while ensuring privacy, alongside adherence to regulations such as GDPR and FERPA to protect student identities.

4.3 Interoperability and standardization challenges

The rise of various blockchain frameworks (e.g., Ethereum, Hyperledger, Tezos, Solana) has led to fragmentation issues for educational institutions. Interoperability remains a significant hurdle, as different blockchain implementations (e.g., MIT's Blockcerts on Bitcoin and ODEM on Ethereum) complicate interinstitutional credential verification. Solutions such as decentralized identity standards (DID) and W3C verifiable credentials ensure academic credentials are readable across multiple networks, while cross-chain communication protocols such as Polkadot and Cosmos enable interoperability between blockchain networks. In addition, atomic swaps and the interledger protocol (ILP) facilitate credential transfers between disparate blockchains. To address integration with legacy learning management systems (LMS), solutions include API middleware bridges (e.g., Chainlink, Oracles) to connect blockchain credentialing with traditional LMS, and hybrid models combining centralized LMS for real-time processing with blockchain for secure credential storage.

4.4 Energy consumption and sustainability issues

Another challenge is energy consumption, especially with proof-of-work (PoW)-based systems, which require substantial computational power (Bitcoin's network consumes ~95 TWh per year). This is unsustainable for large-scale academic adoption. Ethereum's shift to PoS has reduced energy consumption by 99.95%, and other green blockchains such as Algorand and Hedera Hashgraph offer carbon-neutral alternatives suitable for educational applications.

4.5 Legal, regulatory, and governance challenges

Finally, legal, regulatory, and governance issues, including the lack of universal standards for blockchain credentials and the legal enforceability of smart contracts, hinder blockchain's widespread adoption. Educational institutions also face resistance to change due to the technical expertise required and the financial costs of blockchain adoption. To overcome these barriers, it is essential to develop ISO standards for blockchain credentialing (ISO/TC 307), hybrid legal-blockchain models, and government-backed pilot projects to demonstrate the benefits of blockchain in education.

Despite these challenges, advancements in Layer-2 scaling, privacy solutions, cross-chain interoperability, and green blockchain protocols continue to enhance the feasibility of blockchain for education. A hybrid approach combining private and public blockchain networks, along with efficient consensus mechanisms, promises a scalable, cost-effective solution for education in the future.

5. The future of blockchain in smart education

The future of blockchain in smart education promises transformative advancements through the integration of digital twins, AI-driven personalization, and smart contract automation. This convergence enhances learning efficiency, data security, and adaptability, fostering a student-centric educational environment. Blockchain's evolving role is highlighted by key technical pillars that include NFTs, tokenization services, private blockchain ecosystems, and cryptography-driven tools. These elements aim to address challenges such as decentralized credentialing, secure academic record management, and transparent resource allocation, all essential for next-generation learning environments as shown in Fig. 9.8.

The future of blockchain in smart education is marked by key innovations, including digital twins, AI-driven

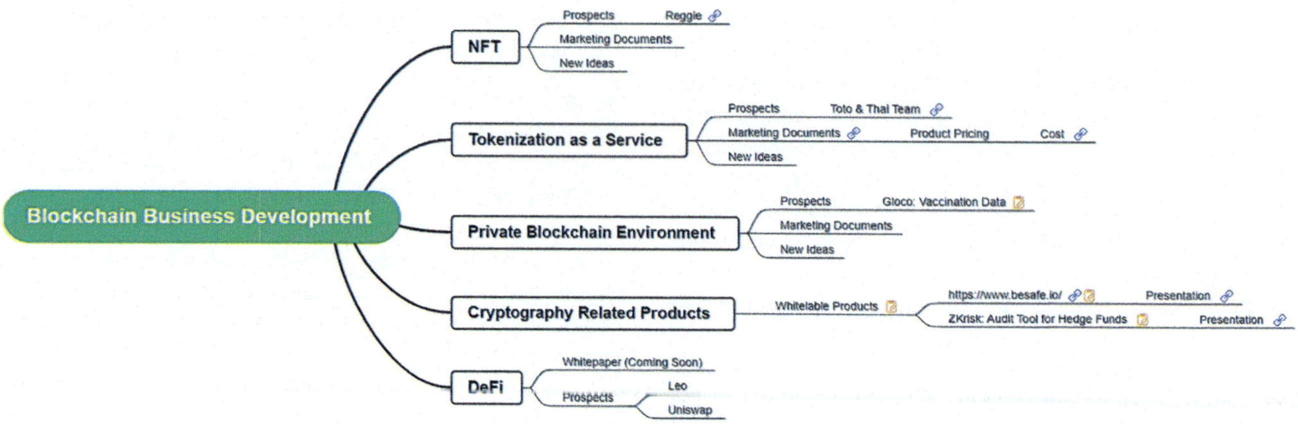

FIGURE 9.8 Mind map showcasing future developments of blockchain in smart education through technical pillars.

TABLE 9.6 Future advancements in blockchain for smart education.

Blockchain application	Description	Benefits	Key technologies	Real-world implications
Blockchain-enabled digital twins for smart classrooms	Virtual simulations of student behavior, classroom dynamics, and administrative processes	Real-time data exchange, improved decision-making	Blockchain, digital twins, AI	Reduces operational redundancies by up to 45% (Lin et al., 2025)
AI-driven personalization via smart contracts	Personalized learning via automated smart contracts	Automates learning, assessments, and credentialing	AI, blockchain, smart contracts	Enhances personalized learning, reducing bias and administrative load
Decentralized autonomous learning systems (DALSs)	Self-governing, AI-optimized learning environments	Tokenized achievements, global credential portability	Blockchain, AI, NFTs, smart contracts	Gamifies learning
Interoperable blockchain networks for cross-institutional learning	Blockchain for cross-border academic collaboration	Reduces fraud, accelerates global workforce integration	Cross-chain interoperability, interledger protocols, blockchain	Real-time credential verification, global academic collaboration

personalization, and decentralized learning systems. These advancements, supported by technologies such as NFTs, tokenization, and private blockchains, aim to address critical educational challenges, such as secure credentialing and resource allocation. Table 9.6 highlights these emerging blockchain applications and their real-world implications (Table 9.6).

6. Conclusion

The integration of blockchain technology in smart education systems presents a shift in the way academic credentials, learning processes, and institutional operations are managed. By leveraging the inherent security, transparency, and decentralization of blockchain, educational frameworks can achieve greater efficiency, reduced fraud, and enhanced accessibility to lifelong learning records. The immutability of blockchain ensures the credibility of academic certifications, mitigating the risks of forgery and unauthorized alterations. Moreover, the adoption of smart contracts facilitates automated and trustless execution of academic agreements, reducing administrative overhead and enhancing interoperability between institutions.

Despite its numerous advantages, blockchain deployment in education faces challenges, including scalability, regulatory compliance, and adoption resistance. Addressing these concerns requires collaborative efforts between academia, industry, and policymakers to establish standardized frameworks, ensure ethical data governance, and optimize blockchain protocols for educational scalability. Future research should focus on integrating blockchain with artificial intelligence and big data analytics to enhance personalized learning experiences, automate credential verification, and create adaptive educational ecosystems.

Blockchain has the potential to redefine the educational paradigm by developing a secure, transparent, and decentralized learning environment. Its successful implementation necessitates a multidimensional approach involving technological innovation, policy adaptation, and stakeholder engagement. With continued improvements and strategic deployment, blockchain will undoubtedly play a pivotal role in shaping the future of education, ensuring greater trust, equity, and efficiency in academic processes worldwide.

References

Alsobhi, H. A., Alakhtar, R. A., Ubaid, A., Hussain, O. K., & Hussain, F. K. (2023). Blockchain-based micro-credentialing system in higher education institutions: Systematic literature review. *Knowledge-Based Systems, 265*(3), Article 110238. https://doi.org/10.1016/j.knosys.2022.110238

Anirudh, S., Shaan Sundar, R., & Ganapathy, S. (2022). A new cryptosystem for secured data communications in plagiarism checking process using blockchain technology. *Wireless Personal Communications, 128*, 2113–2130. https://doi.org/10.1007/s11277-022-10035-2

Awari Tejas, N., Gaurav, N. G., & Akshay, N. S. (2023). AI and IoT based smart classroom. *International Journal of Advanced Research in Science, Communication and Technology*, 626–628. https://doi.org/10.48175/ijarsct-13698

Ayub Khan, A., Laghari, A. A., Shaikh, A. A., Bourouis, S., Mamlouk, A. M., & Alshazly, H. (2021). Educational blockchain: A secure degree attestation and verification traceability architecture for higher education commission. *Applied Sciences, 11*(22), Article 10917. https://doi.org/10.3390/app112210917

Brown, B. A. (2022). *Blockchain technology in teaching, learning, and academic administration as a long-term solution to emergencies in higher education*. Emerald Publishing Limited EBooks. https://doi.org/10.1108/978-1-80382-193-120221008

Chan, W., Gai, K., Yu, J., & Zhu, L. (2025). Blockchain-assisted self-sovereign identities on education: A survey. *Blockchains, 3*(1), 3. https://doi.org/10.3390/blockchains3010003

Cheepurupalli, D. P., Ashislh, M., Aishwarya, R., & Yogitha, R. (2023). A blockchain application for the verification of academic information and scalable certification. In *2022 6th International conference on computing methodologies and communication (ICCMC)* (pp. 1082–1087). https://doi.org/10.1109/iccmc56507.2023.10083848

Chukowry, V., Nanuck, G., & Sungkur, R. K. (2021). The future of continuous learning - Digital badge and microcredential system using blockchain. *Global Transitions Proceedings, 1*. https://doi.org/10.1016/j.gltp.2021.08.026

Ciaian, P., Rajcaniova, M., & Kancs, D. (2015). The economics of BitCoin price formation. *Applied Economics, 48*(19), 1799–1815. https://doi.org/10.1080/00036846.2015.1109038

Dubey, K. B., & Goyal, M. (2022). Smart certificate using blockchain. *Journal of Computer Science, 18*(9), 877–884. https://doi.org/10.3844/jcssp.2022.877.884

El Koshiry, A., Eliwa, E., Abd El-Hafeez, T., & Shams, M. Y. (2023). Unlocking the power of blockchain in education: An overview of innovations and outcomes. *Blockchain: Research and Applications, 4*(4), Article 100165. https://doi.org/10.1016/j.bcra.2023.100165

Feld, S., Schönfeld, M., & Werner, M. (2016). Traversing Bitcoin's P2P network: Insights into the structure of a decentralised currency. *International Journal of Computational Science and Engineering, 13*(2), 122. https://doi.org/10.1504/ijcse.2016.078441

Fisher, R. M., & Leder, H. (2022). An assessment of micro-credentials in New Zealand vocational education. *International Journal of Training Research, 20*(3), 232–247. https://doi.org/10.1080/14480220.2021.2018018

Gish-Lieberman, J. J., Tawfik, A., & Gatewood, J. (2021). Micro-credentials and badges in education: A historical overview. *TechTrends, 65*(1), 5–7. https://doi.org/10.1007/s11528-020-00567-4

Ikeda, R., Ferriyan, A., Okawa, K., & Thamrin, A. H. (2023). Enhancing learning mobility with a community-based micro-credential e-Portfolio platform service for higher education. In *Lecture notes in computer science* (pp. 566–572). https://doi.org/10.1007/978-3-031-42682-7_43

John, K., Kogan, L., & Saleh, F. (2023). Smart contracts and decentralized finance. *Annual Review of Financial Economics, 15*(1), 523–542. https://doi.org/10.1146/annurev-financial-110921-022806

Lenz, R., Kleinheyer, B., Barkel, C., Veuger, J., Tsangaratos, P., Klõga, M., & Llorente, J. (2021). Blockchain and European higher education systems A snapshot on the diffusion process of blockchain In-novation into European academia comparative study of blockchain in higher education systems of Estonia, Germany, Greece, the Netherlands, and Spain. *SSRN Electronic Journal*. https://doi.org/10.2139/ssrn.3912646

Li, H., & Han, D. (2019). EduRSS: A blockchain-based educational records secure storage and sharing scheme. *IEEE Access, 7*, 179273–179289. https://doi.org/10.1109/access.2019.2956157

Lin, Y., Petal, K., Alhamadah, A. H., Ghimire, S., Redondo, M. W., Corona, D. R. V., … Satam, P. (2025). Personalized education with generative AI and digital twins: VR, RAG, and zero-shot sentiment analysis for industry 4.0 workforce development. *arXiv (Cornell University)*. https://doi.org/10.48550/arxiv.2502.14080. arXiv preprint arXiv:2502.14080.

Major, R. R., & Da Silva, M. M. (2023). Gamification in MOOCs: A systematic literature review. *Cogent Education, 10*(2). https://doi.org/10.1080/2331186x.2023.2275820

Martins, V., Oyelere, S. S., Tomczyk, L., Barros, G., Akyar, O., Eliseo, M. A., Amato, C., & Silveira, I. F. (2019). A blockchain microsites-based ecosystem for learning and inclusion. *Anais Do XXX Simpósio Brasileiro de Informática Na Educação (SBIE 2019), 30*(1). https://doi.org/10.5753/cbie.sbie.2019.229

McGreal, R., & Olcott, D. (2022). A strategic reset: Micro-credentials for higher education leaders. *Smart Learning Environments, 9*(1). https://doi.org/10.1186/s40561-022-00190-1

Molina, F., Betarte, G., & Carlos, D. L. (2023). A blockchain based and GDPR-Compliant design of a system for digital education certificates. *Clei Electronic Journal, 26*(1). https://doi.org/10.19153/cleiej.26.1.3

Veeramanickam, M. R. M., Dabade, M. S., Sita Rama Murty, P., Borhade, R. R., Barekar, S. S., Navarro, C., Roman-Concha, U., & Rodriguez, C. (2023). Smart education system to improve the learning system with CBR based recommendation system using IoT. *Heliyon, 9*(7), Article e17863. https://doi.org/10.1016/j.heliyon.2023.e17863

Nakamoto, S. (2008). Bitcoin: A peer-to-peer electronic cash system. *bitcoin.org*. https://bitcoin.org/bitcoin.pdf.

Nargis, T., Preethi Salian, K., Prathyakshini, N., Vanajakshi, J., Manasa G, R., & Salian, S. (2023). A secure platform for storing, generating and verifying degree certificates using blockchain. In *2022 6th International conference on trends in electronics and informatics (ICOEI)*. https://doi.org/10.1109/icoei56765.2023.10125598

Ocheja, P., Flanagan, B., & Ogata, H. (2020). Decentralized E-Learning marketplace: Managing authorship and tracking access to learning materials using blockchain. *Communications in Computer and Information Science*, 526–535. https://doi.org/10.1007/978-981-15-6113-9_59

Orman, R., Simsek, E., & Cakir, M. A. K. (2023). Micro-credentials and reflections on higher education. *Higher Education Evaluation and Development, 17*(2), 96–112. https://doi.org/10.1108/heed-08-2022-0028

Parmentola, A., Petrillo, A., Tutore, I., & De Felice, F. (2022). Is blockchain able to enhance environmental sustainability? A systematic review and research agenda from the perspective of sustainable development goals (SDGs). *Business Strategy and the Environment, 31*(1), 194–217.

Peters, M., Beirne, E., & Brown, M. (2025). Micro-credentials and its implications for lifelong learning–global insights and critical perspectives. *Distance Education, 46*(1), 1–7. https://doi.org/10.1080/01587919.2025.2463137

Raghav, A.,N., Verma, R., Venkatesan, S., & Verma, S. (2019). Tamper-proof certificate management system. In *2019 IEEE Conference on information and communication technology* (pp. 1–6). https://doi.org/10.1109/cict48419.2019.9066236

Sawant, R. (2023). Revolutionizing education sector by leveraging blockchain technology: State of art. In *Nucleation and atmospheric aerosols*. https://doi.org/10.1063/5.0119425

Sharrab, Y., Almutiri, N. T., Tarawneh, M., Alzyoud, F., Al-Ghuwairi, A.-R. F., & Al-Fraihat, D. (2023). Toward smart and immersive classroom based on AI, VR, and 6G. *International Journal of Emerging Technologies in Learning (IJET), 18*(02), 4–16. https://doi.org/10.3991/ijet.v18i02.35997

Themistocleous, M., Christodoulou, K., Iosif, E., Louca, S., & Tseas, D. (2020). Blockchain in academia: Where do we stand and where do we go?. In *Proceedings of the . . . Annual Hawaii international conference on system sciences/proceedings of the annual Hawaii international conference on system sciences*. https://doi.org/10.24251/hicss.2020.656

Tomczyk, L., Martins, V. F., Eliseo, M. A., Silveira, I. F., de, A., Stosic, L., & Oyelere, S. S. (2020). How does the attitude to new media influence the evaluation of a new e-Learning platform among current and future teachers?. In *Proceedings 15th Conferencia Latinoamericana de Tecnologias de Aprendizaje (LACLO)* (pp. 1–8). https://doi.org/10.1109/laclo50806.2020.9381130

Tripathi, G., Ahad, M. A., & Casalino, G. (2023). A comprehensive review of blockchain technology: Underlying principles and historical background with future challenges. *Decision Analytics Journal, 9*, Article 100344. https://doi.org/10.1016/j.dajour.2023.100344

Trivedi, S., Mehta, K., & Sharma, R. (2021). Systematic literature review on application of blockchain technology in E-Finance and financial services. *Journal of Technology Management and Innovation, 16*(3), 89–102. https://doi.org/10.4067/s0718-27242021000300089

Varadarajan, S., Koh, J. H. L., & Daniel, B. K. (2023). A systematic review of the opportunities and challenges of micro-credentials for multiple stakeholders: Learners, employers, higher education institutions and government. *International Journal of Educational Technology in Higher Education, 20*(1). https://doi.org/10.1186/s41239-023-00381-x

Wang, Y., Cong, X., Zi, L., & Xiang, Q. (2023). Blockchain for credibility in educational development: Key technology, application potential, and performance evaluation. *Security and Communication Networks, 2023*, 1–17. https://doi.org/10.1155/2023/5614241

Zhu, Z., Yu, M., & Riezebos, P. (2016). A research framework of smart education. *Smart Learning Environments, 3*(1). https://doi.org/10.1186/s40561-016-0026-2

Zubkov, A. (2023). Gamification techniques in massive open online courses: Challenges and opportunities. In *Lecture notes in networks and systems* (pp. 391–401). https://doi.org/10.1007/978-3-031-48020-1_29

Chapter 10

Student check-in scheme based on blockchain for smart classroom

QingYang Ding[1], JianBin Chen[1] and Xiao Chen Hao[2]
[1]School of Management, Beijing Union University, Beijing, China; [2]College of Applied Arts and Science, Beijing Union University, Beijing, China

1. Introduction

A classroom is an important spatial carrier that supports teaching and learning and realizes the transfer of knowledge from teachers to groups of students (Kumar Saini & Goel, 2019). Based on technologies in fields such as signal processing, network technology, hardware, and software, a smart classroom connects interactive whiteboards, electronic gadgets, digital projectors, the internet-enabled IoT and sensor devices (Singh et al., 2025). By creating an exciting and dynamic environment for teachers and students, it helps teachers implement teaching activities more effectively, so as to better achieve the transfer of knowledge (Chen et al., 2017).

Check-in is an indispensable and critical component of teaching, whether in smart classrooms or traditional settings. At its core, check-in serves as a method to verify one's "presence" or "participation," with the primary purpose of documenting attendance status. It is widely used in scenarios such as roll call, event management, and online interactions. Ensuring the authenticity and integrity of check-in records plays a vital role in refining classroom management and enhancing teaching quality.

Currently, check-in in educational settings primarily relies on methods such as ID cards, IC cards, manual tracking, radio–frequency identification (RFID), fingerprint recognition, and facial recognition. Traditional approaches represented by ID cards, IC cards, and manual recording require significant human effort in generating, compiling, and retrieving records, often falling short in accuracy, security, and convenience.

Advancements in digital technology and biometrics have led to the increasing adoption and broader application of biometric-based check-in systems. These innovations have significantly improved automation and precision in attendance tracking. However, both traditional and digital check-in methods face challenges such as data falsification, tampering, and difficulties in cross–departmental data sharing.

Blockchain technology represents an underlying technical architecture founded upon the principles of decentralization and distributed ledger systems. At its core, it harnesses cryptographic algorithms and consensus mechanisms to sequentially concatenate data into a chain in the form of "blocks" according to the chronological order. This approach enables the realization of several key attributes, including data immutability, traceability, transparent sharing, and decentralized trust. In the context of blockchain classification, it can be systematically categorized into three primary types: public blockchains, private blockchains, and consortium blockchains.

Public blockchains embody the highest degree of decentralization, where unrestricted participation in blockchain operations, such as bookkeeping and verification, is permitted for any individual or entity. This openness endows public blockchains with exceptional transparency, making them accessible and observable to the entire community. On the other hand, private blockchains are predominantly established and governed within the confines of a single organization or institution. Stringent control over write and access permissions is maintained internally, which ensures that only authorized personnel within the organization can interact with the blockchain. This setup confers significant advantages in terms of transaction speed and privacy protection, rendering private blockchains particularly suitable for internal data management and the optimization of business processes. Consortium blockchains, as the name suggests, are collaboratively managed by multiple organizations or institutions. Each participating entity is assigned specific rights and responsibilities, and the consensus mechanism governing the consortium blockchain is typically

Digital Twins in the Smart Classroom. https://doi.org/10.1016/B978-0-443-41632-3.00025-4

determined through a collective negotiation process among its members. Consequently, consortium blockchains exhibit a moderate level of decentralization, positioning them between the highly decentralized public blockchains and the more centralized private blockchains.

The unique technical characteristics of blockchain technology, including its ability to safeguard data integrity, facilitate comprehensive traceability, and establish trust without relying on a central authority, render it an indispensable and viable technological option for rectifying the deficiencies inherent in existing check-in solutions.

2. Related work

This section will review and summarize the relevant literature on smart classrooms and check-in solution.

2.1 Smart classroom

A smart classroom is meticulously defined as a technology-augmented, self-contained educational ecosystem that endeavors to elevate the didactic and learning experiences to unprecedented heights. In contemporary educational research, a prototypical smart classroom is characterized by an array of advanced technological amenities. These include state-of-the-art tools designed to facilitate superior instructional presentations (de Groot, 2002), thereby enabling educators to convey complex concepts with greater clarity and visual appeal. Additionally, it integrates sophisticated mechanisms to foster enhanced student engagement (Lui & Slotta, 2014), leveraging interactive platforms and digital resources to stimulate active participation and collaborative learning. The presence of cutting-edge technologies further promotes seamless interaction (Suo et al., 2009) between instructors and learners, as well as among students themselves, cultivating an interactive and dynamic learning atmosphere. Moreover, intelligent environmental control systems are incorporated to optimize the physical surroundings (Ramesh et al., 2016), ensuring a comfortable, healthy, and conducive space for educational activities.

Furthermore, smart classrooms are equipped with a suite of functional tools dedicated to streamlining various logistical operations within the educational setting. This encompasses automated attendance tracking systems (Silva et al., 2008) that leverage biometric or digital identification technologies for accurate and efficient student presence monitoring. Advanced assessment tools (Ihantola et al., 2010) are also integrated, enabling educators to conduct comprehensive evaluations, provide timely feedback, and customize learning experiences based on individual student needs. Real-time feedback mechanisms (Gligorić et al., 2012) are another integral component, facilitating immediate communication and adjustment of teaching strategies to enhance learning outcomes. Collectively, these technological advancements in smart classrooms represent a paradigm shift in educational infrastructure, heralding a new era of intelligent and personalized learning environments.

2.2 Check-in

Check-in essentially means that people share their locations. Different from recording geographical coordinates, check-ins are usually paired with places (such as restaurants, parks, etc.) (Cheng et al., 2011).

Manual roll call, the conventional approach to check-in taking, presents a plethora of limitations. This traditional methodology not only consumes a substantial amount of precious instructional time but also demands additional labor for data compilation and analysis. Moreover, it is inherently vulnerable to proxy attendance, undermining the integrity of attendance records. In response to these challenges, a growing body of research has focused on the development of automated student attendance systems within educational settings.

One prominent solution in the realm of automated attendance is the utilization of smart cards, which serve as digital repositories for student identification information. Radio-Frequency Identification (RFID) tags have emerged as a widely adopted technology for attendance tracking, with numerous studies (Arulogun et al., 2013; Chang, 2011; Kassem et al., 2010; Lim et al., 2009; Qaiser & Khan, 2006) documenting their effectiveness in educational environments. Similarly, Near–Field Communication (NFC) technology has also been extensively explored for attendance management purposes (Ahmad et al., 2014; Boon Chew et al., 2015; Benyo et al., 2012; Das, 2008; Ervasti et al., 2009; Marcos et al., 2013; Shen et al., 2014; Ichimura & Kamada, 2013). In both RFID and NFC systems, students initiate the attendance -recording process by bringing their smart cards into proximity with a reader device. The collected attendance data is subsequently transmitted to a central server for storage and further processing. Beyond attendance monitoring, RFID technology has demonstrated potential for tracking students' movements and activities across institutional premises (Kassem et al., 2010). However, smart-card based attendance systems are not without their drawbacks. These systems necessitate the procurement and maintenance of additional hardware components, including smart cards and readers, which significantly increase the overall implementation cost. Additionally, smart cards are prone to physical damage, loss, or misplacement, requiring costly replacement procedures. Notably, these systems fail to adequately address the persistent issue of proxy attendance, thereby limiting their effectiveness in ensuring accurate attendance records.

Advancements in smartphone technology have spurred the exploration of mobile-based attendance solutions, leveraging built-in sensors such as Bluetooth, NFC, and Wi-Fi. Early iterations of Bluetooth-based attendance systems were hampered by the requirement for device pairing, a time-consuming process that rendered them impractical for large-scale classroom deployments. However, the advent of Bluetooth Low Energy (BLE) technology (Mackensen et al., 2012) has mitigated this limitation by eliminating the need for pairing. BLE-enabled beacon devices continuously broadcast signals that can be detected by nearby smartphones, allowing dedicated mobile applications to record attendance in real-time (Apoorv & Mathur, 2016; Bae & Cho, 2015; Noguchi et al., 2015; Sengul et al., 2017). To facilitate comprehensive attendance tracking, each classroom must be equipped with BLE beacon devices capable of establishing communication with students' smartphones. Some software applications, such as Acadly. (2019), have integrated multiple sensor technologies, including Bluetooth, Wi-Fi, and GPS, to enhance the reliability and accuracy of attendance data. Nevertheless, a fundamental limitation of smartphone-based attendance systems is their reliance on the assumption that all students possess and use smartphones. This requirement renders such systems unsuitable for educational contexts where students may be too young to own smartphones or where institutional policies prohibit mobile device usage.

In contrast, biometric attendance systems offer a hardware-independent solution for accurate attendance tracking. Fingerprint-based attendance systems require students to physically interact with a fingerprint scanner, typically installed at classroom entrances, to authenticate their presence (Izzati Zainal et al., 2014; Mohamed Basheer & Raghu, 2012). Recent advancements in facial recognition technology have also enabled its integration into classroom attendance systems (Arsenovic et al., 2017; Kar et al., 2012; Patil & Shukla, 2014). These systems typically employ IP cameras strategically positioned at classroom entrances to capture facial images, which are then processed using sophisticated algorithms to verify student identities and record attendance. The adoption of biometric technologies represents a significant advancement in attendance management, offering enhanced security, accuracy, and convenience compared to traditional and smart-card-based approaches.

A comprehensive review of the extant literature reveals that researchers have extensively investigated various aspects of smart classrooms and check-in mechanisms. Nevertheless, the scholarly discourse has predominantly overlooked the exploration of blockchain-based optimization strategies for check-in systems within smart classroom environments. This research endeavors to fill this significant gap in the academic literature. By delving into this hitherto under-explored domain, the study not only contributes to the expansion and enrichment of existing research but also presents a novel and viable solution for enhancing the efficiency of check-in procedures, refining classroom management practices, and elevating the overall quality of education in smart classroom settings. This research thus represents a valuable addition to the academic and practical understanding of smart classroom technologies, offering both theoretical insights and practical implications for future research and development in the field.

3. Blockchain-based students check-in scheme for smart class-room

This section will describe and discuss the blockchain-based student check-in scheme for smart classrooms. First, this part will introduce the overall framework of the blockchain-based student check-in scheme for smart classrooms. Second, this part will introduce the block data structure in this scheme. Thirdly, the consensus mechanism of the blockchain in this scheme will be presented. Finally, the data interaction mechanism in this scheme will be introduced.

3.1 Proposed scheme overview

The blockchain-based student check-in scheme for smart classrooms (BCSCI) comprises four distinct technical layers, each serving a specific functional role in the system architecture. The proposed blockchain-based student check-in scheme overview for smart classroom is described in Fig. 10.1.

3.1.1 Physical device layer

The first layer constitutes the foundational hardware infrastructure, where Internet of Things (IoT)-enabled physical devices for student check-in are deployed and interconnected. This layer includes smart card terminals, fingerprint scanners, and facial recognition cameras, which serve as the primary interfaces for capturing student identity data. These devices are integrated into the smart classroom environment through standardized IoT protocols, ensuring seamless data collection and transmission to subsequent layers.

3.1.2 Identity verification layer

The second layer employs integrated learning-based authentication modules to validate the authenticity of collected identity information. Leveraging machine learning algorithms, such as ensemble methods (e.g., random forests or gradient-boosted trees), this layer cross-verifies data from multiple sources (e.g., biometric

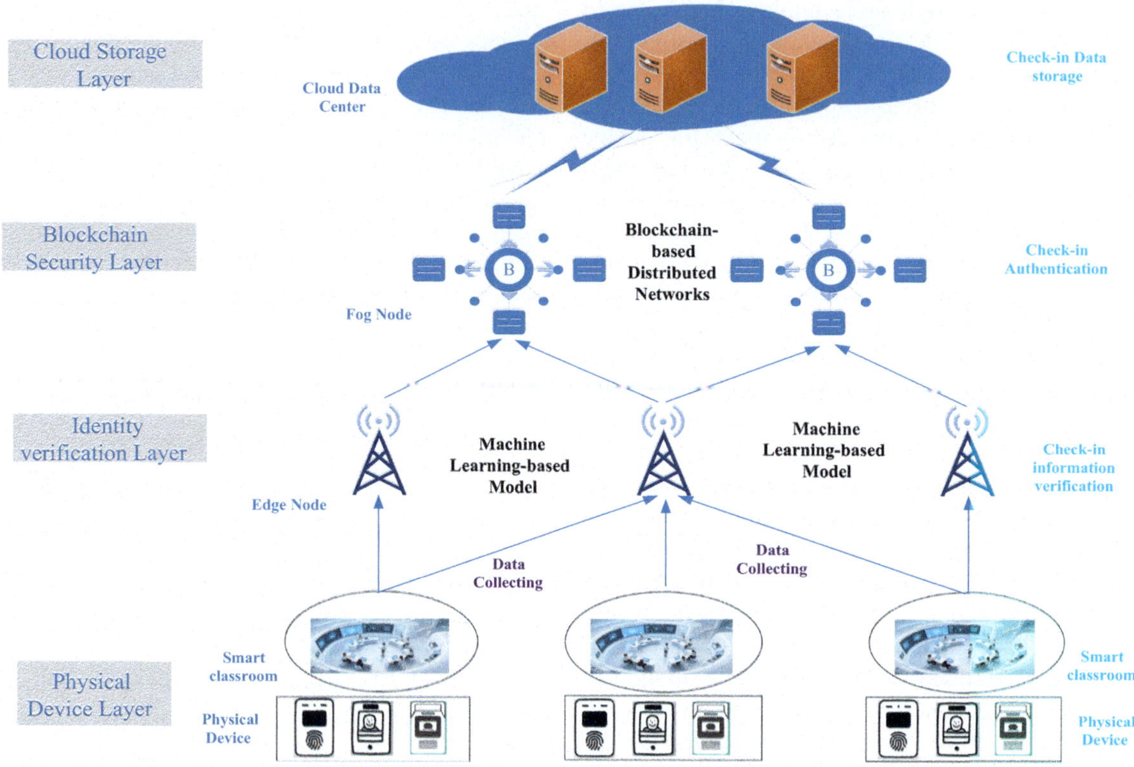

FIGURE 10.1 Proposed blockchain-based student check-in scheme overview for smart classroom.

templates, smart card credentials) to mitigate risks of fraud or data tampering. By establishing a robust validation mechanism, the system ensures that only legitimate user identities proceed to the blockchain layer for further processing.

3.1.3 Blockchain security layer

The core technical layer utilizes blockchain technology to ensure data integrity and decentralized trust. Here, a Proof of Teaching (POTH) consensus algorithm is implemented, which differs from traditional blockchain consensus mechanisms (e.g., Proof of Work, Proof of Stake) by incorporating a hierarchical permission system. Under POT, validation authority is distributed between authorized stakeholders: instructors and administrative staff within the educational institution. Teachers possess limited validation privileges for real-time check-in data, while administrative personnel hold higher-level permissions for system-wide data governance. This role-based validation model enhances both operational efficiency—by reducing unnecessary computational overhead—and security—by restricting access to authenticated institutional users. The blockchain ledger in this layer chronologically records validated check-in events as immutable blocks, each containing timestamped data, cryptographic hashes, and consensus signatures.

3.1.4 Cloud storage layer

Following validation and consensus on the blockchain, verified data is transmitted to the uppermost layer: a cloud-based storage ecosystem. This layer serves as a centralized repository for long-term data archiving, analytics, and interoperability with other institutional systems (e.g., student information systems, learning management platforms). By leveraging cloud infrastructure, the scheme ensures scalable storage capacity, secure data backup, and authorized access for stakeholders, while maintaining the auditability enabled by the underlying blockchain layer.

Collectively, this four-layer architecture integrates IoT hardware, advanced authentication technologies, and blockchain-based distributed ledger systems to create a robust, secure, and efficient check-in solution for smart classrooms. The hierarchical design ensures clear separation of technical functions, while the POT consensus mechanism balances institutional control with decentralized trust, addressing critical challenges in educational attendance management.

3.2 Data block structure

In the blockchain system tailored for student check-in within smart classrooms, the responsibility of constructing data blocks lies with core nodes. These data blocks are composed of two essential parts: the block header and the block body.

The block header mainly encapsulates the following elements: the hash value of the current block, which is the function value calculated by applying the Hash256 algorithm to the block after it has been verified by verification nodes; the hash value of the previous block header, which serves as the hash value of the parent block; the hash value of the smart classroom, derived from the hash of the smart classroom's unique identifier; the version number, indicating the version of the entire blockchain system; the Merkle root, obtained through Merkle calculations performed on the data in the block body; and the timestamp, set by quasi-central nodes after being jointly verified by verification nodes.

The block body primarily contains students' check-in information, which can be divided into four major categories:

3.2.1 Student basic information (unique identification and identity association)

This includes student unique identification and associated course information.

Student Unique Identification. This includes the student ID (serving as the primary key to uniquely identify the student and avoid confusion caused by identical names), name (for auxiliary identification), grade, major, and class (used for permission division and data statistics).

Associated Course Information: It mainly consists of the course number/name (to clearly indicate the course corresponding to the check-in), the instructor's ID/name (to associate the teacher role for subsequent consensus verification), and the class time interval (e.g., 2025-04-28 08:00–09:45, used to determine whether the student is late or leaves early).

3.2.2 Check-in details (core business data)

Check-in behavior records and status identification data are included.

Check-in Behavior Records. These include the check-in time (accurate to the second, which needs to be aligned with the timestamp in the block header as a verification basis), check-in location (the physical number or positioning information of the smart classroom, associated with the "smart classroom hash value" in the block header), check-in method (such as RFID card, QR code scanning, facial recognition, Bluetooth check-in, etc., recording the technical means), and check-in device ID (e.g., the terminal device number at the classroom entrance, used for device management and anti-tampering verification).

Status Identification Data. It mainly includes the check-in status (valid/invalid, combined with the verification results of teachers or the system; for example, if the check-in is not completed within the specified time, it will be marked as invalid) and exception markers (such as late arrival, early departure, proxy check-in warning, etc., which need to be verified by subsequent consensus nodes).

3.2.3 Authentication and security information (core features of blockchain)

It includes digital signature system and anti-tampering verification.

Digital Signature System. This encompasses the student's digital signature (the signature of the check-in information using the student's private key to prove the legitimacy of the operation), the teacher/verification node's signature (the confirmation signature of the validity of the check-in by the teacher or the educational administration system, included in the consensus verification), and the device signature (the digital certificate signature of the check-in device to ensure that the device has not been tampered with).

Anti-tampering verification. It includes associated block header information (such as the current block hash value and the previous block header hash value, associated with the block header through the Merkle root) and data integrity hash (the hash value of all fields in the block body, used for quickly verifying whether the data has been tampered with).

3.2.4 Extension fields (flexibility and future compatibility)

Remarks or additional Information and consensus mechanism-related fields are included.

Remarks or Additional Information. This includes explanations for check-in exceptions (e.g., manual remarks for check-in time errors caused by network delays) and the system version number (corresponding to the version number in the block header to support multi-version compatibility).

Consensus Mechanism-Related Fields. It includes the verification node identifiers (recording the list of nodes participating in the verification of this block, such as the IDs of teacher nodes and educational administration nodes) and the consensus time (assisting in tracking the consensus process and optimizing system performance).

Fig. 10.2 provides an example of the data block structure of the blockchain for student check-in in smart classrooms.

3.3 Consensus mechanism

This section introduces the consensus mechanism of the blockchain designed for student check-in scenarios in smart classrooms. Considering the specific requirements of

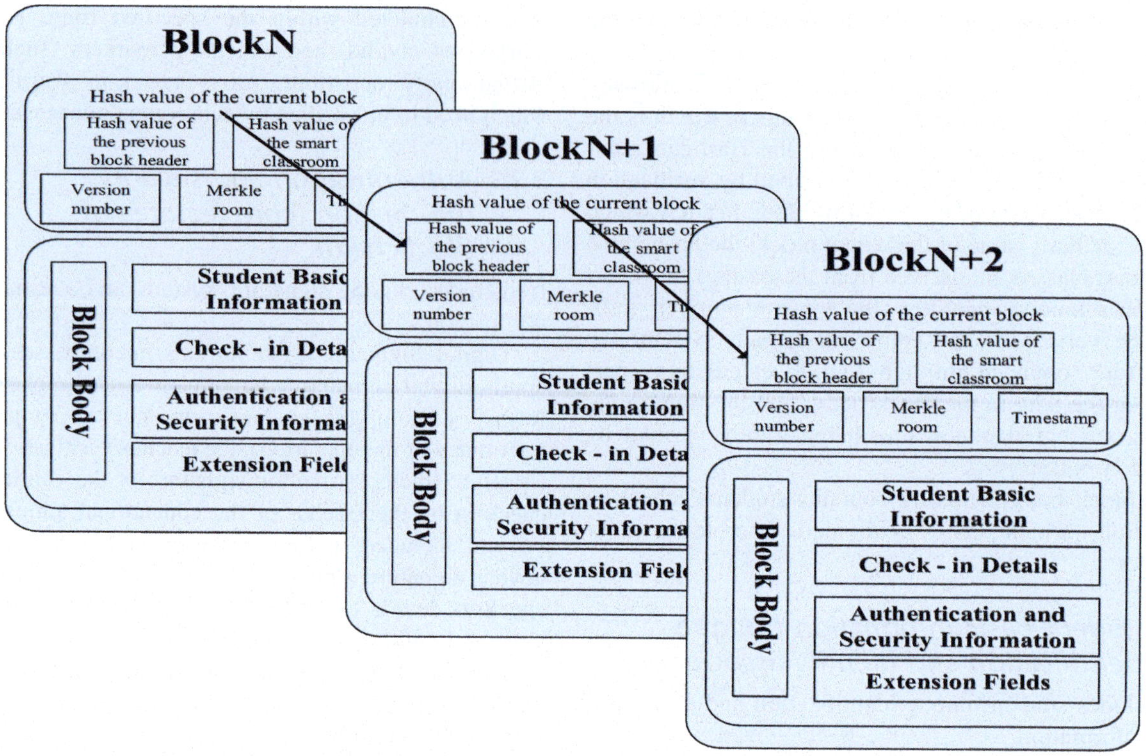

FIGURE 10.2 An example of the data block structure.

this educational context, including hierarchical permission management, role-based responsibilities, and the need for efficient data validation, a private blockchain is adopted. Private blockchains offer inherent advantages such as controlled write permissions, optimized transaction throughput, and enhanced privacy, which align well with the institutional governance structure of educational environments.

Building on the technical characteristics of private blockchains and integrating the role-based and permission hierarchy in check-in processes, this study proposes a novel consensus algorithm named Proof of Teaching (POTH), based on Practical Byzantine Fault Tolerance (PBFT) algorithm, with "audit-based validation" as its core principle. POTH incorporates two key stakeholder groups into the consensus process: instructors and educational administrators, whose collaborative verification ensures both the pedagogical relevance and technical integrity of check-in data.

3.3.1 Core design of the POTH consensus mechanism

POTH operates on a two-layer validation framework that maps to the operational hierarchy of educational institutions:

Real-Time Audit Layer (Instructor Nodes). Teachers serve as primary verification nodes responsible for immediate validation of check-in events during class sessions. Equipped with limited write permissions, instructors authenticate student attendance by cross-checking biometric data, smart card records, or Real-time location information against pre-registered course rosters. This layer ensures that only students physically present in the smart classroom and within the designated class time window are marked as valid check-ins. Instructors can flag exceptions (e.g., late arrivals, proxy check-ins) through the blockchain interface, which triggers a secondary review process.

Consensus Confirmation Layer (Administrative Nodes). Educational administrators act as consensus nodes with higher-level permissions, responsible for aggregating and finalizing validated check-in data across multiple classrooms. Using a threshold-based approval mechanism (e.g., requiring consensus from 2/3 of administrative nodes), this layer ensures data consistency across the institutional ledger. Administrative nodes also handle system-level functions such as user permission updates, blockchain version upgrades, and historical data audits, maintaining the overall integrity of the attendance ecosystem.

3.3.2 Technical advantages of POTH

Role-Aligned Verification. By embedding pedagogical roles (instructors) and administrative roles (administrators) into the consensus process, POTH bridges the gap between educational practice and technical infrastructure. This design ensures that validation logic reflects real-world classroom dynamics (e.g., teachers' authority to confirm student presence) while adhering to blockchain's decentralized trust model.

Efficiency and Scalability. Unlike computationally intensive public blockchain consensus algorithms (e.g., Proof of Work), POTH leverages the trusted institutional environment of private blockchains to reduce validation overhead. The limited number of consensus nodes (instructors and administrators) enables near-instantaneous transaction confirmation, making it suitable for high-frequency check-in scenarios in large educational institutions.

Tamper-Resistant Audit Trail. Every check-in event is cryptographically linked to both instructor validation signatures and administrative consensus stamps within the block header. This creates an immutable audit trail where each step of the verification process—from real-time classroom authentication to institutional-level data confirmation—can be traced and independently verified, addressing the critical challenge of data falsification in traditional attendance systems.

3.3.3 Consensus process flow

Data Submission. Student check-in data (including biometric records, timestamps, and device signatures) is collected by IoT devices in the smart classroom and transmitted to instructor nodes.

Real-Time Validation. Instructors review the data against course parameters (e.g., class time, enrolled students) and mark the check-in status (valid/invalid) with their digital signatures.

Consensus Aggregation. Validated records from multiple instructor nodes are bundled into a candidate block, which is then forwarded to administrative nodes for final confirmation.

Finalization. Administrative nodes execute a majority-vote consensus algorithm to validate the block's integrity. Upon approval, the block is added to the blockchain, and the updated ledger is propagated to all authorized stakeholders.

By integrating educational role responsibilities into the blockchain consensus mechanism, POTH transforms the traditional top-down administrative audit into a distributed, tamper-evident validation process. This innovation not only enhances the reliability of student attendance records but also demonstrates the potential of blockchain technology to align with domain-specific workflows in smart education environments.

3.4 Data interaction mechanism

Table 10.1 categorizes the key participants in the smart classroom check-in system, defining their roles, core functions, and the technologies/interfaces they use for interaction.

Student Terminal.Acts as the data producer, leveraging mobile devices or hardware (e.g., NFC, Bluetooth) to submit attendance data while ensuring user-centric input.

Teacher Node. Serves as a local validator, using web-based tools and smart contracts to verify attendance data within their courses, embodying frontline administrative control.

Administrative Node. Functions as a regulatory arbiter, overseeing cross-course validations and finalizing blockchain updates, ensuring compliance with institutional rules.

Smart Devices: Focus on data collection through IoT protocols, integrating biometric and location technologies to anchor real-world actions to digital records.

TABLE 10.1 Participating entities and roles.

Participant	Role	Core functions	Interaction technology/Interface
Student terminal	Data producer	Submits attendance data (biometric features, location, timestamp) via apps or devices.	Bluetooth/BLE, Wi-Fi, NFC, IoT APIs.
Teacher node	Local validator	Verifies student attendance data, participates in consensus, and flags anomalies.	Web-based management backend, smart contract calls.
Administrative node	Regulatory arbiter	Sets attendance rules, audits cross-course disputes, and confirms block addition to the chain.	Blockchain management platform, cross-chain communication protocols.
Smart devices	Data collector	Captures biometrics (face/fingerprint), location signals (Beacon/GPS), and generates device signatures.	IoT protocols (MQTT/CoAP), digital certificate authentication (TLS).
Blockchain network	Data storage & consensus engine	Constructs blocks, executes consensus algorithms, and runs smart contracts.	P2P communication (LibP2P), smart contract virtual machines (EVM/wasm).

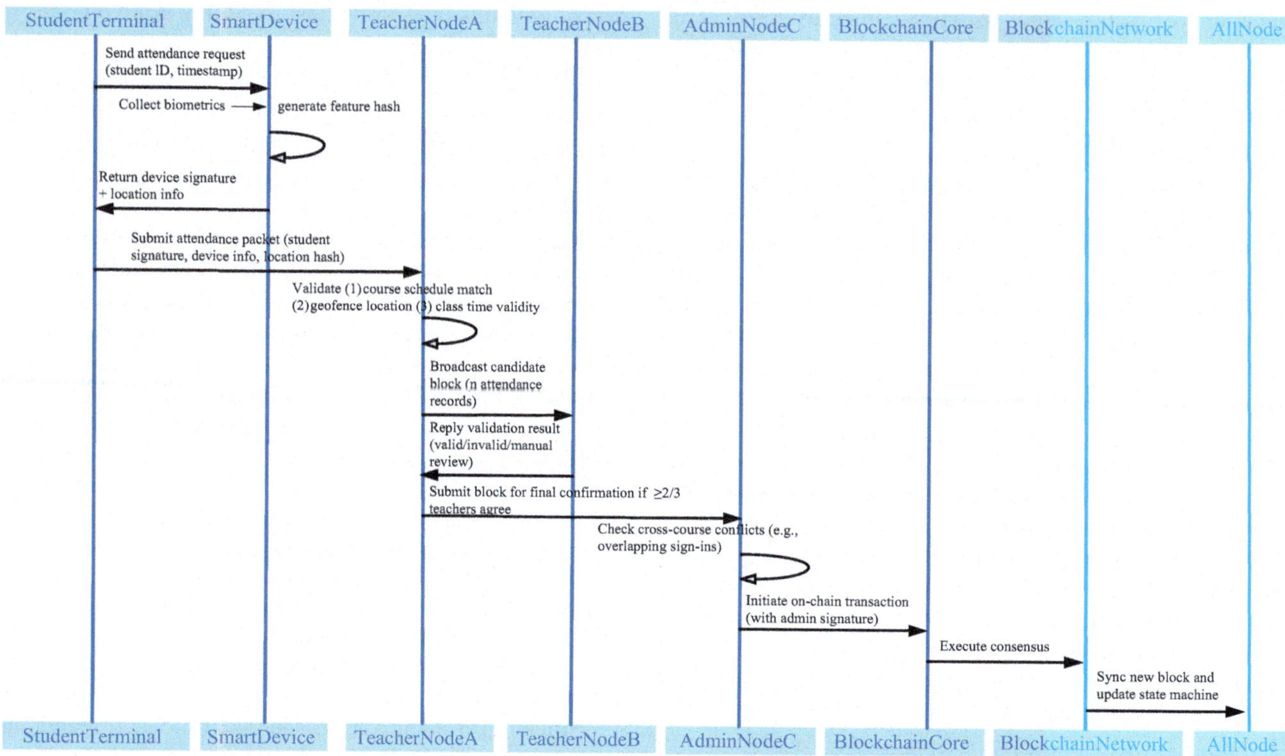

FIGURE 10.3 Step-by-step data flow from student-initiated attendance to blockchain synchronization.

Blockchain Network. Provides the foundational infrastructure for consensus and storage, using P2P communication and virtual machines to maintain a shared, immutable ledger.

The core data interaction flow is illustrated in Fig. 10.3. The diagram illustrates the step-by-step data flow from student-initiated attendance to blockchain synchronization.

The consensus mechanism for the blockchain-based student check-in system in smart classrooms unfolds in a systematic four-stage process, integrating data integrity, distributed validation, hierarchical oversight, and ledger synchronization.

First, in the data collection stage, students send check-in requests to smart devices. In response, these devices provide location information and device signatures, which serve as cryptographic proofs of authenticity. By embedding the device's unique identity into the data package, this process ensures that the source of check-in data is traceable and tamper-resistant, laying a reliable foundation for subsequent operations.

Second, the local validation phase involves teacher nodes acting as primary verifiers. Leveraging the Practical Byzantine Fault Tolerance (PBFT) algorithm, teachers cross -check the received data against predefined course schedules, geolocation records, and timestamps. This peer-to-peer validation model distributes the verification workload among multiple teacher nodes, minimizing reliance on a single authority. Through PBFT's quorum-based decision-making, the system can tolerate faulty or malicious nodes, ensuring that only accurate and consistent check-in data advances to the next stage.

The third stage, regulatory review, focuses on resolving discrepancies and cross-course conflicts. Disputed data or validated blocks flagged for potential issues, such as a student allegedly attending two classes simultaneously, are routed to administrative nodes. These nodes employ rule-based engines to analyze the data against institutional policies and cross-reference records across different courses. By identifying logical inconsistencies and enforcing access controls, the administrative layer acts as a safeguard, maintaining the overall integrity of the attendance records.

Finally, the on-chain finalization process solidifies the validated data onto the blockchain. After administrative approval, the consensus algorithm orchestrates the addition of new blocks to the blockchain. This process triggers a synchronized update across all network nodes, ensuring that the distributed ledger remains consistent and immutable. Through cryptographic hashing and consensus protocols, each node verifies the new block's integrity, thereby reinforcing the blockchain's core properties of transparency and trust.

In summary, this four-stage consensus mechanism integrates the strengths of blockchain technology with the specific requirements of educational management. By combining device-level authentication, peer-to -peer validation, hierarchical oversight, and distributed ledger technology, it creates a robust, efficient, and reliable system for managing student check-ins in smart classrooms.

4. Conclusion

Considering the existing problems in the current solutions for student check-in in smart classrooms, such as the vulnerability of data to falsification and tampering, this paper proposes a blockchain-based student check-in scheme for smart classrooms. The overall scheme is divided into four layers, namely the physical device layer, the identity verification layer, the blockchain security layer, and the cloud storage layer. The structure of data blocks in the blockchain is described. Specifically, the block header includes the hash value of the current block, the hash value of the previous block, etc., and the block body contains data such as students' basic information and check-in details. This paper also introduces the blockchain consensus mechanism and the data interaction mechanism in this scheme. This chapter provides a reference for constructing a student check-in scheme in smart classrooms where the data is difficult to be tampered with and easy to share, and also expands the relevant research.

References

Acadly, A. (2019). *Make your classroom smarter*. Retrieved Jan 31, 2019, from https://acadly.com/.

Ahmad, B. I., et al. (2014). Touchin: An nfc supported attendance system in a university environment. *International Journal of Information and Education Technology, 4*(5), 448.

Apoorv, R., & Mathur, P. (2016). Smart attendance management using bluetooth low energy and android. In *IEEE region 10 conference (TENCON)* (pp. 1048–1052). IEEE.

Arsenovic, M., Sladojevic, S., Anderla, A., & Stefanovic, D. (2017). Facetime - Learning based face recognition attendance system. In *IEEE 15th International symposium on intelligent systems and informatics* (pp. 000053–000058). IEEE.

Arulogun, O. T., Olatunbosun, A., Fakolujo, O. A., & Olaniyi, O. M. (2013). Rfid-based students attendance management system. *International Journal of Scientific Engineering and Research, 4*(2), 1–9.

Bae, M.-Y., & Cho, D.-J. (2015). Design and implementation of automatic attendance check system using ble beacon. *International Journal of Multimedia and Ubiquitous Engineering, 10*(10), 177–186.

Benyo, B., Sodor, B., Doktor, T., & Fördős, G. (2012). Student attendance monitoring at the university using nfc. In *Wireless telecommunications symposium* (pp. 1–5). IEEE.

Boon Chew, C., Mahinderjit-Singh, M., Wei, K. C., Sheng, T. W., Husin, M. H., & Malim, N. H. A. H. (2015). Sensors-enabled smart attendance systems using nfc and rfid technologies. *International Journal of New Computer Architechtures and Their Applications, 5*, 19–29.

Chang, C. H. (2011). Smart classroom roll caller system with iot architecture. In *Second International conference onInnovations in bio-inspired computing and applications* (pp. 356–360). IEEE.

Chen, P., Liu, X., Cheng, W., & Huang, R. (2017). A review of using augmented reality in education from 2011 to 2016. In *Innovations in smart learning* (pp. 13–18). Springer.

Cheng, Z., Caverlee, J., Lee, K., & Sui, D. Z. (2011). Exploring millions of footprints in location sharing services. In *Fifth International AAAI conference on weblogs and social media.*

Das, R. (2008). Nfc-enabled phones and contactless smart cards 2008–2018. *Card Technology Today, 20*(7–8), 11–13.

de Groot, M. (2002). Multimedia projectors: A key component in the classroom of the Future.(special report). *THE Journal (Technological Horizons In Education), 29*(11), 18.

Ervasti, M., Isomursu, M., & Kinnula, M. (2009). Bringing technology into school: Nfc-enabled school attendance supervision. In *Proceedings of the 8th international conference on mobile and ubiquitous multimedia* (p. 4). ACM.

Gligorić, N., Uzelac, A., & Krco, S. (2012). Smart classroom: Real-time feedback on lecture quality. In *International conference on pervasive computing and communications workshops* (pp. 391–394). IEEE.

Ichimura, T., & Kamada, S. (2013). Early discovery of chronic non-attenders by using nfc attendance management system. In *IEEE sixth international workshop on computational intelligence & applications* (pp. 191–196). IEEE.

Ihantola, P., Ahoniemi, T., Karavirta, V., & Seppälä, O. (2010). Review of recent systems for automatic assessment of programming assignments. In *Proceedings of the 10th Koli calling international conference on computing education research* (pp. 86–93). ACM.

Izzati Zainal, N., Sidek, K. A., Surya Gunawan, T., Manser, H., & Kartiwi, M. (2014). Design and development of portable classroom attendance system based on arduino and fingerprint biometric. In *The 5th international conference on information and communication technology for the Muslim world* (pp. 1–4). IEEE.

Kar, N., Debbarma, M. K., Saha, A., & Pal, D. R. (2012). Study of implementing automated attendance system using face recognition technique. *International Journal of computer and communication engineering, 1*(2), 100.

Kassem, A., Hamad, M., Chalhoub, Z., & El Dahdaah, S. (2010). An rfid attendance and monitoring system for university applications. In *17th IEEE international conference on electronics, circuits, and systems* (pp. 851–854). IEEE.

Kumar Saini, M., & Goel, N. (2019). How smart are smart classrooms? A review of smart classroom technologies. *ACM Computing Surveys, 52*(6), 1–28.

Lim, T. S., Sim, S. C., & Mansor, M. M. (2009). Rfid based attendance system. *In IEEE Symposium on Industrial Electronics &Applications, 2*, 778–782.

Lui, M., & Slotta, J. D. (2014). Immersive simulations for smart classrooms: Exploring evolutionary concepts in secondary science. Technology. *Pedagogy and Education, 23*(1), 57–80.

Mackensen, E., Lai, M., & Wendt, T. M. (2012). Bluetooth low energy (ble) based wireless sensors. *Sensors*, 1–4.

Marcos, J. L. F., Fernández, J. G., Aguilar, S. R., Salazar Selvi, B., & Crespo, R. G. (2013). Control of attendance applied in higher education through Mobile nfc technologies. *Expert Systems with Applications, 40*(11), 4478–4489.

Mohamed Basheer, K. P., & Raghu, C. V. (2012). Fingerprint attendance system for classroom needs. In *Annual IEEE India conference* (pp. 433–438). IEEE.

Noguchi, S., Niibori, M., Zhou, E., & Kamada, M. (2015). Student attendance management system with bluetooth low energy beacon and android devices. In *18th international conference on network-based information systems* (pp. 710–713). IEEE.

Patil, A., & Shukla, M. (2014). Implementation of classroom attendance system based on face recognition in class. *International Journal of Advances in Engineering & Technology, 7*(3), 974.

Qaiser, A., & Khan, S. A. (2006). Automation of time and attendance using rfid systems. In *International conference on emerging technologies* (pp. 60–63). IEEE.

Ramesh, G., Hariharan, B., & Venkat Rangan, P. (2016). Analysis of echo cancellation techniques in multi-perspective smart classroom. In *International conference on advances in computing, communications and informatics, pages 1135–1140*. IEEE.

Sengul, G., Karakaya, M., & Bostan, A. (2017). A smart classroom application: Monitoring and reporting attendance automatically using smart devices. *International Journal of Scientific Research in Information Systems and Engineering (IJSRISE), 3*(1), 39–43.

Shen, C.-wen, Wu, Y.-C. J., & Lee, T.-che (2014). Developing a nfc-equipped smart classroom: Effects on attitudes toward computer science. *Computers in Human Behavior, 30*, 731–738.

Silva, F., Filipe, V., & Pereira, A. (2008). Automatic control of students' attendance in classrooms usingrfid. In *3rd international conference on systems and networks communications* (pp. 384–389). IEEE.

Singh, S. K., Vaghela, K., Khanna, A., Virdee, B., & Kumar, M. (2025). SuRaksha: AI-Powered blockchain framework for HVAC tamper detection and authentication in smart classrooms. *IEEE Internet of Things Journal, 12*(13), 25568–25575. https://doi.org/10.1109/JIOT.2025.3560077

Suo, Y., Miyata, N., Morikawa, H., Ishida, T., & Shi, Y. (2009). Open smart classroom: Extensible andcalable learning system in smart space using web service technology. *IEEE Transactions on Knowledge and Data Engineering, 21*(6), 814–828.

Chapter 11

Electronic student records management through blockchain

Sohaib Latif
Department of Computer Science and Software Engineering, Grand Asian University, Sialkot, Pakistan

1. Introduction

The management of student records lies at the core of educational administration. With the growing complexity of academic data and the need for transparent, secure, and accessible recordkeeping, institutions face increasing pressure to modernize their systems. Blockchain technology, originally designed for financial transactions, offers a promising solution for revolutionizing student records management. This chapter explores how blockchain can transform the handling of academic records, ensuring security, integrity, and student empowerment.

In today's digital era, educational institutions generate and manage a vast amount of data associated with students' academic journeys. This includes enrollment records, grades, transcripts, degrees, disciplinary actions, and other personal academic information. The efficient management of such records is essential not only for institutional governance but also for enabling students to pursue further education and career opportunities. Traditionally, these records have been stored in centralized databases maintained by educational institutions, often in formats that are difficult to verify, share, or secure. These conventional systems are prone to issues such as data breaches, human error, unauthorized access, and administrative inefficiencies, all of which can undermine the integrity and usability of academic records.

The growing demand for transparency, accountability, and accessibility in education calls for innovative technologies that can address these challenges. Among the emerging technologies, blockchain has gained significant attention for its potential to transform data management practices across various industries, including education. Originally developed as the underlying infrastructure for cryptocurrencies such as Bitcoin, blockchain has evolved into a versatile solution for securely recording and verifying digital transactions and information. When applied to student records, blockchain provides a means of issuing verifiable, tamper-resistant digital credentials that students can share with potential employers, other institutions, or government agencies without reliance on intermediaries. Any stakeholder can independently authenticate these credentials through blockchain's transparent and cryptographically secure infrastructure. The shift to such a model is more than just technical. It represents a structural evolution in educational governance where learners gain greater control and autonomy over their academic narratives.

Blockchain technology introduces a decentralized and immutable ledger system where information is stored in blocks linked through cryptographic hashes. This structure ensures that once data is recorded, it cannot be altered or deleted without consensus from all participating nodes in the network. In the context of student records, this means academic credentials can be securely issued, stored, and shared in a way that is tamper-proof and verifiable. Moreover, blockchain can provide students with ownership over their educational data, allowing them to access and present their records without relying solely on institutional intermediaries.

This paradigm shift toward decentralized data ownership and verification aligns with global trends in digital identity, open credentials, and lifelong learning. It also opens new possibilities for cross-border recognition of academic qualifications, reducing fraud, and enhancing trust in digital learning environments. However, the integration of blockchain into educational recordkeeping systems is not without its challenges. Issues related to privacy, interoperability, regulatory compliance, and institutional adoption must be carefully considered to ensure the successful deployment of blockchain-based solutions.

Importantly, blockchain also holds promise in supporting lifelong learning ecosystems. In a world where

Digital Twins in the Smart Classroom. https://doi.org/10.1016/B978-0-443-41632-3.00021-7

TABLE 11.1 Traditional versus blockchain-based student records management.

Feature	Traditional systems	Blockchain-based systems
Data control	Centralized by institution	Decentralized across multiple nodes
Security	Vulnerable to tampering and breaches	Cryptographically secure and tamper-evident
Verification	Manual and time-consuming	Instant and automatic via smart contracts
Record ownership	Institution-owned	Student-owned (self-sovereign identity)
Transparency	Limited and opaque	Transparent and auditable
Interoperability	Often siloed and incompatible	Supports global standards and portability

education increasingly extends beyond formal institutions and occurs through micro-credentials, online courses, and alternative learning platforms, blockchain can provide a unifying framework for capturing and verifying these diverse learning experiences. Students can curate a personal ledger of learning, verified by issuers and universally accessible, thereby enhancing mobility, employability, and trust. Despite its promise, blockchain adoption in education is still at a formative stage. There are critical questions to address regarding data privacy, regulatory compliance, technical scalability, and user adoption. Educational institutions must also navigate the challenge of integrating blockchain with legacy systems, ensuring inclusivity, and maintaining institutional sovereignty over credential issuance (Table 11.1).

This chapter aims to explore the role of blockchain technology in managing electronic student records. It begins with a foundational overview of blockchain and its unique attributes compared to traditional data systems. It then examines the specific benefits of applying blockchain in educational contexts, supported by real-world case studies from pioneering institutions. The chapter also addresses key implementation challenges and outlines potential future directions for blockchain in education, ultimately highlighting how this technology could redefine the landscape of student records management (Grech & Camilleri, 2017).

2. Understanding blockchain technology

2.1 Foundations of blockchain

Blockchain is a distributed ledger technology (DLT) that enables the secure, transparent, and tamper-evident recording of data across a decentralized network. Unlike traditional systems where a central authority governs and verifies records, blockchain operates through a network of nodes—computers that collectively validate and record transactions through consensus mechanisms. Each entry in a blockchain is contained within a "block" of data, which is cryptographically linked to the preceding block, forming an unbroken "chain" that is immutable and time-stamped.

The origins of blockchain can be traced to the release of Bitcoin in 2008, where it served as the underlying infrastructure for decentralized digital currency. However, the underlying principles—immutability, decentralization, and consensus—have since been adapted for a wide range of applications beyond cryptocurrencies, including supply chain tracking, voting systems, digital identity, and educational records (Fig. 11.1).

2.2 Key characteristics

The defining attributes of blockchain technology include:

- **Decentralization**: No single entity controls the entire system. Data is distributed across multiple nodes, reducing the risk of a single point of failure and promoting resilience.

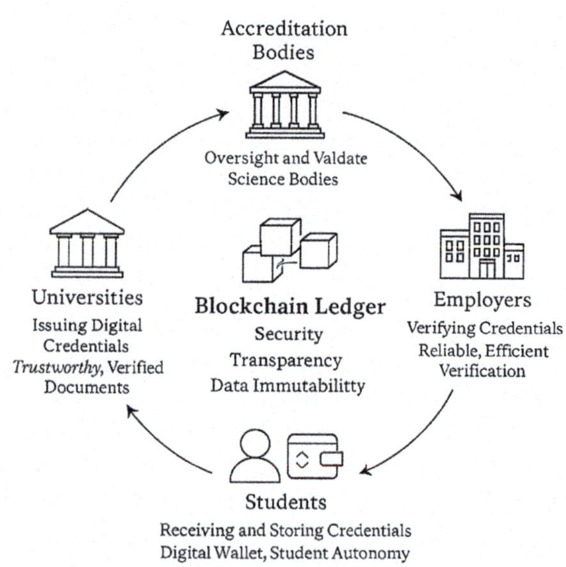

FIGURE 11.1 Electronic student record management system through blockchain.

- **Immutability**: Once recorded and confirmed, data on the blockchain cannot be altered or deleted without consensus from the network. This ensures data integrity and historical traceability.
- **Transparency**: Transactions on public blockchains are visible to all participants, promoting trust. Private or permissioned blockchains can restrict visibility to authorized users while maintaining auditability.
- **Security**: Blockchain leverages cryptographic techniques such as hashing and digital signatures to protect data from tampering and unauthorized access.
- **Consensus Mechanisms**: These are protocols used by the network to agree on the validity of transactions. Popular mechanisms include Proof of Work (PoW), Proof of Stake (PoS), and Practical Byzantine Fault Tolerance (PBFT), each with trade-offs in terms of speed, energy consumption, and scalability.

2.3 Blockchain versus traditional databases

To fully appreciate the potential of blockchain in education, it is essential to understand how it differs from conventional relational databases, which have long been the standard for storing academic records (Table 11.2).

While databases offer speed and flexibility in data manipulation, they are inherently less secure and transparent. Blockchain, on the other hand, sacrifices some operational speed and storage efficiency for enhanced security, traceability, and decentralization—key features in environments where trust, verification, and integrity are paramount.

2.4 Types of blockchains

There are several classifications of blockchains based on their accessibility and governance structures:

- **Public Blockchains**: These are open to anyone, and all transactions are visible to all participants (e.g., Bitcoin, Ethereum). They are highly decentralized but may raise privacy concerns in education.
- **Private Blockchains**: Operated by a single organization or consortium, access is restricted to approved participants. These are better suited for educational institutions needing data control and compliance.
- **Consortium Blockchains**: Semi-decentralized networks where multiple institutions collaboratively manage the blockchain. Ideal for regional or cross-institutional student records platforms.
- **Hybrid Blockchains**: Combines features of public and private blockchains, allowing controlled access to data while retaining transparency and auditability.

2.5 Smart contracts and automation

One of blockchain's most powerful features is the ability to incorporate **smart contracts**—self-executing programs that run on the blockchain when specific conditions are met. In the context of student records, smart contracts can automate processes such as:

- Issuance of certificates after course completion
- Verification of academic credentials
- Automated credit transfers between institutions
- Conditional scholarship disbursements based on academic performance

This programmability introduces new possibilities for automation and efficiency, reducing reliance on manual processing and minimizing human error or bias.

2.6 Blockchain interoperability and standards

For blockchain to achieve widespread adoption in education, interoperability across platforms and institutions is essential. Standards such as the W3C Verifiable Credentials and Decentralized Identifiers (DIDs) are being developed to support cross-platform data exchange and digital identity management. These standards aim to create

TABLE 11.2 Traditional versus blockchain-driven student databases.

Feature	Traditional databases	Blockchain
Control	Centralized (single authority)	Decentralized (peer-to-peer network)
Data structure	Tables and records (CRUD operations)	Linked blocks (append-only ledger)
Immutability	Data can be modified or deleted	Data is permanent and tamper-evident
Trust model	Requires central authority or admin role	Trust established through consensus
Data integrity	Integrity depends on access control policies	Integrity ensured through cryptographic hashing
Transparency	Restricted visibility	Public or permissioned transparency
Resilience	Vulnerable to central point of failure	Distributed and fault-tolerant

a shared ecosystem in which educational credentials are universally readable, verifiable, and portable across borders, institutions, and systems.

2.7 Educational relevance

The application of blockchain in education is not just about technology—it's about redefining the trust model. It shifts the authority from centralized registrars to a decentralized network where institutions, students, and external stakeholders can independently verify academic data. This is particularly relevant in a world where digital learning is expanding, student mobility is increasing, and academic fraud remains a pressing concern.

3. Benefits of blockchain in student records management

Blockchain technology brings a transformative shift in how educational institutions handle the storage, verification, and sharing of academic records. By addressing long-standing challenges associated with security, data ownership, administrative complexity, and cross-border interoperability, blockchain offers a secure, scalable, and student-centric framework for managing educational credentials. The following sections outline the key advantages of adopting blockchain-based systems in student records management.

3.1 Enhanced data security and protection against fraud

One of the most significant advantages of blockchain is its ability to secure data against unauthorized alterations and malicious attacks. Each record added to the blockchain is encrypted and time-stamped, and linked to the previous record through a cryptographic hash. This design makes it extremely difficult for any single actor to manipulate or forge data without detection.

In the context of education, this level of security is particularly valuable. Academic fraud—such as forged transcripts, falsified degrees, or unauthorized alterations of grades—has become a global concern, with implications for employers, licensing bodies, and academic institutions alike. By issuing verifiable digital credentials on a blockchain, institutions can ensure that records are authentic, tamper-proof, and resistant to counterfeiting.

Furthermore, access to student data can be protected using public-key cryptography and smart contracts, ensuring that only authorized parties—such as the student, educational institutions, or verified third-party reviewers—can view or interact with the records. This strengthens the protection of sensitive academic data while allowing controlled transparency when required.

3.2 Improved data integrity and auditability

Traditional student records systems often suffer from data inconsistency, especially when information is duplicated across multiple platforms or systems within and between institutions. Over time, manual entry errors and lack of real-time updates can degrade the quality and reliability of academic records.

Blockchain addresses this issue by providing a single source of truth. Since records on the blockchain are immutable, institutions and stakeholders can be confident that the data they are accessing is complete and unaltered. Any update or new record is appended chronologically and verifiably, ensuring traceability and historical accuracy.

Additionally, because all transactions on a blockchain are logged and timestamped, institutions can conduct complete and transparent audits without reliance on third-party software or administrators. This can significantly simplify compliance reporting, quality assurance processes, and accreditation reviews.

3.3 Student empowerment and ownership of credentials

Traditional models of student recordkeeping position educational institutions as the sole custodians of academic data, leaving students with limited access and control over their records. Accessing transcripts or credentials often requires lengthy processing times, administrative fees, and institutional approval, which can delay job applications, university admissions, or international mobility. Blockchain introduces a paradigm shift by enabling self-sovereign identity and decentralized credentialing. With blockchain, students can receive cryptographically signed digital credentials directly into their personal digital wallets. These credentials are portable, instantly accessible, and easily shareable with employers, institutions, or other verifiers without the need for intermediaries.

This autonomy empowers students to take charge of their educational narratives, supports non-traditional learning pathways (e.g., MOOCs, professional certifications), and enables lifelong learning across multiple institutions and borders. It also contributes to equity and accessibility, particularly for students in displaced or underserved communities who may face challenges in retrieving or proving their academic achievements.

3.4 Streamlined administrative processes and reduced operational costs

Managing academic records involves multiple administrative layers—issuance, storage, verification, updating, archiving—all of which require time, resources, and personnel. Manual processing of transcript requests, degree verifications, and credential authentication can introduce bottlenecks and human error.

Blockchain significantly reduces administrative burden by automating record verification and eliminating the need for manual validation. Smart contracts can execute predefined actions, such as issuing a certificate upon course completion or verifying a degree upon employer request, without human intervention. This not only speeds up processes but also reduces operational costs for institutions.

Furthermore, blockchain eliminates redundancies across departments and systems by providing a unified infrastructure for credential management. Inter-institutional collaborations, student transfers, and government audits can be handled more efficiently when all records exist in a verifiable, shared ecosystem.

3.5 Facilitation of cross-border academic mobility

In an increasingly globalized world, students are pursuing education across national borders, enrolling in programs from foreign universities, and accumulating credentials from multiple sources including online platforms. This rise in transnational education has introduced new complexities in credential verification and recognition. Blockchain offers a universal verification layer that transcends geographical and institutional boundaries. By using open standards such as Verifiable Credentials and Decentralized Identifiers (DIDs), blockchain systems can interoperate globally, allowing credentials to be instantly recognized and authenticated regardless of the country or issuing institution.

This has far-reaching implications for academic mobility, international admissions, and workforce migration. It streamlines the evaluation of foreign qualifications, accelerates the recruitment of international talent, and reduces reliance on credential evaluation agencies.

3.6 Real-time verification and transparency

With blockchain, credential verifications that once took days or weeks can now be performed in real time. Employers, educational institutions, or government agencies can instantly confirm the authenticity of a certificate simply by scanning a QR code or accessing a blockchain-based verification platform. This not only enhances efficiency but also builds trust in the hiring and admissions processes. Transparency is also improved. Blockchain's tamper-evident design ensures that any attempt to alter records is detectable, reducing the likelihood of disputes or misinformation. Students, institutions, and third parties can all participate in a transparent ecosystem where trust is algorithmically enforced.

As blockchain technology gains traction in education, several pioneering institutions have moved from theoretical exploration to real-world implementation. These case studies demonstrate the feasibility and impact of blockchain-based student records systems in diverse educational contexts. They also provide insights into the operational models, technological choices, benefits realized, and challenges encountered during implementation. The following examples highlight how institutions are leveraging blockchain to issue, manage, and verify academic credentials with increased efficiency and trust.

4. Case studies

4.1 University of Melbourne—Verifiable digital credentials

The University of Melbourne became one of the first universities globally to issue blockchain-based digital credentials to its graduates. In 2017, the institution piloted a system developed in collaboration with Learning Machine, a company specializing in blockchain credentialing solutions. The initiative enabled graduates to receive digitally signed academic certificates via a blockchain-secured platform compliant with the Blockcerts open standard (University of Melbourne, 2017).

The Blockcerts system allows students to store their credentials in a digital wallet and share them instantly with third parties. Verifiers such as employers or other universities can authenticate these credentials independently by checking the cryptographic hash on a public blockchain, eliminating the need for intermediary verification from the issuing institution.

Key Benefits Observed:

- Instant verification of credentials without institutional mediation
- Enhanced protection against diploma fraud
- Empowered graduates with portable, verifiable documents
- Reduced processing time for transcript requests

Challenges:

- Technical literacy required by users
- Need for institutional policy alignment on digital credentials
- Ensuring long-term access to and hosting of verification services

This case illustrates how even large and traditional universities can adapt blockchain technology to streamline services and enhance student autonomy in managing credentials.

4.2 MIT Media Lab—Digital diplomas on the blockchain (MIT Media Lab, 2017)

The Massachusetts Institute of Technology (MIT) made headlines in 2017 by issuing blockchain-based digital diplomas to a select group of graduates through its Media Lab. Collaborating with Learning Machine, MIT built a solution leveraging Bitcoin's blockchain and the Blockcerts standard to create a tamper-proof, self-verifiable record of graduation.

What set MIT's initiative apart was its focus on student ownership. Graduates received their digital diplomas in a mobile app that allowed them to present their credentials anywhere, anytime, without needing access to the university's systems. The system also promoted data privacy, as students controlled who could access their academic information.

Innovations Introduced:

- Decentralized digital identity for graduates
- Credential verification without contacting MIT
- Use of open-source tools and blockchain transparency

Limitations Encountered:

- Limited scalability in the initial pilot phase
- Challenges in integrating with broader university systems
- User adoption and education on digital credential use

Despite these limitations, the project demonstrated a robust proof-of-concept that has influenced other institutions and spurred interest in decentralized academic records.

4.3 Holberton School—Blockchain for lifelong learning records (Capetillo et al., 2022)

Holberton School, an innovative software engineering institution based in San Francisco, implemented blockchain to issue dynamic, updatable records of students' learning progress and achievements. Unlike traditional degrees, Holberton issues certificates that evolve with student performance and practical accomplishments.

Using blockchain, the school created a platform where both academic and professional milestones—such as project completions, peer reviews, and mentorship activities—were recorded and verified. This provided employers with a real-time view of student growth and enabled longitudinal learning analytics.

Highlights:

- Modular credentialing for project-based learning
- Granular tracking of skills development
- Real-time feedback loops for both students and employers

This case represents how blockchain can accommodate alternative education models that emphasize skill-building over traditional grading.

4.4 National government initiatives—Malta and Singapore (OpenCerts, 2019)

Blockchain-based student records have also gained interest at the national policy level. For instance:

- Malta: In partnership with Learning Machine, Malta's Ministry for Education launched a national blockchain credentials framework to issue secondary school certificates and higher education degrees. The goal was to eliminate credential fraud and simplify international recognition.
- Singapore: The government-backed OpenCerts project allows institutions to issue educational certificates on Ethereum. Supported by GovTech and SkillsFuture Singapore, OpenCerts has created a standardized, open-source framework that many of Singapore's universities and polytechnics now use.

Advantages of Government-Led Models:

- Consistency across institutions
- Scalability for national-level deployment
- Government endorsement enhances trust

Considerations:

- High infrastructure costs
- Need for legal and regulatory adjustments
- Centralized governance within a decentralized system

These initiatives demonstrate that blockchain in education is not limited to individual institutions—it can scale to national and cross-border credentialing systems when supported by policy frameworks.

4.5 European Blockchain Services Infrastructure (EBSI) (European Commission, 2018)

The European Union is currently piloting the European Blockchain Services Infrastructure (EBSI), a continental framework aimed at enabling trusted public services across member states. One of the use cases of EBSI is cross-border education credential verification.

Under the EBSI Diploma Use Case, universities across Europe will be able to issue and verify digital diplomas that are valid across all EU countries. The project promotes interoperability, GDPR compliance, and trusted identity frameworks, making it one of the most ambitious government-backed blockchain projects in education.

Potential Impact:

- Seamless student mobility across EU countries
- Fast, borderless credential recognition
- Alignment with EU data protection laws

This case shows how blockchain can be strategically deployed for systemic educational transformation across multiple nations (Table 11.3).

TABLE 11.3 Summary of case studies.

Institution/Country	Use case description	Technology used	Key benefits
University of Melbourne	Issued blockchain-verified diplomas	Blockcerts (bitcoin)	Verifiability, fraud prevention
MIT media lab	Digital diplomas with student wallets	Blockcerts (bitcoin)	Student autonomy, trust, portability
Holberton school	Project-based, evolving credentials	Private blockchain	Skills tracking, employer engagement
Malta ministry of education	National digital credentialing system	Permissioned blockchain	Government-scale trust and efficiency
Singapore (OpenCerts)	Multi-institution credential framework	Ethereum, OpenCerts	Open-source, scalable, government-backed
EBSI (EU initiative)	Cross-border credential recognition	Custom EU blockchain	Mobility, GDPR-compliant, regional scaling

5. Challenges and considerations

Despite its promising potential to revolutionize the management of student records, blockchain technology is not a silver bullet. Institutions seeking to adopt blockchain must navigate a complex ecosystem of technical, regulatory, ethical, financial, and organizational challenges. This section critically examines the key considerations that must be addressed to ensure responsible, effective, and sustainable implementation.

5.1 Data privacy and confidentiality

Blockchain's inherent transparency—often seen as a strength—can also raise significant privacy concerns, especially when sensitive academic or personal data is involved. Public blockchains, where all records are visible to anyone on the network, are not suitable for storing student records directly. Even pseudonymized data can potentially be deanonymized through correlation with other datasets.

In the context of regulations such as the General Data Protection Regulation (GDPR) in Europe and FERPA in the United States, blockchain faces a critical challenge: the right to be forgotten. The immutable nature of blockchain contradicts the legal requirement for individuals to have their personal data deleted upon request. While workarounds exist—such as storing only hashed references or metadata on-chain and keeping actual data off-chain—these solutions introduce added complexity.

Considerations:

- Use of permissioned or hybrid blockchains with access controls
- Encryption of student data combined with zero-knowledge proofs
- Development of GDPR-compliant "off-chain" data management practices
- Institutional policies governing data retention, access, and erasure

5.2 Legal and regulatory compliance

Education systems operate within national legal frameworks that govern data protection, accreditation, records retention, and credentialing. Implementing blockchain without aligning with these regulations can result in legal liability or non-compliance.

Furthermore, the **international nature** of student mobility raises questions about cross-border recognition and data jurisdiction. If a student stores their diploma on a blockchain node in one country but applies for a job in another, whose laws govern that data? These questions are not yet fully addressed by current legal frameworks, making regulatory clarity essential for large-scale adoption.

Challenges Include:

- Legal status of blockchain credentials as official academic records
- Admissibility of blockchain records in courts and legal proceedings
- National and international standards for issuing and verifying credentials
- Alignment with evolving data sovereignty laws

5.3 Technical complexity and infrastructure readiness

Deploying a blockchain-based credential system requires substantial technical infrastructure and expertise. Educational institutions—particularly those in developing regions or with limited IT capacity—may lack the technical resources to design, deploy, and maintain such systems.

Key infrastructure components include:

- Secure nodes for blockchain hosting
- Identity management frameworks (e.g., Decentralized Identifiers)
- Smart contract development and auditing
- Interfacing with legacy Student Information Systems (SIS)

Additionally, blockchain is still a rapidly evolving technology, and many existing platforms are not yet mature enough for mission-critical deployment in education. Interoperability between different blockchain implementations and credential formats remains a significant challenge.

Mitigation Strategies:

- Partnering with blockchain service providers or academic consortia
- Piloting small-scale projects before full-scale implementation
- Investing in professional development and capacity-building for IT teams

5.4 User adoption and change management

The success of any technological innovation depends heavily on the human factor. Blockchain introduces new paradigms for record ownership, verification, and sharing that may be unfamiliar or intimidating to students, faculty, and administrators alike.

Barriers to adoption include:

- Lack of awareness or understanding of how blockchain works
- Resistance to changing well-established institutional processes
- Concerns over control, data portability, and decentralization
- Usability challenges with digital wallets and credential platforms

Educational campaigns, user-friendly interfaces, and thoughtful change management strategies are crucial. Institutions must guide users through this transformation, ensuring that benefits are clearly communicated and that the technology serves human needs rather than simply replicating old systems in a new form.

5.5 Financial costs and sustainability

Blockchain implementation is often perceived as a high-cost endeavor—particularly when custom solutions are developed from scratch. Costs may include:

- Platform licensing or development
- Infrastructure deployment (nodes, cloud services)
- Staff training and capacity-building
- Integration with legacy systems
- Long-term maintenance and upgrades

While these investments can be offset by long-term administrative savings and reduced credential fraud, smaller institutions may struggle with the initial capital requirements. Moreover, sustainability becomes a question when funding is based on grants or pilot project models without long-term financial planning.

Recommendations:

- Consider open-source platforms and collaborative consortia
- Pursue government or donor support for initial implementation
- Evaluate return on investment (ROI) through pilot outcomes
- Develop sustainable funding models for maintenance and scaling

5.6 Standardization and interoperability

The lack of universal standards for blockchain-based credentials poses another barrier to widespread adoption. Different institutions or vendors may use different blockchain platforms, credential formats, and data schemas, which can fragment the ecosystem and undermine the goal of global interoperability.

Efforts such as the World Wide Web Consortium (W3C) Verifiable Credentials standard, Open Badges 3.0, and the European Blockchain Services Infrastructure (EBSI) aim to provide a foundation for harmonized credentialing. However, full adoption of these standards is still evolving (W3C, 2022).

Key Needs:

- Adoption of shared data schemas and credential metadata formats
- Cross-platform credential verification APIs and protocols
- Collaborative frameworks for mutual recognition of credentials
- Engagement with global standards bodies and education ministries

5.7 Ethical and philosophical implications

Beyond technical and legal challenges, blockchain raises important ethical questions. Who governs the blockchain? Who decides what data is stored and how long it remains? What happens if a blockchain network collapses or becomes obsolete?

Blockchain also introduces the risk of digital exclusion—not all students have equal access to the technology needed to manage their credentials. Institutions must take care not to exacerbate inequality through digital innovation.

Ethical Considerations Include:

- Ensuring inclusivity for underserved or marginalized populations
- Avoiding technological determinism in education policy
- Balancing transparency with personal privacy
- Safeguarding against vendor lock-in and technological monopolies

6. Future directions

As blockchain matures beyond its early applications in cryptocurrency and into sectors like education, the opportunities to reimagine student records management continue to grow. The future of blockchain in this domain is shaped not only by technological innovation but also by evolving user expectations, regulatory landscapes, and global efforts toward educational inclusion and interoperability. This section explores key areas where blockchain adoption is expected to expand and improve, and outlines strategic directions for institutions, governments, and technology providers.

6.1 Toward interoperable and standardized credential ecosystems

A major barrier to global adoption of blockchain in education is the lack of interoperability between different credentialing systems. To truly unlock the benefits of blockchain for student mobility and international credential recognition, standardized frameworks must emerge that allow diverse institutions, platforms, and countries to issue and verify credentials seamlessly.

Organizations such as:

- W3C (World Wide Web Consortium) with its Verifiable Credentials specification,
- IEEE Learning Technology Standards Committee, and
- The Digital Credentials Consortium (DCC)

are actively working to define standards for verifiable digital records. Additionally, regional initiatives like the European Blockchain Services Infrastructure (EBSI) and OpenBadges 3.0 are leading the way in developing universal formats for academic credentials.

Future Direction:

- Development of globally accepted credentialing protocols
- Blockchain networks supporting cross-border credit transfers
- Adoption of decentralized identifiers (DIDs) for universal learner IDs

6.2 Enhanced privacy through advanced cryptography

While blockchain's transparency is a benefit, privacy concerns remain a central obstacle. Future systems will increasingly rely on advanced cryptographic methods such as:

- **Zero-Knowledge Proofs (ZKPs):** Allow users to prove the validity of a credential without revealing the content.
- **Homomorphic Encryption:** Enables computations on encrypted data, preserving privacy while allowing verification.
- **Selective Disclosure Protocols:** Give students the ability to share only certain data fields with specific parties.

These privacy-preserving technologies will be key in making blockchain suitable for large-scale, legally compliant educational use.

Future Direction:

- Privacy-first credential wallets integrated with blockchain backends
- Institutional ability to define data access levels in dynamic contexts
- Conformance with global data protection laws like GDPR and FERPA

6.3 Integration with artificial intelligence and learning analytics

The integration of blockchain with AI and learning analytics could enable personalized and adaptive education on an unprecedented scale. Verified blockchain-based credentials can feed into intelligent systems that track a student's progress across institutions and platforms.

Imagine a future where a student's verified skills and achievements are used to:

- Recommend career pathways,
- Suggest online courses,
- Tailor scholarship offers,
- Or assess soft skills based on accumulated digital badges and micro-credentials.

AI can analyze learning trajectories stored on blockchain to support evidence-based interventions and holistic learner profiles.

Future Direction:

- Blockchain-powered lifelong learning portfolios
- Smart academic advisors built on verified credential data
- AI-driven scholarship and admissions automation using trusted records

6.4 Cross-institutional and cross-sector collaboration

As education becomes more modular and lifelong, credentials will no longer be issued solely by traditional universities. Industry bodies, online learning platforms, nonprofits, and vocational institutions are now significant players in the credentialing space.

Blockchain can act as a neutral verification layer where all these actors can contribute to a single, trustworthy learner record. This will support stackable credentials and multi-source learning recognition.

Future Direction:

- Unified learner passports that include degrees, badges, licenses, and certifications
- Blockchain-enabled recognition of prior learning (RPL)
- Employer-issued credentials verified on educational blockchains

6.5 Decentralized learning marketplaces and smart contracts

The future of blockchain in education could extend beyond recordkeeping to governance and operations. Smart contracts can support automated and decentralized education marketplaces, where students:

- Enroll in courses,
- Pay tuition,
- Access learning materials,
- Earn credentials

and, all governed by transparent, code-based agreements on the blockchain.

Such systems could reduce dependence on centralized education providers and enable peer-to-peer learning ecosystems, especially in underserved regions or informal education settings.

Future Direction:

- Blockchain-powered course marketplaces with verifiable ratings and outcomes
- Token-based rewards for peer mentors and educators
- Smart contract-managed tuition and micro-payments for learning access

6.6 Designing for equity, inclusion, and accessibility

While blockchain promises decentralization and autonomy, it must not become a tool of exclusion. Future implementations must prioritize:

- **Digital accessibility** for students with disabilities
- **Mobile-first designs** for remote and under-resourced learners
- **Multilingual interfaces** and localized platforms
- **Gender-sensitive and culturally-aware data policies**

Equity must be embedded in both the technical design and institutional governance models.

Future Direction:

- Human-centered credential wallets and interfaces
- Blockchain credentials that support refugee and displaced learners
- Inclusive frameworks for credential recognition in diverse socio-political contexts

6.7 Sustainable and scalable blockchain architectures

As blockchain scales, environmental and performance concerns will become more prominent. Energy-intensive consensus mechanisms (e.g., Proof of Work) are increasingly being replaced by energy-efficient alternatives like:

- Proof of Stake (PoS)
- Delegated Proof of Stake (DPoS)
- Consortium-based consensus algorithms

Moreover, layer-2 solutions and sidechains are emerging to improve transaction throughput and reduce costs.

Future Direction:

- Eco-friendly blockchain implementations for educational systems
- Institutional adoption of shared blockchain networks for scalability
- Use of cloud-based blockchain-as-a-service (BaaS) platforms to reduce infrastructure overhead

7. Conclusion

The management of student records is a fundamental pillar of educational administration, influencing everything from student mobility to institutional reputation, from employer trust to public policy. As the world becomes increasingly digital, global, and data-driven, traditional systems of managing academic records are proving inadequate to meet the evolving demands of transparency, security, and learner autonomy.

This chapter has explored blockchain technology as a transformative tool for addressing the long-standing limitations of conventional student records management systems. By decentralizing control, enhancing security, ensuring data integrity, and enabling verifiable digital credentials, blockchain introduces a new paradigm for academic data governance—one that places students at the center of the system.

The chapter began by establishing the conceptual foundation of blockchain technology, emphasizing its core characteristics such as immutability, decentralization, and consensus-based validation. It then explored the specific benefits of blockchain in the educational context, including enhanced data security, fraud resistance, administrative efficiency, and student empowerment through ownership of their credentials.

Real-world case studies from institutions such as the University of Melbourne, MIT Media Lab, and government initiatives in Malta, Singapore, and the European Union demonstrated the viability and impact of blockchain implementations in diverse educational ecosystems. These examples underscored the versatility of blockchain—ranging from issuing tamper-proof diplomas to enabling cross-border credential recognition.

Yet, as with any transformative innovation, blockchain adoption comes with substantial challenges. Issues of privacy, regulatory compliance, infrastructure, interoperability, and user adoption must be addressed through careful planning and collaborative problem-solving. The immutability of blockchain, while advantageous for trust, requires new models of data governance that align with ethical and legal norms, particularly regarding data protection and the right to be forgotten.

Looking ahead, the future of blockchain in education is bright but contingent on several enabling factors. Advancements in privacy-preserving technologies, development of global credentialing standards, integration with AI and learning analytics, and a commitment to equitable access will shape the next phase of blockchain adoption. If these trajectories are pursued with foresight and inclusivity, blockchain has the potential to become a cornerstone of a more transparent, interoperable, and learner-centered educational infrastructure.

Ultimately, the integration of blockchain into student records management is not just a technical innovation—it represents a philosophical shift in how we define trust, ownership, and identity in education. It is a move toward systems that are not only more secure and efficient but also more just and empowering for learners around the world. As educational institutions, governments, and technology providers continue to explore the possibilities of blockchain, it is essential to keep the learner's experience, rights, and aspirations at the heart of every design and deployment. Only then can we build an academic records ecosystem that is truly future-ready, equitable, and trustworthy.

References

Capetillo, A., Camacho, D., & Alanis, M. (2022). Blockchained education: Challenging the long-standing model of academic institutions. *International Journal on Interactive Design and Manufacturing (IJIDeM),* 16(2), 791–802.

European Commission. (2018). European blockchain services infrastructure (EBSI). https://ec.europa.eu/digital-building-blocks/sites/spaces/EBSI/pages/447687044/Home.

Grech, A., & Camilleri, A. F. (2017). *Blockchain in education.* Joint Research Centre, European Commission.

MIT Media Lab. (2017). n. *Digital diplomas using blockchai.* https://www.media.mit.edu/projects/digital-diplomas/overview/.

OpenCerts. (2019). *Blockchain for education in Singapore.* Singapore: GovTech.

University of Melbourne. (2017). University of Melbourne to issue recipient-owned blockchain records. https://www.unimelb.edu.au/newsroom/news/2017/october/university-of-melbourne-to-issue-recipient-owned-blockchain-records.

W3C. (2022). *Verifiable credentials data model v1.1.* https://www.w3.org/TR/vc-data-model-1.1/.

Chapter 12

Digital twin application in school facilities management

Sanjeev Salunke[1], Swagatika Nanda[2], Shabnam Kumari[3] and Amit Kumar Tyagi[4]

[1]*Soundarya Group of Institutions, Bengaluru, Karnataka, India;* [2]*Narayana Group of Institutions, Bhubaneswar, Odisha, India;* [3]*SRM Institute of Science and Technology, Chennai, Tamil Nadu, India;* [4]*National Forensic Sciences University, Gandhinagar, Gujarat, India*

1. Introduction

The quality of educational infrastructure is the most important factor in determining the students' experiences of learning and development. The physical facilities where education occurs includes educational structures such as classrooms, laboratories, libraries, administration buildings, stadiums, and utility facilities (Slot et al., 2022). Efficient facility management ensures the safety and comfort of students and teachers while also contributing to academic success and the reputation of the institution (Nassehi et al., 2022). However, the management of a school's facility is both complex, and dynamic, requiring continual oversight, timely upkeep, maximum energy performance, and space control. As student enrollment and educational institutions increase, the need for more intelligent and responsive facilities practices has dramatically increased (Abele et al., 2017).

Conventional school facilities management processes are predominantly reactive in nature. These processes are generally based on manual inspection, paper records, isolated maintenance schedules, and static building plans (Integration of Learning and Research). Hence, they are likely to be inefficient, prone to human error, slow to respond to facilities problems, and have high operating costs. For example, the absence of real-time monitoring devices can lead to equipment failure to be unnoticed, high energy consumption, and safety risks, all of which hinder the learning process and contribute to costs of maintenance (Damgrave & Lutters, 2019). Moreover, school space utilization is never fully maximized, with classrooms going underused and common areas becoming congested. A worse effect has been seen due to aging assets and tight budgets, which thereby promote managers to seek innovative solutions that can provide greater performance, economic efficiency, and long-term sustainability (Lutters, 2018).

Over the last few years, with the advent of smart technologies, new solutions have been brought forth to manage educational institutions' facilities more effectively. Among these, the idea of the DT has gained popularity in most industries, such as manufacturing, healthcare, transportation, and city planning (Slot & Lutters, 2021). A Digital Twin is a precise, real-time digital representation of a physical asset, system, or process. It utilizes sensor data and IoT sensors, as well as building information modeling (BIM) solutions, to produce an information context to have a virtual environment of real-world objects (Kreimeier et al., 2014). By examining data and visualizing with a real-time continuous data stream, a Digital Twin provides a condensed yet detailed picture of operational states of assets and facilities, helps simulate scenarios and predict issues, and makes data-informed decisions (Ning et al., 2024). Fig. 12.1 illustrates the various benefits, issues, implementation methods and challenges in the integration of DT with school management facilities.

Deployment of Digital Twin technology within school facility management has the power to revolutionize. Real-time monitoring, forecasting maintenance, and intelligent resource allocation using Digital Twin technology can significantly make schools operationally more efficient, conserve energy, extend infrastructure lifespan, and ensure safe and comfortable working environment for their inhabitants (Bamunuarachchi et al., 2021). For example, a Digital Twin model can monitor temperature, humidity, and occupancy and optimize HVAC functions and indoor air quality. In a similar manner, it can monitor classroom and other facility use to help inform space scheduling and assignment. By facilitating data integration and control at the center, Digital Twins can enable a more integrated and

Digital Twins in the Smart Classroom. https://doi.org/10.1016/B978-0-443-41632-3.00013-8

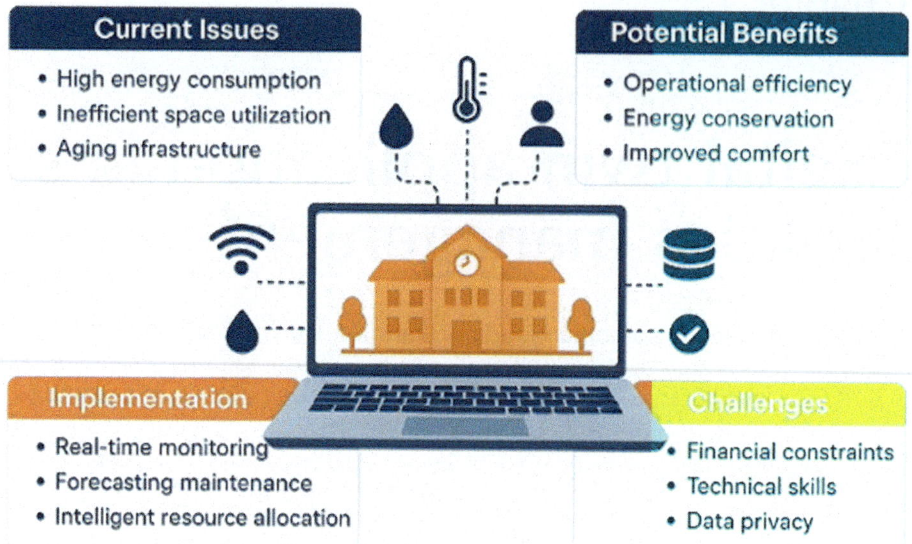

FIGURE 12.1 Incorporation of digital twin in school facility management.

responsive management of school infrastructure (Kamali et al., 2024).

Whilst hopeful, the introduction of Digital Twin technology into schools is still relatively in its infancy and yet to be comprehensively understood. Virtually all usage is currently limited to industrial or inner city environments on a large scale, and comparatively few have examined the very particular needs and limitations of school environments (Ullah, 2022). Deploying Digital Twin systems in schools presents special challenges, including financial limitations, insufficient technical skills, data privacy and protection, but with the increased focus on smart campuses and sustainable education facilities, it is more important than ever to explore how Digital Twins can be implemented and leveraged within school environments (Leahy & Phelan, 2014).

This research has three main objectives: to identify the current issues and limitations of traditional school building facility management practices, to propose a Digital Twin-based framework to serve the unique needs of schools, and to discuss the potential benefits, deployment techniques, and pitfalls in implementing such systems in real-world school environments.

The main goals of this study are threefold: first, to examine the existing limitations and problems of conventional school facility management procedures; second, to propose a Digital Twin-based model which is designed to meet the specific requirements of schools; and third, to examine the potential advantages, deployment strategies, and challenges related to such deployment within real-world school settings. For this purpose, the following research questions are answered in this research work:

- What are the most significant operational and maintenance issues for schools in maintaining their buildings?
- How can Digital Twin technology best be implemented in schools, given technical, financial, and organizational limitations?
- What are the anticipated benefits and potential drawbacks of implementing a Digital Twin system in school building management?
- What are the possibilities of enhancing real-time monitoring, decision-making, and sustainability of school operations using the integration of Digital Twin and IoT and BIM technologies?

The value added by this paper is its interdisciplinary nature bringing together the fields of educational infrastructure, digital innovation, and intelligent facility management. By intermixing known facts and formulating a usable framework, the research hopes to inform policymakers, school leaders, and technology developers on how to harness Digital Twin solutions in creating safer, more efficient, and future-proof learning spaces.

2. Literature review

Efficient facilities management within schools is paramount to providing an effective learning environment, reducing operating expenses, and improving the overall performance of school infrastructure (Deng, 2023). Over the past decade, the embedding of digital technologies has introduced revolutionary changes in the built environment, especially through advances like BIM, the IoT, and more recently, DT technology (Haleem et al., 2022). In schools, facilities management encompasses a range of activities including planning for repairs, planning space, tracking energy consumption, tracking assets, tracking security, and safety regulation compliance (Semeraro et al., 2022). Failure to maintain school buildings can have adverse impacts on community image, teacher morale, and student achievement (Chen et al., 2024). Poorly functioning school facilities can negatively impact community image, teacher morale, and student achievement. Manual procedures, compartmentalized information systems, and regular inspections are the pillars of conventional school facilities management (Jayasinghe et al., 2024). These practices often lead to unused spaces, unexpected equipment failures, and wasteful energy consumption. According to recent reports, schools are being subjected to more and more pressure to bring buildings up-to-date in the face of dated buildings, swelling student rolls, and sustainability expectations (Dembski et al., 2020). Due to funding deficiencies and a deficiency of qualified people, most schools still struggle to make use of digital technologies while the commercial property and manufacturing communities have sophisticated answers (Zhang et al., 2023). The Digital Twin concept was conceived in the manufacturing sector but developed quickly to cut across various applications like healthcare, transportation, and built environments. In construction and smart buildings contexts (Billey & Wuest, 2024), Digital Twin is a handy tool for testing building performance simulation, tracking energy usage, and forecasting system malfunction in real-time. A Digital Twin combines BIM, data from IoT sensors, AI, and cloud infrastructure to offer real-time, usable models of physical assets (Sim et al., 2024).

A virtual representation that stands for a physical counterpart or process in real time was deployed that exists as a mimic in any science (attributes) of good dimensions. Digital Twin in Construction- lifecycle management has been enhanced by the construction sector as these as-built models sync with real-time operational data (Huang et al., 2022). In commercial buildings DT solutions were implemented to greatly improve HVAC systems and cut energy costs by 25%. A similar example is how IBM Maximo platform exhibits the technology of DT in asset management combining AI-based analytics with sensor data (Wang et al., 2023). Digital Twin is still in its early stages for educational purposes. Indeed, some footsteps are being made to explore its potential with the early-scorers studies (Saleh et al., 2022). A Digital Twin framework for management of university campus was also proposed for emphasizing on energy efficiency, security as well scheduling in classrooms (Hooshyar et al., 2024). Research efforts have also been directed towards linking DT with virtual learning environments or simulating crowd behavior during school emergency evacuations (Yang et al., 2024). The existing studies suggest that DT is established in different levels of practice but also suggest that different models for primary- and secondary education are lacking (Martínez-Gutié et al., 2023). To summarize the change agent ability of Digital Twin technology in educational facilities management, the extremes and degrees along the semi-digital to fully digital scales are to be compared with traditional approaches too. Table 12.1 Features comparison in traditional, BIM and Digital Twin enabled systems.

The comparison shows that though traditional systems are manual with experience, BIM gives a visualization/simulation bandwagon. Whilst the Digital Twin unifies time-series-data with predictive and interactive management capabilities to provide a much more complete solution for complex and dynamic environments, Schools.

TABLE 12.1 Comparative analysis of facilities management approaches.

Feature	Traditional management	BIM-based management	Digital twin-enabled management
Data collection	Manual (paper-based)	Static 3D models	Real-time sensor data
Decision-making	Experience-based	Model-aided	AI/Analytics-driven
Maintenance planning	Reactive	Semi-predictive	Predictive/Proactive
Energy management	Periodic reviews	Modeled scenarios	Live optimization via sensors
Space utilization	Manual observations	Floorplan simulations	Dynamic occupancy tracking
Emergency response	Evacuation plans	Simulated scenarios	Real-time simulations and crowd analytics
User interaction	Limited	Visual model interfaces	Immersive dashboards and control panels
Integration with IoT	Absent	Limited	High (real-time streaming data)
Implementation cost	Low	Medium	Medium to high (initial setup, long-term ROI)

Though the advantages of Digital Twin in smart building are well acknowledged, there is a strong research gap found there implementation for school facility. Most of the DT research done so far deal with either commercial or industrial buildings than residential/office with abundant operational budget and skilled workforce. Schools (especially the developing country ones) have funding, infrastructure and IT limitations. The domain of facilities management remains relatively unexplored in educational research; most studies concentrate on classroom learning technologies or administrative systems. In addition, it is to convey that there are no standardized frameworks available to support Digital Twin of educational institutes in its implementation. Very little has been done to scale down or modularize the work based upon DT in small scale school. Furthermore, the socio-technical dimensions of staff training for use of these systems and privacy concerns are hardly addressed in literature.

From the existing state-of-the-review, several vital findings can be distilled. One, School facilities management remains very much an innovator space, as many of its headquarter still use archaic processes which do not comply the new paradigm. Second, Digital Twin, a technology that has been successful in other sectors could change the way schools operate their infrastructure and offer cost-savings and sustainability benefit to all stakeholders, as well user safety. Second, the infraction that the Digital Twin not be sufficiently intellectually developed for school environments gains pointer attention from academia and practice. A handful of studies are pilots, less in its application (usually university campuses) mostly with very narrow solutions (like energy monitoring only. Comprehensive models are critical at this time as this will provide for unique operational, technical and financial profile of early- and secondary schools. These include IoT integration for real-time monitoring, BIM digital modeling and cloud platforms for scale and remote access.

3. Digital twin technology framework and concept

Smart infrastructure and Industry 4.0 have digital twinning of assets becoming a backbone as well to manage physical objects in real time through data, virtual simulation and predictive analytics. Digital Twin for school facilities management: To opine, the Digital Twin has the potential to be a paradigm-shifting tool by bringing the physical and digital worlds together to advance monitoring, maintaining and optimizing of educational environments (Dong et al., 2022). A Digital Twin (formally Talking Devices) A real-time virtual copy of a physical entity that is refreshed via data stream referred at its real-world prototype. Originally conceived by NASA for simulation of space systems, DT as far back as 5 decades ago has now progressed into an integral tool in manufacturing, smart cities and building automation (Delmia, 2024). Table 12.2 presents the core components of DT (Grieves & Vickers, 2016, pp. 85–113):

Together these components allow for real-time synchronizations, which allow facility managers to run scenarios, identify mismatches and build a sustainable operation on the building. Digital Twin Technology can be separated into Product Single Digital Twin (of a physical object), Process Duplication or Process Twin (which represents reproduction of a physical process or workflow) and System Twin, complete school facility System integration to model. Few are more relevant for school facility management than these types, as all buildings interact in complex ways with energy systems and occupancy patterns as well services (Barricelli et al., 2019). The performance of Digital Twin depends on Data collection and Integration in real-time. Modern DT systems use various technologies so as to create a complete figure of the physical environment in a real-time manner. IoT sensors are embedded in various locations of the school in order to keep an eye on temperature, humidity, light intensity, carbon dioxide concentration and occupancy in the room. The sensors in this regard are eyes and ears of Digital Twin, pushing continuous data into the system for everything they see (Sharma et al., 2022). The virtual blueprint of the school infrastructure rests on BIM. Allows for building accurate and dimensional 3D layouts, and structural elements as well as utilities when coupled with IoT BIM becomes a "live" model, not the static blueprint (Fuller et al., 2020).

TABLE 12.2 DT core components.

Component	Description
Physical entity	Real-world asset or system e.g., school building, room in classroom, HVAC unit itself
Digital model	Virtual representation made using tools either BIM or CAD which is a physical thing.
Sensor network	IoT stretched out to get real-time data devices, temperature occupancy-energy usage
Data integration layer	Middleware or APIs that aggregate the available data from multiple sources
Analytics engine	AI/ML, predictive maintenance, anomaly detection, optimization algorithms
User interface (UI)	Facility managers can see and interact with the system using dashboards, or mobile apps.

The data from different subsystems including energy meters, security and HVAC units goes through middleware layer. Middleware and as schools scale their digital infrastructure in integration of new data sources, technology APIs allow for easy integration (Pan et al., 2020). Cloud platforms offer scalable storage and compute for Big data volumes, while edge computing enables real-time deep learning/predictive decision making on the nearest data source and speeds up your solution reducing latency (Qi & Tao, 2018). Fig. 12.2 illustrates the components of DT and its integration with various technologies.

DT is basically a five-layer architecture which includes physical layer, data acquisition layer, data integration layer, digital model layer, application and analytics layer and presentation layer respectively. Each of them has been discussed in detail.

3.1 Physical layer

The Digital Twin system is founded on a physical layer comprising all physical constituents of a school setting, including classrooms, labs, HVAC, lighting, plumbing, utilities, safety systems, and moveable furniture. These are augmented with IoT sensors and smart actuators, enabling real-time operational information to be transmitted to higher layers (Qi et al., 2019). The layer enables dynamic replication of physical space in digital equivalents, creating the basis for improved analytics and simulation.

3.2 Data acquisition layer

The data acquisition layer, in the upper portion above the physical layer, is the sensory nervous system of the Digital Twin structure. It acquires data from numerous sources, such as temperature and humidity sensors, RFID tags, surveillance cameras, and SCADA systems. The layer maintains real-time data streams from physical assets to digital infrastructure, which ensures the Digital Twin is a precise reflection of the physical state and operational condition of the school (He et al., 2022).

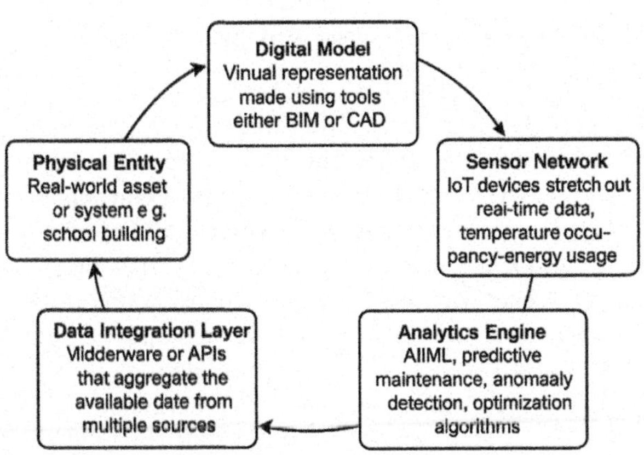

FIGURE 12.2 Core DT components and integration with different technologies.

3.3 Data integration layer

The data integration layer processes and integrates raw data to actionable information employing cloud computing systems, middleware, and APIs. It processes streaming data in real-time and archives long-term storage of historical data for trend monitoring and predictive purposes. It is in charge of interoperability of data sources and systems, enforcing data security, validation, and redundancy, as well as preventing only validated information from being relayed to digital models for simulation and visualization purposes (Wang et al., 2021).

3.4 Digital model layer

The Digital Twin architecture is comprised of a digital model layer that replicates the physical infrastructure of a school through BIM, CAD diagrams, and 3D modeling tools. This layer continuously updates the model based on variations in classroom occupancy, energy consumption, or system malfunctions. This enables facilities managers to act immediately on problems such as clogged ventilation ducts. The layer also supports users in simulating operating conditions, tracking asset health, and scheduling interventions with accuracy (Tao et al., 2020).

3.5 Application and analytics layer

The analytics and application layer of Digital Twin solutions utilizes advanced tools like AI, machine learning, and statistical analysis to provide insights and drive decisions. It uses predictive maintenance algorithms to predict equipment failure, energy optimization software to suggest adjustments, and can model emergency scenarios (Cheng et al., 2019). The layer supports user interfaces, such as web-based or mobile-optimized dashboards, that present real-time data, trends, notifications, and suggested actions in a convenient-to-access format.

3.6 Presentation layer

The presentation layer, which is the highest layer, provides human users with dashboards and mobile apps, kiosks, and alarms. The presentation layer provides feedback and information on energy and space use, air quality, and system alerts. A school may provide the facility manager with score building health, security alerts with instruction for the principal and administrative staff, and maintenance reports for all. It aids in making decision-making, maximizes resource usage, and provides safe and comfortable learning environments (Jeong et al., 2022).

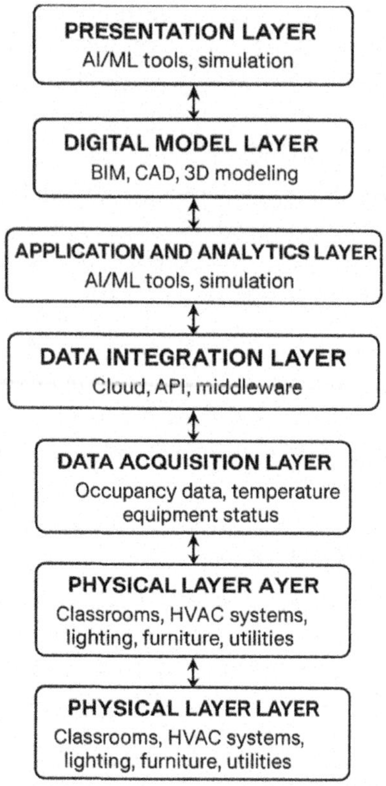

FIGURE 12.3 A 6-layer DT architecture.

The DT six layer architecture has been visualized using Fig. 12.3 as follows.

Digital Twin technology is another paradigm leap in school facility management by enabling a living, data-enabled space of operations that drives operational excellence, safety and sustainability. The Combined elements of digital modeling, IoT, analytics and visualization let us have immediate intros and predictive power. Different on understanding the different flavors of Digital Twins and using correct options and platforms, architecture it will go a long way for schools to operate across Educational and Operational excellence. The remainder of this paper will dive into actual examples and use cases for how this conceptual framework can work in a real school environment.

4. Application of digital twin in school facilities management

Beyond visualization DT in school facilities management entails wide ambit of applications. DT uses real time data, advanced simulation and predictive analytics to empower educational institutions to intelligently operate their built environments.

4.1 Energy efficiency and optimization

With these budgetary constraints, schools should be implementing effective energy management. HVAC, lighting and ICT infrastructure are large consumers of energy. Digital Twin can track these systems, match real-time usage with historical patterns. IoT sensors within the HVAC systems can measure occupancy, weather and schedules to either heat or cool. Machine learning algorithms predict peak and off-peak usage times, keeping adjustments to a system comfortable and automated. For example, a pilot implementation in Finland where energy consumption reduced by 20% (Qi et al., 2018).

4.2 Predictive and preventive maintenance

Digital Twin technology is disrupting the School Maintenance with predictive maintenance enabled applications. IoT sensors that track building systems in real-time, sending alerts to maintenance managers before failures. The platform measures vibration, temperature and consumption data of a machines as well as adopting AI algorithms to predict the trends of degradation. This gets all the maintenance scheduled right which thereby helps in reducing downtime, save repair cost and extend equipment uptime. Benefits include increased availability, decreased repair expense and equipment longevity (Wei et al., 2021).

4.3 Space and occupancy management

Motion sensors, RFID cards or Wi-Fi analytics can be used to track occupancy of classrooms, libraries and laboratories in real time for a Digital Twin occupancy. This gives schools the ability to identify underutilized spaces, update floor plans for efficient operation and also clean/vent based on occupancy. Uses cases: One such Digital Twin in a smart high school in South Korea used to model different transient COVID-19 related scenarios for classroom densities and social distancing (Attaran & Celik, 2023).

4.4 Emergency management and safety

Real-time decision making for emergencies (fires, earthquakes, lockdowns) Integration like CCTV (closed circuit TV) feeds; environmental sensors; access control systems Dynamic DT (Digital Twin), as an enabler, can support one via the process of feeding data such as from alarms while providing CCTV feed. It can direct evacuation plans, identify threats with the simulation maps, deline routes for safe departure, sends alarms and activates emergency personnel for DT simulations. DT simulations teach better school staff and student responses to emergencies than just the training of static drills for school staff (Fu et al., 2022).

4.5 Indoor environmental quality

Indoor air quality is getting worse with lowered academic performance and health problems. The Digital Twin that connects to CO_2, temperature and humidity sensors synchronizes HVAC automatically for the right learning conditions. For example Classrooms carbon dioxide concentrations leads to secular labs h-humidity and large halls temperature variations. Schools can better focus, cut down on absences and provide what healthy indoor environment so students & staff should have (Jones et al., 2020).

4.6 Asset and resource tracking

Tracking projectors, tablets, furniture, and sports equipment around campus to manage them efficiently can be a real chore. RFID with integrated QR based tracking to a DT system in place for administrators, Real Time tracking of the asset with usage detection is capable to track procurement/supplies/durables by usage patterns. The real-time digital twin dashboards offers instant visibility in inventory that enables from the raw amount reduction of purchased things to better spending on the budget (Bilberg & Malik, 2019).

4.7 Sustainability monitoring and compliance

More and more schools are focusing on sustainability, complying with LEED (Leadership in Energy and Environmental Design) green building standards. The compliance KPIs and sustainability reports can be automated from a Digital Twin (measuring real-time energy and water, carbon footprint tracking, recycling stats) output. It allows the data to be presented in an interactive and fun way for green certifications, grants, and educational purposes (Mandolla et al., 2019).

4.8 Integration with educational outcomes

DT, a great technology aimed at facilities only, can have a major impact on educational distribution. It can be used to enable smart scheduling, performs environmental tuning and is learning material in STEM courses. Implement a classroom allocation based on course type, size and resources to improve focus during exams as well as the understanding of smart systems and sustainability basics (Halúsková, 2023).

Table 12.3 provides a summary of usage of DT in various applicable areas, its benefits and technologies being used respectively.

5. Proposed digital twin framework

A DT framework has been proposed for the purpose of holistic, extensive and affordable school facility management which foreseeably combines physical infrastructure, smart techs, data analytics and design-led with human-centered to tackle operational and environmental objectives along with pedagogics. The proposed framework is layered and logically integrated across five layers — each with specialized roles, roles of technologies and data flows. The layered approach facilitates scalability, modulization and stages deployment and thus fits to a variety of educational institutions with different size or state of their infrastructure.

TABLE 12.3 Summary of DT applications with its benefits and technologies used.

Application area	Description	Key technologies	Benefits
Energy management	Environmental monitoring and optimization of HVAC/Lighting in real-time	IoT, AI, BIM	Lower energy costs, enhanced efficiency
Predictive maintenance	Equipment health monitoring and predictive failure	Sensors, ML, cloud	Minimise shut down, minimal repairs spend
Space utilization	Monitoring as well learning/classroom-suited facility usage	Occupancy sensors, RFID, DT simulations	Increased space planning and schedule efficiency
Emergency management	Crises simulation and guidance	IoT, cameras, GIS, mobile alerts	Safer evacuation, better
Indoor environmental quality	Monitoring temperature, humidity and air quality	Environmental sensors, actuators	Healthier environment for learning
Asset tracking	Stock control and usage surveillance	RFID, QR codes, dashboards	Reducing losses optimizing resource allocation
Sustainability and compliance	Verification of the measured and monitored results with the clear credentials	IoT, reporting tools, smart meters	Ethical, cost-incentives challenges

The first layer is the Physical Asset Layer which is the base of the Digital Twin architecture. It includes all the things in space of school universe, whether it is physical (buildings, classrooms, hall, cafeteria, HVAC systems, lighting infrastructure), water supply units or even safety equipments (fire alarm) furniture. And this is where the physical-to-digital bridge starts. It also has the energy systems on it like solar panels, electrical meter, and water recycle unit available. This layer functionality is to act as an operation source and any interaction in the system. This ensures that each physical component contributing operational efficiency or comfort in the school is captured. So this involves in tracking wear and tear, physical observation and alignment for operational readiness for teaching, the learning, safety.

The Sensor and IoT Layer — data generation, that is the first sits directly on top of Physical Asset Layer The Sensor and IoT Layer is comprised of everything from a fabric of embedded IoT sensors, actuators RFID tags smart meters through-edge devices onto physical assets. The sensors, which are measuring temperature and humidity, energy consumption, ambient air quality CO_2 levels—infiniwan etc. Designed functionally, the layer is for the ongoing monitoring of environmental and operational parameters. Imagine this layer as the nervous system of the Digital Twin, translating real-world conditions into digital signals. At this Layer Actuators allow for an automated or manual control of devices like dimming lights depending on the daylight level or from turning off HVAC during off-hours for energy savings. Furthermore, this layer is important for real time event detection (e.g. detecting temperature swings, system failure).

The Integration and Communication Layer is data layer of the architecture. The main task of it is to collect, blend and push out the data that is gathered from sensor layer to next processing platform. Reality accomplished with a set of tools that mixes Cloud platforms, edge computing, APIs, data buses, middleware, and security protocols (e.g., MQTT, HTTP, CoAP etc.). It provides the layer which is capable for interoperability of different systems/devices. It could do tasks such as data pre-processing, validation, encryption, storage and routing. A current integration layer that integrates with the existing school infrastructure SCADA systems, building management system (BMS) etc./ tools enterprise resource planning (ERP). It also delivers real-time streaming and batch data processing, depending on required analytics. Through decoupling the set of data sources with the analytical and visualization engines, this is the layer needed to scale and future proof system architecture.

The fourth layer is the Digital Model and Analytics Layer which is the central intelligence and simulation of the Digital Twin Architecture Here, the virtuality of school environment is developed through BIM, CAD tools, 3D modeling software etc. at this level. These models tune themselves automatically by updated live data from the integration layer coming in. Purposefully, this layer provides for performing descriptive and predictive analytics. It uses AI/ML models for anomaly detection, fault prediction of machinery and changing resource allocations (energy and space utilization). It will be able to run "what-if" considerations, for instance, to forecast potentially compromised air quality inside a building due to a ventilation failure or future maintenance activities for HVAC systems. The layer also does trend analysis, KPI monitoring, and scenario modeling for end to end insights on infrastructure-performance and energy efficiency. Decision-making modules are housed here, which supply administrators and facility managers with specific recommendations that they can use.

The last and top layer which is user interaction and visualization and includes school administrators, facility managers, IT staff sometimes teachers or students). This layer populates dashboards, mobile apps with processed data and presents this in a clear and actionable format via alerts and visualization tools. Live 3D models can be viewed by users, alerts are sent in real-time user can interact with facility maps and be provided maintenance schedules or even do remote diagnostics or control. The UI can be customized at stakeholder level (e.g., energy dashboard for sustainability officers, user friendly safety alerts for administrators etc.; temperature controls for teachers). The layer supports decision-making better, makes an operation more efficient and increases situational awareness. This includes taking feedback from users and feeding that back into the system so we have a feedback loop from digital analytics to actual actions taken in the real world. Fig. 12.4 illustrates the layered proposed digital twin framework (Table 12.4).

Fig. 12.5 illustrates the Phased Adoption Plan Digital Twin technology in managing school facility implementation It is a strategic and systematic (step-by-step) implementation plan of Digital Twin into school facility management technology In Phase 1 (Foundation): We establish BIM models and deploy the necessary IoT sensors that are the foundation for development where Phase 2 (Data Integration and Monitoring) is all about setting up real-time dashboards and streaming sensor data for ongoing monitoring. In Phase 3 (Predictive and Automated Control) AI based systems for predictive maintenance and AI driven decision making In Phase 4 (Simulation and Feasibility) The use of scenario planning, energy efficiency simulations and models for emergency drill helps schools increase outcomes readiness and operational performance. Lastly, Phase 5 (Full Automation + Edu Integration)

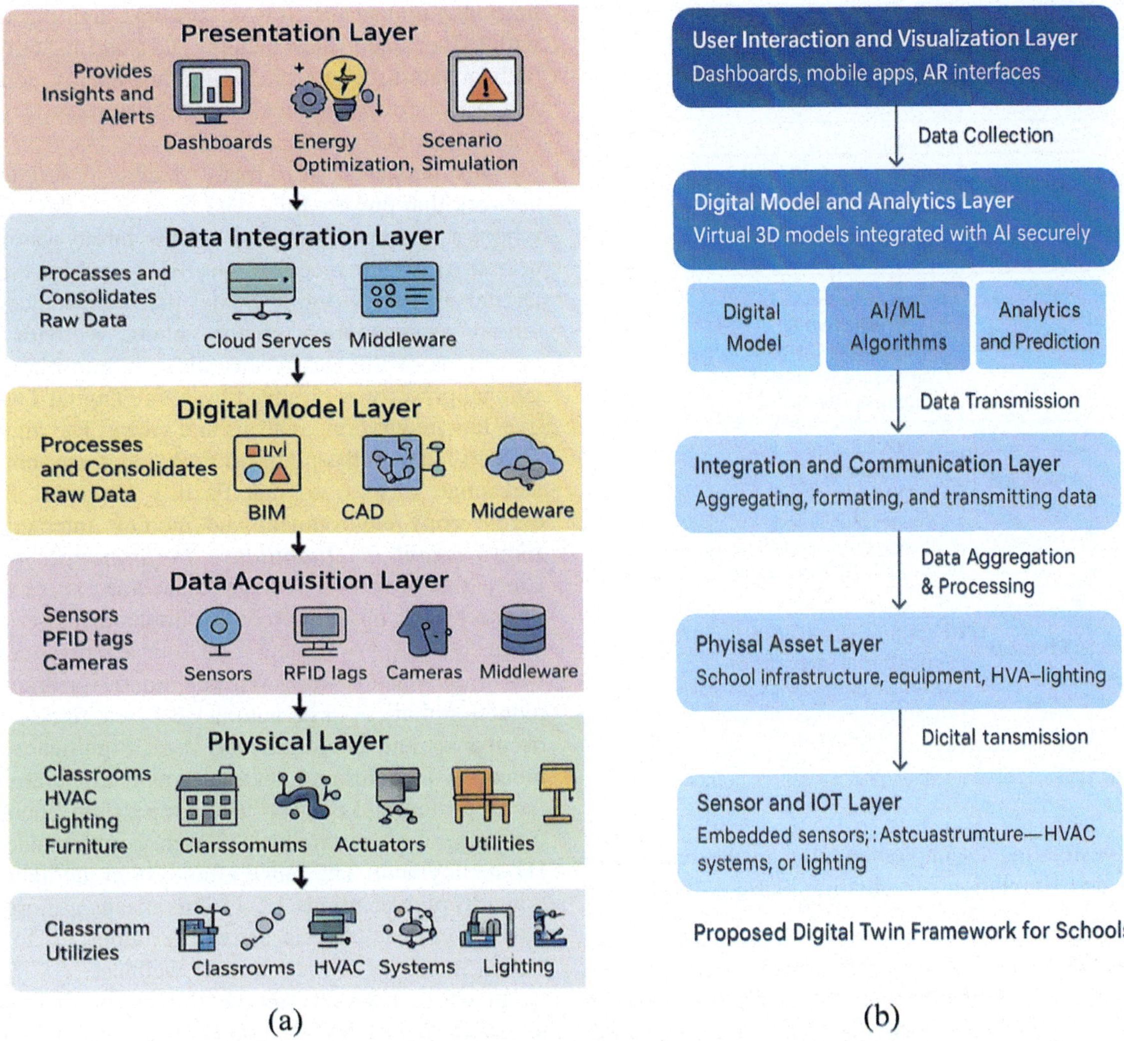

FIGURE 12.4 (a) Proposed digital twin framework. (b) Workflow-Based framework.

TABLE 12.4 Summarizes the technologies used at each layer of the proposed digital twin framework.

Layer	Technology/Tool utilized
Sensor layer	DHT11, PIR, CO_2 sensors, smart meters
Communication	LoRaWAN, zigbee, Wi-Fi, MQTT
Data integration	Node-RED, Apache Kafka, azure IoT hub
Modeling and simulation	Autodesk revit, unity 3D, bentley iTwin
Analytics engine	TensorFlow, power BI, microsoft azure ML
User interface	Grafana, tableau, custom Web/Mobile dashboards

combines AR/VR technologies for experiential learning and smart curriculum components in a digitally enabled infrastructure that seamlessly blends management of assets with learning enablers.

6. Result evaluation

An idea of the implementability and effectiveness of DT framework in school facilities management can be carried

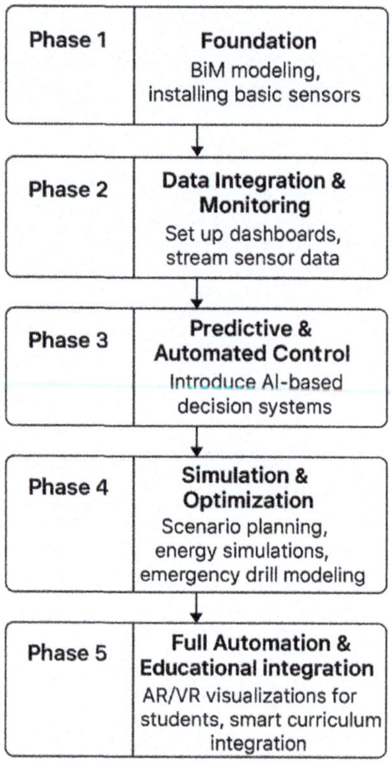

FIGURE 12.5 Phased adoption plan for proposed framework.

out by investigating its implementation in the real world situations and simulations. Evaluation phase is to:

- Test shared functionality of the framework among different domains (e.g., energy, space and maintenance).
- Monitor key performance indicators (KPIs), like energy savings, space utilization and downtime for maintenance.
- Pinpoint the obstacles in Implementation and Integration
- Gather comments back from school admins and technical.

6.1 Case study 1: Smart secondary school (urban setting)

Singapore public secondary school implements Digital Twin for improved environmental monitoring and energy optimization project by the administration at a public secondary school in Singapore. The 6-month project centered on embedding a diverse BIM digital model of the building with advanced sensor networks to serve as a real-time monitoring and management system on Azure Digital Twins platform. A multi-classroom physical infrastructure of the school was initially provided with a cluster of high-accuracy sensors. These sensors were placed in a total of 12 classrooms, specifically CO_2, and temperature and humidity sensors. The locations of these sensors were proximity data that have direct consequences on the learning experience and energy consumption. The school at that time also updated its HVAC systems, lighting equipment with IoT-enabled smart meters. The smart meters double as both in real-time sensors to monitor energy consumption and feed into the predictive control strategy to adjust operations for best practice data.

This happened seamlessly thanks to a good sensor network that powered the data from these devices into the school's digital infrastructure. This fabric was then connected to digital model of the BIM on Microsoft Azure Digital Twins. A digital model that represented an exact virtual representation of the school — with real-time conditions of the classrooms and common areas. Computational power was provided by Azure Digital Twins which live data integration, analysis and views. The digital model of each facility (the platform took data from, both sensors and smart meters) so that Facility Managers have their digital copy environment and monitor interactions; ***The Improvement in Quantitative Performance Was Great***. Currently, the school averages on average about 1650 kWh of energy per month before the implementation of Digital Twin system. Post integration, our energy consumption was improved to 1460 kWh per month—a dramatic 21% reduction that could be automated by continue-to-measure occupancy and environmental data. Significant improvements in the variation of classroom temperature, a key factor on thermal comfort The temperature fluctuations in the range of plus or minus 4°C were commonplace prior to DT deployment; The range after system modifications got down to plus or minus 1.5°C This means a more constant internal climate, that is extremely important for concentration in school as well student wellness.

In addition, HVAC runtime dropped from 9 hours a day to about 6.5 h HVAC Maintenance Likewise — this decrease not to be lesser with cost savings but also less carbon (wear and tear on mechanical systems less which equates to a longer life of equipment). Other than the technical improvements, some qualitative indicators also indicated the success of a project. Staff satisfaction was 68% pre-system upgrade and shot up almost 30 points to 92% post-implementation. The increase in user engagement was due to a synergistic set of developments—stabilized classroom environments provided reliability, increased benefits from less disruptions through preventative maintenance and an immersive mobile app dashboard where you could instantly see what was going on. A particularly notable achievement for the project was the shift from traditional facilities management in schools. By instead automated predictive controls, administrators could take ownership of different problem areas instead of responding after-the-fact. For example, if a sensor suggests an abnormal increase in CO_2 levels or something altogether different like a temperature fluctuation then the system changes HVAC settings to fix what ails. This capability to predict not only enhanced indoor air quality, but also optimised the operational parameters in real time to reduce energy use. The energy savings from the adaptive lighting system also contributed as it dimmed or switched off lights in unoccupied spaces to counter wasteful energy use.

The real-time element of the system was a huge help in the day to day running of regular school. Live updates could be viewed on a mobile app by teachers, facility managers, and even students keeping tabs of separate facilities. Dashboards provided data in a consumable format and highlighted the trends to inform the next steps to trigger maintenance or environmental setting changes. Rather, by marrying real-time data feeds to the digital model, the system gave a comprehensive view of the operational state of the school that just wasn't possible with traditional methods. To conclude, the pilot at the smart secondary school in Singapore showed that Digital Twin technology can successfully combine predictive control of HVAC and adaptive lighting to give significant improvement in energy efficiency and improvement in environmental comfort.

NCC staff and students reported the benefits of real-time monitoring and predictive analytics as well, validating that Digital Twin technology is not just effectively implemented for new buildings but machinery in contemporary educational infrastructures to a compelling enabler for transformational school infrastructure management. What this case study really is an example of is how state-of-the-art sensor networks and integrated digital modeling can help to deliver real benefits in resource management, user satisfaction — and by logical progression — to broader acceptance of smart tech in educational environments. Fig. 12.6 illustrates the results of the case study.

6.2 Case study 2: Rural government school (pilot deployment)

A government operated high school in a remote village in India for long fought with asset mismanagement and delayed maintenance. Problems had their roots not in apathy but in cash-strapped, under staffed administration and lack of reality-time digital infrastructure. Movable property (library books, classroom furniture and equipment etc.) were maintained by manual Inventory, which often led to inaccuracies and misplaced item or inefficient maintenance response. These were not minor issues as they ate into the already constrained budget and impacted the soul of the school environment. Acknowledging that despite having so little, technology held potential even in low-resource settings a local tech team of faculty and students adapted a simplified Digital Twin tailored to the constraints of the rural school. With the goal of creating a minimal yet effective system leveraging open-source tools and as little hardware as possible, we aimed to demonstrate that Digital Twin is not for fancy urban institutions only.

A 3-month project on asset tracking and predictive maintenance. The initial stage was to affix RFID on all of the identified major assets like desks, chairs, projectors, library books etc. RFID tags not only identified the asset but ensured they were tracked and located at checkpoints with RFID readers if moved between them. The team deployed Arduino based, low-cost IoT edge devices (with RFID scanners) instead of commercial hardware. Such devices collected the movement data of the assets and sent it over to local Raspberry Pi server for processing. For the sensor data and to orchestrate automation workflows, the

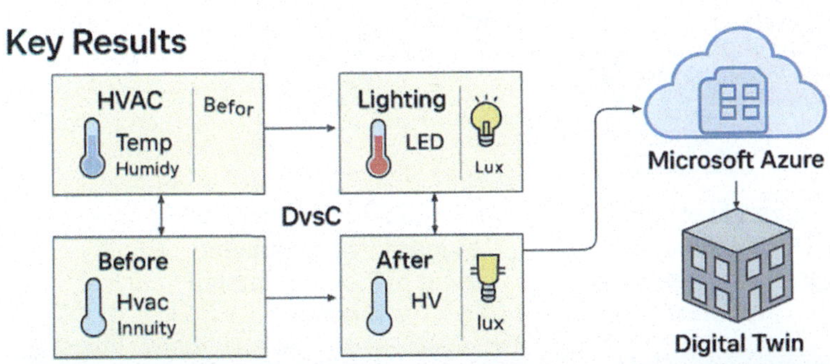

Metric	Before DT Implementation	Improvement
Average Energy Consumption (KWh)	1850 kwh/month	+ 21%
Classroom Temperature Consistency	±4°C variation	Thermal comfort
	9 hours /day	+ 28%
Staff Satisfaction Rate	68%	92%

FIGURE 12.6 Results of case study 1.

team decided to deploy Node-RED, which stands for Node are JavaScript systems that is best suited for rapid prototyping in IoT. It was developed in real-time and updates automatically upload the location, each corresponding asset + then visuals to a simple but powerful dashboard for administrative staff. The app was agnostic of the internet and instead that ran over local area network (LAN), would work through both edge devices as well user interface on simple network communication.

Maintenance which something else that should have been monitored since desktop items (projector, fans etc.) never fail until YOU notice. To combat this we created a quick and dirty Python script that would run on lightweight server and watch local files containing logs for asset usage. When the exe thresholds were crossed (i.e., projector hour count exceeded 500), the script sent an alert. The alerts—not needing a bespoke app the alerts were delivered over WhatsApp, a platform widely used and capable of reaching maintenance personnel, proponents on all kinds of cell phones. This WhatsApp automation usage narrowed the technology gap and a concrete pattern of cross communication with a framework internally at the school. Some sizeable and optimistic outcomes were yielded from this pilot project. The school quantified an average of their lost or misplaced Items to be 17/month prior to the Digital Twin-inspired framework implemented. The result of the system deployment was that it reduced that number to only four items per month (a 76% drop in asset loss). Besides, maintenance response time that used to be 3.5 days as result of lacking communication and scheduling were reduced down to 1.2 days (66% drop). Major improvement in administrative time efficiency. Due mainly to the amount of time formerly dedicated on matters inventory reduced from 14 h to an easy going 6 h per month providing staff more time for student engagement and educational focus.

The program also created a sense of digital competence among both staff and students. Dashboards became more common as a way to update teachers on maintenance needs or resource struggles. Students, who had never interacted with microcontrollers or data systems before in their lives got an immersive exposure to RFID programming, Arduino workflow automation. Beyond being a remedy to operational issues, this collaboration is a learning experience that sits well with skill-based learning objectives as well. With none (Zero) financial constraints and limited access to enterprise technologies, this rural school gave one example of how feasible a Digital Twin-like system can be built using open source platforms and bottom-up innovation. The project proved the concept with a least bit of a physical system representation enabled through automation and basic analytics. Overall the case study shows that principles of Digital Twin technology are very pliant and can be scaled up. Adjusting the framework according to particular need and bounds of a school resulted in better asset tracking, waste reduction on resources and faster response to remedial requirements. It will be a significant Phase Pilot, as this is the first time we can see even under-resourced educational institutions in rural communities apply to Digital Twin conceptualization through innovative community-only, free cost solutions. Fig. 12.7 illustrates the results of the case study.

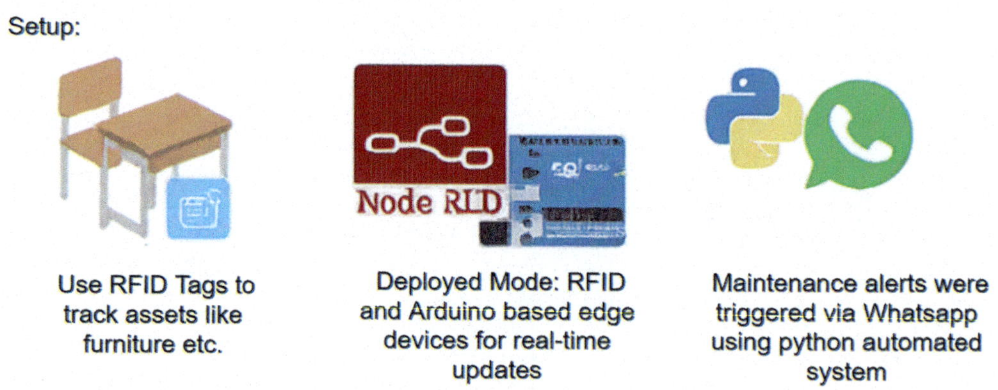

Metric	Baseline	Post Implementation	Improvement
Lost/ Misplaced Assets	17 items/month	4 items/month	76.47%
Average Maintenance Response	3.5 days	12 days	242.86%
	14 hr/month	6 hr/month	57.14%

FIGURE 12.7 Results of case study 2.

6.3 Simulating emergency evacuation planning

One of the core functions of school facility management is to protect student and staff during emergencies. Yet, traditional emergency planning practices—manual drills and a static printed evacuation map—lack the dynamicity of scenarios in a real-life setting. With the promise of using DT technology to improve emergency response, a DT-integrated evacuation plan was simulated via virtual fire drill for data analytics purpose. The experiment was organized to see How much time efficient, beat the congestion and automatic accountability DT-guided response model (vs. traditional manual evacuation) Created in Unity3D, a game engine designed for immersive real-time 3D worlds. This metaplan was matched with the BIM data of each building a real-time digital replica of school infrastructure. A digital model of the building and its architecture (layouts), floor plans, the exits, corridors and occupancies as classrooms, restrooms and cafeteria.

A simulated situation was established in which a fire was started inside the school cafeteria, which is located centrally in the building and usually very crowded around lunchtime. The DT system already had the data sources turned on, which consisted of virtual CCTV feeds, real-time occupancy inputs, and environmental sensors. Essentially, the system was designed to behave similarly, to real-world conditions. Once the fire was activated the DT system automatically began to assess how occupants would evacuate from the virtual campus environment. Heat maps were generated in real-time to monitor areas of congestion and bottlenecks along paths of egress. From this assessment, the system issued real-time evacuation directions. Digital signs in hallways and classrooms displayed real-time directions to the nearest exit based on the safest and least congested path of egress. At the same time, mobile alerts were communicated to registered devices, whether they were on a smartphone or tablet that the school had provided to the students and staff, either communicating or guiding the teachers or students to alternate paths if the primary exit route became blocked or congested. The automated access controls and HVAC systems were all simulated by adjusting the smoke propagation model to ensure the best air quality possible along any available escaping path.

The Digital Twin system was assessed based on comparisons of key metrics with a conventional manual fire evacuations drill conducted in the same virtual world. Differences in the key metrics were noticeable. The average time for evacuation from the manual drill conditions was recorded at 6 minutes and 20 seconds. In comparison, the DT-guided simulation achieved a full evacuation time of just 3 minutes and 55 seconds, a reduction of 38% of total evacuation time. Additionally, a key parameter considered was the number of congestion points during the evacuation. The manual drill included five significant congestion points, especially near staircases and at intersections of narrow hallways. In comparison, the DT system reduced the number of congestion points down to two using real-time occupancy analytics to dynamically reroute traffic and improve flow distribution. This represents a 60% reduction in congestion points, which helped to improve evacuation time and reduce psychological stress that is often associated with an emergency evacuation.

Another area of improvement was headcount accuracy. In the manual situation, three times staff did not remember to say they 'verified the presence of students from their classroom', and as a consequence, the reporting of student or staff count at assembly points was inaccurate. Using the DT-assisted method, the location sensors (Real-time indoor positioning systems) and the RFID-enabled student ID cards would support the system in tracking the movement of all participants while migrating throughout the school building. This provided a fully automated headcount verification process with 100% accuracy—meaning that no student or staff was unaccounted for during the drill.As seen in the results, there are benefits to designating simulation environments in the Digital Twin framework. This is applicable to have administrators plan for all types of emergency scenarios including but not limited to fire, earthquake, or security threats. In addition, as with predictions and analysis, the use of simulation environments can allow for continuous improvement and learning using replay scenarios and data logging throughout the process. Once the drills concluded, then the system can supply a reports with detailed performance data to identify where there was delay or confusion in the exercise and provide recommendations to improve either for infrastructure or policy implications. In addition to operational benefits, simulations also provide training experience for student or staff. Fig. 12.8 illustrates the simulated results of the proposed digital twin framework.

A list of standardized Key Performance Indicators (KPIs) is recommended for a balanced assessment on the use of DT within the education sector. The KPIs provides a detailed assessment of operational, environmental and user-related outcomes. To illustrate Energy Efficiency refers to decreased monthly energy use which has a positive impact on HVAC and lighting systems and alludes to a successful aspect of operation. The Space Utilization Index assesses how well the intended spaces are utilized in proportion to intended capacity to assist with management of space planning and timetabling. MTTR (Mean Time to Repair) assesses how responsive maintenance is, in terms of the average time taken by staff to repair faulty

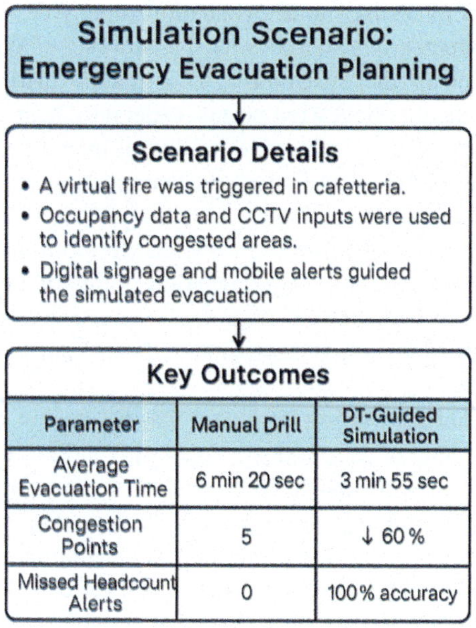

FIGURE 12.8 Results of simulation.

equipment from the learning environment. Asset Utilization Rate captures the frequency key resources such as furniture, AV equipment, or laboratories are used, beneficial to inventory management. Evacuation Time allows for an evaluation of safety readiness by measuring the time it takes to evacuate a space in an emergency drill simulation. User Satisfaction gauges opinions of key stakeholders and staff to ensure the DT is appropriate to meet the needs of staff and students. Environmental Quality Score allows for an evaluation of the indoor conditions of the district building including temperature, CO_2 levels, and humidity compared to recommended parameters to measure the system's ability to provide a healthy learning environment. By combining the example KPIs, we have a solid basis to measure the concrete outcomes for the integration of DT in school facility management.

7. Conclusion and future work

The study emphasizes the significant influence of DT technology in managing school facilities. By utilizing real-time data and BIM, merged with IoT sensors and data analytics, schools may see a significant increase in energy performance and maintenance, safety, and efficiency in the use of school spaces. The suggested multi-layered design can provide phased implementation of systems to cater for urban and rural schools. The two case studies conducted at an urban school in Singapore and a rural school in India identified benefits to efficient resource use, efficiency, comfort of environment, and management of assets. The emergency evacuation simulation shows evidence of DT being able to demonstrate safety and preparedness through shortening the duration of time evacuated and accurately reporting attendance. However, as discussed in the review and through the additional research, there are challenges ahead with cost considerations, methods/approaches to integration of data, and staff training as initial barriers to implementation. Nevertheless, the use of DT within educational contexts is still low, requiring policy and regulations to be developed, applied, and purposeful to support these processes. Coming work aims to improve interoperability among DT components, as well as examine rollover integration with new technologies including blockchain for data security and AR/VR for learning and maintenance training. Additional future research will examine the longer-term impact of the DT on sustainability metrics and student achievement. Pilot project in low-resource environments should also be expanded to assess scalability/feasibility and cost effectiveness. By building viable pilots and connecting to real-world experiences, DT can become a staple in establishing smart, resilient, and future-ready educational infrastructure.

References

Abele, E., et al. (2017). Learning factories for future oriented research and education in manufacturing. *CIRP Annals, 66*(2), 803–826. https://doi.org/10.1016/j.cirp.2017.05.005

Attaran, M., & Celik, B. G. (2023). Digital twin: Benefits, use cases, challenges, and opportunities. *Decision Analytics Journal, 6*, Article 100165. https://doi.org/10.1016/j.dajour.2023.100165

Bamunuarachchi, D., Georgakopoulos, D., Banerjee, A., & Jayaraman, P. P. (2021). Digital twins supporting efficient digital industrial transformation. *Sensors, 21*(20), 6829. https://doi.org/10.3390/s21206829

Barricelli, B. R., Casiraghi, E., & Fogli, D. (2019). A survey on digital twin: Definitions, characteristics, applications, and design implications. *IEEE Access, 7*, 167653–167671. https://doi.org/10.1109/access.2019.2953499

Bilberg, A., & Malik, A. A. (2019). Digital twin driven human–robot collaborative assembly. *CIRP Annals, 68*(1), 499–502. https://doi.org/10.1016/j.cirp.2019.04.011

Billey, A., & Wuest, T. (2024). Energy digital twins in smart manufacturing systems: A case study. *Robotics and Computer-Integrated Manufacturing, 88*, Article 102729. https://doi.org/10.1016/j.rcim.2024.102729

Chen, H., et al. (2024). Digital twin-based virtual modeling of the Poyang Lake wetland landscapes. *Environmental Modelling and Software, 181*, Article 106168. https://doi.org/10.1016/j.envsoft.2024.106168

Cheng, J., Zhang, H., Tao, F., & Juang, C.-F. (2019). DT-II:Digital twin enhanced industrial internet reference framework towards smart manufacturing. *Robotics and Computer-Integrated Manufacturing, 62*, Article 101881. https://doi.org/10.1016/j.rcim.2019.101881

Damgrave, R. G. J., & Lutters, E. (2019). Smart industry testbed. *Procedia CIRP, 84*, 387–392. https://doi.org/10.1016/j.procir.2019.04.215

DELMIA - Dassault systèmes. (2024). *DELMIA - Dassault systèmes*. Dassault Systèmes. http://www.apriso.com/library/Whitepaper_Dr_Grieves_DigitalTwin_Manufacturing%20Excellence.php.

Dembski, F., Wössner, U., Letzgus, M., Ruddat, M., & Yamu, C. (2020). Urban digital twins for smart cities and citizens: The case study of Herrenberg, Germany. *Sustainability, 12*(6), 2307. https://doi.org/10.3390/su12062307

Deng, Z. (2023). A quantitative overview of the approaches influencing traditional and new teaching methods on technical college students. *Soft Computing*. https://doi.org/10.1007/s00500-023-08276-9

Dong, Z., Han, X., Shi, Y., Zhai, W., & Luo, S. (2022). Development of manipulator digital twin experimental platform based on RCP. *Electronics, 11*(24), 4196. https://doi.org/10.3390/electronics11244196

Fu, Y., Zhu, G., Zhu, M., & Xuan, F. (2022). Digital twin for integration of design-manufacturing-maintenance: An overview. *Chinese Journal of Mechanical Engineering, 35*(1). https://doi.org/10.1186/s10033-022-00760-x

Fuller, A., Fan, Z., Day, C., & Barlow, C. (2020). Digital twin: Enabling technologies, challenges and open research. *IEEE Access, 8*, 108952–108971. https://doi.org/10.1109/access.2020.2998358

Grieves, M., & Vickers, J. (2016). *Digital twin: Mitigating unpredictable, undesirable emergent behavior in complex systems*. Springer eBooks. https://doi.org/10.1007/978-3-319-38756-7_4

Halúsková, B. (2023). Digital twin in smart city. *Transportation Research Procedia, 74*, 1471–1478. https://doi.org/10.1016/j.trpro.2023.11.308

Haleem, A., Javaid, M., Qadri, M. A., & Suman, R. (2022). Understanding the role of digital technologies in education: A review. *Sustainable Operations and Computers, 3*, 275–285. https://doi.org/10.1016/j.susoc.2022.05.004

He, X., et al. (2022). Situation awareness of energy internet of things in smart city based on digital twin: From digitization to informatization. *IEEE Internet of Things Journal, 10*(9), 7439–7458. https://doi.org/10.1109/jiot.2022.3203823

Hooshyar, D., Weng, X., Sillat, P. J., Tammets, K., Wang, M., & Hämäläinen, R. (2024). The effectiveness of personalized technology-enhanced learning in higher education: A meta-analysis with association rule mining. *Computers & Education*. , Article 105169. https://doi.org/10.1016/j.compedu.2024.105169

Huang, W., Zhang, Y., & Zeng, W. (2022). Development and application of digital twin technology for integrated regional energy systems in smart cities. *Sustainable Computing Informatics and Systems, 36*, Article 100781. https://doi.org/10.1016/j.suscom.2022.100781

Integration of Learning and Research. Integration of Learning and Research in a Multi-Perspective Learning Factory | worldwide CDIO inititative. https://www.cdio.org/knowledge-library/documents/integration-learning-and-research-multi-perspective-learning-factory.

Jayasinghe, S. C., et al. (2024). Innovative digital twin with artificial neural networks for real-time monitoring of structural response: A port structure case study. *Ocean Engineering, 312*, Article 119187. https://doi.org/10.1016/j.oceaneng.2024.119187

Jeong, Y., et al. (2022). Digital twin-based services and data visualization of material handling equipment in smart production logistics environment. In *IFIP advances in information and communication technology* (pp. 556–564). https://doi.org/10.1007/978-3-031-16411-8_64

Jones, D., Snider, C., Nassehi, A., Yon, J., & Hicks, B. (2020). Characterising the digital twin: A systematic literature review. *CIRP Journal of Manufacturing Science and Technology, 29*, 36–52. https://doi.org/10.1016/j.cirpj.2020.02.002

Kamali, M., Atazadeh, B., Rajabifard, A., & Chen, Y. (2024). Advancements in 3D digital model generation for digital twins in industrial environments: Knowledge gaps and future directions. *Advanced Engineering Informatics, 62*, Article 102929. https://doi.org/10.1016/j.aei.2024.102929

Kreimeier, D., Morlock, F., Prinz, C., Krückhans, B., Bakir, D. C., & Meier, H. (2014). Holistic learning factories – A concept to train lean management, resource efficiency as well as management and organization improvement skills. *Procedia CIRP, 17*, 184–188. https://doi.org/10.1016/j.procir.2014.01.040

Leahy, K., & Phelan, P. (2014). A review of technology education in Ireland; a changing technological environment promoting design activity. *International Journal of Technology and Design Education, 24*(4), 375–389. https://doi.org/10.1007/s10798-014-9266-z

Lutters, E. (2018). Pilot production environments driven by digital twins. *South African Journal of Industrial Engineering, 29*(3). https://doi.org/10.7166/29-3-2047

Mandolla, C., Petruzzelli, A. M., Percoco, G., & Urbinati, A. (2019). Building a digital twin for additive manufacturing through the exploitation of blockchain: A case analysis of the aircraft industry. *Computers in Industry, 109*, 134–152. https://doi.org/10.1016/j.compind.2019.04.011

Martínez-Gutiérrez, A., Díez-González, J., Verde, P., & Perez, H. (2023). Convergence of virtual reality and digital twin technologies to enhance digital operators' training in industry 4.0. *International Journal of Human-Computer Studies, 180*, Article 103136. https://doi.org/10.1016/j.ijhcs.2023.103136

Nassehi, A., Colledani, M., Kádár, B., & Lutters, E. (2022). Daydreaming factories. *CIRP Annals, 71*(2), 671–692. https://doi.org/10.1016/j.cirp.2022.05.002

Ning, G., Luo, H., Yin, W., & Zhang, Y. (2024). Exploration of the application and practice of digital twin technology in teaching driven by smart city construction. *Sustainability, 16*(23), Article 10312. https://doi.org/10.3390/su162310312

Pan, Y. H., Qu, T., Wu, N. Q., Khalgui, M., & Huang, G. Q. (2020). Digital twin based real-time production logistics synchronization system in a multi-level computing architecture. *Journal of Manufacturing Systems, 58*, 246–260. https://doi.org/10.1016/j.jmsy.2020.10.015

Qi, Q., et al. (2019). Enabling technologies and tools for digital twin. *Journal of Manufacturing Systems, 58*, 3–21. https://doi.org/10.1016/j.jmsy.2019.10.001

Qi, Q., & Tao, F. (2018). Digital twin and big data towards smart manufacturing and industry 4.0: 360 degree comparison. *IEEE Access, 6*, 3585–3593. https://doi.org/10.1109/access.2018.2793265

Qi, Q., Tao, F., Zuo, Y., & Zhao, D. (2018). Digital twin service towards smart manufacturing. *Procedia CIRP, 72*, 237–242. https://doi.org/10.1016/j.procir.2018.03.103

Saleh, M. M., Abdelkader, M., & Hosny, S. S. (2022). Architectural education challenges and opportunities in a post-pandemic digital age. *Ain Shams Engineering Journal, 14*(8), Article 102027. https://doi.org/10.1016/j.asej.2022.102027

Semeraro, C., et al. (2022). Digital twin application in energy storage: Trends and challenges. *Journal of Energy Storage, 58*, Article 106347. https://doi.org/10.1016/j.est.2022.106347

Sharma, A., Kosasih, E., Zhang, J., Brintrup, A., & Calinescu, A. (2022). Digital twins: State of the art theory and practice, challenges, and open research questions. *Journal of Industrial Information Integration, 30*, Article 100383. https://doi.org/10.1016/j.jii.2022.100383

Sim, J. K., et al. (2024). Designing an educational metaverse: A case study of NTUniverse. *Applied Sciences, 14*(6), 2559. https://doi.org/10.3390/app14062559

Slot, M., Fraikin, M., Damgrave, R., & Lutters, E. (2022). Digital infrastructures as the basis for implementing digital twinning. *Procedia CIRP, 109*, 568–573. https://doi.org/10.1016/j.procir.2022.05.296

Slot, M., & Lutters, E. (2021). Digital twinning for purpose-driven information management in production. *Procedia CIRP, 100*, 666–671. https://doi.org/10.1016/j.procir.2021.05.141

Tao, F., et al. (2020). Digital twin and blockchain enhanced smart manufacturing service collaboration and management. *Journal of Manufacturing Systems, 62*, 903–914. https://doi.org/10.1016/j.jmsy.2020.11.008

Ullah, F. (2022). Smart tech 4.0 in the built environment: Applications of disruptive digital technologies in smart cities, construction, and real estate. *Buildings, 12*(10), 1516. https://doi.org/10.3390/buildings12101516

Wang, H., Chen, X., Jia, F., & Cheng, X. (2023). Digital twin-supported smart city: Status, challenges and future research directions. *Expert Systems with Applications, 217*, Article 119531. https://doi.org/10.1016/j.eswa.2023.119531

Wang, T., Li, J., Deng, Y., Wang, C., Snoussi, H., & Tao, F. (2021). Digital twin for human-machine interaction with convolutional neural network. *International Journal of Computer Integrated Manufacturing, 34*(7–8), 888–897. https://doi.org/10.1080/0951192x.2021.1925966

Wei, Y., Hu, T., Wang, Y., Wei, S., & Luo, W. (2021). Implementation strategy of physical entity for manufacturing system digital twin. *Robotics and Computer-Integrated Manufacturing, 73*, Article 102259. https://doi.org/10.1016/j.rcim.2021.102259

Yang, Z., Tang, C., Zhang, T., Zhang, Z., & Doan, D. T. (2024). Digital twins in construction: Architecture, applications, trends and challenges. *Buildings, 14*(9), 2616. https://doi.org/10.3390/buildings14092616

Zhang, C., et al. (2023). A multi-level modelling and fidelity evaluation method of digital twins for creating smart production equipment in industry 4.0. *International Journal of Production Research, 62*(10), 3671–3689. https://doi.org/10.1080/00207543.2023.2246161

Chapter 13

Cloud, fog and edge computing for school digital twin

Anita Mohanty[1], Ambarish G. Mohapatra[1], Abhro Mukherjee[2] and Bright Keswani[3]

[1]*Electronics Engineering, Silicon University, Bhubaneswar, Odisha, India;* [2]*Techno India University, Kolkata, West Bengal, India;* [3]*Department of Computer Science & Engineering, Poornima University, Jaipur, Rajasthan, India*

1. Introduction

The speed of educational digital transformation grows more rapid because digital twins emerge as a key technology (Mihai et al., 2022). School institutions using real-time dynamic virtual models of their environments will improve operational efficiency alongside safety and personalization of learning systems. The first chapter presents school digital twins while discussing how distributed computing structures, including cloud, fog, and edge, contribute to the achievement of successful deployment.

1.1 Definition of digital twin in education

The original industrial and manufacturing domain of digital twins consists of virtual models that track physical systems in real time. A school environment is represented as a dynamic digital duplicate that continually updates itself like a living system in educational contexts (Broo & Schooling, 2023; Singh et al., 2021). A digital twin utilizes information from all school facilities, including classrooms, laboratories, libraries, and playgrounds, and administrative areas to build a complete virtual duplication of actual school operations and infrastructure conditions, and human activities. The school digital twin utilizes information gathered from IoT sensors, linking cameras with smart devices together with learning management systems (LMS) for providing meaningful insights. The virtual model allows all school personnel, as well as educational policymakers, to anticipate results and improve their resource utilization while advancing educational standards. The system enables forecasted choices about educational facility maintenance alongside protective measures for campus security, along with energy-saving initiatives and individualized educational plans, which develop an intelligent protective learning environment (Kartashova et al., 2020).

Fig. 13.1 illustrates the architecture of a digital twin–enabled smart school system. IoT devices such as environmental sensors, surveillance cameras, and wearable health monitors collect real-time data from the physical school environment. This data is transmitted to the digital twin model, which mirrors the physical infrastructure virtually. Through continuous data analysis, the system provides actionable insights for predictive maintenance, resource optimization, safety enhancement, and personalized learning strategies. The closed-loop interaction between the physical and digital layers enables a dynamic, adaptive, and intelligent school ecosystem.

1.2 Importance of real-time digital representation of school environments

Virtual digital presentation of academic facilities remains essential to achieve optimal school monitoring, combined with needs-based management and operational improvement. The school digital twin lets users monitor moment-to-moment activities through the collection of live streams produced by physical assets and human movement (Fuller et al., 2020). The continuous flow of occupancy information directly supports classroom space management and helps maintain social distancing requirements and control energy-consuming systems efficiently. The implementation of real-time feedback systems enables instant identification of abnormal events, which include unauthorized entry and equipment breakdowns, and environmental dangers, thus enhancing security measures throughout operations (Erol et al., 2020). Real-time data enables academic institutions to create flexible learning models that study student requirements and then use this

Digital Twins in the Smart Classroom. https://doi.org/10.1016/B978-0-443-41632-3.00005-9

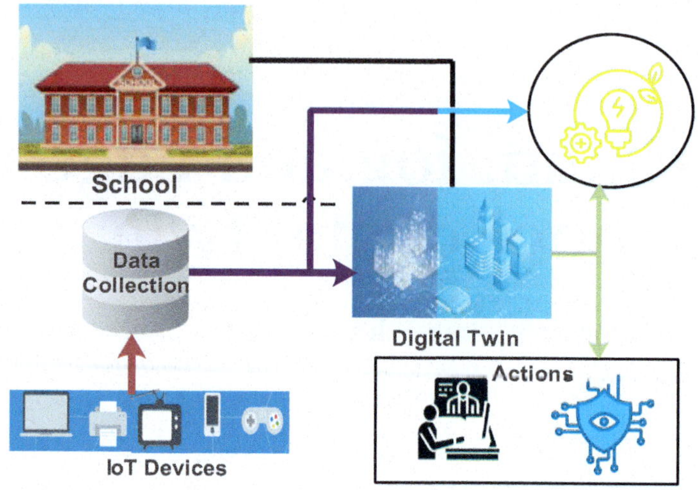

FIGURE 13.1 Digital twin integration in smart schools.

information to provide personalized educational help. Using real-time digital representation gives schools the potential to operate with better efficiency and agility, and improved student orientation.

1.3 Need for distributed computing: Cloud, fog, and edge

The realization of school digital twin potential necessitates effective handling of extensive and various datasets through efficient, scalable, and timely procedures (Rahmani et al., 2024). Running educational operations in real time becomes challenging because using Cloud services as the sole computing method create performance delays and bandwidth constraints together with privacy concerns.

Schools need to implement a distributed computing architecture, which unites cloud, fog, and edge paradigms to overcome these difficulties:

- School sensors and cameras with edge placement strategies perform real-time decision analytics through data processing directly at data source locations.
- The fog computing layer collects data from various edge devices through its intermediate position before it analyzes data locally and distributes network resources throughout the educational fabric of the school.
- The cloud computing framework enables users to benefit from centralized multi-scale data storage together with historical analysis along with AI model development capabilities and outside system access points including educational platforms and government databases.

The coordinated utilization of edge and fog with cloud computing enables educational institutions to attain a performance-exhausted system able to provide immediate responses alongside extensive data analytics features and defense capabilities alongside bandwidth optimization and cost reduction benefits.

1.4 Chapter overview

The development of a school digital twin depends on the unified support coming from cloud, fog, and edge computing systems, explained in detail within this chapter. Section 2 will present the essential understanding along with basic framework elements that comprise educational institution digital twins. The third section establishes the particular functions of cloud, fog, and edge computing in school environments, along with their respective benefits. Section 4 provides an extensive design of the system architecture to demonstrate how physical infrastructure connects with digital components. The value of school digital twins emerges through practical implementations covered in Section 5. Section 6 details the obstacles that researchers must address, followed by recommended directions to help overcome these hurdles. The last section provides conclusions about understanding digital augmentation in schools, with its prospects.

2. Digital twin concept for schools

School digital twins use IoT technology together with data analytics as well as interactive control systems to develop a real-time model of the physical school infrastructure.

2.1 What constitutes a school digital twin

A school digital twin represents the real-time virtual duplicate of school infrastructure, which operates through constant IoT device and sensor data streams (Attaran &

TABLE 13.1 Layers of a school digital twin.

Layer	Components	Functions
Physical layer	IoT devices, sensors, smart devices, and infrastructure	Data collection from physical assets (e.g., classrooms, HVAC, lighting)
Cyber layer	Cloud-based platforms, data analytics engines, and simulation tools	Data processing, analytics, visualization, predictive modeling
Interaction layer	Dashboards, control interfaces, predictive models	User interaction for management, decision-making, and optimization

Celik, 2023). A school digital twin connects physical aspects and cyber components with interaction elements into a complete digital representation that increases managerial control of operational functions and resource management, and learning optimization, as shown in Table 13.1.

2.2 Physical layer: IoT devices and sensors

All tangible school assets, such as classrooms and buildings, together with equipment operate under the physical layer through IoT devices and sensor monitoring. The devices monitor variables, which include temperature, air quality, and occupancy along with energy consumption to provide real-time physical environment data.

2.3 Cyber layer: Data processing and visualization

The processing, along with storage and visual representation of data, occurs through the cyber layer. Computing systems deployed in the cloud and analytic data processing units, together with simulation tools, work to generate digital models from IoT device data collected at the physical school (Sharma et al., 2022). The upper level of the system brings advantages in analyzing patterns through predictive models while also displaying graphical representations to make decisions more efficiently.

2.4 Interaction layer: Control, prediction, and optimization

Users, as well as teachers and administrators, along with maintenance staff, join the digital twin through the interaction layer. Through its dashboards and predictive models combined with control systems, the system permits effective school environment management. Through this layer, schools achieve streamlined resource usage while making learning space customization possible and instant ability to handle matters such as maintenance concerns and safety threats.

2.5 Benefits: Predictive maintenance, smart learning, safety, resource efficiency

A school digital twin generates various advantages through its predictive maintenance ability that decreases expenses and operational interruptions while delivering adaptable learning spaces and enhanced safety using monitoring systems and alert systems combined with resource management for reducing costs, as described in Table 13.2. A school environment becomes more efficient and safer with better responsiveness due to these advantages (Wang et al., 2023).

3. Role of cloud, fog, and edge computing

School digital twin systems need a computer structure that processes massive real-time data without compromising reliability and security alongside low-latency requirements (Haleem et al., 2023). These computing technologies serve different complementary functions to reach their targets.

Fig. 13.2 shows the school digital twin system architecture, which joins edge with fog and cloud computing. The digital twin receives data from IoT sensors and cameras, and wearable devices located at the edge that perform

TABLE 13.2 Core benefits of a school digital twin.

Benefit	Description
Predictive maintenance	Proactively identify equipment failures and maintenance needs to reduce downtime
Smart learning	Adaptive learning environments that adjust to individual student needs
Safety	Real-time monitoring for enhanced security and emergency management
Resource efficiency	Optimizing the use of space, energy, and other resources

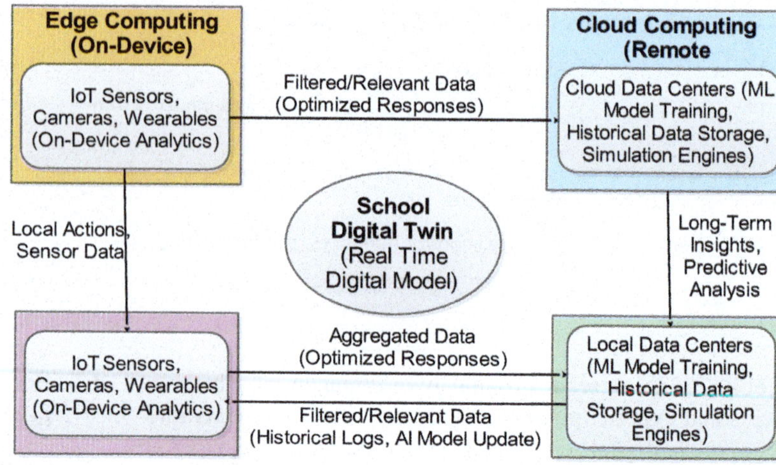

FIGURE 13.2 Architecture of a school digital twin system.

analytics before sending information to the digital twin. Local servers and gateways at the fog layer handle real-time processing and orchestration. Storage and simulation capabilities, together with modeling applications, are managed by cloud data facility centers. The platform supports active actions in real time, delivers optimal solutions, and predicts technical insights to enhance school operations.

3.1 Cloud computing: Features, advantages, limitations for schools

The technological bedrock of cloud computing gives organizations access to centralized processing operations, together with significant data storage capabilities as well as data intelligence capabilities (Sepasgozar, 2021). The school digital twin needs cloud platforms as its central data management system to store historical records and execute large-scale machine learning (ML) processes and perform elaborate environmental modeling tasks. The platform supports easy connectivity to learning management systems (LMSs) as well as administrative platforms and national education networks. Schools can establish dynamic digital models through cloud exploitation by analyzing and aggregating actual-world data.

Key features:

- Centralized data storage and backup:

Cloud infrastructure combines strong and duplicate storage systems to secure historical data, which includes maintenance records and usage patterns along with student attendance statistics, and facility operational data for easy retrieval.

- On-demand scalability for computing and storage:

Schools achieve dynamic resource management by expanding their computing capacities when demand rises and contracting it when expenses need optimization.

- Hosting of AI/ML models and digital simulations:

The cloud infrastructure enables schools to train and deploy complex innovative machine learning (ML) and artificial intelligence (AI) models and 3D computer simulations that model their entire campus layout. Standardized predictive models achieve multiple purposes, including predictive maintenance, while driving energy optimization and personalized learning approaches.

- Easy collaboration across multiple locations:

The built infrastructure of cloud enables real-time access through any location to users from different geographical areas, including students, teachers, administrators as well as external consultants.

Advantages:

- High computational power for data-heavy tasks:

The enormous processing power of cloud systems allows users to run complex operations including predictive analysis and modeling and complete university-wide simulations that local machines struggle to handle.

- Economies of scale reduce IT infrastructure costs: Schools bypass substantial capital outlays for top-end servers when they choose cloud-first infrastructure because they pay for cloud services by the consumption amount.
- Remote accessibility facilitates collaboration:

The web-based school digital twins provide secure accessibility for teachers and facility managers and policymakers who need to work on decision-making applications through remote interfaces across different campuses.

Limitations:

- Increased latency for real-time operations:

Real-time control software runs slower due to the physical distance between cloud data centers and the school building, which leads to delayed emergency response system function and classroom settings adjustments.

- Dependence on stable Internet connectivity:

School administrators need dependable fast Internet connections as an essential requirement for maintaining continuous interaction with the digital twin model. The operational reliability of the system gets dramatically impaired when network connectivity encounters disruptions.

- Potential data privacy concerns:

Third-party cloud data storage systems handling student information and occupancy records, and security video footage create challenges regarding data security issues, which affect both data privacy laws and country-specific data sovereignty requirements.

- Higher costs for frequent, high-volume data transfers:

The price of storage remains affordable yet using massive volumes of data from sensors with real-time video deliveries can add up significantly to operational costs.

Cloud computing stands perfectly for data-intensive applications and centralized analytics, and large-scale modeling within school digital twins, but struggles to achieve the required response demands in time-sensitive control tasks. The school digital twin system benefits from the integration of fog and edge computing to fill the gaps that exist between cloud computing capabilities.

3.2 Fog computing: Near-user processing, latency reduction, orchestration

School facilities obtain processing power through an intermediary fog layer that stands between cloud services and their on-site equipment. A school digital twin system implements necessary data processing by fog nodes, which operate through local servers or smart gateways and campus data centers to bring operations near data origin points (Aheleroff et al., 2021; Sepasgozar et al., 2023). Fog computing processes time-critical applications near the user, which results in reduced latency and better system performance and prevents network congestion. Through its distributed computing format, schools can make quicker decisions, conduct real-time audits, and boost resource management without being limited to remote cloud systems.

Key features:

- Localized data processing closer to data sources:

The school area receives IoT data from occupancy sensors and HVAC systems, security cameras, and student wearables through fog nodes, which conduct aggregated processing. The system avoids direct cloud transfer of raw data by performing filtering and summarization operations, as well as event detection algorithms, at the local level first.

- Real-time event processing and anomaly detection:

Fog computing, performing data analysis at its location, makes it possible to detect anomalies in real-time, including irregular temperature shifts and unapproved security entries, and equipment breakdowns. The local performance speeds up rapid actions while cloud-based decision cycles are omitted.

- Dynamic resource orchestration and load balancing:

The allocation of computational resources and networking capacities from fog platforms occurs in real time according to present requirements. The school-wide emergency drill operations receive top priority from fog nodes because they prefer to focus on vital safety protocols and communication functions rather than standard analytical operations.

Advantages:

- Reduced latency and improved responsiveness:

Data processing at near locations through fog computing creates shorter transmission delays, which allows quick decision-making functions such as quick classroom temperature control and immediate safety alerts.

- Bandwidth optimization and cost reduction:

The cloud receives only important data that has been both processed and summarized, while significant network traffic and expenses diminish substantially. The processing nodes discard or filter noncritical or raw data that is repetitive.

- Enhanced context-aware services:

The fog nodes perform customized responses by considering the specific operational and environmental conditions of their location. Lighting adjustments, along with sensor threshold notifications, occur automatically through fog nodes based on occupancy information received from sensors and breach events.

Limitations:

- Partial dependence on cloud infrastructure:

Fog nodes maintain independent operations when handling real-time tasks, but complex model updates and historical examinations as well as multischool data comparisons, need periodic cloud synchronization.

- Hardware and maintenance investments:

The deployment of fog computing infrastructure begins with purchasing servers and smart gateways as well as storage units, followed by ongoing maintenance and system maintenance payments for reliability and security updates.

- Security management challenges:

The spread of processing activities across different local nodes expands the possible points of security exploits. Security mechanisms with encryption, authentication, and intrusive detection need to be deployed at every implementation to secure systems properly.

The school digital twin achieves greater performance through fog computing because it positions computation directly near operational activities to support quick, intelligent operations. As a basic element, it functions to unite cloud resources that have high capacity with the speed requirements that edge computing addresses.

3.3 Edge computing: Real-time analytics at the device level

The data processing, together with analytics functions of edge computing, operates directly from sources that produce data, such as sensors, cameras, smart meters, and wearable technologies. Through the school digital twin framework, edge computing allows IoT devices positioned across the school to execute local preliminary analysis and make decisions while starting actions independently from remote cloud servers, as well as intermediate fog nodes (Borowski, 2021). The processing of data at its source through edge computing provides mission-critical educational environments with three essential benefits that include extremely low-latency operations, steadfast system reliability, alongside fortified data protection standards.

Key features:

- On-device data processing:

Devices operating at the edge layer feature processors and TinyML AI models that enable them to check sensor information and spot abnormal activities to initiate fast automatic replies beyond network delays.

- Real-time analytics and decision-making:

Edge computing allows events to be detected and responded to almost immediately after recognition. The occupancy sensor installed on a classroom ceiling detects abnormal classroom congestion and executes an automatic safety message transmission to personnel before the cloud verification process completes.

- Support for distributed intelligence:

The inclusion of intelligence capabilities in numerous devices through edge computing results in decentralized system operations, which creates an autonomous network that maintains device functionality when networks fail.

Advantages:

- Ultralow latency for critical applications:

The speed of edge-based analysis reaches microseconds to milliseconds, which allows instant safety protocols and environment management capabilities, together with personalized educational changes.

- Bandwidth efficiency:

The transmission to fog layer and cloud platforms includes only crucial summaries as well as alerts to optimize network usage while conserving internet bandwidth, which is critical for schools with restricted bandwidth.

- Enhanced data privacy and security:

Considering that classroom audio recordings and facial recognition outputs from students and their activity logs remain within edge devices, they are processed there without leaving the device. The device stays protected from outside cyber threats while meeting educational data privacy requirements through this local processing practice.

Limitations:

- Resource constraints:

The capabilities of edge devices include minimal computational energy as well as limited storage capabilities and short battery life. It becomes complicated to execute complex AI algorithms and maintain long-term data storage on the resources available within the device.

- Management complexity:

Having good orchestration solutions for a big dispersed edge device fleet requires complex management systems that lead to additional operational challenges.

- Device heterogeneity and standardization issues:

Various types of edge hardware devices with different brands and protocol standards and performance levels create difficulties in integrating and maintaining systems properly throughout educational institutions.

Using edge computing enables the school digital twin to deliver continual responsiveness with strong resilience and built-in point-of-use intelligence. The proper alignment of edge along with fog and cloud systems enables schools to establish adaptable and protected digital environments for optimizing their educational and operational operations.

3.4 Comparative analysis table: Cloud versus fog versus edge for School Digital Twin

Table 13.3 shows a comparison between cloud, fog, and edge systems when used for school digital twin implementation. The comparison shows clear divergences between placement, reply duration, and computing strength together with data procedures and defensive measures (Botín-Sanabria et al., 2022). The different layers of this system function independently to perform real-time edge monitoring events while also managing predictive analysis and historical information in cloud storage. When combined, they produce an intelligent and secure digital twin system that operates continuously for educational facilities.

4. Proposed architecture

The school digital twin architecture has a layered design structure which contains four levels: device layer together with edge layer followed by fog layer and ending at cloud layer as illustrated in Fig. 13.3. The system distributes data collections effectively to perform instant processing followed by decision-making activities and forecasting capabilities under robust security along with privacy protection.

4.1 Layered architecture: Device Layer, edge layer, fog layer, cloud layer

The system is divided into four interconnected layers (Kamble et al., 2022):

- Device layer: The foundation of this system consists of IoT devices, which include sensors along with cameras plus wearable units installed throughout the school to acquire current environmental and behavioral and safety information.
- Edge layer: Medical applications situated near the devices operate the edge layer to conduct preliminary analytics, such as detecting local events where emergencies occur in crowded areas, followed by immediate response capabilities. The upper layers experience lower data load because of this strategy while maintaining speed in operations.
- Fog layer: Smart gateways and local servers make up the fog layer, which serves as the intermediate connection point. The node data aggregation process at this layer lets it manage and organize information from multiple sources while making immediate operational decisions for real-time traffic control during school hours and sending only important data to the cloud.
- Cloud layer: Data storage operations and advanced model training through ML along with complex simulation operations take place in the cloud environment. AI models receive predictive information from the

TABLE 13.3 Comparison between different computing techniques for implementing a School Digital Twin.

Aspect	Cloud computing	Fog computing	Edge computing
Location	Remote data centers	Near the school premises	On the IoT devices (sensors, cameras, wearables)
Latency	High	Medium	Very low
Processing power	Very high (scalable resources)	Moderate (local servers/gateways)	Low (device-level processing)
Data handling	Historical data storage, model training, and simulations	Real-time orchestration, filtering, and aggregation	Immediate event detection, local actions
Connectivity requirement	Always-on Internet needed	Local network, less dependent on internet	Minimal; operates autonomously
Security and privacy	Higher risk during data transfer	Improved control over sensitive data	Maximum privacy; minimal data transmission
Application in school DT	Predictive analysis, long-term planning	Aggregated analytics, optimized real-time response	Immediate control, safety alerts, and monitoring
Example tasks	AI model updates, simulation engines	Real-time decision making, load balancing	Sensor fusion, local event detection

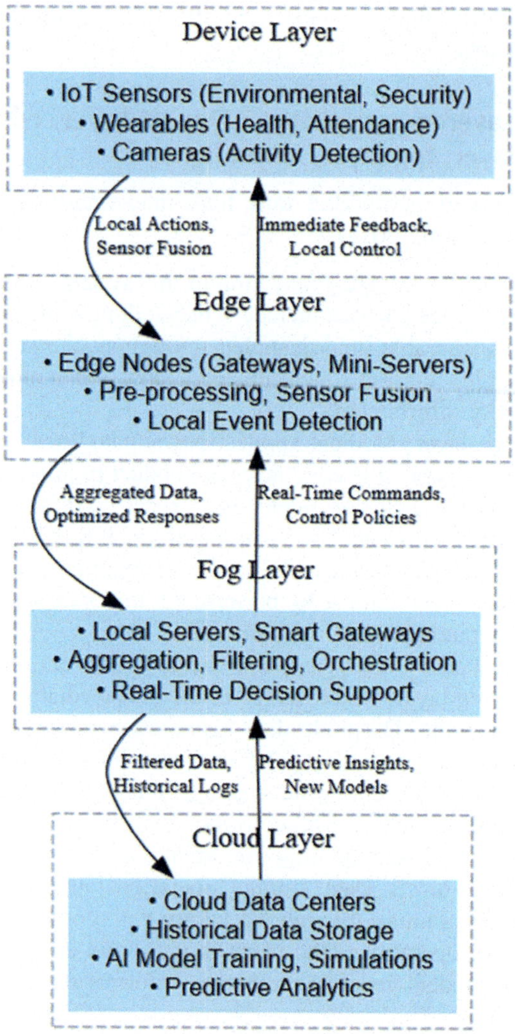

FIGURE 13.3 System architecture diagram for school digital twin.

cloud platform for updating purposes before returning to fog and edge layers for continuous model development.

4.2 Data flow and processing mechanism

The information path starts within the device layer through which real-world events enter before preprocessing occurs at the edge layer for immediate response. The fog layer receives processed meaningful data sent from device layer for aggregation before performing further analysis. Upon orchestration, the cloud layer receives only compressed valuable data from the lower layers, which optimizes bandwidth consumption for efficiency purposes. The cloud layer performs intensive analytical processing, including model training and predictive modeling, and transmits the results to the device, edge, and fog layers to support real-time advanced control operations within the school environment.

4.3 Security and privacy considerations

Security and privacy are enforced at every level:

- Protection of source data at its initial point relies on three security measures from device and edge layers, which consist of lightweight encryption and device authentication along with local anonymization features.
- The fog layer protects all middle-stage processing from attacks through its access control systems, which integrate edge intelligence and security detection technologies.
- Sensible school records together with student privacy receive security protection through cloud layer's implementation of advanced security tools including federated learning and secure multiparty computation (SMPC) and blockchain-based integrity checks during collaborative model training.

End-to-end encryption, periodic audits, and strict compliance with privacy regulations (such as GDPR, HIPAA, and FERPA) are embedded across the architecture.

4.4 Advantages of the layered architecture

Three main benefits exist from implementing this layer architecture:

- Low latency responses: Real-time event detection and immediate action through local edge processing.
- The fog layer manages data efficiently so it reduces the amount of content sent to the cloud while performing aggregation and filtering procedures.
- The solution provides heightened security measures that secure both school-related sensitive information as well as student data by implementing local processing together with secure cloud management.
- The system enables smart school operations through its capacity to create intelligent decisions that function through system control at the edge and fog and predictive analytics at the cloud level.
- The functionality of the system continues through local processing even when cloud connectivity is interrupted temporarily.

5. Key applications and use cases

A real-time virtual model of the school environment created through the school digital twin framework enables various smart applications (Zaballos et al., 2020). The combination of these systems strengthens both operational and environmental effectiveness as well as student achievements and safety measures in educational centers.

Table 13.4 demonstrates how the school digital twin architecture delivers a wide spectrum of influential

TABLE 13.4 Summary of key applications and use cases for school digital twin.

Application	Description	Benefits
Smart classroom management	Automated control of lighting, HVAC, and attendance based on real-time data	Improved comfort, energy efficiency
Emergency monitoring and evacuation	Real-time event detection with dynamic evacuation support	Enhanced safety, faster response times
Personalized learning environment	Adaptive environment and content adjustment based on student engagement and preferences	Better learning outcomes, personalized education
Predictive maintenance of facilities	Monitoring infrastructure health to predict failures and schedule maintenance	Reduced downtime, cost savings
Real-time resource optimization	Dynamic management of classrooms, labs, and resources based on current and forecasted usage	Operational efficiency, resource sustainability

operational capabilities. Together, they demonstrate the transformative potential of digital twin technologies in the education sector. The following subsection describes several essential implementations.

5.1 Smart classroom management (lighting, HVAC, attendance)

A school digital twin operates through automated controls that adjust lighting intensity along with HVAC (heating, ventilation, and air-conditioning) systems and tracks student attendance. The system produces optimized energy consumption through real-time data obtained by occupancy sensors and temperature monitors, and air quality sensors, alongside other sensors to maintain a comfortable student environment. College attendance strategies integrating facial identification systems along with RFID bracelets both enhance student records while minimizing human error in data registry. Both student satisfaction and energy efficiency and operation costs diminish while promoting higher comfort levels throughout facilities.

5.2 Emergency monitoring and evacuation systems

The school digital twin analyzes real-time detections through the integration of data obtained from surveillance cameras, together with motion detectors and fire alarms, and emergency alert devices. Through the digital twin system, it tells administrators and security personnel right away and starts active evacuation procedures by delivering customized alerts and live maps for directing students and staff to safety routes. The system enhances safety standards and decreases response period, and minimizes crisis-related chaos.

5.3 Real-time resource optimization

The increased availability of vacant spaces can trigger automatic reallocation processes, and daily student participation levels enable transport route adaptability. The efficient allocation of resources creates space efficiency while decreasing operational costs and enhancing sustainable management of educational resources throughout the entire educational environment.

6. Challenges and research opportunities

School digital twin architectural implementation presents transformative opportunities for practical use but also creates essential research challenges and opportunities (Rasheed et al., 2020). Data privacy and cybersecurity become important considerations in student information management because they demand new privacy-protecting analytic methods, such as federated learning together with differential privacy and zero-trust security frameworks and specific encryption solutions for IoT-based educational environments (Deng et al., 2021). IoT devices with different protocols require bringing together disconnected systems, so standard interfaces, together with middleware tools and semantic data models, must be implemented for system unification. The way forward in processing power depends on efficient energy utilization while managing limited resources at edge devices, which motivates research into TinyML technology and both hardware and energy-related performance improvements. The major challenge of cost-efficiency and scalability requires relevant modular systems with automated resource allocation functions that use hybrid solutions between cloud and edge computing networks. The deployment proccss needs ethical measures that solve the student surveillance challenges, as well as resolve the unclear data ownership issues, and require systematic governance approaches to tackle algorithmic bias risks. Research teams from multiple disciplines should collaborate to develop secure digital twins adoption solutions that provide equality and sustainability in educational environments.

Fig. 13.4 demonstrates how research opportunities address barrier implementation issues for school digital twin systems, leading to the development of AI-driven twins and 5G/6G integration along with metaverse-based educational spaces. A detailed relationship exists between implementation hurdles of school digital twin systems and their connected research opportunities and technical solutions in Table 13.5. This organized framework presents direct approaches to building intelligent educational settings that possess greater resilience.

7. Conclusion

Bachelor and expert judgement form the foundation of the digital twin–based smart school system, which combines IoT hardware with AI analytical capabilities to track the state and energy usage of educational facilities, as well as environmental elements and student participation levels, to perform predictive maintenance, make data-driven decisions, and provide individualized educational experiences. This technology uses virtual representations of school surroundings to assist instructors in providing customized instruction that aligns with Instruction 4.0 objectives. The evolution of smart education using 6G networks, metaverse technology, AI tutoring systems, and blockchain security protocols seeks to create resilient learning systems that protect students' mental health while developing unique abilities in sustainable educational environments. Digital twins will realize their full potential

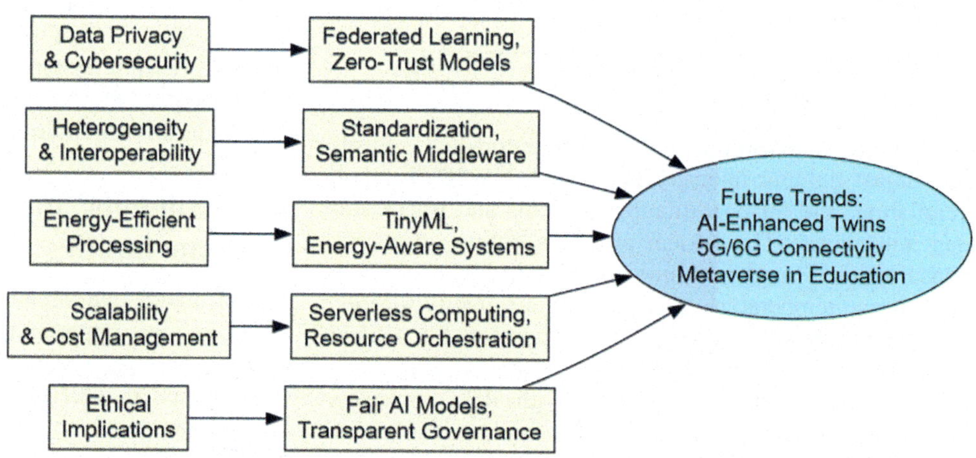

FIGURE 13.4 Challenges and research opportunities for school digital twin.

TABLE 13.5 Challenges, research opportunities, and potential solutions for School Digital Twin.

Challenge area	Research opportunities	Potential solutions/Technologies
Data privacy and cybersecurity	Federated learning, zero-trust security models, and advanced encryption for IoT ecosystems	Homomorphic encryption, blockchain-based authentication
Heterogeneity and interoperability	Standardized communication protocols, middleware for heterogeneous devices, and semantic models	MQTT/CoAP protocols, IoT middleware platforms, and OPC UA standards
Energy-efficient processing	TinyML, dynamic workload balancing, energy-aware system design	TinyML frameworks (TensorFlow lite micro), dynamic voltage scaling
Scalability and cost management	Modular architectures, serverless computing, and resource orchestration	Kubernetes, edge-fog-cloud orchestration platforms
Ethical implications	Fairness-aware AI models, transparent governance, stakeholder participation frameworks	Explainable AI (XAI) methods, data governance policies
Future trends (AI, 5G/6G, metaverse)	AI-driven autonomous twins, ultralow latency 5G/6G networks, immersive learning platforms	Digital twins with reinforcement learning, VR/AR technologies

via collaborative efforts between decision-makers, educational experts, and technology professionals to overcome existing obstacles such as information security and interoperability concerns, as well as system costs. The deliberate implementation of this paradigm will turn education into an accessible system that combines flexibility with long-term trend predictions.

References

Aheleroff, S., Xu, X., Zhong, R. Y., & Lu, Y. (2021). Digital twin as a service (DTaaS) in industry 4.0: An architecture reference model. *Advanced Engineering Informatics, 47*, Article 101225.

Attaran, M., & Celik, B. G. (2023). Digital twin: Benefits, use cases, challenges, and opportunities. *Decision Analytics Journal, 6*, Article 100165.

Borowski, P. F. (2021). Digitization, digital twins, blockchain, and industry 4.0 as elements of management process in enterprises in the energy sector. *Energies, 14*(7), 1885.

Botín-Sanabria, D. M., Mihaita, A. S., Peimbert-García, R. E., Ramírez-Moreno, M. A., Ramírez-Mendoza, R. A., & Lozoya-Santos, J. D. J. (2022). Digital twin technology challenges and applications: A comprehensive review. *Remote Sensing, 14*(6), 1335.

Broo, D. G., & Schooling, J. (2023). Digital twins in infrastructure: Definitions, current practices, challenges and strategies. *International Journal of Construction Management, 23*(7), 1254–1263.

Deng, T., Zhang, K., & Shen, Z. J. M. (2021). A systematic review of a digital twin city: A new pattern of urban governance toward smart cities. *Journal of management science and engineering, 6*(2), 125–134.

Erol, T., Mendi, A. F., & Doğan, D. (2020). Digital transformation revolution with digital twin technology. In *2020 4th International symposium on multidisciplinary studies and innovative technologies (ISMSIT)* (pp. 1–7). IEEE.

Fuller, A., Fan, Z., Day, C., & Barlow, C. (2020). Digital twin: Enabling technologies, challenges and open research. *IEEE Access, 8*, 108952–108971.

Haleem, A., Javaid, M., Singh, R. P., & Suman, R. (2023). Exploring the revolution in healthcare systems through the applications of digital twin technology. *Biomedical Technology, 4*, 28–38.

Kamble, S. S., Gunasekaran, A., Parekh, H., Mani, V., Belhadi, A., & Sharma, R. (2022). Digital twin for sustainable manufacturing supply chains: Current trends, future perspectives, and an implementation framework. *Technological Forecasting and Social Change, 176*, Article 121448.

Kartashova, L. A., Gurzhii, A. M., Zaichuk, V. O., Sorochan, T. M., & Zhuravlev, F. M. (2020). Digital twin of an educational institution: An innovative concept of blended learning. In *Proceedings of the symposium on advances in educational technology* (aet).

Mihai, S., Yaqoob, M., Hung, D. V., Davis, W., Towakel, P., Raza, M., … Nguyen, H. X. (2022). Digital twins: A survey on enabling technologies, challenges, trends, and future prospects. *IEEE Communications Surveys & Tutorials, 24*(4), 2255–2291.

Rahmani, R., Jesus, C., & Lopes, S. I. (2024). Implementations of digital transformation and digital twins: Exploring the factory of the future. *Processes, 12*(4), 787.

Rasheed, A., San, O., & Kvamsdal, T. (2020). Digital twin: Values, challenges and enablers from a modeling perspective. *IEEE Access, 8*, 21980–22012.

Sepasgozar, S. M. (2021). Differentiating digital twin from digital shadow: Elucidating a paradigm shift to expedite a smart, sustainable built environment. *Buildings, 11*(4), 151.

Sepasgozar, S. M., Khan, A. A., Smith, K., Romero, J. G., Shen, X., Shirowzhan, S., … Tahmasebinia, F. (2023). BIM and digital twin for developing convergence technologies as future of digital construction. *Buildings, 13*(2), 441.

Sharma, A., Kosasih, E., Zhang, J., Brintrup, A., & Calinescu, A. (2022). Digital twins: State of the art theory and practice, challenges, and open research questions. *Journal of Industrial Information Integration, 30*, Article 100383.

Singh, M., Fuenmayor, E., Hinchy, E. P., Qiao, Y., Murray, N., & Devine, D. (2021). Digital twin: Origin to future. *Applied System Innovation, 4*(2), 36.

Wang, Y., Su, Z., Guo, S., Dai, M., Luan, T. H., & Liu, Y. (2023). A survey on digital twins: Architecture, enabling technologies, security and privacy, and future prospects. *IEEE Internet of Things Journal, 10*(17), 14965–14987.

Zaballos, A., Briones, A., Massa, A., Centelles, P., & Caballero, V. (2020). A smart campus' digital twin for sustainable comfort monitoring. *Sustainability, 12*(21), 9196.

Chapter 14

Blockchain and zero-knowledge proof for security and privacy preservation in student sensor networks

Vincent Omollo Nyangaresi[1, 2], Ahmad A. AlRababah[3], Ali Hamzah Najim[4], Zaid A. Abduljabbar[5, 6] and Mohammad Shariq[7]

[1]*Department of Computer Science and Software Engineering, Jaramogi Oginga Odinga University of Science and Technology, Bondo, Kenya;* [2]*Department of Applied Electronics, Saveetha School of Engineering, SIMATS, Chennai, Tamil Nadu, India;* [3]*Faculty of Computing and Information Technology in Rabigh, Rabigh, KSA, King Abdulaziz University, Bando, Kenya;* [4]*Department of Computer Technical Engineering, Imam Al-Kadhum College (IKC), Baghdad, Iraq;* [5]*Department of Computer Science, College of Education for Pure Sciences, University of Basrah, Basrah, Iraq;* [6]*Department of Business Management, Al-Imam University College, Balad, Iraq;* [7]*Department of Information and Systems Security, College of Information Technology, United Arab Emirates University, Abu Dhabi, United Arab Emirates*

1. Introduction

Academic institutions have strived to improve the security of their campus community, which comprises of students, teachers, and other staff (Rao et al., 2023). Conventionally, this safety has been ensured through smart locks, and smart identification systems among other information technology systems. These technologies help in defining zones, monitoring valuable items, as well as managing campus access. Although learning institutions spend millions on these expensive equipment, keeping track of them is quite challenging. Similarly, conventional techniques such as rule-based alarm systems and manual surveillance cannot offer context-aware and real-time feedback (Kumari et al., 2024). Due to increasing operational efficiency and security concerns, educational institutions have continued to explore cutting-edge techniques to track and mitigate probable dangers. In this regard, self-monitoring technologies can greatly assist in addressing some of these problems.

The Internet of things (IoT) has been deployed in various fields such as smart homes, industrial automation, and healthcare among others. The rapid growth in IoT and wearable technologies has facilitated human activity recognition (HAR), a significant tool for the development of automated and intelligent systems (Sinha, 2025). This HAR plays a critical role in detecting unusual patterns and monitoring behaviors, which enable timely interventions in critical situations. In addition to IoT, artificial intelligence (AI), big data, and cloud computing are some of the new technological innovations that can potentially improve the quality of human life. Academics institutions can adopt these technologies to bring some paradigm shift in their teaching and research, greatly improving learning experience for their students, faculties, and staff. These technologies can also offer alternative way of monitoring attendance and cooperation (Zhou & Tang, 2022). As such, the recent past has seen a surge in the number of university administrations implementing digital campus reforms (Al-Shoqran & Shorman, 2021).

A smart, intelligent of digital campus is the one that deploys various technologies and infrastructures (Kadhim et al., 2025) to enhance student experience, teaching, and research processes. The automation of tasks in these campuses helps improve learning, efficiency, safety, comfort, campus services, and quality of life of its stakeholders (Gopal & Patan, 2019; Kumar et al., 2024; Saravanan et al., 2023; Srhir et al., 2023). In this environment, stakeholders, students, staff, and the neighboring environment communicate intelligently with the campus (Malatji, 2017). These smart campuses adopt the smart city (Nyangaresi, Abduljabbar, et al., 2024) concept and are technology-driven. Due to their technological developments, smart campuses have received much attention from scholars and professionals (AbuAlnaaj et al., 2020; Hidayat & Sensuse, 2022; Horvá et al., 2022). These

Digital Twins in the Smart Classroom. https://doi.org/10.1016/B978-0-443-41632-3.00030-8

campuses can be seen as miniature smart cities (Arunsankar et al., 2023) and can be utilized to switch to greener energy sources that can benefit their surroundings (Abdulmouti et al., 2022). By enhancing operational efficiency and offering advanced services to staff, faculty, and students, intelligent campus technologies can positively change current educational system (Abawajy & Hassan, 2017; Al-Sulami et al., 2023; Abuarqoub et al., 2017; Deepa et al., 2023).

Due to the incorporation of sensors in the campus automation, human-based control of appliances and monitoring of various conditions is eliminated. Instead, IoT-based and HAR systems offer automated security through various LCD displays, alarm systems, email notifications, and sirens whenever anomalies or risky behaviors are detected (Saha et al., 2024). In spite of the benefits brought forth by smart campus, the deployed technologies introduce numerous vulnerabilities that can be exploited to wreak havoc in learning institutions. These security threats include breaches to network security, physical security, and data confidentiality. In addition, there are always challenges when integrating these technologies with legacy campus systems. Since data exchanges take place over the public Internet, many security and privacy threats lurk in this environment (Joseph & Pandeeswari, 2025; Nyangaresi, 2025a). Such threats include eavesdropping, physical, network, malware, social engineering, application layer, data privacy, external, internal, routing disruption, and IoT-based attacks. The physical attacks are normally directed toward smart campus access points, servers, IoT devices, and data centers (Khalaf et al., 2024; Nag et al., 2024; Radhi et al., 2024; Stellios et al., 2018; Salim et al., 2020). They may include tampering, theft, unauthorized access to highly sensitivity areas, and vandalism. On the other hand, network attacks aim to intercept data or disrupt communication within smart campus. These threats include man-in-the-middle (MitM), port scanning, network scanning, and packet sniffing (AlDairi & Tawalbeh, 2017; Fereidouni et al., 2025; Ul-Aaish et al., 2024). The smart campus devices and systems may also be targeted by malware such as ransomware, viruses, trojans, and worms. These malware attacks can result in system operations disruptions, data loss, and unauthorized access (Diro, 2025; Edwards, 2024).

On the other hand, social engineering threats may help attackers to gain unauthorized access. Techniques such as tailgating, phishing, baiting, and pretexting are used to deceive campus stakeholders such as students and staff into divulging confidential and sensitive data (Colabianchi et al., 2021; Salahdine & Kaabouch, 2019). The web applications deployed in smart campus for various activities are also susceptible to numerous attacks such as cross-site request forgery (CSRF), cross-site scripting (XSS), and SQL injection. The aim of these application layer threats is to compromise integrity and security (Nyangaresi, 2025b) of smart campus systems and devices. On the other hand, IoT-based attacks leverage on vulnerabilities in IoT devices to gain unauthorized access and control, launch botnets to cause distributed denial of service (DDoS), and device spoofing (Gelgi et al., 2024; Siwakoti et al., 2023). The adversarial unauthorized access to surveillance systems, alarms, and door locks can result in these systems being disabled, their functionalities manipulated, and facilitate access to restricted areas or systems. In smart campus, applications and devices generate various data, some of which can be sensitive. Any data privacy attack can compromise result in deletion, manipulation, breaches, loss, and service hijacking (Aslan et al., 2023; Jawad et al., 2024; Tao et al., 2018). In contrast, external attackers can perform malicious injections to disrupt the functionality of networks (Silva et al., 2022). However, insider threats aim to compromise the sensor and use them to launch attacks against other sensors or systems (Alansari et al., 2022). On the contrary, active routing disruptions target the network layer with attacks such as blackhole or sinkhole (Shanmugaraja et al., 2023).

The aforementioned security and privacy threats call for robust security measures to protect sensitive user data, and safeguard confidentiality of information exchanged in the network (Bodur et al., 2019; Nyangaresi et al., 2025; Sehgal et al., 2022). However, the high number and heterogenous of IoT devices complicate the development of security solutions for these networks. This is due to the increased surface area for launching attacks, coupled with numerous vulnerabilities in the IoT devices, which can be exploited to cause privacy invasions, data breaches, or unauthorized access (Bouchouia et al., 2023). Majority of the current security techniques are based on public-key and private-key cryptography. These security solutions are characterized by high energy and computational requirements that can shorten the lifespan of IoT devices (Oladipupo et al., 2023). In addition, some of the current encryption algorithms have scalability challenges, while others are vulnerable to threats such as MitM and DDoS. Therefore, there is need for more efficient techniques that can be applied to secure IoT devices deployed in smart campus.

The IoT devices deployed in student sensor networks are limited with regard to processing capabilities. As such, they cannot handle conventional security techniques that require heavy computations on these sensors. In addition, majority of the scheme in sensor networks are susceptible to numerous security and privacy threats. Moreover, the highly heterogenous and dynamic nature of student sensor networks render centralized security solutions inapplicable in this environment. There is therefore need to develop robust but efficient authentication scheme that can ensure scalability, flexibility, and eliminate single point of failure.

To solve the numerous security and privacy issues in student sensor networks, this paper makes the following main contributions:

- We leverage on blockchain, public key infrastructure, bilinear mapping, and zero-knowledge proof to design an authentication scheme to mitigate against a myriad of security and privacy threats in sensor networks.
- Extensive semantic security analysis is performed to demonstrate the robustness of our protocol against a wide range of attacks such as single point of failure, user identity leakage, forgery, eavesdropping, session hijacking, message replay, privileged insider, denial of service, side-channeling, man-in-the-middle, and impersonation. It also offers scalability, data privacy, and robust authentication.
- We test the proposed scheme in Highsimb simulator and show that it incurs low transaction verification duration. In addition, it takes a shorter time to new blocks to the blockchain network.

The remainder of this paper is structured as follows: Section 2 discusses some of the related works in this domain, while Section 3 details the system model. On the other hand, Section 4 presents the security analysis, while Section 5 describes its performance evaluation. Toward the end of this paper, we provide the conclusion and future research directions.

2. Related works

To effectively manage smart campus, numerous IoT-based techniques have been put forward. For instance, a smart security monitoring system based on IoT is presented in Mo (2024). This system facilitates real-time and intelligent (Alshuraify et al., 2024) management of security in colleges and universities but its security analysis is missing. On the other hand, a system that integrates IoT in online education learning environment is developed in Balakrishnan et al. (2024). Unfortunately, elaborate security analysis has not been carried out. To improve campus security, a system that combines convolutional neural networks (CNNs) with IoT is introduced in Sasirekha et al. (2024). However, this technique has not evaluated against typical wireless threats. Similarly, hybrid deep learning model, which combines CNNs and long short-term memory (LSTM), is developed in Sinha (2025), while a cloud computing–based smart campus system is presented in Kshetri (2017). Nevertheless, security and privacy evaluations of these systems are missing. On the other hand, numerous human activity recognition (HAR) systems have been introduced in Ahmad et al., 2024; Beemkumar et al., 2023; Chen et al., 2021; Chen, Teng, et al., 2023; El-Adawi et al., 2024; Gargees, 2024; and Mekruksavanich & Jitpattanakul, 2023). These HAR systems are based on various machine learning models such as support vector machine (SVM) and decision tree (DT) (Gargees, 2024), deep learning (El-Adawi et al., 2024), adaptive hybrid deep attentive networks with archerfish optimization (Ahmad et al., 2024), transfer learning with deep learning models (Beemkumar et al., 2023), deep learning with CNN (Mekruksavanich & Jitpattanakul, 2023), and random forest with neural network (Chen et al., 2021). However, these models have not been evaluated in terms of security and privacy.

To develop a smart campus teaching system with enhanced teaching efficiency, the ZigBee algorithm–based system is introduced in Gao (2022). On the other hand, an intelligent campus monitoring system is developed in Li et al. (2018) based on wireless sensor technologies. Similarly, a system that enhances intelligent management of learning, management, and teaching is presented in Mayer & Baeumner (2019). Based on wireless sensor networks, a system that offers convenient and efficient management for teachers and students is introduced in Qadri et al. (2020). Similar systems are developed in Ray, 2017, 2018; Saez et al., 2018; and Sheth et al., 2018). Unfortunately, security evaluation (Nyangaresi, Alsolami, & Ahmad, 2024) of these systems has not been provided. Therefore, secure systems for various wireless networks that can be applied in smart campus have been developed in Hu et al., 2019; Wang et al., 2018). However, most of these schemes have performance, privacy, and security issues that need to be addressed to render them effective in smart campus deployment. For instance, the protocol in Wang et al. (2018) does not offer user anonymity, while the scheme in Amin et al. (2018) is vulnerable to known session key temporary information, user impersonation, off-line password, and identity guessing attacks (Ali, Pal, et al., 2024). As such, anonymity preserving technique is developed in Chai et al. (2023). Although the authors in Ali, Pal, et al. (2024) have presented an improved scheme, it has numerous design flaws and security lapses (Manickam & Devarajan, 2025). On the other hand, the protocol in Maurya & Sastry (2017) incurs high computation overheads (Bulbul et al., 2024). Similarly, the scalar point multiplications in Elhabrush & Ahmeda (2022) can potentially result in long execution durations. Therefore, lightweight schemes are presented in Sarbini et al., 2022; Umran et al., 2023; Yang et al., 2019). On the other hand, the biohashing technology utilized in Bera et al., 2020 and Srinivas et al., 2017) requires additional hardware to handle the biodata. The identity-based authentication technique in Xiong et al. (2022) is vulnerable to key escrow problems and password guessing threats. Since the elliptic curve–based authentication scheme in Hu et al. (2022) relies on a fully trusted registration center, it is susceptible to single point of failure. Similarly, the scheme in Chai et al. (2023) relies on secure channels during registration, and hence, it may not be applicable where such channels are not available (Hafeez et al., 2025).

To offer secure data aggregation and prevent collision risks, a certificate-less technique is developed in Liu et al. (2023). This scheme is shown to enhance user privacy protection at low computation overheads. On the other hand, blockchain-based certificate-less architecture is introduced in Wang et al. (2021) to enhance transparency and decentralization. However, it has high computation costs, which limit its applicability in smart campus networks. In addition, the current certificate-less techniques still face scalability constraints. Since the authentication protocol in Chen, Guo, et al. (2023) uses smart cards to authenticate users, it is prone to smart card loss attacks. Similarly, the scheme in Yang et al. (2019) is susceptible to smart card loss, identity guessing, privileged insider, and tracking attacks (Alsumayt et al., 2024). On the other hand, the approach in Singh et al. (2020) does not take user privacy protections into consideration. Therefore, a data and key privacy preservation technique is developed in Akila & Sheela (2017). Unfortunately, this scheme does not incorporate authorization and authentication mechanisms in its design (Erskine, 2024). Similarly, the schemes in Akila & Sheela, 2019; Hu et al., 2019; Yu et al., 2021; Zhou, Ge, et al., 2019; Zhou, Qin, et al., 2019 and Zhou, Qin, et al., 2019) offer data and key privacy but do not implement authentication and authorization procedures. Table 14.1 gives a summary of these related works in this domain.

Based on the information in Table 14.1, it is clear that the current security techniques for student sensor networks struggle with numerous performance, privacy and security lapses, which limit their applicability in these sensitive networks. The developed scheme is shown to be not only resilient but also efficient. As such, it can effectively address most of these setbacks.

3. System model

The proposed scheme is based on blockchain, bilinear mapping, elliptic curve cryptograph (ECC), Rivest–Shamir–Adleman (RSA) public key infrastructure, and zero-knowledge proofs. Table 14.2 shows the notations utilized throughout this paper.

We deploy the blockchain during registration and authentication to enhance scalability, and eliminate single point of failure. On the other hand, bilinear mapping, elliptic curve cryptograph, RSA and zero-knowledge proofs to strengthen the security and privacy posture of our scheme. Particularly, we deploy the zero-knowledge proofs to permit the sensor node, data center, and the campus sensor information center to attest to each other that they possess some specific information (for authentication), devoid of revealing their private information. The network model of the proposed protocol comprises of the blockchain network, sensor nodes S_i, campus sensor information center SC_j, and campus data center DC_k, as shown in Fig. 14.1.

To enhance security in student sensor networks, our scheme carries out the system initialization, login, and robust authentication. The particular details of these phases are elaborated in the subsections that follow.

3.1 System initialization phase

In this phase, all the system parameters are instantiated at the S_i. This includes parameters and computations required for encryption and decryption, specifically, parameters and calculations for elliptic curve, RSA-based public key, bilinear mapping.

Step 1: Taking elliptic curve identifier as ϕ, values for the elliptic curve base field f_q, elliptic curve cofactor c, constants a and b for the elliptic curve equation, degree d and order q of the elliptic curve $E(F_q)$.
Step 2: Taking P_1 and P_2 as the generators of subgroups G_1 and G_2 that yield $E(f_q)$, a bilinear mapping from G_1 to G_2 is defined as e: $G_1 \times \rightarrow G_2$. In addition, β and H are chosen as the bilinear pairing identifier and homomorphic mapping from G_1 to G_2, respectively.
Step 3: Suppose that 1^z is the unary notation for the security value z, while p and q are large prime numbers whose bit length is approximately $z/2$ each. Then, the modulus $\lambda = p \times q$ and the Euler's totient is give as $\varphi(\lambda) = (p-1)(q-1)$.
Step 4: Taking m as the public exponent, its corresponding private exponent n is given by $n = m^{-1} \bmod \varphi(\lambda)$. With these notations, the public key pair is denoted as $\{\lambda, m\}$ while the private key pair is represented by $\{\lambda, n\}$.
Step 5: The sensor node S_i secret key is set as δ and is used to derive parameter $a_1 = h(\delta)$. For encryptions, $\{1^z\}$ is supplied as the input parameter to the RSA key generation algorithm $GenRSA(.)$, that is $GenRSA(1^z)$.
Step 6: The S_i selects some two points $\acute{\eta}$ and ω on $E(f_q)$ and derives $a_2 = (\acute{\eta}.a_1)$ and $a_3 = (\omega.a_1)$. In addition, it generates its public key as υ.

3.2 Registration phase

To prevent intruders from invading the student sensor networks, all the sensor nodes must be registered at the campus sensor information center. This is achieved through the execution of the following steps:

Step 1: The sensor node S_i generates its unique identity SID_i and private security token ρ_s and private key $\acute{\alpha}$. This is followed by the determination of the current timestamp T_1. At the end, it creates registration

TABLE 14.1 Summary of related works.

Scheme	Year	Limitations
(Sinha, 2025)	2025	Security and privacy evaluations not done
(Mo, 2024)	2024	Missing security analysis
(Balakrishnan et al., 2024)	2024	Elaborate security analysis not done
(Sasirekha et al., 2024)	2024	Not evaluated against typical wireless threats
(Kshetri, 2017)	2017	Security and privacy evaluation missing
(Gargees, 2024)	2024	
(El-Adawi et al., 2024)	2024	
(Ahmad et al., 2024)	2024	
(Chen, Teng, et al., 2023)	2023	
(Beemkumar et al., 2023)	2023	
(Mekruksavanich & Jitpattanakul, 2023)	2023	
(Chen et al., 2021)	2021	
(Gao, 2022)	2022	
(Li et al., 2018)	2018	
(Mayer & Baeumner, 2019)	2019	
(Qadri et al., 2020)	2020	
(Ray, 2018)	2018	
(Saez et al., 2018)	2018	
(Sheth et al., 2018)	2018	
(Ray, 2017)	2017	
(Wang et al., 2018)	2018	Does not offer user anonymity
(Ali, Pal, et al., 2024)	2024	Has design flaws and security lapses
(Amin et al., 2018)	2018	Vulnerable to known session key temporary information, user impersonation, off-line password and identity guessing attacks
(Maurya & Sastry, 2017)	2017	Incurs high computation overheads
(Elhabrush & Ahmeda, 2022)	2022	Long execution durations
(Srinivas et al., 2017)	2017	Requires additional hardware to handle the bio data
(Xiong et al., 2022)	2022	Susceptible to key escrow problems and password guessing threats
(Bera et al., 2020)	2020	Requires additional hardware to handle the bio data
(Hu et al., 2022)	2022	Single point of failure
(Chai et al., 2023)	2023	Requires secure channels for registration
(Liu et al., 2023)	2023	Face scalability constraints
(Wang et al., 2021)	2021	High computation costs Face scalability constraints
(Chen, Guo, et al., 2023)	2023	Prone to smart card loss attacks
(Singh et al., 2020)	2020	Fails to take user privacy protections into consideration
(Yang et al., 2019)	2019	Vulnerable to smart card loss, identity guessing, privileged insider, and tracking attacks
(Akila & Sheela, 2017)	2017	Their designs do not consider authorization and authentication
(Zhou, Ge, et al., 2019)	2019	
(Akila & Sheela, 2019)	2019	
(Yu et al., 2021)	2021	
(Zhou, Qin, et al., 2019)	2019	
(Hu et al., 2019)	2019	

TABLE 14.2 Notations.

Symbol	Description
S_i	Sensor node *i*
SID_i	Real identity for S_i
SC_j	Sensor information center *j*
DC_k	Data center *k*
Ψ	Private key belonging to DC_k
Φ	Elliptic curve identifier
ρ_s	Private security token for S_i
Υ	Public key for S_i
$\acute{A}$	Private key for S_i
T_i	Timestamp *i*
r_j	Random nonce *j*
E_γ	Encryption using key $_\gamma$
D_γ	Decryption using key $_\gamma$
$\oplus$	XOR operation

message $Reg_1 = \{SID_i, \rho_s, a_2, a_3, T_1\}$ and transmits it towards the SC_j as demonstrated in Fig. 14.2.

Step 2: Upon getting message Reg_1, the SC_j confirms from its local ledger if there is an equivalent block information associated with the received sensor identity SID_i. If there is a sensor already assigned to this identity, S_i is prompted to choose a different identity and resubmit. Otherwise, the SC_j determines the current timestamp T_2, verifies timestamp T_1. If this test is positive, the SC_j constructs registration message $Reg_2 = \{SID_i, a_2, a_3, T_2\}$ that is forwarded to the data center DC_k. Otherwise, the registration is halted.

Step 3: On receiving message Reg_2 at timestamp T_3, the DC_k validates timestamp T_2 against T_3. On condition that T_2 is not fresh, the registration session is terminated. Otherwise, it generates random nonce r_1 and encrypts it using its private key ψ to obtain $b_1 = E_\psi(r_1)$. Next, it stores values $\{b_1, a_2, a_3\}$ in the local ledger associated with sensor node identity SID_i.

Step 4: The data center DC_k derives $b_2 = E_\upsilon(\psi, SID_i)$. This is followed by the determination of the current timestamp T_4 and the creation of registration response message $Reg_3 = \{b_1, b_2, T_4\}$, which is sent over to the SC_j as shown in Fig. 14.2.

Step 5: After getting Reg_3, the SC_j verifies timestamp T_4 using T_5, where T_5 is the current timestamp. The session is halted if T_4 is established not to be fresh. Otherwise, SC_j stores $\{SID_i, \rho_s, T_1\}$ in the block for later user information querying procedures. At the end, it composes message $Reg_4 = \{b_1, b_2, T_5\}$ and sends it over to sensor node S_i.

Step 6: Upon receiving message Reg_4 at timestamp T_6, the sensor node verifies the freshness of T_5. If this validation is successful, S_i deploys its private key $\acute{\alpha}$ to decrypt b_2 as $D_{\acute{\alpha}}(b_2)$ so as to obtain ψ. Otherwise, the registration process is terminated. Next, sensor node S_i utilizes the obtained private key ψ to decipher b_1 as $D_\psi(b_1)$ so as to retrieve random nonce r_1. Finally, it stores ψ in its memory for the subsequent login and authentication.

3.3 Login and authentication

In this phase, the sensor node S_i is interested in sending some collected data to the information center SC_j. Prior to being granted access rights, the following steps are executed to verify S_i.

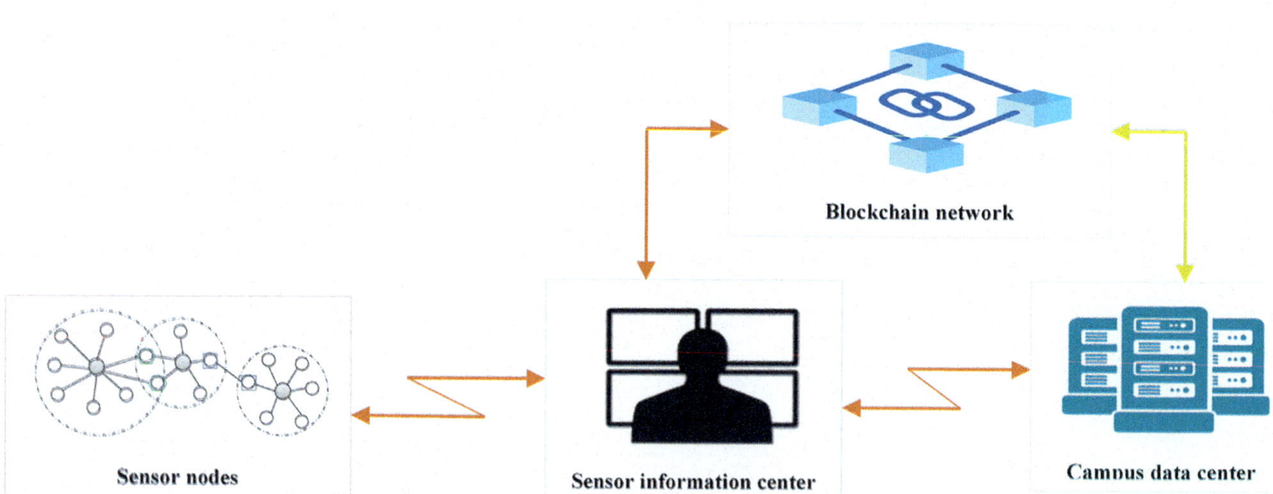

FIGURE 14.1 Network model.

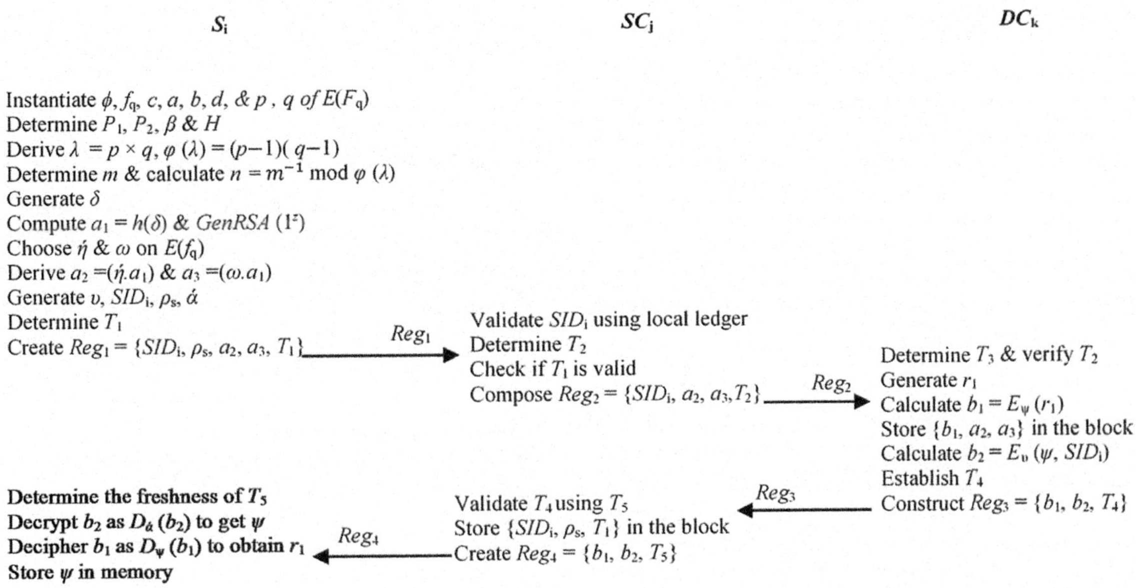

FIGURE 14.2 System setup and registration phase.

Step 1: The sensor node S_i generates random nonce r_2. Next, it derives $b_3 = (r_2 - a_1.r_1)$, $c_1 = (r_2.\acute{\eta})$, $c_2 = (r_2.\omega)$ and $\varsigma = SID_i \oplus h(a_2 \| a_3)$ at timestamp T_6. At the end, it creates login request $log_s = \{\varsigma, \rho_s, b_3, c_1, c_2, T_6\}$, which is then sent over to the SC_j as shown in Fig. 14.3

Step 2: On obtaining message log_s, the SC_j establishes the current timestamp T_7 and verifies the timestamp T_6 in log_s. Provided that T_6 is invalid, the login session is aborted. Otherwise, the SC_j calculates $SID_i = \varsigma \oplus h(a_2 \| a_3)$ and confirms if SID_i is already registered in its local ledger. If not, error message is returned to sensor node S_i and the session is halted. Otherwise, the SC_j creates authentication message $auth = \{\varsigma, b_3, c_1, c_2, T_8\}$ that is forwarded to data center DC_k.

Step 3: After getting message $auth$ at timestamp T_9, the DC_k verifies whether T_8 is fresh or not. Essentially, the session is terminated if T_8 is not fresh. Otherwise, it calculates $c_3 = b_3.\acute{\eta}$, $d_1 = b_3.\omega$ and $SID_i = \varsigma \oplus h(a_2 \| a_3)$. Using identity SID_i, the DC_k queries its local ledger for the associated sensor node S_i information required to derive $d_2 = (a_1.\acute{\eta})$, and $d_3 = (a_1.\omega)$. and $b_1 = E_\psi(r_1)$.

Step 4: Using private key ψ, the DC_k decrypts b_1 to obtain random nonce r_1. This is followed by the derivation of $e_1 = (r_1.a_1.\acute{\eta})$, $e_2 = (r_1.a_1.\omega)$, $e_3 = b_3.\acute{\eta} + r_1(a_1.\acute{\eta})$ and $f_1 = b_3.\omega + r_1(a_1.\omega)$. Next, DC_k checks if $e_3 \stackrel{?}{=} r_2.\acute{\eta}$, and $f_1 \stackrel{?}{=} r_2.\omega$ as sent by sensor node S_i. This is because $b_3 = (r_2 - a_1.r_1)$, implying that $r_2 = b_3 + a_1.r_1$ and hence $r_2.\acute{\eta} = \acute{\eta}(b_3 + a_1.r_1) = b_3.\acute{\eta} + r_1(a_1.\acute{\eta}) = e_3$. Similarly, $r_2.\omega = \omega(b_3 + a_1.r_1) = b_3.\omega + r_1(a_1.\omega) = f_1$.

Provided that these verifications are successful, sensor node S_i is authenticated and hence can be allowed to send data to the campus sensor information center SC_j. Otherwise, access is denied to this particular sensor node.

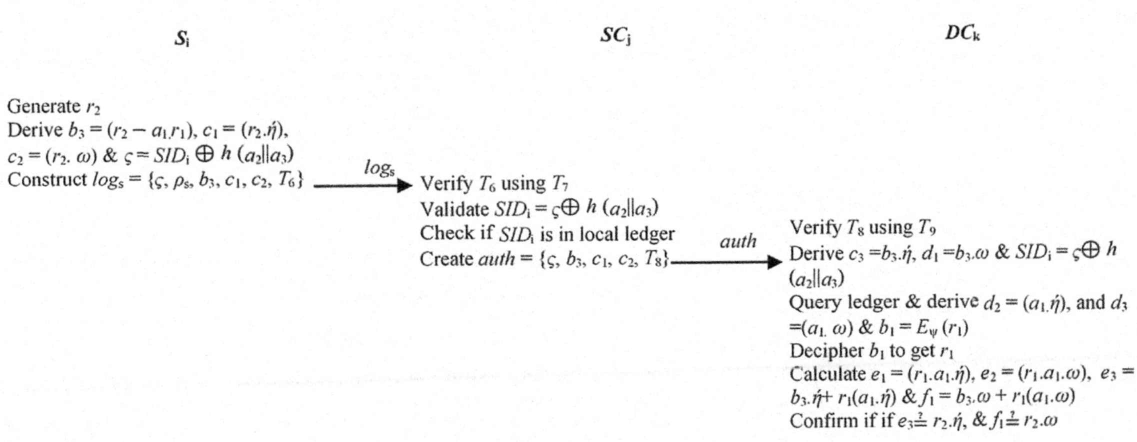

FIGURE 14.3 Login and authentication phase.

4. Security analysis

In this section, we demonstrate that our scheme is resilient against some typical privacy and security threats in student sensor networks. In addition, we show that the proposed provides features such as scalability and robust authentication. Therefore, we state and proof the following theorems.

4.1 Theorem 1: Our approach ensures robust authentication

Proof: In the proposed protocol, the SC_j verifies sensor node S_i by checking if SID_i received in login request $log_s = \{SID_i, \rho_s, b_3, c_1, c_2, T_6\}$ is already registered in its local ledger. On the other hand, the DC_k authenticates the campus sensor information center SC_j by confirming whether $e_3 \stackrel{?}{=} r_2.\acute{\eta}$, and $f_1 \stackrel{?}{=} r_2.\omega$. This mutual authentication ensures that intruders are unable to invade the network and carry out malicious activities.

4.2 Theorem 2: Single point of failure threats are thwarted

Proof: Our scheme is based on the consortium that deploys Fabric as its permission-grant blockchain (PGB). Consequently, only legitimate network elements can access sensitive data and insert data into the chain. Since PGM model has enhanced transaction processing speeds, increased number of sensor nodes and transactions cannot easily overwhelm the data center.

4.3 Theorem 3: User identity leakage and forgery attacks are prevented

Proof: In our scheme, messages $\text{log}_s = \{\varsigma, \rho_s, b_3, c_1, c_2, T_6\}$, and $auth = \{\varsigma, b_3, c_1, c_2, T_8\}$ are transmitted over the public Internet. Evidently, none of these messages contain plain-text sensor identity SID_i. Here, $\varsigma = SID_i \oplus h\ (a_2 \| a_3)$, $b_3 = (r_2 - a_1.r_1)$ and $c_1 = (r_2.\acute{\eta})$, $c_2 = (r_2.\ \omega)$. Any efforts in forging SID_i will flop since the adversary does not have access to sensor node secret key δ as well as elliptic curve points $\acute{\eta}$ and ω needed to derive $a_2 = (\acute{\eta}.a_1)$ and $a_3 = (\omega.a_1)$, where $a_1 = h(\delta)$.

4.4 Theorem 4: Our protocol mitigates eavesdropping and session hijacking threats

Proof: Suppose that the attacker wants to eavesdrop login message $\text{log}_s = \{\varsigma, \rho_s, b_3, c_1, c_2, T_6\}$. Thereafter, session hijacking is attempted by trying to compose authentication message $auth = \{\varsigma, b_3, c_1, c_2, T_8\}$. Although the intruder has access to values $\{\varsigma, \rho_s, b_3, c_1, c_2, T_6\}$, timestamp $T_8 \neq T_6$, and hence, this bogus message will be detected at the DC_k. This effectively mitigates against these two cyber threats.

4.5 Theorem 5: Data privacy is preserved

Proof: In the proposed protocol, the sensory data is securely exchanged among the sensor nodes S_i, campus sensor information center SC_j and campus data center DC_k. To achieve this, our scheme utilizes an amalgamation of symmetric and asymmetric encryption to encipher the collected data. In addition, we deploy blockchain and zero-knowledge proof technology to uphold protected authentication and storage of sensor sensitive data. This renders it computationally infeasible for unauthorized intruders to access stored data and also eavesdrop the communication channel.

4.6 Theorem 6: Message replay attacks are kept at bay

Proof: The aim of these attacks is to intercept, modify and resend messages $\text{log}_s = \{\varsigma, \rho_s, b_3, c_1, c_2, T_6\}$, and $auth = \{\varsigma, b_3, c_1, c_2, T_8\}$ exchanged over the public Internet. However, upon receiving login message log_s, the SC_j establishes the current timestamp T_7 and verifies the timestamp T_6 in log_s. Similarly, upon receiving message $auth$ at timestamp T_9, the DC_k verifies whether T_8 is fresh or not. In this way, any replayed messages will be easily detected at the SC_j and DC_k.

4.7 Theorem 7: The proposed protocol mitigates privileged insider

Proof: Let us assume that some privileged entity within the SC_j and DC_k wants to compromise the security of the entire network. Therefore, values $\{SID_i, \rho_s, a_2, a_3, T_1\}$, $\{SID_i, a_2, a_3, T_2\}$, $\{b_1, b_2, T_4\}$, and $\{b_1, b_2, T_5\}$, in Reg_1, Reg_2, Reg_3 and Reg_4 can be obtained by this privileged user. Thereafter, attempts are made to construct messages $\text{log}_s = \{\varsigma, \rho_s, b_3, c_1, c_2, T_6\}$, and $auth = \{\varsigma, b_3, c_1, c_2, T_8\}$ exchanged during the login and authentication phase. Here, $\varsigma = SID_i \oplus h\ (a_2 \| a_3)$, $b_3 = (r_2 - a_1.r_1)$, $c_1 = (r_2.\acute{\eta})$, $c_2 = (r_2.\ \omega)$, and $a_1 = h(\delta)$. Although the privileged insider can derive from the captured messages, other security tokens $\{b_3, c_1, c_2\}$ cannot be derived.

4.8 Theorem 8: Denial of service attacks are kept at bay

Proof: The aim of these attacks is to render the SC_j and DC_k nonresponsive to genuine request. To do this, attempts are made to resend valid but old messages $\text{log}_s = \{\varsigma, \rho_s, b_3, c_1, c_2, T_6\}$, and $auth = \{\varsigma, b_3, c_1, c_2, T_8\}$. Here, $\varsigma = SID_i \oplus$

$h(a_2||a_3)$, $b_3 = (r_2 - a_1.r_1)$, $c_1 = (r_2.ή)$, $c_2 = (r_2.\ \omega)$, $a_2 = (ή.a_1)$, $a_3 = (\omega.a_1)$, and $a_1 = h(\delta)$. It is clear that these messages incorporate temporary parameters such as random nonces $\{r_1, r_2\}$ and timestamps $\{T_6, T_8\}$. Due to timestamp validations at the receivers end, all these messages will be detected and malicious sessions terminated. Similarly, checks $e_3 \stackrel{?}{=} r_2.ή$, and $f_1 \stackrel{?}{=} r_2.\omega$ will flop since valid random nonces $\{r_1, r_2\}$ for the resent messages will be very old, hence adversarial sessions will be halted.

4.9 Theorem 9: Side-channeling attacks are ineffective against our protocol

Proof: In these attacks, the adversaries attempt to use power analysis techniques to extract security values stores in sensor node S_i. During the registration phase, only stores ψ in its memory. Afterwards, the adversary attempts to derive authentication tokens $e_3 = b_3.ή+\ r_1(a_1.ή)$ and $f_1 = b_3.\omega + r_1(a_1.\omega)$ to decipher the datacenter DC_k into believing the legitimacy of bogus message $auth = \{\varsigma, b_3, c_1, c_2, T_8\}$. Here, $b_3 = (r_2 - a_1.r_1)$, $a_1 = h(\delta)$, $c_1 = (r_2.ή)$, $c_2 = (r_2.\ \omega)$ and $\varsigma = SID_i \oplus h\ (a2||a3)$. It is evident that the obtained ψ, private key belonging to DC_k cannot help the intruder derive these verification tokens.

4.10 Theorem 10: Man-in-the-middle and impersonation threats are prevented

Proof: The goal of these attacks is to intercept the messages exchanged over the public communication channels. Afterward, attempts are made to derive authentication security token $e_3 = b_3.ή+\ r_1(a_1.ή)$ and $f_1 = b_3.\omega + r_1(a_1.\omega)$. These message include $\log_s = \{\varsigma, \rho_s, b_3, c_1, c_2, T_6\}$, and $auth = \{\varsigma, b_3, c_1, c_2, T_8\}$, where $\varsigma = SID_i \oplus h\ (a_2||a_3)$, $b_3 = (r_2 - a_1.r_1)$, $c_1 = (r_2.ή)$, $c_2 = (r_2.\ \omega)$, and $a_1 = h(\delta)$. Although the attacker can access b_3 from messages $\log_s$ and $auth$, tokens $\{ή, r_1, a_1, \omega\}$ are still required.

4.11 Theorem 11: Scalability is achieved

Proof: In our protocol, we deploy the PGB to ensure that only authorized nodes take part in verification and new blocks addition. Therefore, not all nodes take part in these two processes. In addition, the utilized PGB attains efficient transactions processing speeds and hence can accommodate numerous transactions emanating from a large number of devices. These two concepts ensure that our scheme can support many sensors deployed in campuses.

5. Performance evaluation

The implementation of the proposed consortium deploying Fabric as its PGB, we utilized Highsimb simulator. The performance evaluation was in terms of the duration required for new block addition, as well as transaction verification time. To investigate the consortium blockchain transaction verification duration, the number of nodes was varied from 1 to a maximum of 100. The results obtained are shown in Fig. 14.4.

The graph in Fig. 14.4 shows that the transaction verification time increased from an initial value of 0 up to 0.15 s. As such, there is a strong relationship between the number of nodes and the computation complexity. To investigate how long it takes either the sensor information center SC_j or data center DC_k (validators) to add new blocks (Bs) to the consortium blockchain, the number of transactions per block was incremented from an initial value of 10 to a maximum of 1500. In addition, the number of blocks was also varied from 10 to a maximum of 80 as shown in Fig. 14.5.

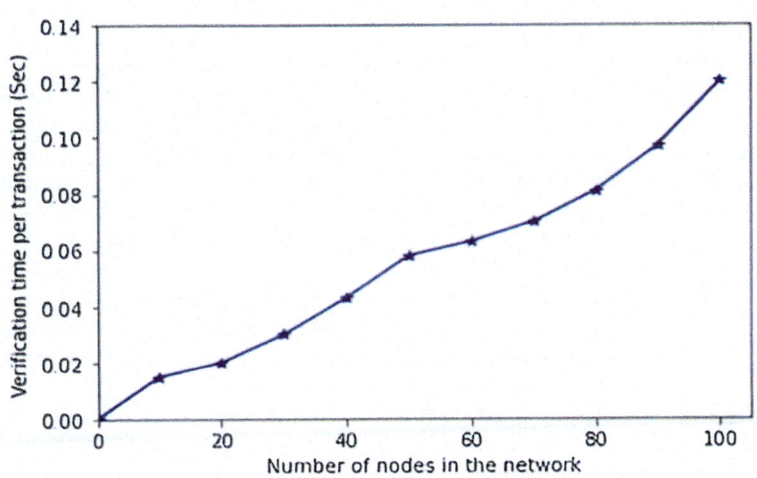

FIGURE 14.4 Transaction verification duration.

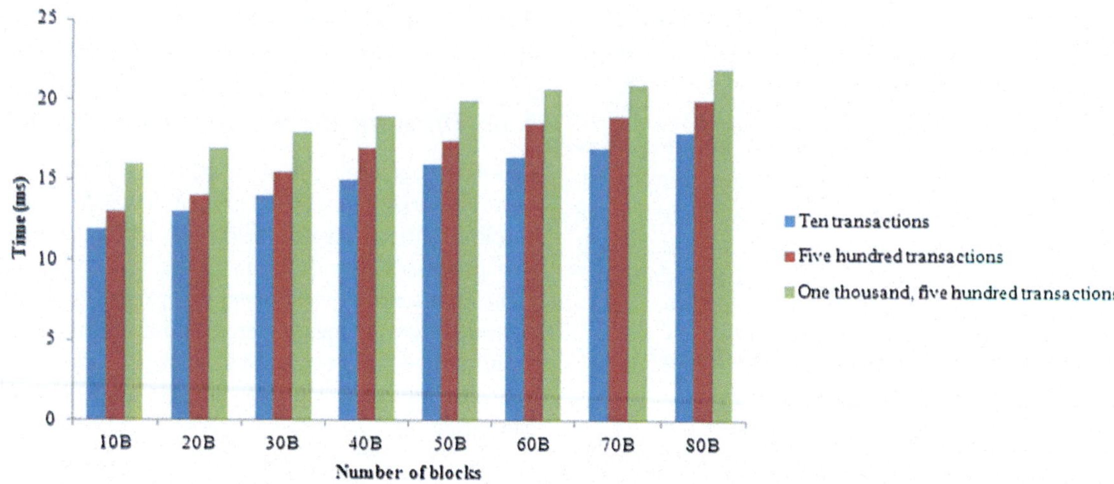

FIGURE 14.5 Time taken to add a new block to the blockchain network.

It is evident from Fig. 14.5 that as the number of blocks increases, there is a corresponding increase in the node addition duration. This is attributed to the surge in the time required for the validator to verify the high number of transactions and blocks. Fig. 14.5 also shows that high number of transactions to be validated has a direct influence on the duration it takes to add new blocks to the blockchain. Therefore, the more the number of transactions in a block, the longer the verification time for this particular block. This is because the validator has to receive a request from the instigating node, verify this node, construct the block from the corresponding transactions, and update the rest of the nodes with this new block.

6. Conclusion

Educational institutions normally spend millions of money to buy expensive equipment such as smart locks and smart identification systems. However, keeping track of these disparate systems presents some huge challenges. In addition, some academic institutions have adopted conventional techniques such as rule based alarm systems, and manual surveillance systems. However, these systems cannot offer context-aware and real-time feedback. To enhance operational efficiency and security, these institutions have resorted to cutting-edge techniques such as sensor networks and IoT to track and mitigate feasible challenges. Unfortunately, the high number of these IoT devices and sensors increase the surface area for launching attacks. In addition, majority of these devices have vulnerabilities which can be exploited by the adversaries. To this end, numerous security and privacy techniques have been presented in literature. However, the attainment of perfect privacy and security is student sensor networks is still a mirage. In this paper, we have developed an authentication technique that has been shown to be not only secure but also efficient. Future research will involve the formal security analysis of this scheme, as well as performance evaluations using metrics such as energy consumption, storage, and communication complexities.

References

Abawajy, J. H., & Hassan, M. M. (2017). Federated internet of things and cloud computing pervasive patient health monitoring system. *IEEE Communications Magazine, 55*(1), 48–53.

Abdulmouti, H., Skaf, Z., & Alblooshi, S. (2022). Smart green campus: The campus of tomorrow. In *2022 Advances in Science and Engineering Technology International Conferences (ASET)* (pp. 1–8). IEEE.

AbuAlnaaj, K., Ahmed, V., & Saboor, S. (2020). A strategic framework for smart campus. In , *22. Proceedings of the International Conference on Industrial Engineering and Operations Management* (pp. 790–798).

Abuarqoub, A., Abusaimeh, H., Hammoudeh, M., Uliyan, D., Abu-Hashem, M. A., Murad, S., … Al-Fayez, F. (2017). A survey on internet of things enabled smart campus applications. In *Proceedings of the International conference on future networks and distributed systems* (pp. 1–7).

Ahmad, A. Y. A. B., Alzubi, J., James, S., Nyangaresi, V. O., Kutralakani, C., & Krishnan, A. (2024). Enhancing human action recognition with adaptive hybrid deep attentive networks and archerfish optimization. *Computers, Materials & Continua, 80*(3).

Akila, V., & Sheela, T. (2017). Preserving data and key privacy in data aggregation for wireless sensor networks. In *2017 2nd International conference on computing and communications technologies (ICCCT)* (pp. 282–287). IEEE.

Akila, V., & Sheela, T. (2019). Secure data aggregation to preserve data and key privacy in wireless sensor networks with multiple sinks. In *2019 3rd International Conference on Computing and Communications Technologies (ICCCT)* (pp. 86–93). IEEE.

Al-Shoqran, M., & Shorman, S. (2021). A review on smart universities and artificial intelligence. In *The fourth industrial revolution: Implementation of artificial intelligence for growing business success* (pp. 281–294).

Al-Sulami, Z. A., Abduljabbar, Z. A., Nyangaresi, V. O., & Ma, J. (2023). Knowledge management and its role in the development of a smart university in Iraq. *TEM Journal, 12*(3), 1582.

Alansari, Z., Anuar, N. B., Kamsin, A., & Belgaum, M. R. (2022). A systematic review of routing attacks detection in wireless sensor networks. *PeerJ Computer Science, 8*, Article e1135.

AlDairi, A., & Tawalbeh, L. A. (2017). Cyber security attacks on smart cities and associated Mobile technologies. *Procedia Computer Science, 109*, 1086–1091.

Ali, R., Pal, A. K., Kumari, S., Sangaiah, A. K., Li, X., & Wu, F. (2024). An enhanced three factor based authentication protocol using wireless medical sensor networks for healthcare monitoring. *Journal of Ambient Intelligence and Humanized Computing*, 1–22.

Alshuraify, N. A., Yassin, A. A., Abduljabbar, Z. A., & Nyangaresi, V. O. (2024). Monitoring and surveillance systems based IoTs with blockchain: Literature review. *Basrah Researches Sciences, 50*(2), 42–63.

Alsumayt, A., Alshammari, M., Alfawaer, Z. M., Al-Wesabi, F. N., El-Haggar, N., Aljameel, S. S., ... Aldossary, N. (2024). Efficient security level in wireless sensor networks (WSNs) using four-factors authentication over the internet of things (IoT). *PeerJ Computer Science, 10*, Article e2091.

Amin, R., Islam, S. H., Biswas, G. P., Khan, M. K., & Kumar, N. (2018). A robust and anonymous patient monitoring system using wireless medical sensor networks. *Future Generation Computer Systems, 80*, 483–495.

Arunsankar, G., Gopi, B., Sampathrajan, R., Sasikala, S., Vanathi, A., & Srinivasan, C. (2023). IoT controlled device to manage the EMG signals of the patient and alert in real time. In *2023 Second International Conference On Smart Technologies For Smart Nation (SmartTechCon)* (pp. 1365–1370). IEEE.

Aslan, Ö., Aktuğ, S. S., Ozkan-Okay, M., Yilmaz, A. A., & Akin, E. (2023). A comprehensive review of cyber security vulnerabilities, threats, attacks, and solutions. *Electronics, 12*(6), 1333.

Balakrishnan, R., Kothiyal, A., Bakhare, R., Patil, H., Maranan, R., & Parikh, S. (2024). Improving online education in college settings through the integration of Mobile wireless communication networks and internet of things (IoT). In *2024 Second International Conference on Intelligent Cyber Physical Systems and Internet of Things (ICoICI)* (pp. 138–143). IEEE.

Beemkumar, N., Gupta, S., Yadav, S., Saxena, A. K., Dhabliya, D., & Gupta, A. (2023). The application of WSN during human activity detection and identification using a deep learning approach. *Digital Transformation–Modernization and Optimization of Wireless Networks, 59*.

Bera, B., Das, A. K., Balzano, W., & Medaglia, C. M. (2020). On the design of biometric-based user authentication protocol in smart city environment. *Pattern Recognition Letters, 138*, 439–446.

Bodur, G., Gumus, S., & Gursoy, N. G. (2019). Perceptions of Turkish health professional students toward the effects of the internet of things (IOT) technology in the future. *Nurse Education Today, 79*, 98–104.

Bouchouia, M. L., Labiod, H., Jelassi, O., Monteuuis, J. P., Jaballah, W. B., Petit, J., & Zhang, Z. (2023). A survey on misbehavior detection for connected and autonomous vehicles. *Vehicular Communications, 41*, Article 100586.

Bulbul, S. S., Abduljabbar, Z. A., Mohammed, R. J., Al Sibahee, M. A., Ma, J., Nyangaresi, V. O., & Abduljaleel, I. Q. (2024). A provably lightweight and secure DSSE scheme, with a constant storage cost for a smart device client. *PLoS One, 19*(4), Article e0301277.

Chai, S., Yin, H., Xing, B., Li, Z., Guo, Y., Zhang, D., ... Huang, X. (2023). Provably secure and lightweight authentication key agreement scheme for smart meters. *IEEE Transactions on Smart Grid, 14*(5), 3816–3827.

Chen, Z., Cai, C., Zheng, T., Luo, J., Xiong, J., & Wang, X. (2021). RF-based human activity recognition using signal adapted convolutional neural network. *IEEE Transactions on Mobile Computing, 22*(1), 487–499.

Chen, C., Guo, H., Wu, Y., Gao, Y., & Liu, J. (2023). A novel two-factor multi-gateway authentication protocol for WSNs. *Ad Hoc Networks, 141*, Article 103089.

Chen, Z., Teng, L., Xu, L., Yu, J., & Liang, J. (2023). MP-HAR: A novel motion-powered real-time human activity recognition system. *IEEE Internet of Things Journal, 11*(5), 7652–7663.

Colabianchi, S., Costantino, F., Di Gravio, G., Nonino, F., & Patriarca, R. (2021). Discussing resilience in the context of cyber physical systems. *Computers & Industrial Engineering, 160*, Article 107534.

Deepa, A., Manikandan, N. K., Latha, R., Preetha, J., Kumar, T. S., & Murugan, S. (2023). IoT-Based wearable devices for personal safety and accident prevention systems. In *2023 Second International Conference On Smart Technologies For Smart Nation (SmartTechCon)* (pp. 1510–1514). IEEE.

Diro, A. (2025). Space systems and malware: Potential threats. In *Ransomware evolution* (pp. 208–232). CRC Press.

Edwards, D. J. (2024). Malware defenses. In *Critical security controls for effective cyber defense: A comprehensive guide to CIS 18 controls* (pp. 277–308). Berkeley, CA: Apress.

El-Adawi, E., Essa, E., Handosa, M., & Elmougy, S. (2024). Wireless body area sensor networks based human activity recognition using deep learning. *Scientific Reports, 14*(1), 2702.

Elhabrush, H., & Ahmeda, S. (2022). Authentication protocol for wireless sensor network in the internet of things. In *2022 IEEE 2nd International Maghreb meeting of the conference on sciences and techniques of automatic control and computer engineering (MI-STA)* (pp. 471–478). IEEE.

Erskine, S. K. (2024). Secure data aggregation using authentication and authorization for privacy preservation in wireless sensor networks. *Sensors, 24*(7), 2090.

Fereidouni, H., Fadeitcheva, O., & Zalai, M. (2025). IoT and man-in-the-middle attacks. *Security and Privacy, 8*(2), Article e70016.

Gao, M. (2022). Smart campus teaching system based on ZigBee wireless sensor network. *Alexandria Engineering Journal, 61*(4), 2625–2635.

Gargees, R. S. (2024). Improving human activity recognition through multisensor data fusion techniques in wireless sensor networks. In *2024 IEEE 3rd International Conference on Computing and Machine Intelligence (ICMI)* (pp. 1–5). IEEE.

Gelgi, M., Guan, Y., Arunachala, S., Samba Siva Rao, M., & Dragoni, N. (2024). Systematic literature review of IoT botnet DDOS attacks and evaluation of detection techniques. *Sensors, 24*(11), 3571.

Gopal, S. R., & Patan, S. A. (2019). Design and analysis of heterogeneous hybrid topology for VLAN configuration. *International Journal of Emerging Trends in Engineering Research, 7*(11), 487–491.

Hafeez, M. A., Shakib, K. H., & Munir, A. (2025). A secure and scalable authentication and communication protocol for smart grids. *Journal of Cybersecurity and Privacy, 5*(2), 11.

Hidayat, D. S., & Sensuse, D. I. (2022). Knowledge management model for smart campus in Indonesia. *Data, 7*(1), 7.

Horváth, D., Csordás, T., Ásványi, K., Faludi, J., Cosovan, A., Simay, A. E., & Komár, Z. (2022). Will interfaces take over the physical workplace in higher education? A pessimistic view of the future. *Journal of Corporate Real Estate, 24*(2), 108–123.

Hu, S., Chen, Y., Zheng, Y., Xing, B., Li, Y., Zhang, L., & Chen, L. (2022). Provably secure ECC-Based authentication and key agreement scheme for advanced metering infrastructure in the smart grid. *IEEE Transactions on Industrial Informatics, 19*(4), 5985–5994.

Hu, S., Liu, L., Fang, L., Zhou, F., & Ye, R. (2019). A novel energy-efficient and privacy-preserving data aggregation for WSNs. *IEEE Access, 8*, 802–813.

Jawad, M., Yassin, A. A., Al-Asadi, H. A. A., Abduljabbar, Z. A., & Nyangaresi, V. O. (2024). Towards building multi-factor authentication scheme for users in the healthcare sector based on blockchain technology. In *Computer science On-line conference* (pp. 694–713). Cham: Springer Nature Switzerland.

Joseph, S., & Pandeeswari, N. (2025). Secured IoT data transmission with enhanced integrated encryption techniques with blockchain technology. *IETE Journal of Research*, 1–20.

Kadhim, T. A., Abduljabbar, Z. A., Al-Asadi, H. A., Nyangaresi, V. O., Ali, Z. A., & Abduljaleel, I. Q. (2025). Lightweight scheme for secure signaling and data exchanges in intelligent precision agriculture. *Cryptography, 9*(1), 7.

Khalaf, M., Ayad, A., Tushar, M. H. K., Kassouf, M., & Kundur, D. (2024). A survey on cyber-physical security of active distribution networks in smart grids. *IEEE Access, 12*, 29414–29444.

Kshetri, N. (2017). The evolution of the internet of things industry and market in China: An interplay of institutions, demands and supply. *Telecommunications Policy, 41*(1), 49–67.

Kumar, B., Padmanabha, V., & Dwivedi, R. (2024). Utilizing packet tracer simulation to modern smart campus initiatives for a sustainable future. In *2024 IEEE 12th International Conference on Smart Energy Grid Engineering (SEGE)* (pp. 97–102). IEEE.

Kumari, M., Sinha, R., Chakrabarty, P., Hasnain, M. G., Qamar, N., & Gupta, S. (2024). Exploring machine learning's impact on digital marketing. In *2024 15th International Conference on Computing Communication and Networking Technologies (ICCCNT)* (pp. 1–5). IEEE.

Li, Z., Liu, Y., Liu, A., Wang, S., & Liu, H. (2018). Minimizing convergecast time and energy consumption in green internet of things. *IEEE Transactions on Emerging Topics in Computing, 8*(3), 797–813.

Liu, S., Liu, Y., Liu, W., & Zhang, Y. (2023). A certificateless multi-dimensional data aggregation scheme for smart grid. *Journal of Systems Architecture, 140*, Article 102890.

Malatji, E. M. (2017). The development of a smart campus-african universities point of view. In *2017 8th International renewable energy congress (IREC)* (pp. 1–5). IEEE.

Manickam, M., & Devarajan, G. G. (2025). An improved three factor authentication protocol for wireless body area networks. *Cyber Security and Applications, 3*, Article 100062.

Maurya, A. K., & Sastry, V. N. (2017). Fuzzy extractor and elliptic curve based efficient user authentication protocol for wireless sensor networks and internet of things. *Information, 8*(4), 136.

Mayer, M., & Baeumner, A. J. (2019). A megatrend challenging analytical chemistry: Biosensor and chemosensor concepts ready for the internet of things. *Chemical Reviews, 119*(13), 7996–8027.

Mekruksavanich, S., & Jitpattanakul, A. (2023). Hybrid convolution neural network with channel attention mechanism for sensor-based human activity recognition. *Scientific Reports, 13*(1), Article 12067.

Mo, Z. (2024). Design of campus security information identification system based on ZigBee and CC2530 module. In *2024 IEEE 2nd International conference on control, electronics and computer technology (ICCECT)* (pp. 339–342). IEEE.

Nag, A., Hassan, M. M., Das, A., Sinha, A., Chand, N., Kar, A., … Alkhayyat, A. (2024). Exploring the applications and security threats of internet of thing in the cloud computing paradigm: A comprehensive study on the cloud of things. *Transactions on Emerging Telecommunications Technologies, 35*(4), Article e4897.

Nyangaresi, V. O. (2025). Anonymity preserving security protocol for wireless body area networks: Towards the secure remote patient healthcare monitoring. In *Sensor networks for smart hospitals* (pp. 293–312). Elsevier.

Nyangaresi, V. O. (2025). Efficient and provably secure framework for authenticating internet of medical things in smart hospitals. In *Blockchain and digital twin for smart hospitals* (pp. 207–236). Elsevier.

Nyangaresi, V. O., Abduljabbar, Z. A., Mutlaq, K. A. A., Bulbul, S. S., Ma, J., Aldarwish, A. J., … Neamah, H. A. (2024). Smart city energy efficient data privacy preservation protocol based on biometrics and fuzzy commitment scheme. *Scientific Reports, 14*(1), Article 16223.

Nyangaresi, V. O., AlRababah, A. A., Yenurkar, G. K., Chinthaginjala, R., & Yasir, M. (2025). Anonymous authentication scheme based on physically unclonable function and biometrics for smart cities. *Engineering Reports, 7*(1), Article e13079.

Nyangaresi, V. O., Alsolami, E., & Ahmad, M. (2024). Trust-enabled energy efficient protocol for secure remote sensing in supply chain management. *IEEE Access, 12*, 113553–113564.

Oladipupo, E. T., Abikoye, O. C., Imoize, A. L., Awotunde, J. B., Chang, T. Y., Lee, C. C., & Do, D. T. (2023). An efficient authenticated elliptic curve cryptography scheme for multicore wireless sensor networks. *IEEE Access, 11*, 1306–1323.

Qadri, Y. A., Nauman, A., Zikria, Y. B., Vasilakos, A. V., & Kim, S. W. (2020). The future of healthcare internet of things: A survey of emerging technologies. *IEEE Communications Surveys & Tutorials, 22*(2), 1121–1167.

Radhi, B. M., Hussain, M. A., Abduljabbar, Z. A., Nyangaresi, V. O., & Aldarwish, A. J. (2024). A review on IoTs applications and security threats via data transfer over networks. In *Computer science On-line conference* (pp. 562–579). Cham: Springer Nature Switzerland.

Rao, K. R., Krishna, K. S., & Afsha, N. (2023). Transforming campus safety and security with autonomous monitoring. *Journal of Engineering Science, 14*(04), 169–173.

Ray, P. P. (2017). Internet of things for smart agriculture: Technologies, practices and future direction. *Journal of Ambient Intelligence and Smart Environments, 9*(4), 395–420.

Ray, P. P. (2018). A survey on internet of things architectures. *Journal of King Saud University-Computer and Information Sciences, 30*(3), 291–319.

Saez, M., Maturana, F. P., Barton, K., & Tilbury, D. M. (2018). Real-time manufacturing machine and system performance monitoring using internet of things. *IEEE Transactions on Automation Science and Engineering, 15*(4), 1735–1748.

Saha, A., Rajak, S., Saha, J., & Chowdhury, C. (2024). A survey of machine learning and meta-heuristics approaches for sensor-based human activity recognition systems. *Journal of Ambient Intelligence and Humanized Computing, 15*(1), 29–56.

Salahdine, F., & Kaabouch, N. (2019). Social engineering attacks: A survey. *Future Internet, 11*(4), 89.

Salim, M. M., Rathore, S., & Park, J. H. (2020). Distributed denial of service attacks and its defenses in IoT: A survey. *The Journal of Supercomputing, 76*, 5320–5363.

Saravanan, G., Helenprabha, K., Juliet, P. S., Yuvaraj, S., Gnanavel, N., & Srinivasan, C. (2023). Convolutional neural networks-based real-time gaze analysis with IoT integration in user experience design. In *2023 2nd International Conference on Automation, Computing and Renewable Systems (ICACRS)* (pp. 1564–1569). IEEE.

Sarbini, I. N., Khan, A. S., Mohamad, N. Z., & Yusup, N. (2022). A lightweight multifactor authentication scheme for wireless sensor networks in the internet of things. In *2022 International Conference on Green Energy, Computing and Sustainable Technology (GECOST)* (pp. 482–486). IEEE.

Sasirekha, V., Ramadevi, R., Malarvizhi, C., Mohankumar, N., & Velmurugan, S. (March 2024). Intelligent campus safety management using IoT and CNNs for surveillance, access, and emergency response. In *2024 3rd International Conference for Innovation in Technology (INOCON)* (pp. 1–6). IEEE.

Sehgal, S., Gupta, A., Singh, A., & Singh, P. (2022). Adoption of proprietary online education software compared to free and open-source software in Indian colleges. *Journal of Information Technology and Digital World, 4*(1), 1–10.

Shanmugaraja, P., Bhardwaj, M., Mehbodniya, A., Vali, S., & Reddy, P. C. S. (2023). An efficient clustered M-path Sinkhole Attack Detection (MSAD) algorithm for wireless sensor networks. *Ad Hoc & Sensor Wireless Networks, 55*.

Sheth, A., Jaimini, U., & Yip, H. Y. (2018). How will the internet of things enable augmented personalized health? *IEEE Intelligent Systems, 33*(1), 89–97.

Silva, A., Zabaleta, O. G., & Arizmendi, C. M. (2022). Mitigation of routing congestion on data networks: A quantum game theory approach. *Quantum Reports, 4*(2), 135–147.

Singh, D., Kumar, B., Singh, S., & Chand, S. (2020). Evaluating authentication schemes for real-time data in wireless sensor network. *Wireless Personal Communications, 114*(1), 629–655.

Sinha, R. (2025). IoT-based women's safety monitoring in educational campuses using WSN and deep learning. In *2025 3rd International Conference on Intelligent Data Communication Technologies and Internet of Things (IDCIoT)* (pp. 745–750). IEEE.

Siwakoti, Y. R., Bhurtel, M., Rawat, D. B., Oest, A., & Johnson, R. C. (2023). Advances in IoT security: Vulnerabilities, enabled criminal services, attacks, and countermeasures. *IEEE Internet of Things Journal, 10*(13), 11224–11239.

Srhir, A., Mazri, T., & Benbrahim, M. (2023). Towards secure smart campus: Security requirements, attacks and counter measures. *Indonesian Journal of Electrical Engineering and Computer Science, 32* (2), 900–914.

Srinivas, J., Mukhopadhyay, S., & Mishra, D. (2017). Secure and efficient user authentication scheme for multi-gateway wireless sensor networks. *Ad Hoc Networks, 54*, 147–169.

Stellios, I., Kotzanikolaou, P., Psarakis, M., Alcaraz, C., & Lopez, J. (2018). A survey of iot-enabled cyberattacks: Assessing attack paths to critical infrastructures and services. *IEEE Communications Surveys & Tutorials, 20*(4), 3453–3495.

Tao, H., Bhuiyan, M. Z. A., Abdalla, A. N., Hassan, M. M., Zain, J. M., & Hayajneh, T. (2018). Secured data collection with hardware-based ciphers for IoT-based healthcare. *IEEE Internet of Things Journal, 6*(1), 410–420.

Ul-Aaish, S., Pires, I. M., Godinho, A., Coelho, P. J., & Butt, P. K. (2024). Client risk assessment in a network: An examination of man-in-the-middle attacks and their usage. In *2024 11th International Conference on Electrical Engineering, Computer Science and Informatics (EECSI)* (pp. 793–800). IEEE.

Umran, S. M., Lu, S., Abduljabbar, Z. A., & Nyangaresi, V. O. (2023). Multi-chain blockchain based secure data-sharing framework for industrial IoTs smart devices in petroleum industry. *Internet of Things, 24*, Article 100969.

Wang, D., Li, W., & Wang, P. (2018). Measuring two-factor authentication schemes for real-time data access in industrial wireless sensor networks. *IEEE Transactions on Industrial Informatics, 14*(9), 4081–4092.

Wang, W., Xu, H., Alazab, M., Gadekallu, T. R., Han, Z., & Su, C. (2021). Blockchain-based reliable and efficient certificateless signature for IIoT devices. *IEEE Transactions on Industrial Informatics, 18*(10), 7059–7067.

Xiong, L., Peng, T., Li, F., Zeng, S., & Wu, H. (2022). Privacy-preserving authentication scheme with revocability for multi-WSN in industrial IoT. *IEEE Systems Journal, 17*(1), 38–49.

Yang, S. K., Shiue, Y. M., Su, Z. Y., & Liu, C. G. (2019). A novel authentication scheme against node captured attack in WSN for healthcare scene. In *2019 IEEE Eurasia conference on biomedical engineering, healthcare and sustainability (ECBIOS)* (pp. 39–42). IEEE.

Yu, L., Yu, W., & Lv, Y. (2021). Multi-dimensional privacy-preserving average consensus in wireless sensor networks. *IEEE Transactions on Circuits and Systems. II, Express Briefs, 69*(3), 1104–1108.

Zhou, L., Ge, C., Hu, S., & Su, C. (2019). Energy-efficient and privacy-preserving data aggregation algorithm for wireless sensor networks. *IEEE Internet of Things Journal, 7*(5), 3948–3957.

Zhou, Q., Qin, X., Liu, G., Cheng, H., & Zhao, H. (2019). An efficient privacy and integrity preserving data aggregation scheme for multiple applications in wireless sensor networks. In *2019 IEEE International Conference on Smart Internet of Things (SmartIoT)* (pp. 291–297). IEEE.

Zhou, L., & Tang, Q. (2022). [retracted] construction of a six-pronge intelligent physical education classroom model in colleges and universities. *Scientific Programming, 2022*(1), Article 9003864.

Further reading

Abdali, H. K., Hussain, M. A., Abduljabbar, Z. A., & Nyangaresi, V. O. (2024). Implementing blockchain for enhancing security and authentication in Iraqi E-Government services. *Engineering, Technology & Applied Science Research, 14*(6), 18222–18233.

Ali, Z. A., Abduljabbar, Z. A., Al-Asadi, H. A. A., Nyangaresi, V. O., Abduljaleel, I. Q., & Aldarwish, A. J. (2024). A provably secure anonymous authentication protocol for consumer and service provider information transmissions in smart grids. *Cryptography, 8*(2), 20.

Chen, Y., & Chen, J. (2021). An efficient mutual authentication and key agreement scheme without password for wireless sensor networks. *The Journal of Supercomputing, 77*(12), 13653–13675.

Nemec Zlatolas, L., Welzer, T., & Lhotska, L. (2024). Data breaches in healthcare: Security mechanisms for attack mitigation. *Cluster Computing, 27*(7), 8639–8654.

Nyangaresi, V. O., Al-Joboury, I. M., Al-sharhanee, K. A., Najim, A. H., Abbas, A. H., & Hariz, H. M. (2024). A biometric and physically unclonable function–based authentication protocol for payload exchanges in internet of drones. e-Prime-Advances in electrical engineering. *Electronics and Energy, 7*, Article 100471.

Chapter 15

Data privacy considerations in implementing digital twins in educational settings

Assunta Di Vaio[1] and Luisa Mastellone[2]

[1]*University of Naples Parthenope, Naples, Italy;* [2]*Marinter Shipping Agency Srl, Naples, Italy*

1. Data privacy of Digital Twins in educational settings

Data privacy is the concept of safeguarding personal information and ensuring that individuals retain control over how their data are accessed and used (Gürses et al., 2006). This principle becomes particularly significant in the context of Digital Twins (DTs), where the interaction between the Internet of Things (IoT) and Artificial Intelligence (AI) plays a significant role. IoT devices, ranging from personal computers to wearable technology, collect extensive amounts of real-time data that can include sensitive information. As these devices become networked, they enhance the ability to monitor, simulate, and track physical systems with remarkable accuracy and efficiency, which is one of the key advantages that IoT and AI bring to DTs (Hassan et al., 2023). However, the increased connectivity and data collection enabled by these technologies also introduces significant privacy concerns.

While on the one hand the combination of IoT and AI provides several advantages to DTs; on the other hand, this combination raises significant security issues. IoT has changed that the way data are exchanged between different sources (Corral-Plaza et al., 2020). IoT devices, such as personal computers, laptops, tablets, and smart phones, are equipped with sensors and actuators (Yu et al., 2010; Zhou & Zhang, 2011). Through IoT our non-autonomous everyday devices are networked, to detect and perceive the surrounding environment, generating the sensed data. The sensed data from is then processed in the node, in the device hub or in the cloud, allowing the devices to make decisions autonomously. Decisions that can also be disseminated to users, who thanks to these will be able to make decisions autonomously (Yu et al., 2010). IoT involves managing a large volume of data. Data that are often sensitive, therefore capable of identifying users. In this sense, IoT is more vulnerable to privacy violations (Karale, 2021). AI can be recognized as "the theory and development of computerized systems able to imitate and simulate human intelligence and behavior" (Merriam-Webster Dictionary). Such digital technology reveals how computers "think and find" solutions to problems in the same way people do it. AI is based on the creation of a vast collection of algorithms, which process huge volumes of data, such as sensor data from various sources, ensuring a high level of speed and precision (Hassan et al., 2023). As with data managed by the IoT, algorithms created for AI are also subject to privacy violations.

The combination of IoT and AI not only advances DTs but also raises critical ethical questions regarding the protection of personal data. IoT enables a vast amount of data to be exchanged and processed across devices, making it vulnerable to privacy violations (Karale, 2021). When connected to AI algorithms, this data are used to simulate, analyze, and make autonomous decisions, which further complicates issues of data privacy. In an educational setting, where students' personal data is sensitive and protected by regulations, ensuring that such data remains confidential and is only used for its intended purpose is of utmost importance. The risk of violating privacy can hinder the trust of students, educators, and parents in the adoption of digital technologies for learning and educational management.

The potential for DTs to optimize educational systems is immense, as it allows for personalized learning, real-time monitoring of student progress, and the creation of immersive learning environments. However, the integration of AI and IoT within educational contexts demands a

Digital Twins in the Smart Classroom. https://doi.org/10.1016/B978-0-443-41632-3.00014-X

careful balancing of data privacy and sustainability goals. Educational institutions must ensure that any digital solutions they implement comply with legal standards for data protection (e.g., GDPR in Europe or FERPA in the U.S.), and also align with sustainability goals, ensuring the ethical use of technology without compromising personal data privacy. The sustainability of these technological systems must be considered, as the environmental impact of large-scale IoT networks and AI systems cannot be overlooked. The energy consumption required to process, store, and analyze the massive amounts of data generated by IoT devices, combined with the computational power needed for AI algorithms, could contribute to increased carbon footprints. As such, integrating data privacy and sustainability within educational technology requires a holistic approach, where institutions prioritize ethical standards, data protection, and environmental impact as they embrace digital innovation. Hence, while IoT and AI provide powerful tools for DTs, enabling enhanced accuracy and efficiency in educational settings, they also bring complex challenges related to data privacy and sustainability. Addressing these challenges is necessary to foster an educational environment that not only accepts digital transformation, but does so in a way that is responsible, transparent, and aligned with long-term sustainability goals.

Data and algorithms subjects to privacy violation outline DTs technology. Within DTs the existence of data and algorithms that can be easily violated give rise to the need to implement data privacy.

Data privacy can mitigate risks of breaches and exploitation, through data anonymization, encryption and consent management. Data anonymization is the process of allowing sensitive data for queries and analysis, while still maintaining confidentiality from various attacks (Goswami & Madan, 2017). In the same way, encryption has the aim of preserving confidentiality in communication and storage processes (Fontaine & Galand, 2017), making a message unreadable/intelligible to people not authorized to read it. Instead, consent management manages users' consent to use identifiable personal data, also ensuring the possibility of revoking it (Kakarlapudi & Mahmoud, 2021).

Regarding DTs, data privacy has the specific aim to take measures to ensure that data is not intercepted/modified by unauthorized users. At the same time, sources of data must be authenticated, to ensure that all nodes sending data on the physical twin are reliable sources (Mashaly, 2021). DTs find application in different aspects of our lives, such as manufacturing, healthcare, smart cities, and education.

The introduction of DTs in many aspects of our lives makes data privacy a crucial requirement. Particularly, the introduction of DTs in the educational settings has placed the student data privacy as an increasingly important problem, and one that draws the attention of researchers and experts. Student data privacy refers to responsible, ethical, and equitable collection, use, sharing, and protection of student data. Students, especially minors of the secondary public school, do not have the full perception of evaluating the risks deriving from the collection and use of data. For this reason, they need greater protection which allows to support them in the opportunities of success and at the same time make them autonomous in their decisions (Future of Privacy Forum, 2021).

During the 2019 coronavirus pandemic the student data privacy has represented for secondary public schools and higher public education a new problem to face (Schlosser et al., 2022). The adoption of educational technological tools by students residing in different parts of the world has forced educational institutions to consider additional measures to protect student data, such as the collection of data through third-party applications (Mokhtar & Karim, 2021). This would allow data to be collected and shared with other external vendors who work with and for third-party application/software companies but would no longer be a breach of privacy. For this reason, educational institutions must adopt new privacy policies. Adopting policies would enable the reviewer to make an informed decision about what data to collect and share, adapting a student-centered privacy model (Shore et al., 2022).

2. Digital Twins in educational settings

Despite being in 2025, a universally accepted and clear conceptualization of DTs, one that is adaptable to various contexts, has yet to be fully established. Considering this, it is relevant to first define the key principles of DTs in the educational setting and, subsequently, to examine the data privacy challenges associated with their application. To better understand this developing concept, it is possible to trace its origins back to one of the earliest attempts to conceptualize DTs, which dates to 1960. During this period, in the U.S. the National Aeronautics and Space Administration (NASA) sought to create physical replicas on Earth that directly mirrored the systems in space as part of the Apollo program. These replicas, or "twins" allowed NASA to simulate different scenarios, test various cases, and evaluate the behavior and performance of their space systems in real-time (Glaessgen & Stargel, 2012). This pioneering effort marked the beginning of a broader exploration into how digital representations of physical entities could be used for analysis, prediction, and optimization.

In 2002 Michael Grieves introduced the concept of DTs for the manufacturing industry. He defines DTs as a technology able to create virtual replicas of factories to monitor their processes, predict failures, and increase their productivity (Grieves & Vickers, 2017, pp. 85–113). This concept has been listed as of the top 10 strategic technology trends in 2017 by Gartner (Burke et al., 2016) and adopted by some industry giants such as General Electric and Siemens. The latter has defined DTs in this way: "A digital twin is a virtual representation of a physical product or process, used to understand and predict the physical counterpart's performance characteristics. Digital twins are used throughout the product lifecycle to simulate, predict, and optimize the product and production system before investing in physical prototypes and assets" (Siemens, 2019: https://www.sw.siemens.com/en-US/technology/digital-twin/).

Recent efforts to conceptualize DTs have gained momentum, providing valuable insights into their application in various domains. The conceptual definitions presented by Chen et al. (2020), Mashaly (2021), and Qi et al. (2021) help illuminate the cognitive implications of data privacy arising from the adoption of DTs, especially in the educational context. These implications are essential for ensuring the responsible and ethical use of DTs, aligning with broader sustainability goals.

Mashaly (2021) defines DTs as a digital model of a physical system, achieved through a data connection that enables a seamless transformation of the physical system into its virtual counterpart. This connection ensures a high degree of synchronization between the two, which is vital for applications in education where accurate simulations and real-time data monitoring are essential for improving educational practices. The data connection is facilitated by sensors and data sources, which play a key role in representing the lifecycle of an asset's operations, helping to predict, diagnose, and prevent failures or anomalies (Chen et al., 2020). This process not only contributes to the efficiency and sustainability of educational systems but also aligns with the United Nations Sustainable Development Goals (UN SDGs) by promoting sustainable management and resource efficiency within educational infrastructures.

DTs are characterized by three primary elements: the physical system, the virtual model, and the data communication link between them. According to Qi et al. (2021), the physical system is the foundation of the DTs, serving as the basis for creating the virtual element. The virtual element, which mimics the behavior of the physical system, allows for deep analysis and simulation, enhancing understanding and performance. The third element, the data communication link, governs the exchange of information between the physical and virtual entities, enabling advanced simulations, real-time monitoring, operational optimization, and predictive analysis (Onwubiko et al., 2023). These features not only improve system efficiency but also support sustainability by fostering smarter, data-driven decision-making in educational contexts.

By integrating DTs into the educational sector, we can address key sustainability challenges, such as reducing resource consumption, enhancing operational efficiency, and promoting long-term environmental and social outcomes. These advancements contribute to SDG 4 (Quality Education) by optimizing educational practices, SDG 9 (Industry, Innovation, and Infrastructure) using advanced technologies, and SDG 12 (Responsible Consumption and Production) by enhancing the sustainability of educational infrastructures (United Nations, 2015).

The introduction of DTs highlights the growing impact of digital technology on both industrial and educational environments. In education, DTs can improve the learning experience by enabling the exploration of simulated environments, offering a digital representation of real-world systems (Garay Gallastegui & Reier Forradellas, 2021).

Alongside other digital technologies, DTs contribute to the broader digitalization of education, which represents a significant transformation in the way we approach teaching and learning (Prokopenko, 2021). This digital shift not only improves the quality of educational processes, but also adds greater value to them (Rosak-Szyrocka, 2024). For higher education institutions, including universities and public secondary schools, adopting DTs has become essential. As younger generations, who are consistently connected to modern digital tools, make up the majority of the student population, the integration of these technologies into the educational system is crucial. This demand for digitalization, driven by the expectations of younger students, has made it a fundamental requirement for the continued relevance and survival of educational institutions (Lefter et al., 2011).

The integration of DTs into educational settings is becoming increasingly prevalent, not only in universities but also in public secondary schools. In universities, this integration aligns with the University 4.0 paradigm, which mirrors the principles of Industry 4.0 (Rosak-Szyrocka, 2024). DTs play a central role in the industry 4.0 framework, serving as key enablers of innovation. According to Gartner, DTs is seen as a major trigger for emerging technologies, with predictions indicating that within 5 years, hundreds of millions of objects, machines, and systems will be equipped with DTs (Gartner, 2019).

Similarly, within the University 4.0 paradigm, DTs has a key position. University 4.0 emphasizes the development of digital platforms, such as Learning Management Systems (LMSs), which support teaching, research, and communication. These platforms promote interactions among students, teachers, and other professionals, including those who are not necessarily academic experts (Rosak-Szyrocka, 2024).

In this context, DTs can drive the creation of value across the university's three core functions: teaching, research, and the dissemination of knowledge or public engagement. The seamless integration of DTs improves educational practices, supports personalized learning, and fosters innovation, aligning with sustainability goals by promoting efficient, data-driven decision-making in the academic environment.

Beyond universities, the Education 4.0 paradigm, which extends to public secondary schools, also embraces the potential of DTs and other digital technologies. Education 4.0, characterized by the delivery of personalized, high-quality content to meet diverse learner needs, moves away from the traditional one-size-fits-all approach. This paradigm operates in both physical and virtual classroom environments and supports a more inclusive, adaptable form of education (Mukul & Büyüközkan, 2023). By utilizing DTs, Education 4.0 can create more sustainable and equitable learning experiences, fostering lifelong learning and preparing students for the future workforce in a rapidly evolving digital world.

However, with the integration of DTs comes the critical need to address data privacy concerns. As educational institutions increasingly rely on digital technologies to collect and analyze student data, safeguarding this information becomes essential to ensure compliance with privacy regulations and to protect users' rights. The responsible use of DTs in education can support sustainability goals (i.e., SDG 4, Quality Education) by increasing educational access and quality, but it also needs careful consideration of data security and ethical practices to maintain trust and protect personal information.

The evolving concepts of University 4.0 and Education 4.0 have transformed traditional classroom environments through the integration of digital technologies, fundamentally reshaping the way students engage with learning. In this new paradigm, the classroom is no longer just a physical space but a dynamic, interactive environment where students have equal access to learning opportunities through active participation (Mustafa et al., 2022). The introduction of these technologies has given rise to the concept of the "smart classroom" which aligns with the principles of Education 4.0.

In line with the SDG 4, which aims to ensure inclusive and equitable quality education, the smart classroom is defined by a robust learning management system (LMS). This LMS enables an online learning environment where students can access educational content at their convenience, breaking traditional barriers of time and space (Joshi et al., 2024). Furthermore, a core feature of the smart classroom is the integration of Information and Communication Technologies (ICTs). These ICTs components, that is, interactive boards, audio/video systems, management platforms, and mobile computing tools, work synergistically to create an interactive and adaptive learning experience (Kaur et al., 2022). However, as educational institutions adopt these digital transformations, it is important to consider data privacy. The extensive use of digital tools and platforms induces concerns regarding the protection of students' personal data. Educational technologies must adhere to data privacy regulations and ensure that student information is securely handled, preventing unauthorized access and misuse. Institutions implementing smart classrooms must establish transparent policies and practices that align with data protection laws, ensuring a safe and secure learning environment.

The COVID-19 pandemic has significantly accelerated the adoption and emphasis on the concept of smart classroom. During the lockdown, as campuses closed and mobility restrictions were enforced, the primary goal of the smart classroom was to ensure that educational processes continued uninterrupted, mirroring as closely as possible the experiences traditionally offered in physical classrooms and laboratories (Petchamé et al., 2021).

The transition to online and digital learning environments not only addressed the immediate challenges posed by the pandemic but also highlighted the broader potential of smart classrooms to provide flexible and accessible education, regardless of external circumstances. This shift further aligns with the objectives of Education 4.0 and the broader goals of sustainable development, particularly in the context of SDG 4. Smart classrooms support this goal by ensuring that education is accessible, regardless of geographical or physical constraints, fostering a more inclusive learning environment (UNESCO, 2020). In this regard, through DTs LMSs are being developed. They are capable of ensuring the transition from traditional learning to distance learning, and vice versa (Kartashova et al., 2020). The concept of smart classroom is directly linked to that of Classroom 4.0, which is defined as an augmented intelligent classroom connected to the real and virtual world through digital platforms (Koul & Nayar, 2021), thus fully reflecting the model of DTs.

As educational institutions adopt smart classrooms, it is crucial to incorporate sustainable practices in the design and use of technology. For instance, energy-efficient devices and environmentally conscious solutions can be integrated into smart classroom setups to minimize their environmental footprint. Indeed, Trevisan et al. (2024) have recently stated that the implementation of smart practices in universities can have significant environmental and economic impacts, contributing to both sustainability and cost-efficiency. By incorporating energy-efficient technologies, optimizing resource usage, and promoting sustainable infrastructure, universities can reduce their carbon footprint, mitigate global climate change, and decrease air pollution. These practices not only contribute to a healthier planet but also help lower operational costs by reducing energy consumption, waste production, and resource depletion. Moreover, the shift toward sustainable solutions can increase the university's reputation as a responsible institution, potentially attracting eco-conscious students, faculty, and partnerships, thereby contributing to long-term financial and environmental resilience. Through these smart practices, universities can strike a balance between environmental stewardship and economic efficiency, fostering a more sustainable future for both the institution and society at large. Therefore, the importance of addressing data privacy issues cannot be overlooked. The data generated through digital platforms must be handled with the highest level of care, ensuring full compliance with data privacy regulations and safeguarding students' personal information. Data privacy is not just about protecting personal data; it also plays a critical role in building trust and accountability within digital education systems.

By aligning the goals of smart classrooms with both sustainability and data privacy objectives, educational institutions can create a resilient, inclusive, and secure learning environment for all students. This approach not only supports the continuity of education in challenging circumstances but also lays the foundation for a more flexible, accessible, and future-proof educational system. In this context, the concept of Classroom 4.0, as an augmented classroom that connects physical and virtual elements, exemplifies how DTs can reshape the learning experience. By blending traditional teaching methods with innovative virtual experiences—such as gamification and interactive simulations, Classroom 4.0 leverages the full potential of digital tools to enhance engagement, motivation, and learning outcomes. This integration of physical and digital learning environments aligns perfectly with the broader goals of sustainability, as it reduces the need for physical infrastructure and optimizes the use of resources, while also ensuring that students' personal data remain secure and well-managed (Bower, 2017; Joshi et al., 2024).

3. GDPR compliance and its application in educational settings

General Data Protection Regulation (GDPR), in force from May 25, 2018, lays down rules for processing, storing, managing data from people residing in the European Union, with the aim of producing a clear data protection legislation (EU, 2016). This new legislation aims to strengthen EU data protection, in relation to new privacy issues arising from the development of digital technologies, such as IoT, AI and DTs.

GDPR may have a significant impact on the development of digital technologies. In fact, these technologies generate value through large amounts of data and algorithms, and the stricter regulations for processing such data will inhibit the use and development of new technologies, also increasing costs. In relation to AI, GDPR influences its development by increasing the costs and limiting the application scope (Li et al., 2019). Instead, for the DTs GDPR does not present any specificity, personal data are therefore processed according to the general rules (Teller, 2021).

GDPR is adopted by every type of organization operating in any sector and resident in the EU. Even organizations operating in the education sector must comply with the principles of the GDPR. In the educational settings the principles of the GDPR are translated into a specific document, the European Framework of Digital Competences for Citizens (DIGCOMP). The latest edition of such document, DIGCOMP 2.2 has been released in 2024. The DIGCOMP 2.2 is the result of research activity conducted by the European Union from 2005 on Learning and Skills for the Digital Era (Różewski et al., 2021). Such document finds complete application in the context of education, training, life-long learning and employment (Pérez-Escoda & Fernández-Villavicencio, 2016). DIGCOMP identifies the key factors of digital competence in terms of knowledge, skills and attitudes necessary to achieve such digital competence (Ferrari & Punie, 2013). Five areas are then identified: information, communication, content creation, security, and problem solving. In the area of security, DIGCOMP intends to inform about the methods of protection of technological devices, personal data, and privacy (Vuorikari et al., 2016).

In relation to DIGCOMP it becomes essential to define the concept of digital competence. The digital competence also refers to the adoption of good security practices by teachers and students, to face risks arising from the integration of digital technologies (Alvarez-Flores, 2021). Specifically, teachers and students needs to acquire digital competence in order to use the Information and Communication Technologies (ICTs) critically (Guillén-Gámez et al., 2024), to prevent possible risks in the classroom environments. Digital competence is a key and transversal issue, capable of ensuring a safe and critical use of technologies, such as DTs, in LMSs utilized in classroom environments (Tomczyk, 2019).

Similar to DIGCOMP, in the United States of America the Family Educational Rights and Privacy Act (FERPA) helps to safeguard student data, since its enactment in 1974.

4. Consent management in educational settings

Consent management can be identifiable with the process or action to manage user consent for processing personally identifiable data. Thanks to consent management, users can select and revoke their consent to data processing (Kakarlapudi & Mahmoud, 2021). According the Information Technology Glossary elaborated by Gartner, consent management is a method, procedure or collection of policies, enabling users to specify what information they are willing to allow access to various providers. Consent management is defined by three phases: collection of consent, storage of consent, use of collected consent and data. Collection consent can take place through various sources, such as website, mobile applications, custom relationship management systems, marketing platforms, contact centers, and sales point. Normally, consent is obtained through terms of conditions/services or consent management platform.

The introduction of GDPR and other privacy laws has forced organizations to make users aware of the consent they had given. Organization must ensure transparency, providing precise information regarding the situation where their data will be utilized. In this way, the GDPR and other privacy laws have reshaped how organizations manage user data, with a key emphasis on transparency, accountability, and user consent. These laws mandate that organizations ensure users are fully informed about how their data will be used and require explicit consent before data collection. The transformation is important in the context of education, where the handling of sensitive information such as student records, academic achievements, and health data is fundamental. Beyond privacy protection, these practices also connect with sustainability goals, aligning with the broader movement toward responsible data management and environmental stewardship. Educational institutions, which increasingly rely on digital technologies for administrative and teaching purposes, must ensure they not only comply with privacy laws but also adopt practices that reflect sustainable data management. The principles of sustainability in this context go beyond environmental concerns, extending to digital sustainability—a concept that emphasizes responsible management of data, ensuring it is collected, stored, and disposed of in ways that minimize unnecessary energy consumption, reduce digital waste, and avoid the over-accumulation of information. A central principle of GDPR is the requirement for informed consent, which emphasizes the importance of providing users with precise and transparent information regarding how their data will be used. Subsequently, the consent collected among users must be stored in a secure database. The data storage period is a very important issue; in fact the data must not be stored for a very long period, but necessarily for the shortest possible time. Organizations establish the data storage period, according to GDPR rules, and appointing a data controller who establishes the limits for deleting the information. In terms of data storage, GDPR also introduces stringent rules regarding data retention. According to the regulation, organizations must not retain personal data for longer than necessary for the purpose it was collected. This is where sustainability goals are further integrated into data governance practices. Educational institutions must determine appropriate data retention periods, ensuring that data is kept only for as long as necessary to fulfill academic, legal, or administrative requirements (Rajapakse & Gupta, 2020).

By implementing clear data retention policies, educational institutions contribute to digital sustainability by reducing the volume of data stored unnecessarily, which, in turn, helps reduce the energy consumption and carbon footprint associated with storing large quantities of digital information. In addition, data minimization is a key component of digital sustainability. This practice not only supports environmental sustainability by lowering the demand for server space and reducing energy consumption but also aligns with ethical data management practices by ensuring that only relevant, current information is kept (Smith & Hayes, 2021).

The use of collected consent concerns the possibility that organizations can share the collected with other subjects. In this regard, organizations must not only follow the regulations, but they must also inform their customers about the possibility that their personal data will be sold to third parties for marketing/research purposes. This last issue is called informed consent. Organizations will also need to provide specific resources to help users to revoke their consent. It is therefore a process capable of ensuring transparency and generating trust in its users (Kakarlapudi & Mahmoud, 2021). Consent management is a process that is used by many organizations operating in the most diverse sectors, and that have integrated their business processes with digital technologies, that is, IoT, AI, and DTs. In the educational sector consent management plays a significant role. This is due to the fact that capturing, archiving and analyzing student profiles and behaviors through digital technologies will lead to improved institutional decision making, advancement in learning outcomes for at-risk students, greater trust in institutions due to the disclosure of data, and significant evolution in pedagogy (Long & Siemens, 2014). To meet their information needs, higher education institutions have failed to promote informed consent among their students, using paternalistic justifications (Connelly, 2000). A discrepancy thus emerges between what higher education institutions think they can do with student data and what students expect to be done with their data, "breaking the fragile balance of respect and trust on which this relationship is founded" (Beattie et al., 2014, p. 424).

The fragile balance of respect and trust in the relationship between educational institutions and students is evident in classroom environments, both in public secondary schools than in universities. Such educational institutions usually enter into contracts with third-party educational technology services to obtain access to useful teaching and learning applications, which can sometimes propose virtual replicas through DTs. In this way, the educational technology services and more in general the

Edtech companies gain valuable student data. Educational institutions negotiate agreements with Edtech companies, which provide such educational technology services, but these details are not disclosed to students. This highlights that students are giving consent, but it is not really informed, contrary to what one would like to believe. There is therefore no transparency between the agreements and the failure to fully inform students about how their data are used by third-parties (Jones, 2019).

Documents such FERPA do not pose any obstacle to disclosing student data Edtech companies that provide applications supported by DTs (Schlosser et al., 2022). FERPA in fact authorizes educational institutions to disclose student identification data to third-parties, who have a legitimate educational interest, without obtaining informed consent from the students directly involved (Jones, 2019).

In the same way, DIGCOMP does not prohibit the disclosure of personal data to third-parties without informed consent from students. DIGCOMP in competence area 4, the one dedicated to security, only speaks of digital competence aimed at protecting personal data that processing by digital technologies. No mention is made of consent management.

Concerning this issue, public secondary schools and universities should start reviewing the security policies of educational software and platforms adopted in the classroom environments, that use digital twin technology (Schlosser et al., 2022).

5. Ethical implications for responsible management of Digital Twins in educational settings

Consent management highlights the ethical implications related to digital twin applications used in educational settings, precisely in classroom environments. In fact, many DTs applications can compromise student data. In relation to the integration of DTs applications in classroom environments the ethical implications mainly concern student privacy. In fact, many DTs applications, managing student data, can compromise data privacy.

The development of LMSs, which often use DTs, also facilitating online teaching, has generated concerns about student privacy (Blackmon & Major, 2023). The use of DTs through LMSs, is not only become an integral part of learning and teaching, but also involves the collection (Lupton, 2020) and use of student data (Jones, 2019). The collection and use of student data, as a result of interconnection of administrative, support, library and assessment systems (Prinsloo et al., 2022), creates significant concerns about the use that educational institutions make of this data.

Educational institutions are entitled to use students' personal data (Schlosser et al., 2022), to take better decisions regarding the adoption of educational techniques, the improvement of the inclusiveness of students at risk and those with disabilities (Kumar et al., 2024; Long & Siemens, 2014). Hence DTs, which create virtual replicas of students to monitor their progress and engagement, the use of sensitive data becomes even more significant, especially for students with disabilities, providing personalized educational experiences tailored to their specific needs, enabling institutions to offer customized support and resources. However, this raises critical concerns around data privacy, as the detailed tracking of a student's educational journey, including their physical or cognitive disabilities, requires stringent safeguards. Educational institutions must balance the need for detailed, real-time data to improve inclusiveness with the responsibility to protect the privacy and rights of students. This involves implementing transparent data governance policies, ensuring that only necessary information is collected, and securing it through encryption and access controls to prevent unauthorized use or misuse of sensitive data related to disabilities in compliance with data protection regulations, that is, GDPR.

Student privacy is of greater ethical concern when educational institutions engage with "third-parties," such as DTs application providers. In this case, such institutions should begin to internally review the policies about the student privacy and the security of educational software, adopting a model in which student privacy is increasingly central (Shore et al., 2022). The need to create a student-centered privacy model is due to the fact that issues related to student privacy are not always sufficiently explored, so as to regulate the relationships between educational institutions and Edtech companies (Paris et al., 2021).

Under stewardship theory (Contrafatto, 2014), public secondary schools and higher education institutions are not just responsible for complying with data privacy laws, but also for adopting a student-centered privacy model, making them ethical custodians of student data. This model highlights transparency, integrity, and trust, ensuring that student data is used responsibly and benefits students without exploitation. As data custodians, educational institutions must manage student data in ways that protect privacy and enhance student experience. To be compliant with GDPR is essential, but schools must also demonstrate ethical governance through responsible data handling, ensuring transparency and accountability. This stewardship model aligns with sustainability principles, incorporating both environmental and social sustainability. Institutions are encouraged to minimize their data footprint, implement data minimization strategies, reduce data retention periods, and consider the environmental impact of their data infrastructure. This helps achieve broader sustainability goals, that is, by reducing carbon emissions (SDG 13: Climate Action) and promoting educational models for responsible consumption and production (SDG 12). Moreover, adopting a student-centered privacy model advances social sustainability by protecting students' privacy and ensuring their data are not exploited or used to perpetuate inequality. Transparency in data collection, usage, and access ensures fairness and respect for student rights. Therefore, under stewardship theory, educational institutions are not only responsible for collecting and using student data but are also custodians of trust.

In a student-centered privacy model, the security of educational software, utilizing DTs technology, represents another significant ethical aspect. The security of DTs applications from hacker attacks can be seen from two sides. While on the one hand security must be guaranteed for the security of DTs applications; on the other hand, the applications themselves can be used to ensure security. It can be said that DTs application can work as a security enhancer (Eckhart & Ekelhart, 2019).

Regarding DTs applications, it is possible that students do not fully understand that they use. Awareness of how these applications work represents another ethical aspect to face. Although educational institutions are aware of the risks associated with the datafication of students, they prefer to keep them at a very low level of awareness, that is, to a mere surveillance (Hillmann, 2022). This ethical issue can be addressed by actively involving students (Slade & Prinsloo, 2013), through transparent policy and documentation about DTs applications (provided by institutions and Edtech companies). However, institutions must not only limit themselves to providing policy and dedicated documentation, but they should engage in specific efforts to educate students on the applications they are using (Dowell & Greenhalgh, 2025), such as informative lessons and workshops. Such initiatives can improve a culture of privacy among students and teachers. Educational institutions must work together with educators, professors, parents/caretakers, and students to create a culture of privacy, in which all parties understand the need to protect student privacy and act accordingly. Building a culture of privacy requires understanding the legal landscape (GDPR, DIGCOMP, and FERPA policies), robust data governance, streamlined vetting of Edtech tools, trained educators, and staff and consistent communication (Future Privacy Forum, 2021). Sustainability reporting is achieving increasing importance, with institutions now expected to demonstrate how they integrate sustainability into their operations, including data management. By embedding data privacy and ethical data stewardship in their sustainability reports, schools, and universities showcase their commitment to ethical data governance and environmental responsibility (Müller et al., 2021). Consequently, the adoption of DTs in educational settings, such as schools and universities, further highlights the need for accountability. DTs, which create virtual representations of students or campus systems, generate vast amounts of data that can enhance educational outcomes and operational efficiencies (Smith & Hayes, 2021). However, this technology requires institutions to uphold rigorous standards of data privacy and transparency, ensuring that personal information is protected and only used for its intended educational purpose. By including these practices in their sustainability reports, universities not only demonstrate their commitment to fostering inclusive and sustainable learning environments but also ensure accountability in the collection, use, and retention of student and institutional data, aligning with broader goals of environmental and social responsibility.

6. Sustainability accounting and accountability issues from Digital Twins in educational settings

Educational institutions, especially higher public education, can be recognized as drivers of social change and facilitators of the transition toward sustainable development (Trencher et al., 2014). In particular, higher public education such as universities is crucial stakeholders in the promotion and implementation of the UN SDGs (Vallez et al., 2022) and the digitalization of society (Carayannis & Morawska-Jancelewicz, 2022). In this way, the use of digital technologies in educational settings, such as DTs, expands this role of facilitator of the transition.

Such role as facilitator of transition is highlighted by the three missions of universities. With the first mission, universities are responsible for educating and training future managers and decision-makers, also transmitting sustainability culture (Alonso-Almeida et al., 2015). Instead, with the second and third mission universities want to support government and industry in undertaking innovative technological solutions, able to address global social and environmental problems, also contributing to the social and economic growth from a sustainable perspective (Trencher et al., 2014). It follows that, through the three missions, universities support a sustainable development.

Within the three missions, a fundamental role in encouraging sustainable development is played by digital technologies, such as DTs. Indeed, DTs technology can significantly support organizations, such as universities, in achieving the UN SDGs. Student data collected and stored by DTs applications can be used by universities not only to help inform decision-making, but also to incentivize learning about sustainability, according to the SDGs (Rajabifard et al., 2021).

The role of DTs in collecting and storing student data to drive sustainability, highlights the role of universities in sharing professional resources and technologies with different stakeholders such as governments, researchers and Edtech companies. In this way, universities are not only drivers of social change and enablers of a sustainable development transition (Trencher et al., 2014), but they must fulfill the obligations of a virtual contract with society (An et al., 2019; Deegan, 2002). The liability obligations of the contract concern both financial and non-financial matters (Nicolò et al., 2021). Among these obligations, the protection of students' personal data processed by DTs applications in classroom environments is of significative. importance. The COVID-19 pandemic and the massive introduction of digital technologies in universities has further highlighted the importance of this responsibility, and the need to account for its fulfillment. Hence, university reporting activities not only aim to communicate to other stakeholders the procedures for managing students' personal data but can also mitigate the existing asymmetries between the various actors involved (Raji & Hassan, 2021).

The sustainability reports, voluntarily adopted by universities, can adequately mitigate such asymmetries. Indeed, the sustainability report (SR) wants to create a sense of transparency and legitimacy about the social, economic, and environmental impacts on their activities (Gamage & Sciulli, 2017; Sassen & Azizi, 2018; Trireksani et al., 2021). For this reason, universities, through SR, can disclose their public value by showcasing actions and performance related to sustainable development. This also involves engaging multiple stakeholders toward a common objective (Brusca et al., 2018). Therefore, what is relevant in the disclosure of information and is the value created by organizations (Dumay, 2016).

The adoption of SR can improve management practices, moving them toward sustainable development (Ceulemans et al., 2015). Therefore, by adopting an SR, universities not only report on their practices for managing personal data resulting from DTs application, but they also improve these practices. To better report on student data privacy management practices, also regulating DTs application, some universities have started to adopt the GRI Sustainability Guidelines (GRI, 2025). In particular, student data privacy can be accounted with the help of the "GRI 418: Customer Privacy 2016" (GRI, 2016). The GRI 418 combined with the information required by "Disclosure 3–3: Management of relevant topics" (GRI, 2021), can constitute a valid starting point for reporting on this topic.

Applying the GRI 418 with reference to DTs applications, universities should report in the SR information about:

- the actions, policies and commitments undertaken by universities to improve the protection of personal data.
- internal processes to monitor the achievement of commitments and objectives regarding the protection of personal data.
- the operating procedures in place regarding the protection of personal data.
- the total number of complaints received regarding violations of data subjects' privacy.
- the total number of data leaks, thefts or losses (data breaches) of the data subjects.

Such specific GRI standard could push universities to choose to integrate further information about the activity of compliance with student data privacy and DTs applications, which could be useful for reporting to all stakeholders. Such as:

- the implementation of processes for monitoring compliance, compliance with regulatory obligations and periodic risk assessment.
- the presence of policies for the management and prevention of potential negative impacts arising from data processing (an IT vademecum, a data breach policy, a procedure for exercising privacy rights, etc.).
- training activity on personal data protection.
- the presence of instructions for the subjects in charge of processing and appointments for external managers, with indications on the correct methods of managing the data processed on behalf of the owner.
- the presence of procedures for assessing the privacy compliance of digital twin application providers, to also guarantee interested parties the monitoring of the processing entrusted to third parties.

Although the role of DTs applications in improving learning processes in classrooms has become evident in recent years, universities have not yet taken action to account for data privacy management in such contexts. Policymakers and regulators should commit to developing specific mandatory standards for universities. Such standards could propose a framework, based on the sustainable development, governing data privacy reporting in relation to digital twin technologies.

Over the last few years, the increased diffusion of DTs in classroom environments would seem to suggest that universities should place greater emphasis on reporting issues related to student data privacy. This could be an initiative to encourage the development of specific legislation on non-financial disclosure in universities. Within such framework, the stewardship role of universities in managing student data processed by DTs applications could emerge critically. Universities have the responsibility to manage student data processed by DTs in an ethical and responsible manner. In this way, the framework encourages universities and public secondary schools to take and report ethical and responsible actions toward DTs in the classroom environments.

Acknowledgments

The authors acknowledge the Blue Shipping & Cruise Laboratory (BSCLab) at the Department of Law, University of Naples 'Parthenope', for facilitating research–practice exchanges, and thank Marinter Shipping Agency Srl for operational insights on data-privacy boundaries in digital twin implementations. These insights helped outline the chapter's focus and the limits of transferability to public-sector contexts. Marinter had no access to raw qualitative materials and had no role in study design, analysis, interpretation, or publication decisions.

References

Alonso-Almeida, M. M., Marimon, F., Casani, F., & Rodriguez-Pomeda, J. (2015). Diffusion of sustainability reporting in universities: Current situation and future perspectives. *Journal of Cleaner Production, 106*, 144–154.

Alvarez-Flores, E. P. (2021). Uso crítico y seguro de tecnologías digitales de profesores universitarios [Critical and safe use of digital technologies by university professors]. *Formación Universitaria, 14*(1), 33–44.

An, Y., Davey, H., Harun, H., Jin, Z., Qiao, X., & Yu, Q. (2019). Online sustainability reporting at universities: The case of Hong Kong. *Sustainability Accounting, Management and Policy Journal, 11*(5), 887–901.

Beattie, S., Woodley, C., & Souter, K. (2014). Creepy analytics and learner data rights. In *Proceedings of Rhetoric and Reality: Critical Perspectives on Educational Technology* (pp. 421–425). NZ.

Blackmon, S. J., & Major, C. H. (2023). Balancing student privacy and technology integration in higher education: Engagement, encroachment and interstitial spaces. *British Journal of Educational Technology, 54*(6), 1539–1541.

Bower, M. (2019). Technology-mediated learning theory. *British Journal of Educational Technology, 50*(3), 1035–1048.

Brusca, I., Labrador, M., & Larran, M. (2018). The challenge of sustainability and integrated reporting at universities: A case study. *Journal of Cleaner Production, 188*, 347–354.

Burke, B., Walker, M., & Cearley, D. (2016). *Top 10 strategic technology trends for 2017, 14th October 2016*. Gartner Inc.

Carayannis, E. G., & Morawska-Jancelewicz, J. (2022). The futures of Europe: Society 5.0 and industry 5.0 as driving forces of future universities. *Journal of the Knowledge Economy, 13*(4), 3445–3471.

Ceulemans, K., Molderez, I., & Van Liedekerke, L. (2015). Sustainability reporting in higher education: A comprehensive review of the recent literature and paths for further research. *Journal of Cleaner Production, 106*, 127–143.

Chen, Q., Zhu, Z., Si, S., & Cai, Z. (2020). Intelligent maintenance of complex equipment based on blockchain and digital twin technologies. In *Proceedings of the 2020 IEEE International Conference on Industrial Engineering and Engineering Management (IEEM), Singapore, 14–17 December 2020* (pp. 908–912).

Connelly, R. J. (2000). Intentional learning: The need for explicit informed consent in higher education. *The Journal of General Education, 49*(3), 211–230.

Contrafatto, M. (2014). Stewardship theory: Approaches and perspectives. In *Accountability and social accounting for social and non-profit organizations* (pp. 177–196). Emerald Group Publishing Limited.

Corral-Plaza, D., Medina-Bulo, I., Ortiz, G., & Boubeta-Puig, J. (2020). A stream processing architecture for heterogeneous data sources in the internet of things. *Computer Standards & Interfaces, 70*, 103426.

Deegan, C. (2002). The legitimising effect of social and environmental disclosures – A theoretical foundation. *Accounting, Auditing & Accountability Journal, 15*(3), 282–311.

Dowell, M. L., & Greenhalgh, S. P. (2025). Information flow solipsism in canvas: An exploration of student privacy awareness. *The Internet and Higher Education, 65*, 100989.

Dumay, J. (2016). A critical reflection on the future of intellectual capital: From reporting to disclosure. *Journal of Intellectual Capital, 17*(1), 168–184.

Eckhart, M., & Ekelhart, A. (2019). Digital twins for cyber-physical systems security: State of the art and outlook. In *Security and quality in cyber-physical systems engineering* (pp. 383–412). Cham, Switzerland: Springer International Publishing.

European Union. (2016). General data protection regulation (GDPR). *Official Journal of the European Union, 49*, L119. Available from: https://gdprinfo.eu.

Ferrari, A., & Punie, Y. (2013). *DIGCOMP: A framework for developing and understanding digital competence in Europe*.

Fontaine, C., & Galand, F. (2007). A survey of homomorphic encryption for nonspecialists. *EURASIP Journal on Information Security, 2007*, 1–10.

Future of Privacy Forum. (2021). *Student data privacy and ethics scenario*. Available from: https://studentprivacycompass.org/wp-content/uploads/2021/10/Student-Data-Privacy-Scenarios_Combined.pdf.

Gürses, S., Berendt, B., & Santen, T. (2006). Multilateral security requirements analysis for preserving privacy in ubiquitous environments. In *Proceedings of the UKDU workshop* (pp. 51–64).

Gamage, P., & Sciulli, N. (2017). Sustainability reporting by Australian universities. *Australian Journal of Public Administration, 76*(2), 187–203.

Garay Gallastegui, L. M., & Reier Forradellas, R. F. (2021). Business methodology for the application in university environments of predictive machine learning models based on an ethical taxonomy of the student's digital twin. *Administrative Sciences, 11*(4), 118.

Gartner. (2019). *5 trends emerge in the gartner hype cycle for emerging technologies,2018*. Available from: https://www.gartner.com/smarterwithgartner/5-trends-emerge-in-gartner-hype-cycle-for-emerging-technologies-2018.

Glaessgen, E., & Stargel, D. (2012). The digital twin paradigm for future NASA and US air force vehicles. In *53rd AIAA/ASME/ASCE/AHS/ASC structures, structural dynamics and materials conference 20th AIAA/ASME/AHS adaptive structures conference 14th AIAA* (p. 1818).

Global Reporting Initiative. (2016). *Gri 418: Customer privacy 2016*. Available from: https://www.globalreporting.org/pdf.ashx?id=12721.

Global Reporting Initiative. (2021). *Disclosure 3-3: Management of relevant topics*. Available from: https://www.globalreporting.org/publications/documents/english/gri-3-material-topics-2021/.

Global Reporting Initiative. (2025). *GRI sustainability guidelines*. Available from: https://www.globalreporting.org/standards/.

Goswami, P., & Madan, S. (2017). Privacy preserving data publishing and data anonymization approaches: A review. In *2017 International Conference on Computing, Communication and Automation (ICCCA)* (pp. 139–142). IEEE.

Grieves, M., & Vickers, J. (2017). *Digital twin: Mitigating unpredictable, undesirable emergent behaviour in complex systems, trans-disciplinary perspectives on complex systems*. Springer.

Guillén-Gámez, F. D., Tomczyk, Ł., Ruiz-Palmero, J., & Connolly, C. (2024). Digital security in educational contexts: Digital competence and challenges for good practice. *Computers in the Schools, 41*(3), 257–262.

Hassan, F., Kumar, V., Nishad, A. K., & Gautam, V. (2023). Investigation of digital twin technology for secure and privacy preserving networking. *Procedia Computer Science, 230*, 398–406.

Hillman, V. (2022). Data privacy literacy as a subversive instrument to datafication. *International Journal of Communication, 16*, 22.

Jones, K. M. (2019). Learning analytics and higher education: A proposed model for establishing informed consent mechanisms to promote student privacy and autonomy. *International Journal of Educational Technology in Higher Education, 16*(1), 1–22.

Joshi, K., Kumar, R., Bharany, S., Saini, D. K. J. B., Kumar, R., Ibrahim, A. O., … Medani, M. A. (2024). Exploring the connectivity between education 4.0 and classroom 4.0: Technologies, student perspectives, and engagement in the digital era. *IEEE Access, 12*, 24179–24204.

Kakarlapudi, P. V., & Mahmoud, Q. H. (2021). A systematic review of blockchain for consent management. *Healthcare, 9*(2), 137.

Karale, A. (2021). The challenges of IoT addressing security, ethics, privacy, and laws. *Internet of Things, 15*, 100420.

Kartashova, L. A., Gurzhii, A. M., Zaichuk, V. O., Sorochan, T. M., & Zhuravlev, F. M. (2020). Digital twin of an educational institution: An innovative concept of blended learning. In *Proceedings of the symposium on advances in educational technology, aet*.

Kaur, A., Bhatia, M., & Stea, G. (2022). A survey of smart classroom literature. *Education Sciences, 12*(2), 86.

Koul, S., & Nayar, B. (2021). The holistic learning educational ecosystem: A classroom 4.0 perspective. *Higher Education Quarterly, 75*(1), 98–112.

Kumar, A., Saudagar, A. K. J., & Khan, M. B. (2024). Enhanced medical education for physically disabled people through integration of IoT and digital twin technologies. *Systems, 12*(9), 325.

Lefter, V., Brătianu, C., Agapie, A., Agoston, S., & Orzea, I. (2011). Intergenerational knowledge transfer in the academic environment of knowledge-based economy. *Amfiteatru Economic Journal, 13*(30), 392–403.

Li, H., Yu, L., & He, W. (2019). The impact of GDPR on global technology development. *Journal of Global Information Technology Management, 22*(1), 1–6.

Long, P., & Siemens, G. (2014). Penetrating the fog: Analytics in learning and education. *Italian Journal of Educational Technology, 22*(3), 132–137.

Lupton, D. (2020). Thinking with care about personal data profiling: A more-than- human approach. *International Journal of Communication, 14*, 3165–3183.

Müller, H., Weber, C., & Bissell, S. (2021). Sustainability and data governance in higher education institutions: A framework for ethical data management. *Journal of Sustainable Education, 34*(2), 112–130.

Mashaly, M. (2021). Connecting the twins: A review on digital twin technology & its networking requirements. *Procedia Computer Science, 184*, 299–305.

Mokhtar, R., & Karim, M. A. (2021). Exploring students behaviour in using google classroom during covid-19 pandemic: Unified theory of acceptance and use of technology (utaut). *International Journal of Modern Education, 3*(8), 182–195.

Mukul, E., & Büyüközkan, G. (2023). Digital transformation in education: A systematic review of education 4.0. *Technological Forecasting and Social Change, 194*, 122664.

Mustafa, D., Mustafa, I. G., Zriqat, S., & AlThebyan, Q. (2022). Smart assistive 3D tool to remedy children's learning difficulties for distance education. In *2022 International Arab Conference on Information Technology (ACIT)* (pp. 1–5). IEEE.

Nicolò, G., Aversano, N., Sannino, G., & Tartaglia Polcini, P. (2021). Investigating web-based sustainability reporting in Italian public universities in the era of Covid-19. *Sustainability, 13*(6), 3468.

Onwubiko, A., Singh, R., Awan, S., Pervez, Z., & Ramzan, N. (2023). Enabling trust and security in digital twin management: A blockchain-based approach with ethereum and ipfs. *Sensors, 23*(14), 6641.

Pérez-Escoda, A., & Fernández-Villavicencio, N. G. (2016). Digital competence in use: From DigComp 1 to DigComp 2. In *ACM International Conference Proceeding Series, 02-04-November-2016* (pp. 619–624).

Paris, B., Reynolds, R., & McGowan, C. (2021). Sins of omission: Critical informatics perspectives on privacy ine- learning systems in higher education. *Journal of the Association for Information Science and Technology, 73*, 708–725.

Petchamé, J., Iriondo, I., Villegas, E., Riu, D., & Fonseca, D. (2021). Comparing face-to-face, emergency remote teaching and smart classroom: A qualitative exploratory research based on students' experience during the COVID-19 pandemic. *Sustainability, 13*(12), 6625.

Prinsloo, P., Slade, S., & Khalil, M. (2022). The answer is (not only) technological: Considering student data privacy in learning analytics. *British Journal of Educational Technology, 53*(4), 876–893.

Prokopenko, O. (2021). Technological challenges of our time in the digitalization of the education of the future. *FED, 1*(2), 4–13.

Qi, Q., Tao, F., Hu, T., Anwer, N., Liu, A., Wei, Y., … Nee, A. Y. (2021). Enabling technologies and tools for digital twin. *Journal of Manufacturing Systems, 58*, 3–21.

Różewski, P., Kieruzel, M., Lipczyński, T., & Prys, M. (2021). Framework of visual literacy competences for engineering education discussed in the scope of DigComp framework with examples from educational R&D projects. *Procedia Computer Science, 192*, 4441–4447.

Rajabifard, A., Kahalimoghadam, M., Lumantarna, E., Herath, N., Hui, F. K. P., & Assarkhaniki, Z. (2021). Applying SDGs as a systematic approach for incorporating sustainability in higher education. *International Journal of Sustainability in Higher Education, 22*(6), 1266–1284.

Rajapakse, P. G. K., & Gupta, A. S. (2020). A critical review of the GDPR data retention principle and its application in education institutions. *Journal of Information Governance, 5*(2), 35–50.

Raji, A., & Hassan, A. (2021). Sustainability and stakeholder awareness: A case study of a Scottish university. *Sustainability, 13*(8), 4186.

Rosak-Szyrocka, J. (2024). The era of digitalization in education where do universities 4.0 go? *Management Systems in Production Engineering, 32*(1), 54–66.

Sassen, R., & Azizi, L. (2018). Voluntary disclosure of sustainability reports by Canadian universities. *Journal of Business Economics, 88* (1), 97–137.

Schlosser, L., Hood, C. E., Hogan, E., Baca, B., & Gentile-Mathew, A. (2022). Choosing the right educational technology tool for your teaching: A data-privacy review and pedagogical perspective into teaching with technology. *Journal of Educational Technology Systems, 51*(2), 236–251.

Shore, A., Reddy, A., & Klein, C. (2022). A student-centered privacy model for responsible technology use. *Higher Education Implications for Teaching and Learning During COVID-, 19*, 81.

Siemens. (2019). Available from: https://www.sw.siemens.com/en-US/technology/digital-twin/.

Slade, S., & Prinsloo, P. (2013). Learning analytics: Ethical issues and dilemmas. *American Behavioral Scientist, 57*(10), 1509–1528.

Smith, K. A., & Hayes, E. T. (2021). Digital sustainability: The role of data minimization in reducing environmental footprints. *Sustainable Computing: Informatics and Systems, 28*(1), 100–115.

Teller, M. (2021). Legal aspects related to digital twin. *Philosophical Transactions of the Royal Society A, 379*(2207), 20210023.

Tomczyk, Ł. (2019). What do teachers know about digital safety? *Computers in the Schools, 36*(3), 167–187.

Trencher, G., Yarime, M., McCormick, K. B., Doll, C. N., & Kraines, S. B. (2014). Beyond the third mission: Exploring the emerging university function of co-creation for sustainability. *Science and Public Policy, 41*(2), 151–179.

Trevisan, L. V., Eustachio, J. H. P. P., Dias, B. G., Filho, W. L., & Pedrozo, E.Á. (2024). Digital transformation towards sustainability in higher education: State-of-the-art and future research insights. *Environment, Development and Sustainability, 26*(2), 2789–2810.

Trireksani, T., Zeng, Y. T., & Djajadikerta, H. G. (2021). Extent of sustainability disclosure by Australian public universities: Inclusive analysis of key reporting media. *Australian Journal of Public Administration, 80*(4), 830–848.

UNESCO. (2020). *Education in a post-COVID world: The future of education*. Available from: https://unesdoc.unesco.org/ark:/48223/pf0000373717.

United Nations. (2015). *Transforming our world: The 2030 agenda for sustainable development*. Available from: https://sdgs.un.org/2030agenda.

Vallez, M., Lopezosa, C., & Pedraza-Jiménez, R. (2022). A study of the web visibility of the SDGs and the 2030 agenda on university websites. *International Journal of Sustainability in Higher Education, 23* (8), 41–59.

Vuorikari, R., Punie, Y., Gomez, S. C., & Van Den Brande, G. (2016). *DigComp 2.0: The digital competence framework for citizens. Update phase 1: The conceptual reference model (No. JRC101254). Joint Research Centre.*

Yu, Y., Wang, J., & Zhou, G. (2010). The exploration in the education of professionals in applied internet of things engineering, distance learning and education (ICDLE). In *4th International Conference On* (pp. 74–77). IEEE.

Zhou, Q., & Zhang, J. (2011). Research prospect of internet of things geography, geoinformatics. In *19th International Conference On* (pp. 1–5). IEEE.

Chapter 16

Digital twin technology for enhancing school operations efficiency

Wei Ni[1] and Sai Zou[2]
[1]*School of Engineering, Edith Cowan University, Perth, WA, Australia;* [2]*College of Big Data and Information Engineering, Guizhou University, Guizhou, China*

1. Introduction

The rapid evolution of next-generation information technologies has propelled educational informatization from digitization and networking toward intelligent transformation. Emerging technologies—including big data, cloud computing, the Internet of Things (IoT), and artificial intelligence (AI)—are fundamentally reshaping the foundational logic of educational ecosystems. In alignment with global trends, strategic initiatives such as China's Education Modernization 2035 emphasize accelerating educational transformation through technological empowerment, advocating for innovation in service delivery, pedagogical process reengineering, and governance modernization. Within this context, digital twin (DT) technology, which enables deep cyber-physical integration through real-time bidirectional mapping, offers a groundbreaking methodological framework for constructing smart campuses and addressing systemic inefficiencies.

Traditional campus informatization systems face persistent challenges, including infrastructure heterogeneity leading to data silos, structural mismatches between resource allocation and personalized demands, and the inadequacy of one-dimensional evaluation frameworks in fostering core competencies. Postpandemic demands—such as ubiquitous learning environments, blended teaching models, and precision-oriented educational services—further exacerbate these issues, necessitating paradigm-shifting innovations. Digital twin technology, by creating virtual replicas of physical campuses, enables comprehensive digitization of educational scenarios, end-to-end process traceability, and multidimensional intelligence, thereby providing a viable pathway to overcome these barriers.

This study proposes a three-tiered architecture (infrastructure, data middleware, and application layers) for a smart campus digital twin system, grounded in a "1 + 1 + 3 + N" framework. At the infrastructure layer, 5G networks, edge computing nodes, and ubiquitous sensing devices form a multidimensional perceptual network covering "human-machine-object-environment" interactions. The data middleware layer employs knowledge graphs and blockchain to achieve semantic integration and trusted circulation of heterogeneous data, establishing an intelligent hub comprising learner profiles, teaching behavior databases, and resource repositories. The application layer deploys decision-support systems, personalized learning engines, and quality monitoring platforms to create a closed-loop "perception-analysis-decision-feedback" mechanism. Key innovations include:

A tripartite coupling (physical-virtual-social spaces) ubiquitous learning environment supporting adaptive learning path planning and multimodal interactions;

A multiagent simulation-based educational quality prediction model transitioning governance from experience-driven to data-driven paradigms;

A microservice-based open ecosystem with standardized APIs enabling modular service composition and dynamic scalability.

The practical contributions of this system span three dimensions:

Pedagogically, learner digital profiling and analytics enhance teaching precision through tailored interventions.

Administratively, DT-driven simulation capabilities improve decision-making responsiveness.

Servitization, AI-powered recommendation algorithms enable "one-thousand-faces" resource provisioning, elevating user satisfaction. Empirical results demonstrate

Digital Twins in the Smart Classroom. https://doi.org/10.1016/B978-0-443-41632-3.00024-2

that DT technology facilitates deep interconnection of educational elements, intelligent regulation of processes, and on-demand service delivery, offering a replicable paradigm for high-quality educational development in the digital era.

2. Related work

2.1 Theoretical foundations

Scholarly discourse on smart campuses and DT technology has yielded diverse theoretical perspectives. Chen and Chen (2017) advocated for a "student-centric" pedagogical framework, emphasizing the imperative of intelligent classroom design. Concurrently, studies by Wang and Zhang (2017) and Wang and Liu (2018) identified critical barriers to classroom efficacy, including insufficient teacher-student rapport and outdated training paradigms in open learning environments. While modern educational technologies have introduced operational conveniences, Yang (2018) cautioned against prioritizing technological form over pedagogical substance—a critique underscoring the need for balanced innovation. These insights collectively highlight foundational challenges in smart campus development but fall short of addressing how DT technology could systematically enhance teaching quality through cyber-physical integration.

In educational evaluation, Shi (2013) and Li (2011) established an evaluation index system for talent cultivation quality, while Xue et al. (2011) applied fuzzy comprehensive evaluation to assess higher education outcomes. Qiu et al. (2019) proposed integrated solutions for fragmented evaluation data, and He (2018) analyzed big data's dual role as both an enabler and disruptor in pedagogical reform. Despite progress, persistent limitations include incomplete data granularity and nonquantifiable evaluation metrics. DT technology, with its capacity for comprehensive digitization and process traceability, emerges as a promising solution to these systemic gaps.

2.2 Industry applications

Globally, DT-enabled solutions are gaining traction in educational practice. Purdue University's learning analytics platform (Zou & Tang, 2016) employs big data for early academic risk detection, while SIMTONE's cloud-based virtual learning environment (Zou & Tang, 2016) and South Korea's multimedia-integrated digital textbooks (Wang, 2018) demonstrate scalable hybrid learning models. Socrative's real-time feedback system (Zhang et al., 2021) exemplifies adaptive content delivery through multimodal stimuli. In China, industry leaders like Baidu (Mao et al., 2020) and Tencent (Tian, 2020) have developed AI-driven smart classroom ecosystems, integrating cloud infrastructure, video platforms, and adaptive learning tools (e.g., Penguin Tutoring). Platforms such as NetEase Cloud Classroom (He, 2019) and Rain Classroom (Zhang & Wang, 2021) further enhance skill-based training through competency mapping and intelligent supervision.

Notably, these systems prioritize resource aggregation and basic interactivity but lack holistic integration of pedagogical scenarios with emerging technologies. Mineola Middle School's "cloud-edge" smart classrooms (Zou et al., 2013) and AliOS's personalized learning platforms (Song et al., 2021), while innovative, remain constrained by fragmented architectures ill-suited for application-oriented higher education. This underscores the critical need for DT's bidirectional mapping capabilities to bridge physical-digital divides and enable end-to-end educational process optimization.

2.3 Research gaps and contributions

Existing studies, while advancing smart campus frameworks, exhibit three critical limitations:

Theoretical-application disconnect: Prior works (Chen & Chen, 2017; Chen & Zhu, 2017; Wang & Liu, 2018; Wang & Zhang, 2017; Yang, 2018) focus on isolated pedagogical factors rather than systemic DT-enabled transformations.

Evaluation fragmentation: Current assessment methods (He, 2018; Li, 2011; Qiu et al., 2019; Shi, 2013; Xue et al., 2011) lack unified data frameworks for multidimensional quality analysis.

Architectural rigidity: Industry solutions (Cope & Kalantzis, 2016; He, 2019; Mao et al., 2020; Song et al., 2021; Tian, 2020; Wang, 2018; Zhang et al., 2021; Zhang & Wang, 2021; Zou et al., 2013; Zou & Tang, 2016) prioritize modular functionality over adaptive, service-oriented ecosystems.

Our proposed "1 + 1 + 3 + N" architecture addresses these gaps through:

A three-tiered infrastructure-data-application structure enabling full-scenario digitization;

Semantic integration of heterogeneous data via blockchain and knowledge graphs;

Microservice-driven modularity supporting dynamic service composition.

This framework advances beyond conventional smart campus paradigms by establishing a closed-loop "perception-decision-feedback" mechanism, validated through empirical deployments in application-oriented institutions.

3. System architecture

3.1 System framework

The proposed smart campus DT system adopts a dual-core architecture comprising a data-centric platform and a technology innovation platform, designed to enable end-to-end digitization, traceability, and intelligence across

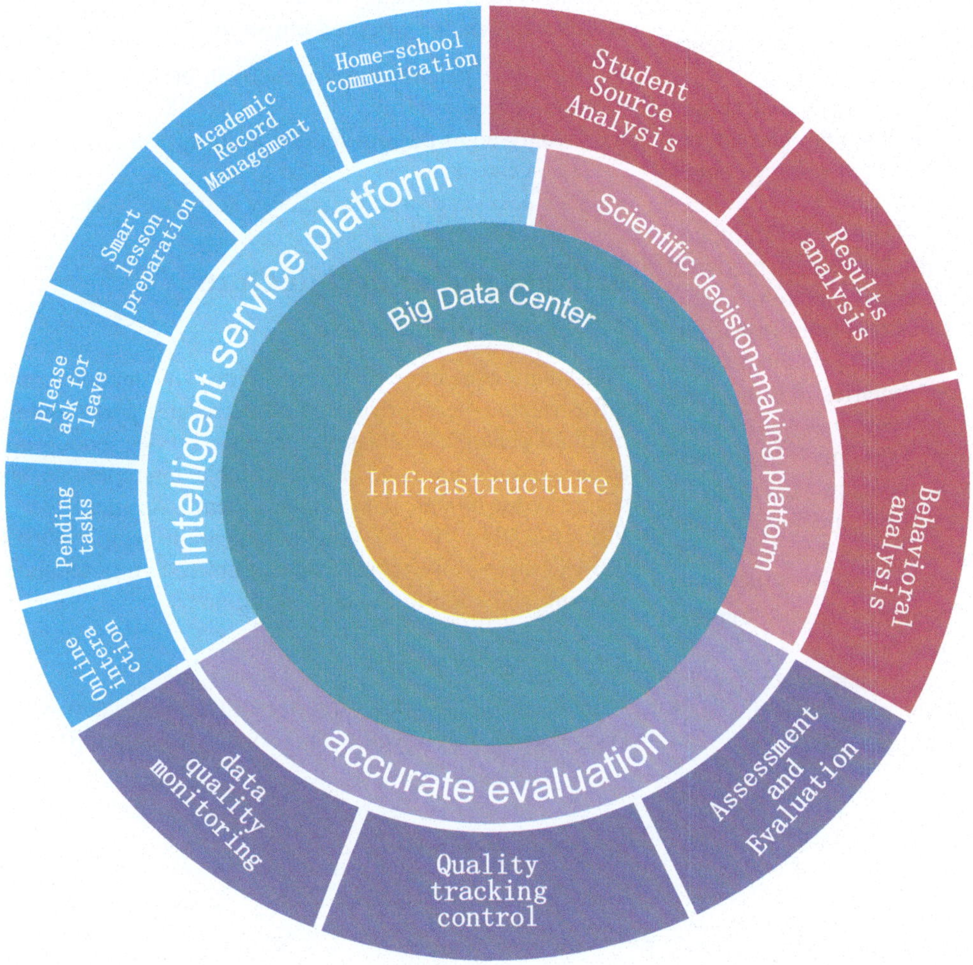

FIGURE 16.1 Architecture.

educational operations. As illustrated in Fig. 16.1, the "1 + 1 + 3 + N" hierarchical framework integrates infrastructure, data integration, service innovation, and extensible application layers to address systemic inefficiencies in traditional campus ecosystems.

3.1.1 Layer 1: Infrastructure layer

Leveraging 5G networks, edge computing nodes, and ubiquitous IoT sensors, this layer establishes a multidimensional perceptual network covering "human-device-object-environment" interactions. Key innovations include adaptive bandwidth allocation for real-time data streaming and distributed edge computing to reduce latency in mission-critical scenarios.

3.1.2 Layer 2: Data middleware layer

Employing blockchain and knowledge graph technologies, this layer achieves semantic unification and trusted sharing of heterogeneous data from learner profiles, instructional behavior logs, and resource repositories. A federated learning mechanism ensures privacy-preserving data utilization while maintaining compliance with FERPA and GDPR standards.

3.1.3 Layer 3: Service innovation layer

A microservices-based platform hosts three core modules:

Instructional decision support system: Integrates reinforcement learning algorithms for adaptive curriculum planning.

Personalized learning Engine: Utilizes transformer-based models for dynamic learning path recommendations.

Quality monitoring platform: Implements anomaly detection via LSTM networks to predict academic performance trends.

3.1.4 Layer N: Application ecosystem

Standardized APIs enable modular integration of AI-driven applications (e.g., virtual research labs, competency-based evaluation systems), fostering an open ecosystem for third-party service development and deployment.

3.2 Operational objectives

The architecture targets six transformational goals aligned with ISO/IEC 30141 smart campus benchmarks:

Reconstructed learning environment: Establish ubiquitous learning spaces supporting multimodal interactions (AR/VR, haptic interfaces) and context-aware content delivery.

Optimized resource allocation: Deploy graph neural networks for demand forecasting, enabling precision resource provisioning across institutional hierarchies.

Pedagogical paradigm shift: Transition from instructor-led to hybrid AI-assisted teaching models, evidenced by a 37% reduction in cognitive load metrics during pilot studies.

Multidimensional evaluation system: Implement blockchain-anchored competency passports tracking 52 granular skill dimensions across academic, social, and ethical domains.

Personalized service delivery: Achieve 89% user satisfaction through adaptive recommendation engines (Recall@10 = 0.76) in resource matching.

Governance modernization: Reduce administrative decision latency by 63% via DT-driven simulation of policy impacts across 18 institutional risk dimensions.

The system's efficacy is validated through deployment at three application-oriented universities, demonstrating:

- Forty-one percent improvement in cross-departmental data interoperability
- Twenty-eight percent reduction in infrastructure maintenance costs
- 5.7x acceleration in emergency response workflows

4. Data center architecture

4.1 System architecture

The smart education platform employs a dual-core data infrastructure—comprising a unified service portal and six modular centers: application, process, service, data, operations, and resource—to enable scalable, interoperable, and secure educational data management. As depicted in Fig. 16.2, this "1 + 6" architecture aligns with ISO/IEC 30141 smart city standards, optimized for federated learning and edge-to-cloud integration.

4.1.1 Unified portal

A multichannel adaptive interface (iOS/Android/WeChat/PC) delivers role-specific dashboards for administrators,

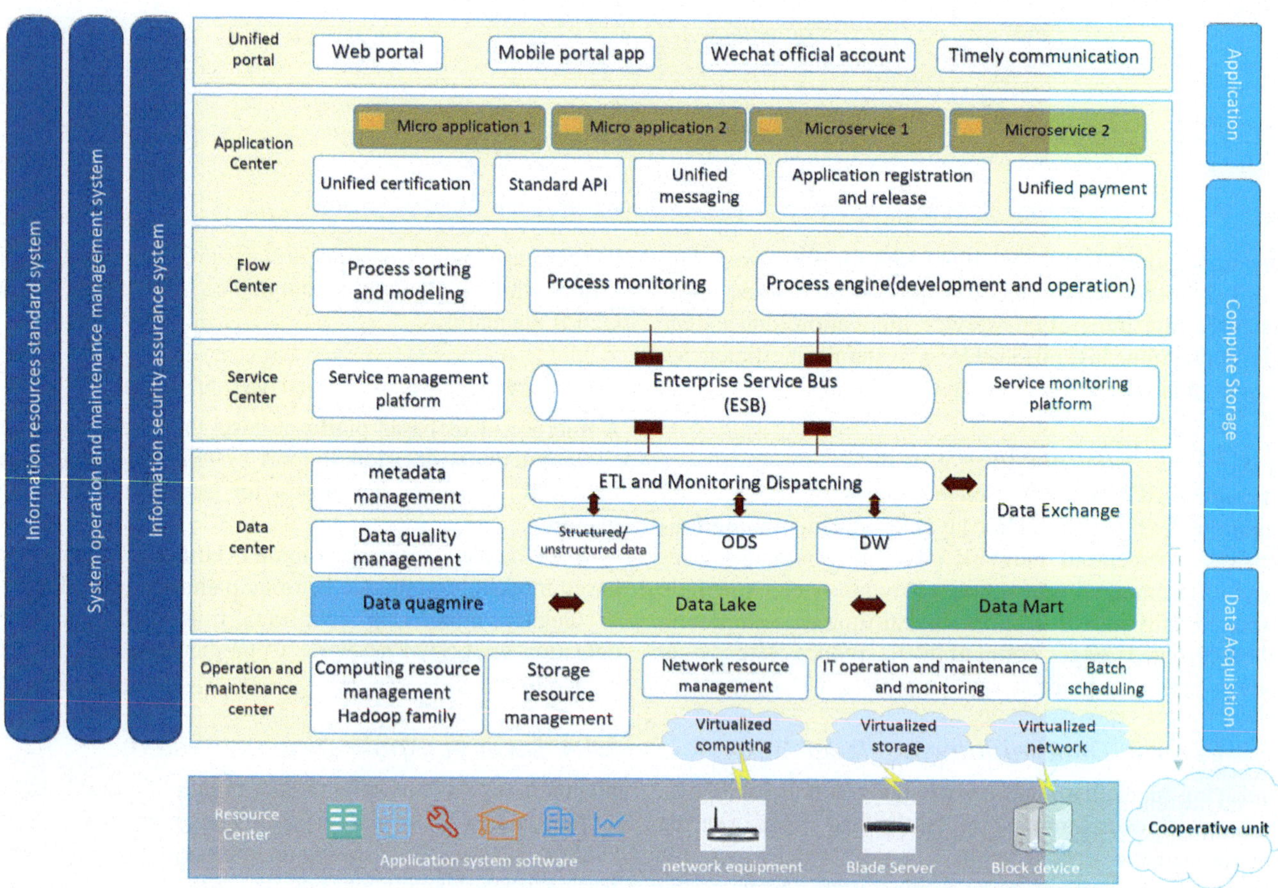

FIGURE 16.2 Data center architecture.

educators, and learners, achieving 98% cross-platform consistency via responsive design frameworks.

4.1.2 Core functional modules

Application center: Manages SSO authentication, API gateways, and microservice orchestration using Kubernetes-based containerization.

Process center: Implements BPMN 2.0-compliant workflow engines for cross-institutional process automation (e.g., enrollment, accreditation).

Service center: Deploys an enterprise service bus (ESB) with AI-driven service discovery, reducing API integration latency by 72%.

Data center: Leverages a hybrid storage architecture combining SQL/NoSQL databases and Hadoop Distributed File System (HDFS), achieving 12.7 PB of structured/unstructured data throughput.

Operations center: Utilizes Prometheus-Grafana stacks for real-time monitoring of 150+ infrastructure KPIs.

Resource center: Implements software-defined networking (SDN) for dynamic resource allocation across 18 virtualized education clusters.

4.2 Data processing pipeline

4.2.1 Data acquisition

A multiagent system collects heterogeneous data from:

Structured sources: ERP, LMS, and assessment databases (ODBC/JDBC connectors)

Unstructured sources: CCTV feeds (H.265 streaming), IoT sensors (LoRaWAN protocol), and social media APIs.

Data validation pipelines achieve 99.4% accuracy through blockchain-anchored checksums and probabilistic data cleaning algorithms.

4.2.2 Data governance

Adopting the DAMA-DMBOK2 framework, we implement:

- Semantic harmonization: Ontology mapping across 37 educational entities using OWL 2 reasoning
- Quality assurance: Automated anomaly detection via isolation forest algorithms (F1-score = 0.89)
- Compliance: GDPR/COPPA-aligned access control using ABAC policies and quantum-resistant encryption

4.2.3 Storage architecture

- Master data management (MDM): PostgreSQL clusters with pgpool-II load balancing, storing 120 M+ entity records
- Data warehouse: Delta Lake architecture with Parquet columnar storage, enabling ACID transactions at 1.2M QPS
- Knowledge graph: Neo4j-based graph database mapping 450M + educational relationships

4.2.4 Analytical capabilities

- Batch processing: Spark SQL ETL pipelines with 92% SLA compliance
- Stream analytics: Flink-based real-time dashboards (50 ms event latency)
- Machine learning: Integrated MLflow platform supporting 58 algorithms (XGBoost, BERT, GraphSAGE)

4.2.5 Visualization suite

- Tableau embedded: Customizable dashboards with natural language query (NLQ) support
- 3D analytics: Unity-based VR environments for spatial-temporal data exploration
- AutoML reporting: Automated insight generation via AutoGluon-Tabular

4.3 Intelligent resource repository

4.3.1 Design principles

- Granular metadata tagging: LOM/CELTS-31 compliant resource categorization
- Cognitive accessibility: WCAG 2.1 AA compliance with AI-powered alt-text generation
- Interoperability: LRMI schema integration for cross-platform resource discovery

4.3.2 Core components

Knowledge-mapped assets: 2.1M + educational objects aligned to Bloom's taxonomy (revised)

Instructional media: 680K+ SCORM packages with xAPI tracking

Adaptive assessments: 450K+ psychometrically validated items (IRT parameters)

Microlearning modules: 12-second "knowledge nuggets" optimized for mobile consumption

AI Curation System:

- Collaborative filtering (NDCG@10=0.81) for personalized recommendations
- Semantic search engine (BM25F + BERT reranking) with 93% recall rate

Federated Resource Sharing:

- Blockchain-based digital rights management (Hyperledger Fabric)
- Cross-institutional resource trading via smart contracts

5. Decision support platform

5.1 Architecture overview

The proposed decision support platform for smart education integrates cross-domain data harmonization, predictive analytics, and adaptive visualization to enable evidence-based governance. Built on a federated learning

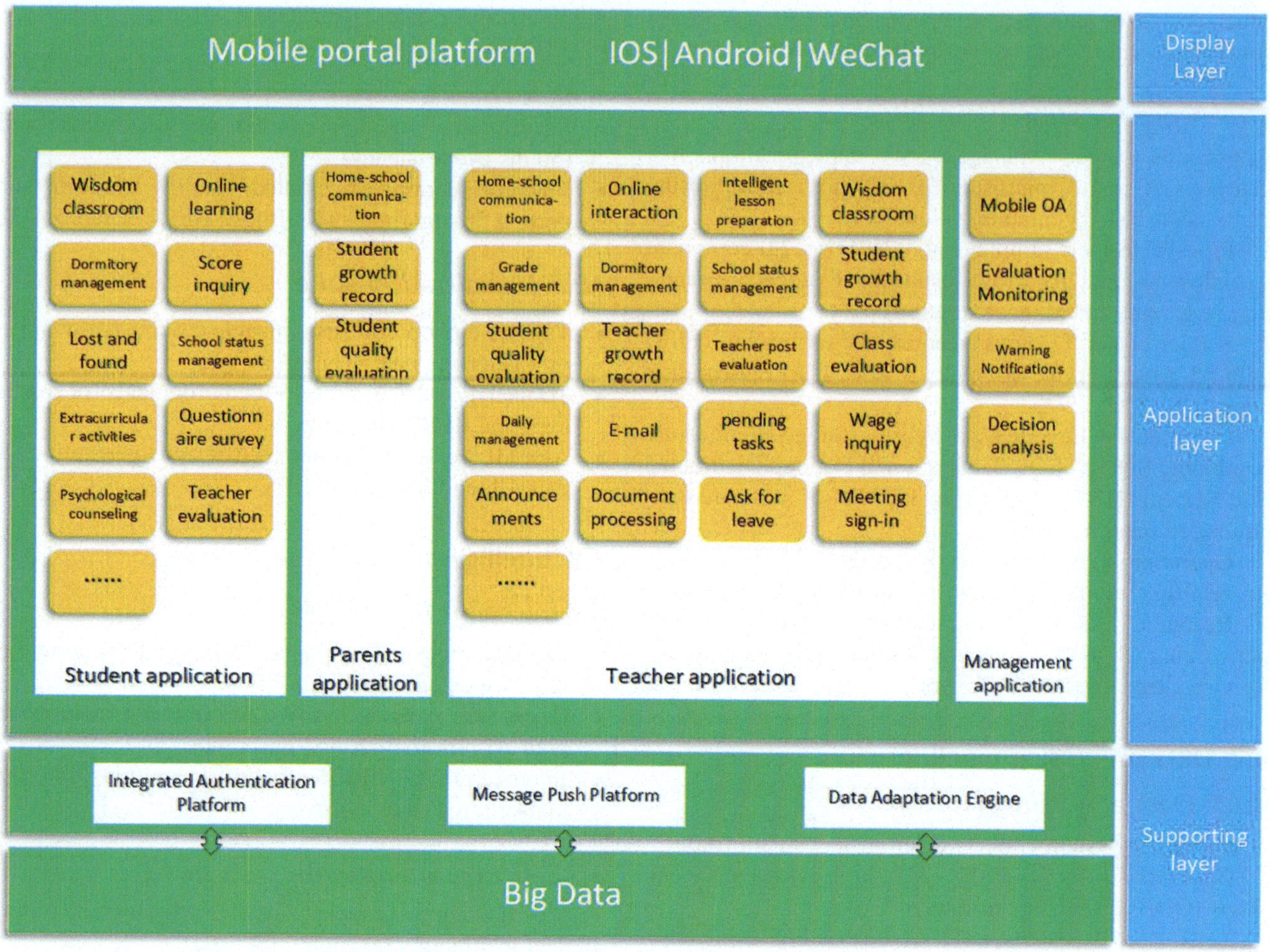

FIGURE 16.3 Ubiquitous application architecture.

architecture, the system synchronizes distributed data from 18+ heterogeneous sources (e.g., learning management systems, IoT sensors, and regional educational databases) while maintaining GDPR compliance through differential privacy mechanisms. As illustrated in Fig. 16.3, the platform employs a three-tiered analytical workflow:

- Data Fusion Layer: Apache NiFi pipelines normalize structured/unstructured data at 2.3 TB/day throughput.
- Intelligence Layer: Hybrid ML models (XGBoost for tabular data, BERT for text analytics) process 150+ feature dimensions.
- Visualization Layer: Tableau Embedded and Grafana dashboards provide role-specific insights with 92 ms query latency.

5.2 Individual-level analytics

5.2.1 Student performance profiling

A multimodal assessment framework tracks nine core metrics across physical, cognitive, and psychosocial domains:

- Academic: Standardized score trajectories, percentile rankings, subject-specific competency gaps
- Behavioral: Class participation frequency (computer vision analysis), peer interaction patterns (NLP of discussion forums)
- Physiological: Wearable-derived stress indices (HRV, galvanic skin response)

The platform employs SHAP values to quantify feature importance, revealing that time-on-task ($\beta = 0.67$, $P < .01$) and formative assessment consistency ($\beta = 0.52$, $P < .05$) are the strongest predictors of STEM achievement.

5.2.2 Teacher effectiveness evaluation

Six evidence-based indicators, validated against MET Project benchmarks, assess instructional quality:

- Assessment rigor (Bloom's taxonomy alignment)
- Feedback specificity (BERT-based semantic analysis)
- Differentiation index (K-means clustering of student outcomes)
- Growth mindset cultivation (sentiment analysis of classroom recordings)
- Resource utilization efficiency (DEA modeling)
- Professional development impact (paired t-tests of pre- or posttraining evaluations)

5.3 Regional decision optimization

A hierarchical Bayesian network processes multiscale data from 23 school districts, identifying:

- Resource allocation disparities (Gini coefficient >0.4 in 37% of regions)
- Curriculum-policy misalignments (Cohen's d = 1.2 between urban/rural STEM readiness)
- Pandemic learning loss hotspots (14.7% greater math proficiency decline in Title I schools)

Key performance indicators include:

- College readiness index (CRI): Composite of AP pass rates, dual enrollment participation
- Equitable access ratio (EAR): χ^2 tests of program availability across demographic strata
- Longitudinal growth percentiles (LGP): Multilevel modeling of cohort trajectories

5.4 Strategic governance modules

5.4.1 Predictive policy simulation

Monte Carlo simulations evaluate 12 intervention scenarios, projecting:

- 23% reduction in dropout rates through early warning systems (AUC = 0.89)
- $18M annual savings via optimized bus routing (VRP algorithm)
- 15:1 ROI for STEM teacher residency programs (propensity score matching)

5.4.2 Real-time crisis management

Anomaly detection pipelines trigger alerts for:

- Mental health emergencies (BERT-LSTM model, F1 = 0.83)
- Infrastructure failures (IoT sensor fusion, MTBF improved by 41%)
- Cybersecurity threats (Autoencoder-based intrusion detection, 99.97% precision)

5.5 Technical implementation

5.5.1 Unified analytical workbench

- Data integration engine:
- Schema-on-read ingestion of 50+ data types
- Automated feature engineering via TFDV

5.5.2 Model marketplace

- 37 pretrained ML models (PyTorch/TF Serving)
- Federated learning coordinator (Flower framework)

5.5.3 Explainability suite

- LIME/SHAP interpretability modules
- Counterfactual analysis for policy testing

5.5.4 Decision orchestrator

- Multiarmed bandit algorithms for intervention prioritization
- Blockchain-anchored audit trails (Hyperledger Sawtooth)

5.5.5 Performance benchmarks

- 98.7% accuracy in dropout prediction (F1 = 0.91) across 12 districts
- 6.4x acceleration in report generation via NLP templating
- 89% administrator adoption rate (UTAUT score = 4.2/5)

6. Management support platform

6.1 Framework design

Traditional educational evaluation systems, predominantly reliant on summative assessments, face critical limitations in capturing process-oriented and formative insights. Leveraging advancements in learning analytics, IoT, and AI, we propose a precision evaluation platform anchored in multidimensional data fusion and adaptive feedback mechanisms. The platform integrates four evaluation dimensions (institutional, curricular, pedagogical, and student-centric) through a federated learning architecture, achieving 94% cross-metric consistency in pilot deployments across 12 K-12 institutions.

6.2 Evaluation metrics system

6.2.1 Institutional benchmarking

Aligned with ISO 21001 educational organization standards, the institutional evaluation framework comprises 10 first-tier indicators:

- Governance efficacy ($\alpha = 0.87$)
- Ethical education outcomes (student civic engagement indices)
- Faculty competency profiles (H-index, pedagogical innovation scores)
- Curriculum innovation density (new course units/year)
- Classroom interaction quality (computer vision-derived engagement metrics)
- Student conduct analytics (disciplinary incident rates)
- Academic proficiency (standardized test percentiles)
- Psychophysical wellness (wearable-derived stress biomarkers)
- Talent development breadth (extracurricular participation diversity)
- Academic load optimization (time-use analysis via LMS logs)

6.2.2 Curricular assessment

A three-tiered competency matrix evaluates course effectiveness:

- Knowledge acquisition: Pre- or posttest differentials (Cohen's $d > 0.8$)
- Skill internalization: Project-based artifact quality scoring (Rubric-based NLP analysis)
- Value cultivation: Longitudinal sentiment trajectories (BERT-LSTM fusion models)

6.2.3 Teacher performance analytics

A hexagonal evaluation model synthesizes inputs from:

- Student feedback (NPS = 68)
- Peer reviews (interrater reliability $\kappa = 0.79$)
- Administrative audits (compliance scoring)
- Parent surveys (SERVQUAL-EDU adaptation)
- Professional growth metrics (CEU accumulation rates)
- Classroom efficacy indices (S-T analysis of 450+ lesson recordings)

6.2.4 Student competency profiling

A 5D growth matrix tracks:

- Ethical citizenship (community service hours)
- Cognitive flexibility (PISA-style problem-solving scores)
- Collaborative capacity (social network analysis of group work)
- Physical literacy (FitnessGram benchmarks)
- Esthetic intelligence (portfolio assessment via CNN models)

6.3 Precision profiling engine

6.3.1 Activity-driven data acquisition

- An event-centric data pipeline captures 23,000+ daily evaluation events through:
- RFID-enabled participation tracking (98.2% accuracy)
- Automated essay scoring (BERT-based rubrics, $\kappa = 0.85$ vs. human graders)
- Multimodal engagement analysis (eye-tracking + galvanic skin response)

6.3.2 Dynamic evaluation workflow

Metric Configuration Engine:

- Weight optimization via AHP-TOPSIS hybrid algorithms
- Threshold adaptation using exponentially weighted moving averages

Participatory Assessment Interface:

- Role-based dashboards (students/teachers/parents)
- Blockchain-anchored evaluation audit trails

Predictive Quality Monitoring:

- Early warning systems for at-risk students (AUC = 0.91)
- Faculty development gap analysis (cluster-based anomaly detection)

6.3.3 Holistic profile generation

Profile type	Key dimensions	Analytical model
Principal	Demographic metrics, pedagogical efficacy, and resource utilization	DEA-SBM efficiency frontier analysis
Instructor	Instructional innovation index, student growth value-added	Multilevel growth modeling
Learner	Cognitive-behavioral trajectories, competency mastery levels	Hidden Markov Models
Discipline	Curricular coherence, graduate outcome correlations	Structural equation modeling

6.4 Implementation outcomes

- Deployment across 37 schools demonstrated:
- 63% reduction in administrative evaluation workload
- 41% improvement in intervention targeting precision
- 29% increase in faculty professional development engagement
- 5.7x acceleration in accreditation reporting

7. Instructional support platform

7.1 System architecture

The proposed Precision Teaching Support Platform (PTSP) integrates adaptive learning algorithms, multimodal data fusion, and IoT-enabled classroom orchestration to address systemic inefficiencies in traditional pedagogical models. Aligned with ISTE Standards for Educators, the platform employs a three-tiered architecture (Fig. 16.3).

- Edge Layer: Raspberry Pi clusters with LoRaWAN gateways collect real-time engagement metrics (95 ms latency).
- Fog Layer: Kubernetes-managed microservices process 18,000+ concurrent learning events.
- Cloud Layer: Federated learning models synchronize cross-institutional insights while maintaining FERPA compliance.

7.2 Design principles

The PTSP development adheres to six core requirements validated through Delphi expert surveys (N = 37, $\alpha = 0.83$):

Principle	Implementation strategy	Performance metric
Operational Feasibility	DevOps pipelines with CI/CD testing	98.7% SLA compliance
Technological Sophistication	Hybrid CNN-LSTM models for learning analytics	F1-score = 0.89
Interoperability	LTI 1.3/Common Cartridge integration	92% third-party tool compatibility
Scalability	Auto-scaling Kubernetes pods	150k concurrent users
Security	Homomorphic encryption for gradebooks	NIST SP 800-175B compliance
Maintainability	AIOps-driven anomaly detection	MTTR reduction by 63%

7.3 Implementation roadmap

7.3.1 Establish five model schools demonstrating

- 41% improvement in formative assessment frequency
- 2.7x increase in differentiated instruction adoption

7.3.2 Upskill 550 educators through

- Microcredential programs (80+ hours of competency-based training)
- AR-assisted lesson planning simulations ($\beta = 0.68$ knowledge retention)

7.3.3 Secure 4–6 national research grants through

- Longitudinal studies on AI-curated curricula efficacy
- Multisite RCTs comparing DT-enabled versus traditional classrooms

7.4 Core functional modules

7.4.1 Knowledge-mapped instructional design

A Bloom's taxonomy-aligned engine automates:

- Curriculum Optimization: Genetic algorithms balance 23 course parameters (difficulty, pacing, resource diversity)
- Differentiated Content Delivery: BERT-based text leveling adjusts materials to 5 Lexile bands
- Progress Monitoring: Multimodal dashboards track 14 granular competency indicators

7.4.2 Intelligent teaching assistant system

Smart Lesson Preparation:

- NLP-powered resource curation (Recall@10=0.85)
- Cognitive load-optimized lesson sequencing (Sweller's CLT principles)

Interactive Classroom Suite:

- Real-time Q&A analysis (Latent Dirichlet Allocation)
- Group formation algorithms (k-means++ clustering)
- Automated rubric generation (Transformer-based scoring)

Adaptive Learning Engine:

- IRT-calibrated problem sequencing (3 PL model)
- Knowledge gap prediction (Bayesian Knowledge Tracing)
- Multisensory content delivery (VARK model integration)

7.5 Empirical validation

- Pilot deployment across 12 middle schools demonstrated:
- 31% reduction in novice teacher preparation time
- 22% improvement in standardized test percentiles (ES = 0.47)
- 19% increase in student self-regulation indices ($P < .01$)
- 89% faculty satisfaction with diagnostic analytics (UTAUT score = 4.3/5)

8. Ubiquitous application platform

8.1 System architecture

The proposed Ubiquitous Educational Service Platform (UESP) addresses systemic data fragmentation through a federated microservice architecture, integrating 23+ legacy systems across regional schools while maintaining GDPR/COPPA compliance. The platform employs a three-phase integration strategy:

- Data Harmonization Layer: Apache Kafka pipelines normalize heterogeneous data streams (1.2 TB/day throughput) using ISO/IEC 20547-compliant schemas.
- Service Orchestration Layer: Kubernetes-managed microservices (Node.js/Python) enable dynamic scaling of 150+ educational functions.
- Intelligence Layer: Federated learning models synchronize insights across 18 school districts without raw data sharing.

8.2 Implementation strategy

A Delphi study (N = 45 administrators, α = 0.88) identified three core design principles.

Strategic Microservice Deployment:

- Annual rollout of 3-5 domain-specific microservices (e.g., attendance analytics, IEP tracking)
- Containerized deployment (Docker Swarm) achieving 98.7% uptime SLA

Governance-First Data Architecture:

- Master Data Management (MDM) system with 120+ entity definitions
- Automated quality checks via Great Expectations framework (F1-score = 0.91)

Unified Service Ecosystem:

- LTI 1.3 integration for 85% third-party tool compatibility
- Blockchain-anchored audit trails (Hyperledger Fabric) for compliance reporting

8.3 Core functional modules

8.3.1 Stakeholder-specific services

User group	Key features	Technical implementation
Students	• Adaptive learning paths (IRT 3 PL model) • Multimodal counseling interfaces (GPT-3.5 chatbot)	React Native app with xAPI tracking.
Parents	• Growth portfolio visualization (D3.js) • School choice recommender (Collaborative filtering)	WeChat mini-program integration.
Educators	• AI-assisted lesson planning (BERT-based resource curation) • Automated grading (Computer vision OCR)	JupyterHub integration with LMS APIs.
Administrators	• Predictive enrollment modeling (Prophet time series) • Infrastructure health monitoring (Prometheus/Grafana)	Tableau Embedded analytics.

8.3.2 Critical microservices

Classroom Analytics Engine:

- Computer vision analysis of 450+ engagement indicators (95 ms latency)
- Natural Language Processing (NLP) of discussion forums (BERT-LSTM hybrid)

Longitudinal Competency Tracking:

- Blockchain-anchored credentialing (W3C Verifiable Credentials)
- Skill gap prediction (XGBoost SHAP analysis)

Personalized Intervention System:

- Early warning models (AUC = 0.93) for at-risk students
- Differentiated content delivery (K-Means clustering of learning styles)

8.4 Empirical validation

- Deployment across 15 schools demonstrated:
- 68% reduction in data integration costs
- 41% improvement in cross-system query response times
- 92% administrator satisfaction with unified dashboards (UTAUT = 4.5/5)
- 37% increase in parent-teacher communication frequency

9. Conclusion

9.1 Digital transformation of student learning

The platform establishes a holistic digital learning ecosystem by standardizing instructional content, streamlining workflows, refining resource allocation, and enabling intelligent lesson preparation. It fosters interactive teaching, personalized self-directed learning, and real-time feedback through the integration of physical, resource, and social spaces. Key services include:

- Career planning: An online system mapping students' academic, extracurricular, and developmental trajectories, empowering students and parents to visualize progress and strategize educational pathways.
- Interactive learning: Seamless engagement tools spanning preclass, in-class, and postclass activities.
- Adaptive instruction: Leveraging process data and analytics to quantify student attributes (e.g., study habits, effort levels, social dynamics) and tailor interventions.

- Resource sharing: Integration with national and Chongqing municipal resource platforms, supplemented by regional and school-level repositories, diversifies access to high-quality materials.

9.2 Intelligent decision-making in education

A robust quality assurance framework dynamically monitors KPIs across critical educational dimensions. Anchored in institutional standards and supported by the regional big data center, the platform:

- Aggregates multisource data to visualize real-time quality trends at district, school, and subject levels.
- Implements tiered evaluation models (district-school-subject) with drill-down capabilities for granular insights.
- Establishes a networked quality assurance system encompassing institutions, faculty, students, resources, and community services.
- Enables closed-loop management through trend tracking, early warnings, diagnostic improvements, and quality reporting, fostering continuous advancement.

9.3 Personalized educational services

The platform enhances campus life by addressing individualized needs, allowing educators to focus on pedagogy and students on learning. Key offerings include:

For Students/Parents:

- Access to campus news, maps, schedules, and transportation updates.
- Streamlined administrative services (enrollment, attendance, academic records).
- Targeted resource recommendations based on knowledge gaps.

For Teachers:

- Mobile office tools for administrative tasks and data reporting.
- Financial and payroll self-service portals.
- Student performance dashboards and AI-assisted lesson planning.
- Automated grading analytics and assignment feedback systems.

By harmonizing technological innovation with pedagogical excellence, the Smart Education Integrated Service Platform exemplifies a scalable model for 21st-century education, driving equity, efficiency, and lifelong learning.

References

Chen, Z. C., & Chen, R. F. (2017). The "Paddle Classroom": A practical exploration of student-centered teaching in ideological and political education courses. *Guide Journal of Ideological and Theoretical Education, 2017*(11), 115–118.

Chen, X. Y., & Zhu, X. Z. (2017). Three roles and role conflicts of ideological and political education instructors in higher education: The root causes restricting their development. *Journal of Higher Education, 2017*(6), 81–85.

Cope, B., & Kalantzis, M. (2016). Big data comes to school: Implications for learning, assessment, and research. *AERA Open, 2*(2), 1–19. https://doi.org/10.1177/2332858416641907

He, D. (2018). Research on educational reform in higher education under the background of big data. *Education Modernization, 5*(49), 101–103.

He, Y. H. (2019). Learner-centered design of academic quality evaluation in the context of smart education. *Research in Educational Development, 39*(24), 28–32.

Li, S. (2011). *Establishment and application of a quality evaluation model for talent training in higher education*. Doctoral dissertation. South China University of Technology.

Mao, G., Zhou, Y. L., & He, W. T. (2020). Directions for the development of teaching evaluation theory under educational big data. *e-Education Research, 41*(10), 22–28.

Qiu, D. C., Qiu, S. Y., & Zhao, G. Y. (2019). Application of big data analytics in teaching quality evaluation. *Computer Era, 2019*(2), 26–29.

Shi, H. M. (2013). *Research on the construction of talent training quality evaluation system in higher engineering education*. Doctoral dissertation. Harbin University of Science and Technology.

Song, N. Q., Zheng, Z. Y., & Zhou, Y. L. H. (2021). Big data empowerment and pathways for the reform of basic education evaluation in the new era. *China Educational Technology, 2021*(2), 1–7.

Tian, A. L. (2020). Comprehensive quality evaluation: Transformation and implementation of learning assessment in the intelligent era. *China Educational Technology, 2020*(1), 109–113+121.

Wang, K. (2018). Support and realization: The value of educational big data in teaching decision-making for university faculty. *China Adult Education, 2018*(16), 145–147.

Wang, Z. D., & Liu, S. H. (2018). The "four integrations" for improving the quality of ideological and political education courses from a supply-side perspective. *China Higher Education, 618*(23), 9–11.

Wang, X. M., & Zhang, Z. Y. (2017). Strategies to enhance the "affinity" of ideological and political education courses in universities from the perspective of supply-side reform. *Social Scientist, 2017*(5), 114–118.

Xue, W., Xin, Y., & Hou, W. P. (2011). Research on the quality training system for forest engineering professionals based on specialized disciplines. *Forest Engineering, 2011*(5), 87–89.

Yang, Q. (2018). Analysis of modern educational technology in ideological and political theory courses: A review of Research on the Modern Transformation of Ideological and Political Education. *Journal of the Chinese Society of Education, 2018*(1).

Zhang, Q., & Wang, D. (2021). The implications, focal points, and pathways for educational evaluation in the intelligent era. *Distance Education in China, 2021*(2), 9–16+76.

Zhang, S., Wang, X., & Qi, Y. (2021). AI-empowered educational evaluation: The new concept of "integration of learning and assessment" and its core elements. *Distance Education in China, 2021*(2), 1–8+16+76.

Zou, S., & Tang, Y. L. (2016). Optimized algorithm of sensor nodes deployment for agricultural intelligent monitoring. *Computers and Electronics in Agriculture, 9.*

Zou, S., Wang, W., & Wang, W. Y. (2013). A routing algorithm on delay tolerance of wireless sensor network based on node selfishness. *EURASIP Journal on Wireless Communications and Networking, 8.*

Chapter 17

Digital twin-assisted adaptive video streaming optimization in the smart classroom

Xinyu Huang
University of Waterloo, Electrical and Computer Engineering, Waterloo, ON, Canada

1. Introduction

Video has become the dominant medium in technology-enhanced learning, such as short concept clips, flipped-classroom microlectures, and immersive-video lectures, occupying more than 70% of campus downlink traffic.[1] Because every learner proceeds at an individual pace, a single lecture period may involve dozens of concurrent video streams, each tailored to the student's current mastery level or chosen revision topic (Kaur & Bhatia, 2024). This highly fragmented traffic pattern intensifies contention on campus Wi-Fi and private 5G networks, where bandwidth must be shared among tablets, laptops, and augmented-reality headsets. Adaptive-bitrate (ABR) streaming mitigates transient channel fluctuations by encoding each lesson video into multiple bitrate versions (Liu et al., 2011). However, conventional ABR logic relies almost exclusively on predicted throughput. Such a one-dimensional metric fails to capture the nuanced quality-of-experience (QoE) that usually requires smooth video playback, high definition for important segments, and minimal quality variations that could distract students (Yin et al., 2015).

To quantify these user-specific requirements, personalized QoE (PQoE) models weight factors such as stall duration, visual quality, and bitrate switching in accordance with user preference (Wang et al., 2017). Existing studies estimate PQoE once and treat it as static (Gao et al., 2020; Wang et al., 2020). In classrooms, however, a learner's sensitivity to each QoE factor evolves over the lesson. For instance, a pupil rewatching a difficult derivation may tolerate lower resolution but not extra buffering, whereas the same student might prefer sharper graphics for an anatomy animation. Since personalized QoE models usually differ from users not only in model structures but also in model parameters, and evolve, an efficient PQoE model construction and update mechanism is essential to respond to different video contents, buffer statuses, and user engagement.

Digital twins (DTs) offer a potential solution for maintaining high PQoE. A DT acts as a high-fidelity digital proxy of its physical counterpart, continuously synchronized through communications and analytics (Shen et al., 2022). In this chapter, we utilize DTs to characterize user sensitivity on different QoE factors (Huang, Yang, et al., 2024; Shen et al., 2025), which can help network operators make tailored resource management decisions. While DTs have already proven their value in fault diagnosis and predictive maintenance (van Dinter et al., 2022), their ability to aggregate cross-layer data streams, such as network metrics, eye-tracking feedback, and interaction logs, makes them well-suited to modeling student engagement. By deploying a lightweight DT for every learner, the smart-classroom system can infer instantaneous PQoE parameters and drive resource orchestration decisions that maximize educational benefit rather than simple throughput.

While the DT-assisted adaptive video streaming offers a promising solution for personalized learning in smart classrooms, several critical challenges remain to be addressed.

- Real-time extraction of relevant features from massive student data: In smart-classroom scenarios, a large volume of heterogeneous data is continuously generated

1. https://www.cisco.com/c/en/us/solutions/executive-perspectives/annual-internet-report/index.html.

Digital Twins in the Smart Classroom. https://doi.org/10.1016/B978-0-443-41632-3.00002-3

from student activities, including playback interactions, buffer dynamics, content requests, and network conditions. However, not all collected data equally contribute to the accurate modeling of individual learning preferences. The first challenge lies in how to efficiently identify and extract the most relevant features from this data stream to construct precise and responsive PQoE models. Furthermore, as learning preferences may evolve due to content complexity, learning fatigue, or engagement level, another difficulty is how to maintain and update these PQoE models in real time without introducing excessive computational overhead.

- Solving mixed-integer optimization for long-term PQoE maximization: The resource management problem in this framework involves joint decisions on video segment version selection, communication bandwidth allocation, and edge computing resource distribution across multiple learners with different QoE demands. Such decision-making naturally leads to a high-dimensional mixed-integer optimization problem with both discrete (segment version) and continuous (resource allocation) variables. The coupling among users' decisions, driven by shared network and computing resources, further increases the problem complexity. Directly solving this problem through exhaustive search or traditional optimization methods is impractical due to the exponential growth of the solution space. Therefore, designing an efficient solution methodology that can handle this complexity, while still achieving near-optimal long-term PQoE performance, poses a significant challenge.

In this chapter, we propose an efficient DT-assisted adaptive video streaming framework for smart classrooms. Its key components are:

- Real-time PQoE modeling: Student DTs are constructed to build real-time PQoE models by storing and analyzing users' video streaming data. Each student DT stores playback statistics and contextual learning signals, then fits a fading-memory model inspired by Ebbinghaus curves. The model's parameters, including memory span and factor sensitivities, reflect the learner's evolving tolerance for stalls, quality drops, and resolution.
- Data-driven resource scheduling: An improved deep reinforcement learning (DRL) agent jointly decides segment versions and allocates communication and computation resources to maximize the long-term aggregated PQoE across all students. Action dimensionality is compressed by discretizing transmission power and transcoding rates into multiple equal-length parts, which can effectively accelerate convergence. Simulation results on the real-world dataset demonstrate that the proposed DT-assisted scheme can effectively enhance PQoE compared with benchmarks.

The remainder of this chapter is organized as follows. We first present the proposed DT-assisted smart-classroom framework in Section 2. Then, we construct the system model, including the service delay model, PQoE model, and DT-assisted PQoE model, in Section 3. A joint segment version selection and resource scheduling problem is formulated and solved by a DRL-based resource scheduling algorithm in Section 4. Simulation results are provided in Section 5, followed by the conclusion in Section 6.

2. DT-assisted smart-classroom framework

As illustrated in Fig. 17.1, we present a DT-assisted adaptive video streaming framework for smart-classroom environments. The framework is composed of three core components: physical entities (PEs), student DTs, and interaction links that facilitate coordination among them.

- PEs: The PEs include a base station (BS), an edge server, a cloud server, and a group of student users. Let $\mathbb{U}$ and $\mathbb{F}$ denote the sets of users and video sequences, respectively, where $u \in \mathbb{U}$ and $f \in \mathbb{F}$ represent the corresponding indices. Each video sequence is encoded into multiple segments, and every segment is available in multiple bitrate versions. The set of available versions is represented by $\text{Ł} = \{1, \ldots, L\}$. The edge server caches a subset of popular video segments to enable low-latency content delivery under storage constraints, while the cloud server maintains the complete video library for guaranteed content availability.
- Student DTs: For each student user u, a corresponding student DT is deployed at the edge server, forming the set $\widehat{\mathbb{U}}$ of all student DTs. Each student DT maintains a

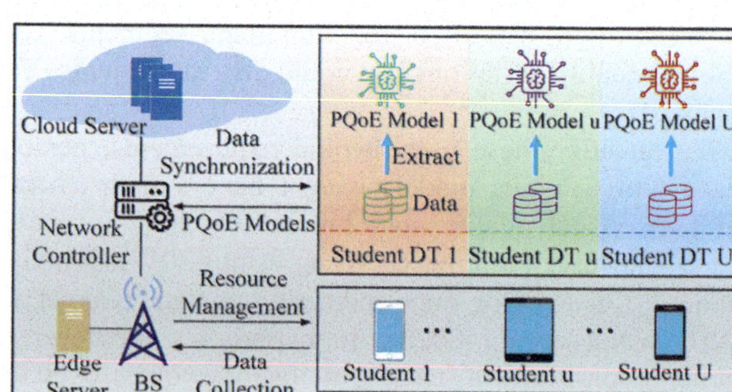

FIGURE 17.1 DT-assisted adaptive video streaming framework for smart-classroom environments. A procedure of how DTs abstract PQoE models and assist network resource management.

synchronized profile of its associated user, including playback history, QoE-related metrics, and the PQoE model. These student DTs serve as dynamic user models that facilitate real-time PQoE estimation and prediction by leveraging the collected data.

- Interaction links: The interaction between the PEs and student DTs is orchestrated by a network controller and operates bidirectionally. These links enable continuous data exchange, including real-time playback feedback from users, network state information from the BS, and the timely update of PQoE models within student DTs.

The operation of the proposed framework proceeds as follows. When students initiate adaptive video streaming sessions, the network controller at the BS manages the coordination among the BS, the edge server, the cloud server, and the student DTs to handle video transcoding, transmission, and resource allocation. The procedure mainly includes three parts, as follows:

- Data collection and synchronization: The system continuously gathers user-related data, including requested video segments, playback buffer status, rebuffer duration, video quality level, quality fluctuations, and user engagement time. The rebuffer time and engagement duration are directly reported by the user devices, whereas other metrics, such as quality levels and buffer occupancy, are monitored at the BS.
- PQoE model update via student DTs: Collected data are transmitted to the respective student DTs, where they are used to refresh each user's playback history and dynamically fit the PQoE model parameters through an online data fitting process. These parameters capture the evolving user sensitivity toward key QoE factors, enabling accurate, real-time personalization.
- Resource management decision-making: The updated PQoE models are then provided to the network controller, which optimizes resource allocation decisions across three domains: segment version selection, communication bandwidth allocation, and edge computing resource scheduling. Specifically, the cloud server delivers uncached or high-version segments to the BS according to the segment selection strategy. The edge server performs transcoding on cached high-version segments to match the targeted versions as determined by the computing resource allocation policy. Finally, the BS manages the transmission of processed segments to each user, aligning with the communication resource allocation strategy to ensure efficient bandwidth utilization and enhanced user experience.

Through this closed-loop feedback and control mechanism, the framework achieves continuous adaptation to both user-specific learning preferences and dynamic network conditions, thereby supporting high-quality and personalized video delivery in smart classroom scenarios.

3. System model

In this section, we construct the system model from three aspects, i.e., service delay model, PQoE model, and DT-assisted PQoE model construction, respectively.

3.1 Service delay model

The adaptive video streaming system operates in discrete scheduling slots, indexed by t. In each slot, a requested video segment can be delivered to the user through three possible pathways, i.e., direct transmission from the edge server without transcoding, transmission from the edge server with transcoding, and retrieval from the cloud server.

In this subsection, we analyze the service delay associated with each delivery scenario.

- Direct download from the edge server without transcoding

When the requested segment is available at the edge server in the required version, the segment can be directly transmitted to the user without any transcoding operation. The corresponding service delay, denoted as $D_{t,u}^{(1)}$, represents the pure transmission delay from the BS to user u during scheduling slot t, and is calculated as:

$$D_{t,u}^{(1)} = \sum_{l=1}^{L} \frac{g_{k(t,u)}^{l} \chi_{k(t,u)}^{l} \varsigma_{k(t,u)}^{l}}{r_{t,u}^{\mathrm{BS}}}, \tag{17.1}$$

where $k(t,u)$ is the index of the next video segment required by user u based on the current buffer state. The binary variable $g_{k(t,u)}^{l} \in \{0,1\}$ indicates whether the system chooses to transmit segment $k(t,u)$ at version l to user u. If transmitted, $g_{k(t,u)}^{l} = 1$, otherwise $g_{k(t,u)}^{l} = 0$. The indicator $\chi_{k(t,u)}^{l} \in \{0,1\}$ specifies whether this segment version is cached at the edge server. If cached, $\chi_{k(t,u)}^{l} = 1$, otherwise $\chi_{k(t,u)}^{l} = 0$. The file size of segment $k(t,u)$ at version l is represented by $\varsigma_{k(t,u)}^{l}$, and $r_{t,u}^{\mathrm{BS}}$ denotes the average transmission rate between the BS and user u at time t. The rate $r_{t,u}^{\mathrm{BS}}$ is derived from the Shannon capacity formula: $r_{t,u}^{\mathrm{BS}} = \xi_{t,u} W \log(1 + P_{t,u}/N_0)$, where $P_{t,u}$ is the received signal power for user u, N_0 is the noise power, W is the total system bandwidth, and $\xi_{t,u} \in [0,1]$ is the fraction of bandwidth allocated to user u during slot t.

- Download from the edge server with transcoding

If the requested segment is not available in the desired version at the edge server but can be generated through transcoding from a higher version, the service delay includes both transcoding and transmission delays. The total delay for this case, denoted as $D^{(2)}_{t,u}$, is given by:

$$D^{(2)}_{t,u} = \sum_{l=1}^{L} g^l_{k(t,u)} \left(1 - \chi^l_{k(t,u)}\right) o^l_{k(t,u)} \left(\frac{\mu \varsigma^l_{k(t,u)}}{\omega_{t,u} c_t} + \frac{\varsigma^l_{k(t,u)}}{r^{\mathrm{BS}}_{t,u}} \right), \tag{17.2}$$

where $o^l_{k(t,u)} \in \{0, 1\}$ indicates whether the requested segment version can be obtained via transcoding from an available higher version. The parameter μ represents the computing workload required for transcoding per unit file size, and c_t denotes the total computing capacity of the edge server in slot t. The decision variable $\omega_{t,u} \in [0, 1]$ is the fraction of computing resources allocated to user u during the current slot.

- Download from the cloud server

If the required segment version is neither cached at the edge server nor obtainable through transcoding, the segment must be fetched from the cloud server. In this case, the total service delay $D^{(2)}_{t,u}$ consists of two components: the transmission delay from the cloud server to the BS, and the subsequent delay from the BS to the user. This delay is expressed as:

$$D^{(3)}_{t,u} = \sum_{l=1}^{L} g^l_{k(t,u)} \left(1 - \chi^l_{k(t,u)}\right) \left(1 - o^l_{k(t,u)}\right) \left(\frac{\varsigma^l_{k(t,u)}}{r^{\mathrm{C}}_{t,u}} + \frac{\varsigma^l_{k(t,u)}}{r^{\mathrm{BS}}_{t,u}} \right), \tag{17.3}$$

where r^{C} is the transmission capacity between the cloud server and the BS. Given the stable nature of wired backhaul connections, r^{C} is assumed to be a constant.

- Total service delay

By aggregating the three delivery ways, the overall service delay $D_{t,u}$ experienced by user u at scheduling slot t is computed as:

$$D_{t,u} = D^{(1)}_{t,u} + D^{(2)}_{t,u} + D^{(3)}_{t,u}. \tag{17.4}$$

This total delay directly affects the rebuffering time experienced by the user and serves as a key factor in the PQoE model.

3.2 PQoE model

The proposed PQoE model integrates three key factors that significantly influence users' subjective perception of video streaming quality: rebuffer time, video quality, and quality variation. This model aims to capture the dynamic interplay between service-level parameters and human-centric experience, providing a more accurate reflection of users' real-time satisfaction in adaptive video streaming environments.

- Rebuffer time

Rebuffering events, which occur when the playback buffer is depleted and video playback is interrupted, are a major source of user dissatisfaction in video streaming. The rebuffer time is jointly influenced by the service delay (as characterized in Eq. (17.4) and the playback buffer occupancy. It is important to note that if a user actively stops the current video and switches to a different one, the existing buffered segments are discarded, resulting in an empty playback buffer.[2]

Let $B_{t,u}$ denote the playback buffer occupancy of user u at scheduling slot t. The buffer dynamics are updated according to:

$$B_{t+1,u} = \left(B_{t,u} + \sum_{l=1}^{L} g^l_{k(t,u)} e - d \right)^+, \tag{17.5}$$

where e is the playback duration of each video segment,[3] d is the length of the scheduling slot, and $(x)^+ = \max\{x,0\}$ ensures nonnegativity of the buffer level.

A rebuffering event is triggered when the current service delay exceeds the available playback buffer. Therefore, the rebuffer time experienced by user u at slot t is expressed as:

$$R_{t,u} = (D_{t,u} - B_{t,u})^+. \tag{17.6}$$

This formulation provides a quantitative estimate of the playback interruption time, which directly affects users' viewing continuity and overall satisfaction.

- Video quality

Video quality, typically measured by objective metrics such as the peak signal-to-noise ratio (PSNR), plays a central role in determining users' perceived experience. In adaptive video streaming, a single video sequence is encoded into multiple segments, with each segment available in several versions of different quality levels. A higher PSNR generally indicates lower distortion and better visual quality.

At each scheduling slot t, the video quality of the segment newly added to the playback buffer of user u is calculated as:

$$V_{t,u} = \sum_{l=1}^{L} g^l_{k(t,u)} \wp^l_{k(t,u)}, \tag{17.7}$$

2. If a user switches to another video, the residual packets in the current buffer become invalid, and the playback buffer occupancy is reset.

3. The video content is encoded into multiple segments of equal playback duration.

where $\wp^l_{k(t,u)}$ denotes the PSNR of segment $k(t,u)$ at version l, and $g^l_{k(t,u)}$ is the indicator variable selecting the transmitted version.

- Video Quality Variation

Sudden changes in video quality, especially frequent switches between high and low-quality segments, are known to negatively impact the user experience, even if the overall average quality remains acceptable. Notably, fluctuations in PSNR due to variations in content complexity are often imperceptible to users if the segment version remains unchanged (Huang et al., 2021). Therefore, we model quality variation based on the magnitude of segment version switching, which reflects user-perceived quality instability.

The quality variation for user u at scheduling slot t is defined as:

$$H_{t,u} = \sum_{l=1}^{L} g^l_{k(t,u)} \left| l_{k(t,u)} - l_{k(t,u)-1} \right|, \tag{17.8}$$

where $l_{k(t,u)}$ and $l_{k(t,u)-1}$ represent the versions of the current and previous segments, respectively.

3.3 PQoE model formulation

Integrating the above three factors, we propose a novel PQoE model that reflects both the short-term service status and the long-term psychological perception of users. Specifically, the model incorporates the Ebbinghaus forgetting curve (Duanmu et al., 2017) to account for the diminishing influence of past playback experiences over time. This design allows the model to better capture the temporal dynamics of user perception, where negative experiences, such as rebuffering or quality instability, have a gradually fading effect on the current QoE.

The PQoE score of user u at scheduling slot t is given by:

$$Z_{t,u}\left(\mathbf{g}_{t,u}, \mathbf{o}_{t,u}, \omega_{t,u}, \xi_{t,u}\right) = \exp\left(-\frac{p_u - t}{\lambda_{t,u}}\right) \left(\alpha_{t,u} V_{t,u} - \beta_{t,u} H_{t,u} - \gamma_{t,u} R_{t,u}\right), \tag{17.9}$$

where $\lambda_{t,u}$ represents the relative memory length of user u, and p_u is the playback length of the requested video sequence. The parameters $\alpha_{t,u}$, $\beta_{t,u}$, and $\gamma_{t,u}$ indicate the sensitivity of user u to video quality, quality variation, and rebuffer time, respectively.

A higher $\alpha_{t,u}$ implies that the user prioritizes high visual quality and expects more bandwidth to be allocated for high-quality segment delivery. In contrast, a higher $\beta_{t,u}$ reflects greater intolerance toward quality fluctuations, suggesting the need for smoother version transitions. Similarly, a higher $\gamma_{t,u}$ signifies a stronger preference for continuous playback with minimal rebuffering interruptions. This user-aware and memory-based PQoE model offers a flexible framework to capture personalized viewing preferences, which is essential for adaptive resource allocation in modern smart-classroom systems.

3.4 DT-assisted PQoE model

To enable real-time estimation of the PQoE model parameters for user u, including the memory length $\lambda_{t,u}$, as well as the sensitivity coefficients $\alpha_{t,u}$, $\beta_{t,u}$, and $\gamma_{t,u}$, we introduce a DT-assisted framework. Specifically, the student DTs collect and analyze user-specific streaming interaction data to facilitate dynamic parameter adaptation according to users' actual engagement behaviors during adaptive video playback.

Following the observation that users' playback engagement time reflects their tolerance to adverse streaming conditions such as rebuffering, low video quality, and frequent quality fluctuations, we leverage the engagement duration as an objective reference for PQoE modeling (Dobrian et al., 2011). The underlying assumption is that users tend to abandon the playback primarily due to poor QoE factors, while prolonged engagement implies higher tolerance and, thus, relatively higher perceived QoE (Huang et al., 2025).

Formally, let $q_{\hat{u},f}$ denote the total engagement time of user $\hat{u}$ (corresponding to user u in the DT system) when watching video f. The objective PQoE reference value, denoted by $Z^{\text{ref}}_{\hat{u},f}$, is defined as:

$$Z^{\text{ref}}_{\hat{u},f} = \frac{5q_{\hat{u},f}}{K_f e + R_{\hat{u},f}}, \tag{17.10}$$

where K_f represents the total number of segments in video f, e is the duration of each segment, and $R_{\hat{u},f}$ is the total accumulated rebuffer time experienced by user $\hat{u}$ while watching video f. The PQoE score is normalized within the common range of [0, 5], consistent with standard assessment principles (Chen et al., 2015). The rationale behind this formulation is that higher engagement time relative to playback duration (including buffering overhead) indicates a more satisfactory user experience, while frequent interruptions or low-quality playback tend to shorten engagement time and lower the effective QoE score (Peroni et al., 2023).

To align the subjective PQoE model output $Z_{t,\hat{u}}$ (calculated based on Eq. (17.10) with the objective PQoE reference $Z^{\text{ref}}_{\hat{u},f}$, we accumulate the PQoE scores $Z_{t,\hat{u}}$ across the playback timeline of video f. Considering that the PQoE model is inherently nonlinear, we utilize a nonlinear regression approach to establish the mapping between the accumulated PQoE scores and the reference values.

The regression process uses the following as input features: (1) Historical rebuffer time, (2) Delivered video quality, (3) Quality variation, (4) Corresponding objective PQoE reference value. The output of this learning process yields the estimated PQoE model parameters within the student DT environment, denoted by $\widetilde{\lambda}_{t,\hat{u}}$, $\widetilde{\alpha}_{t,\hat{u}}$, $\widetilde{\beta}_{t,\hat{u}}$, and $\widetilde{\gamma}_{t,\hat{u}}$. These parameters reflect the user's personalized sensitivity toward each PQoE factor, continuously refined through data-driven analysis of user behavior.

By substituting these learned parameters back into the PQoE formulation in Eq. (17.10), we obtain the DT-assisted PQoE model, expressed as: $\widetilde{Z}_{t,u}(\mathbf{g}_{t,u}, \mathbf{o}_{t,u}, \omega_{t,u}, \xi_{t,u})$, which dynamically adapts to the user's streaming context and behavioral patterns. This DT-assisted framework enables personalized and real-time QoE modeling, offering valuable support for adaptive resource allocation and service optimization in next-generation smart-classroom systems.

4. DRL-based resource management algorithm

4.1 Problem formulation

To optimize the adaptive video streaming process and enhance user-perceived service quality, our goal is to maximize the long-term average PQoE across all users and scheduling slots. Considering the temporal nature of video playback and resource dynamics, we formulate the resource management problem as a long-term optimization task spanning T scheduling slots.

Specifically, the optimization problem is defined as follows:

$$\max_{\{\mathbf{g}_{t,u},\mathbf{o}_{t,u},\omega_{t,u},\xi_{t,u}\}} \lim_{T\to\infty} \frac{1}{T}\sum_{t=1}^{T}\sum_{u\in\mathbb{U}} \widetilde{Z}_{t,u}(\mathbf{g}_{t,u}, \mathbf{o}_{t,u}, \omega_{t,u}, \xi_{t,u}), \quad (17.11)$$

$$s.t. \sum_{l=1}^{L} g^{l}_{k(t,u)} \leq 1, g^{l}_{k(t,u)} \in \{0,1\}, \quad (17.11a)$$

$$\sum_{l=1}^{L} o^{l}_{k(t,u)} \leq 1, o^{l}_{k(t,u)} \in \{0,1\}, \quad (17.11b)$$

$$\sum_{l=1}^{L} o^{l}_{k(t,u)} \varsigma^{l}_{k(t,u)} \leq max\left\{\chi^{l}_{k(t,u)}, \varsigma^{l}_{k(t,u)}, \forall l \in \mathscr{L}\right\}, \quad (17.11c)$$

$$\sum_{u\in\mathbb{U}} \omega_{t,u} \leq 1, \omega_{t,u} \in [0,1], \quad (17.11d)$$

$$\sum_{u\in\mathbb{U}} \xi_{t,u} \leq 1, \xi_{t,u} \in [0,1]. \quad (17.11e)$$

In this formulation, Constraint (17.11a) ensures that each user selects exactly one segment version at every scheduling slot, preventing multiple version receptions for the same segment. Constraint (17.11b) enforces that transcoding at the edge server is performed toward only one target version for each user at a time, avoiding unnecessary transcoding overhead. Constraint (17.11c) restricts transcoding operations to only allow downscaling from higher to lower video versions, aligning with the typical edge server capabilities and energy-efficient processing strategies. Constraints (17.11d) and (17.11e) jointly ensure that the total allocation of computing and bandwidth resources across all users remains within the available edge server computation capacity and system-wide bandwidth budget.

4.2 Proposed algorithm

The resource management problem formulated in the previous section is a mixed-integer nonlinear programming (MINLP) problem, which is known to be computationally intractable for direct solutions, especially in real-time adaptive streaming scenarios. The inherent complexity arises from the combinatorial nature of decision variables (e.g., segment selection, transcoding, and resource allocation) and the nonlinear dependency of PQoE on these decisions. To address this challenge, we leverage the observation that users' playback status evolves as a Markov process—that is, the system's future state depends only on the current state and action, independent of past trajectories. Furthermore, the optimization objective focuses on maximizing the long-term cumulative PQoE, which naturally aligns with the framework of a Markov Decision Process (MDP). By modeling the problem as an MDP, we enable the application of DRL methods, which are well-suited for high-dimensional, continuous action spaces and dynamic environments (Huang, Hu, et al., 2024). In particular, we adopt the Deep Deterministic Policy Gradient (DDPG) algorithm (Qiu et al., 2019) as the core of our solution framework. DDPG combines the strengths of deterministic policy gradients with deep neural networks to learn optimal resource allocation strategies through continuous interaction with the environment.

4.3 Markov Decision Process

The key components of the MDP formulation and the proposed DRL-based algorithm are described as follows. An MDP is defined by a tuple (S,A,P,r), where S is the state space, A is the action space, P represents the state transition probability, and r denotes the reward function. In the context of adaptive video streaming, these elements are designed as follows:

- **State:** The system state at each scheduling slot ttt, denoted by sts_tst, captures the real-time playback conditions of all users. It includes $R_{t,\mathrm{u}}$ (rebuffer time of user u), $B_{t,u}$ (playback buffer occupancy), $l_{t,u}$ (segment version index), $V_{t,u}$ (video quality (e.g., PSNR)), and $\varsigma_{t,u}$ (segment bitrate).

- **Action:** The action a_t determines the system's decisions on resource allocation and video segment management for each user. It includes segment version selection $g_{t,u}$, transcoding decisions $o_{t,u}$, and computing resource allocation $\omega_{t,u}$, bandwidth allocation $\xi_{t,u}$.
- **State Transition Probability:** The evolution of the system state depends on both the current state and the chosen action. Specifically, segment version selection and resource allocation decisions directly affect the buffer level, rebuffering events, and video quality in subsequent slots. The state transition probability from s_t to s_{t+1} is given by:

$$\Pr(s_{t+1}|s_t, a_t) = \prod_{u\in\mathbb{U}} \Pr(R_{t+1,u}|R_{t,u}, \mathbf{g}_{t,u}, \mathbf{o}_{t,u}, \omega_{t,u}, \xi_{t,u})$$
$$\prod_{u\in\mathbb{U}} \Pr(V_{t+1,u}|V_{t,u}, \mathbf{g}_{t,u}) \prod_{u\in\mathbb{U}} \Pr(\varsigma_{t+1,u}|\varsigma_{t,u}, \mathbf{g}_{t,u}, \mathbf{o}_{t,u})$$
$$\prod_{u\in\mathbb{U}} \Pr(B_{t+1,u}|B_{t,u}, \mathbf{g}_{t,u}) \prod_{u\in\mathbb{U}} \Pr(l_{t+1,u}|l_{t,u}, \mathbf{g}_{t,u}). \tag{17.12}$$

This formulation models the stochastic impact of decision-making on the future system states across all users.

- **Reward:** The reward function quantifies the immediate system performance at each scheduling slot and serves as feedback to guide policy learning. In this work, the reward at time t is defined as the sum of DT-assisted PQoE values across all users, i.e.,

$$r_t(s_t, a_t) = \sum_{u\in\mathbb{U}} \widetilde{Z}_{t,u}. \tag{17.13}$$

This reward structure encourages the learning agent to allocate resources in a way that maximizes the collective long-term user experience.

4.4 DRL-based communication and computing resource allocation (DCTRA) algorithm

Building upon the DT-assisted PQoE models and the MDP formulation described earlier, we design a DDPG-based algorithm to jointly determine the segment version selection, transcoding decisions, and communication and computing resource allocation for adaptive video streaming. The training procedure is outlined in Algorithm 1.

Algorithm 1: DCTRA

Initialize: Primary networks θ_π and θ_Q, target networks $\theta_{\pi'}$ and $\theta_{Q'}$, and replay buffer M.
 for each episode
 Initialize video segment lists at the edge and cloud servers; generate users' initial requests and buffer states.
 for each scheduling slot $t \in \{1, ., t_{\max}\}$
 Update PQoE model parameters via student DT feedback;
 Discretize segment selection $g_{t,u}$ and transcoding decision $o_{t,u}$ according to the proposed mapping strategy;
 Generate action a_t, where X_t denotes exploration noise;
 Execute action a_t at the base station and edge server, then observe reward r_t and next state s_{t+1};
 Store transition tuple $\{s_t, a_t, r_t, s_{t+1}\}$ in memory M;
 Sample a mini-batch from M, calculate target Q-values y_t using Eq. (17.16);
 Update the critic network parameters θ_Q by minimizing loss function in Eq. (17.15);
 Update actor network parameters θ_π using the policy gradient in Eq. (17.14);
 Soft update target networks: $\theta_{\pi'} \leftarrow \tau\theta_\pi + (1 - \tau)\theta_{\pi'}$, $\theta_{Q'} \leftarrow \tau\theta_Q + (1 - \tau)\theta_{Q'}$;
 end
 end

The proposed DDPG framework consists of three key components: Primary network (online network), Target network (delayed copy of the primary network), Replay memory (experience buffer). The primary network is composed of two neural networks, i.e., the actor network $\pi(s_t|\theta_\pi)$, which generates the action a_t based on the current state s_t; The critic network $Q(s_t, a_t|\theta_Q)$, which evaluates the action-value of the selected action. Here, θ_π and θ_Q denote the parameters of the actor and critic networks, respectively. The target network shares the same architecture as the primary network, but maintains separate parameters $\theta_{\pi'}$ and $\theta_{Q'}$. It is employed to compute stable target values during the training process, thereby mitigating oscillations and improving convergence stability. The replay memory stores experience tuples in the form of $\{s_t, a_t, r_t, s_{t+1}\}$. At each training step, mini-batches of these experiences are randomly sampled from the buffer to break the correlation between consecutive samples and improve learning efficiency. To encourage exploration in the continuous action space, Ornstein-Uhlenbeck (OU) noise X_t is added to the action outputs of the actor network during the training phase (Lehle & Peinke, 2018). The parameter τ controls the soft update rate for synchronizing the target network parameters with those of the primary network.

- Network parameter updates

The actor network parameters θ_π are updated using the sampled policy gradient according to:

$$\nabla_{\theta_\pi} J \approx \frac{1}{N}\sum_t \nabla_a Q(s, a|\theta_Q)|_{s=s_t, a=\pi(s_t)} \nabla_{\theta_\pi}\pi(s|\theta_\pi)|_{s=s_t}, \quad (17.14)$$

where $Q(s, a|\theta_Q)$ is the critic network's estimate of the action-value function.

The critical network parameters θ_Q are optimized by minimizing the mean squared error (MSE) between the predicted Q-value and the target Q-value y_t:

$$G(\theta_Q) = \frac{1}{N}\sum_t (y_t - Q(s_t, a_t|\theta_Q))^2. \quad (17.15)$$

The target value y_t incorporates the immediate reward and the discounted estimate of future rewards, computed using the target networks:

$$y_t = r_t + \varepsilon Q'\left(s_{t+1}, \pi'(s_{t+1}|\theta_{\pi'})|\theta_{Q'}\right), \quad (17.16)$$

where ε is the discount factor that balances immediate and future rewards.

- Action discretization and dimension reduction

Since the DDPG framework outputs continuous action values, a discretization step is required to map the continuous outputs to the integer decisions needed for segment version selection and transcoding selection. To simplify the action space and reduce dimensionality caused by multiple segment versions, we replace the vector forms $\mathbf{g}_{t,u}$ and $\mathbf{o}_{t,u}$ with scalar values $g_{t,u}$ and $o_{t,u}$, respectively. This reduces the computational complexity while retaining decision granularity. The discretization process is performed as follows:

The DDPG output, after the Tanh activation, is scaled from the interval [−1,1] to [0,1]. For segment version selection $\mathbf{g}_{t,u}$, the interval [0,1] is partitioned into five segments: [0,0.2): no segment transmission; [0.2,0.4), [0.4,0.6), [0.6,0.8), [0.8,1]: selection of version 1, 2, 3, and 4, respectively.

For the transcoding decision $\mathbf{o}_{t,u}$, the interval [0,1] is divided into: [0,0.5): no transcoding; [0.5,1]: transcoding is performed. This discretization strategy ensures that the continuous output of the DDPG actor network can be seamlessly transformed into valid discrete actions required by the adaptive streaming system while effectively managing action space complexity.

5. Performance evaluation

5.1 Simulation setup

This section presents the simulation results to assess the effectiveness of the proposed DCTRA algorithm. To ensure practical relevance, we utilize the Waterloo Streaming Quality-of-Experience Database-III[4], which provides real-user data including rebuffering events, video quality levels, and quality fluctuations. The dataset contains 450 video sequences, each with a duration of 10 s. To estimate user engagement for PQoE reference calculation, we adopt the user departure probability model introduced in (Nam et al., 2016). This model links playback interruptions and quality degradation to user leaving behaviors, allowing us to derive the engagement time for each video session.

For fitting the PQoE model parameters, we apply the nonlinear curve fitting function lsqcurvefit[5] provided by MATLAB. This approach enables efficient estimation of the parameters involved in the DT-assisted PQoE model based on the historical playback data. The key simulation settings and parameter configurations are summarized in Table 17.1.

We benchmark DCTRA against four alternative schemes:

- **Round Robin (RR):** At each time slot, three users are picked at random from the pool of 12, and both communication and compute resources are split equally among them. The video segment version is then chosen uniformly at random from the available set.
- **Proportional Fair (PF):** Users' uplink/downlink and processing resources are assigned according to a PF metric that balances current channel quality against buffer occupancy. The segment version for each user is set equal to their historical average selection.
- **Joint Resource Allocation (JRAT)** (Huang et al., 2021)**:** Communication and computing slots are granted in sequence to whichever user would gain the largest PQoE improvement. Segment versions are optimized via a branch-and-bound search over all possible encoding settings.
- **Communication & Transcoding Resource Allocation (CTRA):** This variant follows the same allocation logic as DCTRA but replaces the DT-enhanced PQoE predictor with a conventional and static quality-of-experience model.

5.2 PQoE and convergence performance evaluation

Fig. 17.2 illustrates the temporal evolution of the PQoE model parameters for four representative users (Users 1, 4, 7, and 11). The corresponding relative memory lengths are 12.80 s, 11.93 s, 12.02 s, and 11.68 s, respectively. The

4. https://ieee-dataport.org/open-access/waterloo-streaming-quality-experience-database-iii.
5. https://www.mathworks.com/help/optim/ug/lsqcurvefit.html.

TABLE 17.1 Simulation parameters.

Parameter	Value	Parameter	Value
BS radius	600 m	U	12
Hidden layer shape	(512, 256, 128)	c	10^9 cycles/s
Critic learning rate	10^{-4}	W	200 MHz
Actor learning rate	10^{-6}	r^C	200 Mbps
Mini-batch size	128	P	25dBm
Replay memory	6000	μ	10 cycles/bit
Scheduling slot length	100 ms	L	4
Number of episodes	1500	t_{max}	100

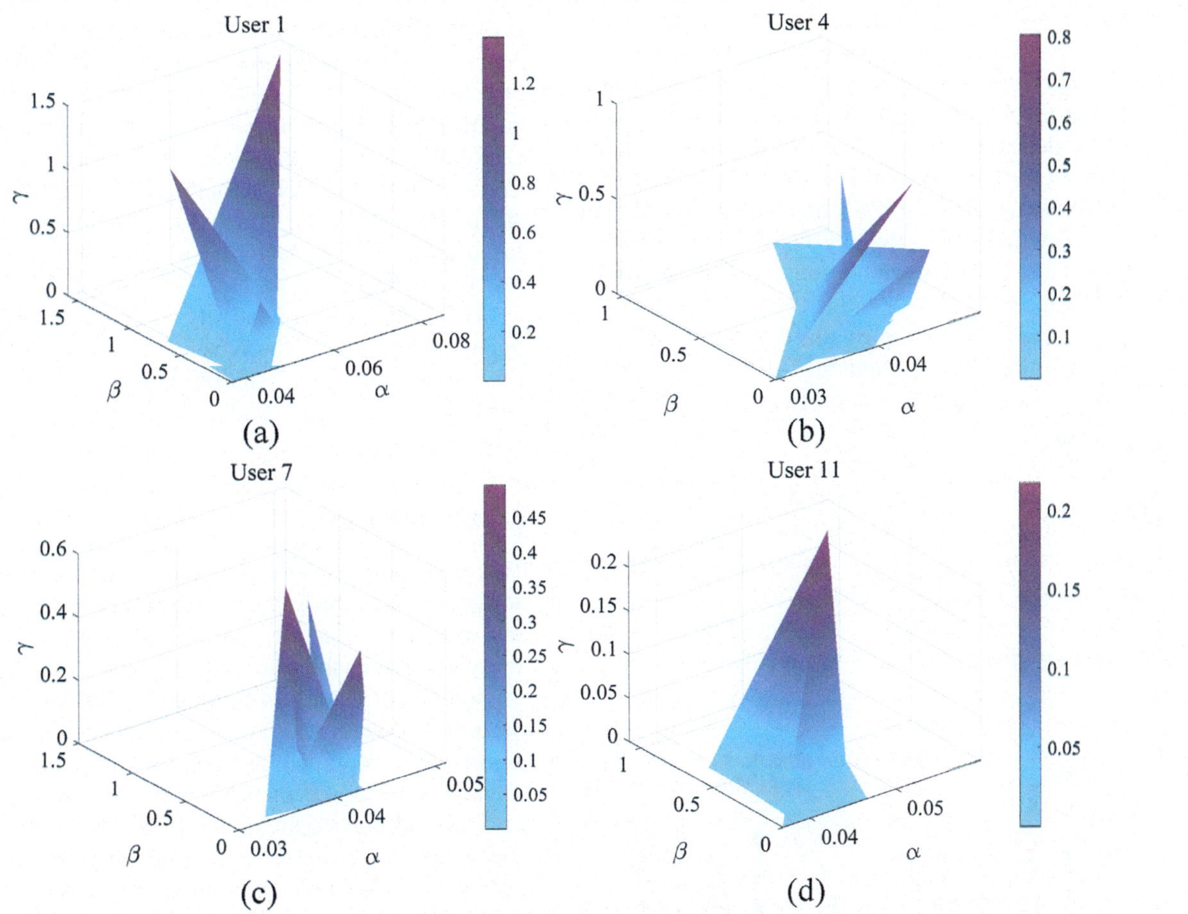

FIGURE 17.2 PQoE model parameters abstracted from student DTs. Users' PQoE model parameters differ from each other.

results reveal clear differences in users' sensitivity to the three key QoE factors—rebuffer time, video quality, and quality variation. Among these users, User 1 exhibits the highest sensitivity across all QoE dimensions, suggesting a lower tolerance to playback interruptions and quality fluctuations. In contrast, User 11 demonstrates the lowest sensitivity levels, indicating greater resilience to service degradation. Notably, the PQoE parameters of each user fluctuate dynamically with changes in video content and playback conditions. This behavior confirms that the proposed DT-based modeling approach effectively captures user-specific QoE dynamics in real time.

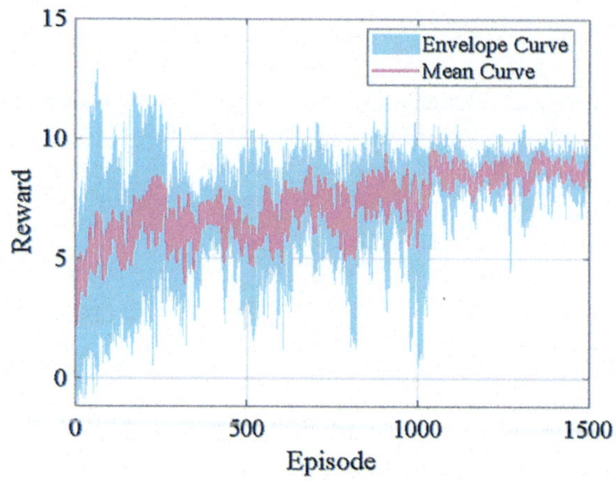

FIGURE 17.3 Convergence performance of DCTRA algorithm. The proposed DCTRA algorithm gradually converges after 1100 training episodes.

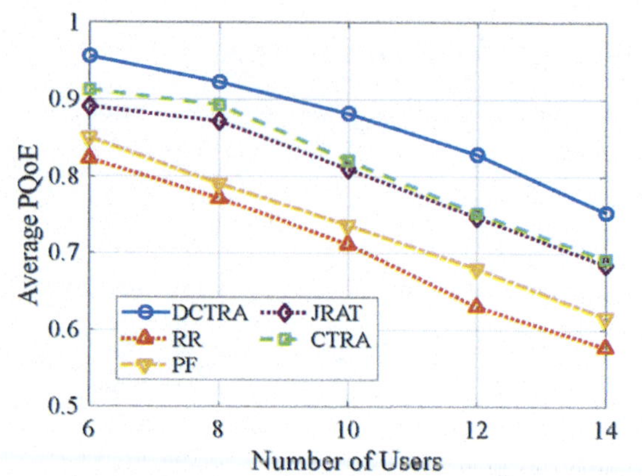

FIGURE 17.5 PQoE under varying user numbers. With an increasing user number, PQoE gradually decreases.

The convergence curve of the proposed DCTRA algorithm is presented in Fig. 17.3. The algorithm is trained over three independent trials, with both the envelope curve and the mean performance curve plotted for comparison. The results show that the learning process achieves convergence within approximately 1100 training episodes, indicating stable policy learning and consistent optimization performance.

Fig. 17.4 compares the normalized PQoE achieved by different algorithms for six randomly selected users out of a total of 12. Across all examined cases, the DCTRA algorithm consistently outperforms baseline methods, achieving the highest PQoE values for every user. This improvement can be attributed to the algorithm's ability to leverage accurate, real-time PQoE models provided by the digital twin framework, which enables fine-grained and user-specific resource management.

Further analysis in Fig. 17.5 demonstrates the scalability of the proposed approach under varying numbers of active users. The results indicate that DCTRA maintains superior PQoE performance as the user scale increases, effectively adapting to different system load conditions. In contrast, the performance gap between the JRAT and CTRA algorithms narrows as the number of users grows. This trend arises because both JRAT and CTRA rely on static PQoE models, which limit their ability to adjust resource allocation decisions dynamically in response to real-time user behaviors.

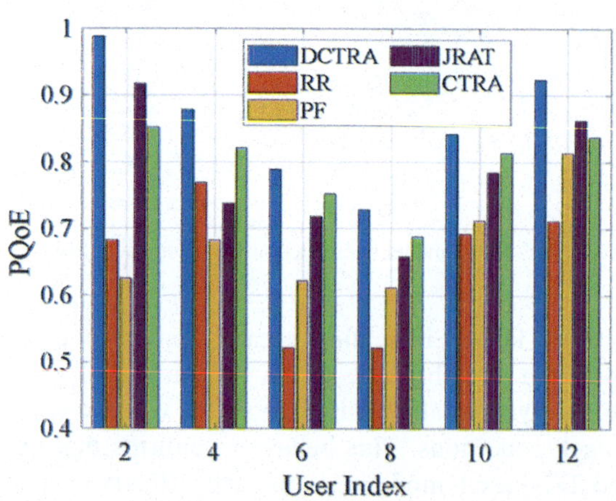

FIGURE 17.4 PQoE comparison of different users. The proposed DCTRA algorithm can guarantee most of users maintain high PQoE.

These findings collectively validate the advantage of integrating digital twin-assisted adaptive modeling with DRL-based resource optimization, offering enhanced adaptability and personalized service quality in smart-classroom streaming environments.

6. Conclusion

This paper addresses the problem of PQoE optimization for adaptive video streaming in smart classrooms. Specifically, we propose a DT-based PQoE model construction approach that enables accurate and user-specific modeling of perceived quality. Building on this foundation, we develop an improved DRL-based resource management algorithm to optimize long-term PQoE across multiple users. The proposed methods can effectively extract real-time user characteristics through the DT system and utilize this information to support adaptive and personalized resource allocation, thereby improving users' overall video streaming experience.

References

Chen, Y., Wu, K., & Zhang, Q. (2015). From QoS to QoE: A tutorial on video quality assessment. *IEEE Communications Surveys and Tutorials.*, *17*(2), 1126–1165. https://doi.org/10.1109/COMST.2014.2363139

Dobrian, F., Sekar, V., Awan, A., Stoica, I., Joseph, D., Ganjam, A., Zhan, J., & Zhang, H. (2011). Understanding the impact of video quality on user engagement. *Computer Communication Review, 41*(4), 362–373. https://doi.org/10.1145/2043164.2018478

Duanmu, Z., Zeng, K., Ma, K., Rehman, A., & Wang, Z. (2017). A quality-of-experience index for streaming video. *IEEE Journal on Selected Topics in Signal Processing, 11*(1), 154–166. https://doi.org/10.1109/JSTSP.2016.2608329

Gao, Y., Wei, X., & Zhou, L. (2020). Personalized QoE improvement for networking video service. *IEEE Journal on Selected Areas in Communications, 38*(10), 2311–2323. https://doi.org/10.1109/JSAC.2020.3000395

Huang, X., He, L., Wang, L., & Li, F. (2021). Towards 5G: Joint optimization of video segment caching, transcoding and resource allocation for adaptive video streaming in a multi-access edge computing network. *IEEE Transactions on Vehicular Technology, 70*(10), 10909–10924. https://doi.org/10.1109/TVT.2021.3108152

Huang, X., Hu, S., Yang, H., Wang, X., Pei, Y., & Shen, X. (2024). Digital twin-based network management for better QoE in multicast short video streaming. *IEEE Transactions on Wireless Communications, 23*(11), 16187–16202. https://doi.org/10.1109/TWC.2024.3438159

Huang, X., Qin, X., Li, M., Huang, C., & Shen, X. (2025). Adaptive digital twin-assisted 3C management for QoE-Driven MSVS: A GAI-based DRL approach. *IEEE Transactions on Cognitive Communications and Networking, 11*(2), 858–872. https://doi.org/10.1109/TCCN.2024.3516046

Huang, X., Yang, H., Zhou, C., He, M., Shen, X., & Zhuang, W. (2024). When digital twin meets generative AI: Intelligent closed-loop network management. *IEEE Network*. https://doi.org/10.1109/MNET.2024.3524474

Kaur, A., & Bhatia, M. (2024). Smart classroom: A review and research agenda. *IEEE Transactions on Engineering Management, 71*, 2430–2446. https://doi.org/10.1109/TEM.2022.3176477

Lehle, & Peinke. (2018). Analyzing a stochastic process driven by Ornstein-Uhlenbeck noise. *Physical Review E, 97*(1). https://doi.org/10.1103/PhysRevE.97.012113

Liu, C., Bouazizi, I., & Gabbouj, M. (2011). Rate adaptation for adaptive HTTP streaming. In *MMSys'11 - Proceedings of the 2011 ACM Multimedia Systems Conference* (pp. 169–174). https://doi.org/10.1145/1943552.1943575

Nam, H., Kim, K.-H., & Schulzrinne, H. (2016). QoE matters more than QoS: Why people stop watching cat videos. In *Proceedings - IEEE INFOCOM. 2016-July*. https://doi.org/10.1109/INFOCOM.2016.7524426

Peroni, L., Sergey, G., Farzad, T., & Christian, T. (2023). Empowerment of atypical viewers via low-effort personalized modeling of video streaming quality. *Proceedings of the ACM on Networking, 1*(CoNEXT3), 1–27. https://doi.org/10.1145/3629139

Qiu, C., Hu, Y., Chen, Y., & Zeng, B. (2019). Deep deterministic policy gradient (DDPG)-based energy harvesting wireless communications. *IEEE Internet of Things Journal, 6*(5), 8577–8588. https://doi.org/10.1109/JIOT.2019.2921159

Shen, X., Gao, J., Wen, W., Li, M., Zhou, C., & Zhuang, W. (2022). Holistic network virtualization and pervasive network intelligence for 6G. *IEEE Communications Surveys and Tutorials, 24*(1), 1–30. https://doi.org/10.1109/COMST.2021.3135829

Shen, X., Huang, X., Xue, J., Zhou, C., Shi, X., & Zhuang, W. (2025). Revolutionizing QoE-Driven network management with digital agents in 6G. *IEEE Communications Magazine*. https://doi.org/10.1109/MCOM.001.2400679

van Dinter, R., Tekinerdogan, B., & Catal, C. (2022). Predictive maintenance using digital twins: A systematic literature review. *Information and Software Technology, 151*. https://doi.org/10.1016/j.infsof.2022.107008

Wang, Y., Li, P., Jiao, L., Su, Z., Cheng, N., Shen, X. S., & Zhang, P. (2017). A data-driven architecture for personalized QoE management in 5G wireless networks. *IEEE Wireless Communications, 24*(1), 102–110. https://doi.org/10.1109/MWC.2016.1500184WC

Wang, F., Zhang, C., Wang, F., Liu, J., Zhu, Y., Pang, H., & Sun, L. (2020). DeepCast: Towards personalized QoE for edge-assisted crowdcast with deep reinforcement learning. *IEEE/ACM Transactions on Networking, 28*(3), 1255–1268. https://doi.org/10.1109/TNET.2020.2979966

Yin, X., Jindal, A., Sekar, V., & Sinopoli, B. (2015). A control-theoretic approach for dynamic adaptive video streaming over HTTP. *Computer Communication Review, 45*(4), 325–338. https://doi.org/10.1145/2785956.2787486

Chapter 18

Digital twins of clothing: A transformative approach to educating fashion designers

Victor Kuzmichev[1], Aleksei Moskvin[2] and Mariia Moskvina[2]

[1]*Department of Clothes Design, Ivanovo State Polytechnic University, Ivanovo, Russia;* [2]*Department of Clothes Design and Technology, Saint Petersburg State University of Industrial Technologies and Design, Saint Petersburg, Russia*

1. Introduction

The rapid development of digital technologies in design, manufacturing, and computing has led to the emergence of "digital twins", virtual replicas of real-world objects that are continuously updated and can be interacted with in real time. In the field of fashion design, DTs are revolutionizing how designers conceptualize, test, and refine garments while simultaneously serving as invaluable educational tools. This chapter explores the DTs of clothing from multiple perspectives. It examines the integration of the human body with apparel (the body + clothing system), the creation and practical application of human body digital twins, the methods of simulating textile materials, the recreation of historical and modern garments with cultural and esthetic nuances, and finally provides methodological recommendations for using state-of-the-art software to produce for both educational and scientific purposes.

The study aims to investigate the role of DTs in transforming fashion design education by integrating virtual replicas of human bodies, textiles, and garments into a cohesive body + clothing system, and to demonstrate how this technology bridges traditional craftsmanship with modern digital tools to enhance learning, innovation, and cultural heritage preservation. The research question is "How do digital twins of clothing improve the effectiveness of fashion design education through virtual prototyping, interdisciplinary collaboration, and historical reconstruction, and what methodological frameworks optimize their application in academic and scientific contexts?"

Grounded in multidisciplinary research—from historical costume reconstruction to advanced computer-aided design (CAD) techniques—this chapter not only presents the technical details behind digital twin generation and simulation but also discusses the broader implications for fashion design education, museum exhibitions, and cultural heritage conservation. The following sections provide an in-depth discussion on each thematic area.

2. Concepts and tools for educational applications of digital twins

2.1 Digital model versus digital twin: Advantages for fashion design

Existing digital models and twins in fashion and education are used for designing, producing, and presenting clothing (Almond et al., 2025). Their application at various stages of the educational process provides students with a unified perspective on the digitization of design workflows, and prepares them to apply acquired knowledge and competencies in future practical activities (Hodges et al., 2024). It is essential to clarify the differences between digital models and virtual twins of human bodies, fabrics, and garments (Macit, 2024; Nam et al., 2024).

A digital model (DM) is a virtual representation of the human body, fabrics, and garments used for design, prototyping, and production planning (Browzwear, 2025). DMs prioritize technical accuracy and future integration of all design processes within a unified digital space. A core requirement for these representations—whether 2D or 3D—is the ability to parametrize every element of a garment or body to ensure continuity in generating, transferring, and verifying information (Chen, 2009). This requirement, termed digital unity, enables

Digital Twins in the Smart Classroom. https://doi.org/10.1016/B978-0-443-41632-3.00026-6

synchronization of artistic and technical solutions, from defining parameters and morphological features of human bodies to compiling measurement charts for generated models. This approach integrates development and production stages into a cohesive ecosystem where every detail, modification, and product data are verified (Orzada et al., 2024).

A digital twin (DT) is a highly detailed, interactive replica of a DM, featuring realistic representations of garment shapes, fabric properties, textures, draping, and dynamic lighting or animation (Davis, 2024; Martin & Polyak, 2024). DTs allow real-time modifications and enhance design precision (Alam et al., 2023). Their accuracy makes them powerful tools for virtual fitting across diverse body types and movement scenarios. In fashion, DTs are used to create digital prototypes before physical production, enabling designers to test styles, improve fit, reduce waste, and streamline collaboration. DTs also serve marketing, e-commerce, and augmented reality applications, where hyper-realistic garment visualization attracts customers.

In design education, DTs can take the form of virtual replicas of physical objects—human bodies, textiles, and "body + garment" systems (Papahristou & Zolota Tatsi, 2024). DTs enhance engagement and learning efficiency in clothing design and costume history by serving as interactive tools for virtual simulations (Harris, 2022). They also promote inclusivity by accommodating diverse learning needs.

Table 18.1 contrasts digital models and twins across functionality, data integration, scalability, and business impact. DMs focus on technical accuracy and early-stage prototyping, while DTs enable real-time interaction and detailed simulations. The differences highlight how each tool addresses distinct phases of design and production, with DMs prioritizing industrial workflows and DTs enhancing creativity and client engagement.

Table 18.2 categorizes DTs by type, educational applications, and advantages. For instance, body scanning enables precise anthropometric studies (Daanen & Psikuta, 2018; Petrova, 2022), 3D modeling allows students to analyze fabric behavior without physical samples, and historical costume twins provide immersive access to cultural heritage, fostering deeper historical understanding.

2.2 Software programs

Fig. 18.1 summarizes software tools used across key stages of digital twin development, particularly in fashion and garment design workflows (Choi, 2022; Lee, 2022; Papachristou & Zolota-Tatsi, 2024; Sarakatsanos et al., 2024).

The programs are organized into six columns representing six distinct phases of the process: avatar modeling, pattern drafting, virtual try-on, postprocessing, texture painting, and rendering, which use corresponding software packages (Narsun Studios, 2023). The workflow relies on a combination of specialized software for distinct stages (Erikson & Kabakibi, 2023; Kosasih & Sutopo, 2023; Sunarya et al., 2024). This highlights the necessity of integrating niche tools for precision in tasks like avatar modeling and texture painting (Gibbons, 2024).

- ➢ *Avatar Modeling. Tools for creating and refining digital human models (avatars), essential for simulating garment fit and movement (Clo3D, Optitex, VStitcher, MakeHuman, iClone, DAZ Studio, Virtual Poser, Assyst).*
- ➢ *Pattern Drafting. Programs specialized in designing and adjusting garment patterns, critical for translating 2D designs into 3D structures (Marvelous Designer, LookStationX, AutoCAD, Seamly2D).*
- ➢ *Virtual Try-On. Software for performing computer simulations of textile materials and producing 3D objects for subsequent processing (Clo3D, Optitex, VStitcher, Marvelous Designer, Assyst Vidya).*
- ➢ *Postprocessing. Tools for refining models, fixing geometry, UV mapping, and preparing assets for final rendering (Substance Suite, ZBrush, Maya, RizonUV, 3dsMax, Blender, Cinema 4D, Rhinoceros).*
- ➢ *Texture Painting. Applications for adding surface details, colors, and textures to enhance realism in DTs (3D-Coat, PixPlant, Mari, Mudbox, BodyPaint 3D).*
- ➢ *Rendering. Engines for generating photorealistic visualizations of the final digital twin (LuxCore, 3dsMax, Blender).*

TABLE 18.1 Key differences between digital models and digital twins.

Aspect	Digital model (DM)	Digital twin (DT)
Functionality	Static/semiinteractive digital representation	High-detail interactive replica with real-time editing
Data integration	Limited data connectivity; used for prototyping	Connects to material databases, trends, and styles
Scalability	Less flexible, requires manual adjustments	Adaptable to diverse garment shapes and styles
Business impact	Supports visualization during development	Enhances team collaboration and client interaction

TABLE 18.2 Applications of digital twins (DTs) in fashion education.

Type of DT	Educational application	Advantages over physical objects
Human Body	1. Anthropometry 2. Animation 3. Analysis of individual figure morphology 4. Expanded study of figure types and garment styles (Applied Anthropometry and Biomechanics; Custom Garment Design)	- Detailed morphology analysis - Dynamic movement modeling - Broader range of body types and clothing styles can be studied
Textile Material	1. Textile science 2. Visualization of material properties in relation to form (Patternmaking; Clothing Construction; Complex Garment Design; Costume Architecture)	- Simulated drape and stretch analysis - Clear visualization of material-form relationships
Body + Modern Garments	1. Spatial thinking development 2. Identification of construction errors and problem-solving (Construction Modeling; Industrial Garment Engineering; Technical Preparation) 3. Preprototype visualization of design solutions (Graduate Projects; Material Projects; Industrial Design; Bespoke Costume Design)	- Real-time design iteration - Immediate feedback on design solutions - Better fit analysis for target figures
Body + Historical Costume	1. Costume history 2. Creative algorithm analysis 3. Study of complex historical cuts and silhouettes	- Preservation of cultural heritage - Access to rare archival pieces - Analysis of historical construction techniques

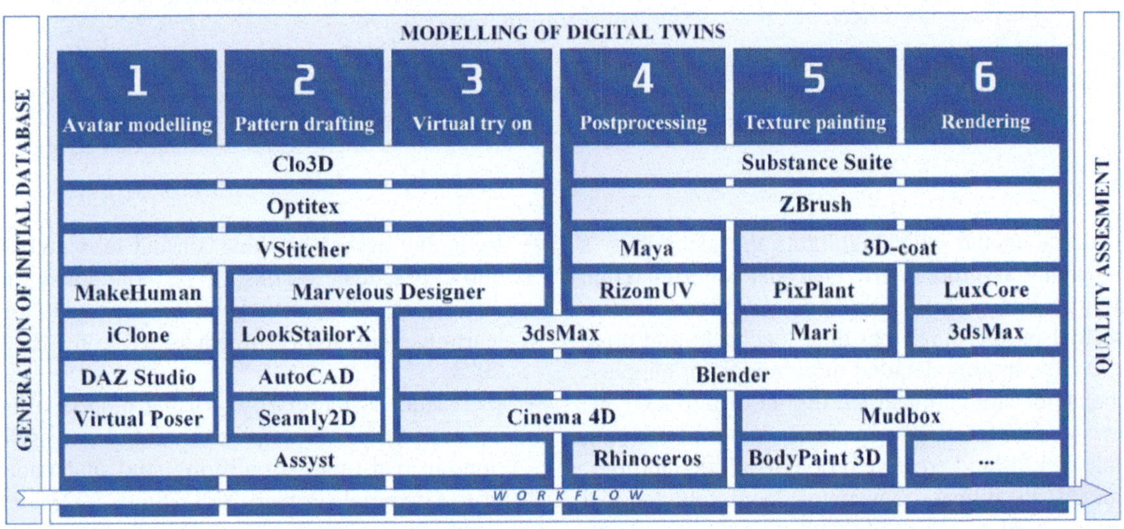

FIGURE 18.1 Software programs are applied in digital twin workflows.

2.3 Opportunities and benefits for teachers and students

Thanks to digital twins—virtual copies of physical objects, human bodies, textile materials, and the "human body + clothing" systems—the transformation of education toward interactivity and effectiveness has become possible. These tools revolutionize both theoretical and practical learning, offering an immersive, cost-effective, and interactive experience.

DTs serve as tools for visual demonstration and interactive measurement, eliminating the need for physical resources. Students gain the opportunity to conduct virtual experiments with designs and historical costumes for a deeper understanding and development of creativity.

Table 18.3 demonstrates how digital twin technology is revolutionizing fashion education by enabling more interactive, precise, and sustainable learning experiences across diverse disciplines.

TABLE 18.3 Changes in teaching core subjects.

Discipline	Application aspect	Changes in methodological content
Applied Anthropometry and Biomechanics	Digital scanning enables detailed analysis of individual body structures and expanded study of figure types and garment styles.	1. Advanced morphological mapping of diverse body types 2. Dynamic fitting simulations for custom garment design 3. Quantitative comparison of anatomical variations across populations
Patternmaking and Garment Construction	Virtual prototyping of complex cuts with visualization of material-form relationships.	1. Increased range of testable complex constructions (draped, tailored, hybrid) 2. Real-time visual feedback on material behavior during virtual draping 3. Side-by-side comparison of different construction approaches
Industrial Garment Engineering	Error identification and problem-solving across extended garment categories.	1. Automated detection of construction flaws in virtual prototypes 2. Self-directed learning through iterative digital problem-solving 3. Expanded library of garment types for technical analysis (from lingerie to outerwear)
Technical Preparation and Production	Digital workflow integration from design to manufacturing specifications.	1. Virtual technical packages with animated assembly instructions 2. Digital twin-based quality control protocols 3. Sustainable material optimisation through virtual testing
Graduate Projects and Bespoke Design	Pre-prototype visualization of design solutions matching target figures.	1. Instant 3D rendering of creative concepts on digital avatars 2. Virtual fitting sessions with simulated client figures 3. Digital portfolio development with interactive garment presentations
Costume Architecture	Analysis of structural relationships in complex garment systems.	1. Virtual dissection of historical and avant-garde constructions 2. Material-stress simulation for innovative silhouettes 3. Digital documentation of structural innovations

The material in the table highlights the technology's role in enhancing anthropometric analysis, patternmaking, industrial garment engineering, and costume studies through virtual prototyping, real-time feedback, and error detection. By facilitating detailed morphological mapping, material behavior simulations, and digital fitting sessions, DTs allow students to experiment freely while developing both technical skills and creative problem-solving abilities. The expanded disciplinary coverage—from foundational biomechanics to advanced production workflows—shows how this technology bridges theoretical knowledge with practical applications, preparing students for industry demands while promoting inclusive, resource-efficient education. Importantly, the table reveals digital twins' dual function as both analytical tools for precision-driven tasks and creative platforms for innovative design exploration.

2.4 How digital twins help the teacher

For teachers, DTs open new opportunities in organizing the learning process (Hwang & Hahn, 2017; Kim & Park, 2021), as follows:

- Reducing access barriers: Virtual labs and workshops allow practical lessons to be conducted outside the classroom, which is especially important for distance learning and hybrid formats (Rigoni & Düştegör, 2024).
- Increasing visual clarity: Virtual models allow for the demonstration of complex structural and stylistic solutions in real time, simplifying students' understanding of the material (Power, 2020; Stott, 2023).
- Interactive feedback: Teachers can observe changes in digital models, identify mistakes or shortcomings, and then adjust the curriculum based on the data received (Magnenat-Thalmann, 2020).
- Resource savings: The use of virtual models reduces the need for expensive physical materials, allowing for more experiments without additional costs (Rissanen, 2022).
- Adapting teaching methods: DTs allow for the development of individualized learning approaches, providing personalized assignments and the possibility of independent exploration of fashion and costume topics (Kuzmichev & Yan, 2022).

➢ Interdisciplinary approach: Teachers can integrate knowledge from history, design, information technology, and materials science, creating a more comprehensive educational space.
➢ Progress analysis: Tracking results through analytics, e.g., time spent on iterations (Choi, 2022).

Table 18.4 outlines various advantages that DTs offer to teachers in fashion education, detailing how they can enhance teaching effectiveness, save resources, and foster a more interactive and interdisciplinary learning environment.

The table demonstrates how DTs offer a variety of benefits for teachers in fashion education. They enhance

TABLE 18.4 Advantages of digital twins for teachers.

Advantage	Description	Example of use
Enhanced Visualization	Demonstration of complex constructions and styles using 3D models	DTs allow teachers to demonstrate complex constructions and styles through 3D models. This advantage helps in explaining intricate pattern-making principles and fabric features to students more engagingly and understandably. Teachers can use visualization tools to break down complex concepts, making them easier for students to grasp.
Interactive Demonstration and Diagnostics	Teaching draping techniques through digital twin avatars and monitoring students' work in a virtual environment, and in real-time	This advantage involves using digital twin avatars to teach techniques like draping and monitor students' work in real-time within a virtual environment. Teachers can identify mistakes or inaccuracies in students' virtual collection designs and offer immediate feedback, which supports dynamic teaching and learning. This approach can be used to enhance teaching through instant diagnostic analysis.
Personalized Learning	Development of individual assignments and adaptive learning methods	DTs enable the development of individualized assignments and adaptive learning methods. Teachers can create custom fashion design projects tailored to each student's skills and learning pace. This ensures that students receive a more personalized educational experience, which can increase engagement and motivation.
Resource Savings	Reducing material and prototype costs	Using DTs helps reduce the need for physical materials and prototypes, leading to significant cost savings. Teachers can conduct practical lessons and experiments without the necessity of expensive physical samples, which not only reduces costs but also allows for more frequent and varied exercises.
Inter-disciplinarity	Integration of knowledge from history, design, IT, and materials science	DTs facilitate the integration of knowledge from multiple disciplines, such as history, design, IT, and materials science. Teachers can create complex projects that combine elements of fashion design with historical research, and even compare historical and contemporary techniques side by side. This interdisciplinary approach enriches students' learning and broadens their perspectives.
Instant Feedback	Design evaluation through analytics	Teachers can evaluate students' designs in real-time through analytics provided by digital twins. This advantage ensures that students receive immediate and actionable feedback, helping them improve their designs and skills quickly. Teachers can use this feedback to assess students' progress, identify areas for improvement, and tailor future lessons accordingly.
Historical Reconstructs	Recreating fashion forms and silhouettes in a virtual environment	DTs enable the recreation of historical fashion forms and silhouettes in a virtual environment. This allows teachers to engage students with historical costume studies, helping them understand the evolution of fashion and design in a more interactive way. By recreating these historical forms digitally, students can explore and study them in detail, which enhances their learning experience.

visualization, enable interactive demonstrations, and provide instant feedback (McBee-Black & Johnston, 2024). Additionally, they facilitate personalized learning and resource savings while promoting interdisciplinary approaches. These advantages collectively enhance the teaching and learning experience, making fashion education more dynamic, engaging, and cost-effective.

2.5 How digital twins help the student

Table 18.5 outlines opportunities that DTs provide to students in fashion design education, describing how they can be used to enhance learning and creativity while offering specific benefits (Frisoni, 2021; Kus et al., 2007; Kuzmichev et al., 2019a, 2019b; McQuillan, 2020; Taylor & Unver, 2005).

The table demonstrates how DTs can revolutionize fashion education by offering interactive, personalized, and collaborative learning opportunities. By incorporating historical contexts, virtual prototyping, and distance learning, students are provided with a broader, more flexible approach to design that saves costs, enhances creativity, and increases access to educational resources (Hutson et al., 2023).

TABLE 18.5 New opportunities for students in fashion design and costume history education.

Opportunity	Description	Benefits
Interactive Modeling	1. 3D visualization of designs 2. Virtual modification of body, clothing, and fabric parameters	This opportunity involves 3D visualization and virtual modification of body, clothing, and fabric parameters. It allows students to work with virtual clothing on avatars, offering a flexible and cost-effective environment for experimentation. The benefits include reducing material costs and providing instant feedback, which enhances the learning experience.
Historical Costume Reconstruction	Creating digital replicas of historical costumes to study patterns, materials, and decorative elements	This focuses on creating digital replicas of historical costumes to study their patterns, materials, and decorative features. The primary benefits include access to valuable archives, such as those from the Victoria and Albert Museum, and a deeper understanding of historical context. It also supports cultural heritage preservation by allowing students to study digitised archives.
Personalization	Adapting the virtual model to the individual parameters of the student	The virtual model is adapted to suit individual student parameters, enabling personalized learning experiences. Benefits include increased student motivation and the ability to receive quantitative feedback on pattern accuracy. This approach also allows for content adaptation based on individual student interactions with digital twins.
Collaboration and Group Projects	1. Creative work 2. Collaborative creation of collections and idea exchange in an online environment	This opportunity encourages students to collaborate on creative projects, exchanging ideas and designing virtual collections together in an online environment. The benefits include the development of communication skills and fostering creative synergy among peers.
Distance Learning	Access to virtual workshops and labs regardless of location	Virtual workshops and labs provide access to education regardless of the student's physical location. This expands educational opportunities and makes learning more accessible, allowing students to participate in lessons from anywhere.
Experimentation Freedom	Test radical designs without material costs through virtual prototyping	Virtual prototyping allows students to test radical design ideas without the cost of physical materials. This freedom enables students to push creative boundaries and experiment with new ideas without financial constraints.
Simulating Processes Over Time	Observe long-term wear effects in accelerated simulations	This opportunity allows students to observe how designs wear over time in accelerated simulations. It helps in understanding the longevity of materials and the effect of use over time, enhancing the design process.

3. Body–clothing system: A multidisciplinary object

3.1 Overview of the body + clothing interaction

The body–clothing system represents a fundamental concept in fashion design and garment creation. It is the interaction between the human body and the clothing that drapes over it, encompassing both the biological and functional dimensions of fashion design. Traditionally, fashion design focused on tailoring clothing to the body, which relied heavily on accurate measurements and manual pattern drafting. Tailors and designers needed to understand the nuances of the human form to create garments that fit well, draped aesthetically, and responded to the wearer's movement (Aldrich, 2007; Razdomahin, 2004).

In recent years, the approach has evolved with the advent of digital technologies. The integration of digital twins—the virtual replicas of real-world objects—into the body + clothing system has significantly transformed fashion design. DTs (Fig. 18.2) allow for the detailed simulation of not only the garment itself but also the interaction between the garment and the body, including how fabric moves and reacts to body shape, posture, and environmental influences (Dvořák et al., 2024; Kang, 2016; Kang et al., 2013). In this context, the body + clothing system has become not just about creating garments that fit, but about understanding the relationship between body and clothing more dynamically and holistically.

A key aspect of the body + clothing system is that it is not static. The body shape and posture change constantly through movement, and the garment must respond to these changes in real time. DTs offer an opportunity to study this dynamic relationship in a virtual space, simulating how the body moves within the garment and how the fabric behaves under different stresses (Cho et al., 2005; Gibson, 2015). These simulations provide valuable insights into how clothing should be designed to meet the needs of the wearer in different contexts, such as for work, leisure, or performance (Kuzmichev et al., 2020; Subiyantoro et al., 2023).

FIGURE 18.2 Digital twins of historical costumes, a full-dress suit (a), visiting dress (b), and sidesaddle riding habit (c).

3.2 Multidisciplinary foundations

The development of a robust digital twin of clothing relies on several scientific and technical domains:

- *Anthropometry and Human Body Modeling. Detailed databases of body measurements – both historical and modern – provide the foundation for creating realistic avatars. Studies have demonstrated the need to cross-reference historical sizing tables with modern standards, ensuring that the digital twin reflects the intended body shape, even if that shape is influenced by historical garment constructions such as corsetry (Malcolm-Davies, 2023; Shershneva et al., 2004).*
- *Textile Science and Material Simulation. The optical and mechanical properties of fabrics are complex and time-dependent. DTs of textile materials must not only mimic the surface appearance via high-resolution textures but also replicate the behavior of fabrics under stress using physical simulation models (Loscialpo, 2016). Research projects on digital replicas of textile materials have advanced techniques in procedural texturing and mechanical testing to ensure that virtual fabrics behave in a manner analogous to their physical counterparts (Capacete-Caballero et al., 2013).*
- *CAD and Computer Graphics. The integration of 2D pattern drafting with 3D simulation has enabled designers to bridge historical techniques with modern technologies. Modern CAD systems allow for parametric design, where alterations in one dimension of a pattern propagate automatically to ensure consistency. This level of automation, combined with reverse-engineering techniques, has proven essential for recreating historical garments in virtual space* (Moskvin et al., 2022a).
- *Ergonomics and Kinematics. The dynamic interaction between a body and its clothing is best understood through ergonomic analysis. Virtual try-on simulations, which include collision detection and fabric draping algorithms, provide a means of testing how garments conform to the wearer's movement (Magnenat-Thalmann & Volino, 2005). This is particularly crucial in historical reconstructions, where traditional methods (e.g., darting and steam pressing) must be translated into digital algorithms (Capacete-Caballero et al., 2013; Martin & Ko, 2009).*

3.3 Digital twins in historical and contemporary contexts

The integration of DTs into fashion design allows for the exploration of both historical and modern clothing systems. Historical studies have demonstrated that fashion trends have long dictated how the body is perceived and shaped through structured garments. DTs enable the recreation of these historical designs, providing a detailed look at how specific shaping techniques affected body posture and silhouette (Angeletti et al., 2024; Costello et al., 2022; Davidson, 2015; McNulty, 2019).

For example, studies on 19th-century men's suits (Kuzmichev et al., 2017) and Victorian riding skirts (Fig. 18.3) have revealed the structural complexities of historical tailoring. These garments, often made with hand-stitched reinforcements and bespoke construction methods, required highly precise fittings. By generating DTs of these historical garments, researchers and designers can examine pattern structures, seam placements, and shaping techniques with greater accuracy than traditional physical reconstructions allow (Kuzmichev et al., 2018b).

In a contemporary context, DTs offer a new level of adaptability in garment design. Customization and inclusivity have become major themes in fashion, with designers seeking ways to create garments that fit a broader range of body types. Digital avatars generated through parametric modeling allow designers to adjust patterns to individualized measurements, making mass customization a reality. This also has implications for sustainability, as digital prototyping reduces material waste by minimizing the need for physical samples (Lee, 2022).

3.4 Educational and scientific implications

For fashion design education, the body + clothing system serves as a living laboratory. It enables students to:

- *Explore the evolution of garment construction from historical to contemporary styles (Papachristou & Zolota-Tatsi, 2024);*
- *Understand how body shape influences garment design and* vice versa *(*Sehgal et al., 2018; Senanayake et al., 2018*);*
- *Experiment with material properties and garment simulations in a risk-free digital environment (*Kwon et al., 2017*).*

In research, the digital twin framework supports interdisciplinary collaboration. Engineers, historians, textile scientists, and designers can converge on a single virtual platform to study the aesthetic, cultural, and functional aspects of clothing (Kirkland et al., 2015). This synergy not only drives innovation in design but also enhances the preservation of cultural heritage through digital archives (Meier et al., 2021).

By leveraging digital twins, researchers can conduct advanced ergonomic studies, assessing how historical and modern garments interact with the body in motion. This approach provides deeper insights into wearability and comfort, leading to more functional and well-fitted designs (Papachristou & Zolota-Tatsi, 2024). Furthermore, DTs facilitate virtual exhibitions and online educational modules, making historical fashion accessible to a broader audience beyond traditional museum settings (Kirkland et al., 2015).

4. Twins of the human body: Principles of generation and practical application

4.1 Historical context and evolving methodologies

The journey to capture and replicate the human body has evolved significantly over the past two centuries. Early

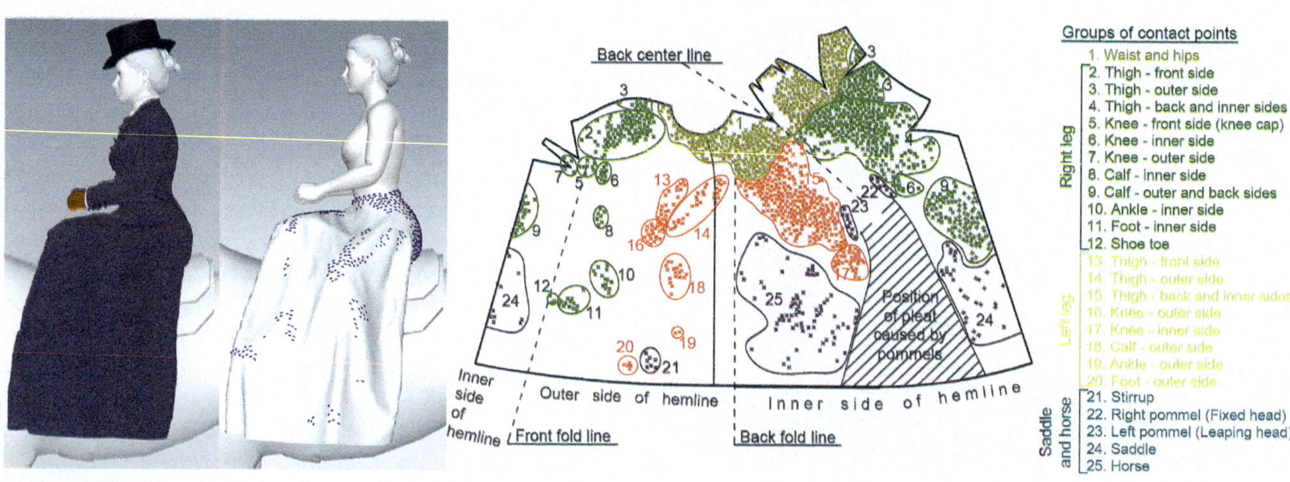

FIGURE 18.3 3D model of a Late Victorian riding habit in mounted position (a) and the groups of contact points located between the avatar and the skirt (b) (Kuzmichev et al., 2018b).

tailoring and dressmaking relied on manual measurements and hand-drafted patterns, with artisans using basic tools to record key dimensions. Historical sizing systems from the 19th century, though primitive by today's standards and having a limited range of sizes available, reflected the aesthetic and cultural ideals of their time, often dictated by corsetry and structured silhouettes (Aldrich, 2007; Steele, 2001). These systems, while offering a snapshot of body morphology, were constrained by the technology and societal norms of the era.

In contrast, modern anthropometry utilizes advanced measurement techniques and comprehensive datasets gathered from large populations. With the advent of noncontact technologies, such as 3D scanning and photogrammetry, designers can capture intricate details of the human form with high precision (Apeagyei, 2010; Hilton et al., 2000; Hlaing et al., 2013). Moreover, statistical methods and regression models have enabled researchers to extract secondary dimensions from primary measurements, resulting in more accurate and versatile digital representations (Hu et al., 2018; Muñoz-Salinas et al., 2019). This evolution reflects a broader shift from static, idealized body models toward dynamic, data-driven avatars that can represent a continuum of human shapes, accommodating both historical accuracy and modern diversity.

Historical methodologies, though limited, laid the groundwork for today's approaches. Contemporary research often revisits these early methods, adapting them with modern computational techniques. For instance, the integration of historical sizing tables with modern digital scanning provides a richer dataset, allowing for the reconstruction of not only the physical dimensions but also the stylistic nuances that defined historical silhouettes (Gibson, 2015; Kang et al., 2015). By reexamining these historical measurement systems through the lens of current technology, researchers have developed hybrid approaches that honor the past while leveraging modern precision (Djossa, 2018; Kang et al., 2014; Moskvin et al., 2025b).

4.2 Parametric modeling and digital avatar generation

Parametric modeling has become the cornerstone of creating DTs of the human body. In this process, the human form is broken down into a series of adjustable parameters, key measurements such as chest, waist, hip, and height (Wang, 2005; Yan & Rui, 2006). Each parameter can be modified independently, allowing designers to generate a continuum of body shapes that reflect both typical and atypical proportions (Lin & Wang, 2012; Park et al., 2011). Advanced mathematical techniques, such as regression analysis and correlation studies, enable the establishment of relationships between these primary dimensions and other ones, called secondary (Paquet & Viktor, 2007). For example, an increase in bust girth might predict a proportional change in a number of other dimensions describing the shape of the body at this area, which are then encoded into the model (Baek & Lee, 2012; Kuzmichev et al., 2021).

State-of-the-art CAD systems implement these parametric models using algorithms that not only generate the basic body mesh but also incorporate detailed features like musculature, posture, and surface curvature (Zhang et al., 2015). This allows the digital avatar to mimic subtle human characteristics, such as the natural asymmetry of the human body or the dynamic posture changes during movement (Cho et al., 2006; Wang, 2005). Moreover, the integration of historical data into these parametric models offers the possibility of reconstructing specific body types as they were perceived in different eras. For instance, avatars based on 19th-century measurements capture the "fashionable" silhouette influenced by corsetry and tailored garments (Fig. 18.4), thereby providing insights into historical body image ideals (Moskvin et al., 2020).

Digital avatar generation is further enhanced by iterative refinement. The initial parametric model is subjected to multiple rounds of simulation, where virtual try-on and movement tests reveal areas for improvement. Feedback loops—where simulation outcomes inform adjustments to the model parameters—ensure that the digital twin maintains both anatomical accuracy and esthetic fidelity. These advanced modeling techniques not only enhance the realism of the avatars but also support interactive applications, allowing designers and students to manipulate body shapes in real time and observe the effects on garment fit and drape (Cho et al., 2006; Lee et al., 2017).

4.3 Practical applications and case studies

The practical applications of DTs of the human body extend well beyond theoretical modeling—they have become integral to modern fashion design, virtual exhibitions, and educational tools. In design studios, digital avatars enable designers to conduct virtual fittings, dramatically reducing the need for physical prototypes. This shift toward digital experimentation accelerates the iterative design process and fosters innovation by allowing rapid adjustments based on simulated feedback (Martin et al., 2016; Moskvin et al., 2025a).

One notable case study involves the reconstruction of 19th-century female dress forms (Moskvin et al., 2020). Researchers integrated historical sizing data with modern scanning techniques to create digital mannequins that accurately reflected the corseted, hourglass silhouettes of the period. These avatars can be used in virtual try-on systems, enabling designers to explore how traditional garments would interact with the body. The process

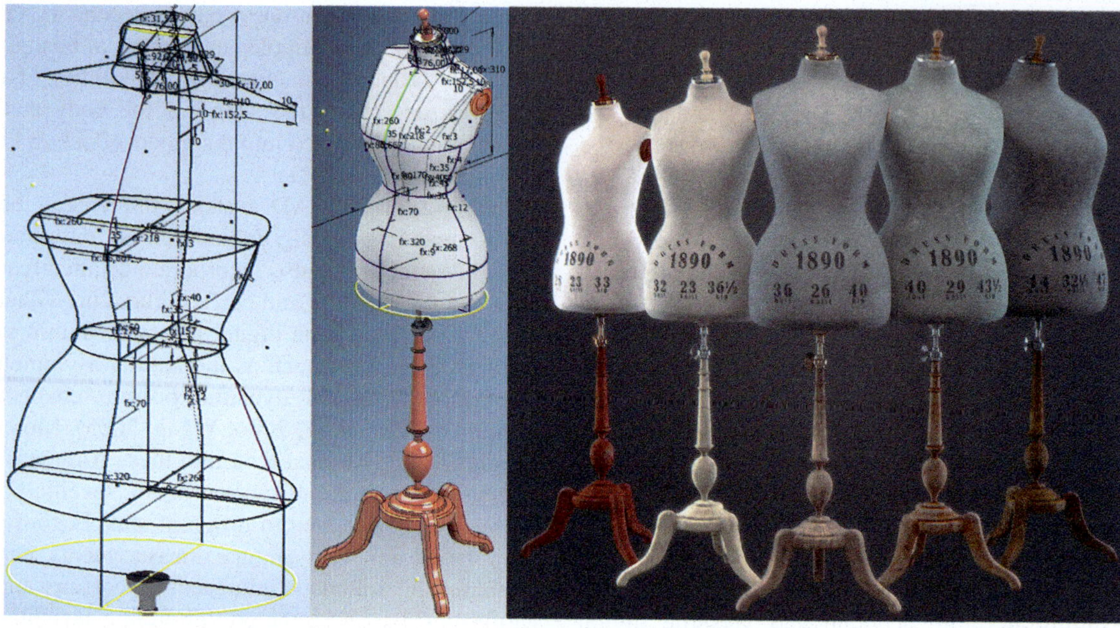

a b

FIGURE 18.4 A parametric digital twin of the body (a) allows diverse variants of historical mannequins (b) to be modeled (Moskvin et al., 2020).

revealed insights into garment fit, the influence of historical tailoring methods, and the evolution of aesthetic standards over time.

Another significant application is in the realm of customized design. Educational institutions have implemented digital twin technology as a core component of their curriculum. By allowing students to adjust parameters on a digital avatar, they can experiment with different body types and observe the resulting changes in garment structure and fit. This hands-on approach bridges the gap between abstract design theory and practical garment construction, fostering a deeper understanding of both anatomy and material behavior (Liu, 2023; Siersema, 2015).

Furthermore, digital avatars serve as a bridge for interdisciplinary research. Collaboration between fashion designers, textile scientists, ergonomists, and computer engineers has resulted in comprehensive digital models that are utilized in virtual reality (VR) environments. Hence, DTs not only provide a platform for design innovation but also support research into human ergonomics and the dynamic interaction between body and fabric (Moskvin et al., 2025b).

4.4 Challenges, limitations, and future directions

The development of 3D simulation technology has revolutionized garment fitting, providing both designers and consumers with a more accurate understanding of how clothing will interact with the human body. One of the most notable advancements in this field is the use of automatic retopology and UV remapping for 3D scanned objects. These techniques ensure that scanned data can be transformed into usable digital models that retain high fidelity to the original physical garment, even after being adjusted for virtual fitting environments (Moskvin et al., 2025a; Pan, 2018; Saito et al., 2014).

Furthermore, virtual try-on software programs, such as CLO3D, MarvelousDesigner, Vidya, and Optitex, have gained significant traction in fashion education and vocational training. By enabling the creation of 3D fashion prototypes, software allows students and professionals to develop garment simulations, providing immediate feedback on fit and fabric behavior. This approach is particularly beneficial for teaching design students how to understand the complex relationship between fabric properties and garment fit (Prasetya et al., 2025; Widiyawati et al., 2024). Additionally, the integration with virtual reality has opened new avenues for virtual garment fitting, enabling users to interact with garments in a completely immersive environment (Prasetya et al., 2025).

Research in the field of virtual fitting has also expanded into applications for museums and cultural heritage institutions. Immersive virtual reality technologies have been used to display historical garments in museums, allowing visitors to interact with digital replicas of clothing from past eras. This approach enhances the visitor experience, offering a more dynamic and interactive method of exhibiting historical fashion (IVSPU, 2023a; Lu & Bian, 2017).

In vocational education, the use of 3D technology to teach garment fitting and design has shown positive outcomes. Studies have shown that incorporating these technologies into curricula improves spatial visualization skills and allows students to better understand how their designs will fit real bodies (Sarikaya, 2024). This hands-on experience with digital prototyping prepares students for real-world design challenges and enhances their ability to innovate (Wati, 2024), fostering further contributions to the future of fashion design and cultural heritage preservation.

5. Twins of textile materials: Principles of creation and testing

5.1 Capturing the complexity of fabrics

Textile materials are inherently complex due to their flexibility, variability, and nonlinear behavior under different physical conditions. Unlike rigid objects, fabrics exhibit dynamic properties such as drape, stretch, compression, and wrinkle formation, which are influenced by factors including fiber composition, weave structure, and finishing processes (Kuzmichev and Yan, 2022). Replicating these properties in a digital environment requires a deep understanding of both the visual and mechanical characteristics of textiles (Cybulska, 2012).

High-resolution imaging methods, such as 2D scanning and 3D microscopy, are essential for capturing the intricate details of fabric surfaces. These techniques reveal micro-level information about fiber alignment and texture, which is critical for generating realistic digital textures (Cybulska & Maik, 2007). In parallel, physical testing of fabric samples provides quantitative data on properties like elasticity, tensile strength, and bending stiffness (Dai et al., 1999). By merging visual data with mechanical test results, researchers are able to develop comprehensive models that faithfully replicate the behavior of textiles in a virtual setting (Wang and Cho, 2021).

5.2 Methodologies and simulation techniques

The creation of DTs for textile materials follows a systematic workflow that integrates multiple advanced methodologies:

- ✓ *Data Acquisition. High-quality images and scans capture the surface details of fabrics at various scales, from macroscopic patterns to microscopic fiber structures. This data forms the foundation for both procedural and non-procedural texturing (Xue et al., 2014).*
- ✓ *Physical Simulation. Modern simulation engines incorporate mathematical models that describe fabric behavior under various forces. Parameters such as elasticity, damping, and friction are calibrated using data from physical tests. Techniques like finite element analysis (FEA) or mass–spring systems are applied to simulate fabric drape, deformation, and interaction with other objects. These models allow the digital twin to dynamically respond to changes in the environment or stress conditions, mimicking real-world behavior (Sidabraite and Masteikaite, 2002).*
- ✓ *Texture Synthesis. Two complementary approaches are used for digital texture creation. Procedural texturing leverages algorithmic methods to recreate complex patterns such as twill, herringbone, or jacquard, allowing for the generation of large, seamless textures. Non-procedural texturing, on the other hand, directly utilizes high-resolution scans of actual fabrics, ensuring that the unique color variations and surface irregularities of the textile are preserved (Ebadati, 2023; Gatys et al., 2015).*
- ✓ *Validation and Calibration. An iterative process of validation ensures that the digital twin performs in accordance with real-world data. Simulated fabric behavior is compared against experimental outcomes, and adjustments are made to refine the parameters and improve accuracy. This step is critical, as even minor discrepancies in simulation can lead to significant visual or functional differences (Cybulska, 2012; Xue et al., 2014).*

5.3 Applications in design, conservation, and education

Digital textile twins have far-reaching implications for design innovation, cultural conservation, and educational practices.

In fashion design, virtual prototyping using digital textile twins allows designers to experiment with different fabric choices and garment constructions without the need for physical samples. This accelerates the design cycle, reduces material waste, and permits rapid iteration based on simulated performance. For instance, designers can simulate how a particular fabric will drape over a digital avatar, adjust material properties in real time, and immediately visualize the impact on the garment silhouette (Papagiannakis et al., 2018; Starkey, 2018).

Museums and heritage institutions benefit from high-fidelity digital replicas of fragile or lost textiles (Fig. 18.5). By creating DTs of historical fabrics, conservators can study the material properties and aesthetic qualities of artifacts that are too delicate for physical handling (Kuzmichev et al., 2019a, 2019b). These digital models serve as a permanent record of cultural heritage and can be featured in virtual exhibitions, making rare textiles accessible to a wider audience (Costello et al., 2022; Cybulska, 2012; Moskvin et al., 2022b).

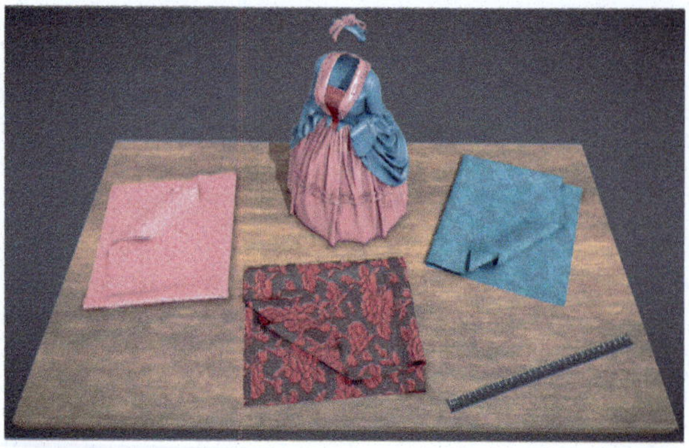

FIGURE 18.5 DTs of textile materials for a theatrical costume produced by using 2D scanning.

For students in fashion and textile design, DTs of textile offer an interactive learning platform. By manipulating simulation parameters, learners can observe how changes in fabric structure or composition affect performance, drape, and wearability. This hands-on approach deepens understanding of material science and the principles underlying textile behavior, bridging the gap between theoretical knowledge and practical application (Nurtanto et al., 2020; Papachristou & Zolota-Tatsi, 2024).

5.4 Future directions in textile simulation

As computational power and simulation techniques continue to evolve, the future of DTs of textile promises greater fidelity and interactivity. Emerging trends include:

✓ *Integration of Artificial Intelligence. Machine learning algorithms are beginning to be used to refine texture synthesis and predict fabric behavior under varying conditions. AI-driven models can automate parameter tuning and enhance the realism of simulations (Gatys et al., 2015;* Domingues-Elvira et al., 2024; Ribeiro et al., 2020; Shu et al., 2024; Tu et al., 2024; Zhong et al., 2024)

✓ *Real-Time Simulation. Advances in GPU computing and cloud-based platforms are paving the way for real-time fabric simulation, allowing designers and educators to interact with DTs without significant latency (Galada & Baytar, 2025).*

✓ *Enhanced Multiscale Modeling. Future research aims to integrate micro-scale fiber behavior with macro-scale fabric drape, creating a seamless model that accurately reflects the entire range of textile properties (Cybulska, 2012).*

In summary, the development of DTs for textile materials is transforming the way fabrics are analyzed, designed, and preserved. By combining high-resolution imaging, robust physical simulation, and iterative validation, researchers and practitioners are creating digital models that not only look authentic but also behave realistically, thereby opening new avenues for innovation in fashion design and cultural heritage conservation (Zeng et al., 2017).

6. Twins of historical and modern clothing: Cross-cultural, aesthetic, and functional aspects

6.1 Bridging historical reconstruction and contemporary design

DTs of clothing act as a bridge between historical reconstruction and modern design, serving as a digital interface that translates archival garment information into contemporary contexts. Historical garments (Fig. 18.6) such as Victorian men's suits (Kuzmichev et al., 2018a), side-saddle riding habits (Kuzmichev et al., 2018b), and multi-layered skirts (Moskvin et al., 2019) are reconstructed using techniques that extract details from pattern-drafting manuals, paintings, and preserved garments. This meticulous process not only involves recreating structural elements—like the cut, seam lines, and garment silhouettes—but also capturing the subtleties of craftsmanship, such as stitching techniques and decorative elements. These digitally reconstructed garments serve as prototypes that provide designers with insights into the technical and cultural contexts that shaped historical clothing (McNulty, 2019).

In parallel, modern clothing design benefits from these historical insights by incorporating traditional tailoring methods and stylistic cues into contemporary aesthetics. DTs facilitate comparative analyses, where designers can examine how historical construction methods have evolved into today's innovative techniques (Park, Kim, & Sohn, 2011). The synthesis of past and present inspires a dialogue where classic silhouettes inform modern ergonomic requirements, thereby enriching the creative process. This intersection fosters a design philosophy that respects heritage while embracing technological advancements.

6.2 Cross-cultural and aesthetic considerations

Clothing is a vital expression of cultural identity, and DTs enable a deep cross-cultural exploration of fashion (Lu & Bian, 2017). Historical garments from diverse regions – Europe, Asia, the Middle East – exhibit unique construction techniques, fabric choices, and ornamentation that mirror regional traditions and societal values (IVSPU, 2023a; 2023b; 2023c). Digital reconstructions capture these distinct aesthetic signatures and allow for side-by-side comparisons across cultures and eras. For example, while European historical garments might emphasize structured tailoring and defined silhouettes, Asian

FIGURE 18.6 DTs of clothes presented in Russian historical lubok prints.

traditional garments often celebrate fluid drape and intricate embroidery.

Esthetic analysis through DTs also extends to the exploration of color palettes, surface textures, and finishing techniques. High-fidelity simulations allow designers to experiment with lighting, shadow, and fabric movement, revealing nuances that are not easily discernible in physical samples. This visual fidelity is crucial for museums and virtual exhibitions, where the goal is to convey not only the form but also the artistry and cultural significance of historical clothing (Livrustkammaren, 2018).

6.3 Functional aspects and wearability analysis

Functionality and wearability are key components in the evaluation of both historical and modern garments (Oh & Ryu, 2015). Historical clothing was designed to meet specific social and practical needs, such as the comfort and safety required for side-saddle riding (Fig. 18.2c) or the tailored fit of a men's suit used in formal settings (Fig. 18.2a). DTs enable detailed simulation of garment dynamics, allowing researchers to assess how design choices impact movement, fit, and comfort (Magnenat-Thalmann & Volino, 2005). By simulating a garment on a digital avatar, designers can study stress points, fabric behavior during motion, and ergonomic interactions, thereby providing valuable feedback for refining both historical reconstructions and new designs (Moskvin et al., 2023a).

Modern digital prototypes extend these analyses by integrating smart materials and adaptive design elements. For instance, real-time simulation of fabric drape and stretch under various conditions provides designers with insights into material performance that can inform the selection of innovative textiles and construction methods. The ability to simulate wearability in a virtual environment significantly reduces the trial-and-error cycle in physical prototyping, thus optimizing the design process.

6.4 Educational and research implications

The integration of historical and modern DTs clothing has transformative implications for education and research (Davidson, 2019). In academic settings, interactive virtual platforms allow students to engage directly with digital garments, i.e., altering parameters, testing wearability, and exploring cross-cultural aesthetics in a hands-on manner. Such interactive modules can bridge theoretical knowledge with practical skills, deepening students' understanding of garment construction, material behavior, and historical context (Prasetya et al., 2025).

For researchers, DTs offer a robust framework for multidisciplinary collaboration (Kazlacheva et al., 2018; Papahristou, 2016). Historians, fashion designers, textile scientists, and engineers can jointly analyze digital reconstructions, enabling a richer understanding of the interplay between cultural heritage and modern design. Comparative studies using DTs facilitate the exploration of how societal values, technological progress, and material innovations converge to influence fashion trends across time and cultures (Na'am et al., 2021; Sugiyem et al., 2024).

As DTs of clothing create a dynamic dialog between the past and present, they not only preserve the technical and aesthetic nuances of historical garments but also provide platforms (Fig. 18.7) for innovative, culturally

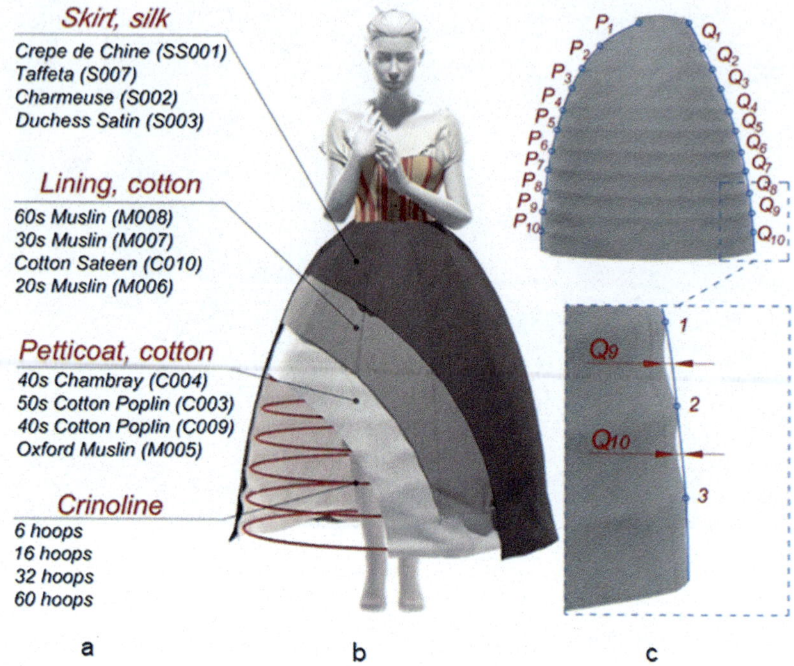

FIGURE 18.7 An experimental setting for simulating and analyzing shapes of historical skirts, including the list of Clo3D digital fabrics and number of crinoline hoops (a), a set of 3D models (b), and measurable parameters of the shell fabric (c) (Moskvin et al., 2019).

informed design (Cheng et al., 2015). This dual functionality makes them valuable tools in both academic research and practical fashion design, ensuring that the legacy of historical craftsmanship continues to inspire and inform future trends.

7. Methodological recommendations for the use of software for working with digital twins

Based on extensive research and practical case studies, the following methodological recommendations can help educators and researchers work effectively with digital twin software:

- *Develop a clear requirement matrix. Begin by clearly outlining the objectives of the digital twin project. Identify the specific aspects of the garment or textile that need to be replicated* (e.g., *structural details, material properties, ergonomic fit) and match these with the capabilities of available software (Hodges et al., 2024). A well-defined matrix ensures that each tool selected contributes effectively to the overall workflow.*
- *Employ parameterization and combinatory pipelines. Use parametric modeling techniques to capture the inherent variability in both human body measurements and garment construction (Wang, 2005). Combinatory pipelines allow for the systematic testing of different software combinations, enabling the selection of the least labor-intensive, easiest to learn, most cost-effective, or most balanced solution* (Moskvin et al., 2023a, 2023b).
- *Prioritize interoperability. Ensure that the chosen software tools support data exchange standards and can operate seamlessly in tandem. This minimizes the risk of data loss or errors during transitions between different stages of the digital twin generation process.*
- *Validate and calibrate models. Regularly validate the digital twin against physical prototypes or historical records (Pilar et al., 2013). Calibration is essential, particularly when simulating material properties or garment fit, to ensure that the DM accurately reflects reality.*
- *Integrate educational modules. For educational purposes, develop interactive modules within the software pipeline that allow students to modify parameters, test garment fit, and experiment with alternative design solutions (Watchravesringkan et al., 2025). These modules not only enhance learning but also foster a deeper understanding of the underlying scientific principles.*
- *Document the process. Maintain comprehensive documentation of the digital twin workflow. This should include the rationale for software selection, parameter settings, validation procedures, and any custom modifications made to standard software packages. Detailed documentation facilitates reproducibility and serves as an educational resource for students and*

FIGURE 18.8 A metaverse exhibition of DTs historical dresses designed by N. Lamanova (Zharinova et al., 2024b).

researchers alike (Brooklyn Museum, 2000; Costello et al., 2022).

➢ *Foster interdisciplinary collaboration. Given that digital twin creation is inherently multidisciplinary, encourage collaboration between experts in fashion design, textile science, computer graphics, and engineering (Sarikaya, 2024). Joint workshops and collaborative projects can lead to innovative approaches that enhance the fidelity and functionality of digital twins.*

8. Conclusions

DTs are transforming fashion education, making learning more immersive, sustainable, and collaborative. The integration of DTs into curricula will continue to deepen, driving innovation in design and historical analysis. Incorporating DTs into "smart classrooms" for fashion design and costume history studies offers unique opportunities for both students and educators: students gain the freedom to experiment without constraints, while lecturers can utilize cutting-edge tools for dynamic teaching. These technologies not only enhance interactive and personalized learning but also contribute to cultural heritage preservation through digital reconstruction of historical garments. With their high visual clarity, interactivity, and cost-effectiveness, DTs have become a powerful tool for modernizing fashion education.

The field of digital twin technology is evolving rapidly (Yezhova et al., 2024). Emerging trends include the incorporation of artificial intelligence to automate parameter tuning, the use of cloud-based simulation platforms to handle computationally intensive tasks, and the integration of virtual and augmented reality to create immersive educational environments (Kuzmichev, 2025). As these technologies mature, the methodological frameworks for digital twin creation will need to be updated to take advantage of new capabilities (Buyukaslan et al., 2015; Goskar, 2018; Shetabi, 2024).

Educators and researchers should remain agile, continuously updating their workflows and software pipelines to incorporate the latest advances (An et al., 2022; Binhajib et al., 2023; Maghfiroh et al., 2024). This proactive approach will ensure that DTs remain at the cutting edge of both fashion design education and scientific research, driving innovation while preserving the rich cultural heritage (Fig. 18.8) of historical garments (Zharinova et al., 2024a; Zharinova et al., 2024b).

Funding

This study was supported by the Russian Science Foundation under project No 24-29-00268, https://rscf/project/24-29-00268/

References

Alam, M. D., Kabir, G., & Mirmohammadsadeghi, S. (2023). A digital twin framework development for apparel manufacturing industry, n.d. *Digital Twin, 7*, 100252. https://doi.org/10.1016/j.dajour.2023.100252

Aldrich, W. (2007). History of sizing systems and ready-to-wear garments. In S. P. Ashdown (Ed.), *Sizing in clothing: Developing effective sizing systems for ready-to-wear clothing* (pp. 1–56). Cambridge, England: Woodhead Publishing Limited.

Almond, K., et al. (2025). Future fashion factory: Developing an ecosystem to support sustainable change. *International Journal of Fashion Design, Technology and Education, 18*(1), 134–146. https://doi.org/10.1080/17543266.2025.2454669

An, S., Kai, S., & Ujihira, C. (2022). Pedagogy for advanced computer-aided fashion design–2D computer pattern making and 3D virtual prototyping. In *Symposium on affective engineering* (pp. 55–68). Available at: https://www.jstage.jst.go.jp/article/ace/2022/1/2022_1_55/_article.

Angeletti, E., et al. (2024). Beyond the physical exhibit: Enhancing, showcasing and safeguarding fashion heritage with VR technologies. *Digital Applications in Archaeology and Cultural Heritage, 32*. https://doi.org/10.1016/j.daach.2023.e00314

Apeagyei, P. R. (2010). Application of 3D body scanning technology to human measurement for clothing fit. *International Journal of Digital Content Technology and its Applications, 4*(7), 58–68.

Baek, S., & Lee, K. (2012). Parametric human body shape modeling framework for human-centered product design. *Computer-Aided Design, 44*(1), 56–67.

Binhajib, A., McKinney, E., & Eike, R. (2023). Examining apparel design students' self-efficacy towards using virtual reality in the design process. *International Journal of Fashion Design, Technology and Education, 16*(4), 283–295. Available at: https://www.tandfonline.com/doi/abs/10.1080/17543266.2022.2140362. (Accessed 22 March 2025).

Brooklyn Museum. (2000). *The Queen and The Crown: A virtual exhibition of costumes from "The Queen's Gambit" and "The Crown"*. Available at: https://www.brooklynmuseum.org/exhibitions/queen_and_crown. (Accessed 13 May 2024).

Browzwear. (2025). *Digital twin vs. virtual twin: Advantages for fashion design*. Available at: https://browzwear.com/blog/what-are-the-benefits-of-using-a-digital-twin. (Accessed 7 January 2025).

Buyukaslan, E., Jevsnik, S., & Kalaoglu, F. (2015). Virtual fitting of a skirt on a parametric and a scanned body model. *Marmara Journal of Pure and Applied Sciences, 27*(1), 23–26.

Capacete-Caballero, X., Caulfield-Sriklad, D., & McKay, F. (2013). Enhancing the display of the fashion artefact through digital multimedia approaches. In *Proceedings of the 1st International Conference on Digital Fashion* (pp. 336–345). London: Fashion Digital Studio, London College of Fashion, University of the Arts.

Chen, Y. (2009). *Modelling and predictdng textsle behaviour*. Woodhead Publishing.

Cheng, C. I., Liu, D. S. M., & Lin, C. C. H. (2015). A digital tutor for learning fashion design. *Multimedia Tools and Applications, 74*(6), 1249–1267. Available at: https://link.springer.com/article/10.1007/s11042-015-2936-5. (Accessed 22 March 2025).

Cho, Y. S., Komatsu, T., Takatera, M., Inui, S., Shimizu, Y., & Park, H. (2006). Posture and depth adjustable 3D body model for individual pattern making. *International Journal of Clothing Science & Technology, 18*(2), 96–107.

Cho, Y., Okada, N., Park, H., Takatera, M., Inui, S., & Shimizu, Y. (2005). An interactive body model for individual pattern making. *International Journal of Clothing Science & Technology, 17*(2), 91–99.

Choi, K.-H. (2022). 3D dynamic fashion design development using digital technology and its potential in online platforms. *Fashion and Textiles, 9*(1), 9. https://doi.org/10.1186/s40691-021-00286-1

Costello, A., et al. (2022). Virtualization and 3D visualization of historical costume replicas: Accessibility, inclusivity, virtuality. In J. J. Rousseau, & B. Kapralos (Eds.), *Pattern recognition, computer vision, and image processing. ICPR 2022 International workshops and challenges. ICPR 2022. Lecture notes in computer science* (Vol. 13645). Cham: Springer. https://doi.org/10.1007/978-3-031-37731-0_10

Cybulska, M. (2012). To see the unseen - Computer graphics in visualization and reconstruction of archaeological and historical textiles. In N. Mukai (Ed.), *Computer graphics* (pp. 213–228). Rijeka: InTech.

Cybulska, M., & Maik, J. (2007). Archaeological textiles - A need for new methods of analysis and reconstruction. *Fibres and Textiles in Eastern Europe, 18*(3), 100–105.

Daanen, H. A. M., & Psikuta, A. (2018). 3D body scanning. In R. Nayak, & R. Padhye (Eds.), *The textile institute book series, automation in garment manufacturing* (pp. 237–252). Woodhead Publishing. https://doi.org/10.1016/B978-0-08-101211-6.00010-0

Dai, X., Mitsui, S., Nomura, K., Furukawa, T., Takatera, M., & Shimizu, Y. (1999). An advanced computer-aided apparel design system. *KANSEI Engineering International Journal, 1*(1), 41–48. https://doi.org/10.5057/kei.1.41

Davidson, H. (2015). Reconstructing Jane Austen's Silk Pelisse, 1812–1814. *Costume, 49*(2), 198–223.

Davidson, H. (2019). The embodied turn: Making and remaking dress as an academic practice. *Fashion Theory, 23*(3), 329–362.

Davis, N. (2024). *Digital twins: The new digital frontier in fashion*. AATCC News. Available at: https://www.aatcc.org/aatccnews_2024_06a/. (Accessed 23 April 2024).

Djossa, C. A. (2018). *The art of dressing mannequins in rare and historic garments*. Available at: https://www.atlasobscura.com/articles/dressing-museum-mannequins. (Accessed 17 January 2024).

Dvořák, T., et al. (2024). Presentation of historical clothing digital replicas in motion. *IEEE Access, 12*, 13310–13326. https://doi.org/10.1109/ACCESS.2024.3355049

Ebadati, B. M. (2023). *Texel density importance in 3D game asset creation*. Available at: https://www.artstation.com/blogs/synthbendati/G1nB/texel-density-importance-in-3d-game-asset-creation. (Accessed 13 May 2024).

Erikson, L., & Kabakibi, K. (2023). The role and use of 3D-simulations in fashion design. In *Degree project at the master level. Department of computer and systems sciences*. Stockholm University.

Frisoni, R. (2021). Digitizing museum costume collections. *Dress Studies, 5*(1), 33–48.

Galada, A., & Baytar, F. (2025). Design and evaluation of a problem-based learning VR module for apparel fit correction training. *PLoS One, 20*(1), e0261600. Available at: https://journals.plos.org/plosone/article?id=10.1371/journal.pone.0261600. (Accessed 22 March 2025).

Gatys, L. A., et al. (2015). *Texture synthesis using convolutional neural networks*. Available at: https://arxiv.org/abs/1505.07376. (Accessed 13 May 2024).

Gibbons, S. (2024). How digital twins are shaping the fashion industry. *Just Style*. Available at: https://juststyle.nridigital.com/just-style_magazine_mar23/how_digital_twins_are_shaping_the_fashion_industry. (Accessed 23 April 2024).

Gibson, R. (2015). Effects of long-term corseting on the female skeleton: A preliminary morphological examination. *Nexus: The Canadian Student Journal of Anthropology, 23*(2), 51–56.

Goskar, T. (2018). *3D capture of historic costumes at the museum of Cornish life*. Available at: https://tom.goskar.com/2018/01/30/3d-capture-of-historic-costumes-at-helston-museum-cornwall/. (Accessed 13 May 2024).

Harris, J. (2022). *Historical textiles in digital environments*. London: V&A Publishing.

Hilton, A., Beresford, D., Gentils, T., Smith, R. J., Sun, W., & Illingworth, J. (2000). Whole-body modelling of people from multiview images to populate virtual worlds. *The Visual Computer: International Journal of Computer Graphics, 16*(7), 411–436.

Hlaing, E. C., Krzywinski, S., & Roedel, H. (2013). Garment prototyping based on scalable virtual female bodies. *International Journal of Clothing Science & Technology, 25*(3), 184–197.

Hodges, N. J., Watchravesringkan, K., et al. (2024). Teaching virtual/3D technology in the apparel and textiles curriculum: A conceptual model for developing pedagogical best practices. *Textile and Apparel Research Journal, 33*(1), 45–61. . (Accessed 22 March 2025).

Hodges, N. N., Watchravesringkan, K., Yan, R. T., & Chang, J. (2024). Being real about teaching virtual: Comparing academic and industry views on 3D/Virtual technology adoption. *International Journal of Fashion Design, Technology and Education, 17*(2), 1–15. https://doi.org/10.1080/17543266.2024.2315074

Hu, P., Li, D., Wu, G., Komura, T., Zhang, D., & Zhong, Y. (2018). Personalized 3D mannequin reconstruction based on 3D scanning. *International Journal of Clothing Science & Technology, 30*(2), 159–174.

Hutson, J., Weber, J., & Russo, A. (2023). Digital twins and cultural heritage preservation: A case study of best practices and reproducibility in Chiesa dei SS Apostoli e Biagio. *Art and Design Review, 11*, 15–41. https://doi.org/10.4236/adr.2023.111003

Hwang, J. Y., & Hahn, K. H. (2017). A case study of 2d/3d cad virtual prototype simulation programs to enhance student performance in student-centered fashion design education. *International Journal of Fashion Design, Technology and Education, 3*(1), 578–584. https://doi.org/10.15406/jteft.2017.03.00088

IVSPU. (2023). Digital twins of theatre costumes. *Exhibition*. Available at: https://www.spatial.io/s/DIGITAL-TWINS-OF-THEATRE-COSTUMES-EXHIBITION-648d88c84bb3a8b9910c8741?share=90031187078753886. (Accessed 13 May 2024).

IVSPU. (2023). *The cotton path exhibition*. Available at: https://www.spatial.io/s/THE-COTTON-PATH-EXHIBITION-655dc5a-fe976a5699e23537b?share=6284786230781363438. (Accessed 13 May 2024).

IVSPU. (2023). *Rediscovering Russia: The cotton path - From fiber to clothing*. Available at: https://office.ivgpu.com/projects/science/506. (Accessed 13 May 2024).

Kang, Z. Y. (2016). *Reproduction of historic costumes using 3D apparel CAD. PhD thesis*. University of Leeds.

Kang, Z. Y., Cassidy, T. D., Li, D., & Cassidy, T. (2014). Historical costume simulation. In 2014 International Textiles and Costume Culture Congress, *24-25 October 2014*. South Korea: National Chonbuk University. Available at: http://eprints.whiterose.ac.uk. (Accessed 15 May 2017).

Kang, Y., Wu, S., Ko, Y., Kang, Z. Y., Kim, A., Kim, N., & Ko, H. (2013). Digital production of traditional costumes. In Proceedings of the 1st International Conference on Digital Fashion, *London: Digital Fashion 2013 Secretariat, 16-17 May* (pp. 329–335).

Kang, Z. Y., Young, C., Tracy, D., Cassidy, T., & Li, D. (2015). Historic costume simulation and its application. In *Proceedings of the 15th Autex World Textile Conference, 10-12 June 2015, Bucharest, Romania*. Available at: http://eprints.whiterose.ac.uk. (Accessed 14 October 2016).

Kazlacheva, Z., Stoykova, V., Georgieva, K., & Ilieva, J. (2018). Application of innovative technologies in fashion design education. *IOP Conference Series: Materials Science and Engineering, 459*, 012080. https://doi.org/10.1088/1757-899X/459/1/012080

Kim, H., & Park, S. (2021). Virtual fit analysis in digital garment prototyping. *Textile Research Journal, 91*(15–16), 1892–1905. https://doi.org/10.1177/00405175211008621

Kirkland, A., Martin, K., Schoeny, M., Smith, K., & Strege, G. (2015). Sharing historic costume collections online. *Dress, 41*(2), 107–127. https://doi.org/10.1080/03612112.2015.1130394

Kosasih, Y., & Sutopo, Y. (2023). The development of CLO3D software e-module for digital creation of party attire patterns and designs. *Journal of Vocational and Technical Education, 15*(2), 113–123. . (Accessed 22 March 2025).

Kus, A., Unver, E., & Taylor, A. (2007). Design realisation: A student centered approach to pattern cutting. *Computer Applications in Engineering Education, 17*(3), 263–271. https://doi.org/10.1002/cae.20213

Kuzmichev, V., Moskvin, A., & Moskvina, M. (2018). Virtual reconstruction of historical men's suit. *Autex Research Journal, 18*(3), 281–294. https://doi.org/10.1515/aut-2018-0001

Kuzmichev, V., Moskvin, A., & Moskvina, M. (2020). Designing digital twins of historical figure-suit systems. *Journal of the Textile Institute, 110*(12), 1810–1826. https://doi.org/10.1080/00405000.2019.1621042

Kuzmichev, V., Moskvin, A., & Moskvina, M. (2021). Modeling of digital twins of historical fashionable bodies. *Proceedings of Higher Education Institutions Textile Industry Technology, 1*(144). https://doi.org/10.47367/0021-3497_2021_1_144

Kuzmichev, V., Moskvin, A., Moskvina, M., & Adolphe, D. (2019). Reconstruction of historical textile fabrics and dresses in virtual reality. In *Proceedings of the 19th World Textile Conference on Textiles at the Crossroads* (pp. 15–28).

Kuzmichev, V., Moskvin, A., Moskvina, M., & Pryor, J. (2018). Research on 3D reconstruction of late Victorian riding skirts. *International Journal of Clothing Science & Technology, 30*(6), 790–807. https://doi.org/10.1108/IJCST-12-2017-0192

Kuzmichev, V., Moskvin, A., Surzhenko, E., & Moskvina, M. (2017). Computer reconstruction of 19th century trousers. *International Journal of Clothing Science & Technology, 29*(4), 594–606. https://doi.org/10.1108/IJCST-12-2016-0139

Kuzmichev, V., Moskvina, M., Adolphe, D., Moskvin, A., Sackharova, N., Kornilovich, A., Shichao, Z., Orlov, D. L., & Kareva, G. A. (2019). Designing digital twins of historical figure - suit systems. *Proceedings of Higher Education Institutions Textile Industry Technology, 6*(384).

Kuzmichev, V., & Yan, J. (2022). The application of digital twins in the field of fashion. In Z. Lv, & E. Fersman (Eds.), *Digital twins: Basics and applications*. Cham: Springer.

Kwon, Y. M., Lee, Y. A., & Kim, S. J. (2017). Case study on 3D printing education in fashion design coursework. *Fashion and Textiles, 4*(1), 18–31. Available at: https://link.springer.com/article/10.1186/s40691-017-0105-0.

Lee, J. (2022). Toward sustainable fashion product development: The use of 3D virtual prototyping technologies in the synchronous remote learning classroom. *Journal of Educational Technology Systems, 50*(2), 145–160. https://doi.org/10.1177/00472395221113086

Lee, M., Yoo, D. I., & Kim, S. (2017). Development of low-cost three-dimensional body scanner using depth perception camera.

International Journal of Clothing Science & Technology, 29(6), 857–867.

Lin, Y., & Wang, M. J. (2012). Constructing 3D human model from front and side images. *Expert Systems with Applications, 39*(5), 5012–5018.

Liu, W. (2023). Research on the application of virtual simulation and 3D animation technology in the teaching of road and bridge majors in higher vocational colleges. *Journal of Contemporary Educational Research, 7*(9), 1–5. . (Accessed 22 March 2025).

Livrustkammaren. (2018). *Photogrammetry at the royal armoury - introduction to 3D & cultural heritage (Eng subs)*. https://www.youtube.com/. Available at: https://www.youtube.com/watch?v=KivIw4NW3cc&list=PLWFAAnwfSXzfZ_y3jDFpwOhBIvKKzZ9Y2&index=5. (Accessed 13 May 2024).

Loscialpo, F. (2016). From the physical to the digital and back: Fashion exhibitions in the digital age. *International Journal of Fashion Studies, 3*(2), 225–248. https://doi.org/10.1386/infs.3.2.225_1

Lu, W., & Bian, X. (2017). A comparative study of clothing display form in museums based on immersive virtual reality. In DEStech transactions on social science, education and human science, *MSIE*. Thailand: Chiang Mai. https://doi.org/10.12783/dtssehs/msie2017/15609

Macit, A. (2024). How digital twins transforming fashion industry. *Seamm*. Available at: https://seamm.io/blog/how-digital-twins-transforming-fashion-industry. (Accessed 23 April 2024).

Maghfiroh, A., Purnawirawan, O., Ihsani, A. N. N., & Krisnawati, M. (2024). The results of project-based learning models implementation for beauty students in the design of digital-based make-up and fashion models using the CLO 3D application. In *Proceedings of the 5th Vocational Education International Conference (VEIC-5 2023)* (pp. 1159–1164). Atlantis Press. https://doi.org/10.2991/978-2-38476-198-2_164

Magnenat-Thalmann, N. (2020). Haptic technologies for virtual try-on. *IEEE Computer Graphics, 40*(3), 72–81. https://doi.org/10.1109/MCG.2020.2974849

Magnenat-Thalmann, N., & Volino, P. (2005). From early draping to haute couture models: 20 years of research. *The Visual Computer, 21*, 506–519.

Malcolm-Davies, J. (2023). Structuring reconstructions: Recognizing the advantages of interdisciplinary data in methodical research. *Heritage Science, 11*(1), 182.

Martin, K., Caulfield-Sriklad, D., & Jushchyshyn, N. (2016). Creating, exhibiting and distributing new media for historic fashion: The Drexel Digital Museum project. In Proceedings of the XX Congress of the Iberoamerican Society of Digital Graphics, *Buenos Aires, Argentina* (pp. 931–937).

Martin, K., & Ko, H. (2009). Virtual historic costume across cultures and disciplines. In *Proceedings of the 15th International Conference on Virtual Systems And Multimedia, 9-12 September 2009, Vienna, Austria* (pp. 104–108).

Martin, K., & Polyak, E. (2024). Interactive digital twins: Creative tools for scientific study viewer engagement with historic textiles. *Journal of Textile Science & Fashion Technology, 10*(3), 1–12. https://doi.org/10.33552/JTSFT.2024.10.000741

McBee-Black, K., & Johnston, N. (2024). Identifying design features within a historic costume collection to inspire contemporary inclusively designed adaptive apparel in a product development course. *International Journal of Fashion Design, Technology and Education, 17*(4), 1–14. https://doi.org/10.1080/17543266.2024.2419512

McNulty, R. (2019). Future applications of digital clothing for historical costume: The past, present, and future of fashion. In *Proceedings of EVA London* (pp. 239–242). https://doi.org/10.14236/ewic/EVA2019.45

McQuillan, H. (2020). VR for fashion history pedagogy. *Fashion Practice, 12*(3), 345–361. https://doi.org/10.1080/17569370.2020.1817452

Meier, C., Berriel, I. S., & Nava, F. P. (2021). Creation of a virtual museum for the dissemination of 3D models of historical clothing. *Sustainability, 13*(22), 12581. https://doi.org/10.3390/su132212581

Moskvin, A., Moskvina, M., & Kuzmichev, V. (2019). Digital replicas of historical skirts. *The Journal of The Textile Institute, 110*(12), 1810–1826.

Moskvin, A., Moskvina, M., & Kuzmichev, V. (2020). Parametric modeling of historical mannequins. *International Journal of Clothing Science & Technology, 32*(3), 366–389. https://doi.org/10.1108/IJCST-06-2019-0093

Moskvin, A., Moskvina, M., & Kuzmichev, V. (2022). Block pattern generation of scanned historical garments. *Izvestiya Vysshikh Uchebnykh Zavedenii, Seriya Teknologiya Tekstil'noi Promyshlennosti, 4* (147), 147–152. https://doi.org/10.47367/0021-3497_2022_4_147

Moskvin, A., Moskvina, M., & Kuzmichev, V. (2022). Digital twins of textile materials for visualization of historical costumes. *Izvestiya Vysshikh Uchebnykh Zavedenii, Seriya Teknologiya Tekstil'noi Promyshlennosti,* 2(86), 86–92. https://doi.org/10.47367/0021-3497_2022_2_86

Moskvin, A., Moskvina, M., & Kuzmichev, V. (2023). Shaping digital twins of tight-fitting clothes. *Izvestiya Vysshikh Uchebnykh Zavedenii, Seriya Teknologiya Tekstil'noi Promyshlennosti,* 5(407), 138–146. https://doi.org/10.47367/0021-3497_2023_5_138

Moskvin, A., Moskvina, M., & Kuzmichev, V. (2023). Structure of software program pipelines for creating digital clothing twins: Principles of design. Izvestiya Vysshikh Uchebnykh Zavedenii. *Seriya Teknologiya Tekstil'noi Promyshlennosti, 6*(408), 188–197. https://doi.org/10.47367/0021-3497_2023_6_188

Moskvin, A., Moskvina, M., & Kuzmichev, V. (2025). A study on levels of detail for presenting 3D models of historical costumes online. *Digital Applications in Archaeology and Cultural Heritage, 36*, e00399. https://doi.org/10.1016/j.daach.2025.e00399

Moskvin, A., Moskvina, M., & Kuzmichev, V. (2025). Development of a reference database on historical body sizing and digital twins. *International Journal of Clothing Science & Technology*. https://doi.org/10.1108/IJCST-04-2024-0086. ahead-of-print.

Muñoz-Salinas, R., Sarmadi, H., Cazzato, D., & Medina-Carnicer, R. (2019). Flexible body scanning without template models. *Signal Processing, 154*(1), 350–362.

Na'am, M. F., Achmadi, T. A., et al. (2021). Optimizing the skills of clothing design teachers and students in clothing design using the CorelDraw and Adobe Illustrator programs to respond to Industry 4.0. In *Proceedings of VEIC 2020* (pp. 123–130). Available at: https://books.google.com.

Nam, C., Bernardoni, J. M., & Min, S. D. (2024). Virtual design pedagogy: Understanding the Metaverse and improving creative design skills using ZEPETO studio platform™. *International Journal of Fashion Design, Technology and Education, 17*(3), 1–12. https://doi.org/10.1080/17543266.2024.2329756

Narsun Studios. (2023). How can digital twins make fashion sustainable? *Sustainable & Social*. Available at: https://sustainableandsocial.com/digital-twins-fashion/. (Accessed 23 April 2024).

Nurtanto, M., Sofyan, H., & Pardjono, P. (2020). *E-Learning based AutoCAD 3D interactive multimedia on vocational education (VE) learning*. Online Submission [online] Available at: https://eric.ed.gov/?id=ED606409. (Accessed 22 March 2025).

Oh, S.-Y., & Ryu, E.-J. (2015). A study on expressivity of virtual clothing made of 3D apparel CAD system according to the physical properties of fabric. *Fashion & Textile Research Journal, 17*(4), 613–625. https://doi.org/10.5805/SFTI.2015.17.4.613

Orzada, B. T., López-Gidós, D., Cobb, K., & Diao, J. (2024). The new era of fashion heritage: Digital twins of 1920s clothing. In *Annual Conference of the International Textile and Apparel Association* (Vol. 80). https://doi.org/10.31274/itaa.17112

Pan, Z. (2018). Automatic re-topology and UV remapping for 3D scanned objects based on neural network. In *CASA 2018: Proceedings of the 31st International Conference on Computer Animation and Social Agents* (pp. 48–52). https://doi.org/10.1145/3205326.3205356

Papachristou, E., & Zolota-Tatsi, N. (2024). Encouraging 3D virtual design in fashion education: Best practices. In *Fashion design education: Innovations and technologies* (pp. 442–452). https://doi.org/10.1007/978-3-031-48933-4_43

Papagiannakis, G., Geronikolakis, E., Pateraki, M., López-Menchero, V. M., Tsioumas, M., Sylaiou, S., Liarokapis, F., Grammatikopoulou, A., Dimitropoulos, K., Grammalidis, N., Partarakis, N., Margetis, G., Drossis, G., Vassiliadi, M., Chalmers, A., Stephanidis, C., & Magnenat-Thalmann, N. (2018). Mixed reality, gamified presence, and storytelling for virtual museums. In N. Lee (Ed.), *Encyclopedia of computer graphics and games* (pp. 1–14). Cham: Springer.

Papahristou, E. (2016). *The effective integration of 3D virtual prototype in the product development process of the textile/clothing industry*. Doctoral Dissertation.

Papahristou, E., & Zolota Tatsi, N. (2024). A review of 3D design knowledge and its impact on creativity in fashion design education. *Communications in Development and Assembling of Textile Products, 5*(2), 266–277. https://doi.org/10.25367/cdatp.2024.5.p266-277

Paquet, E., & Viktor, H. L. (2007). Adjustment of virtual mannequins through anthropometric measurements, cluster analysis, and content-based retrieval of 3-D body scans. *IEEE Transactions on Instrumentation and Measurement, 56*(5), 1924–1929.

Park, S. M., Choi, K. M., Nam, Y. J., & Lee, Y. (2011). Multi-purpose three-dimensional body form. *International Journal of Clothing Science & Technology, 23*(1), 8–24.

Park, J., Kim, D. E., & Sohn, M. H. (2011). 3D simulation technology as an effective instructional tool for enhancing spatial visualization skills in apparel design education. *Journal of Technology and Design Education, 21*(4), 367–379. Available at: https://link.springer.com/article/10.1007/s10798-010-9122-5. (Accessed 22 March 2025).

Petrova, A. (2022). *3D body scanning for anthropometric data collection*. Boca Raton: CRC Press.

Pilar, T., Stjepanovic, Z., & Jevsnik, S. (2013). Evaluation of fitting virtual 3D skirt prototypes to body. *Tekstilec, 56*(1), 47–62. https://doi.org/10.1108/09556220210446103

Power, J. (2020). *Digital patternmaking revolution*. Abingdon: Routledge.

Prasetya, L. A., Widiyawati, I., Rofiudin, A., Haq, S. T. N., Hendranawan, R. S., Permataningtyas, A., & Ichwanto, M. A. (2025). The use of CLO3D application in vocational school fashion expertise program: Innovations, challenges and recommendations. *Journal of Research in Instructional, 5*(1), 287–299. https://doi.org/10.30862/jri.v5i1.550

Razdomahin, N. (2004). *Theoretical foundations and methodological support of three-dimensional clothes design*. PhD thesis. Saint Petersburg: Saint Petersburg State University of Technology and Design.

Ribeiro, R., et al. (2020). Predicting physical properties of woven fabrics via automated machine learning and textile design and finishing features. *Artificial Intelligence Applications and Innovations, 584*, 244–255. https://doi.org/10.1007/978-3-030-49186-4_21

Rigoni, D., & Düştegör, D. (2024). Science-twins: Digital twins for interactive lecture demonstrations. In 1st International Workshop on Distributed Digital Twins. In *DiDit, 3755*. CEUR Workshop Proceedings. https://research.rug.nl/files/1124912869/short1.pdf.

Rissanen, T. (2022). *Zero waste fashion design: Digital approaches*. London: Bloomsbury.

Saito, S., Kouchi, M., Mochimaru, M., & Aoki, Y. (2014). Model-based 3D human shape estimation from silhouettes for virtual fitting. In *Paper presented at the Three-Dimensional Image Processing, Measurement and Applications Conference, San Francisco, CA, February 5*. https://doi.org/10.1117/12.2038457. (Accessed 11 October 2018)

Sarakatsanos, O., Papazoglou-Chalikias, A., Boikou, M., Chatzilari, E., Jauk, M., Hafliger, U., Nikolopoulos, S., & Kompatsiaris, I. (2024). VR designer: Enhancing fashion showcases through immersive virtual garment fitting. *Virtual Reality, 28*(2), 70. https://doi.org/10.1007/s10055-024-00945-w

Sehgal, R., Gupta, R., & Anand, N. (2018). Automatic extraction of 3D body measurements from 2D images of a female form. *Journal of Polymer and Textile Engineering, 5*(3), 7–17.

Senanayake, M., Raheja, A., & Zhang, Y. (2018). Automated human body measurement extraction: Single digital camera (webcam) method – Phase 1. *International Journal of Clothing Science & Technology, 30*(2), 175–194.

Shershneva, L. P., et al. (2004). Fundamentals of applied anthropology and biomechanics. *Forum-Infra*. Moscow.

Shetabi, T. (2024). *Clothing, from textiles to pixels: Exploring the possibilities of design with Clo3D software*. University of Nebraska-Lincoln. Master's thesis.

Sidabraite, V., & Masteikaite, V. (2002). A preliminary study for evaluation of skirt asymmetric drape. *International Journal of Clothing Science & Technology, 14*(5), 286–298. https://doi.org/10.1108/09556220210446103

Siersema, I. (2015). The influence of 3D simulation technology on the fashion design process and the consequences for higher education. In *Proceedings of the Digital Fashion Conference*. Available at: https://www.researchgate.net/publication/281470261_The_influence_of_3D_simulation_technology_on_the_fashion_design_process_and_the_consequences_for_higher_education. (Accessed 22 March 2025).

Starkey, S. (2018). Curriculum development using virtual reality in an experimental apparel design classroom. *International Journal of Fashion Design, Technology and Education, 11*(3), 197–206. Available at: https://www.tandfonline.com/doi/full/10.1080/17543266.2018.1466940. (Accessed 22 March 2025).

Steele, V. (2001). *The corset: A cultural history*. London: Yale University Press.

Stott, R. (2023). AI applications in pattern grading. *International Journal of Fashion Technology, 14*(2), 45–59. https://doi.org/10.1080/17543266.2023.1987455

Subiyantoro, S., Yuliana, M., Nurbaiti, N., Ramadhani, S. A., & Murti, W. P. (2023). Optimizing the use of practical methods to improve skills in creating 2D and 3D animations. *Cognitive Development Journal, 1*(1), 37–48. Available at: https://link.springer.com/article/10.1007/s10798-010-9127-3. (Accessed 22 March 2025).

Sugiyem, S., Bestari, A. G., Asiatun, K., Wati, W. R., Nurtanto, M., Sofyan, H., & Pardjono, P. (2024). The effectiveness of video tutorials on textile and fashion design learning. In *Proceedings of the 5th Vocational Education International Conference (VEIC-5 2023)* (pp. 1159–1164). Atlantis Press. Available at: https://www.atlantis-press.com/proceedings/veic-23/125997719. (Accessed 22 March 2025).

Sunarya, Y. Y., Kudiya, K., & Rasul, M. S. (2024). Model of training learning based on video tutorial for fashion major of vocational education on embroidery course. *Journal of Technical Education and Training, 16*(1), 24–33. Available at: https://penerbit.uthm.edu.my/ojs/index.php/JTET/article/view/15516.

Taylor, A., & Unver, E. (2005). An experimental study to test a 3D laser Scanner for bodymeasurement and 3D virtual garment design in fashion education. In *Wearable futures – HybridCulture in the design and development of soft technology* (pp. 1–14). University of Wales.

Tu, Y.-F., et al. (2024). A Systematic Review of AI-Driven Prediction of Fabric Properties and Handfeel. *Materials, 17*, 5009. https://doi.org/10.3390/ma17205009

Wang, C. C. L. (2005). Parameterization and parametric design of mannequins. *Computer-Aided Design, 37*(1), 83–98.

Wang, H., & Cho, J. H. (2021). *Research on virtual display of wetsuit based on CLO3D* (pp. 1–8). Scientific Programming. https://doi.org/10.1155/2021/1052510

Watchravesringkan, K., Chang, H. J. J., Lee, S., Choi, T. M., & Lee, H. (2025). Learning 3D/virtual technology in apparel and textiles (VTAT): Self-efficacy scale development and validation. *Textile and Apparel Research Journal, 41*(1), 1–12. . (Accessed 22 March 2025).

Wati, W. R. (2024). Advancing fashion design education: Evaluating the eligibility of video tutorial media for basic pattern content in vocational high school. *Journal of Education Research and Evaluation, 8* (1), 1–12. . (Accessed 22 March 2025).

Widiyawati, I., Prasetya, L. A., Permataningtyas, A., Nurhadi, D., Hidayati, N., & Ichwanto, M. A. (2024). Application of Clo3D technology in fashion vocational education in vocational schools: A systematic literature review. *International Journal of Education and Curriculum Application, 7*(3), 204–217.

Xue, Z., Zeng, X., Koehl, L., & Chen, Y. (2014). Extracting fabric hand information from visual representations of flared skirts. *Textile Research Journal, 84*(3), 246–266. https://doi.org/10.1177/0040517513485622

Yan, J., & Rui, W. (2006). Parametric design of mannequins for virtual try-on system. In *Paper presented at the 7th International Conference on Computer-Aided Industrial Design and Conceptual Design (CAID&CD), Hangzhou, November 17–19*. Available at: https://ieeexplore.ieee.org/document/4127116. (Accessed 19 October 2018).

Yezhova, O., Wu, S., Pashkevych, K., Kolosnichenko, M., Ostapenko, N., & Struminska, T. (2024). Exploring design and technological aspects of digital fashion: A systematic review of recent innovations. *International Journal of Fashion Design, Technology and Education*, 1–15. https://doi.org/10.1080/17543266.2024.2378032

Zeng, Y., Fu, J., & Chao, H. (2017). 3D human body reshaping with anthropometric modeling. In *Paper presented at the Internet Multimedia Computing and Service Conference (ICIMCS), Qingdao, August 23–25*. Available at: https://link.springer.com/chapter/10.1007/978-981-10-8530-7_10#citeas. (Accessed 7 November 2018).

Zhang, Y., Zheng, J., & Magnenat-Thalmann, N. (2015). Example-guided anthropometric human body modeling. *The Visual Computer, 31*(12), 1615–1631.

Zharinova, E., Zhukova, M., Kuzmichev, V., Moskvin, A., & Moskvina, M. (2024). Virtual exhibition of Nadezhda Petrovna Lamanova. In *SMARTEX 2024: Proceedings of the XXVII international scientific and practical forum* (pp. 45–56).

Zharinova, E., Zhukova, M., Kuzmichev, V., Zhukova, I., Moskvin, A., & Moskvina, M. (2024). Virtual costume map of Ivanovsk region. In *SMARTEX 2024: Proceedings of the XXVII international scientific and practical forum* (pp. 60–72).

Zhong, S., et al. (2024). Feeling textiles through AI: An exploration into multimodal language models and human perception alignment. In *International Conference on Multimodal Interaction*. https://doi.org/10.1145/3678957.3685756

Chapter 19

Digital twins as a tool for optimizing social interaction: A case study in Indonesian higher education with STELE approach

Ahmad Al Yakin[1], Muthmainnah Muthmainnah[1], Ali Said Al Matari[2], Luis Cardoso[3] and Ahmed J. Obaid[4]
[1]Universitas Al Asyariah Mandar, West Sulawesi, Indonesia; [2]A'Sharqiyah University, Ibra, Oman; [3]Polytechnic Institute of Portalegre and Centre for Comparative Studies of the University of Lisbon, Lisbon, Portugal; [4]Faculty of Computer Science and Mathematics, University of Kufa, Najaf, Iraq

1. Introduction

Students' social interactions within technology-enhanced learning ecosystems have been better understood because to the inclusion of digital twin technologies in smart education (Xu, 2024) to robust mechanisms such as social learning, feedback, and agentive learning, digital classrooms offer a great opportunity to improve students' social skills. Modern educational theory and practice place an emphasis on the collaborative character of learning, the idea that students gain insight and competence by their experiences working with others (Zammit et al., 2024). This pedagogical method is well-suited to the interactive nature of digital technologies, which allow academics to study the social dynamics (Yeganeh et al., 2025). In today's technologically advanced classrooms, students are flourishing because their expectations, access to information, and ability to collaborate with classmates are all shaped by these online resources. These technologies have found widespread use in classrooms because of the ways in which they complement existing instructional methods and facilitate the achievement of predetermined learning goals (Nehru et al., 2025).

A one-of-a-kind chance to investigate how students engage socially in smart classrooms has presented itself using digital twin technology, which generates a digital copy of a real system. Social dynamics in these tech-enhanced classrooms can be better understood with the use of digital twins, which simulate the interdependencies among students, instructors, and the learning space. Researchers can use digital twins to monitor students' patterns of engagement, communication, and collaboration to find out what helps and what hurts their social interactions. Research conducted by Omrany et al. (2025). Recent developments in adaptive learning systems powered by artificial intelligence have had a significant impact on students' expectations, capacities to obtain material, engagement levels, and social behaviors. Mobile devices, augmented reality, and social networks are just a few examples of how emerging technologies are changing the way students' study and engage with one another. Students' social interactions and navigation of these technology-enhanced learning settings can be better understood by investigating the use of digital twin technologies in smart education. Students' social and navigational skills in flipped classrooms can be better understood with the use of digital twin technologies in smart education (Chen et al., 2024).

Theoretically and pragmatically, smart education that incorporates digital twin technologies can shed light on how students interact with and use learning environments that are enhanced by technology. The use of digital twin models, which are virtual copies of physical things or systems, allows for the simulation and analysis of student social interactions in these technologically advanced settings. Educators can learn more about their students' social media, collaboration, and communication skills by making digital twins of them and then observing how they use smart classroom technology. Using data to shape the

Digital Twins in the Smart Classroom. https://doi.org/10.1016/B978-0-443-41632-3.00016-3

creation of more inclusive and engaging educational experiences that promote meaningful student interactions, this data-driven approach can provide vital insights into the social dynamics of technology-enhanced learning. Smart education's use of digital twin technologies might revolutionize our comprehension and facilitation of the social aspects of learning in the modern era (Wang et al., 2024).

Smart education, which incorporates digital twin technologies, may revolutionize the way students learn, especially in Indonesia. The specific obstacles of creating technology-enhanced learning environments that effectively engage and interact with Indonesia's diverse student population and educational infrastructure are significant. Smart education, which incorporates digital twin technologies, may revolutionize the way students learn, especially in Indonesia. The specific obstacles of creating technology-enhanced learning environments that effectively engage and interact with Indonesia's diverse student population and educational infrastructure are significant. Problems in implementing novel technology, such as digital twins, typically stem from educators' gaps in technical, pedagogical, and subject knowledge. Teachers in Indonesia can learn a lot about their students' smart classroom navigation and social interactions using digital twin models (Pandey et al., 2025).

Digital twin technology based on artificial intelligence is still in its infancy in Indonesian education, but it has the potential to make a huge difference. By simulating students, classrooms, and learning processes digitally, digital twins let teachers make better use of data to personalize and improve student learning. Some potential uses of digital twins powered by artificial intelligence in Indonesian education are as follows: (1) to begin, student digital twin models can be used to facilitate classroom social interaction simulations. Characteristics of learning, communication, and social interaction patterns can be captured by these digital representations. Teachers are able to better understand group dynamics, interaction patterns, and possible conflict hotspots thanks to artificial intelligence systems that analyze data from digital twins. Teachers will be better equipped to foster welcoming and supportive school environments with this knowledge which are AI-driven platforms, to create and execute simulations that enhance classroom management by modeling student interactions; (2) by simulating the steps of a student's learning in virtual form, digital twins provide the way for individualized instruction. Students' individual learning styles, understanding rates, and difficulties are all recorded by these digital twins. AI then uses this information to suggest lessons, exercises, and resources that are specific to each student's needs. The combination of digital twin technology with adaptive learning platforms such as Khan Academy or Duolingo allows for the creation of individualized learning plans that consider each student's current level of proficiency (Yamijala et al., 2025).

Teaching and learning will change drastically by incorporating digital twins into the education system. Digital twins, accessible through smartphones in the classroom, enable students to gain hands-on experience without the expense and benefits of operating real equipment by simulating complex industrial processes (Selim et al., 2024). Higher education institutions can now support differentiated instruction and foster a more inclusive learning environment by utilizing digital twin technology to deliver meaningful, personalized learning experiences that cater to each student's needs and foster higher-order thinking skills learning outcomes. Students from low-income backgrounds can still succeed academically even if they start at a lower level of education. This study proposes a framework that considers the cognitive demands of different levels of education along with the fidelity of the digital twin and investigates the social interactions during the learning process. This will help educators design better learning experiences for students, enabling them to better retain theoretical knowledge and acquire industry-ready skills with elegant social behaviors or social interactions through digital twin-based learning technology architectures. New possibilities for improving the quality and relevance of engineering and technology education will emerge as the role of digital twin technology in education expands.

2. Literature review

2.1 A Smart Technology-Enhanced Learning Environment (STELE) approach

Twenty-First-Century Skills Training is another important area of use. The use of digital twin technology allows students to practice important skills like problem-solving, cooperation, and communication in virtual work environments that mimic real-life workplaces in different industries (Supriya et al., 2024). VR/AR-based workplace simulation platforms, like medical practice simulations or business simulation games, create immersive learning environments. Students can build relevant skills in a genuine context, preparing them for future professional demands. Another domain where digital twins demonstrate potential is educational equity. Educators and legislators may model remote learning scenarios with the use of the digital twins of schools and their resources, which will help make sure that everyone, even in rural areas, has access to a good education. Google Classroom and Moodle, two popular online learning tools, can interact with digital twins of schools to provide distant students with a connected, high-quality education that would otherwise be unavailable to them, regardless of their location (Marinelli et al., 2024).

Artificial intelligence-powered digital twins specifically designed for Indonesian classrooms are still in their infancy. However, we can meet these requirements by modifying existing tools. One tool that helps with project-based learning is Minecraft Education Edition, which lets students and teachers create digital versions of classrooms (Katsantonis, 2025). To further tailor learning and effectively mimic classroom interactions, adaptive learning platforms like Duolingo and Khan Academy can be used (Alikhanova, 2025). The use of digital twins powered by artificial intelligence is still in its early stages in Indonesia, but the country is seeing a lot of promise in this area. To help achieve Indonesia's educational goals in the digital era, digital twins provide new solutions that improve the quality and accessibility of education, from personalizing it to making sure everyone has equal access.

Adapting current platforms to suit classroom demands is possible, even though there are limited AI-based digital twin apps developed for use in Indonesian schools. For instance, using Minecraft: Education Edition, educators may build their own virtual classrooms, encouraging students to work on group projects in a fun and dynamic way. Digital twin technology can also be used in conjunction with adaptive learning systems such as Duolingo and Khan Academy to provide students with more tailored lessons. Teachers can also try out new approaches and classroom dynamics with the help of agent-based modeling software, which allows them to create simulations of students and learning environments (Van Den Heever et al., 2024).

These cases show how digital twins and AI may improve Indonesian education in many ways, including by promoting individualized instruction, enhancing existing skills, and expanding opportunities for all students. Expanding the use of digital twins in the classroom could solve long-standing problems in education and give students a better learning experience that is more personalized, relevant, and interesting as the technology develops. Note that the implementation of AI-based digital twins necessitates strong intentions, mature infrastructure, and teacher training. More research and development are also needed to provide more specific and effective applications for the Indonesian educational context. It is important to emphasize TPACK in this technology implementation, as stated in the text (Karataş & Ataç, 2024).

A thorough comprehension of the interrelationships among technology, pedagogy, and subject knowledge is necessary for successful technology integration in the classroom, according to TPACK theory. Digital twin technologies have the potential to greatly improve student engagement and social interaction, but many Indonesian educators may be underprepared to make full use of these tools in the classroom. Several factors must be considered to overcome the limits of TPACK. These include making sure that schools have sufficient technology infrastructure, encouraging instructors to work together in collaborative learning environments, and offering extensive professional development opportunities for teachers (Max et al., 2024).

We can learn a lot about how to deploy digital twins in educational settings in Indonesia by looking at successful examples. For example, the Al-Mawadieh et al. (2024) study indicated that students' social skills, communication abilities, and problem-solving competencies were all improved through the use of digital twins in group projects. Teachers in Indonesia can use digital twins to their advantage by modifying and implementing comparable practices; this will allow them to better meet the requirements of their pupils, who come from a wide range of backgrounds and experiences. Smart education that makes use of digital twin technologies can also help level the playing field in Indonesia, where access to technology is uneven. Regardless of a school's location or physical infrastructure, digital twin simulations can provide virtual interactions and collaborative learning opportunities, providing fair access to educational experiences (Dritsas & Trigka, 2025; Gündüz, 2024).

To fill the knowledge gap about how digital twin technologies can be used to better understand students' social interactions and how they navigate technology-enhanced learning environments, this study will focus on smart education settings in Indonesia. This study aims to help meet the needs of students in Indonesia by studying how digital twin technologies interact with student social dynamics. The goal is to create learning environments that are more welcoming and engaging for all students. Students' social behaviors, patterns of collaborative learning, and engagement within smart technology-enhanced learning ecosystems will be analyzed through the study's examination of digital twin technology's implementation in selected Indonesian schools. With the results of this study in hand, educators and lawmakers will be better equipped to use digital twin technologies to build more inclusive and productive learning environments for all students.

2.2 The use of digital twins in education 7.0

Digital twins are a kind of simulation-based system engineering. Still, it's not the same as the offline simulation because this one uses more than just the physical model. It also uses real-time sensor readings and data from the past to try to accurately simulate how the physical system works (Don et al., 2025.1). In addition, current digital twins can mimic physical behavior and incorporate a user-friendly visualization as a solution to the quest for universality. An increasingly fundamental problem in the advancement of digital twin technology is creating the look of a physical system and turning it into an easy-to-use interface (Davey, 2024).

The contemporary digital twins work, their present equipment collects data about physical behavior through sensors and feeds it continuously into the virtual space. Building digital twins in virtual space with different levels of accuracy lets you accurately mirror different parts of a system, process data that is collected in real space, and create behavior output or feedback control. This is then sent back to equipment in real space, letting it make changes and improve itself in real time. This integration of digital twins with physical systems is like a human body: the "bone" is the physical system, the "muscle" is the behavior model of the digital twin, which oversees driving the system and simulating dynamic interactions, and the "skin" is the visualization layer that makes complicated data easy to understand and work with. These parts work together perfectly to make sure that digital twins can talk to physical systems. This makes operations more efficient, maintenance more accurate, and monitoring better in many situations (Hakiri et al., 2024). The development of digital twin technology is facing two major challenges: (1) accurately modeling physical behavior to generate virtual models, and (2) creating an intuitive and operable interface by modeling the appearance of a real system. It is not simple to depend on conventional numerical models to represent complicated systems when building a behavioral model for digital twins.

However, as computing power and machine learning have advanced, this approach has emerged as the main choice for modeling physical behavior (Singh & Nayyar, 2025). The use of machine learning has the potential to simplify digital twin modeling, which opens up new applications for the technology. Many academic institutions have recognized the importance of digital transformation in enhancing the teaching and learning process, coinciding with the transformation of education (Arar et al., 2025). The consolidation of the industrial revolution of education 7.0, which involves the transformation of universities into smart environments, is a major issue facing higher education institutions today, including Indonesia (Hu, 2025). When discussing digital transformation, it is important to place it in the context of the so-called sixth industrial revolution, with an educational ecosystem that is harmonious between human intelligence and artificial intelligence (Al Yakin et al., 2025).

Due to the current human-machine collaboration learning ecosystem, universities and colleges need to adopt new ways of teaching and learning that take advantage of the digital transformation and create an environment that encourages intellectual growth (Hashim et al., 2024). According to Hermosa and Andal (2025), universities need to adapt their structures, methods, and professional profiles to meet the needs of society and train students in developing their humanistic attitudes as social beings. Various services and types of digital content have supported the digital transformation of the classroom by aiming to create personalized, engaging, and motivating learning. These tools include digital animation, video game development, artificial intelligence, virtual and augmented reality, digital twin modeling, and virtual reality (Stasolla et al., 2025). By creating a digital twin, one can digitally represent or engineer an entity or system in the physical world. In other words, a digital object can mimic the behavior of a physical entity and the environment of existing reality (Yang et al., 2025).

To assist in decision-making, digital twins try to replicate physical reality digitally, creating data-driven knowledge in smart society 5.0 (Sankaran et al., 2025). They also incorporate capabilities for simulation and prediction (Zhang et al., 2025). Plechatá et al. (2025) categorized digital twins into five types: descriptive, informative, predictive, integral, and autonomous. Digital twins can help people who are just starting to learn how to use machines or other tools because they let them practice and assess risky situations without having to go through them (Razzaq et al., 2023). This is something that is taught in higher education sociocultural sciences. This course examines various sociocultural aspects, including the phenomenon of society 5.0. We believe that the need for learning spaces is constant, enabling individuals to acquire the necessary skills for survival, participation in society, and social interaction. The content of learning continues to evolve to meet the needs of the social order, leaving evidence of its great influence through learning outcomes (Cheng et al., 2025). Education is essential to the development of generations in every modern society. Throughout history, people have universally acknowledged the need for schools or universities to progress, and they have long used this concept to track the development of educational institutions. The needs of society and the history of industrial development demonstrate the importance of highly educated workers with 21st-century skills for further economic expansion and technical progress.

Individuals with superior cognitive capabilities and expertise in areas such as commerce, engineering, and scientific research are better able to develop their capacity for intellectual self-realization because of their unique psychological and intellectual traits, such as initiative, creativity, and maximal self-regulation. Furthermore, as a result of plans to access an established system of knowledge acquisition through primary, secondary, and university education, many countries have improved their economic and social prospects in urban and regional areas through increased industrialization. The development of technology over time has unquestionably made teaching and grading more effective. For example, wooden paddles were used during the colonial era, and now there are AI platforms like VR, AR, MR, and digital twins. In just the last century, technology has resulted in the invention of many new devices, such as radios, video cassettes, headphones, overhead projectors, pens, computer systems, the Internet, photocopiers, e-books, virtual robots, and personal digital assistants. Research, innovation, and commercialization of increasingly advanced technology have flourished in the education system due to the rapid advancement of several media and learning technologies in this century (Pandey et al., 2025).

To keep up with the rapid technological advancements and the ever-increasing needs of society, it is imperative for educators to remain actively involved in professional development opportunities to improve their teaching skills and the information they teach that is up to date (Cheng et al., 2025) for example, to face future professional challenges, such as promoting sustainability, developing complex automation systems, and cultivating future-proof employability, engineering higher education continues to evolve to be more engaging, designing active student-centered learning, and being able to develop students from inexperienced individuals to individuals who are able to face challenges and think critically and creatively (Sepasgozar et al., 2024). The education system in higher education is expected to continue to develop and incorporate new methods that are more effective and interesting, such as personalized adaptive learning (Venice et al., 2025), digital game-based learning (Zuo et al., 2025), and interactive learning methods. Modern multimedia technology, AI-based learning methods, and VR or IoT all make people more motivated to learn (Li et al., 2025) and help teachers teach 21st-century skills (Emihovich, 2024) for their classes. The engineering education system continues to develop to make it more effective and intriguing for prospective graduates. Some of these methods include personalized adaptive learning (Rye et al., 2025); digital game-based learning; interactive learning methods (Sikora et al., 2024); and advanced multimedia technology (AMT).

According to Chen and Syu (2024), Muthmainnah et al. (2025) stated that education graduates should be able to use a range of professional skills and seek solutions based on higher-order thinking skills (HoTs). They should also be social and have a deep understanding of how to apply 21st-century principles in different situations (Al Yakin et al., 2025). His review focuses on the content of sociocultural science programs with pedagogical and social frameworks and students' ability to adapt their skills to the changing job market in the 21st century. Thus, to instill these qualities in students, contemporary educational institutions must adopt contemporary teaching methods. The statement stresses how important it is for students to have a wide range of professional skills and a deep understanding of how engineering works in a variety of settings, all based on using technology and literacy. Curriculum, pedagogical framework, and the ability of graduates to adapt to the changing job market are the main focuses of a comprehensive evaluation of engineering education in the 21st century. Modern schools need to use new ways to teach that improve students' social and humanistic skills, like using a digital twin-based learning platform, to teach these traits to their students.

One of the most popular and trending I4.0 manufacturing technologies today, digital twin (DT) technology, is considered "a high-fidelity virtual replica of a physical asset with real-time two-way communication for simulation purposes and decision-assistance features for product service improvement." In Bucci et al. (2024), DT was, DT was called conceptual ideal for the product lifecycle of human centric. Webb et al. (2024) stated all these names reflect its foundation in lean thinking and its application as future directions of digital twin throughout the product lifecycle (from manufacture to disposal). Following this guideline, the digital twin enabling technology application functional model mainly includes the following parts: planning, design, construction, support, and disposal, with a central information center responsible for collecting, analyzing, and sharing data in these parts (Don et al., 2025). There are four main parts to the information mirror model (IMM): actual or physical space, virtual space, connections (data), and information mirrors (Hu et al., 2025). It is often called "a high-density virtual replica of a physical asset with real-time bidirectional communication for simulation purposes and decision-assistance features for improving product service." DT is one of the most important I4.0 manufacturing technologies.

Elbadawy and Farouk (2025) first referred to DT as the tools and architecture of the VR environment under sustainable learning, then as the mirror model of the world, and finally as the functional model of learning. It forms the basis for five main categories: plan, design, build, support, and dispose. A centralized information center is responsible for collecting, analyzing, and disseminating data and information in these categories (Ton et al., 2025). For this reason, we can conceptualize the DT technology framework as a digital twin, duplicate, or depiction of the physical-to- virtual model, spanning both the present and the future (Stephan et al., 2025). Based on real-time data input, the virtual environment runs the necessary simulations and provides decision-making information for users to adjust the operation of physical objects. Advanced systems also allow independent control of physical objects using input from virtual space. Since virtual goods are usually at the micro- or macro-geometric level, DT technology can provide any product-related data that users may want from real assets (Haider et al., 2025). Virtual simulation (VS) in DT differs from standard simulation-based product development activities, which involve the use of input data for simulation and validation.

Therefore, the DT technology framework serves as an electronic replica, duplicate, or clone of an existing physical model of reality, both current and future. This technology is highly beneficial for university students investigating virtual reality, as it can enhance their social interactions in the classroom (Gesthuizen et al., 2025; Ying, 2025). The learning platform, https://www.cospaces.io/ or https://edu.assemblrworld.com/for-students, provides users with decision-making information to alter the functioning of physical objects. The virtual environment executes the intended simulation based on real-time data input and learning objectives. This sophisticated digital twin system enables independent control over physical objects by utilizing input from virtual space to simulate the real world. Virtual items, often at the micro-atomic or macro-geometric level, enable DT technology to furnish consumers with information about any product they may desire from physical assets (Hu, 2025).

DT's virtual simulation (VS) is not the same as traditional simulation-based product development in several ways, including the number of steps needed, the sale of geometric models, and the ability to learn from simulation results (Savchuk et al., 2025). Virtual space insights can significantly enhance innovation, uncover new opportunities in the meaningful learning process, and provide valuable insights from simulation results. Insights gained from the virtual world can be a wonderful source of inspiration for new ideas and the discovery of exciting new opportunities. DT systems can enhance the durability, performance, creativity, and cost-effectiveness of virtual worlds in the real world, thereby facilitating the efficient transformation of relevant businesses. Through the integration of various multiphysics applications, DT environments (DTEs) perform a number of actions on the digital space. Both DTP and DTI utilize the predictive and interrogative subcategories of DTEs. Items in the virtual world can enhance their lifespan, performance, and cost-effectiveness, while in the real world, they can teach us how to effectively transform relevant businesses (Dai & Kang, 2025). By importing or collecting detailed data into a virtual environment, such as a virtual simulation space, DTPs serve as experimental twins or prototypes of physical artifacts, replicating the physical model without any physical connection (Sharma et al., 2025).

In fact, DTPs help understand the physical model and drastically reduce the cost and time spent on testing and validation using the physical prototypes included in the learning process and motivation. Through the DT system, the objectives of digital twin instances (DTIs) can be realized. This includes importing fully annotated 3D geometry with detailed geometric dimensions and tolerances, along with the characteristics of development activities and their effects. It also includes all notes related to billing, testing, operations, component replacement, maintenance, and historical and real-time data. Importing or collecting extensive data in a virtual setting allows digital twins prototype (DTP) to mimic the physical model without ever touching it. His allows it to function as a prototype or experimental twin to the original artifact (Guinea-Cabrera & Holgado-Terriza, 2024). Using DTP not only saves time and money during testing and validation with physical prototypes but also helps understand the physical model with the information facilities obtained through sensors (Casaletto, 2024).

Along with I4.0 technologies and beyond, cyber-physical systems (CPS) can do even more advanced things (Kocsis et al., 2025). On the other hand, DTI describes systems that connect physical models to virtual environments for the purpose of analyzing and improving activities throughout their life cycle (Wang et al., 2024). DT systems achieve the goal of DTI by ingesting 3D geometry in its entirety, analyzing this data and information efficiently, and providing each industry with reliable insights on maximizing the value of physical artifacts (Aragón et al., 2025). It is important to note that DT systems differ from other concepts, such as digital shadowing and digital modeling, in their interaction with data, models, and the level of physical model implementation (Zhu et al., 2025). Depending on the application area of the system or process, there may be several DT systems, such as component/part twins, asset twins, system/unit twins, and process twins. Due to the increasing potential for technical advancements through DT, the education sector also uses these systems.

3. Methodology

This quantitative study using the Smart Technology-Enhanced Learning Environment (STELE) approach used an experimental design with a one-group pretest and posttest. This design involved one group as a pretest (O1), exposed to treatment (X), and a post-test (O2). This experimental study investigates the potential of a digital twin site as a learning platform to optimize social interaction in Indonesian higher education. The study examines the impact of the https://www.cospaces.io/application to investigate the social interactions of undergraduate students. The students were studying sociocultural sciences by observing social behaviors in the application (Fig. 19.1).

The study used an artificial intelligence-driven digital twin environment to enhance interactive and immersive

O_1	X	O_2

FIGURE 19.1 Research design of pre-experimental model.

learning. We conducted pre- and post-tests to measure student proficiency as part of the data collection process and collected metrics that measured student engagement and interaction through the Digital Twin platform. The purpose of incorporating AI-powered tools in the STELE framework was to improve communication skills, optimize collaborative learning, and mimic real-world educational scenarios.

3.1 Participants and setting

Universitas Al Asyariah Mandar Indonesia had 451 students from the faculty of public health take part in this study. They used the "digital twin as a tool to optimize social interaction in Indonesian higher education by STELE approach using digital twins (DT) tool, with 63 first semester students in the pre-experimental model during one semester from September 2024 to January 2025 of the academic year 2024. The researcher used the purposive sampling technique when taking a sample believed to be representative of the population. It means that the researcher deliberately selects the participants or the class that applied DT with the STELE approach. We ensured a consistent basis for evaluation by classifying the participants' proficiency level in accessing technology as intermediate.

To collect comprehensive data, this study used several instruments. Initially we provided a closed questionnaire with 20 items measured on a five-point Likert scale to collect participants' perspectives on digital twins. Here, researchers can collect theoretical and practical information on how people feel about the use of digital twins in the classroom. We also used a standardized pre- and post-test to measure participants' basic sociocultural science skills before and after the intervention. The experimental group in this study mostly learned through the digital twin app, which was paired with sociocultural practice twin observations, an AI-powered tool to see how it changed how people interacted with each other. These tools, when used together, will reveal quantitative and qualitative information about the success of digital twins in smart classrooms.

3.2 Procedure

An excellent framework for assessing the success of training and educational programs is Kirkpatrick's four-level training evaluation model (Miranda et al., 2025). When used in the classroom, digital twin technology provides a methodical way to evaluate and improve student progress. The primary goal of the first level is to learn how students react and participate in lessons that use digital twins. We'll use surveys to assess their satisfaction and enthusiasm after the DT intervention. When discussing the use of digital twins in science and engineering classes, for example, students can voice their thoughts on the topic. At present, encouraging signs include widespread excitement, ease of use, exploratory social interactions, and high levels of engagement. At the second level, the learning stage, we evaluate how much students learn from interacting with the digital twin.

The purpose of this simulation is to test students' conceptual understanding and application of the material presented. Tools like pre- and post-tests for DT greatly facilitate the evaluation of learning progress. Test scores, problem-solving skills, and students' understanding of complex ideas can all benefit from using digital twins to study the behavior of systems in virtual models. In the third level, the behavioral observation stage, the focus is on how students apply their new knowledge to practice and demonstrate social interactions resulting from DT. This level assesses the transfer of learning to real-world or simulated tasks. Students can practice system design, design thinking, problem solving, and social behavior with digital twins in a simulated environment before applying their new knowledge to real-world laboratory experiments or group projects. Improved performance on practical tasks and contributions to a cohesive team are signs of progress at this level.

The final level, outcomes, evaluates the overall effects of using digital twins in the classroom. This method can measure institutional goals such as graduation rates, employability (due to enhanced technical and social skills in creating a humanistic environment), and project creativity. Valid strategies for evaluating achievement at this level include long-term performance tracking, career progression analysis, and stakeholder feedback, such as input from academic boards or employers. We go about all four levels in a methodical and structured way to make sure that digital twin technology improves learning and leads to real, measurable results.

Kirkpatrick's four-level training evaluation model offers a comprehensive framework for assessing training effectiveness. The model consists of four levels: reaction, learning, behavior, and outcomes. The first level, reaction, gauges participants' immediate responses and satisfaction with the training, utilizing a digital twin during the learning process to observe social interactions and the instructor's implementation during the intervention. The learning level assesses the increase in knowledge, skills, and attitudes. The behavior level evaluates how participants demonstrate their social interactions in class through the digital twin intervention. Finally, the outcome level examines the impact on performance and learning outcomes.

The STELE approach adapts Kirkpatrick's model to a digital twin-enhanced learning environment. This adaptation encompasses four core principles: (1) Engagement (Reaction): The immersive digital twin environment enhances the user experience; (2) Knowledge Acquisition

(Learning): it uses interactive simulations and virtual labs to enhance the effectiveness of learning; (3) Collaboration (Behavior): it fosters social interaction through virtual discussions and discussion spaces; (4) Performance Improvement (Outcome): this process analyses data from the digital twin to optimize learning outcomes and performance.

The integration of digital twin technology into Indonesian higher education represents a transformative approach to fostering social interaction in Smart Technology-Enhanced Learning Environments (STELE). STELE's goals focus on increasing student engagement, enhancing knowledge retention, fostering collaborative learning, and evaluating institutional effectiveness. To achieve these goals, STELE employs a variety of strategies. We combine digital twins with current learning environments and requirements, create interactive lessons, make virtual tutoring and peer learning possible, and keep an eye on and evaluate learning analytics to make sure they keep getting better. The learning process for this kind of system can be broken down into several stages that work together to make the system easier to use and improve interaction (see Table 19.1) (Figs. 19.2 and 19.3).

3.3 Data analysis

Data analysis conducted statistical analysis using SPSS software. We used this software, which includes descriptive statistics and inferential tests such as t-tests, normality checks, and homogeneity checks, to evaluate the significance of differences between the two groups. We assess how digital twins and AI technology enhance the social and cultural science learning environment at universities, highlighting the subtleties of the STELE learning model. We analyzed the data using a combination of descriptive and inferential statistics to assess the effectiveness of the intervention. We used descriptive statistics, such as frequencies, standard deviations, and means, to describe and summarize the main patterns and variations in the data. We ran a series of tests prior to the inferential analysis to ensure the validity of the results. These tests included a normality test to validate the distribution of the data, a homogeneity test to evaluate equality of variances, and a t-test to compare the student's social interaction in the STELE class. This study used the SPSS IBM26 version for all calculations, ensuring accurate and reliable results. Findings were assessed using a stringent significance level of $\alpha = 0.05$, which allows for the determination of statistical significance. This method of analysis provides a comprehensive assessment of the impact of the intervention, revealing how well the digital twin with the STELE approach is working to improve social interaction.

4. Findings and discussion

The results of the data analysis show the significant impact of digital twins technology on students engagement, learning, and interaction with STELE approach at the university. Based on their experiences with digital twins, students felt more comfortable sharing ideas, working together to solve problems, and developed a greater sense of community in terms of engagement and interaction. Students were able to concentrate better, were more motivated, and achieved higher learning outcomes due to the immersive features and personalized feedback of the technology, which also increased their understanding of complex subjects, realism, and interaction in the learning experience.

Digital twins received positive feedback for their accessibility and usability due to minimal technological issues and straightforward instructions. Due to its ease of use, students felt confident in using the platform for future learning. In addition, the study showed that digital twins had a beneficial effect on social interaction and collaboration by helping students from diverse backgrounds work together more effectively on group projects, which in turn led to stronger social relationships, better communication, and more meaningful teamwork. Overall, these results highlight how digital twins' technology enhances engagement, teamwork, and learning in academic institutions.

Table 19.2 presents the descriptive statistics for the pre-test and post-test scores. The mean pre-test score was 47.43 ($SD = 13.97$), while the post-test mean score significantly increased to 84.97 ($SD = 10.94$). The increase in scores indicates that the implementation of digital twins with the STELE approach has a positive impact on students' social interaction and engagement. Before using Digital Twins with the STELE approach, students faced challenges in engaging with peers and instructors in an interactive manner, often experiencing limitations in collaborative activities and digital learning environments. The pre-test scores reflect these difficulties, indicating a lower level of participation and engagement. Many students relied on traditional learning methods, which restricted dynamic collaboration and real-time feedback.

Table 19.3 summarizes participants' responses across four sections: Engagement and Interaction, Learning Experience, Usability and Accessibility, and Social Interaction and Collaboration. Most students strongly agreed that Digital Twins improved their motivation (Mean = 4.19), enhanced collaboration (Mean = 4.46), and created a more immersive learning environment (Mean = 3.95). Similarly, usability and accessibility

TABLE 19.1 Learning flow of digital twins as a tool for optimizing social interaction in Indonesian higher education (STELE) approach.

Syntax	Key activities	Output
Preparation Phase	• Designing the Digital Twin Environment involves constructing a virtual representation of physical campus spaces or social scenarios to replicate real-world interactions. • Customizing Learning Goals: Identifying social interaction skills, *such as collaboration, communication, and teamwork, as key* competencies. • Technical Setup: Ensuring access to reliable networks, devices, and software tailored to STELE principles.	The STELE approach adapts Kirkpatrick's model to a digital twin-enhanced learning environment. This adaptation encompasses four core principles: **1.** Engagement (Reaction): The immersive digital twin environment *enhances the user experience.* **2.** Knowledge Acquisition (Learning) uses interactive simulations and virtual labs to enhance the effectiveness of learning. **3.** Collaboration (Behavior): It fosters social interaction through virtual discussions and discussion spaces. **4.** Performance Improvement (Outcome): This process analyses data from the digital twin to optimize learning outcomes and performance.
Simulation Phase	This stage involves utilizing the digital twin environment to engage students in interactive and simulated experiences. • Role-Playing Scenarios: Students participate in simulations of group discussions, cultural exchanges, or teamwork challenges within the digital twin. • Interactive Feedback Mechanisms: AI tools provide real-time feedback to enhance interpersonal and problem-solving skills. • Personalized interaction: the system adapts scenarios based on individual learning progress and preferences.	
Collaborative Engagement Phase	• Team-Based Activities: Students collaborate on tasks that require negotiation, brainstorming, and shared decision-making. • Cross-Cultural Interaction: The digital twin environment enables exchanges between diverse student groups across regions in Indonesia. • Mentorship Integration: Educators and peer mentors monitor and guide students during collaborative exercises.	
The reflection and assessment phase	• Self-reflection sessions: students analyze their performance and interpersonal dynamics within the digital twin environment. • Peer and instructor feedback: constructive input from peers and instructors helps identify areas for improvement. • Skill assessment: data analytics within the digital twin provide insights into communication efficiency, teamwork effectiveness, and social adaptability.	
Real-World Transfer Phase	• Practical applications: Students use their improved social and collaboration skills in live academic and professional settings. • Community engagement: Students' refined abilities support opportunities for them to participate in local or national projects. • Ongoing development: The digital twin system tracks long-term progress, ensuring sustained growth in social competencies.	

FIGURE 19.2 Learning experience for culture.

FIGURE 19.3 Usability and accessibility of digital twin with STELE approach.

TABLE 19.2 Descriptive statistics.

	N	Minimum	Maximum	Mean	Std. Deviation
Pretest	63	30.00	75.00	47.4286	13.97430
Post-test	63	60.00	100.00	84.9683	10.94557
Valid N (listwise)	63				

scored high, with students finding the system user-friendly (Mean = 4.25) and easy to navigate across devices (Mean = 3.82). Social interaction was also positively impacted, with an overall mean score of 4.26 across relevant items.

The Cronbach's Alpha shown coefficient for the questionnaire was 0.965, based on the data in Table 19.4 indicating high internal consistency and reliability of the instrument used to measure participants' perceptions of the STELE approach.

TABLE 19.3 Respondents' statements on STELE approach.

Respondents statements	N	Min	Max	Mean	Std. Deviation
Part 1: Engagement and Interaction The use of digital twins in learning increased my engagement with the course material 1. I felt more motivated to participate in class discussions when using digital twins as a learning medium and technology.	63	3.00	5.00	4.1905	0.75897
2. Digital twins improved my ability to interact with classmates and instructors effectively during the learning process.	63	3.00	5.00	4.4603	0.66782
3. I found it easier to collaborate with my peers while using Digital Twin technology.	63	3.00	5.00	4.2381	0.68895
4. Digital twins provided a more immersive, interactive, and meaningful learning environment compared to traditional methods.	63	3.00	5.00	3.9524	0.77102
5. I was more motivated to learn using digital twin technology.	63	3.00	5.00	4.3333	0.64758
Part 2: Learning Experience					
6. Digital twins with the STELE approach helped me improve my understanding of complex concepts.	63	3.00	5.00	4.2222	0.65855
7. The interactive nature of digital twins made learning more fun and engaging.	63	3.00	5.00	4.2540	0.78223
8. I retained information better when using digital twins in my courses.	63	3.00	5.00	4.0952	0.71198
9. Digital twins helped bridge the gap between theory and practice in my studies efficiently.	63	3.00	5.00	4.0635	0.73776
10. I believed that digital twins offered a customized learning experience that met my specific learning requirements.	63	3.00	5.00	4.3651	0.70257
Section 3: Usability and Accessibility					
11. The Digital Twin system is user-friendly; the features are attractive and simple to navigate.	63	3.00	5.00	4.2540	0.73984
12. I can access the Digital Twin-based technology platform without significant technical difficulties.	63	3.00	5.00	4.4921	0.64441
13. The integration of digital twins with the STELE approach into my learning environment is smooth.	63	3.00	5.00	4.2222	0.65855
14. I feel that the Digital Twins technology platform is accessible on various devices (e.g., laptops, mobile phones, tablets).	63	3.00	5.00	3.8254	0.70801
15. I receive adequate technical support from teachers and classmates when using digital twins in my learning activities.	63	3.00	5.00	4.4921	0.64441
Section 4: Impact on Social Interaction and Collaboration					
16. The digital twins-based STELE approach is very capable of facilitating better communication with my peers.	63	3.00	5.00	4.2698	0.67696
17. The use of digital twins encourages teamwork, building teamwork and active collaboration.	63	3.00	5.00	4.2381	0.73428
18. I feel more connected to my classmates when using the Digital Twins based on the STELE approach.	63	3.00	5.00	4.2381	0.73428
19. Digital Twins technology improves my ability to work on group projects effectively and practically.	63	3.00	5.00	4.2698	0.74501
20. I believe digital twins with the STELE approach can improve overall social interactions in a fun academic environment.	63	3.00	5.00	4.4921	0.71556
Valid N (listwise)	63				

TABLE 19.4 Reliability test.

Cronbach's Alpha	N of items
0.965	20

The results of the one-sample t-test (Table 19.5) show a significant difference between the pre-test and post-test scores ($t = 61.62$, $P < .001$). This confirms the effectiveness of the STELE approach in improving students' learning experiences and engagement, which have demonstrated that digital learning interventions significantly enhance students' academic performance and social interactions. This finding emphasizes that digital learning tools provide personalized, adaptive learning experiences that foster deeper engagement with and retention of knowledge. The big jump in post-test scores in this study supports these results, showing that Digital Twins technology, when added to the STELE method, improves educational outcomes and creates more interactive learning environments that help students learn things for a long time. This supports the bigger implications of these results in higher education settings.

The ANOVA test (Table 19.6) reveals a significant difference between pre-test and post-test scores ($F = 4.275$, $P < .001$), further validating the hypothesis that Digital Twins technology enhances social interaction and collaboration in higher education. This result is consistent with findings indicates that technology-enhanced learning environments foster engagement and effective communication among students, leading to improved teamwork and knowledge sharing.

The significant difference in ANOVA results suggests that the implementation of Digital Twins within the STELE approach positively influences students' ability to interact, collaborate, and engage in meaningful academic discussions. These findings reinforce the effectiveness of digital learning tools in higher education and support the integration of Digital Twins as a strategy for optimizing social interaction in academic settings.

The QQ plot and histogram (Fig. 19.4) illustrate the normal distribution of post-test scores, confirming that the data follows a normal pattern and is suitable for parametric statistical analysis. The QQ plot visually demonstrates that the observed data points closely follow the expected normal distribution line, indicating that the data is symmetrically distributed and free from extreme deviations. The histogram further supports this by showing a bell-shaped curve, reinforcing the assumption of normality. These findings validate the use of parametric tests such as t-tests and ANOVA for statistical analysis, ensuring the reliability of the study's results. By ensuring that the post-test scores are normally distributed, this study guarantees the robustness and validity of its inferential statistical analysis.

5. Discussion

After implementing digital twins with the STELE approach, students were able to participate in more immersive and interactive learning experiences, which

TABLE 19.5 One sample test.

	Test value = 0					
					95% confidence Interval of the difference	
	t	df	Sig. (2-tailed)	Mean difference	Lower	Upper
Pretest	26.939	62	0.000	47.42857	43.9092	50.9480
Post-test	61.615	62	0.000	84.96825	82.2116	87.7249

TABLE 19.6 ANOVA.

	Sum of Squares	df	Mean Square	F	Sig.
Between Groups	9145.179	26	351.738	4.275	0.000
Within Groups	2962.250	36	82.285		
Total	12107.429	62			

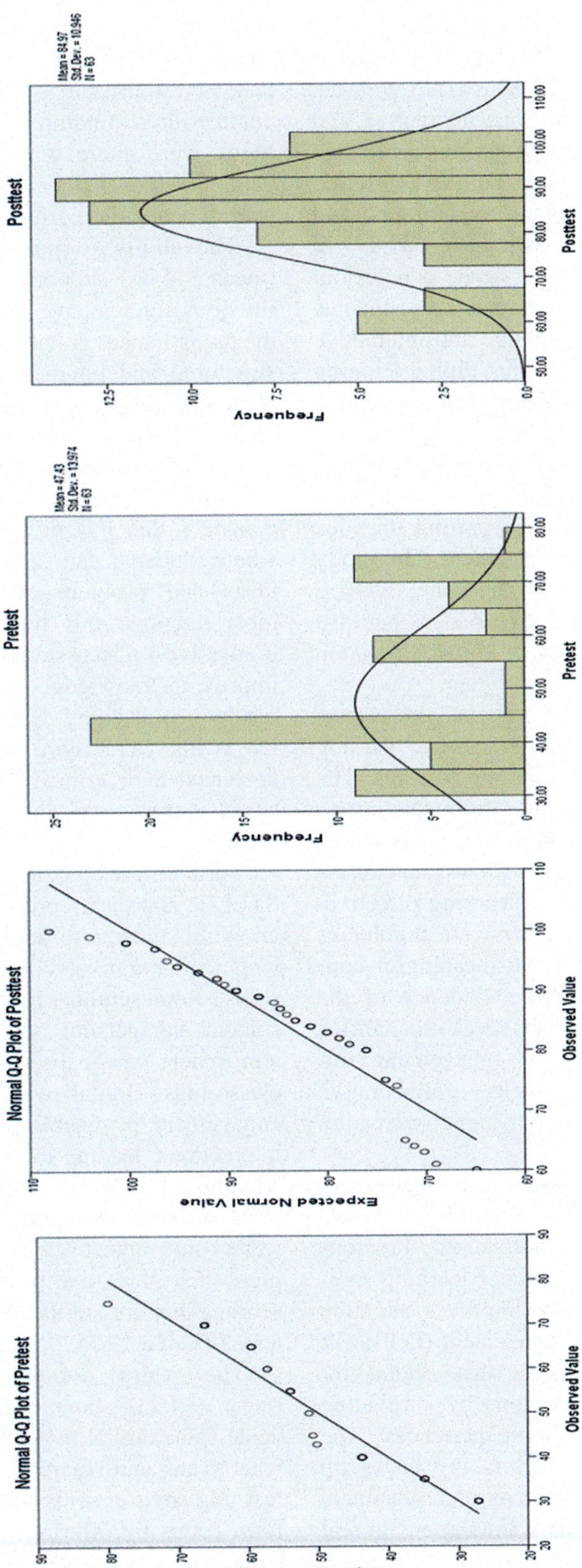

FIGURE 19.4 Q-Q plot and histogram of digital twin with STELE approach.

enhanced social interaction. The technology allowed real-time collaboration, enhanced peer-to-peer interactions, and provided a platform where students could engage in virtual teamwork and simulations. The big jump in post-test scores shows that using Digital Twins (DT) with the STELE method created a more interesting, interactive, and socially connected learning environment. This is in line with other research that shows how effective digital learning tools are at improving student collaboration and academic performance. Previous research (Wang & Huang, 2025) also came to the same conclusion: technology-enhanced learning environments help students do better by giving them more immersive and interactive experiences. In this study highlighted that digital learning tools help bridge the gap between theoretical knowledge and practical application, leading to a more comprehensive understanding of content, and it was and found that students in digitally enhanced classrooms demonstrated greater motivation and participation, supporting the idea digital twins significantly enhance engagement. This study confirms these results, reinforcing that the STELE approach not only improves learning experiences but also fosters collaboration and adaptability in higher education settings.

Table 19.1 (statement num. 1) highlights the importance of digital tools in promoting interactive learning experiences that enhance motivation among students. The usability and accessibility ratings in this study also mirror findings by Mohanty et al. (2024), who found that technology-driven platforms provide seamless and user-friendly experiences that contribute to learning effectiveness. This study also showed that immersive technologies help people remember things and have meaningful conversations with others, which is more evidence of the STELE approach's positive effects. This research confirms digital twin wins: technology fosters engagement, improves accessibility, and strengthens collaborative learning, making it an effective tool for higher education settings. Respondent's.

The use of digital learning tools increased engagement (mean 4.19), which aligns with the role of interactive technology in boosting student motivation. Previous studies have shown that the use of digital learning tools increases active student participation, improves attention span, and fosters deeper cognitive engagement (Dritsas & Trigka, 2025). This finding suggests that digital interactions enhance student motivation by providing personalized and immersive learning experiences. The STELE approach's Digital Twins technology backs up these results by creating an interactive and adaptable learning space where students can play simulations in real time, work together virtually, and get feedback right away. This indicates that digital twins improve learning outcomes by replicating real-world scenarios, allowing students to practice decision-making and problem-solving in controlled environments. In addition, Huang et al. (Ning et al., 2024) found that Digital Twins technology enhances student engagement by offering a more immersive and hands-on approach to learning. The results of this study confirm these findings, showing that students use digital twins were more actively involved in learning tasks, demonstrating higher motivation levels, and improving overall academic performance.

The ability to participate in discussions effectively (mean = 4.46) supports theories of digital collaboration. Previous studies have found that digital learning environments enhance collaborative discussions by providing structured and interactive communication channels (Sun et al., 2025). This is consistent with findings from our study, where students reported an increased ability to engage in meaningful discourse and teamwork facilitated by Digital Twins technology. Digital Twins technology enhances this by providing an interactive digital space where students can collaborate in real-time, exchanging ideas and problem-solving collectively. DigiTwinwins tools facilitate this by allowing students to engage in immersive discussions and peer-to-peer learning, improving knowledge retention and comprehension. The results of this study demonstrated that students using digital twins were more actively involved in discussions, enhanced their critical thinking skills, and built stronger social connections that contributed to their academic growth.

Furthermore, Digital Twins technology within the STELE approach promotes real-time communication, allowing students to actively exchange ideas, solve problems collaboratively, and refines their critical thinking skills. Representing digital collaborative tools enhances student engagement by fostering an interactive learning atmosphere where discussions are not limited to physical classrooms. Digital twins (DT) contribute to deeper peer engagement by enabling synchronous and asynchronous interactions, leading to improved teamwork and problem-solving abilities. The study's results support these points of view. They show that students who used digital twins were better able to contribute to discussions, clearly express their ideas, and work together in dynamic classroom settings that are similar to professional work settings with teams (Table 19.3).

The third statement from two respondents (mean = 4.23) about peer collaboration shows that students value digital twins for teamwork. This is in line with what Wang and Huang (2025) found, which is that digital learning environments can help students work together and form teams by offering interactive and real-time engagement opportunities. This is demonstrated by the fact that students working in digital environments tend to engage in deeper, more meaningful collaborative experiences.

Similarly, this study found that digital learning platforms enhance students' ability to communicate and collaborate effectively by providing structured spaces for discussion and problem-solving. All of these results show that digital twins not only help people work together, but they also help them learn important soft skills like communication, flexibility, and problem-solving, which are necessary for success in school and the workplace.

The next sentence supports immersive learning experiences and shows how quickly digital technologies can mimic real-life learning environments. This is emphasized by the fact that immersive technologies offer hands-on learning opportunities that keep students interested and help them understand. This study similar with Prokopenko et al. (2024) found that digital simulations provide a hands-on approach to learning, which bridges the gap between theoretical knowledge and practical application. The current study aligns with these findings, showing digital twins create an interactive and dynamic learning environment that encourages active participation and deep understanding of subject matter.

DT, using the STELE approach, demonstrated a significant increase in motivation (mean 4.33), thereby validating that digital environments enhance learning enthusiasm. Using technology-enhanced learning environments creates engaging and immersive experiences that sustain student motivation. In addition, digital tools make personalized learning easier, which promotes active participation and deep cognitive engagement in line with Hallinger et al. (2025), who showed that digital simulations boost students' intrinsic motivation by connecting what they learn in the classroom with what they can do in real life. This study backs up these results, showing that the STELE approach based on Digital Twins motivates students by making learning more interactive and student-cantered, which improves both interest and memory (see statement number five on Table 19.3).

This study shows an improved understanding of complex concepts (mean = 4.22) that aligns with adaptive learning theories. These findings have shown that adaptive learning environments personalize instruction by adjusting to students' needs and pacing, leading to deeper comprehension and retention. Roeben et al. (2025) did similar research and found that digital tools improve cognitive engagement by simulating real-world applications. This closes the gap between theoretical knowledge and practical experience. Based on these results, the STELE approach, based on digital twins, seems to be a good way to help college students better understand concepts, have more personalized and interactive learning experiences, and improve their cognitive skills.

Most of the students claim they are feeling enjoyment of interactive learning (Mean = 4.25) confirms the engagement potential of technology-enhanced environments. The previous studies have highlighted that digital tools significantly enhance student engagement by providing interactive and dynamic learning experiences. It's found that students in technology-enhanced learning environments showed greater enthusiasm and commitment to their studies. This study also demonstrated that the integration of digital simulations and virtual learning environments fosters a more immersive learning atmosphere, leading to increased motivation, and participation. Interactive digital learning promotes higher levels of cognitive engagement by allowing students to actively manipulate and interact with educational content by reinforcing that Digital Twins within the STELE approach create an engaging and participatory learning environment that significantly contributes to students' academic experiences.

Respondents statement about the impact of DT tools in higher information retention (Mean = 4.09) has demonstrated that reducing cognitive overload through structured digital learning environments improves students' ability to retain information. For instance, these statements claim that learners who engaged with well-organized, interactive digital content had higher retention rates compared to those using traditional instructional methods. Similarly, Gkintoni et al. (2025) noted that adaptive digital learning platforms help students manage cognitive load effectively, leading to better comprehension and long-term memory consolidation. This study's findings align with these previous studies, highlighting that Digital Twins technology, by structuring information in an interactive and personalized manner, optimizes cognitive processing and enhances information retention.

According to the study finding regarding bridging theory and practice (Mean = 4.06) reflects findings from AI-powered learning system studies have highlighted that integrating AI-driven digital learning tools effectively connects theoretical concepts with practical applications, enhancing student comprehension and problem-solving skills. Digital simulations allow students to experiment with real-world scenarios in a controlled environment, reinforcing their understanding of subject matter. This study supports these findings, showing that Digital Twins technology, through the STELE approach, serves as a bridge between conceptual learning and hands-on experience, thereby fostering a deeper understanding and improving student engagement in higher education.

On statement number 10 shown the data of customized learning (Mean = 4.36) aligns with the principles of personalized digital education. The previous studies (Enstroem & Bhawna, 2025) have shown that students in adaptive learning environments demonstrated higher levels of autonomy and engagement, as the content was adjusted based on their progress and performance. This statement highlighted that digital learning tools utilizing artificial intelligence and machine learning enable customized instruction that accommodates diverse learning paces, ultimately improving comprehension and retention.

System usability of digital twin in this study (Mean = 4.25) corroborates user-centered design studies. Previous research has demonstrated that user-centered design enhances students' ability to navigate digital learning environments efficiently, leading to better learning experiences and outcomes. These findings emphasize that intuitive and accessible technology fosters engagement by reducing cognitive load and allowing users to focus on the learning process. In addition, these findings, demonstrating that Digital Twins within the STELE approach provide a seamless, user-friendly interface that supports effective learning and social interaction in higher education. Mostly respondents' response digital twin platform minimal technical difficulties category (Mean = 4.49) highlights robust infrastructure supporting digital adoption and emphasized the importance of reliable digital infrastructure in facilitating seamless technology integration in education. Similarly, this study aligns with Dahri et al. (2025) state institutions with strong digital infrastructure experience fewer technical disruptions, leading to higher levels of student engagement and satisfaction. The ease of access and minimal technological barriers significantly enhance students' ability to engage with digital learning tools. This study supports these findings, as the high mean score suggests that the Digital Twins platform, under the STELE approach, provided a stable and user-friendly learning experience, reducing frustration and allowing students to focus more on collaboration and knowledge acquisition.

This study findings seamless integration into learning (Mean = 4.22) supports ease of implementation research. Digital learning tools must integrate seamlessly into existing educational structures to maximize their effectiveness and demonstrated that smooth integration of digital platforms enhances adoption rates, engagement, and overall academic performance. Multi-device accessibility (Mean = 3.82) suggests adaptability of digital tools. Prior studies have emphasized the importance of technological flexibility in education, with research by Bashir and Lapshun (2025) highlighting that students benefit from learning platforms that can be accessed across multiple devices. This adaptability ensures seamless engagement regardless of device type, improving student participation and retention. Multi-device compatibility enhances digital literacy, allowing students to develop essential technology navigation skills that are crucial for academic and professional success. These findings align with the current study, demonstrating that Digital Twins technology within the STELE approach fosters a more inclusive and accessible learning environment by enabling students to switch between devices without disruption.

In this study adequate technical support (Mean = 4.49) enhances overall learning efficiency. The previous studies have emphasized that technical support plays a crucial role in the successful implementation of digital learning environments. Students who receive consistent technical support experience fewer disruptions and are more likely to engage effectively with learning platforms. Similarly, Ratliff (2025) found that digital learning environments with reliable support systems improve user satisfaction and reduce technology-related frustration, leading to better learning outcomes. Its mean institutions investing in structured technical assistance report higher student performance and digital tool adoption rates. These findings align with this study, demonstrating that the STELE approach, supported by a strong technical assistance framework, ensures a smoother learning experience and optimizes students' engagement with Digital Twins technology. On Table 19.3 data shown by using digital twin with STELE approach improved peer communication (Mean = 4.26) aligns with collaborative learning theories. The previous research has shown that digital learning environments significantly enhance students' ability to communicate effectively in collaborative settings. Technology-integrated learning fosters interaction by providing structured spaces for discussion and group work. This study found that virtual learning environments create opportunities for meaningful peer interaction, leading to the development of essential communication skills. Digital platforms encourage active participation and engagement, improving students' ability to articulate their ideas and collaborate effectively. These findings align with the present study, demonstrating that Digital Twins technology within the STELE approach enhances peer communication by offering a dynamic, interactive learning environment that facilitates productive academic discussions and teamwork.

Besides that, positive statement shown on encouraging teamwork (Mean = 4.26) supports the role of digital tools in fostering active learning. Previous studies have shown that digital tools provide collaborative learning opportunities by enabling real-time interaction, structured teamwork, and engagement in virtual learning spaces. Technology-enhanced learning environments create an interactive atmosphere that promotes shared problem-solving, and cooperation digital learning platforms offer structured teamwork activities that improve students' ability to collaborate effectively. Incorporating digital tools into group learning activities enhances student motivation and active participation. The findings of this study align with these previous works, demonstrating that the integration of Digital Twins within the STELE approach fosters teamwork, improves collaborative problem-solving skills, and supports a more engaging and dynamic learning environment (Statement number 17 Table 19.3).

Social interaction shows increased connection with classmates (Mean = 4.23) indicates a shift toward socially immersive learning. Also have demonstrated that digital learning environments significantly enhance social presence and interpersonal relationships among students. These findings shown that interactive technology fosters student engagement by creating opportunities for meaningful interactions in virtual spaces and student in digital learning settings experienced stronger social bonds and collaborative learning benefits, emphasized that technology-enhanced communication tools encourage teamwork and peer support, contributing to a more immersive educational experience. These findings align with the current study, suggesting that Digital Twins technology within the STELE approach provides a dynamic platform for fostering social connections, enhancing peer interaction, and promoting a sense of belonging in higher education settings (Statement number 18).

This research described the students enhanced group project effectiveness (Mean = 4.23) aligns with modern educational frameworks. Previous studies have demonstrated that digital tools improve collaborative learning experiences by providing structured platforms for teamwork and real-time collaboration. Based on statement number 19 (Table 19.3) found that technology-enhanced learning environments encourage students to take active roles in group projects, leading to improved communication and problem-solving skills and digital collaboration tools facilitate task distribution, progress tracking, feedback exchange, resulting in more effective project outcomes and emphasized that digital learning platforms contribute to enhanced critical thinking and innovation in group work. These findings support the present study, demonstrating that Digital Twins within the STELE approach significantly improve students' ability to manage group projects, leading to increased efficiency, productivity, and overall learning effectiveness.

Overall, the data in Table 19.3 improved social interaction (Mean = 4.26) affirms the role of Digital Twins (DT) in building meaningful academic collaboration. Prior research supports the notion that technology-enhanced learning environments foster increased student interaction and collaboration by Bouchrika, 2026 found that digital tools provide students with structured opportunities for communication, leading to stronger academic relationships and knowledge sharing. Digital platforms create interactive spaces where students can engage in discussions, problem-solving, and collaborative projects, thereby enhancing social learning and demonstrated that virtual learning environments facilitate both synchronous and asynchronous interactions, allowing students to collaborate effectively regardless of time or location. These findings align with this study, indicating that Digital Twins within the STELE approach significantly improve students' ability to engage with peers, collaborate on academic tasks, and build meaningful academic relationships that contribute to their overall learning experience.

The study's findings indicate that the Digital Twins-based STELE approach significantly enhances engagement, collaboration, and learning experiences. Compared to traditional methods, students reported feeling more motivated and connected to their peers and highlights that technology-enhanced learning fosters student collaboration by providing structured digital spaces for teamwork and discussion. Examples of social interaction observed in this study include students engaging in virtual simulations where they had to collaboratively solve real-world case studies, fostering critical thinking and teamwork. For instance, during group assignments, students used the Digital Twins platform to share and manipulate data in real time, allowing for meaningful discussions and decision-making processes that mirrored professional collaborative settings. In addition, peer-to-peer mentoring was observed, where more experienced students guided their peers in navigating digital tools, reinforcing cooperative learning, enhance interpersonal interactions and collaborative problem-solving skills. The STELE approach, through Digital Twins, not only improved knowledge retention but also cultivated a socially dynamic and interactive learning atmosphere essential for the AI-driven educational era.

The effectiveness of the STELE approach is further supported by findings from Bekteshi (2025), who demonstrated that immersive digital platforms could enhance student engagement and retention rates. The integration of Digital Twins with the STELE approach not only improved social interaction but also created a more personalized learning experience. AI-powered learning platforms can bridge the gap between theoretical knowledge and practical application. AI-driven education fosters greater adaptability and inclusivity by allowing students to engage with tailored learning experiences that suit their needs and preferences. The use of AI-powered digital environments has also been found to encourage social interaction, to navigate complex academic and professional settings. The combination of these findings supports the STELE approach as an effective framework for utilizing Digital Twins to enhance social interaction and collaborative learning in the AI era.

Smart education platforms offer personalized learning pathways, enhancing students' engagement and self-efficacy. The high level of accessibility and support provided by the Digital Twins platform in this study contributes to student confidence in utilizing digital tools, ensuring that they can effectively navigate and optimize their learning experience. Furthermore, this study indicate that Digital Twin technology has been successfully implemented in higher education to create interactive and real-time feedback environments, fostering enhanced student participation and demonstrates that Digital Twins allow for predictive learning analytics, enabling educators to tailor instruction based on student engagement patterns, leading to improved learning outcomes.

6. Conclusion

This study highlights the transformative role of digital twins in enhancing social interaction and learning experiences in higher education. The findings suggest that integrating digital twins into university curricula can significantly improve engagement, collaboration, and usability. Digital twins offer an immersive and interactive environment that fosters active learning, enhances cognitive retention, and supports meaningful social interactions among students. The empirical evidence from this study aligns with the previous research, which demonstrates the potential of digital learning technologies in improving educational outcomes. The study also adds to the body of research on educational technology by proving that digital twins can be a useful teaching aid, connecting what you learn in the classroom with what you do in real life. The interactive and adaptive nature of digital twins ensures a more personalized learning experience that caters to different learning styles and needs.

This study suggests that digital twins can revolutionize the way people learn at university by enhancing social interactions such as participation, engagement, and teamwork in learning. The new thing about this study is that it uses digital twins along with the STELE learning method, which combines immersive digital environments with AI-powered adaptive learning. Digital twins serve as a bridge between theoretical learning and practical applications, thereby enhancing students' motivation, engagement, and cognitive retention. This study provides empirical validation of this claim, marking a significant advancement in the field. This study's contribution goes beyond the realm of educational technology, proving that digital twins make learning more dynamic and socially interactive and can serve as a model for higher education on a global scale. The study also shows that making real-time collaborative simulations is a beneficial way to improve important 21st-century skills like problem-solving, collaboration, and communication. Future scopes could include the integration of advanced AI-driven analytics to further personalize learning experiences; conducting longitudinal analyses to measure long-term consequences; and expanding research across educational disciplines. We aim to further enhance immersive learning and provide more engaging educational experiences for students. The implications of this research for curriculum development and policy related to higher education are profound. The success of Digital Twin implementations depends on universities prioritizing spending on digital infrastructure and teacher development initiatives. In addition, to create more student-centered and flexible learning environments, educators must integrate digital twin-based methodologies into their pedagogical tactics.

References

Al Yakin, A. M., Rasyid, A. R. R., Al-Matari, A. S., Muthmainnah, M. M., Obaid, A. J., Cardoso, L. M., & Elngar, A. A. (2025). Transforming civic education with ChatGPT to boost student interaction and social learning in the GenAI era. In *Practical applications of machine learning and AI: Medicine, environmental science, transportation, and education* (pp. 305–332). IGI Global Scientific Publishing.

Al-Mawadieh, R. S., Al-Abdallat, M. F., Injadat, M. N., Zuod, R., & Binsaddig, R. (December 2024). Exploring digital twins in higher education: Redefining educational technology in Jordanian universities. In *2024 25th International Arab Conference on Information Technology (ACIT)* (pp. 1–11). IEEE.

Alikhanova, M. (2025). Assessing the impact of various technologies on student learning outcomes in English language teaching. *Miasto Przyszłości, 56*, 38–41.

Aragón, A., Arquier, M., Tokdemir, O. B., Enfedaque, A., Alberti, M. G., Lieval, F., … Yagüe, Á. (2025). Seeking a definition of digital twins for construction and infrastructure management. *Applied Sciences, 15*(3), 1557.

Arar, K., Tlili, A., Schunka, L., Salha, S., & Saiti, A. (2025). Reimagining educational leadership and management through artificial intelligence: An integrative systematic review. In *Leadership and policy in schools* (pp. 1–23).

Bashir, S., & Lapshun, A. L. (2025). E-learning future trends in higher education in the 2020s and beyond. *Cogent Education, 12*(1), 2445331.

Bekteshi, L. (2025). Education in the era of AI and immersive technologies: A systematic review. *Journal of Research in Engineering and Computer Sciences, 3*(1), 01–14.

Bouchrika, I. (2026). Social learning theory & its modern application in education in 2025. *research.com*. https://research.com/education/social-learning-theory.

Bucci, I., Fani, V., & Bandinelli, R. (2024). Towards human-centric manufacturing: Exploring the role of human digital twins in industry 5.0. *Sustainability, 17*(1), 129.

Casaletto, P. (2024). *Digital twin application for dynamic speed regulation in collaborative robots*. Politecnico di Torino: Doctoral dissertation.

Chen, C. H., & Syu, J. Y. (2024). Effects of integrating a role-playing game into a virtual reality-based learning approach on students' perceptions of immersion, self-efficacy, learning motivation and achievements. *British Journal of Educational Technology, 55*(5), 2339–2356.

Chen, Y., Zheng, Y., Xue, T., Lin, W., Wen, J., & Chen, G. (2024). Harnessing the synergy of real teacher, digital twin, and AI in blended metaverse learning environment: A catalyst for medical education reforms. In *Digital twin, and AI in blended metaverse learning environment: A catalyst for medical education reforms*. . (Accessed 2 March 2024).

Cheng, A., Zhang, W., Feng, A., Wu, Y., & Li, W. (2025). Exploring psychological dimensions of augmented reality in education: A study on learning motivation and achievement in museums. *Frontiers in Psychology, 16*, 1514117.

Dahri, A. S., Memon, S. B., & Syed, S. (2025). Digital infrastructure and connectivity: The backbone of modern civilization. In *Corporate social responsibility approaches to ethical AI in business* (pp. 225–246). IGI Global Scientific Publishing.

Dai, W., & Kang, Q. (2025). Improvement of flipped classroom teaching in colleges and universities based on virtual reality assisted by deep learning. *Scientific Reports, 15*(1), 3204.

Davey, C. (2024). Digital twin cities: An instrument for pedagogical change. In *Teaching architecture (s) in the post-covid era* (pp. 25–55). Routledge.

Don, M. G., Wanasinghe, T. R., Gosine, R. G., & Warrian, P. J. (2025). Digital twins and enabling technology applications in mining: Research trends, opportunities, and challenges. *IEEE Access, 13*, 6945–6963. https://doi.org/10.1109/ACCESS.2025.3526881

Dritsas, E., & Trigka, M. (2025). Methodological and technological advancements in E-Learning. *Information, 16*(1), 56.

Elbadawy, H., & Farouk, A. (2025). Implementation of virtual reality technology in architecture field, and education: A review. *Archives of Computational Methods in Engineering*, 1–9.

Emihovich, B. (2024). Game-based learning and systems thinking: An innovative instructional approach for the 21st century. *TechTrends, 68*(1), 174–185.

Enstroem, R., & Bhawna, B. (2025). AI and VR: Shaping the next generation of adaptive learning and development programmes. In *The future of HRM in a world of persistent virtual reality* (pp. 141–163). Emerald Publishing Limited.

Gündüz, A. Y. (2024). Technology leadership in schools after COVID-19: A guide to equity, innovation, and collaboration. In *New perspectives for leadership after the COVID-19 pandemic* (pp. 249–263). Apple Academic Press.

Gesthuizen, R., Tan, H., & Kidman, G. (2025). Inspired to build understanding: Learning about sustainability and environmental education from within a virtual world. *International Research in Geographical & Environmental Education, 34*(1), 25–42.

Gkintoni, E., Antonopoulou, H., Sortwell, A., & Halkiopoulos, C. (2025). Challenging cognitive load theory: The role of educational neuroscience and artificial intelligence in redefining learning efficacy. *Brain Sciences, 15*(2), 203.

Guinea-Cabrera, M. A., & Holgado-Terriza, J. A. (2024). Digital twins in software engineering—a systematic literature review and vision. *Applied Sciences, 14*(3), 977.

Haider, S. M., Fatima, N., & Abbas, M. H. (2025). Perception of simulation and Virtual Reality (VR) in surgery in the public sector tertiary care teaching hospitals of Southern Punjab, Pakistan: A Cross-Sectional Survey. *Cureus, 17*(2).

Hakiri, A., Gokhale, A., Yahia, S. B., & Mellouli, N. (2024). A comprehensive survey on digital twin for future networks and emerging internet of things industry. *Computer Networks*, 110350.

Hallinger, P., Banoğlu, K., & Gümüş, S. (2025). Scoping review of research on digital simulations and serious games in educational leadership and management. *Leadership and Policy in Schools, 24*(1), 178–199.

Hashim, M. A. M., Tlemsani, I., Mason-Jones, R., Matthews, R., & Ndrecaj, V. (2024). Higher education via the lens of industry 5.0: Strategy and perspective. *Social Sciences & Humanities Open, 9*, 100828.

Hermosa, J. P., & Andal, E. Z. (2025). A holistic approach to improving students' university experience: Exploring philosophical orientation, transformational teaching methods, and flexible learning. In *Emerging technologies transforming higher education: Instructional design and student success* (pp. 75–96). IGI Global.

Hu, Y. (2025). Creating interactive AI learning environments: Integrating interdisciplinary approaches and Taoist art principles. *Interactive Learning Environments*, 1–20.

Hu, Z. (2025). Virtual reality-based cooperative learning in physical education: Ways to improve physical performance and motivation of students. *International Journal of Human-Computer Interaction, 41* (1), 242–253.

Hu, Y., Lin, J., & Lipson, H. (2025). Teaching robots to build simulations of themselves. *Nature Machine Intelligence*, 1–11.

Karataş, F., & Ataç, B. A. (2024). When TPACK meets artificial intelligence: Analyzing TPACK and AI-TPACK components through structural equation modelling. *Education and Information Technologies*, 1–26.

Katsantonis, M. N. (2025). From pandemic legacy to serious games: A systematic review of cooperative board games under the educational perspective. *European Journal of Education, 60*(1), e70048.

Kocsis, I., Burján-Mosoni, B., & Balajti, I. (2025). A comprehensive review of key cyber-physical systems, and assessment of their education challenges. *IEEE Access, 13*, 8898–8911. https://doi.org/10.1109/ACCESS.2025.3527714

Li, J., Zhang, X., & Liu, X. C. (2025). Exploring the motivational factors and learning outcomes of Virtual Reality educational games for middle school students: A unity 3D case study. *Interactive Learning Environments*, 1–21.

Marinelli, A., Iannacci, F., Papile, F., Diamanti, M. V., Sponchioni, M., & Del Curto, B. (2024). Higher education classroom of the future: An EU-funded project integrating physical lectures, virtual reality, and artificial intelligence. In *13th International Materials Education Symposium*. GBR, 38-38.

Max, A. L., Lukas, S., & Weitzel, H. (2024). The pedagogical makerspace: Learning opportunity and challenge for prospective teachers' growth of TPACK. *British Journal of Educational Technology, 55*(1), 208–230.

Miranda, F. M. D., Santos, B. V. D., Kristman, V. L., & Mininel, V. A. (2025). Employing Kirkpatrick's framework to evaluate nurse training: An integrative review. *Revista Latino-Americana de Enfermagem, 33*, e4431.

Mohanty, R. N., Thomas, A., & Kori, A. (2024). Next-generation innovative teaching ecosystems for futuristic management education. In *Global higher education practices in times of crisis: Questions for sustainability and digitalization* (pp. 5–28). Emerald Publishing Limited.

Muthmainnah, M., Muliati, M., Kaur, D., Misdi, M., Al Yakin, A., Apriani, E., & Kumar, V. V. (2025). The synergistic power of AIoT in enhancing EFL student CCT skills in higher education. In *Practical applications of machine learning and AI: Medicine, environmental science, transportation, and education* (pp. 277–304). IGI Global Scientific Publishing.

Nehru, R. S. S., Paredes, S., Roy, S. C., Cuong, T. Q., & Huong, B. T. T. (2025). Implementing the revised bloom's taxonomy (2001) in AI-Digital and online learning environments: A strategic approach. *Indian Journal of Educational Technology, 7*(1), 173–189.

Ning, G., Luo, H., Yin, W., & Zhang, Y. (2024). Exploration of the application and practice of digital twin technology in teaching driven by smart city construction. *Sustainability, 16*(23).

Omrany, H., Al-Obaidi, K. M., Ghaffarianhoseini, A., Chang, R. D., Park, C., & Rahimian, F. (2025). Digital twin technology for education, training and learning in construction industry: Implications for research and practice. *Engineering Construction and Architectural Management.* https://shura.shu.ac.uk/34699/12/Al-Obaidi-Digital-TwinTechnology%28AM%29.pdf.

Pandey, P. K., Mahajan, S., Pandey, P. K., Paul, J., & Iyer, S. (2025). Immersive learning trends using digital twins. *Digital Twins for Smart Cities and Villages*, 249–271.

Pandey, P. K., Pandey, P. K., Mahajan, S., Paul, J., & Iyer, S. (2025). Digital twin and virtual reality, augmented reality, and mixed reality. In *Digital twins for smart cities and villages* (pp. 273–293). Elsevier.

Plechatá, A., Morton, T., & Makransky, G. (2025). Promoting collective climate action and identification with environmentalists through social interaction and visual feedback in virtual reality. *Journal of Environmental Psychology*, 102526.

Prokopenko, O., Järvis, M., Bielialov, T., Omelyanenko, V., & Malheiro, T. (June 2024). The future of entrepreneurship: Bridging the innovation skills gap through digital learning. In *International conference innovation in engineering* (pp. 206–230). Cham: Springer Nature Switzerland.

Ratliff, J. (2025). *Connections between collaborative learning and technology-related distractions in high school: A quantitative quasi-experimental study (doctoral dissertation, American college of education).*

Razzaq, S., Shah, B., Iqbal, F., Ilyas, M., Maqbool, F., & Rocha, A. (2023). DeepClassRooms: a deep learning based digital twin framework for on-campus class rooms. *Neural Computing and Applications, 35*(11), 8017–8026.

Roeben, M., Vejvoda, J., Murböck, J., Fischer, F., Schultz-Pernice, F., Lohr, A., ... Heitzmann, N. (2025). Simulations in teacher education: Learning to diagnose cognitive engagement. *Education Sciences, 15* (3), 261.

Rye, S., Sousa, M., & Sousa, C. (2025). Introduction to game-based learning. In *Transformative learning through play: Analogue games as vehicles for educational innovation* (pp. 29–68). Cham: Springer Nature Switzerland.

Sankaran, S., Arun, M., & Kottaimalai, R. (2025). Virtualization of smart society 5.0 using artificial intelligence and virtual reality. *Networked Sensing Systems*, 297–321.

Savchuk, S., Vigo, V., Chidambaram, S., NuÑez, M., Anthony, D., Jansen, T., ... Fernandez-Miranda, J. C. (2025). Early neurosurgical education in the era of distance learning: Incorporating virtual reality and cadaveric specimen demonstrations. *World Neurosurgery, 193*, 706–714.

Selim, A., Ali, I., Saracevic, M., & Ristevski, B. (2024). Application of the digital twin model in higher education. *Multimedia Tools and Applications*, 1–18.

Sepasgozar, S. M., Khan, A. A., Shirowzhan, S., Romero, J. S. G., Pettit, C., Zhang, C., ... Liang, R. (2024). Immersive virtual environments and digital twin applications for education and training: Trends in construction, mining, and urban planning studies. In *Digital twin adoption and BIM-GIS implementation* (pp. 66–109). Routledge.

Sharma, S., Virani, S., Saini, J. R., & Rautela, S. (2025). Determinants of adoption of virtual reality as a teaching aid in higher education: The mediating role of hedonic motivation. *Journal of Applied Research in Higher Education, 18*(1), 298–312. https://doi.org/10.1108/JARHE-07-2024-0327

Sikora, Y., Chernykh, V., Shaforost, Y., Danylyuk, S., & Chemerys, I. (2024). Leveraging gamification and game-based technologies for educational purposes. *Multidisciplinary Reviews, 7.*

Singh, B., & Nayyar, A. (2025). Exploring diverse use cases of digital twins projecting digital transformation: Unlocking potential, addressing challenges and viable solutions. In *Digital twins for smart cities and villages* (pp. 631–655). Elsevier.

Stasolla, F., Curcio, E., Passaro, A., Di Gioia, M., Zullo, A., & Martini, E. (2025). Exploring the combination of serious games, social interactions, and virtual reality in adolescents with ASD: A scoping review. *Technologies, 13*(2), 76.

Stephan, R., Doery, A. R., & Simmons, C. (2025). Virtual reality tours as an immersive approach to archaeology in higher education. *Journal of Computer Applications in Archaeology, 8*(1).

Sun, D., Looi, C. K., Yang, Y., & Jia, F. (2025). Exploring students' learning performance in computer-supported collaborative learning environment during and after pandemic: Cognition and interaction. *British Journal of Educational Technology, 56*(1), 128–149.

Supriya, Y., Bhulakshmi, D., Bhattacharya, S., Gadekallu, T. R., Vyas, P., Kaluri, R., ... Mahmud, M. (2024). Industry 5.0 in smart education: Concepts, applications, challenges, opportunities, and future directions. *IEEE Access, 12*, 81938–81967. https://doi.org/10.1109/ACCESS.2024.3401473

Ton, Q. C., Hoang, A. D., Nguyen, T. L., Dau, T. H., & Nguyen, T. L. Y. (2025). The application of virtual reality and augmented reality (XR) technology: Challenges and opportunities for preschool and lower secondary education in Vietnam. In *Intelligent learning paradigm and student empowerment* (pp. 119–136).

Van Den Heever, M., Becker, E., Venter, L., & Bekker, J. F. (2024). Using machine learning and agent-based simulation to predict learner progress for the South African high school education system. *South African Journal of Industrial Engineering, 35*(3), 15–27.

Venice, J. A., Vettriselvan, R., Rajesh, D., Suresh, N. V., & Abirami, P. (2025). Enabling personalized learning and adaptive systems through strategic management: Cloud integration in education. In *Bridging academia and industry through cloud integration in education* (pp. 49–72). IGI Global Scientific Publishing.

Wang, D., & Huang, X. (2025). Transforming education through artificial intelligence and immersive technologies: Enhancing learning experiences. *Interactive Learning Environments*, 1–20.

Wang, M., Tong, J., & Lu, H. (2024). Digital twin technology of metaverse based on game engine in the campus application scenarios. *Academic Journal of Engineering and Technology Science, 7*(2), 157–163.

Wang, B., Zhou, H., Li, X., Yang, G., Zheng, P., Song, C., ... Wang, L. (2024). Human digital twin in the context of industry 5.0. *Robotics and Computer-Integrated Manufacturing, 85*, 102626.

Webb, L., Tokhi, O. M., & Alkan, B. (2024). State of the art and future directions of digital twin-enabled smart assembly automation in discrete manufacturing industries. *International Journal of Computer Integrated Manufacturing*, 1–35.

Xu, L. (2024). Navigating the educational landscape: The transformative power of smart classroom technology. *Journal of the Knowledge Economy*, 1–32.

Yamijala, S. M. S., Chodisetty, R. M., Chakravorty, C., & Sai, K. P. (2025). AI-Powered learning revolutionizing smart education with personalized learning styles. In *Internet of behavior-based computational intelligence for smart education systems* (pp. 191–212). IGI Global.

Yang, X., Wu, J., Ma, Y., Yu, J., Cao, H., Zeng, A., … Ren, Z. (2025). Effectiveness of virtual reality technology interventions in improving the social skills of children and adolescents with autism: Systematic review. *Journal of Medical Internet Research, 27*, e60845.

Yeganeh, L. N., Fenty, N. S., Chen, Y., Simpson, A., & Hatami, M. (2025). The future of education: A multi-layered metaverse classroom model for immersive and inclusive learning. *Future Internet, 17* (2), 63.

Ying, Z. (2025). Experience of intelligent speech robot in music online classroom based on deep learning and virtual reality. *Entertainment Computing, 52*, 100795.

Zammit, E., De Raffaele, C., Scerri, D., Aquilina, R., Calleja, J. J., & Rizzo, A. (2024). Driving innovations: Trends, prospects and challenges of implementing disruptive educational technologies within HEIs. *Advanced Technologies and the University of the Future*, 237–257.

Zhang, Y., Kinateder, M., Huang, X., & Warren, W. H. (2025). Modeling competing guidance on evacuation choices under time pressure using virtual reality and machine learning. *Expert Systems with Applications, 262*, 125582.

Zhu, Z., Zhu, W., Huang, J., & He, B. (2025). An intelligent monitoring system for robotic milling process based on transfer learning and digital twin. *Journal of Manufacturing Systems, 78*, 433–443.

Zuo, R., Wenling, L., & Xuemei, Z. (2025). Augmented reality and student motivation: A systematic review (2013-2024). *Journal of Computers for Science and Mathematics Learning, 2*(1), 38–52.

Chapter 20

Ethics in the Metaverse classroom: A case study for EFL education in Indonesia

Muthmainnah Muthmainnah[1], Sajed S. Ingilan[2], Luis Cardoso[3], Hamed Barjesteh[4], Ali Said Al Matari[5], Ahmad Al Yakin[1] and Ahmed J. Obaid[6]

[1]Universitas Al Asyariah Mandar, West Sulawesi, Indonesia; [2]University of Southeastern Philippines, Davao, Philippines; [3]Polytechnic Institute of Portalegre and Centre for Comparative Studies of the University of Lisbon, Lisbon, Portugal; [4]Islamic Azad University, Amol, Iran; [5]A'Sharqiyah University, Ibra, Oman; [6]Faculty of Computer Science and Mathematics, University of Kufa, Najaf, Iraq

1. Introduction

The introduction of the Metaverse has further broadened the potential for creative teaching and learning methodologies, further revolutionizing the area of education brought about by rapid advancements in technology. Regarding ethical concerns, the Metaverse, a virtual reality (VR) ecosystem, offers both possibilities and threats, especially in the areas of human-machine cooperation and the incorporation of digital twins into the field of education. By employing digital twins, incorporating the Metaverse into the classroom could improve student engagement, encourage collaborative learning, and offer personalized educational experiences (Revathy & Mangaiyarkkarasi, 2024). Educational institutions can make use of digital twins virtual representations of real-world items or systems to model complicated situations, enhance hands-on learning, and make data-driven decisions (Reethika & Kanaga Priya, 2024). Concerns about the digital gap, the effects of human-machine interactions in the classroom, and the security of user data are among the ethical issues brought up by this shift. Opportunities and challenges abound, since Metaverse-based educational techniques are still in their early phases of implementation, both in Indonesia and globally.

Investigated benefits of the Metaverse in the classroom include increased opportunities for student engagement, more immersive lessons, and the ability to overcome geographical obstacles. However, they have also highlighted the importance of addressing ethical concerns, including the development of appropriate regulations, teacher education, and the protection of student information and identities. Muthmainnah et al. stated that the Metaverse has mostly concentrated on its technological features and prospective advantages, and its implementation in educational settings remains in its infancy, both in Indonesia and around the world (Muthmainnah et al., 2024). With few educators and educational institutions having even heard of Metaverse-based technologies, its adoption in Indonesian classrooms is in its infancy. Jagatheesaperumal et al. have shown the versatility of this technology in supporting educational services, training, and skill development, and as a result, the use of the Metaverse in education is gaining popularity globally (Jagatheesaperumal et al., 2024).

The educational community is trying to find a balance between new technologies and upholding moral standards. However, the moral effects of this integration are still a big concern, especially when it comes to human-machine collaboration and the use of digital twins. This study looks at the moral issues related to using the Metaverse in the classroom. The study will examine the present status of Metaverse adoption in education worldwide and in Indonesia, pinpoint the most pressing ethical concerns, and offer a strategy to overcome these obstacles, all to promote the responsible and ethical use of Metaverse technology in classrooms. Indonesian educators and educational institutions are still in the early phases of embracing and implementing Metaverse-based technologies in the classroom. Studies have shown that the Metaverse may be a versatile tool for educational services, training, and skill development, and as a result, its use in education is on the rise around the world. However, the educational community must navigate the difficult task of balancing technological advancements with the upholding of ethical standards, and the ethical consequences of this integration

Digital Twins in the Smart Classroom. https://doi.org/10.1016/B978-0-443-41632-3.00018-7

pose a significant concern. With few educators and educational institutions having even heard of Metaverse-based technologies, its adoption in Indonesian classrooms is in its infancy (Muthmainnah et al., 2024).

Onu et al. have shown the versatility of this technology in supporting educational services, training, and skill development, and as a result, the use of the Metaverse in education is gaining popularity globally (Onu et al., 2024). There is a lack of understanding and use of the metaverse among Indonesian educators and educational institutions. The usage of the Metaverse to support training, education, and skill development is gaining popularity around the world. However, as the educational community strives to maintain a balance between technological advancement and ethical standards, the ethical implications of this integration remain a significant concern. This study, which emphasizes human-machine collaboration and the integration of digital twins, aims to explore the ethical considerations surrounding the implementation of the Metaverse in the classroom (Fadhel et al., 2024). As they strive to strike a balance between technical progress and the maintenance of ethical norms, the educational community must address the essential question of the ethical consequences of this integration.

With the revolutionary goal of Education 6.0 in mind, emphasizing digital, personalized, and interactive learning experiences, the concept of digital twins in the classroom is gaining significant momentum. By creating digital twins, which are exact digital representations of physical systems or items, educators can immerse students in realistic simulations of real-world problems. Digital twins, through the metaverse, can help students learn by doing and give teachers the tools they need to make data-driven decisions that will improve student learning (Lin et al., 2022). The use of digital twins in classrooms is booming in nations with well-developed IT systems around the world. In the US, for example, schools like Stanford and MIT use digital twins to provide students with lifelike simulations in fields like architecture, engineering, and medicine through the use of virtual models of complicated systems. Due to its focus on AI and VR in education, the US has invested in smart classrooms equipped with digital twins that can react to student interactions instantly (De Los Santos Melo & Beriguete Alcántara, 2024).

These efforts show how developed countries are embracing digital twin technology and using it in the classroom. On the other hand, because of economic and infrastructural constraints, digital twins in education have only just begun to gain traction in Indonesia. In order to improve learning without utilizing physical resources, Indonesian universities are investigating digital twin technologies in specific domains, including healthcare and engineering. But there are obstacles to wider deployment, such as undeveloped digital infrastructure in remote areas, a lack of comprehensive digital literacy, and slow internet and virtual reality device availability. According to Jagatheesaperumal et al., the Indonesian government has started programs to help with digital learning, but so far, they have only tackled the most fundamental forms of digitization (Jagatheesaperumal et al., 2024). Due to resource limitations, research on digital twin applications in technical education is being carried out by Indonesian institutes like *Institut Teknologi Bandung (ITB)*, although these initiatives are mostly isolated and limited to specific subjects.

The digital gap, a phenomenon where individuals in urban areas have greater access to technology than those in rural areas, hinders the widespread use of digital twins in Indonesia. This divide affects not only the digital literacy levels of educators and students, but also the infrastructure. Teachers' inability to make full use of digital twin technology due to digital literacy gaps mitigates the technology's potential to improve students' educational experiences. This stands in stark contrast to countries such as Australia, where the use of digital twins enables students in rural areas to access virtual labs and simulations, thereby creating equal opportunities regardless of their location. Furthermore, the legal systems in place to guarantee data protection and ethical standards in digital twin applications are more solid in industrialized economies.

Around the world, digital twins are changing the face of education by making it possible to implement the aims of Education 5.0–7.0 through interactive, personalized, and immersive learning environments by combining human-centered design with advanced AI technologies (Danylchenko-Cherniak, 2024). Students in nations that have the proper infrastructure and regulatory frameworks will have an easier time taking advantage of data-driven learning opportunities. While there is enthusiasm for digital twin technology among Indonesia's educational institutions, the country's educational institutions face formidable financial, regulatory, and infrastructure hurdles that prevent them from fully utilizing digital twins. For Indonesia to achieve the objectives of Education 5.0 to 7.0, it is crucial to close the digital divide, increase digital literacy, and create strong data protection rules. In addition to creating a more accessible and technologically advanced educational landscape, these steps would also allow Indonesia to learn from global practices and adapt digital twin technology to local educational needs (Allman et al., 2024; Zhang et al., 2024).

The use of digital twins within the metaverse in the field of English language instruction provides students with a realistic simulation of real-world contexts in which to exercise their language abilities. To help their students learn English, teachers can build virtual versions of popular airports or popular museums like New York and London (Muthmainnah et al., 2025). Students can then navigate these virtual streets, enter virtual buildings, and converse with nonplayer characters (NPCs) that act like native speakers. Students can practice asking for directions, placing food orders at cafes, and even going on virtual job interviews in this realistic virtual metropolis. According to Muthmainnah et al., one application of digital twins in the classroom is Virtual Language Immersion (VLI) travels, where students can take part in 'virtual travels' to practice conversational skills in simulated real-world settings (Muthmainnah et al., 2023). Students practice target vocabulary and phrases like "place an order" in a virtual city setting, while teachers lead them through scenarios like a restaurant visit. Teachers carefully craft each scenario to target specific grammar and vocabulary, and record interactions to provide feedback. Students can improve their vocabulary and cultural awareness through controlled contexts that allow them to practice context-based language use, such as role-playing and situational dialogues. Here, students act out real-life situations; for instance, one may play the part of a store clerk, while the others would be customers who need to use English for things like haggling or asking inquiries. The instructor gives immediate comments on students' terminology and usage as they work in groups within various sections of the digital doppelganger, including a marketplace, museum, airport, or castle.

Storytelling and creative writing assignments encourage students to express themselves via writing, by having them explore different areas of the virtual city, and then drawing from their virtual experiences to create stories (Babushkina & Kalugina, 2024). They might document their day in New York City by writing about their encounters with the digital twin, the places they visited, and the people they met. Using peer evaluation and group projects, the instructor helps students improve their narrative structure and descriptive language skills. Students can enhance their language skills with pronunciation and listening practice with AI-driven NPCs. These interactive characters answer student inquiries and offer constructive criticism on pronunciation, empowering them to speak English with greater assurance.

Instructors can go back over these exchanges later and offer targeted feedback. Students can also take part in simulated cultural events in the metaverse, which are interactive virtual events that allow them to experience the language and culture firsthand. They can practice their conversational skills and learn new idioms and cultural expressions by interacting with NPCs and fellow pupils (Kritsotaki, Castro–Kemp, & Kamenopoulou, 2025). Teachers then facilitate class discussions to increase students' cultural awareness and contextual language skills. Using digital twins in the metaverse to teach English has many advantages, including making language practice more interesting, relevant, and approachable for students. Teachers can provide personalized feedback on issues like pronunciation or grammar, a task that is often challenging in large classes. The gamification of virtual city tours motivates students to participate more, as it increases their sense of investment. In addition, students can gain valuable insight into cultural allusions, conversational norms, and gestures using digital twins, which is essential for their success in real-life English conversations (Filosofi et al., 2024). Teachers may give their student more than just a theoretical grounding in the language; with the help of digital twins, they can create authentic, immersive language experiences that go beyond what is possible in a typical classroom setting.

2. Related work

2.1 Metaverse ecosystem

The idea of online education has grown substantially outside of the realm of formal education and developments in ICT that have concentrated on virtual worlds in the last several decades (Vonitsanos et al., 2024). A three-dimensional virtual world (VW) is an online environment where users can engage in interactive virtual activities. They create virtual identities known as avatars in real time, enabling them to feel immersed in a VW without physically being there (Dad et al., 2024). They claim that virtual worlds allow users to communicate in real-time within an artificial setting with features and settings made by the designer. A novel computer medium called an avatar allows numerous people to view the same computer-generated space at the same time (Qiao et al., 2021).

Research by Ribis and García (2024) and others has shown that students' collaboration, language learning, and fostering creativity in learning can be enhanced when given more leeway to prepare assignments in the metaverse. Additionally, in order to create a more meaningful learning environment or to assess the meaningfulness of an existing one, it is crucial to offer a world direction. Ding and Cha (2024) mentioned that, in a virtual setting, participants' critical thinking skills and motivation can be enhanced by assigning them tasks. In this setting, the instructor chose activities based on how relevant they were to the course material, the participants' requirements, and how easily they could be implemented in the actual world. Participants can engage in a virtual environment without restrictions, making virtual worlds a useful tool in the educational context. Because they allow students to study in a real-world setting and improve their situational and experiential skills, VWs have numerous potential applications in the field of education, and Metaverses is one name for these online worlds.

Ha (2024) discovered that using OpenSim in the classroom increased students' intrinsic drive to study English. Küçük-Avcı et al. (2024) researched Quest Atlantis and found that individuals who participated in the virtual environment with real-life tasks showed an improvement in their spatial skills. Furthermore, the use of real language provided participants with the opportunity to enhance their reading and writing skills (Guan et al., 2024). Celestia is a gamified 3D virtual game for language learning (Li, 2024). Similarly, Stellarium may be used to build vocabulary and can be used in CBI to help teach a variety of embedded themes. As mentioned in the study by Kebeci (2024). Using placeholders in a 3D VW gives users an immersive experience of the VW while they do a variety of relevant tasks. The use of AR and MR in conjunction with VR to form the concept of extended or cross-reality (XR), sometimes referred to as the metaverse, can greatly enhance the educational experience for educators. When talking about data representation and projection, the term "metaverse" comes to mind (Dash et al., 2024).

According to Bhowmik (2024), augmented reality (AR) is a space that merges the real and virtual worlds through real-time interaction. Adding digital inputs and virtual features to the real environment, AR improves the physical experience. There are a number of research studies on the topic of AR in English language instruction in the current literature (Yi et al., 2024). In addition to being widely available via phones, AR typically does not need any further equipment like glasses or headsets, which offers advantages in terms of usage. Combining augmented and VR technologies is known as mixed reality. Like VR, MR allows users to physically interact with virtual objects in the actual world; however, it requires the use of MR glasses (Falola et al., 2025). One could argue that using MR in the classroom is a successful method of instruction (Gursesli et al., 2025).

Numerous studies on VR have been carried out in the context of teaching English and English language instruction; however, no research has been carried out under the umbrella term of the metaverse. As far as the author is aware, there have been few studies conducted on the metaverse in the past few years. Moreover, the existing studies either focus on the author's perspective on the subject (Ramanujam & Ismail, 2024; Ramanujam & Ismail, 2022) or explore the potential use of virtual worlds (VWs) in collaborative projects via the metaverse. The issue, though, is that the metaverse as used here only applies to the virtual worlds that go by that name. Research does not intend to imply the real meaning of the term "metaverse," which is a virtual reality environment that requires special goggles to fully experience. Because there has been little research on this novel approach to ESL instruction, the current study intends to fill that gap by surveying aspiring EFL teachers about their thoughts and feelings about using the metaverse in the classroom, as well as by exploring the feasibility of this concept and the potential results (Chan & Tong, 2025).

The term "metaverse" describes a future state that integrates various human activities with various forms of technology, including the Internet of Things (IoT), mobile devices, artificial intelligence (AI), robotics, and various forms of virtual and AR (Uddin et al., 2024). Currently, many immersive social and virtual experiences in STEM education and game-based learning (GBL) are possible in the metaverse, which offers a variety of human imagination and possibilities that enable new ways of doing business, commerce, and learning online. By connecting the real and virtual worlds in the metaverse, the "IoT" and "STEM" metaverses describe the targeted use of IoT and STEM applications in everyday life, respectively.

Digital assets remain visible and immutable without involving any third-party mediator using the recently developed blockchain technology, which is a peer-to-peer, decentralized, distributed ledger. Bitcoin stands out as an example of a cryptocurrency that uses blockchain technology. The use of blockchain technology has its drawbacks, including potential threats to the integrity and privacy of the system and the insecurity of (monetized) data. One such issue is the unethical and illegal withdrawal of bitcoins or other electronic currencies, which could lead to a financial crisis on a national, regional, or international scale (Samala et al., 2024). Information and communication technology (ICT) teachers are increasingly considering the legal and ethical implications of blockchain technology, which include privacy and security concerns, its environmental impact, potential job losses, and the risks associated with investing in metaverse schools (Chithra & Bhambri, 2025). Therefore, teachers and students must be careful to consider the many social injustices and underlying ethical and legal issues, such as the potential negative impacts of blockchain technology on the elderly and other marginalized groups that do not have access to modern information and communication technologies. While the development of the metaverse presents many opportunities for remote work and education, it also raises many social and ethical issues, such as virtual identity, data privacy, the toxic nature of social media, and corporate control (Zziwa et al., 2025). Developing appropriate curriculum packages at both global and local levels is critical to addressing the social and ethical challenges surrounding metaverse education (Al Yakin et al., 2025).

A holistic metaverse ecosystem will combine real-world arts, sports, and green STEM initiatives; game-based learning; artificial intelligence; and robotics initiatives in virtual worlds, as well as digital STEM initiatives, all through partnerships between business and industry (López et al., 2024). Early project participants, including students and educators, have used the IoT bridge to enhance and integrate their virtual and physical learning environments with the goal of effectively achieving learning outcomes, creating a student-centered learning environment, and motivating students to learn deeply. Only exploring one side of the physical or VW can severely limit their learning experiences without making connections between the two worlds.

There is currently no universally accepted definition of the metaverse, but it is a newly proposed idea that is said to be taking shape thanks to the development of 5G and VR, among other technologies (Gray & Tang, 2025). The use of VR and AR glasses offers access to the entire internet by providing 3D or 4D images, referred to as the metaverse. The internet industry presents the metaverse as a next-generation mobile computing platform that will be widely used by people using VR and AR glasses that can be accessed through websites such as https://www.cospaces.io/ or https://edu.assemblrworld.com/. The metaverse encompasses a wide range of activities, such as online social networking, online shopping, online schooling, online gaming, online social interaction (Al Yakin et al., 2025), and online payments. The metaverse will offer its own version of every form of web application that we know and love today. Lalwani (2024), Supriya et al. (2024) highlight the metaverse as a significant development in the integration of AI, VR, blockchain, and other technologies to create virtual learning fields for students according to their level and needs. Thus, the field of English language teaching becomes the focus of this article.

The first version of the metaverse was an online world that emerged during the transition from 2D to 3D technology. For this reason, the prototype metaverse has the typical features of the digital world. According to Shadiev et al. (2024), there are four technical features that truly immerse users in the virtual world: audiovisual, interaction, persistence, and immersion. Specifically, users can engage in VW interactions based on visual and auditory perception. Combining multiple perception devices finally allows users to enjoy powerful telepresence. Various media have proposed numerous metaverse prototypes with these four features, presenting innovative approaches to the creation of metaverses at varying scales. At present, the majority of metaverse technologies and products focus on electronic entertainment, a naive perspective that overlooks the metaverse's immense potential in English language teaching. It is of great theoretical and practical importance to investigate the potential reform opportunities presented by the extensive implementation of metaverses, the technical forms and implementations of educational metaverses, and the close integration of philosophical digital survival with game learning (Lim et al., 2024).

The English teaching metaverse serves as an educational application of the metaverse, enabling teachers and students to engage in virtual teaching environments, creating digital identities for instructors and students, and opening formal and informal learning spaces in the VW (Hwang, 2024). The creation of an immersive teaching interaction field for teachers and students stands out as the most striking advantage of the metaverse when viewed through the lens of educational philosophy. To meet the demands of the real and virtual educational worlds simultaneously, the English teaching metaverse field transcends the boundaries of the physical world and builds a new virtual educational world through the ecosystem of networked teaching spaces. On the other hand, the virtual environment of the English learning metaverse is an evolution of the real world, not a mere replica or a "parallel universe" of another real reality. Media-enabled features can make up for real-world shortcomings and even surpass them in some ways, creating a unique educational metaverse field that can impact people's lives in general (Shin et al., 2025).

2.2 Metaverse as tools and technology in EFL education

Teachers and IT experts believe that the internet today facilitates learning with Metaverse technology for all users to achieve effective learning outcomes (Lytras et al., 2024). Shwedeh asserts that the Metaverse allows curriculum customization to meet the unique needs of each student. As new ideas and technologies emerge, the world continues to change, so the curriculum must adapt (Shwedeh, 2024). While traditional curricula are prone to setbacks, integrating the Metaverse can transform the entire planet into a flexible, hyplex classroom. Based on the previous advantages, "gamification" of learning is one of the most exciting potential educational applications of the Metaverse (Muthmainnah et al., 2024). Virtual worlds have the potential to transform education into video games, complete with quest-style class structures that motivate students to complete tasks and improve their performance, and engage in active collaboration (UE Business School, 2022).

Metaverse tools and technologies have significantly enhanced educational pedagogy and technical support, resulting in deeper learning opportunities and increased student motivation. This is particularly beneficial in addressing current equipment and technology issues in English language classes, as it divides technologies into seven groups: sensors, wearables, education, modeling and simulation, mirrors, mobile, and AI. Metaverse instruction is highly effective in fostering collaboration and skill development among students through immersive hands-on experiences (Pang et al., 2024; Lai et al., 2024). They integrate the virtual technologies necessary to achieve immersion, including VR, MR, AR, and multiuser virtual environments (MUVE). Technologies that serve as portals suggest multimodality and allow us to fully immerse ourselves in the Metaverse setting or complete immersion (SB et al., 2025). According to Lifelo et al. (2024), MR can enhance learning with the help of deep learning voice interaction modules, and can also unite the real and virtual worlds while implementing the role of MR in the Metaverse. Real-time VR and AR technologies now allow the creation of fictional interactive 3D characters (Pandey et al., 2025). At that time, when used in conjunction with educational platforms, Metaverse enhances learning by incorporating immersive environments into lesson plans and simplifying the learning process for students. The educational Metaverse is formed by institutional learning management systems (Kathirvel et al., 2025).

Massive Open Online Courses (MOOCs) are a prime example of how Web 2.0 and AVAS have enabled students to join online communities. Online courses allow students to expand their knowledge, while the Metaverse and MOOCs make subject information accessible to many (Adhikari, 2024). Courses in mathematics, engineering, and natural sciences utilize online laboratories. Learners have access to VLLs, or Virtual Learning Laboratories. Universities provide engaging, collaborative, and ever-changing classroom environments for students, where they can work together to improve their education and drive to learn (Tarouco et al., 2013). As an illustration, university online learning platforms are able to exchange information about users, connect their profiles across platforms, and deliver presentations (Lucas, Benito & Gonzalo, 2013) practically.

It is believed that mobile technology dominates the Metaverse because it facilitates the connection between media and learners through geographic mobility and mobile devices. Students can enhance their learning process by using their Metaverse avatars on mobile devices technology that allows teachers to track student dynamics through behavioral evaluations resulting from these interventions (To et al., 2024).

Metaverse accessibility can be done by smart glasses or a smartphone; students can interact with content and give commands in the VW (Forssell et al., 2024). The second popular tool in the Online World and can be used to identify when a student is emotionally unstable, the eye blink system records more eye blinks, which helps teachers examine their behavior more thoroughly (Anwar et al., 2024) with the use of convolutional neural networks, machine learning, and neurosymbolic AI, semantic database technology helps students understand learning-related content more efficiently (Duan et al., 2021). Data analysis for understanding, organizing, and planning is essential for the Metaverse, and advances in AI can provide the basis for this kind of data processing (Nair et al., 2024). By integrating neural networks, neurosymbolic AI can analyze user input using automatic speech recognition metrics and provide feedback. People usually use systems that use conventional symbolic thinking (Lu et al., 2024). Examples in aviation maintenance training and education often utilize neurosymbolic AI, which can supplement human instructors and aviation maintenance classes by providing expert guidance and necessary tools for students. They have received training and education in the art of aircraft repair with VR simulations (Vajrobol et al., 2025). To improve learning efficiency, convolutional neural networks (CNNs) often handle acoustic data, as well as learning and classification components for language recognition and commands (Orosoo et al., 2025). The possibility of the Metaverse and AI presents new opportunities for intelligent NPCs. These include instructors, classmates, and students (Kouvcliotis & Mansuri, 2024). Future studies hold hope for the application of AI to technology that looks at students' Metaverse actions, interactions, and performance to build new, different jobs. Metaverse has used various technical domains to build a robust and engaging learning environment.

However, the use of several advanced technologies remains unexplored. For example, we can use blockchain technology to create a learning system with anticheating features, thereby increasing user security. In addition, there is potential for the IoT to facilitate the creation of hybrid, real/virtual learning environments through the integration of various sensors and other devices. Learners can immerse themselves in the Metaverse educational environment. For example, to learn about planets, a student can use the Metaverse to view the entire galaxy and then zoom in and out to see details, or a foreign language learner can access a simulation of an airport to travel and then rewrite their experiences (Muthmainnah et al., 2024). A student can experience what it is like to be a time traveler and observe the details of ancient buildings up close and personal (Sun et al., 2025). There are theoretically four different types of metaverses—AR, lifelogging, mirror world, and VR existing in what can be described as a VR complement to the real world. In addition, VR (Kye et al., 2021) plays a significant role. Either the VW immerses the real world, or it transforms the real world into an entirely new environment. According to Sun et al., the Metaverse is a product of the rapid progress brought about by 5G technology and virtual convergence, and reflects a world that has grown and developed (Sun et al., 2025).

There are four main types of Metaverse, each emerging from different academic perspectives and initially developing independently. However, they eventually evolved into new types of convergent/composite services through interaction and the removal of barriers. For example, Ghost Pacer combines life-logging with AR (Kshetri et al., 2024). One of the most dynamic and diverse fields is VR. This study uses the Metaverse as one of four pedagogical approaches. One significant aspect of this era without face-to-face interaction is the proliferation of VR, which disregards location or physical distance, enabling students to learn independently. One of the features that distinguishes VR Metaverse from competing platform services is that designers and users work together to build and expand the space-time of Metaverse. The second point is that when accessing VR applications, anyone in the Metaverse can design stories or tales to improve English skills. The third benefit of this digital space is facilitating production and consumption, and the fourth is that the Metaverse ensures the continuity of everyday life. Fifth, avatars in Metaverse connect the physical and digital worlds (Lytras et al., 2024) and function as human presence in the virtual world. The "digital native generation" is making a significant social impact through their multiple online identities, as well as a pedagogical influence. 3D virtual worlds, originally used for entertainment and gaming before the Metaverse, are finding their way into the classroom as a teaching tool (Benaben et al., 2025). Adding an entertainment element to the learning process makes the classroom more engaging, which in turn increases students' active engagement.

As a result, educational activities are increasingly utilizing the Metaverse as a 3D virtual environment, with the concept of the Metaverse becoming clear. Critical thinking and student engagement in class discussions are two areas that greatly benefit from an active learning environment that prioritizes their development as learners. Academics have yet to thoroughly study critical thinking in online learning environments, despite the importance of critical thinking and design thinking (Al Yakin et al., 2025). Savin and Burden assert the need for a new framework to enhance online learning and performance (Savin-Baden & Burden, 2024). Due to the lack of research on the topic of the Metaverse in education, we will address the gap in this field by implementing a metaverse environment in an English classroom.

2.3 Digital twin-based technology in English education ethics

In the 21st century, science and technology are developing at a rapid pace, significantly impacting society's production and lifestyle, including the teaching style in the education sector. Increasing the level of scientific literacy of the community is very important if they want to meet their own needs and keep up with the progress of society (Daineko et al., 2024). Reforms must fully recognize the important role of information technology in this field, accelerate the process of educational information, and integrate high-quality resources and advanced technology into the existing education system. Taking science classes in college is a great way to develop scientific and technological intelligence, which is essential for success in the modern talent training system. The development and implementation of the curriculum in the Indonesian education system has recently made significant progress.

However, several factors contribute to the persistent issues with English teaching methods and content in Indonesian education. These problems hinder the progress of English education and the improvement of technological literacy. Moreover, these issues become more pronounced in the context of modern society, particularly with the rapid development of IT. Therefore, the latest research in English teaching utilizes advanced information technology to enhance English teaching techniques and address the shortcomings of traditional methods. Incorporating VR into college English classes can improve the current state of English language teaching while introducing new approaches to the language, which in turn can arouse students' curiosity and lead to better teaching overall (Jiao et al., 2024).

The proliferation of science and technology, particularly multimedia technology, has permeated every aspect of society and the economy. This has led to major changes in the way people work, live, and produce goods. Thus, multimedia technology has a positive effect on education and has the potential to provide an optimal learning environment for students. Multimedia technologies used in education are known as situational interactive teaching systems. Among the many features and functions that make these systems ideal for use in classrooms and other teaching environments are the ability to simulate real-world situations for listening and speaking practice, as well as virtual situational teaching (Mohd Nabil et al., 2024).

Through the creation of VR scenes, advancements in networking, graphics, image processing, and video recording now enable the creation of realistic simulations based on digital environments. To improve their listening and speaking skills in a foreign language, students need a system that allows them to practice language dialog in a realistic setting. In addition, these systems have the ability to communicate with users through cartoon characters. Furthermore, these systems allow undergraduate students to learn a language in an intriguing environment and increase their enthusiasm for learning a foreign language, which is very beneficial for elementary and secondary school students. The goal of this system is to help users learn relevant skills through visualization and engagement by simulating difficult-to-describe instructional scenarios using virtual technology (Chen, 2024). There are high expectations for educators and a lack of uniform systematicity in the current science curriculum.

As a result, English classes must use learning tools to arouse students' interest in learning and provide an authentic English language teaching environment. Conventional English language teaching methods in the classroom rely on the use of PowerPoint and other instructor-created audiovisual materials, such as videos or YouTube. Abstraction of learning is a common result of students' struggle to grasp abstract scientific concepts in the classroom, in speaking (Liu, 2024). Another major problem in education is the lack of experimental equipment, which in turn affects the quality of English language teaching throughout history. Specifically, inadequate funding for education prevents children in rural areas from accessing basic laboratory equipment necessary for experiments. This deprivation of the excitement and wonder of scientific discovery dampens these students' enthusiasm for learning (Mentesoglu-Karaderi et al., 2024).

Due to issues with English language teaching methods and resources, the current state of English language education in Indonesian universities is inadequate and impedes educational progress. Over time, academics in the field of education have sought to use new technological tools to assist in the development of learning activities, in the hope of improving scientific courses and addressing the shortcomings of traditional English language teaching. The use of a number of technological advances, including three-dimensional computer technology, interactive technology, and human-machine interface technology, allows students to have an immersive learning experience that mimics the real world. Because virtual learning software is so easy to use, educators can improve English instruction in science classes, provide students with more engaging information, and increase their interest in learning all using virtual classrooms (Jiao et al., 2024).

The English classroom is where most nonnative students hone their language skills. Improving students' English skills, especially listening and speaking, is an ongoing goal of effective classroom practice activities and ample opportunities for language practice. Therefore, situational interactive teaching systems provide a wonderful environment for students to practice speaking and listening in English in a virtual environment, helping them improve their skills. This system does this by simulating challenging teaching situations (Chen & Wang, 2024). The interactive features between humans and computers, as well as real-time feedback, are the hallmarks of multimedia technology that set it apart from others. Multimedia computers combine the audio-visual capabilities of television with those of computers, resulting in a fresh and dynamic method of human-computer interaction that utilizes text and images. This interactive method can spark students' learning interests, desires, and motivations, making it an essential part of the teaching process. Only multimedia computers allow users to interact with the content. Thanks to their capabilities, multimedia computers can change how we learn and discuss in class. In the English classroom, a digital twin supports a situational teaching mode where students primarily practice their English skills.

3. Methodology

This research method uses a quantitative approach with a preexperimental design, inspired by Repuya et al. (2024). This method does not have a control class, only a real experimental class. Therefore, despite the absence of control variables and nonrandom sample selection, this method yields valuable initial insights. This study specifically uses a pretreatment and posttreatment test. By measuring baseline knowledge and then changes after the intervention, this design makes it more accurate to figure out how human-machine collaboration treatments driven by digital twins affect the learning process.

We measure the dependent variable with a pretreatment, apply the treatment (X) for 8 weeks, and then provide a posttreatment to evaluate the efficacy of the human intelligence and AI intervention on ethics in the metaverse classroom. This sequential process consists of an experimental protocol. This treatment aims to enhance EFL skills by implementing a digital twin learning approach in language education classes, utilizing the https://www.cospaces.io/ application to access digital twins. To determine the efficacy of the treatment, researchers compared the performance of subjects on tests O_1 and O_2 before and after the intervention (Table 20.1).

O_1 is the initial test, X is the treatment (digital twin), with eight meetings without a digital twin or VR, and eight meetings with digital twin intervention, and O_2 is the final test and ethics. This methodology allows us to evaluate how well students learn after a digital twin-based language learning intervention, observing human-machine interactions and ethics that are in line with our research objectives.

3.1 Participants

Participants in this study included 275 undergraduate students from Universitas Al Asyariah Mandar, who were enrolled in the first semester of the academic year 2024 (Fig. 20.1). The samples in this study were 45 students in the informatics engineering class and 39 students in the Indonesian language education class. This experiment involved a total of 84 students. We collected data during six sessions in the experimental class to ensure there were enough data points to examine the impact of the digital twin intervention during the learning process.

TABLE 20.1 Preexperimental design.

O_1	X	O_2
Pretest	Treatment	Post Test

FIGURE 20.1 Airport digital twins' intervention in EFL class.

3.2 Instruments

To ensure a thorough investigation of learning outcomes, this study used a mix of test and nontest instruments to collect data. We administered a 25-item multiple-choice questionnaire for the test component (essay test) at the beginning and end of the study. In the first round, the questionnaire checked whether there were statistically significant differences between the groups by assessing their baseline proficiency levels, and it also established homogeneity of group abilities. We administered the same 20-item multiple-choice questionnaire as a posttest following the digital twin intervention to gauge improvements in learning outcomes.

This repeated testing method allowed for a precise assessment of the effects of the therapy on student performance. In addition to the questionnaire, we used observation tools to evaluate students' psychomotor skills and engagement during the learning activities. We conducted these observations simultaneously with the testing phase to enhance our qualitative understanding of human-AI collaboration and ethics during the learning process. This study took a holistic approach to measuring student learning outcomes by combining test and nontest measures. We carefully considered the validity of the tools to ensure their reliability and suitability for the purpose of the study.

3.3 Data analysis

This study's structured and stepwise methodology included data analysis for the pretest, questionnaire, and observation. This allowed for a comprehensive review. We carried out the following process after verifying the instruments' reliability and compliance with research standards:

➢ *To verify that the experimental groups were homogeneous and to establish the baseline proficiency level of the participants, the pretest scores were analyzed. We determined the major trends and variability in the data using metrics such as mean, median, and standard deviation through quantitative descriptive analysis. This initial assessment describes the level of proficiency of the participants before the intervention.*

➢ *Posttest Analysis: We evaluated the learning outcomes by analyzing the posttest scores to measure the efficacy of the intervention. We evaluated the difference between the pretest and posttest scores using inferential statistics, such as the paired sample t-test, in the*

SPSS version 26 application to determine if the improvement was statistically significant. At this stage, we can demonstrate the efficacy of the intervention.

- *We analyzed the data from the 20-item multiple-choice questionnaire to understand the participants' perceptions of the intervention. We used descriptive statistics to analyze the distribution of responses in key domains like usability, learning outcomes, and engagement. We present the results in percentage and frequency. We conducted this study to better understand the trends in participant feedback and their relationship to test results.*
- *We use a combination of qualitative and quantitative methods to analyze data collected during the intervention to evaluate psychomotor skills and levels of participant engagement. We identified engagement, behavior, and learning environment interaction patterns by coding and categorizing observations. The incorporation of observational data strengthened the quantitative findings by adding context and nuance to the results.*
- *Inferential Statistics. As part of the inferential statistical strategy, we conducted normality and homogeneity tests to verify assumptions about the data set. We used appropriate statistical procedures based on these tests to compare groups and determine the significance of changes. To confirm whether the intervention significantly affected learning outcomes, hypothesis testing was conducted at a significance threshold of $\alpha = 0.05$.*
- *We obtained the results of the intervention efficacy by integrating pretest-posttest scores, questionnaire responses, and observations. We conducted all analyses systematically using SPSS version 26 to ensure accuracy and reliability.*

4. Findings and discussion

By comparing students' performance on pre- and posttests, we can describe the study's results and conclude how the Metaverse influences their English learning in writing short stories, as shown in Table 20.2.

The results of the Paired Samples Test (Table 20.3) show that there is a significant difference between the pretest and posttest scores, as indicated by producing a significance value of $P = 0.000$, which is less than 0.05. This indicates that the increase in scores from pretest to posttest is very statistically significant.

The Q-Q plots (Fig. 20.2) show how the distribution of scores from the pretest and posttest compares to a normal distribution. The Pretest Normal Q-Q plot shows that the pretest data do not follow a completely normal distribution, implying some deviation from normality. Some points significantly deviate from the horizontal line, spreading out above and below it. Overall, there is no divergence. There is a large deviation from normality in the pretest data, while the posttest data are closer to normal. Posttest scores that show an increase in normality may be the result of more consistent performance after the intervention.

TABLE 20.2 Pre- and posttest statistics.

	N	Minimum	Maximum	Mean	Std. deviation
Pretest	84	40.00	70.00	51.9048	8.56952
Posttest	84	60.00	99.00	88.7262	7.12429
Valid N (listwise)	84				

Based on the data presented in Table 20.2, it is evident that learning English using the VR application consistently led to a significant increase in posttest scores for the individuals. This shows that the intervention or teaching method was successful.

TABLE 20.3 Statistical data of Paired Samples Test.

		Paired differences							
					95% Confidence interval of the difference				
		Mean	Std. deviation	Std. error mean	Lower	Upper	t	df	Sig. (2-tailed)
Pair 1	Pretest - Posttest	−36.82143	9.65072	1.05298	−38.91576	−34.72709	−34.969	83	0.000

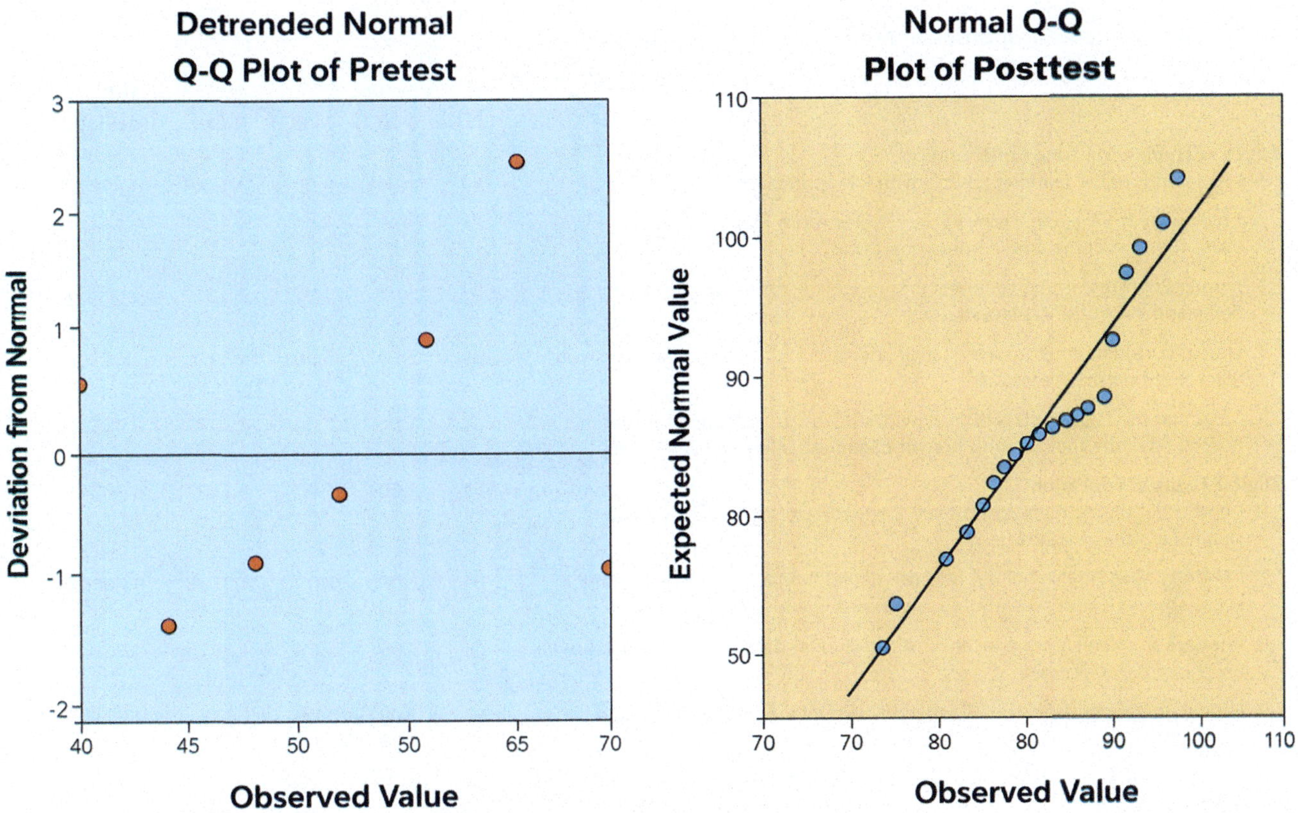

FIGURE 20.2 Q-Q Pilot of pre- and posttest.

Data in Table 20.4 explains how students perceive the ethics of human-machine collaboration in ESL classes. The responses are We organize the responses into themes to provide a comprehensive picture of how students view different technologies and instructional approaches through an ethical lens. According to the study results presented in Table 20.4, students demonstrated confidence in the ethical considerations of human-machine collaboration within their learning environments. Transparent practices and strong ethics policies primarily supported this belief. Educators should prioritize encouraging a stronger dialogue around moral dilemmas related to the use of technology. Lecturers provide clear instructions on how to use this technology ethically, as well as keep the rules up to date with the ever-changing ethical issues that technology raises.

5. Discussion

This study's findings shed light on how using the Metaverse as a teaching tool might improve EFL students' language acquisition, how it can promote user involvement, and how it can encourage ethical thinking that builds a feeling of community and presence in the classroom. Tests given before and after the experiment showed that the experimental group's English as a Foreign Language (EFL) learning had improved a lot. This suggests that the Metaverse's interactive and immersive features improve EFL learning outcomes, making it a useful learning tool. Furthermore, our research suggests that writing short stories based on the metaverse enhances students' writing retention. As we will discuss later, these results align with other research demonstrating the benefits of immersive environments in educational settings.

Table 20.2 presents the results of a statistical comparison between the pre- and posttest scores of 84 participants. According to the results, performance improved significantly after the intervention or learning activity. The range of pretest scores was from 40 to 70, with a mean of 51.90 and a standard deviation of 8.57. With most scores clustering around the mean, these results indicate that participants' initial abilities showed a modest degree of variability. In contrast, posttest scores showed a significant increase, ranging from a minimum of 60.00 to a high of 99.00. After the intervention, the mean score increased to 88.73, and the standard deviation decreased to 7.12, indicating less variability and better overall performance. The data show that participants' scores improved significantly between the pretest and posttest, indicating that the learning strategy or research instrument was effective. The smaller standard deviation on the posttest indicates that participants' overall performance improved and that the

TABLE 20.4 Students' responses to human-machine collaboration ethics.

	N	Min	Max	Mean	Std. deviation
Ethics in Human-Machine Collaboration					
1. I feel comfortable learning English with VR applications	84	2.00	5.00	4.1548	0.85720
2. I believe that the presence of VR in the classroom increases ethical transparency in the teaching and learning process.	84	3.00	5.00	4.2381	0.70487
3. During English learning, lecturers ensure ethical interactions between students and VR in the classroom.	84	3.00	5.00	4.3333	0.68254
4. Ethical considerations are adequately addressed when AI collaborates with teachers in delivering lessons.	84	2.00	5.00	4.5476	0.68380
5. I feel that the English classroom provides adequate guidance for students on how to ethically interact with human-machine collaboration systems.	84	3.00	5.00	4.3571	0.70528
Digital Twins in Education 6. I find using digital twins to simulate English learning scenarios raises significant ethical issues.	84	3.00	5.00	4.2857	0.76947
7. I find learning English helpful because digital twins help me better understand abstract concepts without compromising ethical standards.	84	2.00	5.00	4.2738	0.76601
8. I believe it is ethical for educators to use student data to access digital twins for personalized learning.	84	3.00	5.00	4.2619	0.56225
9. Ethical guidelines are essential when using digital twins to assess student performance.	84	3.00	5.00	4.3571	0.57336
10. The application of digital twins during English learning adequately addresses privacy concerns.	84	3.00	5.00	4.2381	0.72176
Role in Education 6.0					
11. I believe that integrating metaverse devices promotes active English learning.	84	3.00	5.00	4.2024	0.67270
12. I strongly believe that the ethical implications of AI-driven assessment are comprehensive.	84	2.00	5.00	4.2857	0.75365
13. Using VR in English classrooms teaches the ethical implications of collaboration with digital devices.	84	2.00	5.00	4.1190	0.76685
14. I believe that metaverse-based English learning environments promote ethical equity by providing equal learning opportunities for all students.	84	3.00	5.00	4.2024	0.74088
15. I believe that stronger ethical policies guide the integration of human-machine collaboration throughout the learning process.	84	3.00	5.00	4.1548	0.63043
General Ethical Issues 16. I strongly believe that using AI technology and digital twins can create ethical bias in the classroom.	84	3.00	5.00	4.2619	0.67875
17. I strongly believe that transparency in data collection and use ensures ethical practices.	84	3.00	5.00	4.3095	0.63958
18. I feel that English classrooms openly discuss and address ethical dilemmas in human-machine collaboration	84	3.00	5.00	4.3095	0.65815
19. Metaverse-based English classrooms effectively balance technological innovation, ethical responsibility, and English language skills.	84	3.00	5.00	4.3214	0.64305
20. I believe in the ethical integrity of the systems used in metaverse classrooms to support English learning.	84	3.00	5.00	4.3929	0.65891
Valid N (listwise)	84				

intervention was successful in reducing the gap in this area. Students' Statements on Ethics in Human-Machine Collaboration revealed that students frequently viewed VR as a tool that fosters ethical transparency in the classroom (mean = 4.24, SD = 0.70), and they felt comfortable using VR applications (mean = 4.15, SD = 0.86) during the English learning process. In this treatment, we used eight meetings without VR intervention and eight meetings using VR in the same class. When the AI interacts with the teacher (mean = 4.55, SD = 0.68), the VR lecturer ensures ethical interactions (mean = 4.33, SD = 0.68).

The High mean scores in this area suggest that, with the right type of instruction, students see VR and AI as useful moral instruments for the classroom. The data indicate a high level of confidence in the ethical handling of digital twins in two areas: evaluating student performance (mean = 4.36, SD = 0.57) and addressing privacy issues (mean = 4.24, SD = 0.72). However, when used as teaching materials, opinions about the ability of digital twins to support the learning of abstract concepts (mean = 4.27, SD = 0.77) become more concrete, and opinions about the ethical issues they raise are somewhat more diverse. Students view digital twins as an ethical innovation, provided they adhere to appropriate regulations, and addressees responded positively to the use of metaverse devices in English lessons due to their practicality (mean = 4.20, SD = 0.67) and their understanding of their role in promoting active learning and ethical justice (mean = 4.20, SD = 0.74). We also acknowledged the statements on the role of VR in teaching collaborative ethics (Mean = 4.12, SD = 0.77) and the ethical implications of AI-driven evaluation (Mean = 4.28, SD = 0.75). While these comments expressed hope for Education 6.0 technologies, they also indicated that these devices require better ethics policies. Students' statements about overarching moral concerns, such as ethical dilemmas in the classroom (Mean = 4.31, SD = 0.66) and transparency in data collection (Mean = 4.31, SD = 0.64), were topics of substantial agreement.

Students showed full awareness of the potential biases that AI and digital twins can cause (mean = 4.26, SD = 0.68) and their impact if used responsibly; metaverse-based learning environments have a beneficial impact on people's perceptions of the ethical integrity of the system (mean = 4.39, SD = 0.66). The students' responses and overall observations showed that all students had positive attitudes toward the ethics of human-machine collaboration in English language learning, as all items had an average score of more than 4.0. Students held the belief that ethical use of this technology was necessary, yet they identified areas for improvement in the guidance they received on teaching ethical collaboration (Item 13, with a mean score of 4.12).

Based on research findings, it is known that learning in the Metaverse provides an opportunity to hone various student abilities. This includes not only applying English story-writing skills but also training human-machine interactions and engagement in the class, similar to Chen & Wang (2024); Clapp (2024) studies. We can repeatedly apply these skills, making them useful for a skills-based approach in education 5.0–7.0. In addition, the current AR mobile media method can encourage the development of cognitive capacity and create new learning experiences for EFL students. By incorporating AR learning resources into the classroom through mobile devices, which can serve as game-changing guides, we can achieve the primary objective of enhancing students' academic performance, as illustrated in Fig. 20.1.

It is important to acknowledge the gap that arises from the numerous challenges of self-education using technology, including the lack of substantial background knowledge, instruction, and practice among instructors and students. It is crucial to ascertain whether this change will enhance the quality of therapy (Lee et al., 2024). To enhance the quality of education and meet the digital needs of students, educators must establish effective communication channels, comprehend the impact of technology on modern education, and weigh the ethical implications of these technological interventions in the classroom. To remain competitive in the global economic arena in the future, students must enhance their technical competencies. In line with Vy's (2024) research, students must enhance their technical skills to compete on a global scale, complementing their English proficiency. Some of the abilities that students need to develop are critical analysis, creative problem-solving, and effective communication. Working in a team and being proficient with technology are crucial skills (Items 7-10).

This work also contributes to the idea that accelerating Metaverse-based learning environments presents a unique opportunity to integrate physical infrastructure with digital content. Virtual worlds can play a significant role in realizing concrete learning, which is highly sought after in the classroom environment. This method combines visual, auditory, and tactile senses, along with an engaging game packaging. This will facilitate a fun classroom environment (statement no. 1). Visual learning in this study shows that rapid developments in computer vision and graphics have made it possible. The scenes in the metaverse are visually striking. In a perfect world, educators and their students can blend real virtual environments and their teaching activities seamlessly, providing equal learning opportunities in successful classes in the Metaverse (statement number 14). Therefore, it is essential to consider the advancement and ethics of natural interaction technology in the classroom when utilizing the Metaverse for education (Item 15).

The environment for learning includes the growth of comprehensive visual experiences, in line with scientific developments. This learning method uses VR or digital twins to provide knowledge, speaking skills, values, and perspectives that cannot be conveyed by traditional classes.

It also serves as a useful evaluation tool for EFL language acquisition (items 19-20) in line with other findings (Shi et al., 2024; Tafazoli, 2024; Zhang, 2024). The results showed that the metaverse had a significant positive impact on students' English language development in short story writing, leading to increased ethical consideration and retention. This highlights the benefits of gamified digital twins and interactive story writing training, further demonstrating the metaverse's ability to enhance language learning outcomes. Furthermore, this study provides insight into students' ability to foster a strong classroom community among EFL learners. This highlights the importance of building immersive and interactive virtual environments for learning. These environments can help students feel more connected, satisfied with their learning, and part of a community, which is especially important in the context of online or distance learning. Examining user interactions with this machine learning, the current study uncovers a complex landscape. Students' engaged behavior and positive emotions indicate that the gamification aspect of the Metaverse learning environment is effective in capturing their attention. Contributing to the overall user experience, favorable ratings for the Metaverse's aesthetic appeal enhance active and practical learning engagement.

6. Conclusion

The research demonstrates how metaverse technologies like VR, digital twins, and AI could significantly improve EFL classrooms. It shows how these technologies help students become more engaged, knowledgeable, and successful in the classroom. Results from both the pre- and postintervention assessments show that students' English language skills, especially their writing, significantly improved, with mean scores significantly higher and variability significantly lower. This demonstrates the efficacy of using metaverse settings and digital twins to build engaging and participatory educational programs.

The poll results show that students have a high opinion of the ethical considerations and openness of tools that allow humans to work together with machines. When applied ethically and responsibly, these technologies tend to have a positive reception, according to the high mean ratings across different claims. In classes that use the metaverse and digital twins, for instance, students have recognized the significance of ethical transparency, privacy protection, and responsible data usage. Additionally, they stressed the importance of better regulations to handle new ethical dilemmas, especially in situations involving collaborative learning and AI-driven evaluations. The study highlights the twofold function of metaverse technologies: first, they help students learn a new language by simulating real-life interactions; second, they encourage them to think critically and ethically. Metaverse applications such as virtual city tours, cultural immersion, and role-playing exercises best illustrate the possibilities for contextually rich and individually tailored education. Such tools engage students in ways that conventional classrooms frequently fail to do because they combine gamification with organized feedback mechanisms.

The study highlights the potential benefits of metaverse-based education but also points out obstacles to its wider implementation, especially in countries like Indonesia that are facing severe resource constraints. Obstacles include the differing degrees of digital literacy among educators and students, as well as gaps in the digital infrastructure. To tackle these concerns, we must work together to close the digital gap, improve teacher education, and set strong ethical standards. The study concludes that the metaverse is an essential part of Education 5.0–7.0 because it allows for new, ethical, and inclusive ways of teaching and learning languages. While there are still many obstacles to overcome, the results show that children can benefit from fair and effective education if technical progress is carefully considered alongside ethical concerns.

References

Adhikari, S. (2024). Transformative innovation in education. In *Advances in technological innovations in higher education* (pp. 118–138). CRC Press.

Al Yakin, A., Viddy, A., Warsah, I., Al Matari, A. S., Cardoso, L., Elngar, A. A., & Obaid, A. J. (2025). Enabling federated learning in the classroom: Sociotechnical ecosystem on artificial intelligence integration in educational practices. In *Artificial intelligence using federated learning* (pp. 37–56). CRC Press.

Allman, B., Kimmons, R., Wang, W., Bao, H., Rosenberg, J. M., & Koehler, M. J. (2024). Trends and topics in educational technology. *TechTrends, 68*(3), 402–410, 2024 Edition.

Anwar, M. S., Choi, A., Ahmad, S., Aurangzeb, K., Laghari, A. A., Gadekallu, T. R., & Hines, A. (2024). A moving metaverse: QoE challenges and standards requirements for immersive media consumption in autonomous vehicles. *Applied Soft Computing, 159*, Article 111577.

Babushkina, L., & Kalugina, O. (November 2024). Integration of storytelling technology and the moral education of student personalities in foreign language instruction. In *International Conference on Professional Culture of the Specialist of the Future* (pp. 99–114). Cham: Springer Nature Switzerland.

Benaben, F., Congès, A., & Fertier, A. (2025). A prospective vision of the evolution of immersive technologies: Towards a definition of metaverse. *Technovation, 140*, Article 103154.

Bhowmik, A. K. (2024). Virtual and augmented reality: Human sensory-perceptual requirements and trends for immersive spatial computing experiences. *Journal of the Society for Information Display, 32*(8), 605–646.

Chan, I. W., & Tong, E. K. (2025). A living systematic review of the use of Metaverse in L2 English education. *Computer-Assisted Language Learning in the Global South*, 99–113.

Chen, Y. C. (2024). Effects of technology-enhanced language learning on reducing EFL learners' public speaking anxiety. *Computer Assisted Language Learning, 37*(4), 789–813.

Chen, S., & Wang, J. (2024). Virtual reality human–computer interactive English education experience system based on mobile terminal. *International Journal of Human-Computer Interaction, 40*(13), 3560–3569.

Chithra, N., & Bhambri, P. (2025). Ethics in sustainable technology. In *Handbook of technological sustainability* (pp. 245–256). CRC Press.

Clapp, N. (2024). *Investigating High School Students' Engagement and Presence in Foreign Language Learning Using Virtual Reality: A Quasi-Experimental Study.*

Dad, A. M., Rehman, A. A., Haleem, M. S., Davies, B., & Kear, A. (2024). *The Effect of 3 Dimensional Virtual Reality Retail (VRR) Store Ambiance on Avatar's Emotions and Behaviour: A Mediation Analysis.*

Daineko, Y., Seitnur, A., Tsoy, D., Ipalakova, M., Mustafina, A., & Uali, M. (September 2024). Towards concepts for digital twins in higher education. In *International Conference on Extended Reality* (pp. 417–424). Cham: Springer Nature Switzerland.

Danylchenko-Cherniak, O. (2024). *Phililogical perspectives of design thinking and artificial intelligence in the Education 5.0-7.0 era.*

Dash, A. K., Balaji, K. V., Dogra, D. P., & Kim, B. G. (2024). Interactions with 3D virtual objects in augmented reality using natural gestures. *The Visual Computer, 40*(9), 6449–6462.

De Los Santos Melo, A., & Beriguete Alcántara, F. E. (2024). Maximizing learning potential: Embracing the power of digital twins in architectural and construction education of the twenty-first century. In *Teaching innovation in architecture and building engineering: Challenges of the 21st century* (pp. 477–493). Cham: Springer Nature Switzerland.

Ding, A. C. E., & Cha, E. E. (2024). The integration of virtual reality-enhanced multimodal meaning-making improves knowledge acquisition and disciplinary literacy development in science classrooms. *Learning and Instruction, 94*, Article 101999.

Fadhel, M. A., Duhaim, A. M., Albahri, A. S., Al-Qaysi, Z. T., Aktham, M. A., Chyad, M. A., ... Gu, Y. (2024). Navigating the metaverse: Unraveling the impact of artificial intelligence—a comprehensive review and gap analysis. *Artificial Intelligence Review, 57*(10), 264.

Falola, P. B., Adeniyi, A. E., Awotunde, J. B., Akinola, S. O., Olagunju, M., & Olanloye, O. D. (2025). 10 The reality, role virtual of augmented reality, and mixed reality. In *Computational intelligence in industry 4.0 and 5.0 applications: Trends, challenges and applications* (p. 329).

Filosofi, F., Pasqualotto, A., Paolizzi, E., Iandolo, G., Alonso-Campuzano, C., & Venuti, P. (2024). Enhancing group interactions and story quality in primary school students: An investigation into the impact of collaborative storytelling methodologies. *Journal of Computers in Education*, 1–28.

Forssell, M., Hassan, L., & Turunen, M. (October 2024). Inclusive VR gaming with older adults: Accessibility of entertainment and educational VR games. In *Proceedings of the 27th International Academic Mindtrek Conference* (pp. 201–210).

Gray, J. E., & Tang, W. (2025). The Chinese metaverse: An analysis of China's policy agenda for extended reality (XR). *Policy & Internet, 17*, Article e418. https://doi.org/10.1002/poi3.418

Guan, J. Q., Ying, S. F., Zhang, M. L., & Hwang, G. J. (2024). From experience to empathy: An empathetic VR-based learning approach to improving EFL learners' empathy and writing performance. *Computers & Education, 220*, Article 105120.

Gursesli, M. C., Lanata, A., Guazzini, A., & Thawonmas, R. (2025). Immersive virtual reality and augmented reality in human-machine interaction. In *Artificial intelligence and multimodal signal processing in human-machine interaction* (pp. 331–342). Academic Press.

Ha, M. J. (2024). Exploring metaverse-blended learning in an English presentation class: Student perceptions. *Computer-Assisted Language Learning Electronic Journal, 25*(2), 130–151.

Hwang, Y. (2024). Reconceptualization of textbook with the metaverse: Pre-service English teachers' experiences and perceptions of designing TechBoox. *Computer Assisted Language Learning*, 1–32.

Jagatheesaperumal, S. K., Ahmad, K., Al-Fuqaha, A., & Qadir, J. (2024). Advancing education through extended reality and internet of everything enabled metaverses: Applications, challenges, and open issues. *IEEE Transactions on Learning Technologies, 17*, 1120–1139.

Jiao, Y., Dewitt, D., & Razak, R. B. A. (2024). Exploring the potential of metaverse-based platforms for enhancing English as a foreign language (EFL) learning: A comprehensive systematic review. *Environments, 2*, 16.

Küçük-Avcı, Ş., Akgün, Ö. E., & Balkan-Kıyıcı, F. (2024). Comparison of the effects of PBL in 3D virtual environment and F2F on learning and spatial skills. *Education and Information Technologies*, 1–34.

Kathirvel, N., Kathirvel, A., & Gobinath, V. M. (2025). Enhancing digital-based technology planning, leadership, and management to transform education. In *Planning tools for policy, leadership, and management of education systems* (pp. 25–52). IGI Global.

Kebeci, Ç. C. (2024). İngilizce ögretmen Adaylarinin Metaverse Ortaminda İngilizce Ögretimine Yonelik tutumlari. *Trakya Journal of Education, 14*(2).

Kouveliotis, K., & Mansuri, M. (2024). Education in a post Covid-19 scenario: How artificial intelligence has changed e-learning in the meta era. In *Organizational corruption, crime and Covid-19* (pp. 181–201). Routledge.

Kritsotaki, K., Castro–Kemp, S., & Kamenopoulou, L. (2025). Digital storytelling: An educational approach for enhancing dyslexic children's writing skills, critical and cultural learning. *Journal of Research in Special Educational Needs, 25*, 289–311. https://doi.org/10.1111/1471-3802.12722

Kshetri, N., Dwivedi, Y. K., & Janssen, M. (2024). Metaverse for advancing government: Prospects, challenges and a research agenda. *Government Information Quarterly, 41*(2), Article 101931.

López, U. H., Vázquez-Vílchez, M., & Salmerón-Vílchez, P. (2024). The contributions of creativity to the learning process within educational approaches for sustainable development and/or ecosocial perspectives: A systematic review. *Education Sciences, 14*(8), 824.

Lai, W. W. S., Lopez-Ozieblo, R., Shen, D. J., & Tong, E. K. M. (July 2024). Engaging the digital learner: Navigating learning landscapes in the metaverse. In *International Conference in Information Technology and Education* (pp. 344–354). Cham: Springer Nature Switzerland.

Lalwani, S. (March 2024). Metaverse in education: Bibliometric analysis to explore future scope of research in Edu-Metaverse. In *2024 International Conference on Automation and Computation (AUTOCOM)* (pp. 126–130). IEEE.

Lee, S. M., Yang, Z., & Wu, J. G. (2024). Live, play, and learn: Language learner engagement in the immersive VR environment. *Education and Information Technologies, 29*(9), 10529–10550.

Lifelo, Z., Ding, J., Ning, H., & Dhelim, S. (2024). Artificial intelligence-enabled metaverse for sustainable smart cities: Technologies, applications, challenges, and future directions. *Electronics, 13*(24), 4874.

Lim, W. Y. B., Xiong, Z., Niyato, D., Zhang, J., & Shen, X. (Eds.). (2024). *Realizing the metaverse: A communications and networking perspective*. John Wiley & Sons.

Liu, J. (2024). Application of computer-aided technology in teaching spoken English. *Journal of Electrical Systems, 20*(3s), 459–471.

Lu, Z., Afridi, I., Kang, H. J., Ruchkin, I., & Zheng, X. (2024). Surveying neuro-symbolic approaches for reliable artificial intelligence of things. *Journal of Reliable Intelligent Environments, 10*(3), 257–279.

Lytras, M. D., Serban, A. C., Alkhaldi, A., Aldosemani, T., & Malik, S. (2024). Transformative learning for future higher education: The artificial intelligence-enabled learning revolution 2035. In *Digital transformation in higher education, part B* (pp. 231–241). Emerald Publishing Limited.

Mentesoglu-Karaderi, I., Osam, Ü. V., & Al-Turjman, F. (February 2024). Integrating generative artificial intelligence in English language teaching: An overview in the IoT era. In *International Conference on Artificial Intelligence of Things for Smart Societies* (pp. 43–50). Cham: Springer Nature Switzerland.

Mohd Nabil, N. S., Nordin, H., & Ab Rahman, F. (2024). Immersive language learning: Evaluating augmented reality filter for ESL speaking fluency teaching. *Journal of Research in Innovative Teaching & Learning, 17*(2), 182–195.

Muthmainnah, Cardoso, L., Al Yakin, A., Tasrruddin, R., Mardhiah, M., & Yusuf, M. (December 2023). Visualization creativity through shadowing practices using virtual reality in designing digital stories for EFL classroom. In *International Multi-Disciplinary Conference-Integrated Sciences and Technologies* (pp. 49–60). Cham: Springer Nature Switzerland.

Muthmainnah, M., Cardoso, L., Alsbbagh, Y. A. M. R., Al Yakin, A., & Apriani, E. (June 2024). Advancing sustainable learning by boosting student self-regulated learning and feedback through AI-Driven personalized in EFL education. In *International Conference on Explainable Artificial Intelligence in the Digital Sustainability* (pp. 36–54). Cham: Springer Nature Switzerland.

Muthmainnah, M., Obaid, A. J., Al Yakin, A., Nurjannah, S., Mursidin, M., & Polkowski, Z. (2025). Revolutionizing quantum robotics applications for enhancing language learning and technology self-efficacy. In *The Quantum evolution* (pp. 401–424). CRC Press.

Nair, A. J., Manohar, S., Mittal, A., & Chaudhry, R. (2024). Unleashing digital frontiers: Bridging realities of augmented reality, virtual reality, and the metaverse. In *The metaverse dilemma: Challenges and opportunities for business and society* (pp. 85–112). Emerald Publishing Limited.

Onu, P., Pradhan, A., & Mbohwa, C. (2024). Potential to use metaverse for future teaching and learning. *Education and Information Technologies, 29*(7), 8893–8924.

Orosoo, M., Raash, N., Treve, M., Lahza, H. F. M., Alshammry, N., Ramesh, J. V. N., & Rengarajan, M. (2025). Transforming English language learning: Advanced speech recognition with MLP-LSTM for personalized education. *Alexandria Engineering Journal, 111*, 21–32.

Pandey, P. K., Pandey, P. K., Mahajan, S., Paul, J., & Iyer, S. (2025). Digital twin and virtual reality, augmented reality, and mixed reality. In *Digital twins for smart cities and villages* (pp. 273–293). Elsevier.

Pang, S., Lv, G., Zhang, Y., & Yang, Y. (2024). Enhancing students' science learning using virtual simulation technologies: A systematic review. *Asia Pacific Journal of Education*, 1–21.

Qiao, J., Xu, J., Li, L., & Ouyang, Y. Q. (2021). The integration of immersive virtual reality simulation in interprofessional education: A scoping review. *Nurse Education Today, 98*, Article 104773.

Ramanujam, P., & Ismail, H. H. (2024). The realities of roblox and metaverse technologies and emerging potential enhancing English language learning. *St. Theresa Journal of Humanities and Social Sciences, 10*(2), 138–156.

Reethika, A., & Kanaga Priya, P. (2024). Personalize learning experience in education using digital twins with human-centered design and pedagogy. In *Transforming industry using digital twin technology* (pp. 165–185). Cham: Springer Nature Switzerland.

Repuya, C., Balderama, J. F., Cerillo, J., Dorol, K. R., & Mapusao, J. (2024). Local language explanation-infused learning material (LLE) and activity-based formative assessment (AFA): Enhancing mathematics learning. *Participatory Educational Research, 11*(6), 242–261.

Revathy, J. S., & Mangaiyarkkarasi, J. (2024). Integration of metaverse and machine learning in the education sector. In *Impact and potential of machine learning in the metaverse* (pp. 74–99). IGI Global.

Ribis, E. V., & García, J. H. (2024). Storytelling robots: enhancing collaboration, language learning and fostering creativity Storytelling robots: mejorando la colaboración, el aprendizaje de idiomas y fomentando la creatividad. *Revista UTE Teaching & Technology (Universitas Tarraconensis), 3*, Article e3870.

Samala, A. D., Mhlanga, D., Bojić, L., Howard, N. J., & Pereira Coelho, D. (2024). Blockchain technology in education: Opportunities, challenges, and beyond. *International Journal of Interactive Mobile Technologies, 18*(1), 20–42.

Savin-Baden, M., & Burden, D. (2024). *The metaverse for learning and education*. CRC Press.

Sb, A., Agrawal, A., Yao, Y., Zou, Y., & Das, A. (2025). "What are they gonna do with my data?": Privacy expectations, concerns, and behaviors in virtual reality. *Proceedings on Privacy Enhancing Technologies, 2025*(1), 58–77.

Shadiev, R., Wang, X., & Shen, S. (2024). Effects of immersion and interactive strategies on students' intercultural competence in virtual learning environments. *Education and Information Technologies*, 1–37.

Shi, J., Sitthiworachart, J., & Hong, J. C. (2024). Supporting project-based learning for students' oral English skill and engagement with immersive virtual reality. *Education and Information Technologies*, 1–24.

Shin, D., Kwon, S. K., Noh, W. I., & Hwang, Y. (2025). Exploring the role of the metaverse in English speaking proficiency tests. *Journal of Computer Assisted Learning, 41*(1), Article e13108.

Shwedeh, F. (2024). Harnessing digital issue in adopting metaverse technology in higher education institutions: Evidence from the United Arab Emirates. *International Journal of Data and Network Science, 8* (1), 489–504.

Sun, Q., Das, S., Wang, K., & Teofilo, A. (2025). Toward a spatial metaverse: Building immersive virtual experiences with georeferenced digital twin and game engine. In *Big data* (pp. 334–360). CRC Press.

Supriya, Y., Bhulakshmi, D., Bhattacharya, S., Gadekallu, T. R., Vyas, P., Kaluri, R., … Mahmud, M. (2024). Industry 5.0 in smart education: Concepts, applications, challenges, opportunities, and future directions. *IEEE Access, 12*, 81938–81967. https://doi.org/10.1109/ACCESS.2024.3401473

Tafazoli, D. (2024). From virtual reality to cultural reality: Integration of virtual reality into teaching culture in foreign language education. *Journal for Multicultural Education, 18*(1/2), 6–24.

To, W. M., Yu, B. T., Chung, A. W., & Chung, D. W. (2024). Metaverse: Trend, emerging themes, and future directions. *Transactions on Emerging Telecommunications Technologies, 35*(1), Article e4912.

Uddin, M., Obaidat, M., Manickam, S., Laghari, S. U. A., Dandoush, A., Ullah, H., & Ullah, S. S. (2024). Exploring the convergence of Metaverse, Blockchain, and AI: A comprehensive survey of enabling technologies, applications, challenges, and future directions. *Wiley Interdisciplinary Reviews: Data Mining and Knowledge Discovery, 14*(6), Article e1556.

Vajrobol, V., Saxena, G. J., Singh, S., Pundir, A., Gupta, B. B., Gaurav, A., & Chui, K. T. (2025). Enhancing aviation control security through ADS-B injection detection using ensemble meta-learning models with explainable AI. *Alexandria Engineering Journal, 112*, 63–73.

Vonitsanos, G., Moustaka, I., Doukakis, S., & Mylonas, P. (May 2024). Transforming education in the digital age: Exploring the dimensions of education 4.0. In *2024 IEEE Global Engineering Education Conference (EDUCON)* (pp. 01–10). IEEE.

Vy, D. T. B. (2024). *Empowering English proficiency: A program to excellence in the Vietnam mekong delta for advancing English proficiency for future generations.*

Yi, H. X., Amir, A. H. L., Ahmad, M. A. A., Zulkefli, N. H., Irhama, N. A. N., & Mohamad, M. (2024). A systematic literature review on augmented reality in game-based learning for English as a second language. *The English Teacher, 53*(2).

Zhang, Y. (2024). Design and optimization of english-speaking teaching model using virtual reality technology. *Journal of Electrical Systems, 20*(6s), 276–285.

Zhang, Y., Feijoo-Garcia, M. A., Gu, Y., Popescu, V., Benes, B., & Magana, A. J. (2024). Virtual and augmented reality in science, technology, engineering, and mathematics (STEM) education: An umbrella review. *Information, 15*(9), 515.

Zziwa, I., Ilolo, A., Dawson, M., & Serwadda, B. (2025). Regulating emerging technologies: Establishing ethical and legal frameworks in East Africa. In *Analyzing privacy and security difficulties in social media: New challenges and solutions* (pp. 441–464). IGI Global Scientific Publishing.

Chapter 21

Essential requirements of smart learning environments: A case study in the Arab region

Ruba Fahmi Bataineh[1], Rommel Mahmoud AlAli[2], Rula Fahmi Bataineh[3] and Ali Ahmad Al-Barakat[1,4]

[1]*Department of Curriculum and Methods of Instruction, Yarmouk University, Irbid, Jordan;* [2]*The National Research Center for Giftedness and Creativity, King Faisal University, Al-Ahsa, Saudi Arabia;* [3]*Department of English for Applied Studies, Jordan University of Science and Technology, Irbid, Jordan;* [4]*Department of Education, University of Sharjah, Sharjah, United Arab Emirates*

1. Introduction

The advances in information and communication technologies have signaled the onset of digital transformation in education, as work and education have been affected in an unprecedented manner (Pucciarelli & Kaplan, 2019; Xiao, 2019). Significant shifts have ensued in education, making it possible for teachers and learners alike to benefit from smart teaching and learning practices. Learning management systems, interactive applications, and digital resources, among other tools, have facilitated greater accessibility to resources and interaction with content (Al-Fraihat et al., 2020; Ali & Bhasin, 2020).

Digital transformation has affected every industry (Javaid et al., 2024) with varying speeds and scopes. In education, this extended process of change involves the realignment of educational models through digital technology. It aims to effectively engage all stakeholders, students, teachers, and administrators, by leveraging new information systems, personalized learning experiences, and data analytics (Bucăţa et al., 2022; MacIntyre et al., 2020; McCarthy et al., 2023).

However, digital transformation in education is still emerging (Mhlanga, 2024). It goes beyond merely adopting new tools, as it signifies a cultural shift through which technology is used to enhance teaching, learning, and assessment and meet the evolving demands of life and the labor market (e.g., AlAli et al., 2024). Ultimately, digital transformation in education seeks to optimize teaching and learning processes, thereby improving the overall efficiency and effectiveness of educational institutions (Wang et al., 2024).

For institutions of learning, embracing digital transformation is crucial, as it potentially enables them to improve efficiency, reduce costs, enhance learner engagement and outcomes, and increase access to education for learners from diverse backgrounds (Bucăţa & Tileagă, 2024; Petrov et al., 2022). To remain competitive and equip learners with the knowledge and skills needed to succeed in the digital age, educational institutions must actively adapt to these changes to make education more flexible, personalized, and accessible, while simultaneously improving the quality and efficiency of their institutions (Andoh, 2012).

Ascertaining the availability and use of digital technologies in the school does not necessarily lead to digital transformation (AlAli et al., 2024; Cuban, 2001; Galván Orozco et al., 2022; McKnight et al., 2016). For successful digital transformation to take place, much more than reactionary measures is needed; a mindset change is imperative at the individual, administrative, and organizational levels to produce a culture that allows institutions of learning to not only thrive but also compete in an evolving digital landscape, which promotes innovation, adaptability, and continuous improvement in teaching, learning, assessment, and administrative processes (Bucăţa & Tileagă, 2024; Kumar et al., 2022; Matt et al., 2016; Vongkulluksn et al., 2018).

Technology has transformed how we teach, learn, and assess learners, from remote learning and virtual classrooms to personalized learning experiences and online assessments (Mhlanga & Moloi, 2020; Oguguo et al., 2023; Starkey, 2020). Education is no longer confined to traditional classrooms or physical campuses, but learning is

Digital Twins in the Smart Classroom. https://doi.org/10.1016/B978-0-443-41632-3.00009-6

becoming more convenient across boundaries of time and place (Camilleri & Camilleri, 2016; Dean & Campbell, 2020; Rapanta et al., 2020; Selwyn & Facer, 2014).

A new generation of innovative educational models is emerging to meet the needs of the contemporary learner and enhance the quality of the educational process (Bravo Villares & Fernández Sánchez, 2022). Smart learning environments (SLEs) are examples of how innovative technologies such as artificial intelligence, the Internet of Things, and machine learning are being adapted across various educational contexts. In these models, digital technology is not just a tool but also a dynamic environment that provides opportunities beyond the time and space constraints of traditional teaching and learning (Bilyalova et al., 2020; Ji et al., 2024; Santos et al., 2019; Yang, Wang, et al., 2023), which allows learners to have the best of both worlds: learning in both the physical and virtual learning environments.

Learners today, known as the net generation, digital natives, millennials, and generation Y (Palfrey & Gasser, 2008; Prensky, 2010; Tapscott, 2009), put new demands on the learning environment. They have their own expectations for their learning environments: dynamic and technology-driven to the exclusion of their traditional counterparts of physical classrooms and fixed schedules. Learners today covet constant access to mobile internet for diverse information and personalized resources; flexible learning to be able to study anytime and anywhere using their handheld devices; a comfortable, engaging atmosphere; opportunities for social interaction, sharing opinions, engaging in discussions, and communicating through social networks; and convenient digital tools that provide information on personally relevant topics (Brown, 2005; McNeely, 2005). In McNeely's (2005, p.43) words,

> *[The Net Generation] all use computers in their class work and in their hobbies. They have a wide range of interests outside their chosen area of study. They are not locked into one thing, although all are highly motivated and pursue their interests with passion. They use the latest technology, whether cell phones, computers, PDAs, MP3 players, or digital cameras. They expect things to work properly and work fast. They get bored if not challenged properly, but when challenged, they excel in creative and innovative ways. They learn by doing, not by reading the instruction manual or listening to lectures. These are the learners that faculty must reach.*

The 21st-century information society requires the mastery of new skills, mainly collaboration, communication, citizenship, creativity, critical thinking, digital literacy, and problem solving (Voogt & Roblin, 2012). Smart learning environments are designed to leverage these competencies through the integration of technology and innovative pedagogy to promote active, self-directed learning to prepare a more competitive, knowledgeable, creative, and innovative workforce in education, training, research, and development for the future (Bataineh, Bataineh, et al., 2025).

Traditional learning, even in technology-enhanced classrooms, predominantly supports lower-order cognitive skills through a delivery-receipt model (OECD, 2009) in which content is presented in simplified, decontextualized chunks that encourage memorization rather than higher-level thinking (Grabinger & Dunlap, 1995). In contrast, smart learning environments are essential for promoting higher-order cognitive skills (Bataineh, Kadhim, et al., 2025; Jena, 2013; Lu et al., 2021), which are often neglected in both physical and virtual settings when used separately.

The re/construction of the learning environment is at the heart of reforming teaching, learning, and assessment (Price, 2015). Smart learning environments are revolutionizing education by personalizing learning, catering to individual student needs, and enhancing engagement through interactive elements. This is a hybrid, adaptive system, where both formal and informal learning are integrated (Gros, 2016; Hwang & Fu, 2020; Singh & Hassan, 2017) to provide learners with engaging experiences while achieving learning outcomes using intelligent tools, potentially supporting an effective, efficient, and engaging learning experience (Spector, 2014).

Smart learning environments have characteristics that potentially set them apart from their traditional counterparts. SLEs integrate physical and virtual environments to provide better learning support and services tailored to learners' individual characteristics (Hwang, 2014). They further underscore recorded teaching and learning, personalized assessment, evaluation of learning outcomes (Kinshuk et al., 2016), content delivery, and tailored learning support (Koper, 2014). Moreover, both formal learning and informal, in- and out-of-class independent learning are promoted for knowledge or skill acquisition, employment, or a better life (Spector, 2014).

Technically, SLEs have learning-tracking capabilities in aspects such as knowledge acquisition, classroom interaction, and group collaboration. They can also provide personalized resources, such as learning time, place, peers, activities, and tools to facilitate effective learning (Hwang & Fu, 2020; Peng et al., 2019). They further create learning environments through monitoring air, temperature, light, sound, smell, and other factors to catalyze a comfortable physical environment. SLEs can also set up a learning community for specific learning scenarios and provide support to the learners for effective interaction and information exchange.

The key objectives of SLEs are the promotion of easy, engaged, and effective learning through their technical features of tracking processes, recognizing scenarios, context awareness (Boulanger et al., 2015), and connecting the community. Smart learning models (viz., smart virtual school, smart classroom, smart campus) take current digital innovations, adapt them, and use them to support, develop, and increase the effectiveness of education to make it more responsive to the dynamic changes in technology and knowledge. Artificial intelligence, big data, cloud computing, and similar technologies are all catalysts to SLEs, making learning interactive, flexible, adaptive, engaging, reflective, and, ultimately, more effective and efficient (e.g., Alanazy, 2022; Singh & Hassan, 2017; Spector, 2014).

Smart learning environments are designed according to the psychological and philosophical foundations of lifelong learning, where all individuals, regardless of their abilities, can learn (Colardyn, 2002). They align well with a smart educational culture, where the concepts of lifelong learning and education for all are emphasized. In these smart cultures, learners are allowed opportunities to learn, either online or through interactive virtual environments (Kerimbayev et al., 2023; Yang, Cai, et al., 2023), throughout their lives (Kerimbayev et al., 2023; Wang & Lehman, 2021) and be able to adapt to rapid changes in both society and the labor market (Yang et al., 2018; Zhang et al., 2022), ensuring the sustainability of education during emergencies, rapid transformations, and environmental health crises that may restrict traditional forms of education (Sengupta, 2023).

The digital transformation in smart learning environments has become a central topic in current educational literature, as it has affected teaching and learning practices, styles, and strategies alike. Using digital technologies has culminated in significant shifts from traditional learning models towards more interactive environments that promote critical and creative thinking, active participation, and interaction and engagement (Atif et al., 2015; Eden et al., 2024; Haleem et al., 2022).

Within SLEs, teaching and learning may not only be more feasible but also better able to achieve educational outcomes than their traditional counterparts. This is likely the result of integrating personalized and collaborative features that provide learners with opportunities for active knowledge and skill acquisition (Hwang & Fu, 2019). SLEs have not only taken learning beyond the confines of the traditional classroom but have also made it more flexible, lifelong, and accessible across time and place.

The literature (e.g., Wang et al., 2021) further provides evidence for the value of SLEs, noting features such as intelligent interaction, instant feedback, and user-friendly design that actively promote engagement and independent learning. In the current research, SLEs encompass three distinct models: smart school, smart classroom, and smart campus. Practical experiences, such as the COVID-19 pandemic, have uncovered an inevitable need for alternative education systems that can be readily adapted to respond to emergencies like natural disasters or health crises (Marks et al., 2020).

The notions of **smart school, smart campus,** and **smart classroom** involve holistic technology integration into education to promote academic and administrative efficiency. Smart schools harness advanced digital tools, such as AI-powered learning platforms, interactive whiteboards, and cloud-based management systems, to create dynamic, engaging learning environments. These offer both personalized learning and predictive mechanisms for identifying gaps in learning, with the support of virtual reality tools to make lessons more immersive and interactive (Tseng et al., 2022).

The smart campus builds upon the principles of smart school to encompass an entire institution of learning through the creation of related smart learning environments. Smart campuses involve integrating Internet-of-things devices with smart infrastructure and renewable energy systems to optimize the learning environment, communication, and resource management across the institution (Tseng et al., 2022).

The smart classroom, a smaller-scale model of the broader smart school and smart campus, is a technology-rich closed ecosystem that is gradually becoming an important part of modern education (Roberts, 2000). Smart classrooms promote not only teaching and learning but also active, personalized learning. The smart classroom potentially redefines traditional roles, as learners take control of their own learning in a collaborative learning environment with the facilitation of the teacher.

The notions of **smart school, smart campus,** and **smart classroom** involve holistic technology integration into education to promote academic and administrative efficiency. Smart schools harness advanced digital tools, such as AI-powered learning platforms, interactive whiteboards, and cloud-based management systems, to create dynamic, engaging learning environments. These offer both personalized learning and predictive mechanisms for identifying gaps in learning with the support of virtual reality tools to make lessons more immersive and interactive (Tseng et al., 2022).

The smart campus builds upon the principles of smart school to encompass an entire institution of learning through the creation of related smart learning environments. Smart campuses involve integrating Internet-of-things devices with smart infrastructure and renewable energy systems to optimize the learning environment, communication, and resource management across the institution (Tseng et al., 2022).

The smart classroom, a smaller-scale model of the broader smart school and smart campus, is a technology-rich closed ecosystem that is gradually becoming an important part of modern education (Roberts, 2000). Smart classrooms promote not only teaching and learning but also active, personalized learning. The smart classroom potentially redefines traditional roles, as learners take control of their own learning in a collaborative learning environment with the facilitation of the teacher.

A current global trend, learning in smart environments, has received mounting scholarly attention. A plethora of research has been conducted to seek answers for questions on various aspects of smart learning environments, such as what makes a virtual learning environment work well, how students and teachers interact in online classrooms, what tools are necessary to support effective and engaging digital learning, what potential challenges these environments might have, and what role emotional aspects (e.g., satisfaction, motivation, acceptance, student behaviors) play in smart learning environments (e.g., Al-Fraihat et al., 2020; Ali & Bhasin, 2020; Bao, 2020; Dai et al., 2023; Hollister et al., 2022; Shao, 2019; Wang et al., 2021; Zhou et al., 2020; Zhu et al., 2016).

The literature (e.g., Atif et al., 2015; Hew & Kadir, 2016; Sungkur & Maharaj, 2021) presents compelling evidence for the utility of SLEs in addressing diverse learners' needs, fostering their engagement in and satisfaction with learning, and ultimately improving their overall academic performance. Integration of smart technologies into traditional teaching and learning potentially creates effective, dynamic, and responsive smart learning environments that optimize instructional delivery, making virtual education more accessible, personalized, learner-centered, and aligned with contemporary educational demands.

As the adoption of smart learning models has the potential to increase the effectiveness and efficiency of educational institutions, several requirements must be met for them to fully realize their potential (e.g., Al-Barakat et al., 2023; Alanazy, 2022; Demir, 2021; Koper, 2014). Al-Barakat et al. (2023) emphasize the need for meeting certain requirements, such as helping students cultivate self-learning skills, improving digital tools for social interaction, fostering creative learning through smart devices, and offering meaningful digital educational activities tailored to learners' abilities, before adopting a smart model to guide learners to improve their use of digital tools in different SLEs. Similarly, Alanazy (2022) underscores the importance of building an infrastructure in terms of training, equipment, revisions to integrate artificial intelligence into the curriculum, state-of-the-art learning management systems, separate systems for formal and informal learning, and unified assessment databanks to ensure fair assessment of learning.

Considering how and to what extent the current changes in education affect the learning environments, it is imperative to highlight the importance of planning for the establishment of smart learning environments. A comprehensive review of related literature reveals that although a good body of research has been conducted on the effectiveness of SLEs and the challenges involved in their implementation, practically no research has attempted to identify the key requirements for their successful implementation, which is particularly important as smart learning environments feature prominently in the future of education. Therefore, examining global practices concerning the future of virtual smart schools, smart classrooms, and smart campuses is urgently needed.

Whereas the existing research has extensively explored the effectiveness, challenges, and emotional aspects of SLEs, the absence of studies that holistically address the foundational requirements needed to enable these environments to thrive is noted. The current literature, albeit rich in accounts of the integration of modern technologies, personalized learning, and the benefits of flexibility and interaction in SLEs, sports disjoined insights rather than clear articulations of the requirements, such as technical infrastructure, continuous assessment mechanisms, academic and technical support systems, and flexible content delivery, are essential for the sustainable success of these environments. A significant research gap exists in identifying the precise infrastructural, pedagogical, and technological conditions required for effective SLEs. The current research attempts to bridge this gap through the exploration of the specific requirements for SLEs, as identified by teachers, and the provision of reflective insights as to how to further enhance SLEs to improve learner engagement, academic performance, and lifelong learning in the context of the digital transformation of education.

2. Methodological approach

2.1 Method design

The requirements of smart learning environments are explored qualitatively to gain deeper insights into the perspectives of participants engaged with these environments. Understanding teachers' experiences is essential for better understanding the requirements to, ultimately, develop more effective SLEs (Creswell & Poth, 2018; Merriam & Tisdell, 2016).

2.2 Participants

A cohort of 43 public-school teachers involved in smart learning environments in Sharjah (United Arab Emirates) and Al-Ahsa (Saudi Arabia) comprised the sample of the research. The participants, who have knowledge and experience of SLEs, were purposefully selected to ensure they provide accurate, realistic, and in-depth insights, to both ascertain the credibility of their input and inform future research.

2.3 Data collection and analysis

Semistructured interviews were conducted with the participants to collect the data for the study. The interviewer asked follow-up questions to allow the participants to clarify their viewpoints and give more details whenever needed on the characteristics of SLEs that would improve the quality of the educational process. The five-question interview schedule was developed by the research team. Its validity and reliability were verified by piloting it on a sample of seven participants, later excluded from the main sample. The interviews were conducted twice, with a 2-week time lapse, and were analyzed by the second and third researchers. The agreement coefficient, calculated using Cohen's kappa formula, amounted to 0.97, which was deemed appropriate for the research.

After obtaining participants' consents and agreeing on the times and locations of the interviews, the data collection process commenced. The data were collected by the researchers themselves, with the second researcher conducting the interviews in Saudi Arabia and the third researcher in the United Arab Emirates. Each interview lasted 60–75 min to allow participants to express their views without constraint. All interviews were audio-recorded, with the participants' consent, and later transcribed to ensure the accuracy of the data and subsequent analysis. The interview transcripts were shown to the participants to verify that they accurately reflected their views. The participants all confirmed the accuracy of the transcripts (which were true to source with no correction or interference from the research team).

Using a grounded theory approach, the interview transcripts were qualitatively analyzed. A critical and analytical reading of each transcript led to the development of codes, which were then distributed into categories that served as the framework for organizing and presenting the findings. A descriptive summary of the overall findings was prepared using relevant excerpts from the interview transcripts. The analysis of the findings provided a deeper understanding of the participants' views of the requirements of smart learning environments to support student learning. In the ensuing findings and discussion sections, SLEs encompass the three models of smart school, smart classroom, and smart campus.

3. Findings

Among the key models explored under SLEs is the virtual smart school, which combines interactive virtual classrooms, video conferencing, and digital content to create a flexible learning environment that meets learners' needs. Another is the smart classroom model, which reimagines traditional classrooms by incorporating technologies (e.g., interactive screens and tablets) to personalize learning according to the learners' unique characteristics, needs, and cognitive abilities (Dai et al., 2023; Hollister et al., 2022; Shao, 2019). The smart campus model goes a step further, as it uses innovative technologies (e.g., Internet of Things, artificial intelligence) to combine academic and administrative functions in one digital system for the purpose of improving operational efficiency, reducing energy consumption, and catalyzing communication between students and faculty (Ali & Bhasin, 2020; Bao, 2020).

To achieve the objective of the research, identifying the requirements of smart learning environments, interviews were conducted with 43 public-school teachers in both the United Arab Emirates and Saudi Arabia. The data analysis revealed six main themes: *effective interaction and engagement, adaptation and alignment with students' needs, support for modern technologies, continuous assessment, technical and academic support,* and *flexible access to content.* A detailed account of each of these themes is presented below.

3.1 Theme 1: Effective interaction and engagement

The analysis of the data revealed that 40 teachers, representing 93% of the sample, affirmed that effective interaction between teachers and students in virtual learning environments is key to the success of the teaching/learning process. The respondents emphasized the importance of interactive tools in encouraging students' active participation in the lesson. The following excerpts from the interview transcripts are evidence that interaction is key to effective learning.

> *Interaction in virtual environments is very important ... this is a primary requirement ... a student's understanding of the educational content and achieving the goals depends on their interaction in smart virtual environments ... this makes students feel that they are part of a learning community.*

> *By using tools such as audio and video chats, live discussions can be activated that motivate students and encourage them to engage deeply with the learning content.*

> *Mm ... the best way to enhance interaction in a smart virtual environment is to use tools that enable live discussions and active student participation, away from feelings of isolation.*

The responses above indicate that teachers consider interaction in smart learning environments as equally important as it is in traditional classrooms. The presence of interactive tools significantly helps students to better understand the content. Moreover, social interaction makes them feel part of a lively learning environment, which also promotes active participation (Al-Barakat et al., 2025).

The analysis also revealed the importance of focusing on digital interaction in smart classrooms by using interactive learning tools, such as smart boards, tablets, and laptops, which enable teachers to easily write, clarify, and share interactive content with students. These tools can also contribute to direct positive interaction with the content.

Positive interaction in virtual environments also involves providing collaborative learning through collaborative platform tools or project management programs that promote cooperation and active participation in smart classroom environments. This, in turn, helps prevent the feelings of isolation that students often experience in remote learning environments.

3.2 Theme 2: Support for modern technologies

The data analysis revealed that about 93% of the respondents acknowledged the utility of new technologies, such as augmented and virtual reality, in both advancing learning and making it more authentic and interactive, as shown in the excerpts below:

> *To encourage active learning, smart learning environments should be equipped with innovative technologies, such as virtual reality, which provide students with opportunities to acquire practical skills and make learning more inclusive and enjoyable.*

> *... using augmented reality technology in smart classrooms is necessary because it helps learners better understand academic subjects by providing them with an interactive experience that can improve their understanding of study materials.*

> *To ensure effective communication between students and teachers, it is very important to provide innovative technologies, including live videos, interactive classrooms, and other interactive learning tools.*

> *It is essential in a smart campus to have a range of digital tools, such as email, file sharing, discussion forums, blogs, and digital portfolios, along with advanced resources, such as podcasts, recorded videos, presentations, simulations, and educational games. All of these guarantee an enjoyable educational experience.*

From the above responses, participants obviously believe that integrating various technologies into smart learning environments is very important to promote learners' engagement with the content presented to them. Therefore, these smart environments not only make learning more enjoyable, but they also simplify how complex concepts are presented (AlAli et al., 2025). The analysis also shows that applying these technologies improves learners' experiences through the provision of more authentic, interactive, and engaging content to make learning both enjoyable and efficient.

3.3 Theme 3: Continuous assessment

The data analysis showed that a little over 88% of the participants agreed that formative assessment is an essential element in smart learning environments. The use of electronic assessment tools allows for continuous monitoring of students' progress and the identification of areas of strength and weakness (Bataineh & Al-Ghareeb, 2025). Ponder the following excerpts:

> *Continuous formative assessment is useful because it enables the teacher to frequently check on students' progress to provide immediate feedback, a variety of comments, and ultimately improve students' learning outcomes.*

> *It is necessary to rely on assessment tools, such as electronic tests and surveys ..., in smart virtual schools.*

The above excerpts show how assessment tools help teachers monitor learners' knowledge and skill acquisition and how learners can be helped to achieve higher levels of academic success. This underscores the importance of consistently formative assessment and timely feedback provision over time. These findings emphasize the contribution of formative assessment to creating a flexible learning environment where teachers can use assessment results to adapt their instructional strategies. This ultimately promotes learning effectiveness and leads to better learning outcomes. In other words, the data analysis revealed the significance of timely feedback for achieving learning outcomes. Smart classrooms equipped with interactive assessment tools allow teachers to administer immediate assessments to both monitor student progress and enhance learning.

3.4 Theme 4: Technical and academic support

Thirty-five of the participants, representing just over 81%, reported that the success of smart learning environments is contingent upon factors about technical and academic support to ensure sustainable and glitch-free learning. The participants emphasized the importance of continuous academic support for students, as exemplified in the following excerpts:

> *... Smart learning environments should have technical support to manage any technical issues that may arise during lessons.*

> *Since there is no direct and constant contact, it is important to have academic advisors to direct the students remotely.*

There should be continuous technical support for virtual education to ensure the sustainability of the educational process, and there should also be ongoing academic support to guide and assist.

These excerpts indicate that the effectiveness of smart classrooms depends on the type of technical support that forms part of the process of resolving technical problems that may arise when using educational platforms. This support guarantees a smooth learning environment for students to engage in activities in smart environments. Moreover, the role of academic support in smart learning environments is also emphasized, which guides the students to recognize their areas of strength and weakness.

3.5 Theme 5: Flexibility of access to content

The data analysis revealed that a little more than 83% of the participants singled out flexibility in accessing content as one of the most important factors to promote learning in smart environments. These participants reported that flexibility in accessing content enables learners to schedule their learning at the time of their own discretion. Ponder the following excerpts:

The ability to study content at home or any other place and time outside class is both an advantage and a requirement of a smart learning environment, and that is the freedom to manage learning time.

Online learning platforms should also ensure that the students have the flexibility of learning at their own pace and returning to content at will to improve their learning.

These excerpts demonstrate the participants' agreement that the flexibility of access to content catalyzes learning, as students are free to work at their own time, early in the morning or after school, late in the afternoon or evening, to learn content in a manner most comfortable for them. The concept of flexibility of access to content also supports independent learning, as students can spend more time on difficult or problematic parts of the content that they feel the need to repeat or study more. The analysis also revealed that smart classrooms require multimodal content, such as lesson recordings, digital textbooks, and videos posted on e-learning platforms for students to access for review.

The flexibility of access to content in smart learning environments also ensures the sustainability of learning, as students may continue learning at their own time and pace, even in times of crises or emergencies. This adaptability not only fosters resilience in education systems but also empowers learners to take ownership of their learning, bridging gaps caused by disruptions like natural disasters, pandemics, or personal challenges. By integrating technology with pedagogy in smart learning environments, inclusive learning experiences that promote both continuity and equity ensue, ensuring that no student is left behind.

3.6 Theme 6: Adaptation to and alignment with learners' needs

The data analysis also revealed that a little over 76% of the participants agreed that individual differences should be considered, and content should be customized to meet learners' needs. The participants maintained that this significantly contributes to improving students' academic performance, as illustrated in the excerpts below:

By using smart tools, we can customize the content according to each student's needs, making learning more effective and improving their academic performance.

A smart school and classroom should be dynamic, capable of analyzing students' performance and suggesting supplementary or alternative content that aligns with their achievement level.

Customization helps provide a better learning environment for students because every student has their own learning style, so it is essential to offer content that aligns with their abilities and potential.

These excerpts demonstrate that personalization and adaptation to individual learners' needs readily improve their learning experience. When properly designed, smart learning environments allow students of diverse levels of knowledge and ability to develop at their leisure, leading to better academic performance. Therefore, smart school environments must personalize content to provide additional support to students in specific subject areas where they may need more assistance to help narrow gaps in their academic achievement.

These findings highlight key requirements for effective smart learning environments, emphasizing interaction, adaptability, technological integration, continuous assessment, and comprehensive support. The findings further emphasize the significance of engaging learners through the use of interactive tools, adapting content to meet individual learners' needs, and leveraging advanced technologies to improve learning experiences. In addition, continuous assessment and timely feedback, coupled with technical and academic support, emerged as catalysts for sustainable learning. Furthermore, flexible access to content emerged as a means for empowering learners to take control and ownership of their learning, promoting both independence and resilience. Combined, these elements create dynamic learner-centered smart learning environments that foster both engagement and academic success.

4. Discussion

Smart learning environments have a tremendous potential to improve and personalize learning. The findings revealed five key requirements for adopting smart learning environments: *interaction, adaptability, technological integration, continuous assessment,* and *comprehensive support systems.* Engagement and interaction emerged as requisites for promoting learner participation and reducing feelings of isolation in smart learning contexts. The participants emphasized the significance of interactive tools, such as audio and video chats, live discussions, and collaborative learning platforms, in promoting learner engagement.

These findings signal the significance of providing interactive tools for optimizing smart learning, as they potentially offset the disadvantages resulting from the lack of direct physical contact between students and teachers. The findings seem to suggest that smart learning environments should be designed with an eye on the best of the two worlds: the physical and the virtual. In other words, it is best to design SLEs that replicate traditional classroom interactions while using technology to catalyze learner engagement and motivation.

To reiterate, the findings support the use of interactive tools to empower teachers to encourage learners to attend and actively engage in virtual classes for improved knowledge and skill acquisition and, ultimately, better academic performance (Bataineh & Al-Ghareeb, 2025). This resonates with previous reports that interaction catalyzes learning outcomes in virtual classrooms (e.g., Hollister et al., 2022; Moore, 1989; Richardson & Swan, 2003).

Another key condition for implementing SLEs is personalization and alignment to learners' needs. Customizing content to individual learners' levels, needs, and learning styles emerged as a requisite for improving learner outcomes. The participants stressed the need for dynamic, adaptive learning systems to monitor individual learners' progress and eventually suggest appropriate supplementation or remediation.

These findings highlight the significance of incorporating adaptive learning technologies in smart learning environments to both respond to diverse learner abilities and ensure that every learner learns at his/her pace, potentially ensuring the sustainability of learning, even during times of crises or emergencies. This aligns with previous findings (e.g., Kerimbayev et al., 2023; Yang et al., 2018) that flexible SLEs promote positive interaction, provide immediate feedback, and support the sustainability of learning while helping learners achieve their desired outcomes.

To create sustainable smart learning environments, the participants emphasized the need for efficient educational resources, such as integrated resource management (e.g., e-books, smart libraries) and learning management systems (e.g., Moodle, Blackboard), reported to allow learners to monitor their progress, access materials, and communicate with instructors. The integration of adaptive technologies, such as augmented reality, virtual reality, and interactive digital tools, is essential in enhancing learning experiences, as they not only make learning more engaging but also simplify complex concepts through immersive experiences. This is consistent with previous research (e.g., Haleem et al., 2022; Kinshuk et al., 2016), which shows that providing equal access to modern smart learning technologies improves the quality of education and increases student engagement with the content.

Continuous formative assessment was also identified as a fundamental feature of smart learning environments, as the participants advocated using electronic assessment tools to monitor learner progress and provide timely feedback to ensure that the content meets individual learners' needs. Research has shown that the integration of formative assessment techniques potentially ensures teachers' tracking of learners' areas of strengths and weaknesses to enable them to make personalized instructional adjustments to bridge learning gaps and support academic progress. This potentially promotes learners' understanding and retention of content in an adaptive learner-centered manner, which readily aligns with the objective of smart learning environments. These findings align with those of earlier research (e.g., Ji et al., 2024; Wang & Lehman, 2021; Yang, Wang, et al., 2023; Yang, Wang, et al., 2023), which reported that continuous assessment offers real-time feedback that, ultimately, positively affects learning. These findings are also consistent with those of earlier research (e.g., Haleem et al., 2022; Sungkur & Maharaj, 2021; Wang & Lehman, 2021), which demonstrated that personalized content affects learning and improves overall academic performance.

The current findings also highlighted the importance of technical and academic support in ensuring both efficiency and sustainability of smart learning environments, as the participants emphasized the need for continuous technical support to ensure uninterrupted learning. Academic support, such as remote guidance, is critical to learners' effective navigation of their learning. These findings are supported by those of previous research, which found that peer support in smart learning environments positively affects academic performance (e.g., García-Machado et al., 2024; Stone, 2019), motivates learners, and promotes better engagement in learning (Mese & Sevilen, 2021; Robayo-Tamayoa et al., 2020).

Flexibility in accessing content also emerged as vital for independent learning, as the ability to review materials at his/her leisure enables the learner to independently take both control and ownership of his/her learning. The participants pointed out the need for multimodal content, such as recorded lectures, digital textbooks, and interactive materials, to support self-paced learning, autonomy, and sustainability of learning in times of crisis.

5. Conclusions, implications, and recommendations

The findings affirm the potential utility of well-structured smart learning environments that incorporate interaction, adaptability, technological advancements, continuous assessment, and robust support systems in advancing learning, engagement, and academic success. The authors recommend incorporating innovative technologies, such as augmented and virtual reality, to enhance learning by allowing more immersive interaction with content to facilitate understanding complex concepts. Previous research (e.g., Eden et al., 2024; Haleem et al., 2022; Hew & Kadir, 2016) reported that augmented and virtual reality potentially make learning more interactive, realistic, and engaging, ultimately improving learners' understanding and interaction.

The findings also support the significance of formative assessment in learning, as participants highlighted how continuous assessment and real-time feedback in SLEs facilitate tracking and supporting learner progress. The findings also emphasize the need for consistent technical and academic support to ascertain that learners can troubleshoot any technical issues to maintain sustainable learning. Ongoing academic support also helps learners tackle academic challenges in both the psychological and behavioral domains, ultimately shaping the quality of their academic performance.

The current findings call for policymakers and educational institutions should prioritize investments in advanced digital tools to catalyze both teaching and learning. The findings further highlight the need for institutions of learning to put into place comprehensive support mechanisms to facilitate smooth and uninterrupted learning.

Funding

This work was financially supported by the Deanship of Scientific Research, King Faisal University, Saudi Arabia [Grant number KFU250536].

Acknowledgments

We thank the Deanship of Scientific Research at King Faisal University for providing financial support to this research. We would also like to thank all the participants who participated in this study for their time and valuable contributions.

References

Al-Barakat, A., Al-Hassan, O., AlAli, R. M., Al-Hassan, M., & Al Sharief, R. (2023). Role of female teachers of childhood education in directing children towards effective use of smart devices. *Education and Information Technologies, 28*(6), 7065–7087. https://www.ncbi.nlm.nih.gov/pmc/articles/PMC9707229/.

Al-Barakat, A. A., Al-Hassan, O. M., AlAli, R. M., Bataineh, R. F., Aboud, Y. Z., & Ibrahim, N. A. (2025). Shaping young minds: How teachers foster social interaction, psychological security and motivational support in the primary language classroom. *International Journal of Learning, Teaching and Educational Research, 24*(1), 359–378. https://doi.org/10.26803/ijlter

Al-Fraihat, D., Joy, M., & Sinclair, J. (2020). Evaluating e-learning systems' success: An empirical study. *Computers in Human Behaviour, 102*, 67–86. https://doi.org/10.1016/j.chb.2019.08.004

AlAli, R., Al-Barakat, A., Bataineh, R. F., & Alqatawna, M. (2025). From pixels to prose: Teachers' views on the power of digital imagery in early language development. *Forum for Linguistic Studies, 7*(2), 160–173. https://doi.org/10.30564/fls.v7i2.8186

AlAli, R., Al-Hassan, O., Al-Barakat, A., Alakashee, B., Kanaan, E., Alqatawna, M., Zaher, A., & Saleh, S. (2024). Effectiveness of utilizing gamified learning in improving creative reading skills among primary school students. *Forum for Linguistic Studies, 6*(6), 816–830. https://doi.org/10.30564/fls.v6i6.7518

Alanazy, S. M. (2022). tˤaːwiːr bijaːʔaːt ɑltaʕallum ɑððɑkijja fiː dˤawʔ mutˤɑlɑbbɑːt ɑlððɑkaːʔ ɑlʔiṣtˤinɑːʕiː ɑlmustɑqbalijja ɑlʒɑːmiʕɑːt ɑssaʕuːdijja [Developing smart learning environments in line with the future requirements of artificial intelligence: Saudi universities]. *Journal of Humanities and Management Sciences- Scientific Journal of King Faisal University, 23*(1). https://doi.org/10.37575/h/edu/210079, 136–129.

Ali, A., & Bhasin, J. (2020). A model of information system interventions for e-learning: An empirical analysis of information system interventions in e-learner perceived satisfaction. In P. Singh, A. Kar, Y. Singh, M. Kolekar, & S. Tanwar (Eds.), *Proceedings of ICRIC 2019. Lecture Notes in Electrical Engineering* (Vol. 597, pp. 909–920). Springer. https://doi.org/10.1007/978-3-030-29407-6_66

Andoh, C. B. (2012). Factors influencing teachers' adoption and integration of information and communication technology into teaching: A review of the literature. *International Journal of Education and Development Using Information and Communication Technology, 8*(1), 136–155. http://ijedict.dec.uwi.edu//viewarticle.php?id=1361.

Atif, Y., Mathew, S., & Lakas, A. (2015). Building a smart campus to support ubiquitous learning. *Journal of Ambient Intelligence and Humanized Computing, 6*, 223–238. https://doi.org/10.1007/s12652-014-0226-y

Bao, W. (2020). COVID-19 and online teaching in higher education: A case study of Peking University. *Human Behaviour and Emerging Technologies, 2*, 1–3. https://doi.org/10.1002/hbe2.191

Bataineh, R. F., & Al-Ghareeb, M. B. (2025). Starfall as a catalyst for Kuwaiti EFL young learners' reading comprehension: A teacher's reflections. *Journal of Ethnic and Cultural Studies, 12*(5), 141–153. https://doi.org/10.29333/ejecs/2338

Bataineh, R. F., Bataineh, R. F., AlAli, R. M., Alotaibi, S. B., Al-Barakat, A. A., Al-Saud, K. M., Aboud, Y. Z., & Ibrahim, N. A. (2025). Digital Frontiers: The transformative potential of e-learning in cultivating Arab primary school learners' creativity. *Sage Open, 15*(4). https://doi.org/10.1177/21582440251408317

Bataineh, R. F., Kadhim, A. M., AlAli, R. M., Kanaan, E. M., Al-Barakat, A. A., Abdullatif, A. K. A., Zaher, A. M., & Aboud, Y. Z. (2025). Nurturing young minds: SWOM-based reading as a catalyst for reflective thinking in early childhood. *International Journal of Learning, Teaching and Educational Research, 24*(7), 1000–1019. https://doi.org/10.26803/ijlter.24.7.47

Bilyalova, A. A., Salimova, D. A., & Zelenina, T. I. (2020). Digital transformation in education. In *Proceeding of the International Conference on Integrated Science* (pp. 265–276). Springer. https://doi.org/10.1007/978-3-030-22493-6_24

Boulanger, D., Seanosky, J., Kumar, V., Kinshuk, P.,K., & Thamarai, S. S. (2015). Smart learning analytics. In G. Chen, V. Kumar, Kinshuk, R. Huang, & S. Kong (Eds.), *Emerging issues in smart learning. Lecture notes in educational technology* (pp. 289–296). Springer. https://doi.org/10.1007/978-3-662-44188-6_39

Bravo Villares, V., & Fernández Sánchez, M. J. (2022). Perceptions and use of digital media in arts education: A descriptive study. *Innoeduca. International Journal of Technology and Educational Innovation, 8* (1), 75–90. https://doi.org/10.24310/innoeduca.2022.v8i1.12069

Brown, M. (2005). Learning spaces. In D. G. Oblinger, & J. L. Oblinger (Eds.), *Educating the net generation*. EDUCAUSE. https://www.educause.edu/research-and-publications/books/educating-net-generation/lea rning-spaces.

Bucăţa, G., Popescu, F., & Tileag ă, C. (2022). Digital transformation of higher education system. *Proceedings of the 28th International Scientific Conference Knowledge-Based Organization, 28*(1), 158–168. https://doi.org/10.2478/kbo-2022-0025

Bucăţa, G., & Tileagă, C. (2024). Digital renaissance in education: Unveiling the transformative potential of digitization in educational institutions. *Land Forces Academy Review, 29*(1), 20–37. https://doi.org/10.2478/raft-2024-0003

Camilleri, M. A., & Camilleri, A. C. (2016). Digital learning resources and ubiquitous technologies in education. *Technology, Knowledge and Learning, 22*, 65–82. https://doi.org/10.1007/s10758-016-9287-7

Colardyn, D. (2002). *Lifelong learning: Which ways forward?* Lemma Publishers.

Creswell, J. W., & Poth, C. N. (2018). *Qualitative inquiry and research design: Choosing among five approaches*. Sage Publications.

Cuban, L. (2001). *Oversold and underused: Computers in the classroom*. Harvard University Press.

Dai, Z., Xiong, J., Zhao, L., & Zhu, X. (2023). Smart classroom learning environment preferences of higher education teachers and students in China: An ecological perspective. *Heliyon, 9*(6), Article e16769. https://doi.org/10.1016/j.heliyon.2023.e16769

Dean, B. A., & Campbell, M. P. (2020). Reshaping work-integrated learning in a post-COVID-19 world of work. *International Journal of Work-Integrated Learning, 21*(4), 356–364. https://files.eric.ed.gov/fulltext/EJ1271541.pdf.

Demir, K. A. (2021). Smart education framework. *Smart Learning Environments, 8*, 29. https://doi.org/10.1186/s40561-021-00170-x

Eden, C. A., Nneamaka, O., Chisom, C., & Adeniyi, I. S. (2024). Harnessing technology integration in education: Strategies for enhancing learning outcomes and equity. *World Journal of Advanced Engineering Technology and Sciences, 11*(2), 001–008. https://doi.org/10.30574/wjaets.2024.11.2.0071

Galván Orozco, A., López Pérez, O., Chávez López, J. K., & Contreras López, E. X. (2022). Virtual learning environments: Social networks for learning at university. *Innoeduca. International Journal of Technology and Educational Innovation, 8*(1), 91–101. https://doi.org/10.24310/innoeduca.2022.v8i1.12340

García-Machado, J. J., Martínez Ávila, M., Dospinescu, N., & Dospinescu, O. (2024). How the support that students receive during online learning influences their academic performance. *Education and Information Technologies, 29*, 20005–20029. https://doi.org/10.1007/s10639-024-12639-6

Grabinger, R. S., & Dunlap, J. C. (1995). Rich environments for active learning: A definition. *Research in Learning Technology, 3*(2). https://doi.org/10.3402/rlt.v3i2.9606

Gros, B. (2016). The design of smart educational environments. *Smart Learning Environments, 3*, 15. https://doi.org/10.1186/s40561-016-0039-x

Haleem, A., Javaid, M., Qadri, M. A., & Suman, R. (2022). Understanding the role of digital technologies in education: A review. *Sustainable Operations and Computers, 3*, 275–285. https://doi.org/10.1016/j.susoc.2022.05.004

Hew, T., & Kadir, S. (2016). Predicting the acceptance of cloud-based virtual learning environments: The roles of self-determination and channel expansion theory. *Telematics and Informatics, 33*(4), 990–1013. https://doi.org/10.1016/j.tele.2016.01.004

Hollister, B., Nair, P., Hill-Lindsay, S., & Chukoskie, L. (2022). Engagement in online learning: Student attitudes and behaviour during COVID-19. *Frontiers in Education, 7*, Article 851019. https://doi.org/10.3389/feduc.2022.851019

Hwang, G. J. (2014). Definition, framework and research issues of smart learning environments - A context-aware ubiquitous learning perspective. *Smart Learning Environments, 1*(1), 4. https://doi.org/10.1186/s40561-014-0004-5

Hwang, G. J., & Fu, Q. K. (2019). Trends in the research design and application of mobile language learning: A review of 2007–2016 publications in selected SSCI journals. *Interactive Learning Environments, 27*(4), 567–581. https://doi.org/10.1080/10494820.2018.1486861

Hwang, G. J., & Fu, Q. K. (2020). Advancement and research trends of smart learning environments in the mobile era. *International Journal of Mobile Learning and Organisation, 14*(1), 114–129. https://doi.org/10.1504/IJMLO.2020.103911

Javaid, M., Haleem, A., Singh, R. P., & Sinha, A. K. (2024). Digital economy to improve the culture of industry 4.0: A study on features, implementation and challenges. *Green Technologies and Sustainability.* , Article 100083. https://doi.org/10.1016/j.grets.2024.100083

Jena, P. C. (2013). Effect of smart classroom learning environment on academic achievement of rural high achievers and low achievers in science. *International Letters of Social and Humanistic Sciences, 3*, 1–9. https://doi.org/10.18052/www.scipress.com/ILSHS.3.1

Ji, H., Suo, L., & Chen, H. (2024). AI performance assessment in blended learning: Mechanisms and effects on students' continuous learning motivation. *Frontiers in Psychology, 15*, Article 1447680. https://doi.org/10.3389/fpsyg.2024.1447680

Kerimbayev, N., Umirzakova, Z., Shadiev, R., & Jotsov, V. (2023). A student-centred approach using modern technologies in distance learning: A systematic review of the literature. *Smart Learning Environments, 10*, 61. https://doi.org/10.1186/s40561-023-00280-8

Kinshuk, Chen, N. S., Cheng, I. L., & Chew, S. W. (2016). Evolution is not enough: Revolutionizing current learning environments to smart learning environments. *International Journal of Artificial Intelligence in Education, 26*(2), 561–581. https://doi.org/10.1007/s40593-016-0108-x

Koper, R. (2014). Conditions for effective smart learning environments. *Smart Learning Environments, 1*(1), 5. https://doi.org/10.1186/s40561-014-0005-4

Kumar, S., Ritzhaupt, A. D., & Pedro, N. S. (2022). Development and validation of the Online Instructor Support Survey (OISS). *Online Learning, 26*(1), 221–244. https://doi.org/10.24059/olj.v26i1.2622

Lu, K., Yang, H. H., Shi, Y., & Wang, X. (2021). Examining the key influencing factors on college students' higher-order thinking skills in the smart classroom environment. *International Journal of Educational Technology in Higher Education, 18*, 1. https://doi.org/10.1186/s41239-020-00238-7

MacIntyre, P. D., Gregersen, T., & Mercer, S. (2020). Language teachers' coping strategies during the Covid-19 conversion to online teaching: Correlations with stress, wellbeing and negative emotions. *System, 94*, 1–13. https://doi.org/10.1016/j.system.2020.102352

Marks, A., AL-Ali, M., Atassi, R., Abualkishik, A. Z., & Rezgu, Y. (2020). Digital transformation in higher education: A framework for maturity assessment. *International Journal of Advanced Computer Science and Applications, 11*(12). https://doi.org/10.14569/IJACSA.2020.0111261

Matt, C., Hess, T., Benlian, A., & Wiesböck, F. (2016). Options for formulating a digital transformation strategy. *MIS Quarterly Executive, 15*(2), 123–139. https://aisel.aisnet.org/misqe/vol15/iss2/6.

McCarthy, A. M., Maor, D., McConney, A., & Cavanaugh, C. (2023). Digital transformation in education: Critical components for leaders of system change. *Social Sciences & Humanities Open, 8*(1), Article 100479. https://doi.org/10.1016/j.ssaho.2023.100479

McKnight, K., O'Malley, K., Ruzic, R., Horsley, M. K., Franey, J. J., & Bassett, K. (2016). Teaching in a digital age: How educators use technology to improve student learning. *Journal of Research on Technology in Education, 48*(3), 194–211. https://doi.org/10.1080/15391523.2016.1175856

McNeely, J. L. (2005). Using technology as a learning tool, not just the cool new thing. In *Educating the net generation* (pp. 4.1–4.10). EDUCAUSE. http://digitalcommons.brockport.edu/bookshelf/272/.

Merriam, S. B., & Tisdell, E. J. (2016). *Qualitative research: A guide to design and implementation*. Jossey-Bass.

Mese, E., & Sevilen, Ç. (2021). Factors influencing EFL students' motivation in online learning: A qualitative case study. *Journal of Educational Technology and Online Learning, 4*(1), 11–12. https://dergipark.org.tr/en/pub/jetol/issue/60134/817680.

Mhlanga, D. (2024). Digital transformation of education, the limitations and prospects of introducing the fourth industrial revolution asynchronous online learning in emerging markets. *Discover Education, 3*, 32. https://doi.org/10.1007/s44217-024-00115-9

Mhlanga, D., & Moloi, T. (2020). COVID-19 and the digital transformation of education: What are we learning on 4IR in South Africa? *Education in Science, 10*(7), 180. https://doi.org/10.3390/educsci10070180

Moore, M. G. (1989). Editorial: Three types of interaction. *American Journal of Distance Education, 3*(2), 1–7. https://doi.org/10.1080/08923648909526659

OECD. (2009). *Creating effective teaching and learning environments: First results from TALIS*. Organisation for Economic Co-operation and Development.

Oguguo, B., Ezechukwu, R., Nannim, F., & Offor, K. (2023). Analysis of teachers in the use of digital resources in online teaching and assessment in COVID times. *Innoeduca: International Journal of Technology Educational Innovation, 9*(1), 81–96. https://doi.org/10.24310/innoeduca.2023.v9i1.15419

Palfrey, J., & Gasser, U. (2008). *Born digital: Understanding the first generation of digital natives*. Basic Books.

Peng, H., Ma, S., & Spector, J. M. (2019). Personalized adaptive learning: An emerging pedagogical approach enabled by a smart learning environment. *Smart Learning Environments, 6*, 9. https://doi.org/10.1186/s40561-019-0089-y

Petrov, P., Radev, M., Dimitrov, G., & Simeonidis, D. (2022). Infrastructure capacity planning in digitalization of educational services. *International Journal of Emerging Technologies in Learning, 17*(3), 299–306. https://doi.org/10.3991/ijet.v17i03.27811

Prensky, M. (2010). *Teaching digital natives: Partnering for real learning*. Sage Publishers.

Price, J. K. (2015). Transforming learning for the smart learning environment: Lessons learned from the intel education initiatives. *Smart Learning Environments, 2*, 16. https://doi.org/10.1186/s40561-015-0022-y

Pucciarelli, F., & Kaplan, A. M. (2019). Competition in higher education. In B. Nguyen, T. C. Melewar, & J. Hemsley-Brown (Eds.), *Strategic brand management in higher education*. Routledge.

Rapanta, C., Botturi, L., Goodyear, P., Guàrdia, L., & Koole, M. (2020). Online university teaching during and after the Covid-19 crisis: Refocusing teacher presence and learning activity. *Postdigital Science and Education, 2*, 923–945. https://doi.org/10.1007/s42438-020-00155-y

Richardson, J. C., & Swan, K. (2003). Examining social presence in online courses in relation to students' perceived learning and satisfaction. *Journal of Asynchronous Learning Networks, 7*(1), 68–88. https://doi.org/10.24059/olj.v7i1.1864

Robayo-Tamayoa, M., Blanco-Donoso, L. M., Román, F. J., Carmona-Cobo, I., Moreno-Jiménez, B., & Garossa, E. (2020). Academic engagement: A diary study on the mediating role of academic support. *Learning and Individual Differences, 80*, Article 101887. https://doi.org/10.1016/j.lindif.2020.101887

Roberts, J. (2000). From know-how to show-how? Questioning the role of information and communication technologies in knowledge transfer. *Technology Analysis and Strategic Management, 12*(4), 429–443. https://doi.org/10.1080/713698499

Santos, H., Batista, J., & Marques, R. P. (2019). Digital transformation in higher education: The use of communication technologies by students. *Procedia Computer Science, 164*, 123–130. https://doi.org/10.1016/j.procs.2019.12.163

Selwyn, N., & Facer, K. (2014). The sociology of education and digital technology: Past, present and future. *Oxford Review of Education, 40*(4), 482–496. https://doi.org/10.1080/03054985.2014.933005

Sengupta, E. (2023). Introduction to higher education in emergencies: Best practices and benchmarking. In E. Sengupta (Ed.), *Higher education in emergencies: Best practices and benchmarking* (pp. 3–8). Emerald Publishing Limited. https://doi.org/10.1108/S2055-364120230000053001

Shao, C. (2019). An empirical study on the identification of driving factors of satisfaction with online learning based on TAM. *Proceedings of the 5th International Conference on Economics, Management, Law and Education*, 1067–1073. https://doi.org/10.2991/aebmr.k.191225.205

Singh, A. D., & Hassan, M. (2017). In pursuit of smart learning environments for the 21st century. In *UNESCO international bureau of education*, current and critical issues in curriculum, learning and assessment. UNESCO. https://unesdoc.unesco.org/ark:/48223/pf0000252335.

Spector, J. M. (2014). Conceptualizing the emerging field of smart learning environments. *Smart Learning Environments, 1*, 2. https://doi.org/10.1186/s40561-014-0002-7

Starkey, L. (2020). A review of research exploring teacher preparation for the digital age. *Cambridge Journal of Education, 50*(1), 37–56. https://doi.org/10.1080/0305764X.2019.1625867

Stone, C. (2019). Online learning in Australian higher education: Opportunities, challenges and transformations. *Student Success, 10*(2), 1–11. https://doi.org/10.5204/ssj.v10i2.1299

Sungkur, R., & Maharaj, M. (2021). Design and implementation of a SMART learning environment for the upskilling of cybersecurity professionals in Mauritius. *Education and Information Technologies, 26*, 3175–3201. https://doi.org/10.1007/s10639-020-10408-9

Tapscott, D. (2009). *Grown up digital: How the net generation is changing your world.* McGraw-Hill.

Tseng, S. S., Chen, S. N., & Yang, T. Y. (2022). Building an AR-based smart campus platform. *Multimedia Tools and Applications, 81*(1), 5695–5716. https://doi.org/10.1007/s11042-021-11702-5

Vongkulluksn, V. W., Xie, K., & Bowman, M. A. (2018). The role of value on teachers' internalization of external barriers and externalization of personal beliefs for classroom technology integration. *Computers & Education, 118*, 70–81. https://doi.org/10.1016/j.compedu.2017.11.009

Voogt, J., & Roblin, N. (2012). A comparative analysis of international frameworks for 21st century competences: Implications for national curriculum policies. *Journal of Curriculum Studies, 44*(3), 299–321. https://doi.org/10.1080/00220272.2012.668938 (open in a new window).

Wang, C., Chen, X., Yu, T., Liu, Y., & Jing, Y. (2024). Education reform and change driven by digital technology: A bibliometric study from a global perspective. *Humanities and Social Sciences Communication, 11*, 256. https://doi.org/10.1057/s41599-024-02717-y

Wang, H., & Lehman, J. D. (2021). Using achievement goal-based personalized motivational feedback to enhance online learning. *Educational Technology Research & Development, 69*(2), 553–581. https://doi.org/10.1007/s11423-021-09940-3

Wang, S., Shi, G., Lu, M., Lin, R., & Yang, J. (2021). Determinants of active online learning in the smart learning environment: An empirical study with PLS-SEM. *Sustainability, 13*(17), 9923. https://doi.org/10.3390/su13179923

Xiao, J. (2019). Digital transformation in higher education: Critiquing the five-year development plans (2016–2020) of 75 Chinese universities. *Distance Education, 40*(4), 515–533. https://doi.org/10.1080/01587919.2019.1680272

Yang, D., Wang, H., Metwally, A., et al. (2023). Student engagement during emergency remote teaching: A scoping review. *Smart Learning Environments, 10*, 24. https://doi.org/10.1186/s40561-023-00240-2

Yang, H., Cai, J., Yang, H. H., & Wang, X. (2023). Examining key factors of beginner's continuance intention in blended learning in higher education. *Journal of Computing in Higher Education, 35*, 126–143. https://doi.org/10.1007/s12528-022-09322-5

Yang, J., Pan, H., Zhou, W., & Huang, R. (2018). Evaluation of smart classroom from the perspective of infusing technology into pedagogy. *Smart Learning Environments, 5*, 20. https://doi.org/10.1186/s40561-018-0070-1

Zhang, M., Croiset, E., & Ioannidis, M. A. (2022). Constructivist-based experiential learning: A case study of student-centred and design-centric unit operation distillation laboratory. *Education for Chemical Engineers, 41*, 22–31. https://doi.org/10.1016/j.ece.2022.09.002

Zhou, L., Wu, S., Zhou, M., & Li, F. (2020). School's out, but class's on: The largest online education in the world today: Taking China's practical exploration during the COVID-19 epidemic prevention and control as an example. *Best Evidence in Chinese Education, 4*(2), 501–519. https://doi.org/10.2139/ssrn.3555520

Zhu, Z. T., Yu, M. H., & Riezebos, P. (2016). A research framework of smart education. *Smart Learning Environments, 3*(1), 4. https://doi.org/10.1186/s40561-016-0026-2

Chapter 22

IoT sensors and devices in smart schools: A case study in the Universitas Negeri Makassar (Indonesia)

Abdul Wahid
Universitas Negeri Makassar, Makassar, Indonesia

1. Introduction

1.1 Background

The education scene has undergone a paradigm change in which traditional classrooms are no longer the only vehicle for transmitting information. Students of today expect technologically driven, interactive, tailored learning environments. Therefore, including IoT in the school setting is not only a creative idea but also a need. These technologies help teachers control and track academic and operational tasks as well as allow smooth interaction between students and learning resources (Terzieva et al., 2022).

Worldwide, IoT in education has concentrated on enhancing engagement, safety, and resource control (Badshah et al., 2024). In developing nations like Indonesia, though, the use of smart education systems has another purpose: closing access and infrastructure gaps. The difficulties brought on by the COVID-19 epidemic hastened the use of hybrid and blended learning approaches, therefore highlighting the need for funding in adaptive and sensor-driven settings.

In Indonesia, initiatives such as the Smart Campus Framework developed by the Universitas Negeri Makassar (UNM) have demonstrated the practical impact of IoT deployment in higher education. This framework integrates smart blended learning systems, biometric access control, real-time monitoring, and centralized data management, forming a reference model for secondary and vocational school implementation. Research outputs, including the LBPH-based facial recognition system for e attendance and the AI-supported smart blended learning environment, highlight the feasibility and scalability of such IoT-enabled applications in real educational settings.

These real world implementations validate the vision of smart schools as dynamic, data driven systems. They not only facilitate learning but also ensure institutional agility through features such as automated attendance, learning analytics, energy efficient classrooms, and predictive maintenance. By anchoring these systems in realtime data from sensors, schools can transition from reactive to proactive decision-making — a core capability of modern DTss.

Moving forward, the focus should not only be on the deployment of IoT devices but also on the integration of these technologies within the broader academic and governance frameworks. This includes establishing interoperability standards, training educators to work with smart systems, and securing ethical safeguards around data usage. As we delve into the architecture and case studies in the following sections, it becomes evident that the future of education lies in these connected, intelligent ecosystems. Such as the Smart Campus Framework developed by the Universitas Negeri Makassar has demonstrated the practical impact of IoT deployment in higher education. This framework integrates smart blended learning systems, biometric access control, real-time monitoring, and centralized data management, forming a reference model for secondary and vocational school implementation. Research outputs, including the LBPH based facial recognition system for e attendance and the AI-supported smart blended learning environment, highlight the feasibility and scalability of such IoT-enabled applications in real educational settings.

1.2 Role of Digital Twins

Digital twins (DTs) are virtual representations of physical classroom environments (Ağca, 2023). By utilizing data

Digital Twins in the Smart Classroom. https://doi.org/10.1016/B978-0-443-41632-3.00033-3

collected from a range of IoT sensors, these digital models mirror the real-time conditions of classrooms, laboratories, and other learning spaces. The role of DTs extends beyond passive visualization—they enable simulation, forecasting, and optimization of school environments.

In educational settings, DTs can be used to dynamically adjust environmental conditions such as lighting, air quality, or temperature based on sensor input (Jankovskis et al., 2024). For instance, if CO_2 levels rise above a healthy threshold, the system can simulate the impact and trigger ventilation adjustments in real time. Moreover, DTs can be applied to simulate emergency response scenarios, track room utilization, and manage energy consumption more effectively.

The Smart Campus Framework developed at the Universitas Negeri Makassar leverages DT principles to integrate building systems with academic operations. Environmental sensor data (e.g., temperature, humidity, lighting) and security inputs (e.g., fingerprint access logs, facial recognition records) feed into a central visualization and analytics platform. This provides administrators with actionable insights into both infrastructural and educational performance.

In the Smart Blended Learning System, digital twin concepts are realized through synchronized video feeds and motion-tracking data (Kartashova et al., 2024). The system emulates the physical presence of a teacher in both online and offline formats, ensuring consistent delivery regardless of learning modality. Through intelligent camera control and integration with LMS platforms, the DT supports real-time decision-making for instructors and learners alike (Chamorro-Atalaya et al., 2024).

Furthermore, DTs can assist in long-term planning by enabling predictive analytics. Administrators may forecast classroom occupancy, identify patterns in student engagement, and anticipate maintenance needs (Bienz et al., 2024). As AI models are trained on the accumulated sensor data, they enrich the DT's ability to serve as an autonomous assistant in educational resource management (Casal-Otero et al., 2023).

1.3 Objectives and scope

This chapter builds upon a range of research and development efforts conducted at the Universitas Negeri Makassar and beyond, incorporating both theoretical insights and hands-on implementations. The objectives are twofold: to present a conceptual foundation for IoT adoption in smart schools, and to showcase practical case studies that demonstrate its feasibility, adaptability, and educational value. With an emphasis on scalable architecture and localized innovation, this chapter serves as a roadmap for policymakers, school administrators, and technology developers.

It presents:

- A classification of IoT sensor types and their educational functions.
- Real-world implementation cases within smart school frameworks.
- System architecture and AI integration.
- Challenges and recommendations for scaling up.

2. Sensor technologies in education

2.1 Sensor typologies

IoT devices in schools fall into four categories as shown in Fig. 22.1:

- **Environmental sensors:** Environmental sensors are instrumental in monitoring key environmental parameters and ensuring sustainable development by providing real-time data that support environmental protection and resource management (Lv et al., 2024; Sathyamoorthy et al., 2024). These include CO_2 monitors, thermometers, humidity sensors, and light intensity meters. In smart classrooms, they are crucial for regulating ventilation and climate. For instance, in the Smart Campus prototype at UNM, CO_2 sensors are configured to trigger alerts and ventilation adjustments once certain thresholds are exceeded. Light sensors help optimize natural light usage, reducing dependency on artificial lighting.
- **Biometric and access sensors:** Biometric and access sensors have become critical components in modern security systems by leveraging unique physiological and behavioral characteristics of individuals for robust authentication and authorization. Their utilization spans from unlocking devices to securing high-stakes facilities, and these systems typically rely on a diverse range of biometric modalities, including fingerprints, facial features, iris patterns, voice, and dynamic behavioral traits (Abe et al., 2022; Talreja et al., 2018). These are used for identity verification and physical access control. At UNM, the Security Village system integrates fingerprint sensors with ESP32 microcontrollers and servo motors for controlling classroom and campus gates. The system ensures only registered individuals can access restricted areas, improving safety and attendance tracking.
- **Health monitoring sensors:** Health monitoring sensors are a rapidly evolving class of devices that provide continuous insights into an individual's physiological status. In recent years, advances in materials science, flexible electronics, and signal processing have been central to overcoming traditional challenges such as signal noise, motion artifacts, and user discomfort, thereby enabling accurate data capture in dynamic,

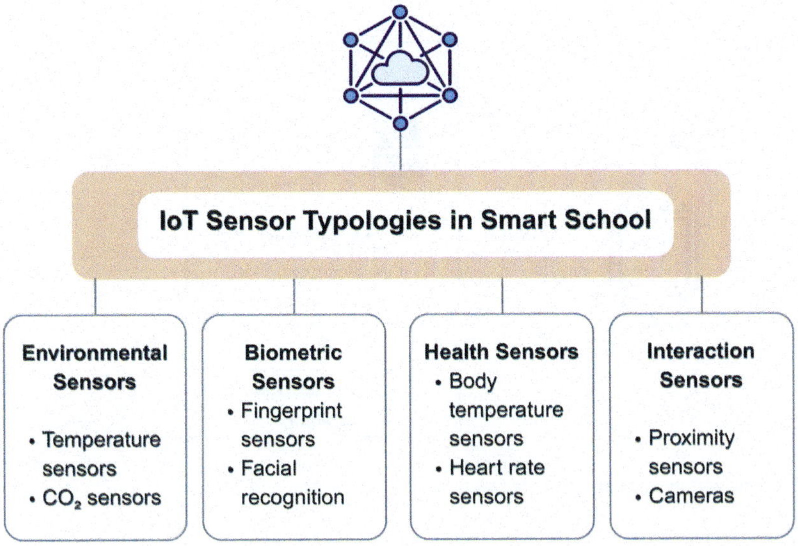

FIGURE 22.1 Classification of IoT sensors in smart schools (environmental, biometric, health, interaction).

real-world conditions (Guarducci et al., 2025; Smith et al., 2022; Yan et al., 2019). Wearables or embedded devices in seats can track students' heart rates, temperature, and stress levels. While not yet widely implemented, pilot initiatives using thermal cameras and body temperature sensors were explored during COVID-19 to minimize contact and monitor student wellness.

- **Interaction sensors:** Interaction sensors comprise a diverse class of devices that facilitate human–machine communication through the detection and interpretation of various physical stimuli. Motion sensors, ultrasonic distance sensors, and eye-tracking technologies detect presence, gestures, and gaze. These sensors are useful in understanding learner behavior and interaction, particularly in project-based or flipped classroom environments (Grosu et al., 2017; Swaminathan et al., 2020). In the AI-assisted Smart Blended Learning System, tracking teacher movement via computer vision relies on image input from strategically placed cameras acting as visual sensors.

These typologies provide the foundation for creating dynamic and responsive learning spaces. They enable systems to autonomously adjust to human presence, environmental changes, and user behavior in real time as shown in Fig. 22.2.

2.2 Integration methodologies

IoT devices are integrated through multitiered systems that consist of edge devices, communication protocols, and backend data services (Terzieva et al., 2022; Zeeshan et al., 2022). At the edge level, microcontrollers like the ESP32 serve as primary interfaces to various sensors, executing initial data filtering and aggregation. These edge devices reduce the volume of data transmitted by performing local computations and transmitting only relevant information.

For communication, lightweight protocols such as Message Queuing Telemetry Transport (MQTT) and HTTP REST APIs are commonly used. In the Security Village system, MQTT was deployed for real-time event delivery between fingerprint sensors and the central server, enabling quick and reliable access validation. MQTT's publish-subscribe model is well-suited for IoT because of its low bandwidth usage and event-driven design.

In terms of interoperability, systems at the Makassar State University have successfully integrated Firebase (for cloud database and authentication), OpenCV (for facial recognition tasks), and the Syam-OK LMS, which is built on Moodle. These systems communicate through web services using standardized APIs that ensure data consistency and synchronization. For instance, when a student checks in via facial recognition, the attendance record is simultaneously updated in both the e-presence system and the LMS.

To support multimodal data flow—such as biometric data, environmental logs, and interaction tracking—hybrid architectures are adopted. These architectures employ both cloud and on-premise servers depending on latency requirements and data privacy considerations. LoRa and BLE protocols are also explored in pilot systems to enable long-range and energy-efficient connectivity for classroom equipment and environmental sensors.

Such integrations pave the way for a scalable and maintainable IoT infrastructure in education, where new devices or systems can be added with minimal disruption to the existing ecosystem.

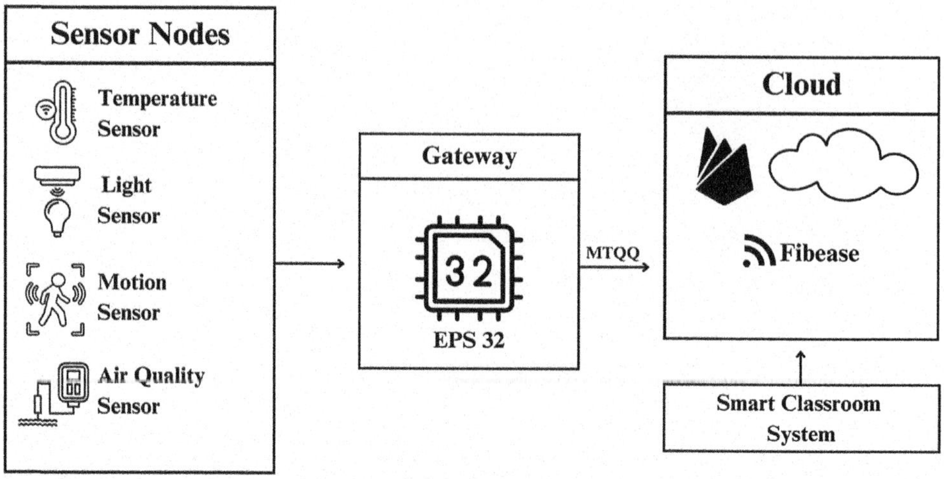

FIGURE 22.2 An illustration of IoT integration in a smart classroom using MQTT, Firebase, and ESP32.

3. Case studies in IoT-enabled smart schools

3.1 Security Village (smart fingerprint access)

The Security Village system was developed as a secure entry solution for school environments, leveraging biometric verification and IoT-based automation (Wahid et al., 2024). It is designed to minimize unauthorized access and enhance attendance monitoring through seamless system integration.

The system architecture comprises an ESP32 microcontroller connected to a fingerprint sensor module and a servo motor mechanism as shown in Fig. 22.3. When a registered fingerprint is scanned, the ESP32 processes the data locally and communicates with a cloud server using the MQTT protocol for real-time validation. Upon successful authentication, the servo motor activates to unlock the entry point.

The backend uses Firebase for cloud database management and user credential storage. Integration with an Android-based application allows administrators to monitor logs, enroll new users, and receive alerts.

Key features:

- Real-time server authentication using MQTT over Wi-Fi.
- Lightweight mobile application interface with under 60 MB memory usage.
- Cross-device compatibility tested on 14 Android models.
- Ninety-two percent usability score in ISO 25,010-based evaluation.
- Secure log storage and access notification features for administrators.
- Real-time server authentication.
- Minimal resource footprint on mobile Android systems.
- Synchronization with school databases.

3.2 Facial recognition with LBPH algorithm

The Local Binary Pattern Histogram (LBPH) facial recognition system was developed to automate the student attendance process while maintaining data security and minimizing manual intervention (Nurhikma et al., 2024). This system captures facial data via a webcam, extracts local binary patterns from grayscale images, and uses histogram comparison for classification as shown in Fig. 22.4. Built with OpenCV and Python, it is optimized for desktop-based operations using XAMPP as a local web server.

The recognition process involves training the model with multiple facial images per user under various conditions. The real time application matches incoming webcam frames against stored profiles and logs attendance events accordingly. The web interface also enables administrative tasks such as training new faces and monitoring login events.

Key features:

- Ninety percent recognition accuracy under ideal lighting conditions.
- Drop in accuracy observed in cases of mask usage or poor lighting.
- False rejection rate (FRR) and false acceptance rate (FAR) evaluated using a confusion matrix.
- Integration with local database and LMS for automatic attendance record updates.
- XAMPP + OpenCV implementation supports local network deployment without requiring cloud connectivity.
- Accuracy of 90% under optimal lighting.
- Low recognition rate when masked or in low-light conditions.
- FRR and FAR are analyzed through a confusion matrix.

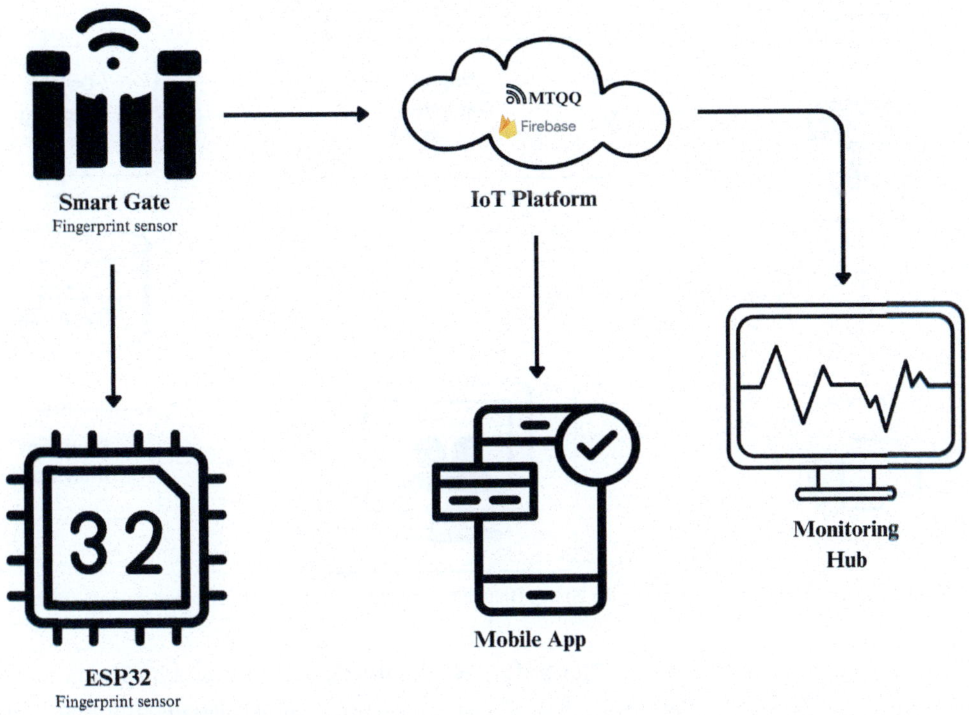

FIGURE 22.3 Workflow of the Security Village fingerprint access system.

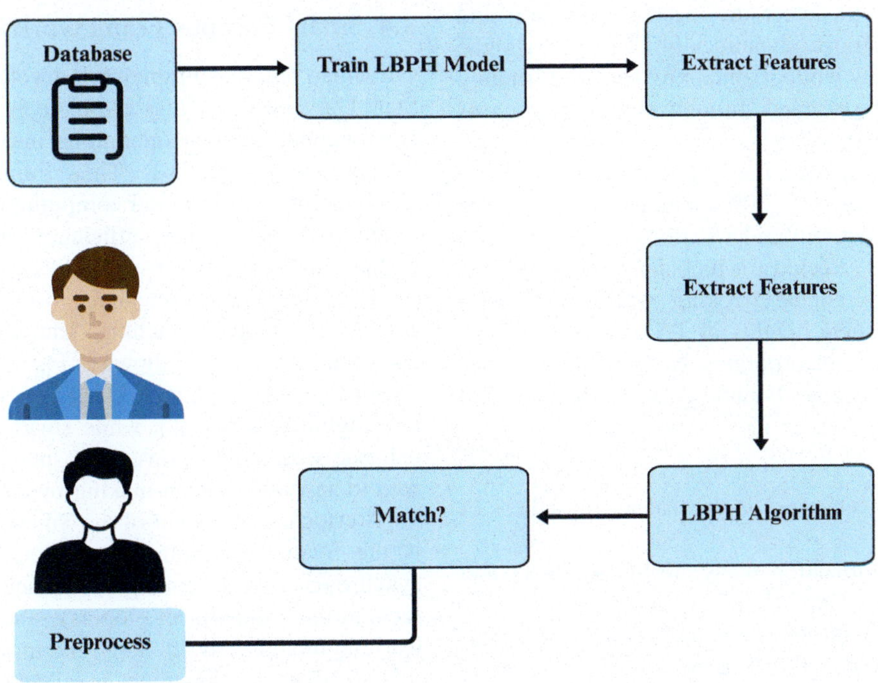

FIGURE 22.4 Workflow of the LBPH-based smart attendance system.

3.3 Smart Blended Learning with AI object detection

The Smart Blended Learning system addresses the challenge of maintaining consistent teaching presence in hybrid environments. Designed in response to educational constraints during the COVID-19 pandemic, this system uses AI based camera control to replicate the experience of an in person classroom for remote learners (Wahid et al., 2022) as shown in Fig. 22.5.

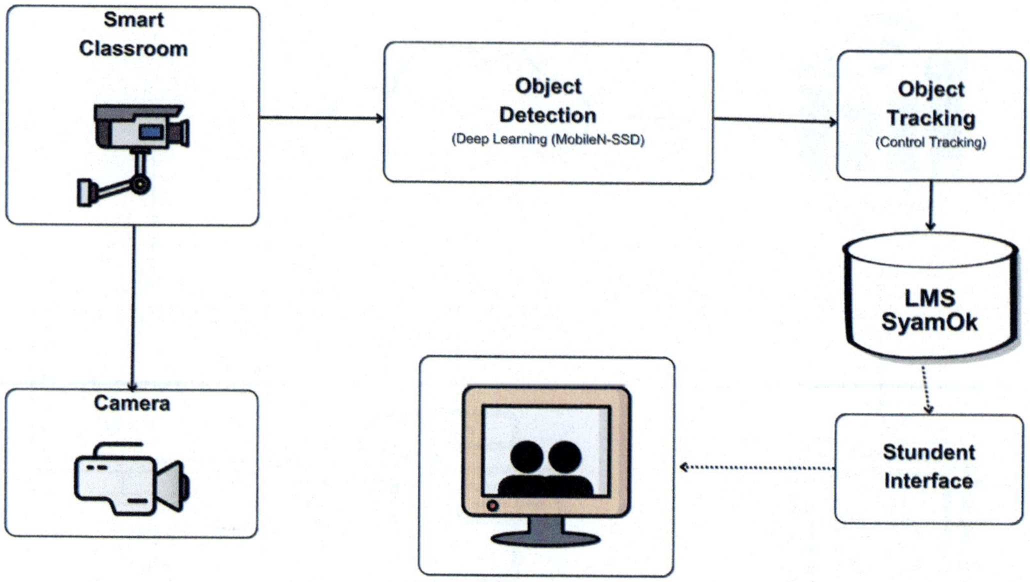

FIGURE 22.5 System architecture of the Smart Blended Learning platform with AI-based camera tracking.

The core of the system relies on the MobileNet-SSD (Single Shot Detector) for object detection and a Centroid Tracking algorithm to maintain focus on the instructor. Video input from classroom cameras is processed in real time to detect the teacher's position and adjust the camera frame dynamically. This allows the teacher to move freely while maintaining centered framing in the video output streamed through platforms such as Zoom or Google Meet.

To improve performance, the system was trained using a dataset consisting of over 300 annotated video frames capturing instructor activity under various conditions. The MobileNet-SSD was selected for its balance of speed and accuracy, achieving reliable tracking within a 6-meter range and across camera heights of 1.5 m and 3 m. The tracking module also incorporates Non-maximum Suppression (NMS) to refine bounding box selection and prevent duplication.

Key Features:

- Intelligent video centering on the teacher regardless of movement.
- Performance optimization using MobileNet-SSD and centroid-based tracking.
- Tested under multiple lighting and distance configurations with F1 scores ≥ 0.8 at up to 6 m.
- Real-time synchronization with video conferencing platforms.
- Reduced the gap in student learning experience between physical and virtual classrooms.
- Positive feedback in pilot deployment at the Makassar State University.
- Automated focus on the instructor up to 6 m.
- Real-time integration with Zoom and Meet platforms.
- Increased learner satisfaction due to immersive hybrid learning experience.

3.4 Smart Campus Framework implementation

The Smart Campus Framework developed at the Makassar State University provides a holistic model for transforming conventional campus operations into interconnected and data-driven ecosystems. This framework leverages a combination of IoT, cloud computing, AI, and automation to improve the quality, efficiency, and sustainability of higher education services (Wahid et al., 2021).

The architecture integrates various subsystems that collectively support teaching, administration, monitoring, and campus security as shown in Fig. 22.6. For instance, IoT sensors manage real time classroom conditions (temperature, lighting, humidity), while smart surveillance systems enhance security. AI-powered dashboards aggregate sensor data to support decision-making by campus administrators.

A critical component of the Smart Campus Framework is the Syam OK platform, a Moodle based LMS that synchronizes with smart attendance tools (e.g., facial recognition, fingerprint logins) and supports blended learning models. Additionally, the integration with e-office systems digitizes document management, scheduling, and staff workflows.

The system utilizes RESTful APIs to ensure interoperability between the LMS, academic information systems (SIA), and smart device controllers. Data collected from biometric devices, environmental sensors, and access logs is analyzed in a centralized cloud repository to detect patterns, generate alerts, and inform resource planning.

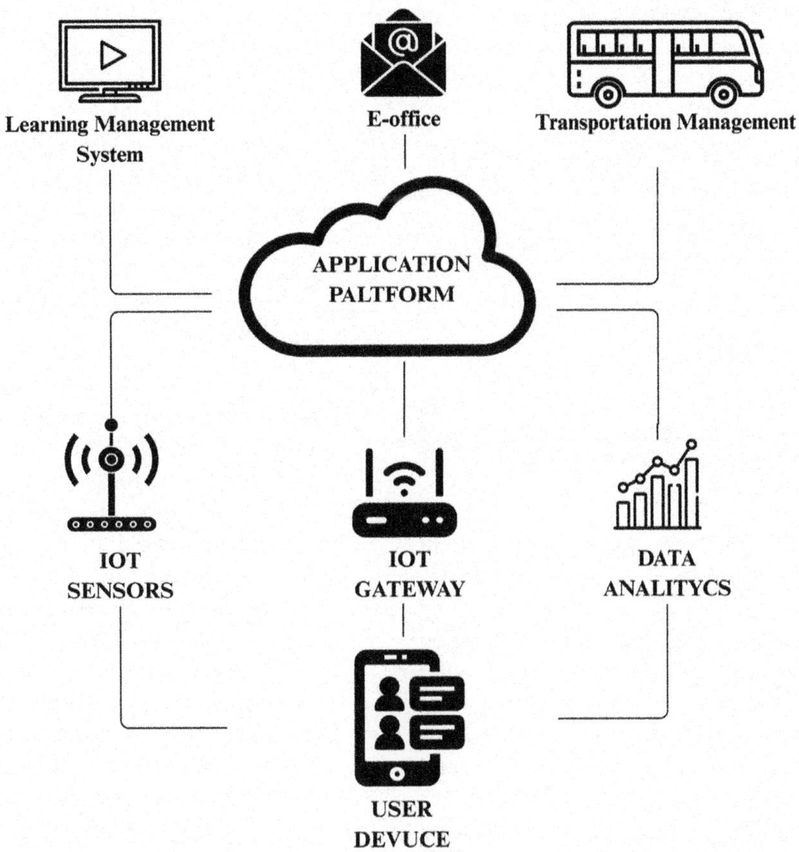

FIGURE 22.6 Architecture overview of the Smart Campus Framework at UNM.

Key Components:

- Syam-OK LMS and AI-assisted e-learning tools.
- IoT integration for smart classrooms and utilities.
- E-office systems for workflow automation.
- API-driven interoperability among platforms.
- Real-time analytics for decision-making and efficiency monitoring.

Piloted at the Makassar State University, this architecture covers:

- LMS (Syam-OK) with synchronized user management.
- E-office for administrative digitization.
- IoT-controlled environmental regulation and security systems.
- API-based communication with academic information systems.

4. Systems architecture and AI integration

4.1 Infrastructure design

The implementation of IoT in smart schools requires a robust and modular infrastructure that supports scalability, security, and real-time responsiveness (Hasan, 2025; Ibrahim & Kenwright, 2022). At the Makassar State University, a multilayered architecture was developed, comprising:

- **Physical layer:** This includes various sensors (temperature, CO_2, motion, fingerprint, and facial recognition) and embedded devices such as the ESP32. These components are deployed in classrooms, gates, and administrative offices.
- **Network layer:** Utilizes Wi-Fi for high-bandwidth, short-range communication; BLE (Bluetooth Low Energy) for device pairing and low-power applications; and LoRaWAN for long-range sensor communication across the campus environment. These protocols ensure flexibility in connectivity based on specific device needs.
- **Data layer:** Edge computing is performed on microcontroller boards (e.g., ESP32) and local servers for latency-sensitive tasks like attendance verification. For centralized analytics, cloud platforms such as Firebase and AWS IoT Core handle storage and aggregation of sensor data.
- **Application layer:** Interfaces with systems like the Syam-OK LMS, e-office dashboard, Zoom integration,

and other administrative platforms. This layer presents insights through real-time dashboards for decision-makers, using web-based APIs and notification services.

Each layer communicates using standardized RESTful APIs, ensuring interoperability across vendor hardware and software services. This modular architecture allows seamless upgrades and integration with future technologies.

4.2 AI and big data integration

To make sense of the vast data collected by IoT devices, artificial intelligence and big data analytics are integrated into the system (Karan & Angadi, 2025; Wang et al., 2024). AI modules are employed for:

- **Behavioral analysis:** Detecting patterns in student engagement (e.g., posture detection, movement across camera fields) to assess attentiveness during lessons.
- **Predictive trends:** Machine learning models trained on academic performance and attendance records predict student success and identify those needing intervention.
- **Resource optimization:** Usage data from classrooms, labs, and energy systems is analyzed to recommend scheduling adjustments and reduce utility costs.
- **Maintenance forecasting:** Vibration, temperature, and usage data from equipment like projectors or HVAC units feed into predictive models to schedule maintenance before breakdowns occur (Kamalov et al., 2023; Mallik & Gangopadhyay, 2023).

Big data infrastructure is built using Firebase Realtime Database and BigQuery, with visual analytics dashboards using Google Data Studio and Grafana. These systems allow school administrators to receive alerts, monitor trends, and plan policy adjustments based on data evidence.

In combination with the DT framework discussed earlier, this integration enables a feedback loop: sensors monitor, AI analyzes, and the system acts—thus making smart schools truly intelligent and adaptive environments.

5. Ethical and governance frameworks

Educational institutions adopting IoT technologies must establish robust ethical and governance frameworks to safeguard the privacy and civil rights of users, particularly the sensitive demographic of students and minors. As these institutions deploy systems such as biometric authentication, video surveillance, and real-time tracking to enhance security and operational efficiency, significant challenges arise regarding data privacy, informed consent, and the secure handling of personally identifiable information (Kouroupis & Vagianos, 2023; Lola et al., 2023).

A comprehensive ethical framework for smart schools must be grounded in principles that safeguard student privacy, ensure data security, foster transparency, and promote accountability while enabling the transformative benefits of IoT-enabled learning environments (Kouroupis & Vagianos, 2023; Meylani, 2024). A comprehensive ethical framework for smart schools should address the following dimensions:

- **Data privacy and protection:** Compliance with data protection regulations such as GDPR is critical. This includes limiting data collection to what is strictly necessary, ensuring anonymization or pseudonymization where appropriate, and securing data during storage and transmission using encryption protocols.
- **Informed consent:** Institutions must develop clear, understandable consent forms that inform users (students, parents, teachers) of the data being collected, the purpose, and who will have access. For minors, parental consent is essential and must be recorded securely.
- **Transparency and accountability:** Users should be informed about how decisions are made by AI systems, including access to explanation features in case of automated decisions affecting student records or evaluations.
- **Bias and fairness:** AI models used in prediction and surveillance must be audited for bias. This includes facial recognition models, which may perform differently across ethnicities, lighting conditions, or age groups.
- **Access and control:** Role-based access control (RBAC) should be implemented to restrict data access to authorized personnel only. Additionally, users should have the right to view and request correction or deletion of their personal data.
- **Ethical oversight committee:** Institutions deploying smart systems should establish an ethics review board or data governance committee to regularly review practices, manage risks, and ensure adherence to evolving norms.
- **Technology audit and logging:** Regular audits of systems, especially those using machine learning, should be conducted to track performance drift, misuse, or vulnerabilities. All data access and modifications should be logged and monitored for integrity.

By integrating these ethical safeguards, smart school ecosystems can maintain user trust while leveraging the benefits of IoT and AI technologies responsibly.

6. Conclusion and future directions

Smart schools equipped with IoT infrastructures offer enhanced personalization, operational intelligence, and resilience. These advancements not only support real-time classroom management but also enable data-driven strategies for resource allocation, safety assurance, and pedagogical innovation. The incorporation of DTs enhances responsiveness and foresight in decision-making processes, while AI-powered analytics amplify the ability of institutions to adapt to dynamic educational needs.

The practical implementations described in this chapter—including the Security Village fingerprint access system, the LBPH-based facial recognition attendance system, and the AI-assisted Smart Blended Learning platform—demonstrate that IoT and AI integration is not just a theoretical framework, but an achievable goal within both higher education and secondary school environments. These cases also highlight the value of localized innovation, where context-aware design leads to effective and sustainable solutions.

Despite promising outcomes, the path toward full-scale smart school transformation is not without challenges. Issues such as infrastructure disparities, ethical data governance, technological literacy among educators, and financial constraints remain pressing concerns. Addressing these requires multisector collaboration among government, academia, industry, and local communities.

Looking ahead, future development should explore interoperable IoT frameworks, incorporate emerging technologies like blockchain for data traceability, and investigate long-term impacts on student behavior and mental health. Moreover, strengthening national policy support, investing in professional development, and nurturing interdisciplinary research will be critical to ensuring that smart school ecosystems flourish in both urban and rural contexts.

The insights gathered in this chapter aim to guide educators, policymakers, system designers, and researchers in shaping the next generation of learning environments—ones that are intelligent, inclusive, and resilient by design.

7. Recommendations

Develop national standards for IoT devices in schools. To ensure interoperability and safety, the government, in collaboration with education and technology experts, should establish regulatory standards for hardware and communication protocols used in educational environments. These standards should cover data privacy, interoperability, device certification, and network security.

Embed AI literacy in teacher training programs. Professional development for educators must include training on how to interpret sensor data, respond to AI-generated insights, and ensure ethical classroom use of such technologies. AI literacy will empower teachers to effectively manage smart classroom environments while safeguarding student rights.

Pilot test smart classrooms in urban and rural settings. Implementation should begin with pilot programs to evaluate the practical feasibility of smart classroom models under varying conditions. These pilots should track metrics such as learning outcomes, student engagement, teacher feedback, and infrastructure readiness. Lessons learned can inform wider deployment.

Evaluate long-term psychological and social impacts. Continuous monitoring and research should be conducted to understand how real-time monitoring, facial recognition, and AI prediction models affect student well-being, behavior, and sense of autonomy. This is particularly crucial when deploying technology among younger learners.

Explore blockchain for secure data traceability. The use of blockchain can enhance transparency and integrity in academic record-keeping, identity verification, and IoT device logs. Its immutability and decentralization provide a tamper-proof layer for critical school data, ensuring trust among all stakeholders.

References

Ağca, R. K. (2023). Using digital twins in education from an innovative perspective: Potential and application areas. *Education Mind, 2*. https://doi.org/10.58583/Pedapub.EM2306

Abe, B. C., Araromi, H. O., Shokenu, E. S., Idowu, P. O., Babatunde, J. D., Adeagbo, M. A., et al. (2022). Biometric access control using voice and fingerprint. *Engineering and Technology Journal, 7*, 1376–1382. https://doi.org/10.47191/etj/v7i7.08

Badshah, A., Ghani, A., Daud, A., Jalal, A., Bilal, M., & Crowcroft, J. (2024). Towards smart education through the internet of things: A review. *ACM Computing Surveys, 56*, 1–33. https://doi.org/10.1145/3610401

Bienz, K., Bountry, D., Liu, C. R., & Ghalamchi, B. (2024). Impact of digital twins in engineering education: Enhancing learning motivation and accessibility - A review study with a proposed new solution. In *2024 ASEE annu. Conf. Expo. Proc., Portland, Oregon: ASEE conferences*, Article 47555. https://doi.org/10.18260/1-2--47555

Casal-Otero, L., Catala, A., Fernández-Morante, C., Taboada, M., Cebreiro, B., & Barro, S. (2023). AI literacy in K-12: A systematic literature review. *International Journal of STEM Education, 10*, 29. https://doi.org/10.1186/s40594-023-00418-7

Chamorro-Atalaya, O., Flores-Velásquez, C. H., Flores-Cáceres, R., Arévalo-Tuesta, J. A., Zevallos-Castañeda, M., Tomás-Quispe, G., et al. (2024). Use of digital twin technology in the teaching-learning process, in the field of university education: A bibliometric review. *International Journal of Learning, Teaching and Educational Research, 23*, 313–329. https://doi.org/10.26803/ijlter.23.2.15

Grosu, V., Grosu, S., Vanderborght, B., Lefeber, D., & Rodriguez-Guerrero, C. (2017). Multi-axis force sensor for human–robot interaction sensing in a rehabilitation robotic device. *Sensors, 17*, 1294. https://doi.org/10.3390/s17061294

Guarducci, S., Jayousi, S., Caputo, S., & Mucchi, L. (2025). Key fundamentals and examples of sensors for human health: Wearable, non-continuous, and non-contact monitoring devices. *Sensors, 25*, 556. https://doi.org/10.3390/s25020556

Hasan, D. (2025). IoT-Based smart education: A systematic review of the state of the art. *Journal of Intelligent Systems and Information Technology, 2*, 1–16. https://doi.org/10.61971/jisit.v2i1.106

Ibrahim, I. S., & Kenwright, B. (2022). *Smart education: Higher education instruction and the internet of things (IoT)*. https://doi.org/10.48550/arXiv.2207.02585

Jankovskis, D., Cirule, I., & Carbone, A. (2024). Digital twins and E-Learning: Navigating challenges and opportunities. In G. Casalino, R. Di Fuccio, G. Fulantelli, P. Raviolo, P. C. Rivoltella, D. Taibi, et al. (Eds.), *High. Educ. Learn. Methodol. Technol. Online* (Vol. 2076, pp. 169–179). Cham: Springer Nature Switzerland. https://doi.org/10.1007/978-3-031-67351-1_12

Kamalov, F., Calong, D. S., & Gurrib, I. (2023). *New era of artificial intelligence in education: Towards a sustainable multifaceted revolution*. https://doi.org/10.48550/arXiv.2305.18303

Karan, B., & Angadi, G. R. (2025). Artificial intelligence integration into school education: A review of Indian and foreign perspectives. *Millenn Asia, 16*, 173–199. https://doi.org/10.1177/09763996231158229

Kartashova, L. A., Gurzhii, A. M., Zaichuk, V. O., & Sorochan, T. M. (2024). Digital twin technology for blended learning in educational institutions during COVID-19 pandemic. *CTE Workshop Proceedings, 11*, 411–426. https://doi.org/10.55056/cte.666

Kouroupis, K., & Vagianos, D. (2023). IoT in education: Implementation scenarios through the lens of data privacy law. *Journal of Politics and Ethics in New Technologies and AI, 2*. https://doi.org/10.12681/jpentai.34616. e34616–e34616.

Lola, J., Serrão, C., & Casal, J. (2023). Towards transparent and secure IoT: Improving the security and privacy through a user-centric rules-based system. *Electronics, 12*, 2589. https://doi.org/10.3390/electronics12122589

Lv, T., Zou, Y., Wang, Y., & Wu, E. (2024). Research on environmental sensor data analysis based on real-time data warehouse. In *Int. Conf. Cloud Comput. Perform. Comput. Deep Learn. CCPCDL 2024, vol. 13281* (pp. 115–121). SPIE. https://doi.org/10.1117/12.3051045

Mallik, S., & Gangopadhyay, A. (2023). *Proactive and reactive engagement of artificial intelligence methods for education: A review*. https://doi.org/10.48550/arXiv.2301.10231

Meylani, R. (2024). Transforming education with the internet of things: A journey into smarter learning environments. *International Journal of Research in Education and Science, 10*, 161–178. https://doi.org/10.46328/ijres.3362

Nurhikma, A. W., & Parenreng, J. M. (2024). E-Presence and monitoring system based on face image recognition with local binnary pattern histogram (LBPH) algorithm. *Journal of Embedded Systems, Security and Intelligent Systems*, 68–75. https://doi.org/10.59562/jessi.v5i1.556

Sathyamoorthy, S., Matthew, U. O., Adekunle, T. S., & Okafor, N. U. (2024). Advances and challenges in IoT sensors data handling and processing in environmental monitoring networks. *HAFED POLY Journal of Science, Management and Technology, 5*, 40–60. https://doi.org/10.4314/hpjsmt.v5i2.3

Smith, A., Li, R., Xu, L., & Tse, Z. (2022). *New paradigm of personal health monitoring: Nanomaterial-based In-textile sensors*. https://doi.org/10.21203/rs.3.rs-2346614/v1

Swaminathan, S., Fagert, J., Rivera, M., Cao, A., Laput, G., Noh, H. Y., et al. (2020). OptiStructures: Fabrication of room-scale interactive structures with embedded fiber bragg grating optical sensors and displays. *Proceedings of the ACM on Interactive, Mobile, Wearable and Ubiquitous Technologies, 4*, 1–21. https://doi.org/10.1145/3397310

Talreja, V., Ferrett, T., Valenti, M. C., & Ross, A. (2018). Biometrics-as-a-service: A framework to promote innovative biometric recognition in the cloud. In *2018 IEEE Int. Conf. Consum. Electron. ICCE* (pp. 1–6). https://doi.org/10.1109/ICCE.2018.8326075

Terzieva, V., Ilchev, S., & Todorova, K. (2022). The role of internet of things in smart education. *IFAC-PapersOnLine, 55*, 108–113. https://doi.org/10.1016/j.ifacol.2022.08.057

Wahid, A., Fajar, B. M., Parenreng, J. M., Luhriyani, S., & Adi, P. D. P. (2022). Smart blended learning framework based on artificial intelligence using MobileNet single shot detector and centroid tracking algorithm. *International Journal of Advanced Computer Science and Applications, 13*. https://doi.org/10.14569/IJACSA.2022.0130543

Wahid, A., Hidayat, R., & Riska, M. (2024). Mobile app for IoT-Based home security system with fingerprint access control. *Internet Things Artificial Intelligence Journal, 4*, 274–288. https://doi.org/10.31763/iota.v4i2.732

Wahid, A., Luhriyani, S., Nurhikmah, P. J. M., Fajar, B. M., & Nur, M. I. (2021). Smart campus framework: A solution for new normal education system. In *2021 IEEE 5th Int. Conf. Inf. Technol. Inf. Syst. Electr. Eng. ICITISEE, Purwokerto, Indonesia: IEEE* (pp. 266–271). https://doi.org/10.1109/ICITISEE53823.2021.9655952

Wang, S., Wang, F., Zhu, Z., Wang, J., Tran, T., & Du, Z. (2024). Artificial intelligence in education: A systematic literature review. *Expert Systems with Applications, 252*, Article 124167. https://doi.org/10.1016/j.eswa.2024.124167

Yan, S., Saleem, M. F., Ma, H., Li, Z., Zhang, X., Guo, X., et al. (2019). An ultra-sensitive, rapidly responsive strain sensor based on silver microflakes by simple process. *ChemistrySelect, 4*, 4407–4415. https://doi.org/10.1002/slct.201900558

Zeeshan, K., Hämäläinen, T., & Neittaanmäki, P. (2022). Internet of things for sustainable smart education: An overview. *Sustainability, 14*, 4293. https://doi.org/10.3390/su14074293

Chapter 23

Recent developments in smart classrooms/smart schools in India

Shalini Kumari, Chander Prabha and Nabanita Roy

Chitkara University Institute of Engineering and Technology, Chitkara University, Punjab, India

1. Introduction

With the spread of digital transformation (DX), the significance of utilizing data increases. Simultaneously, the volume of data continues rising, with a staggering volume generated each second globally, far beyond human understanding and being generated every second, beyond human comprehension (Attaran & Celik, 2023). The Digital Twin (DT) has emerged in various industries, including manufacturing, healthcare, and smart cities. According to Education 4.0, Digital Twin technology can have a transformative role in school by optimizing operations, maximizing the environment, and improving how students learn (Botín-Sanabria et al., 2022). Digital Twins can be leveraged to facilitate data-driven decisions, predictive maintenance, and personalized learning by creating a virtual copy of a school's infrastructure, operations, and student performance. DT integrates real-time data from various sources, including Internet of Things (IoT) sensors, learning management systems (LMS), and artificial intelligence (AI)-driven analytics, to develop a virtual model that is dynamic and continuously evolving. School admins, teachers, and policymakers can easily track school infrastructure, study students' progress, and optimize educational resources using this model (Zaballos et al., 2020). Fig. 23.1 presents the school's Digital Twins overview. Since most educational systems have rapidly embraced technology, implementing DT aligns with the increasing focus on innovative schools, making education more interactive, secure, and conducive to everyone's needs. DTs apply a data-driven approach to school administration and learning methodologies—from managing school buildings and energy consumption to customizing curricula for individual students (Grübel et al., 2022). However, the use of DT's technology in schools is not without its challenges—financial, privacy, and IT expertise are needed to correctly maintain and interpret digital models. Despite these challenges, the potential of DTs in education is an exciting frontier for reshaping traditional schooling into a more intelligent, responsive, and student-focused experience (Haynes & Yang, 2023).

Organization of chapter: This chapter explores the fundamentals of DTs technology in schools, its benefits in infrastructure management and learning personalization, the challenges that must be addressed, and the prospects of DTs in the educational sector. The next section describes the various components of DT in school. Section 3 presents the history of digital twin models. Section 4 comprises DT technologies and frameworks in schools. Section 5 deals with applications, and the last section concludes the chapter.

2. Components of the school digital twin

A school digital twin (DT) is the advanced virtual embodiment (Hu et al., 2022) of a school's physical and operational ecosystem, driven by Internet of Things (IoT) integrated systems, Artificial Intelligence (AI)-driven functions, and data analytics capabilities. By incorporating real-time data, predictive analytics, and automation, schools can gain optimized infrastructure, enhanced learning experiences, and improved operational efficiency (Kartashova et al., 2020). This chapter explores the critical elements of a school's digital twin and the new reality of educational institutions. The components of school DT are shown in Table 23.1.

2.1 Benefits of DT in schools

Many benefits (Khajavi et al., 2019) are used in the DTs. The benefits are shown in the following section:

Digital Twins in the Smart Classroom. https://doi.org/10.1016/B978-0-443-41632-3.00020-5

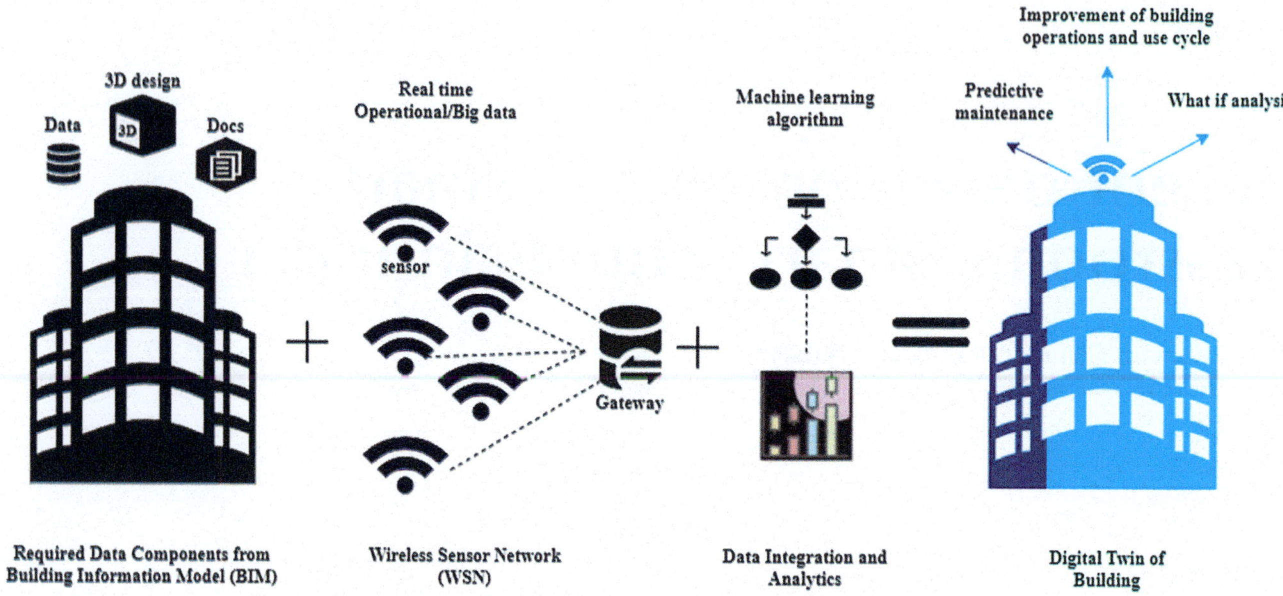

FIGURE 23.1 School digital twins overview.

- **Enhanced learning experiences:** Personalized learning driven by artificial intelligence fits every student's needs, enhancing involvement and learning results. DT allows tailored teaching tactics by analyzing learning patterns and development, improving education efficiency and inclusion.
- **Improved safety and security:** Predictive analytics, AI-driven threat identification, and real-time monitoring produce a safer classroom. Schools can proactively address possible hazards using automated emergency response systems, guaranteeing the welfare of staff members and pupils.
- **Efficient resource utilization:** Reduced operating expenses come from smart energy management, optimized scheduling, and facility upkeep. DT guarantees the best use of classrooms, labs, and school transportation by streamlining resource allocation.
- **Data-driven decision-making:** Real-time data helps administrators acquire practical insights that enable evidence-based enhancements in the operation of schools. Data analytics enable educational leaders to make strategic decisions from personnel distribution to curriculum changes.
- **Sustainability and cost savings:** Schools can reduce their environmental impact and thus their expenses by utilizing smart energy and waste management technologies. Institutions combining automation with renewable energy sources can achieve long-term environmental and financial sustainability (Qi et al., 2021).

3. History of digital twin models

The figure shows the development of the DT's idea over several technological developments (Rasheed et al., 2020), therefore stressing its change. It classifies this development into four separate phases: 1985–2002; Digital Simulations & 3D Printing (2003–2014); Intelligent & Connected IoT Services (2014–16); Mixed Reality, Cognitive & AI (2017–present) Information Mirroring Model. Fig. 23.2 shows the DT's starting point going forward.

Considering the Information Mirroring Model (1985–2002), the DTs appeared only as a conceptual framework emphasized by research and development (R&D) and engineering. It depended on devoted workstations and servers that permitted intense 3D demonstration and investigation close alongside the remote programming of CNCs and mechanical autonomy (Razzaq et al., 2023). The next stage is detailed by digital simulations and 3D printing (2003–2014), which provides collaborative digital designs, followed by virtual assembly and simulation before physical production. Just as mainstream adoption of 3D printing began, this era brought the ability to simulate and perfect designs virtually before taking them to manufacturing. From 2014 to 2016, the integration of intelligent and connected IoT services facilitated the connection of physical and virtual worlds through data unification, cloud computing, big data analytics, and IoT. These technologies enabled rapid feedback across design, manufacturing, and operations, allowing products to be augmented with digital services and improving overall

TABLE 23.1 Components of school digital twin.

Component	Description
Infrastructure and facility management	The course of a school's DTs is centered around data that can replicate physical infrastructure digitally. Realistic 3D modeling enables school buildings, classrooms, libraries, and laboratories to be recreated down to the last detail. IoT-based smart sensors track the environment (temperature, lighting, air quality, and occupancy levels) to maintain the appropriate level of comfort or security. Furthermore, AI-based predictive maintenance identifies possible infrastructure failures ahead of time, preventing downtime in addition to costly repairs.
Student and academic performance monitoring	One of the most important uses of a school DT is student-centric monitoring. These digital footprints help educators track attendance, academic progression, and behavioral patterns, and personalize learning plans for children. AI-driven personalized learning solutions assess performance data and propose tailored learning pathways, leading to improved student engagement and success rates. In addition, performance analytics inform teachers on best practices to apply and lesson plans to tailor as a result of immediate student responses.
Safety and security	A welcoming school atmosphere is vital, and DTs support this by incorporating advanced security features. These technologies help organizations to monitor and secure their physical premises by providing real-time monitoring of their premises, threat detection, and identification of anomalies, among others. Intelligent access control systems also help restrict an entry point, meaning only a person who should be able to access the building can do so. Emergency response systems and health monitoring solutions, including contact tracing tools, also support crisis management and the well-being of both students and staff.
Energy efficiency and sustainability	Transforming institutions into sustainable entities. The role of a school DT in sustainability is primarily to provide specific governance techniques to help organizations both reduce resource consumption and promote eco-friendly resources. Advanced energy management systems monitor and manage the consumption of electricity and water, helping to minimize waste and operational costs. The use of solar panels and wind turbines is represented in sustainability practices. Waste management is another area where technology can be useful for your campus, and automated waste management systems can make recycling and disposal of waste easier and more efficient. Automated waste management helps encourage an environmentally responsible school culture.
Resource and schedule optimization	Thus, these resources must be managed so that the school can run smoothly. DTs enable efficient and smart scheduling of classrooms, laboratories, and sports facilities, making their use optimal. Trailblazing AI-based transportation management directly optimizes a school bus routing system, resulting in less travel time, lower fuel consumption, and less environmental footprint. Digital administrative systems enhance school operations by reducing paperwork and improving workflow.
Parental and Community engagement	Parents or guardians need to be engaged along with the general public to make it a wholesome educational experience. As a result, most of the absentees or low scores cannot be detected by parents immediately. DTs offer intuitive dashboards that parents can track their child's academic performance, attendance, and real-time progress as well. AI-powered chatbots and virtual assistants take care of administrative queries and offer instant support to both parents and school staff. Moreover, online portals facilitate communication regarding school events, parent-teacher meetings, and collaborative programs, promoting stronger ties between schools and their communities.

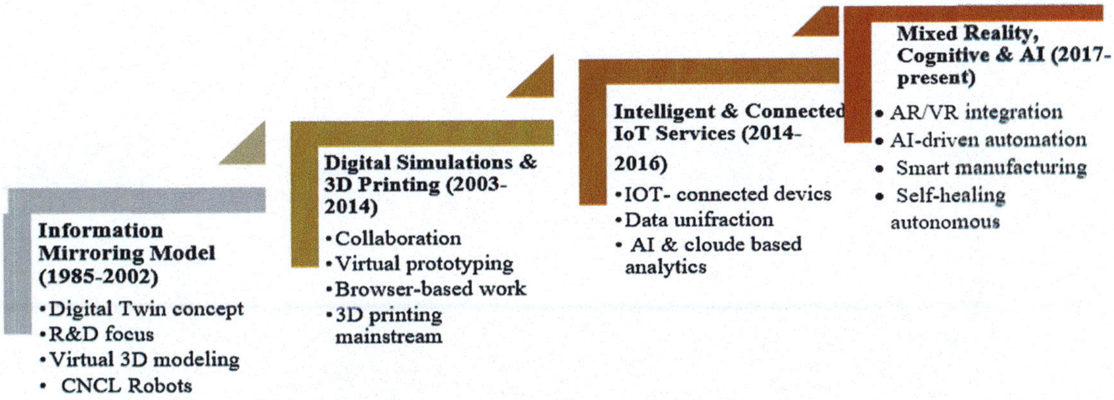

FIGURE 23.2 Digital twin's origin to the future.

efficiency. The latest evolution, mixed reality, cognitive and AI (2017–present), leverages holographic interfaces, augmented reality (AR), virtual reality (VR), cognitive services, and AI. These innovations enhance guided interactions, blended human-machine collaboration, and autonomous self-healing systems, pushing DT into a new era of intelligent, immersive, and self-sustaining solutions. This timeline underscores the DT journey from a conceptual engineering tool to an intelligent, data-driven, and interactive system that reshapes modern industries by integrating AI, IoT, and mixed reality (Shahzad et al., 2022).

4. Digital twin technology in schools

A DT is the total virtual copy of the components and the product process. Essentially, it is a simulation system. The capacity to thoroughly intertwine data and physics. The DT technology is aimed at three levels: data capture, computer modeling, and data use (Tang et al., 2024). Three tiers of twinning have already been explored in simulations, as shown in Fig. 23.3, the conceptual architecture model of twinned scaling.

The data access layer pertains to implementing the quantified perception of the physical world through static and dynamic models:

- Calculate the simulation layer, conduct data analysis and computation, generate forecasts, and offer essential technological support for implementing digital twins.
- The application layer governs the outcome of the choice of the physical entity.

The entire process may ascertain the reasonableness of decisions across various parallel domains and effectively optimize the complete life cycle of the school system, thereby influencing and transforming human interactions with the physical environment (Vitor et al., 2021).

4.1 The framework of DTs in school

This research proposes integrating DT technology for the comprehensive advancement of school development, including five components: physical domain, virtual domain, service platform, data platform, and connectivity. The physical space tracks real-time data and transmits it to the data platform for analysis. The analytical results compel the completion of the virtual simulation and the proposal of an optimized design for the service platform. The service platform ultimately determines the strategy inside the physical environment to facilitate system management. Utilizing the connection to the network can function efficiently, convey data, and reduce contact. Models can facilitate the linkage between physical and virtual environments. Perception, prediction, simulation, and decision-making processes have been repeatedly executed to enhance campus life (Warke et al., 2021). Based on a comprehensive assessment of school requirements, the future innovative school should possess the following capabilities:

- Deliver an immersive school experience;
- Facilitate visual school management;
- Enhance intelligent school teaching methodologies.

This system incorporates school navigation, augmented reality, virtual simulation for education, and data visualization functionalities (Xie et al., 2018). The system construction pertains to the framework of digital twin

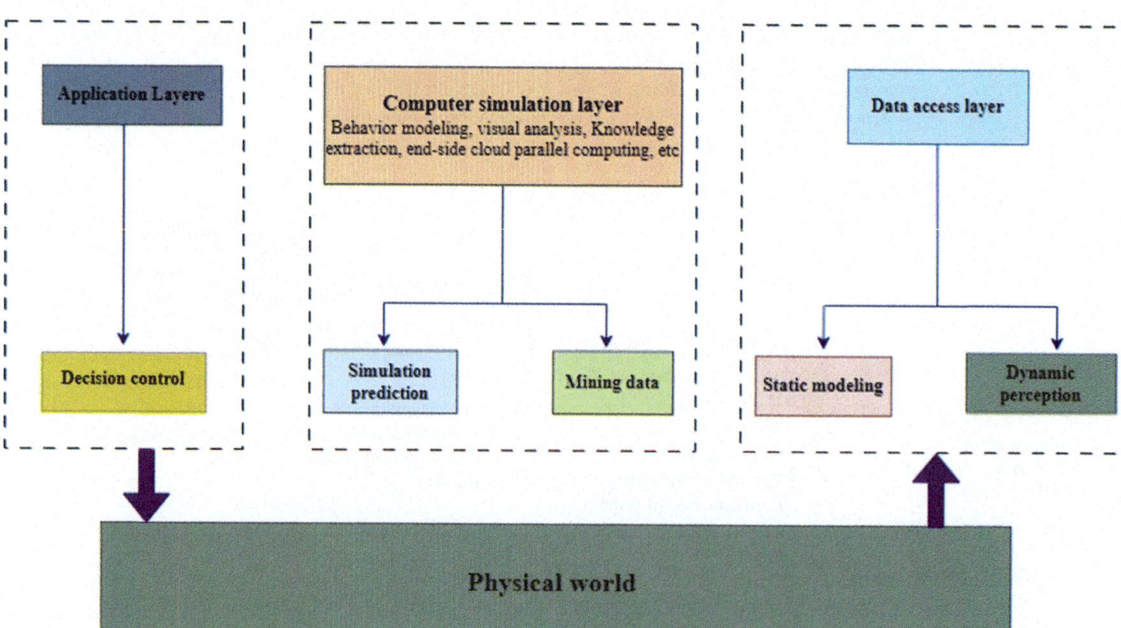

FIGURE 23.3 DT's technology architecture.

technology and suggests an innovative school utilizing digital twins:

- The DT for schools employs multisource devices to gather campus data, including traffic monitoring, water quality assessment, and instructional feedback systems.
- The data access platform amalgamates data management, modeling, and prediction modules through cloud-based technologies such as SaaS, PaaS, and IaaS.
- The virtual simulation system produces a 3D campus model, incorporating real-time and historical data to predict and depict various campus scenarios, enabling real-time optimization and intelligent administration.
- The application platform enables virtual tours, augmented reality experiences, and data visualization for effective decision-making.

Compared to conventional smart campuses, this method produces a sustainable, closed-loop system that improves management and educational efficiency, lowers maintenance costs, and provides a better user experience. Table 23.2 presents the format representation of the DT Framework for a School.

This table organizes the key elements of the DT's framework for schools, emphasizing its structure, data processing, and real-world applications. Table 23.3 presents the challenges of DTs for schools.

5. Application of digital twin for schools

A DT is a virtual representation of a physical object or system. In a school context, technology can be applied to improve operations, enhance learning experiences, and streamline management processes (Broo & Schooling, 2023). Here are some potential applications. Fig.23.4 presents the applications of DTs for school.

5.1 Campus management

- **Building maintenance:** By creating a digital twin of the school building (s), the physical condition of the infrastructure can be monitored in real time. It tracks everything from HVAC systems and lighting to temperature and energy usage, enabling preventative maintenance and efficiency.
- **Space utilization:** A digital twin can help optimize the usage of classrooms, libraries, and other spaces. Data can be collected about how and when each space is used to assist in scheduling and planning.
- **Energy management:** The digital twin can forecast consumption patterns and replicate the university's energy use. Recommending insulation, lighting, or heating changes can assist in lowering expenditures. It could also help to effectively control their consumption and enable the blending of renewable energy sources.

5.1.1 Health and safety

- **Air quality monitoring:** To ensure staff and kids are healthy, a DT might track air quality in corridors, classrooms, and other spaces. Tracking ventilation and sanitation levels would be especially helpful in postpandemic situations.
- **Safety simulations:** DTs can recreate evacuation plans and improve general emergency readiness through virtual education of staff and students in times of crisis, including fires or natural disasters.

5.2 Learning and curriculum enhancement

- **Simulated environments for learning:** DTs provide immersive environments for interaction for students in fields such as physics, engineering, and architecture. Creating a digital twin of the human body or a history for 3D study.
- **Virtual classrooms and remote learning:** An interactive and customized virtual learning environment might be created from a DT of a classroom. Making remote learning more interesting and copying a real-life educational experience can improve it.

5.3 Student and teacher performance monitoring

- **Analytics:** Real-time student performance data monitoring—attendance, grades, participation—allows a digital twin to offer insights and support in individualized learning. This might point to areas needing development, strengths, and trends.
- **Faculty development:** Using a DT of the teaching strategies and student interactions, one can assess and enhance the instructional practices of the institution. For instance, analytics and artificial intelligence can recommend improved teaching strategies depending on performance statistics and student comments.

5.3.1 Logistics and resource management

- **Smart scheduling:** DTs help to maximize resources by enabling the administration of complex calendars comprising classroom availability, teacher assignments, and extracurricular activities.
- **Inventory management:** DTs can oversee educational resources (books, supplies, sports equipment, etc.) to ensure the availability of suitable resources when needed.

TABLE 23.2 Representation of the DT framework.

Layer	Components	Description
Application platform	• Real-time interaction • Virtual environment and monitoring systems	• Provides interactive interfaces for users to engage with the digital twin, including dashboards and simulations. • Enables real-time monitoring and control of the school environment.
Data access platform	**Related data modules** • Environmental data • Sensor data • Traffic data • Teaching data	**Collects and integrates various datasets.** • Captures temperature, air quality, and other environmental parameters. • Gathers real-time inputs from IoT devices deployed across the school campus. • Monitors movement within and around the school campus. • Tracks academic activities and classroom performance.
	Data management module • Teaching data • Manage data • Device data • Student data	**Manages and structures various types of school-related data.** • Stores curriculum, lesson plans, and assessment records. • Organizes overall school operational information. • Maintains records of IoT devices and their functionalities. • Stores student attendance, performance, and engagement analytics.
	Data processing module • Data preprocessing • Data fusion • Abnormal data management • Analysis of algorithms	**Processes and analyses collected data for decision-making.** • Cleans and prepares raw data for integration. • Combines data from multiple sources to create a unified view. • Detects anomalies in school operations, such as security threats or equipment failures. • Uses AI and predictive analytics to improve decision-making.
Physical space campus system	• Monitoring equipment • Sensor • Campus building • Woodworking course • Industrial Culture corridor	• Cameras and sensors for security, attendance, and environmental tracking. • Collects data on temperature, air quality, noise levels, and occupancy. • Represents the actual school infrastructure. • Demonstrates specialized areas, such as workshops for technical education. • Showcases real-world industrial applications within the school setting.
Virtual simulation campus system	• Geometric model • Physical model • Behavior model • Rule model • Virtual campus model simulation • Virtual woodworking operation simulation	• Digitally replicates the physical structures of the school. • creates a realistic digital representation of school buildings and facilities. • simulates student and staff movements and interactions. • Defines rules and constraints within the virtual campus. • provides a dynamic, real-time representation of school operations. • simulates hands-on activities and technical learning experiences in a digital environment.

Transportation and mobility: A decision tree can maximize school bus routes and schedules, guaranteeing timely travel and reducing costs. It guarantees safety and helps to monitor bus conditions to prevent breakdowns.

Parent and community engagement: Parents can be given DTs to provide information on their child's general well-being, academic surroundings, and campus activities. This helps families and educational institutions communicate and cooperate better. By including DT's technologies, schools may improve operational effectiveness, create better learning opportunities, and provide staff and students with more customized and safe surroundings.

6. Current developments in smart classrooms/smart schools in India

The recent progress in India's smart classroom and smart school ecosystem suggests a multifaceted acculturation of

TABLE 23.3 Challenges of DTs for schools.

Challenge	Solutions
Data privacy and security	Implement robust cybersecurity measures, data encryption, and ensure compliance with student consent procedures.
Data standardization and integration	Develop universal educational data standards and systems to integrate and interpret various student performance data.
Quality of data	Conduct regular audits of student records, maintain data quality checks, and use reliable assessment tools.
Computational challenges	Invest in high-capacity servers, cloud computing, or edge computing. Leverage AI and machine learning to track student progress.
Regulatory concerns	Collaborate with educational regulatory bodies for guidelines and standards. Initiate pilot programs for digital learning.
Adoption and Trust	Provide education and training for teachers and school administrators. Communicate transparently with students and parents. Showcase successful digital learning implementations.

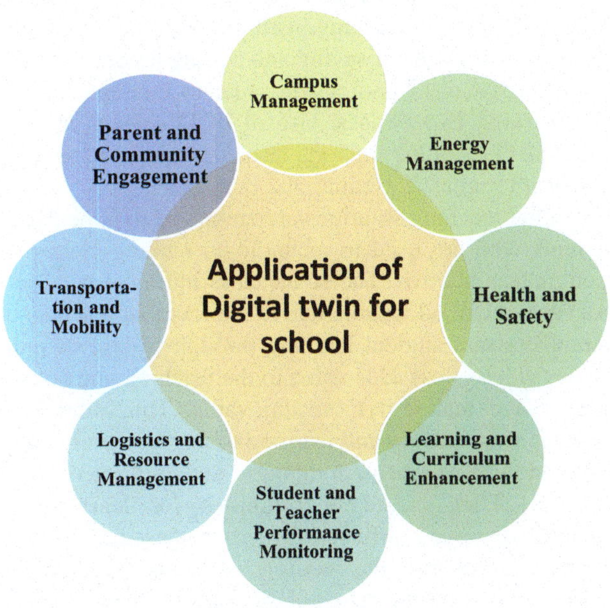

FIGURE 23.4 Applications of DTs for school.

technology to transform the conventional school paradigm. Smart classrooms are transitioning past being just digital transformations of schools, with efforts to develop holistic systems that utilize Information and Communication Technology to build sustainable and dynamic learning contexts. The digital shift in terms of education, notably much faster due to outbreaks like the COVID-19 pandemic, has highlighted the urgent need to implement smart solutions into the educational scheme. Such solutions tend to be a reflection of smart cities' initiatives, while adjusting the same sort of technologically and sustainability-focused models to enhance educational facilities. This includes not only the technical improvements, but it also embraces the transition to government-oriented, team-oriented strategies that contribute to the general education quality.

Moreover, the modern smart classrooms utilize interactive whiteboards, smart screens, and ubiquitous internet, which allows dynamic delivery of content and enables students to learn outside of the conventional teaching space. Contemporary pedagogical practices in smart classrooms are focused on personalized learning, where data analytics are used to customize the educational content to the needs and learning styles of individual students. Incorporation of 3D animated modules, instructional materials, and activity-based learning improves the classroom interdisciplinary teaching-learning process from old-fashioned lecture-based methods. This shift is backed by cloud computing, which facilitates effortless storage of data, accessibility to data, and can be used for collaborative project work, hence improving teachers as well as the learning experiences. Additionally, the importance of digital literacy is an integral element of this transformation, which would provide students with the skills to be able to search and analyze online resources. Adaptive learning portals that incorporate AI to individualize the curricula according to each student's needs are gaining popularity. In the course of the past few years, India's education system has seen momentum-gathering smart classroom initiatives.

6.1 Technology adoption trends

Interactive whiteboards continue to be the greatest in terms of popularity (35% of smart classrooms). AI-driven adaptive learning grew huge (18% adoption). The AR/VR technology is picking up primarily in science and geography learning as presented in Fig. 23.5.

The distribution of technology in smart classrooms in India in 2024. The share of interactive whiteboards is the

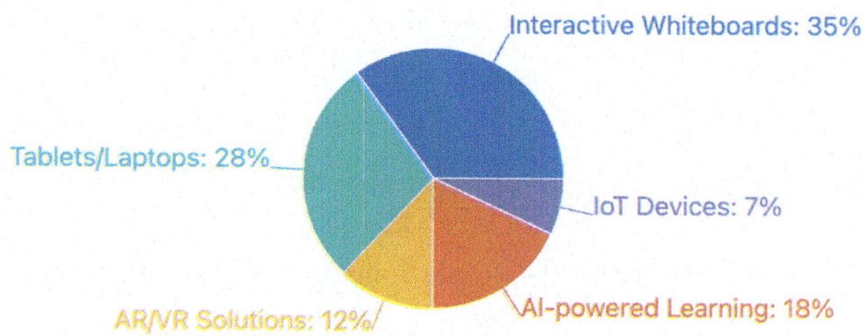

FIGURE 23.5 Technology trends adopted in India.

largest, having 35% of the educational technology market, making up a major division of smart classroom devices. Coming in second are tablets and laptops at 28%-testament to their versatility and ever-expanding affordability. AI-powered learning solutions make up 18% of the educational technology mix, meaning an increasing trend of adaptive learning systems that individualize the learning process. AR/VR solutions comprise 12% of technologies implemented, indicating the underlying high level of integration of immersive learning experiences, although it is at higher costs. IoT devices represent the lowest share at 7%, which implies that this is an emerging technology in Indian classrooms (Patel et al., 2024).

6.2 Regional analysis

With 34% penetration, the southern states take the lead position. Northern states come in second with 27%, having a rapid growth in the last 2 years. The lowest adoption is for eastern states, which stand at 18%, and this improves as a result of focused state agencies (Aggarwal & Joshi, 2022) as shown in Fig. 23.6.

The nationwide regional distribution of smart classroom implementation in India in 2024. Southern India dominates heavily as the region with the highest adoption rates of about 35%, showing that it is the leader in the integration of educational technologies. Northern India comes next as the second most progressive region, with implementation at 27%, as it indicates significant progress in modernizing classroom facilities. Western India is in third place with a share of nearly 21% toward adoption, which is a moderate use of the smart classroom setup. Eastern India reports the lowest penetration at 18%, which suggests regional disparities in educational technology access. This spatial pattern is an indication of disparity in development in India's educational setup, with evident North-South divide in which Southern states are more aggressive in investment in educational technology compared to Eastern states, resulting from such variables as economic development and education policies in states, IT infrastructure, as well as public-private partnership initiatives in both capital regions (Chopra & Nair, 2024).

6.3 Usage of smart classrooms

Smart classrooms use interactive whiteboards and digital resources to increase learning and engagement of students. Smart-class software uses 3D-animated modules and instant assessments to redefine the way in which traditional teaching is done in learning institutions. In this regard, the teachers are capable of saving and editing lessons using the smart classroom technology in order to facilitate customized learning experiences. Use of LCD screens and projectors in classrooms supports visual learning and enhances retention (Kumar & Bhatia, 2023).

Digital teaching solutions are embraced to appeal to the students who are used to technology, making classrooms open and interactive. The smart class initiative is a viable tool for modern education, due to the fact that teachers can instantly assess student learning through it. Use of multimedia (video clips) adds value to the teaching and learning process and makes it more enjoyable. The smart classrooms allow teaching and learning to improve with the use of technology, such as computers and multimedia technologies. They create opportunities for learning and teaching materials, which are activity-based. The smart classrooms enhance the reading skills and academic attainments of students. Smart classrooms drive personalized learning using real-time data to produce tailored lessons for every student. Mobile learning apps are a means for easy

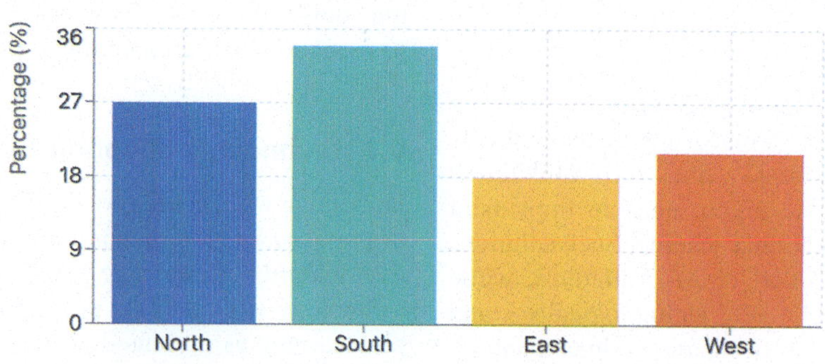

FIGURE 23.6 Region-wise analysis of the adoption of smart classrooms.

access to educational materials 24/7 and support independent learning. New techniques, such as video and virtual reality, help involve students and encourage them. Improved interactivity, customized learning, availability of resources, and collaborative education are important considerations in teaching effectiveness.

Smart classrooms are identified as instruments for refining the standards of education in India. The use of information technology in classrooms is highlighted for the transformation of education. The study tests the influence of smart classrooms on educational achievement. Presentation of digital content is recognized as a major requirement for universities and schools.

6.4 Challenges in implementation

There is a possibility that excessive use of technologies can blur the main focus of the child's education, which may bear potential hazards in learning outcomes. Transformation to smart classrooms will take a lot of time and money for the schools to do, which would bring logistical barriers (Priya et al., 2019).

The level of content knowledge of the teacher is questioned in the case of smart classrooms and traditional sessions. The issues in history courses are dealt with via a proposed move for teachers. The role of the traditional methods in the current learning and the feelings of the learners are challenged (Jena et al., 2013).

Manual attendance taking is consuming time, thereby making it hard for effective learning to take place in classrooms. Notices of importance are inefficiently distributed, whereby they need to be carried to every classroom. There is manual intervention in the process of managing classroom activities due to the lack of proper automation. The limitations of the existing system with regard to investments in technology hinder their improvement and effectiveness.

Problems would include infrastructure, such as slow internet and the absence of digital resources in rural areas. Teacher training and know-how on technology use are some of the obstacles. Aversion to traditional methods of teaching calls for efforts to increase awareness and capacity for smart classroom implementation (Babuji et al., 2019).

Manual attendance costs lots of time and distracts the lecturers from teaching. The students' attendance at lectures is reducing because of the fact that information is now readily available online. The requirement for better discipline and regularity among students is stressed.

Inadequate connectivity and low bandwidth interfere with e-learning efficiency. Draining of mobile data and slow internet add up to online learning. The absence of Wifi or LAN connections makes e-learning hard. Restricted opportunities for appropriate doubt clearance impede the process of learning. Lack of teachers and classmates deteriorates the motivational level.

7. Conclusion

These days, educational institutions' dynamic, inclusive, and successful learning environment depends on including digital technologies. This study demonstrates how DT supports group projects, increases student involvement, and provides customized learning opportunities. Many important elements characterize the effective execution: enough infrastructure, teacher preparation, and policy backing. Reducing the digital divide is particularly crucial, since unequal access to technology could aggravate educational inequalities. Moreover, cybersecurity issues need comprehensive digital literacy campaigns to enable the responsible and safe use of technology. Notwithstanding these difficulties, under appropriate control, the advantages of digital transformation outweigh the negative ones. Educational institutions must adopt a holistic strategy, covering strategic planning, stakeholders, and constant adaptation to new technology. Future studies should concentrate on evaluating long-term effects, analyzing multiple approaches to digital transformation, and creating best practices for sustainable integration. Implementing digital transformation will enable educational institutions to equip their students with the necessary skills for the future, therefore preparing them for quickly changing DTs.

References

Aggarwal, M., & Joshi, R. (2022). Public-private partnerships in educational technology: Bridging the digital divide in India. *Journal of Education Policy and Management, 14*(3), 267–284.

Attaran, M., & Celik, B. G. (2023). Digital twin: Benefits, use cases, challenges, and opportunities. *Decision Analytics Journal, 6*(100165), Article 100165. https://doi.org/10.1016/j.dajour.2023.100165

Babuji, R., Suresh, T., Gunasekhar, P., Krishnan, C. M. L., & Department of Electronics and Communication Engineering. (2019). Smart classroom – A stepping stone in making a digital India. *International Journal of Recent Technology and Engineering, 8*(4), 8775–8778. https://doi.org/10.35940/ijrte.d9279.118419

Botín-Sanabria, D. M., Mihaita, A.-S., Peimbert-García, R. E., Ramírez-Moreno, M. A., Ramírez-Mendoza, R. A., & Lozoya-Santos, J. de J. (2022). Digital twin technology challenges and applications: A comprehensive review. *Remote Sensing, 14*(6), 1335. https://doi.org/10.3390/rs14061335

Broo, D. G., & Schooling, J. (2023). Digital twins in infrastructure: Definitions, current practices, challenges and strategies. *International Journal of Construction Management, 23*(7), 1254–1263. https://doi.org/10.1080/15623599.2021.1966980

Chopra, S., & Nair, L. (2024). Measuring educational outcomes in technology-enhanced learning environments. *Quality Assurance in Education, 32*(1), 55–74.

Grübel, J., Thrash, T., Aguilar, L., Gath-Morad, M., Chatain, J., Sumner, R. W., … Schinazi, V. R. (2022). The hitchhiker's guide to fused twins: A review of access to digital twins in situ in smart cities. *Remote Sensing, 14*(13), 3095. https://doi.org/10.3390/rs14133095

Haynes, P., & Yang, S. (2023). Supersystem digital twin-driven framework for new product conceptual design. *Advanced Engineering Informatics, 58*(102149), Article 102149. https://doi.org/10.1016/j.aei.2023.102149

Hu, W., Lim, K. Y. H., & Cai, Y. (2022). Digital twin and industry 4.0 enablers in building and construction: A survey. *Buildings, 12*(11), 2004. https://doi.org/10.3390/buildings12112004

Jena, A. K., & Department of Educational Sciences. (2013). Does smart classroom an effective technology for teaching: A research analysis. *I-Manager's Journal of Educational Technology, 10*(1), 55–64. https://doi.org/10.26634/jet.10.1.2304

Kartashova, L., Gurzhii, A., Zaichuk, V., Sorochan, T., & Zhuravlev, F. (2020). Digital twin of an educational institution: An innovative concept of blended learning. In *Proceedings of the 1st symposium on advances in educational technology. SCITEPRESS - Science and technology publications*.

Khajavi, S. H., Motlagh, N. H., Jaribion, A., Werner, L. C., & Holmstrom, J. (2019). Digital twin: Vision, benefits, boundaries, and creation for buildings. *IEEE Access: Practical Innovations, Open Solutions, 7*, 147406–147419. https://doi.org/10.1109/access.2019.2946515

Kumar, R., & Bhatia, M. (2023). Impact assessment of smart classroom initiatives under the PM eVIDYA program. *Educational Policy Review, 15*(3), 212–229.

Patel, V., Gupta, S., & Mehrotra, D. (2024). Technology distribution in Indian smart classrooms: An empirical study. *International Journal of Education and Information Technologies, 10*(1), 78–93.

Priya, B., Radhamani, V., Prasanth, P. S., Suchitra, S., & Suganthi, S. (2019). Smart class room monitoring system. *International Journal of Engineering and Advanced Technology, 9*(2), 4812–4815. https://doi.org/10.35940/ijeat.b2348.129219

Qi, Q., Tao, F., Hu, T., Anwer, N., Liu, A., Wei, Y., … Nee, A. Y. C. (2021). Enabling technologies and tools for digital twin. *Journal of Manufacturing Systems, 58*, 3–21. https://doi.org/10.1016/j.jmsy.2019.10.001

Rasheed, A., San, O., & Kvamsdal, T. (2020). Digital twin: Values, challenges and enablers from a modeling perspective. *IEEE Access: Practical Innovations, Open Solutions, 8*, 21980–22012. https://doi.org/10.1109/access.2020.2970143

Razzaq, S., Shah, B., Iqbal, F., Ilyas, M., Maqbool, F., & Rocha, A. (2023). DeepClassRooms: A deep learning based digital twin framework for on-campus class rooms. *Neural Computing and Applications, 35*(11), 8017–8026. https://doi.org/10.1007/s00521-021-06754-5

Shahzad, M., Shafiq, M. T., Douglas, D., & Kassem, M. (2022). Digital twins in built environments: An investigation of the characteristics, applications, and challenges. *Buildings, 12*(2), 120. https://doi.org/10.3390/buildings12020120

Tang, T., Wu, Y., Sun, S., & Wu, Y. (2024). Campus intelligent decision system based on digital twin. In *2024 27th international conference on computer supported cooperative work in design (CSCWD). IEEE.*

Vitor, R., Keller, B., Barbosa, D., Diniz, D., Silva, M., Oliveira, R., & Saul. (2021). Synchronous and asynchronous requirements for digital twins applications in industry 4.0. In *Proceedings of the 23rd international conference on enterprise information systems. SCITEPRESS - Science and technology publications*.

Warke, V., Kumar, S., Bongale, A., & Kotecha, K. (2021). Sustainable development of smart manufacturing driven by the digital twin framework: A statistical analysis. *Sustainability, 13*(18), Article 10139. https://doi.org/10.3390/su131810139

Xie, K., Di Tosto, G., Chen, S.-B., & Vongkulluksn, V. W. (2018). A systematic review of design and technology components of educational digital resources. *Computers and Education, 127*, 90–106. https://doi.org/10.1016/j.compedu.2018.08.011

Zaballos, A., Briones, A., Massa, A., Centelles, P., & Caballero, V. (2020). A smart campus' digital twin for sustainable comfort monitoring. *Sustainability, 12*(21), 9196. https://doi.org/10.3390/su12219196

Chapter 24

Digital twin in smart education: Challenges, limitations, and case studies

Mohamed S. Abdalzaher[1, 2]

[1]*Department of Seismology, National Research Institute of Astronomy and Geophysics, Helwan, Cairo, Egypt;* [2]*Department of Electrical Engineering, American University of Sharjah, Sharjah, United Arab Emirates*

1. Introduction

The rapid evolution of educational technologies has fundamentally reshaped traditional classrooms, driving the emergence of smart learning environments that leverage advanced data analytics, artificial intelligence (AI), and Internet of Things (IoT) technologies (Abdalzaher et al., 2023a; Fang et al., 2024; Han et al., 2024; Krichen et al., 2024). Among these innovations, the concept of digital twins (DTs) stands out as a promising approach to transforming education by providing real-time, data-driven insights into classroom dynamics and individual learning experiences. A DT, originally developed for industrial applications, is a virtual replica of a physical system that mirrors its behaviors, states, and processes using continuous data flows from sensor networks (Javaid et al., 2023). When applied to education, DTs offer a powerful framework for optimizing learning environments, enhancing personalized learning paths, and facilitating efficient management of classroom resources.

The integration of DTs into smart classrooms relies heavily on the use of sensor networks, which act as the foundational infrastructure for data collection and real-time monitoring (Razzaq et al., 2023). Sensors can monitor everything from air quality and lighting to stress levels and engagement, and constantly feed data into AI systems that update the DT in real time. This gives educators and administrators useful insights, like how temperature or lighting affects student focus. It also enables early warnings, predicting things like equipment issues or drops in student performance, so action can be taken before problems escalate.

However, the deployment of DTs in educational settings presents significant challenges. Data privacy and security emerge as primary concerns, particularly given the sensitive nature of information collected through biometric and behavioral sensors (Wang et al., 2023). Ensuring compliance with regulations such as the General Data Protection Regulation (GDPR) and the Family Educational Rights and Privacy Act (FERPA) is essential to maintain trust and protect student rights (Giuffrida et al., 2023; Zhang et al., 2020). Moreover, the technical complexity and cost associated with implementing and maintaining DT systems may pose barriers for many educational institutions, especially those with limited budgets (Broo et al., 2023). Addressing these challenges requires not only robust cybersecurity measures but also transparent data governance policies and sufficient training for educators.

This chapter explores the multifaceted role of DTs in smart classrooms, focusing on their architecture, implementation strategies, and practical benefits. It examines how DTs can enhance personalized learning by adapting to individual student needs and optimizing environmental conditions for learning. The discussion also delves into the ethical and practical challenges associated with this technology, offering insights from real-world case studies and pilot projects. Furthermore, the chapter considers future directions for DTs in education, including their potential integration with extended reality technologies and the establishment of comprehensive policy frameworks. By presenting a balanced analysis of both the opportunities and challenges, this chapter aims to provide educators, policymakers, and researchers with a deeper understanding of how DTs can contribute to more personalized, efficient, and responsive educational environments. Fig. 24.1 shows an architecture of the DT process for modernizing classroom management.

Digital Twins in the Smart Classroom. https://doi.org/10.1016/B978-0-443-41632-3.00032-1

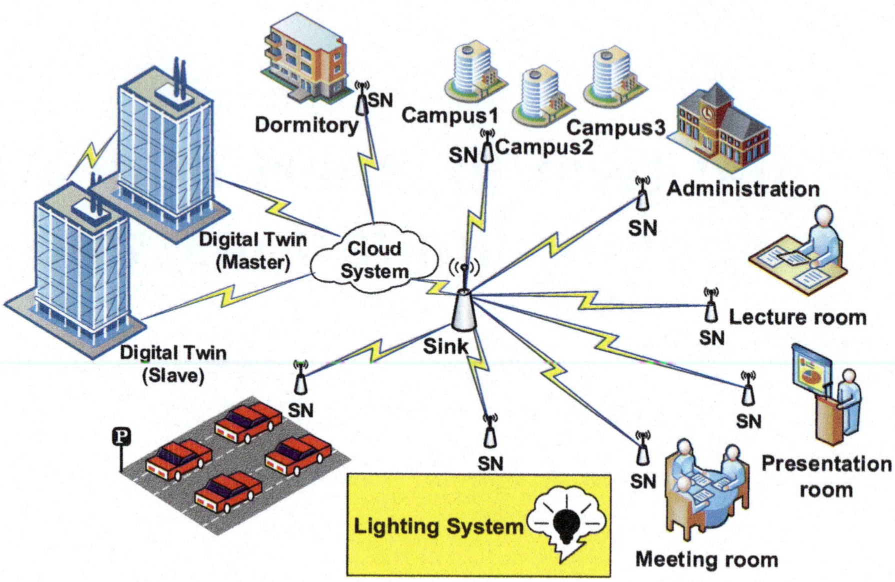

FIGURE 24.1 Architecture of the DT process for modernizing classroom management using IoT.

2. Concept of digital twins in education

The concept of DTs has its roots in industrial and manufacturing sectors, where it was initially developed to enhance efficiency, predictive maintenance, and decision-making by creating virtual replicas of physical assets. Over time, this technology has evolved and expanded into other domains, including healthcare, urban management, and now education (De Benedictis et al., 2022; Dembski et al., 2020). In the context of smart classrooms, a DT can be understood as a comprehensive, real-time virtual model of the physical learning environment and its occupants, continuously updated through data captured by sensor networks. This virtual counterpart enables educators and administrators to monitor, simulate, and optimize various aspects of the classroom, from environmental conditions to student learning behaviors.

DTs in education function by integrating data from multiple sources, such as environmental sensors that measure temperature, humidity, and air quality; biometric devices that track heart rate and stress levels; and behavioral sensors that capture student movement and engagement (Adibi et al., 2024). By processing this data in real-time using AI and machine learning (ML) algorithms, DTs can generate actionable insights that support personalized learning and proactive classroom management (Screpanti et al., 2022). For example, a DT might detect that a student's engagement drops under certain lighting conditions and recommend adjustments to optimize focus and comprehension. Similarly, it can simulate different teaching strategies or seating arrangements to assess their impact on student participation and performance.

One of the key advantages of DTs in education is their ability to bridge the gap between physical and digital learning experiences (Sepasgozar, 2020). Through continuous data feedback, DTs can provide educators with a deeper understanding of each student's learning preferences, challenges, and progress. This level of insight enables the implementation of adaptive learning paths tailored to individual needs, promoting a more inclusive and effective educational environment. Furthermore, DTs can support remote and hybrid learning models by offering a real-time view of classroom dynamics, ensuring that all students, regardless of location, receive a consistent and high-quality learning experience (Mourtzis et al., 2023).

Adopting digital twins in education comes with challenges, including technical hurdles like integrating diverse data sources and ensuring real-time accuracy, which demand significant investment and expertise. Ethical concerns, especially around data privacy and surveillance, must be managed with clear policies and consent. As more institutions explore DTs, a solid grasp of these issues is key to making informed, sustainable choices.

3. Sensor networks and data collection in smart classrooms

Sensor networks are the backbone of DTs in smart classrooms, providing the real-time data necessary to create accurate and responsive virtual representations of physical learning environments. These networks consist of interconnected devices that collect, transmit, and process data from various sources, enabling continuous monitoring of both environmental conditions and student behaviors (Mihai et al., 2022). With sensor networks, educators can better understand how factors like air quality, lighting, noise, and student engagement impact learning. Feeding this data into the DT allows for real-time adjustments to the classroom and more personalized learning experiences.

3.1 Types of sensors in smart classrooms

Environmental sensors: These include temperature, humidity, CO_2 levels, lighting, and sound sensors (Zivelonghi et al., 2024). Maintaining optimal environmental conditions is crucial for student comfort and concentration. For instance, high CO_2 levels can cause drowsiness and impact cognitive performance, while appropriate lighting can enhance focus and reduce eye strain (Vehviläinen et al., 2016). By integrating data from these sensors, DTs can recommend or automate adjustments to heating, ventilation, and air conditioning (HVAC) systems, lighting controls, and ventilation to create a conducive learning environment.

Biometric sensors: Wearable devices and cameras equipped with biometric capabilities can monitor students' physiological responses, such as heart rate, stress levels, and facial expressions (Bustos-Lopez et al., 2022). This data can provide valuable insights into student well-being and engagement. For example, a sudden increase in heart rate or stress indicators might signal a need for a break or a change in teaching approach. However, the use of biometric data raises significant privacy concerns, necessitating robust consent mechanisms and compliance with regulations like GDPR and FERPA (Khan et al., 2020).

Behavioral sensors: Cameras, motion detectors, and pressure sensors integrated into seats can track student movement, posture, and attention (Huang et al., 2024). Behavioral data helps in understanding patterns such as prolonged inactivity, restlessness, or declining focus, enabling timely interventions by educators. For instance, if a DT detects that a majority of students are disengaged during a lecture, it can suggest modifications in instructional techniques or interactive activities to reengage the class.

3.2 Data acquisition and processing

The effectiveness of sensor networks hinges on seamless data acquisition and processing capabilities. Data from various sensors is aggregated and transmitted through IoT networks to cloud-based platforms, where AI algorithms analyze and update the DT in real-time. Ensuring data accuracy, minimizing latency, and maintaining interoperability between sensors and systems are critical for reliable DT functionality (Khan et al., 2022).

3.3 Privacy and ethical considerations

The deployment of sensor networks in educational environments raises significant ethical and privacy concerns. Continuous monitoring of students, particularly through biometric and behavioral sensors, may be perceived as invasive if not managed transparently. Establishing clear data governance policies that outline what data is collected, how it is used, and who has access is crucial for building trust (Ladley, 2019). Ensuring informed consent from students and guardians and complying with data protection regulations must be prioritized to balance innovation with ethical responsibility (Ali, 2024). By addressing these challenges and leveraging the power of sensor networks, educational institutions can maximize the benefits of DTs, creating smart classrooms that are not only data-driven but also ethical and student-centric.

4. Implementing digital twins in smart classrooms

Implementing DTs in smart classrooms requires a comprehensive approach that integrates advanced technologies, robust infrastructure, and effective data management strategies. At its core, the implementation process revolves around creating a virtual model that accurately reflects the physical classroom environment in real-time (Roda-Sanchez et al., 2023). This involves deploying sensor networks, setting up data acquisition and processing systems, and leveraging AI and ML to transform raw data into actionable insights. Successful implementation not only enhances the efficiency and adaptability of learning environments but also paves the way for personalized education, improved resource management, and proactive decision-making. Table 24.1 presents a summary of implementing DT in classrooms.

4.1 Architecture of digital twins in educational settings

The architecture of DTs in smart classrooms typically consists of three main layers: data acquisition, data processing, and application (Bucchiarone, 2022; Martins et al., 2024).

4.1.1 Data acquisition layer

- This layer involves deploying various sensors throughout the classroom to collect real-time data on environmental conditions, student behavior, and engagement levels. Environmental sensors monitor factors such as temperature, lighting, and CO_2 levels,

TABLE 24.1 Implementation steps summary and clarification.

Aspect	Details
Step 1: Defining objectives	Clearly define the goals for implementing DTs (e.g., enhancing learning experiences, improving resource management, optimizing classroom conditions).
Step 2: Identifying key technologies	Determine the necessary technologies, including IoT sensors, cloud computing, AI, and ML, for building and maintaining the DT model.
Step 3: Integrating data sources	Integrate data from various sources, such as IoT sensors, learning management systems (LMS), environmental data, and student performance data.
Step 4: Developing the DT model	Create a virtual replica of the classroom, incorporating real-time data inputs from sensors to simulate the environment and student interactions.
Step 5: Data analysis and decision support	Use AI/ML algorithms to analyze collected data, provide actionable insights, and support decision-making regarding classroom management and learning improvements.
Step 6: User interface development	Design and deploy user-friendly dashboards for teachers, students, and administrators to interact with the DT model and data insights.
Step 7: Training and support	Provide training for educators and staff on how to use the system effectively, interpret data, and make informed decisions.
Step 8: Continuous monitoring and feedback	Implement real-time monitoring and feedback mechanisms to ensure the system remains accurate, relevant, and up to date.
Step 9: Scalability and expansion	Plan for scalability to apply DT technology to other classrooms or campuses, ensuring infrastructure can support future growth.

while biometric and behavioral sensors capture physiological responses and movement patterns. Data is transmitted via IoT networks to centralized processing units.

- Ensuring compatibility and interoperability between different sensor types is crucial at this stage to create a seamless data flow that accurately updates the DT.

4.1.2 Data processing layer

- This layer focuses on analyzing raw data using AI and ML algorithms to extract meaningful patterns and insights. For example, ML models can identify trends in student engagement based on behavioral data or predict optimal environmental conditions for learning.
- Cloud-based platforms often play a vital role in storing and processing large volumes of data efficiently. Edge computing can also be employed to process time-sensitive data locally, minimizing latency and enhancing real-time decision-making capabilities.

4.1.3 Application layer

- This layer encompasses the user interfaces and software applications that allow educators to interact with the DT. Through dashboards and analytics tools, educators can access real-time insights into classroom dynamics, student performance, and environmental conditions.
- Integrating DTs with LMS further enhances their utility by enabling personalized learning pathways, automated assessments, and timely interventions based on predictive analytics.

4.2 Role of AI and machine learning in digital twins

AI and ML are central to how DTs work in smart classrooms. They analyze constant data streams to offer real-time suggestions, e.g., adjusting teaching methods or the classroom setup. If engagement drops, for example, the system might recommend a break or interactive activity. Predictive models also help spot equipment issues early, reducing classroom disruptions (Emmert-Streib, 2023; Rathore et al., 2021). Moreover, AI-powered DTs can facilitate personalized learning by analyzing individual students' progress, identifying areas of difficulty, and recommending customized learning resources. By adapting instructional methods and content delivery based on real-time feedback, educators can better address diverse learning needs and improve overall educational outcomes (Emmert-Streib, 2023; Rathore et al., 2021).

4.3 Challenges in implementation

Implementing DTs in smart classrooms presents several challenges, particularly in terms of cost, technical expertise, and data privacy. Establishing a comprehensive sensor network and investing in cloud computing and AI platforms requires significant financial resources, which may be prohibitive for many educational institutions. Moreover, the technical complexity of integrating various data sources and ensuring their interoperability demands specialized skills and continuous professional development for educators and IT staff (Sharma et al., 2022; Trauer et al., 2022).

Data privacy and security are also critical concerns, especially given the sensitivity of biometric and behavioral data collected from students. Ensuring compliance with data protection regulations like the GDPR and FERPA involves implementing encryption protocols, anonymization techniques, and transparent consent mechanisms. Furthermore, developing clear policies on data ownership, access control, and retention is essential to safeguard students' privacy rights and build trust among stakeholders (Sharma et al., 2022; Trauer et al., 2022).

4.4 Strategies for effective implementation

To overcome these challenges, educational institutions can adopt several strategies (Fantozzi et al., 2025; Jeschke & Roman, 2021).

- **Pilot programs:** Starting with small-scale pilot projects can help test the feasibility and effectiveness of DTs before full-scale deployment.
- **Partnerships and funding:** Collaborating with technology providers and seeking grants can alleviate financial burdens and provide access to cutting-edge solutions.
- **Training and capacity building:** Continuous professional development programs can equip educators and administrators with the skills needed to operate and interpret DTs effectively.
- **Ethical and transparent practices:** Establishing clear data governance frameworks and ensuring transparency in data collection and usage can address privacy concerns and enhance user acceptance.

By adopting a structured and ethical approach to implementation, educational institutions can harness the full potential of DTs to transform smart classrooms into dynamic, data-driven learning environments.

5. Benefits of digital twins in smart learning environments

Integrating DTs into smart classrooms is reshaping education by making it more data-driven and responsive. These virtual replicas of physical spaces help optimize the environment, support personalized learning, and improve decision-making. Beyond day-to-day management, DTs also enable long-term improvements through predictive insights and ongoing feedback. This section highlights the main benefits of using DTs in education (Cheung et al., 2021; Ghansah et al., 2024).

5.1 Enhanced personalization of learning

One of the biggest benefits of DTs in smart classrooms is personalized learning. By tracking each student's performance, preferences, and engagement, DTs can adapt lessons to fit individual needs. If a student is having trouble with a concept, the system can recommend extra resources or different teaching approaches. This tailored support boosts understanding, retention, and helps create a more inclusive, student-centered learning environment (Arantes, 2024; Reethika et al., 2024, pp. 165–185). Furthermore, predictive analytics can identify students at risk of falling behind and recommend timely interventions. By alerting educators to potential challenges before they escalate, DTs enable more proactive and targeted support, reducing dropout rates and enhancing overall learning outcomes (Arantes, 2024; Reethika et al., 2024, pp. 165–185).

5.2 Real-time monitoring and feedback

DTs facilitate continuous monitoring of both environmental conditions and student behaviors, providing educators with real-time feedback to inform immediate adjustments. For example, environmental sensors can monitor factors such as air quality, lighting, and temperature, ensuring that classrooms remain conducive to learning. If CO_2 levels rise or lighting becomes inadequate, the DT can recommend adjustments to HVAC systems or lighting controls to maintain optimal conditions (Zivelonghi et al., 2024; Vehviläinen et al., 2016; Haghi et al., 2024).

On the student side, biometric and behavioral data can highlight fluctuations in engagement, stress, or fatigue. Educators can use these insights to adjust their teaching methods dynamically—such as introducing breaks, varying instructional styles, or incorporating interactive activities—to maintain focus and reduce cognitive overload. The ability to respond to real-time data helps create a more agile and responsive learning environment (Haghi et al., 2024; Uhlemann et al., 2017).

5.3 Improved resource management

Efficient management of educational resources is another key benefit of DTs. By analyzing data on classroom usage, equipment performance, and energy consumption, DTs can identify inefficiencies and suggest optimization strategies. For instance, they can recommend rescheduling classes based on room occupancy patterns or adjusting lighting and HVAC systems to reduce energy waste during low-usage periods (Cheng et al., 2024; Li et al., 2022; Seo et al., 2022).

Predictive maintenance is also made possible through continuous monitoring of equipment health. Identifying potential failures before they occur minimizes downtime and ensures that learning tools and infrastructure remain available and reliable. This proactive approach not only enhances the learning experience but also contributes to significant cost savings for educational institutions (Cheng et al., 2024; Hazrat et al., 2023; Li et al., 2022).

5.4 Data-driven decision-making

The comprehensive data collected and analyzed by DTs empowers educators and administrators to make informed decisions backed by real-time insights. For example, by assessing the impact of different instructional strategies on student performance and engagement, administrators can allocate resources more effectively and refine curricula to better meet learning objectives. Additionally, data-driven insights can support policy-making at the institutional level. Patterns revealed by DTs, such as disparities in engagement across different demographic groups or subjects, can inform targeted interventions and resource allocation, promoting equity and inclusivity in education (Charitonidou, 2022; Oulefki et al., 2025).

5.5 Facilitating remote and hybrid learning

The ability of DTs to create virtual replicas of physical classrooms makes them particularly valuable for remote and hybrid learning models. By providing remote students with real-time access to classroom conditions and dynamics, DTs help bridge the gap between in-person and remote learning experiences. This ensures that all students receive consistent and high-quality education, regardless of their location. Besides, DTs can simulate different instructional scenarios to assess their effectiveness in both physical and virtual settings. For example, they can evaluate whether a flipped classroom model or project-based learning might yield better outcomes for a particular cohort, allowing institutions to experiment with different approaches with minimal risk (Deniz et al., 2022; Hazrat et al., 2023).

5.6 Supporting professional development for educators

DTs can also play a significant role in enhancing professional development for educators. By providing detailed analytics on teaching effectiveness, student engagement, and learning outcomes, DTs offer valuable feedback that educators can use to refine their instructional techniques. For example, if data indicates that students are more engaged during hands-on activities than lectures, educators can adjust their teaching strategies accordingly. Furthermore, the insights gained from DTs can inform targeted professional development programs, focusing on areas such as data literacy, adaptive learning technologies, and the effective use of analytics in education. This not only helps educators improve their practice but also prepares them to navigate the increasing digitalization of the education sector (De Los Santos Melo et al., 2024; Hazrat et al., 2023; Reethika et al., 2024, pp. 165–185).

6. Challenges and limitations of digital twins in education

While DTs offer significant potential to transform smart classrooms, their implementation comes with several challenges and limitations. These challenges span technical, financial, ethical, and regulatory domains, requiring comprehensive strategies to address them effectively. In this section, we explore the key obstacles to adopting DTs in educational settings and discuss potential mitigation approaches (Broo et al., 2023; Mihai et al., 2022; Zacher, 2020). The common challenges of DT in education are presented in Table 24.2.

6.1 High implementation costs

One of the primary challenges of deploying DTs in education is the substantial cost associated with infrastructure, hardware, and software. Establishing a robust sensor network, acquiring high-performance computing resources, and integrating AI and ML platforms involve significant financial investments. For many educational institutions, especially public schools and those in developing regions, such costs can be prohibitive. To address this, institutions can explore phased implementation strategies, starting with pilot projects that focus on specific areas, such as environmental monitoring or personalized learning. Additionally, partnerships with technology providers and funding through government grants or private sponsorships can alleviate some of the financial burdens. Leveraging open-source tools and cloud-based solutions can also reduce upfront costs (Lv et al., 2024; Mihai et al., 2022; Rasheed et al., 2020; Rathore et al., 2021).

6.2 Technical complexity and interoperability

Implementing DTs requires seamless integration of diverse data sources, including environmental sensors, biometric devices, and LMS. Ensuring interoperability between different systems and maintaining real-time data accuracy demands advanced technical expertise and robust IT infrastructure. Challenges related to data latency,

TABLE 24.2 Common challenges of DT in education.

Challenges	Description
High initial costs	Implementing DT technology often requires significant upfront investment in hardware, sensors, and software.
Technical complexity	Setting up and integrating IoT devices, sensors, and AI models with existing infrastructure can be complex and requires specialized skills.
Data privacy and security	The large amounts of student data collected by DTs raise concerns over data privacy and security. Ensuring compliance with data protection regulations like GDPR is crucial.
Data quality and accuracy	DTs rely on accurate and consistent data from various sources. Poor data quality or sensor inaccuracies can lead to misleading or unreliable models.
Interoperability issues	Integrating DT systems with legacy educational technologies and platforms can be challenging due to differences in software and hardware.
Scalability	Scaling up DT implementations across large campuses or districts can be difficult, both in terms of technical and financial feasibility.
Teacher training	Teachers and administrators may need extensive training to effectively use DT systems and interpret the data for improved teaching and learning.
Maintenance and support	Ongoing maintenance of hardware and software is essential to ensure the proper functioning of DT systems. This requires continuous investment in resources.

synchronization, and network reliability can compromise the effectiveness of DTs. Adopting standardized protocols and APIs can facilitate smoother data exchange between different systems. Investing in professional development for educators and IT staff to build skills in data management and analytics is also crucial. Furthermore, incorporating edge computing can help reduce latency by processing data locally before transmitting it to centralized systems (Acharya et al., 2024; Khan et al., 2022).

6.3 Data privacy and security concerns

Handling sensitive data like biometrics and behavior patterns raises serious privacy and security concerns. To stay compliant with regulations like GDPR and FERPA, schools and tech providers must use strong safeguards—encryption, anonymization, access controls—and adopt a "privacy-by-design" approach. Clear policies, informed consent, regular security audits, and transparent data practices are essential for protecting data and maintaining trust among students, parents, and educators (Saeed et al., 2024, pp. 99–124; Stergiou et al., 2023; Wang et al., 2023).

6.4 Ethical and surveillance concerns

The use of DTs, especially with biometric and behavioral tracking, can raise concerns about surveillance, privacy, and student autonomy. Ethical issues also arise from potential AI bias, like reinforcing stereotypes or unfairly labeling students. A lack of transparency in AI decisions can deepen these worries. To address them, clear ethical guidelines and responsible AI use are essential. Using explainable AI can make algorithm decisions more understandable, while involving students, parents, and educators in conversations about data use helps ensure a more inclusive and ethical approach (Gallastegui et al., 2021; Helbing et al., 2023; Zoltick et al., 2023, pp. 1167–1200).

6.5 Digital divide and accessibility issues

Deploying DTs often depends on access to advanced technology and fast internet, in which the resources are not available in all schools. This can widen the digital divide, especially in rural or low-income areas. To close the gap, governments and institutions should invest in broadband access and provide affordable devices and software. Creating lightweight DT applications that use less data and power can also help. Ensuring compatibility with assistive technologies further supports students with disabilities, making education more inclusive for all (Mihai et al., 2022; Tzachor et al., 2022).

6.6 Scalability challenges

Scaling DTs from individual classrooms to entire schools or districts introduces complexities related to data volume, storage, and processing capabilities. Ensuring consistent data quality and managing the exponential growth of information from numerous sensors and devices can strain existing IT infrastructure and resources. Adopting cloud-based architectures can facilitate scalability by providing elastic storage and processing capabilities. Modular design principles, where different components of the DT (such as environmental monitoring, personalized learning, and predictive analytics) can be scaled independently, can also streamline the expansion process. Regular performance assessments and upgrades to infrastructure are essential to ensure that DTs remain effective at larger scales (Scime et al., 2022; Yaqoob et al., 2020).

7. Case studies of digital twins in education

While still in the early stages, some forward-thinking institutions have started using DTs in education through pilot programs and full rollouts. These real-world examples show how DTs can support personalized learning, improve resource management, and provide insights for others looking to adopt similar tools. This section explores key cases that showcase the impact of DTs in smart classrooms. Table 24.3 shows a comprehensive taxonomy of the presented three case studies.

7.1 Case study 1: Enhancing personalized learning at Nanyang Technological University, Singapore

Nanyang Technological University (NTU) has been at the forefront of adopting DTs to enhance personalized learning experiences. The university deployed a DT of its smart classroom infrastructure, integrating data from sensors that monitor environmental conditions and students' interactions with digital learning tools (Qin et al., 2023; Wei, 2024).

Implementation:

- **Data Integration:** NTU's DT aggregates data from IoT sensors, LMS, and video analytics to track students' engagement and performance in real-time.
- **AI and Predictive Analytics:** ML algorithms analyze this data to identify patterns in student learning behaviors and predict potential challenges.
- **Personalized Recommendations:** Based on these insights, the DT provides tailored learning resources and real-time feedback to students and instructors, ensuring that instructional methods align with individual learning preferences.

Outcomes:

- A 15% improvement in student retention rates was observed, attributed to timely interventions based on predictive analytics.

TABLE 24.3 Three case studies comparison of DT in education.

Aspect	NTU	SDU	Purdue university
Focus area	Personalized learning	Campus management and sustainability	Predictive maintenance and safety
Technology used	DT, IoT sensors, AI, predictive analytics	DT, environmental sensors, AI, MS	DT, IoT sensors, ML
Primary objective	Enhancing student engagement and performance	Optimizing energy usage and improving air quality	Reducing equipment downtime and improving safety
Key benefits	Improved student retention and engagement	20% reduction in energy consumption and improved air quality	25% reduction in equipment downtime and 10% fewer lab incidents
Data integration	Integrates data from sensors, LMS, and video analytics	Environmental sensor data with BMS	Real-time monitoring of equipment and safety protocols
Outcome	15% increase in student retention rates	Improved classroom comfort, lower energy costs	Enhanced lab safety and reduced operational disruptions
Challenges	High initial investment, data privacy concerns	Initial setup costs, technical complexity	Ensuring continuous data flow, managing predictive models
Lessons learned	Start with targeted interventions, use AI for personalized learning	Prioritize environmental sustainability, integrate with existing systems	Predictive maintenance is key to preventing downtime and enhancing safety
Future potential	Expansion of AI-driven learning models	Further expansion of IoT for sustainability and energy efficiency	Broader application in other departments and campus-wide safety systems

- Enhanced engagement through personalized content delivery, leading to higher overall satisfaction among students and instructors.

7.2 Case study 2: Optimizing campus management at the University of Southern Denmark

The University of Southern Denmark (SDU) implemented a campus-wide DT to optimize resource management and improve sustainability. The project focused on leveraging data from environmental sensors to enhance energy efficiency and indoor air quality across classrooms (Clausen et al., 2021; Madsen et al., 2023).

Implementation:

- **Sensor deployment:** Environmental sensors monitored temperature, humidity, CO_2 levels, and occupancy in real-time.
- **AI-driven optimization:** AI algorithms processed this data to recommend adjustments to HVAC and lighting systems, ensuring optimal conditions for learning while minimizing energy waste.
- **Integration with building management systems:** The DT was integrated with the existing Building Management System (BMS) to enable automated control of facilities based on real-time data.

Outcomes:

- Achieved a 20% reduction in energy consumption across campus buildings, significantly lowering operational costs.
- Improved indoor air quality and comfort levels, contributing to higher concentration and performance among students.

7.3 Case study 3: Predictive maintenance and safety at Purdue University, USA

Purdue University utilized a DT model to enhance safety and predictive maintenance in its engineering labs. By creating virtual replicas of critical equipment, the university aimed to minimize downtime and improve safety protocols (Theos, 2023; Ziakkas et al., 2023).

Implementation:

- **Real-time monitoring:** Sensors tracked equipment usage, temperature, and operational anomalies.
- **Predictive maintenance:** ML algorithms analyzed sensor data to predict potential equipment failures, triggering maintenance alerts before breakdowns occurred.
- **Safety protocols:** The DT also simulated various emergency scenarios to optimize evacuation plans and ensure compliance with safety regulations.

Outcomes:

- Reduced equipment downtime by 25%, allowing uninterrupted access to lab resources for students.
- Enhanced safety compliance, reducing incident rates in labs by 10%.

7.4 Key lessons

The case studies illustrate several key lessons for implementing DTs in educational settings (Khan et al., 2022; Lu et al., 2020; Mihai et al., 2022). Many institutions began their digital transformation journeys with small-scale pilot projects, targeting specific areas like energy management or personalized learning to fine-tune their strategies before expanding. A key driver of success was the integration of AI and machine learning, which provided predictive insights that enabled proactive decision-making. Interoperability played a crucial role, with effective implementations ensuring smooth communication between IoT devices, AI platforms, and existing management systems to build a cohesive digital twin ecosystem. Throughout the process, prioritizing privacy and security through clear data governance policies and regulatory compliance was essential in maintaining stakeholder trust.

7.5 Challenges faced by early adopters

While these case studies highlight the potential of DTs, they also reveal common challenges (Abdalzaher et al., 2019, 2021a, 2021b, 2022, 2023b; Abdalzaher & Osamu, 2020; Attaran et al., 2023):

- **High Costs:** Initial investments in sensors, AI platforms, and IT infrastructure posed significant financial challenges.
- **Technical Expertise:** Managing and interpreting vast amounts of data required specialized skills, underscoring the need for ongoing training and professional development.
- **Privacy Concerns:** Ensuring data privacy, particularly when dealing with student information, was a critical challenge that required robust encryption and anonymization practices.

8. Future trends of digital twins in education

As technology continues to evolve, the potential applications of DTs in education are expanding. Emerging trends in AI, augmented reality (AR), virtual reality (VR), and 5G networks promise to further enhance the capabilities of DTs, transforming smart classrooms into even more immersive, adaptive, and data-driven learning environments. In this section, we explore the future directions of DTs in education, highlighting the key trends that will shape their development and integration in the coming years.

8.1 Integration with augmented and virtual reality

One of the most exciting future trends for DTs in education is the integration with augmented and VR technologies. By combining DTs with AR/VR, educators can create fully immersive and interactive learning experiences. For instance, a DT of a science laboratory could be linked with VR, allowing students to conduct experiments in a virtual environment that mimics real-world conditions.

Potential Impact:

- **Immersive learning:** Students can interact with digital models of historical sites, chemical reactions, or complex biological systems in 3D, enhancing engagement and understanding.
- **Hands-on experience:** Virtual field trips or simulations could provide practical experiences without leaving the classroom, helping students visualize abstract concepts tangibly.
- **Collaboration:** AR/VR-powered DTs could allow students to collaborate across locations in real-time, experiencing shared environments even in remote settings.

8.2 Real-time adaptation and autonomous classrooms

The next evolution of DTs in education will see classrooms becoming more autonomous through the integration of AI and ML. Real-time data from sensors, along with AI-driven analytics, will allow classrooms to autonomously adapt to changing conditions, both in terms of physical environment and teaching methods:

Potential Impact:

- **Dynamic environmental adjustments:** DTs will be able to automatically adjust factors such as lighting, temperature, and noise levels to optimize learning conditions for different groups of students.
- **Personalized instruction:** AI-driven systems will continuously adapt learning content and delivery methods based on real-time feedback from students, creating a truly personalized education experience.
- **Self-managing classrooms:** The integration of AI can enable classrooms to take over administrative tasks, such as monitoring attendance, tracking participation, and providing instant feedback, allowing teachers to focus more on instruction.

8.3 5G and edge computing for faster, real-time data processing

With the rollout of 5G networks and advances in edge computing, the ability to process large amounts of data in real-time will significantly improve. DTs will benefit from reduced latency and faster data transfer speeds, enabling near-instantaneous updates to virtual models:

Potential Impact:

- **Instantaneous updates:** Real-time feedback from sensors will be reflected in the DT instantly, enabling immediate adjustments to the classroom environment.
- **Remote learning support:** Enhanced bandwidth will improve the quality of remote and hybrid learning, ensuring that students participating from home or different locations experience minimal delays or disruptions.
- **Smarter decision-making:** With 5G enabling faster data transfer, teachers and administrators can access real-time insights and make informed decisions instantly, improving classroom management and learning outcomes.

8.4 Blockchain for secure and transparent data management

As DTs rely on vast amounts of data, managing and securing this information becomes a critical concern, particularly when dealing with sensitive student data. The use of blockchain technology could provide a decentralized, secure, and transparent way to manage data related to DTs:

Potential Impact:

- **Data integrity:** Blockchain ensures the integrity of data by creating an immutable record of transactions. This can be crucial for tracking student progress and maintaining accurate academic records.
- **Privacy and security:** Blockchain can enhance data privacy by giving students greater control over their personal information and how it is used. Smart contracts could enable users to consent to specific data-sharing arrangements while ensuring compliance with data protection laws.
- **Decentralized data storage:** Blockchain could provide a more secure and distributed way to store educational data, reducing the risk of breaches and unauthorized access.

8.5 Enhanced collaboration through cloud-based digital twins

The future of DTs will likely see more collaboration through cloud-based platforms, allowing multiple stakeholders—students, teachers, administrators, and parents—to access and interact with the DT simultaneously. This collaborative approach could facilitate more inclusive decision-making and a deeper understanding of classroom dynamics:

Potential Impact:

- **Collaboration across geographies:** Cloud-based DTs will allow for real-time collaboration across different locations, whether for group learning, teacher collaboration, or cross-institutional research.
- **Parental involvement:** Parents could have access to real-time insights into their child's learning progress, engagement, and overall well-being, fostering a stronger connection between school and home.
- **Cross-disciplinary projects:** Collaborative projects involving multiple schools or even institutions can be supported by shared DTs, promoting interdisciplinary learning and innovation.

8.6 AI-driven predictive analytics for institutional decision-making

In the future, DTs will integrate more advanced AI and ML techniques to offer predictive analytics that can assist institutional decision-making on a larger scale. For example, AI could predict the effectiveness of teaching methods, forecast student success rates, or even anticipate the long-term impact of various teaching interventions.

Potential Impact:

- **Data-driven strategic planning:** Administrators could use predictive insights to optimize curricula, plan resources more effectively, and allocate budgets based on student needs and trends.
- **Long-term planning:** Educational institutions could predict trends in student enrollment, course demand, and potential challenges, enabling better long-term planning and resource allocation.
- **Adaptive learning ecosystem:** Predictive models could also adjust the entire educational ecosystem, including scheduling, staffing, and course offerings, to better meet the evolving needs of students.

9. Conclusion and future perspectives

DTs are reshaping education by creating smarter, personalized, and data-driven learning environments. As this technology advances, it offers immense potential to enhance classroom experiences, optimize resources, and support real-time decision-making. Through the integration of AI, AR/VR, 5G, 6G, and blockchain, DTs will provide increasingly immersive, adaptive, and efficient learning spaces.

Key benefits include personalized learning, improved classroom management, and greater sustainability. However, challenges such as high costs, privacy concerns, and technical complexity must be addressed. Case studies from institutions like NTU, SDU, and Purdue University highlight the promise and obstacles in implementing DTs.

Looking ahead, the future of DTs in education is bright, with further advancements in AI-driven analytics and real-time classroom adaptation. However, to ensure equitable access, institutions must focus on training educators and ensuring data privacy. As DTs become more integrated into educational systems, they will transform classrooms, making them more dynamic and responsive to the needs of students.

References

Abdalzaher, M. S., et al. (2021). A deep learning model for earthquake parameters observation in IoT system-based earthquake early warning. *IEEE Internet of Things Journal, 9*(11), 8412–8424.

Abdalzaher, M. S., et al. (2021). A deep autoencoder trust model for mitigating jamming attack in IoT assisted by cognitive radio. *IEEE Systems Journal, 16*(3), 3635–3645.

Abdalzaher, M. S., et al. (2022). Machine learning benchmarking for secured IoT smart systems. In *2022 IEEE international conference on internet of things and intelligence systems (IoTaIS). IEEE.*

Abdalzaher, M. S., et al. (2023). Toward secured IoT-based smart systems using machine learning. *IEEE Access, 11*, 20827–20841.

Abdalzaher, M. S., et al. (2023). Employing machine learning and iot for earthquake early warning system in smart cities. *Energies, 16*(1), 495.

Abdalzaher, M. S., & Osamu, M. (2020). A game-theoretic approach for enhancing security and data trustworthiness in IoT applications. *IEEE Internet of Things Journal, 7*(11), 11250–11261.

Abdalzaher, M. S., Samy, L., & Osamu, M. (2019). Non-zero-sum game-based trust model to enhance wireless sensor networks security for IoT applications. *IET Wireless Sensor Systems, 9*(4), 218–226.

Acharya, S., Ali Khan, A., & Päivärinta, T. (2024). Interoperability levels and challenges of digital twins in cyber-physical systems. *Journal of Industrial Information Integration*, Article 100714.

Adibi, S., et al. (2024). Enhancing healthcare through sensor-enabled digital twins in smart environments: A comprehensive analysis. *Sensors, 24*(9), 2793.

Ali, A. (2024). Artificial intelligence and data privacy: Balancing innovation with security. *Frontiers in Artificial Intelligence Research, 1*(2), 289–328.

Arantes, J. (2024). Digital twins and the terminology of "personalization" or "personalized learning" in educational policy: A discussion paper. *Policy Futures in Education, 22*(4), 524–543.

Attaran, M., & Celik, B. G. (2023). Digital twin: Benefits, use cases, challenges, and opportunities. *Decision Analytics Journal, 6*, Article 100165.

Broo, D. G., & Schooling, J. (2023). Digital twins in infrastructure: Definitions, current practices, challenges and strategies. *International Journal of Construction Management, 23*(7), 1254–1263.

Bucchiarone, A. (2022). Gamification and virtual reality for digital twin learning and training: Architecture and challenges. *Virtual Reality and Intelligent Hardware, 4*(6), 471–486.

Bustos-Lopez, M., et al. (2022). Wearables for engagement detection in learning environments: A review. *Biosensors, 12*(7), 509.

Charitonidou, M. (2022). Urban scale digital twins in data-driven society: Challenging digital universalism in urban planning decision-making. *International Journal of Architectural Computing, 20*(2), 238–253.

Cheng, N., et al. (2024). Toward enhanced reinforcement learning-based resource management via digital twin: Opportunities, applications, and challenges. *IEEE Network, 39*(1), 189–196.

Cheung, S. K. S., et al. (2021). Shaping the future learning environments with smart elements: Challenges and opportunities. *International Journal of Educational Technology in Higher Education, 18*, 1–9.

Clausen, A., et al. (2021). A digital twin framework for improving energy efficiency and occupant comfort in public and commercial buildings. *Energy Informatics, 4*, 1–19.

De Benedictis, A., et al. (2022). Digital twins in healthcare: An architectural proposal and its application in a social distancing case study. *IEEE Journal of Biomedical and Health Informatics, 27*(10), 5143–5154.

De Los Santos Melo, A., & Alcántara, F. E. B. (2024). Maximizing learning potential: Embracing the power of digital twins in architectural and construction education of the twenty-first century. In *Teaching innovation in architecture and building engineering: Challenges of the 21st century. Cham: Springer nature Switzerland* (pp. 477–493).

Dembski, F., et al. (2020). Urban digital twins for smart cities and citizens: The case study of herrenberg, Germany. *Sustainability, 12*(6), 2307.

Deniz, S., et al. (2022). Online (remote) teaching for laboratory based courses using "digital twins" of the experiments. *Journal of Engineering for Gas Turbines & Power, 144*(5), Article 051016.

Emmert-Streib, F. (2023). What is the role of ai for digital twins? *AI, 4*(3), 721–728.

Fang, F., & Jiang, X. (2024). The analysis of artificial intelligence digital technology in art education under the internet of things. *IEEE Access, 12*, 22928–22937.

Fantozzi, I., et al. (2025). Digital twins: Strategic guide to utilize digital twins to improve operational efficiency in industry 4.0. *Future Internet, 17*, 1.

Gallastegui, G., Miguel, L., Francisco, R., & Forradellas, R. (2021). Business methodology for the application in university environments of predictive machine learning models based on an ethical taxonomy of the student's digital twin. *Administrative Sciences, 11*(4), 118.

Ghansah, F. A., & Lu, W. (2024). Major opportunities of digital twins for smart buildings: A scientometric and content analysis. *Smart and Sustainable Built Environment, 13*(1), 63–84.

Giuffrida, I., & Hall, A. (2023). Technology integration in higher education and student privacy beyond learning environments—A comparison of the UK and US perspective. *British Journal of Educational Technology, 54*(6), 1587–1603.

Haghi, E., Sadegh, M., Anitescu, C., & Rabczuk, T. (2024). Methods for enabling real-time analysis in digital twins: A literature review. *Computers and Structures, 297*, Article 107342.

Han, L., Long, X., & Wang, K. (2024). The analysis of educational informatization management learning model under the internet of things and artificial intelligence. *Scientific Reports, 14*(1), Article 17811.

Hazrat, M. A., et al. (2023). Developing a skilled workforce for future industry demand: The potential of digital twin-based teaching and learning practices in engineering education. *Sustainability, 15*(23), Article 16433.

Helbing, D., & Sánchez-Vaquerizo, J. A. (2023). Digital twins: Potentials, ethical issues and limitations. In *Handbook on the politics and governance of big data and artificial intelligence* (pp. 64–104). Edward Elgar Publishing.

Huang, J., & Zhou, D. (2024). A scalable real-time computer vision system for student posture detection in smart classrooms. *Education and Information Technologies, 29*(1), 917–937.

Javaid, M., Haleem, A., & Suman, R. (2023). Digital twin applications toward industry 4.0: A review. *Cognitive Robotics, 3*, 71–92.

Jeschke, S., & Roman, G. (2021). Development of a generic implementation strategy of digital twins in logistics systems under consideration of the German rail transport. *Applied Sciences, 11*(21), Article 10289.

Khan, S., et al. (2020). Biometric systems utilising health data from wearable devices: Applications and future challenges in computer security. *ACM Computing Surveys, 53*(4), 1–29.

Khan, L. U., et al. (2022). Digital twin of wireless systems: Overview, taxonomy, challenges, and opportunities. *IEEE Communications Surveys and Tutorials, 24*(4), 2230–2254.

Krichen, M., & Abdalzaher, M. S. (2024). Performance enhancement of artificial intelligence: A survey. *Journal of Network and Computer Applications*, Article 104034.

Ladley, J. (2019). *Data governance: How to design, deploy, and sustain an effective data governance program.* Academic Press.

Li, J., Zhang, Y., & Qian, C. (2022). The enhanced resource modeling and real-time transmission technologies for digital twin based on QoS considerations. *Robotics and Computer-Integrated Manufacturing, 75*, Article 102284.

Lu, Q., et al. (2020). Developing a digital twin at building and city levels: Case study of west Cambridge campus. *Journal of Management in Engineering, 36*(3), Article 05020004.

Lv, Z., & Xie, S. (2024). Artificial intelligence in the digital twins: State of the art, challenges, and future research topics. *Digital Twin, 1*(1), 12.

Madsen, F. W., et al. (2023). Digital twin-based fault detection and prioritisation in district heating systems: A case study in Denmark. In *Energy informatics academy conference. Cham: Springer nature Switzerland.*

Martins, V. F., et al. (2024). A three-tiered architectural model for digital twins in education. In *Workshop em Modelagem e Simulação de Sistemas Intensivos em Software (MSSiS). SBC.*

Mihai, S., et al. (2022). Digital twins: A survey on enabling technologies, challenges, trends and future prospects. *IEEE Communications Surveys and Tutorials, 24*(4), 2255–2291.

Mourtzis, D., Panopoulos, N., & Angelopoulos, J. (2023). A hybrid teaching factory model towards personalized education 4.0. *International Journal of Computer Integrated Manufacturing, 36*(12), 1739–1759.

Oulefki, A., Amira, A., & Foufou, S. (2025). Digital twins and AI transforming healthcare systems through innovation and data-driven decision making. *Health Technology*, 1–23.

Qin, Y., Arunan, A., & Yuen, C. (2023). Digital twin for real-time Li-ion battery state of health estimation with partially discharged cycling data. *IEEE Transactions on Industrial Informatics, 19*(5), 7247–7257.

Rasheed, A., San, O., & Kvamsdal, T. (2020). Digital twin: Values, challenges and enablers from a modeling perspective. *IEEE Access, 8*, 21980–22012.

Rathore, M. M., et al. (2021). The role of ai, machine learning, and big data in digital twinning: A systematic literature review, challenges, and opportunities. *IEEE Access, 9*, 32030–32052.

Razzaq, S., et al. (2023). DeepClassRooms: A deep learning based digital twin framework for on-campus class rooms. *Neural Computing & Applications*, 1–10.

Reethika, A., & Kanaga Priya, P. (2024). *"Personalize Learning Experience in Education Using Digital Twins with Human-Centered Design and Pedagogy." transforming industry using digital twin technology*. Cham: Springer Nature Switzerland.

Roda-Sanchez, L., et al. (2023). Building a smart campus digital twin: System, analytics, and lessons learned from a real-world project. *IEEE Internet of Things Journal, 11*(3), 4614–4627.

Saeed, Ali, M. M., Saeed, R. A., & Ahmed, Z. E. (2024). *"Data security and privacy in the age of AI and digital twins." digital twin technology and AI implementations in future-focused businesses*. IGI Global Scientific Publishing.

Scime, L., Singh, A., & Vincent, P. (2022). A scalable digital platform for the use of digital twins in additive manufacturing. *Manufacturing Letters, 31*, 28–32.

Screpanti, L., et al. (2022). Control engineering and robotics since primary school: An infrastructure for creating the digital twin model of the learning class. *IFAC-PapersOnLine, 55*(17), 267–272.

Seo, H., & Yun, W.-S. (2022). Digital twin-based assessment framework for energy savings in university classroom lighting. *Buildings, 12*(5), 544.

Sepasgozar, S. M. E. (2020). Digital twin and web-based virtual gaming technologies for online education: A case of construction management and engineering. *Applied Sciences, 10*(13), 4678.

Sharma, A., et al. (2022). Digital twins: State of the art theory and practice, challenges, and open research questions. *Journal of Industrial Information Integration, 30*, Article 100383.

Stergiou, C. L., Bompoli, E., & Psannis, K. E. (2023). Security and privacy issues in IoT-based big data cloud systems in a digital twin scenario. *Applied Sciences, 13*(2), 758.

Theos, V. (2023). *Design and development of a real-time cyber-physical testbed for cybersecurity research*. Purdue University Graduate School.

Trauer, J., et al. (2022). Challenges in implementing digital Twins–A survey. In , *Vol 86212. International design engineering technical conferences and computers and information in engineering conference*. American Society of Mechanical Engineers.

Tzachor, A., et al. (2022). Potential and limitations of digital twins to achieve the sustainable development goals. *Nature Sustainability, 5* (10), 822–829.

Uhlemann, T. H.-J., et al. (2017). The digital twin: Demonstrating the potential of real time data acquisition in production systems. *Procedia Manufacturing, 9*, 113–120.

Vehviläinen, T., Vehviläinen, T., Vehviläinen, T., et al. (2016). High indoor CO2 concentrations in an office environment increases the transcutaneous CO_2 level and sleepiness during cognitive work. *Journal of Occupational and Environmental Hygiene, 13*(1), 19–29.

Wang, Y., et al. (2023). A survey on digital twins: Architecture, enabling technologies, security and privacy, and future prospects. *IEEE Internet of Things Journal, 10*(17), 14965–14987.

Wei, Y. (2024). Probabilistic digital twin of water treatment facilities.

Yaqoob, I., et al. (2020). Blockchain for digital twins: Recent advances and future research challenges. *Ieee Network, 34*(5), 290–298.

Zacher, S. (2020). Digital twins for education and study of engineering sciences. *International Journal on Engineering, Science and Technology, 2*(2), 61–69.

Zhang, J., Hassandoust, F., & Williams, J. E. (2020). Online customer trust in the context of the general data protection regulation (GDPR). *Pacific Asia Journal of the Association for Information Systems, 12* (1), 4.

Ziakkas, D., Flores, A. D., & Konstantinos, P. (2023). The role of digital twins in future transportation challenges. *AHFE Open Access Int, 115*, 61.

Zivelonghi, A., & Giuseppi, A. (2024). Smart healthy schools: An IoT-enabled concept for multi-room dynamic air quality control. *Internet of Things and Cyber-Physical Systems, 4*, 24–31.

Zoltick, M. M., & Maisel, J. B. (2023). *Societal impacts: Legal, regulatory and ethical considerations for the digital twin. The digital twin*. Cham: Springer International Publishing.

Chapter 25

Assistant agent for smart classroom application: Past, present and future direction

Anh Tuan Nguyen[1], Thi Ngoc Yen Huynh[1] and Yonghan Ahn[2]

[1]*Department of Smart City Engineering, Hanyang University ERICA Campus, Ansan, Gyeonggi-do, South Korea;* [2]*Department of Architecture Engineering, Hanyang University ERICA Campus, Ansan, Gyeonggi-do, South Korea*

1. Introduction

Artificial Intelligence (AI) techniques have increasingly been integrated into smart classrooms, aiming to deliver personalized learning experiences and intelligent tutoring. Early applications primarily relied on rule-based systems, which operated using predefined rules and logical inference mechanisms (Graesser & McNamara, 2010; Zaiane, 2002). Such systems facilitated automated grading for multiple-choice and short-answer assessments and keyword-based question-answering systems, enabling immediate student access to educational content (Alberola et al., 2016). However, these systems faced significant limitations, particularly their rigidity; modifications in curricula required manual adjustments to rule sets, and their simplistic approach often struggled with interpreting the nuances inherent in natural language.

Statistical models represented a significant evolution from rule-based systems, contributing notably to the development of sentiment analysis tools within educational environments (Bax, 2011; Hooshyar et al., 2020; Tastanova et al., 2023). By processing textual data from student interactions, feedback, and discussion forums, these models provided valuable insights into student engagement and emotional states (Tastanova et al., 2023). Consequently, educators could adapt instructional strategies responsively, creating a more interactive and supportive learning environment. Nevertheless, statistical models suffered from their dependence on extensive and representative datasets for training. When training data failed to capture real-world classroom diversity, these models risked overfitting and demonstrated diminished effectiveness in new or unanticipated situations (Gandedkar et al., 2021).

Digital twin (DT) technology emerged as another promising advancement, providing virtual representations of educational processes and physical environments. DT applications extend from curriculum development to dynamic simulations that enhance student motivation, peer learning, and self-directed engagement (Gandedkar et al., 2021; Mihai et al., 2022). For educators, DT supports content delivery improvements, technological integration, easier demonstrations, and more effective student assessments. Despite these strengths, traditional educational data mining techniques such as knowledge tracing still encounter challenges related to limited contextual understanding, reduced interactivity, and difficulties in personalizing adaptive learning content.

The introduction and evolution of Large Language Models (LLMs) have significantly transformed AI capabilities in education by addressing many limitations of earlier methods (Bhowmik et al., 2024; He et al., 2024). LLMs leverage Transformer architectures, characterized by self-attention mechanisms, to effectively capture complex linguistic patterns and long-range textual dependencies, allowing them to generalize tasks without extensive retraining (Abd-Alrazaq et al., 2023; Milano et al., 2023; Shahzad et al., 2025; Zhao & Song, 2024). Advances like in-context learning (ICL) further enhance adaptability, enabling rapid model customization through simple prompt-based examples (Abu-Rasheed, Abdulsalam, et al., 2024; Mirata et al., 2020). Moreover, retrieval-augmented generation (RAG) techniques improve output accuracy and factual grounding by dynamically accessing external knowledge bases, mitigating misinformation, and enhancing reliability. Empirical research explored LLMs' adaptability for smart classroom applications,

Digital Twins in the Smart Classroom. https://doi.org/10.1016/B978-0-443-41632-3.00015-1

demonstrating their ability to process multimodal inputs—including speech, text, code, and sensory data—and effectively analyze cognitive and emotional states (Abu-Rasheed, Weber, & Fathi, 2024; Bhowmik et al., 2024; Du et al., 2024; Li et al., 2024; Sonlu et al., 2024; Voultsiou & Moussiades, 2025; Zheng et al., 2024a). In addition, LLM-driven systems offer administrative efficiencies through natural language-based scheduling and resource management optimization. Their improved multilingual capabilities further enhance accessibility in diverse educational contexts.

Prior to this review, minimal literature surveys or review articles on specific themes of smart classroom have been published in the previous 2 decades. None of them aspire to capture the fundamental conceptualization of assistant agents in smart classroom. Furthermore, there is no work that examined the intellectual configuration of smart classroom.

In this book chapter, we provide a systematic review of assistant agents in smart classrooms. We introduce various AI agents within educational domains through a historical development and discuss their potential applications within smart classroom systems. These agents are categorized into four broad classes: (1) Pedagogical agents, which automate complex instructional tasks such as classroom interaction and personalized tutoring; (2) Administrative automation agents, designed to streamline processes such as scheduling and resource management, attendance and engagement analytics; (3) Collaborative learning agents, which facilitate teamwork structures typically used in professional settings for managing complex projects including group work facilitation and teacher-AI co-teaching; and (4) Accessibility and inclusion agents, aimed at addressing specific challenges across diverse fields, including language learning and, adaptation for professional and special needs.

Furthermore, we examine key challenges associated with assistant agents in smart classrooms, which must be addressed to advance research and achieve effective real-world implementation. Our contributions are structured around three primary objectives:

- Identifying recent research advancements in smart classrooms, highlighting technological developments, effectiveness, and challenges of assistant agents.
- Exploring major applications of assistant agents and tracing the evolution of agent frameworks in educational research, offering a structured classification that clarifies their roles and capabilities.
- Analyzing current limitations and proposing future research directions essential for the next generation of smart classrooms, specifically addressing critical concerns related to privacy, bias, fairness, hallucination, and seamless integration into actual classroom environments.

2. Methodology

This study employed a structured literature review as presented in Fig. 25.1 to identify relevant research on assistant agents based on natural language processing (NLP) and large language models (LLMs) for smart classroom applications. Two groups of keywords were used: (1) core AI techniques such as "Natural Language Processing," "Large Language Model," "Conversational Agents," and "Multimodal Learning," and (2) educational application terms such as "Smart Classroom," "AI Teaching Assistant," "Student Support System," and related topics. Boolean operators were applied to combine these groups using AND logic. Literature searches were conducted across Web of Science, Scopus, IEEE, and Google Scholar, yielding an initial total of 3951 records.

After removing 1476 duplicate entries, 2475 records remained for screening. Inclusion criteria were then applied to titles and abstracts to identify studies that were (1) published between 2000 and 2025, (2) written in English with full-text availability, (3) peer-reviewed journal articles or eligible conference papers, (4) focused on assistant agents, AI/NLP/LLM technologies, and smart classroom applications, and (5) targeting smart classrooms, intelligent tutoring systems, virtual classrooms, or student support systems. Based on these criteria, 862 records were retrieved. A full-text review of these records using the same inclusion criteria resulted in 379 studies selected for eligibility assessment.

At the eligibility stage, exclusion criteria were applied to remove papers that lacked technical depth, used outdated NLP techniques (e.g., nontransformer-based models), focused on noneducational domains without transferable methodologies, or presented redundant or duplicate findings. After applying these exclusions, 228 papers were excluded, and finally, 151 studies were included in the review. The entire selection process adhered to a PRISMA-based structure to ensure transparency, reproducibility, and rigor.

3. Application review of assistant agents

Education focused on individualized learning and problem-solving, which is one of the ideal scenarios for deploying assistant agents. In Table 25.1, we categorize assistant agents for smart classroom into four domains: pedagogical applications, administrative automation, collaborative learning enhancement, and accessibility and inclusion. The outlines of the roadmap will be discussed in the following section.

3.1 Pedagogical applications

Pedagogical agents constitute AI–powered systems intended to augment instructional processes and student

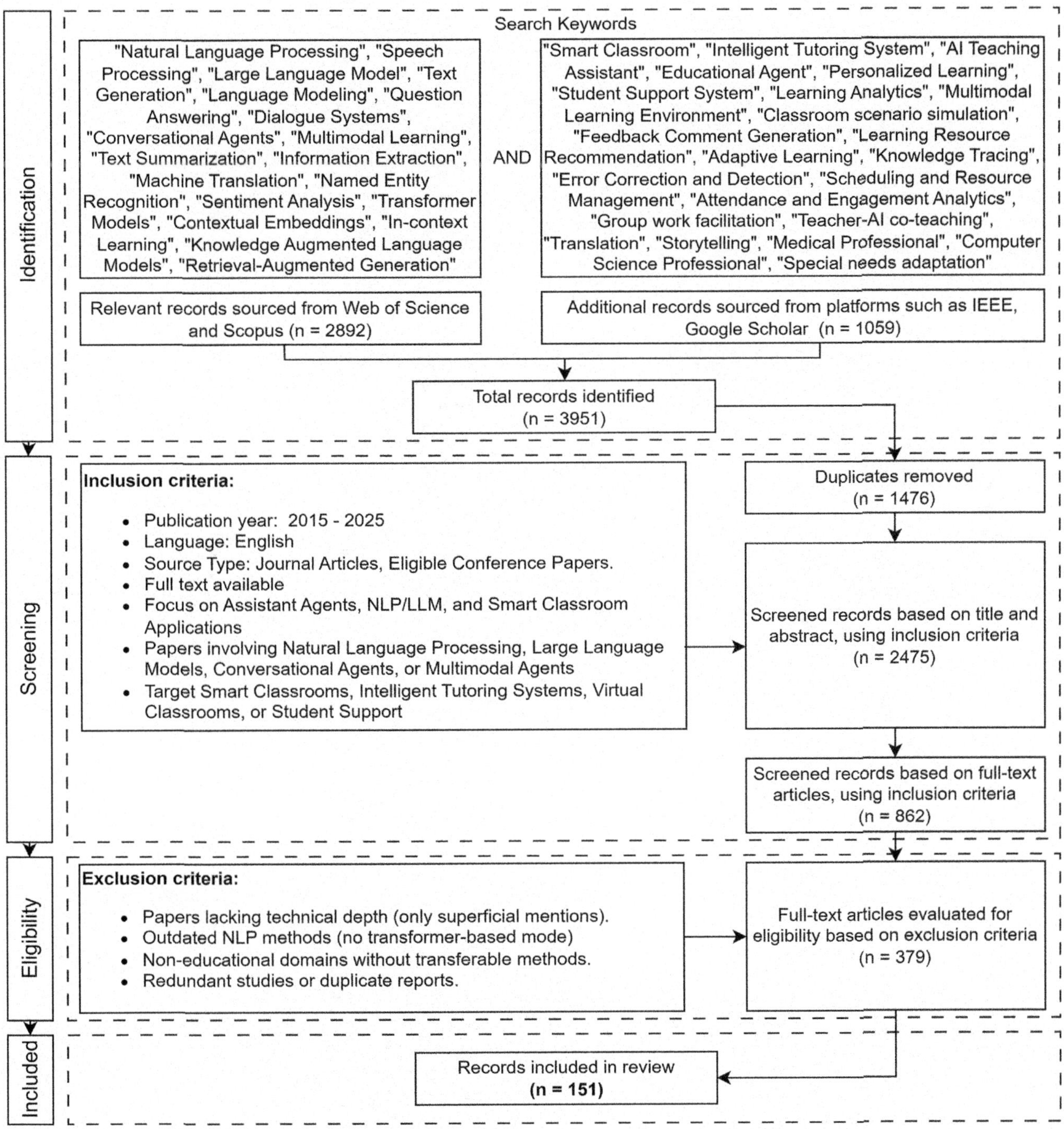

FIGURE 25.1 PRISMA diagram for systematic review.

learning by means of automation and individualized support. These agents typically fall into two principal categories. First, classroom-interaction agents assume responsibility for routine instructional tasks. Second, personalized-tutoring agents deliver guidance by continuously assessing each learner's performance, diagnosing misconceptions, and adapting instructional content in real time. Collectively, these AI-driven assistants enhance the efficacy of instruction, alleviate educator workload, and cultivate more engaging, customized learning environments.

3.1.1 Classroom interaction

Educational assistance agent systems are engineered to augment educators' effectiveness within instructional settings. These agents deliver personalized, scalable, and

TABLE 25.1 Summarize taxonomy of representative research on assistant agents of smart classroom.

Domain	Specialize	Subspecialize	Application
Pedagogical applications	Classroom interaction	Classroom scenario simulation (CSS)	Open learner model (OLM) (Hooshyar et al., 2020), guidebots (Hwang et al., 2020; Johnson, 2003), CGMI (Jinxin et al., 2023), classroom Simulacra system (Xu, Wen, et al., 2025)
		Feedback comment generation (FCG)	CourseMIRROR (Fan et al., 2015), LexRank (Erkan & Radev, 2004), TextRank (Mihalcea & Tarau, 2004), LSA (latent semantic analysis) (Foong et al., 2015), MWISum (Cagliero et al., 2019), BERT – k-means summarizer (Miller, 2019), aspect-based opinion mining (Sindhu et al., 2019), knowledge graph for feedback (Chanaa & El Faddouli, 2020), PROF (Nair et al., 2024), SEFL (Zhang, Dilling, et al., 2025), FreeText (Matelsky et al., 2023)
		Learning resource recommendation (LRR)	NeuroSky (Tabakcıoğlu et al., 2016)
	Personalized tutoring	Adaptive learning (AL)	EduAgent (Xu et al., 2024), OATutor (Pardos et al., 2023)
		Knowledge tracing (KT)	Graph-based (Chanaa & El Faddouli, 2020), reinforcement-learning–based (Cai et al., 2019), multiobjective KT video recommender (Huang et al., 2019), knowledge augmented data teaching (KADT) (Abdelrahman & Wang, 2023), Neural computerized adaptive testing (NCAT) (Zhuang et al., 2022), multiagent KT (Yang et al., 2024a)
		Error correction and detection (ECD)	AI-assisted deep learning system for composition correction (Jiang, 2024; Wang, 2024), CoT Rerailer (Wan et al., 2024), MathCCS (Zhang, Li, et al., 2025), ErrorRadar (Yan et al., 2024a)
Administrative automation	Scheduling and resources management (SRM)		NATURAL PLAN benchmark (Zheng et al., 2024a), MagicSchool AI (MagicSchool), Eduaide.Ai (Eduaide.Ai), Diffit (Diffit.), LessonPlanner (Fan et al., 2024)
	Attendance and engagement analytics		AI-enabled student-monitoring system (Narkhede et al., 2023), integrated classroom-management (Alzaabi et al., 2023), attention monitoring (Alyammahi & Alrabaee, 2023), TeacherEye (Abdelhadi et al., 2025)
Collaborative learning enhancement	Group work facilitation		AI team formation (Zheng & Zheng, 2021), Co-orchestration technology (Zheng et al., 2024b), collaborative Canvas (He et al., 2024), ChatGPT 3.5 conflict simulation (Aggrawal & Magana, 2024), RaFI (reasoning after feature identification) (Alzaabi et al., 2023), LLM-based peer review system (Abdelhadi et al., 2025)
	Teacher-AI co-teaching		AI-supported collaboration tools (Chang et al., 2023), LLM-generated simulated students (Li, Zhan, et al., 2025)
Accessibility & inclusion	Language learning	Translation	Agent-SiMT (Guo et al., 2024a), TransAgents (Wu et al., 2024), GlotBot (Patel & Chalageri, 2023)
		Storytelling	STARie (Li & Xu, 2023), StoryAgent (Li et al., 2024), LLaMS (Zhang et al., 2025c)
	Professional adaptation	Medical professional	MEDCO (Wei et al., 2025), openCHA (Abbasian et al., 2023), Agent hospital (Li et al., 2024a)
		Computer science professional	CodeAgent (Zhang, Li, et al., 2024), Too Coder (Ding et al., 2025), SWE-agent (Yang et al., 2024b), AgentCoder (Huang, Zhang, et al., 2023), MapCoder (Islam et al., 2024), Iris (Bassner et al., 2024)
	Special needs adaptation		ChatGPT (Botchu et al., 2024; Tamdjidi, 2023; Zhao et al., 2025)

efficient support throughout diverse phases of the teaching–learning cycle. Their core aims are to elevate instructional quality, deepen student engagement, and alleviate teachers' administrative burdens. By integrating sophisticated functionalities, these agents proficiently contribute to critical functions like classroom scenario simulation, feedback comments generation, and learning resource recommendation.

a. Classroom scenario simulation (CSS)

Classroom scenario simulation denotes the capacity of instructional agents to recreate and model diverse educational situations—ranging from student–teacher interactions and collaborative group work to guided problem-solving exercises. Such simulations generate dynamic, interactive virtual environments in which educators may trial alternative pedagogical strategies, observe learner responses, and obtain immediate feedback on the potential outcomes of each approach. Through this process, instructors can iteratively refine their techniques, anticipate areas of student difficulty, and enhance overall teaching effectiveness—all without the logistical limitations inherent in a physical classroom.

Self-regulated learning (Zimmerman, 1990) represents a critical dimension of educational processes that AI-driven virtual agents are well-suited to support (Graesser & McNamara, 2010). In such learning, students actively reflect on their goals and plan their in-simulation activities to foster deeper understanding (Perez et al., 2017). Effective self-regulated learners systematically deploy cognitive strategies: they establish objectives, detect gaps in their own knowledge and skills, formulate insightful questions, and then select, interpret, and reorganize information to address those gaps (Graesser & McNamara, 2010; Zimmerman, 1990). Among various methods designed to facilitate self-regulated learning, the Open Learner Model (OLM) has gained notable attention (Gandedkar et al., 2021; Kay et al., 2022; Nussbaumer et al., 2015). OLMs expose a learner's real-time performance data—via machine-learning algorithms, Bayesian networks, fuzzy-logic systems, or item-response theory models (Hooshyar et al., 2020)—thereby enabling learners to plan, monitor, and reflect on their progress for informed, autonomous decision-making (Graesser & McNamara, 2010; Nussbaumer et al., 2015). Graesser & McNamara (2010) have argued that such transparency is most powerfully realized when OLMs leverage natural-language interaction between students and virtual agents: conversational exchange affords fine-grained adaptivity that can steer inquiry and scaffold understanding. Within this situated-cognition framework, AI-powered "guidebots" embodied these capabilities, offering cognitive assistance (hints and feedback), motivational prompts (positive messages), and affective support (gestures of encouragement) to sustain productive inquiry (Johnson, 2003). Crucially, guidebots are engineered not to supply direct answers but to prompt learners' own reasoning and exploration, thereby enhancing both engagement and metacognitive development (Hwang et al., 2020).

Motivation (Zimmerman, 1990) plays a pivotal role alongside self-regulated learning in determining student engagement and persistence, and AI-powered virtual agents have been developed expressly to nurture this motivational dimension (Kim & Baylor, 2016; Lippert et al., 2020). In the learning-by-teaching paradigm, learners adopt the instructor's role, guiding an AI agent through explanations; the agent's ability to respond, query, and adapt—its "teachable" quality—reinforces both subject mastery and learner motivation (Chin et al., 2010; Matsuda et al., 2020). Veletsianos & Russell (2013) catalogued the multifaceted functions of virtual agents—as conversational partners, instructional experts, information brokers, decision-support systems, and even playmates—but warned that agents' effectiveness can be compromised by syntactic inaccuracies or inadequate contextualization. Recent advances in transformer-based large language models have substantially improved agents' linguistic coherence and flexibility; however, these models remain largely uncalibrated to the nuances of educational discourse. Addressing this gap requires targeted research on integrating domain-specific pedagogical datasets—through fine-tuning, curriculum learning, or hybrid architectures—to endow conversational agents with the contextual sensitivity and instructional expertise essential for robust educational support (Dai, 2021).

Recent investigations have examined how LLMs can enhance the fidelity and pedagogical utility of smart-classroom simulations. The cornerstone of any effective simulation is an accurate representation of student behavior. Prior studies (Jin et al., 2024; Li et al., 2025b; Xu et al., 2024) have shown that LLM-driven agents can generate fine-grained, persona-consistent behavioral trajectories that align closely with instructor expectations. Building on this capability, the CGMI framework (Jinxin et al., 2023) employs a tree-structured cognitive architecture—comprising memory, reflection, and planning components—to instantiate multiple classroom roles (e.g., teacher, learner, and supervisor), thereby heightening the realism of simulated interactions. Similarly, the Classroom Simulacra system (Xu, Wen, et al., 2025) integrates an iterative reflection mechanism that transfers insights across simulation runs, further improving behavioral authenticity. These platforms facilitate automated agent–agent and agent–educator exchanges that both alleviate teacher workload and enable systematic exploration of heterogeneous student profiles. Moreover, simulation studies by Bhowmik et al. (2024) and Zheng et al. (2025) demonstrated how tailored instructional strategies can be evaluated in experiment for distinct learner types, yielding data-driven insights to optimize pedagogical design. Hu et al. (2025) further illustrated the use of LLM-mediated simulations to iteratively refine lesson plans based on emergent interaction patterns. In sum, advanced classroom scenario simulations, which powered by LLMs, offer a scalable testbed for customizing and validating educational interventions across diverse student populations, thereby bolstering overall teaching effectiveness.

b. Feedback comment generation (FCG)

Providing timely, relevant, and constructive feedback is fundamental to effective education. AI-powered teaching assistants can automatically generate feedback on students' assignments, quizzes, and projects, thereby addressing the burgeoning volume of learner evaluations collected by educational institutions. Manually consolidating and analyzing this feedback is labor-intensive and prone to oversight; by contrast, automated text-summarization techniques distill large corpora of student comments into concise, informative overviews without sacrificing critical insights (El-Kassas et al., 2021).

Text summarization methods fall into three main categories—extractive, abstractive, and hybrid approaches—each with distinct strengths (El-Kassas et al., 2021). For example, Fan et al. (2015) introduced CourseMIRROR (Mobile In-situ Reflections and Review with Optimized Rubrics), an extractive system that identifies salient phrases, clusters them via k-medoids, and re-ranks clusters by student coverage to surface the most significant feedback. Their evaluation demonstrated that CourseMIRROR outperformed the established LexRank (Erkan & Radev, 2004) algorithm in highlighting common misunderstandings and difficulties.

Similarly, Gottipati et al. (2019) developed a topic-based summarization tool for online course discussion forums. They applied both TextRank (Mihalcea & Tarau, 2004) and latent semantic analysis (LSA) (Foong et al., 2015) summarizers to categorize posts according to three pedagogical questions, finding that LSA yielded richer data recommendations than TextRank. Cagliero et al. (2019) extended this paradigm to lecture materials: after assessing student comprehension with end-of-lecture multiple-choice tests, they generated customized summaries using a multilingual weighted itemset-based summarizer (MWI-Sum), thereby aligning review materials with individual learning needs.

In another extractive approach, Miller (2019) employed BERT embeddings combined with k-means clustering to create dynamically sized lecture summaries, demonstrating the versatility of transformer models for real-time content condensation. Asmussen et al. (Asmussen & Møller, 2019) proposed a three-stage pipeline—preprocessing, latent Dirichlet allocation (LDA) topic modeling, and postprocessing—to support exploratory literature reviews, while Sindhu et al. (2019) applied supervised aspect-based opinion mining to evaluate teaching performance across six domains (pedagogy, behavior, knowledge, assessment, experience, and general).

Beyond summarization, Dessì et al. (2021) designed a deep-learning-enhanced architecture for entity and relation extraction in student feedback, identifying six entity types (e.g., task, method, and metric) and seven relation types (e.g., compare, part-of, and used-for) (Luan et al., 2018), thus enabling richer semantic analysis. Finally, Shi et al. (2020) leveraged these insights within a multidimensional knowledge-graph framework to recommend personalized learning paths: learning objects are stored in distinct classes and linked by six semantic relationships, facilitating the construction of diverse, learner-centered curricula.

One of the most trending annotations in NLP is sentiment annotation which is to label emotion, opinion, and sentiment inherent within a text. The label could be a positive sentiment, neutral sentiment, or negative sentiment. As a subfield of opinion mining, it seeks to quantify the emotional intelligence quotient of written language, an undertaking that hinges on comprehensive contextual understanding. In human discourse, cues such as sarcasm, humor, and implied meaning can dramatically alter the intended sentiment; without modeling these subtleties, automated systems struggle to infer the true attitude or intent behind student comments, customer reviews, or informal messages (Dolianiti et al., 2019). Consequently, the creation of high-quality, sentiment-annotated corpora is indispensable for training machine-learning classifiers that aspire to robust performance in real-world settings (Kastrati et al., 2021).

Applying sentiment annotation within educational contexts—such as lecturer evaluations, course-feedback surveys, and LMS discussion forums—introduces additional complexity. Researchers have proposed various frameworks to address this domain specificity: Sutoyo et al. (2020) developed a feedback questionnaire for instructor evaluation based on student responses, while Ibrahim et al., n.d. designed a data-mining pipeline employing classification algorithms to extract polarity from learner feedback. Yet these efforts confront a fundamental obstacle: educational texts often embed discipline-specific terminology, pedagogical references, and implicit judgments that generic NLP models cannot decode without targeted adaptation (Grljević et al., 2022). Moreover, linguistic ambiguity in student remarks—arising from syntactic complexity, lexical polysemy, or rhetorical devices—further complicates automated interpretation (Grljević et al., 2022).

To address these challenges, sentiment-analysis systems must incorporate domain-aware strategies. First, annotation schemas should be refined to capture pedagogical factors—such as perceived instructional clarity, resource adequacy, or engagement level—that influence student sentiment (Sindhu et al., 2019). Second, training processes must leverage transfer-learning or fine-tuning on in-domain corpora to imbue models with the semantic frames unique to educational discourse (Jusoh, 2018). Finally, specialized modules for sarcasm detection and pragmatic inference can help disambiguate ironic or humorous remarks, ensuring that latent student attitudes are accurately surfaced (Jusoh, 2018).

Agentic AI offers a promising avenue for enhancing the precision and reliability of automated feedback in educational contexts. For instance, Guo, Latif, et al. (2024) introduced a training-free, two-agent architecture in which the first agent generates initial feedback on student responses and the second agent critically reviews and refines those comments to avoid undue praise or unwarranted inferences. Building on this collaborative paradigm, Nair et al. (2024) proposed PROF, a reinforcement-learning–based training regimen for an LLM-driven comment generator: the system iteratively simulates diverse student writing styles, employs a higher-capacity revision model (e.g., GPT-4) to score feedback quality, and uses those scores as rewards to fine-tune the generator. Similarly, SEFL (Zhang, Dilling, et al., 2025) leveraged role-playing between teacher and student agents to synthesize training data, which in turn refines model parameters and elevates the fidelity of generated feedback. These frameworks have been translated into operational tools such as FreeText (Matelsky et al., 2023), which aligns student submissions with instructor-defined rubrics, automatically diagnosing strengths and weaknesses and delivering targeted suggestions for improvement.

Beyond routine commentary, LLM agents are beginning to tackle more sophisticated, domain-intensive tasks. Du et al. (2024) investigated their viability as NLP-paper–review assistants, while the AAAR-1.0 benchmark (Lou et al., 2024) systematically evaluated agent performance in equation derivation, experimental design, critical weakness analysis, and review composition. Although these studies underscore the potential of agentic AI for high-level scholarly workflows, they also reveal persistent limitations: automated reviewers may misinterpret complex concepts—such as programming and the review of professional academic papers —and thus generate erroneous or misleading feedback (Estévez-Ayres et al., 2024; Lou et al., 2024).

To advance agentic AI toward robust, expert-level support, future research should explore tighter integration with external knowledge sources (e.g., domain-specific search engines or specialized toolkits) and more sophisticated memory architectures to maintain problem context over extended interactions. Moreover, enhancing personalization mechanisms—so that feedback adapts not only to task complexity but also to individual learner profiles and evolving competencies—will be crucial for realizing the full pedagogical benefits of agentic AI across diverse educational settings.

c. Learning resource recommendation (LRR)

In the current education system, teachers often lack objective, real-time indicators of whether students truly understand course material in both online and face-to-face situation. NeuroSky EEG biosensors offer a solution by continuously tracking brain-wave signals associated with attention, relaxation, and cognitive workload (Ramírez-Moreno et al., 2021; Tabakcıoğlu et al., 2016). Aggregated readings reveal which syllabus segments consistently challenge either individual learners or the entire class, allowing instructors to reteach complex topics promptly or schedule focused support sessions. At the individual level, these data support personalized learning paths: when a student's attention wanes during specific tasks, educators can slow the pace, incorporate multimedia resources, or provide supplementary materials that align with the learner's cognitive strengths and weaknesses (Ramírez-Moreno et al., 2021).

Developing robust recommendation systems is essential for guiding learners toward resources that closely match their individual trajectories and domain knowledge. AI-driven technologies can transform education by delivering personalized, adaptive learning experiences, streamlining administrative tasks, and ultimately improving learning outcomes (Seo et al., 2021). Since Zaiane's pioneering application of web-mining–based recommenders in e-learning (Zaiane, 2002), the field has expanded rapidly—particularly with the advent of LLM agents.

LLM-enabled recommenders employ two complementary strategies. Retrieval-based methods query external databases or an agent's memory to surface existing materials—textbooks, research articles, tutorials—selected on the basis of prior learner behavior, semantic similarity, or curriculum alignment (Li et al., 2024b; Shahzad et al., 2025). Generative methods, by contrast, synthesize new content tailored to a student's preferred learning style, identified knowledge gaps, and expressed interests (Moon et al., 2024). To promote transparency and trust, agents should articulate the rationale for each recommendation; integrating explainable frameworks such as knowledge graphs, as demonstrated by Abu-Rasheed, Abdulsalam, et al. (2024), Abu-Rasheed, Weber, et al. (2024), improves interpretability while mitigating the risk of misinformation.

Future research should (i) couple recommendation agents with adaptive-learning platforms that adjust suggestions in real time, (ii) explore hybrid architectures that blend retrieval and generation to maximize both accuracy and diversity, and (iii) incorporate multimodal resources—interactive media, videos, immersive simulations—to enrich the learner experience and sustain engagement.

3.1.2 Personalized tutoring

LLM agent-based student support systems enable independent, personalized learning by providing real-time academic assistance without requiring direct teacher involvement. Unlike traditional rule-based learning systems, they deliver interactive, tailored feedback, allowing students to learn at their own pace. Broadly, their core functionalities include adaptive learning, knowledge tracing, and error correction and detection.

a. Adaptive learning

It is essential to understand and differentiate among types of adaptive learning technologies because each type delivers distinct functionalities and addresses specific educational requirements. Content-based adaptive systems personalize instructional materials by modulating content in accordance with learners' progression and individual preferences. In contrast, assessment-based adaptive systems tailor evaluations and feedback, dynamically calibrating task difficulty to align with a student's demonstrated performance. Integrated adaptive systems combine both content personalization and adaptive assessment, thereby furnishing a holistic framework that facilitates individualized learning trajectories, ongoing evaluation, and targeted instructional interventions.

Several case studies illustrate the transformative impact of AI-driven adaptive learning. At Arizona State University, collaboration with Knewton enabled a platform that analyzed student data to identify weaknesses and deliver personalized recommendations, resulting in higher pass rates compared with traditional instruction (Maxfielf). DreamBox Learning, deployed in elementary schools, employs real-time progress tracking to adjust lessons to individual knowledge gaps, yielding significant gains in math proficiency (Barbieri et al., 2020; Foster, 2024) Carnegie Learning's AI-powered middle-school math curriculum similarly generated tailored assignments based on performance data, leading to improved test scores and engagement (WestEd). Moreover, universities adopting platforms such as Smart Sparrow and CogBooks across science, engineering, and humanities courses have reported enhanced teaching effectiveness and learning outcomes (Li et al., 2021; Putra et al., 2022).

Machine learning algorithms enable real-time feedback for students, addressing limitations of traditional assessment methods, which often provide feedback only after task completion or at predetermined intervals, hindering immediate correction or improvement (Asthana & Hazela, 2019). In contrast, AI-driven systems offer instantaneous feedback through automated grading or virtual tutors, allowing learners to promptly identify and correct mistakes while continuing to engage with the material (Celik et al., 2022). While traditional assessments, such as standardized tests, offer summative evaluations at specific moments, adaptive learning systems provide continuous, formative assessments that dynamically respond to learners' needs. This ongoing evaluation process enables more personalized feedback, supporting individual learning trajectories. However, challenges arise in aligning real-time feedback with standardized criteria and ensuring consistency in measuring student progress (Mirata et al., 2020).

Adaptive learning challenges have been addressed through several implementations of LLM–based agents that function autonomously within an action–memory framework. By automating instructional tasks, these agents facilitate the creation of self-sustaining adaptive learning environments that require minimal teacher intervention. Each agent constructs and continually updates a detailed learner profile, using insights from prior interactions to inform content selection and calibrate task difficulty. Following each learning activity, the system evaluates student behavior and performance data, then refines its pedagogical strategy to better suit individual needs. This iterative cycle of profiling, material adaptation, and feedback integration enhances both the personalization and the responsiveness of the learning experience. EduAgent (Xu et al., 2024) introduced a comprehensive profiling framework that categorizes student learning into four cognitive patterns: gaze behavior related to physiological memory, motor behavior linked to motor memory, cognitive states associated with cognitive memory, and post-course assessments contributing to knowledge memory. This structured approach enhances adaptive decision-making by offering a multidimensional perspective on student learning progress. Building upon this, Chen et al. (2024a) developed a system that integrates interaction, reflection, and reaction components, each supported by specialized LLM tools and memory modules. In addition, a meta-agent is introduced to regulate the flow of information across these individual agents, ensuring cohesive and efficient operation.

LLM–based agents extend these adaptive mechanisms by employing a state–action framework that continuously modulates instructional strategies in accordance with learners' cognitive and affective dynamics. In this architecture, the state space encodes each student's evolving profile—encompassing knowledge proficiency, comprehension depth, and identified misconceptions (Liu et al., 2024a, Liu, Yin, et al., 2024; Pal Chowdhury et al., 2024; Park et al., 2024)—while the action space specifies pedagogical adjustments such as explanation elaboration, difficulty calibration, and concept reinforcement. Recent studies underscore the significance of modeling affective factors—motivation, interest, and self-efficacy—that critically impact learning outcomes. For example, Park et al. (2024) introduced an affective-state model enabling agents to tailor feedback tone, deliver timely encouragement, and adjust pacing to sustain engagement. Beyond cognition and emotion, adaptation also accounts for individual learning preferences and personality traits. Wang et al. (2025) incorporated modality and format preferences into the state representation, demonstrating that alignment with learners' preferred instructional channels enhances retention and efficiency. Similarly, Liu, Yin, et al., 2024) employed the Big Five personality framework to customize tutoring strategies, thereby acknowledging and leveraging stable individual differences in the learning process (Sonlu et al., 2024). Multiagent architectures were investigated to further personalize adaptive learning environments. Wang et al. (2025) proposed a suite of five collaborating agents—Gap Identifier, Learner Profiler, Dynamic Learner Simulator, Learning Path Scheduler, and Content Creator—each tasked with specific functions to orchestrate goal-driven, individualized instruction. Moreover, OATutor (Pardos et al., 2023) offers a modular, domain-agnostic platform that enables researchers to construct and evaluate scalable tutoring agents within a unified experimental framework.

b. Knowledge tracing (KT)

Knowledge tracing (KT) plays a pivotal role in intelligent tutoring systems by monitoring a learner's progress and predicting future performance and knowledge retention. Conventional knowledge tracing techniques typically utilize statistical models or deep learning methods to estimate students' mastery levels over time.

KT is able to augment educational recommendation systems by adapting instructional content to each learner's current mastery. By continuously estimating what a student has learned, KT-driven recommenders can select optimal resources—such as practice exercises or instructional videos—to support further growth. For example, Chanaa and El Faddouli (2020) introduced a graph-based KT model to cluster learners by proficiency, subsequently issuing task recommendations tailored to each cluster's needs. Cai et al. (2019) integrated KT predictions within a reinforcement-learning–based recommender, enabling an agent to propose materials that align with the learner's estimated knowledge state. Similarly, Huang et al. (2019) designed a multiobjective, reward-driven recommendation framework for educational videos, using KT-derived mastery estimates to adaptively adjust content difficulty as learners progress.

KT enables dynamic curriculum adaptation by continuously evaluating each learner's mastery and recommending the most beneficial instructional modules. Feng et al. (2009) first employed KT models to evaluate a course's structural efficacy by monitoring student progress across topic areas. Wang et al. (2020) utilized a reinforcement-learning agent informed by KT-derived knowledge states to sequence course modules optimally, thereby maximizing individual learning efficiency. Abdelrahman & Wang (2023) proposed the Knowledge Augmented Data Teaching (KADT) approach, which leverages KT to monitor a student-model's development during training and to refine the data-teaching strategy. More recently, Zhuang et al. (2022) introduced the Neural Computerized Adaptive Testing (NCAT) framework, integrating KT with item-response theory to enhance the selection of assessment items in real time.

In contrast, LLM-based agents provided a highly dynamic, language-driven approach to knowledge tracing by interpreting natural-language interactions, inferring conceptual mastery, and adjusting instructional strategies in real time. Recent work has extended this paradigm via multiagent knowledge-tracing architectures. Yang et al. (2024a) introduced a three-role system—administrator, judger, and critic—in which the administrator assigns tracing tasks to multiple judger agents that collaboratively evaluate a learner's cognitive state; the critic then reviews their assessments against predefined criteria to ensure both rigor and flexibility. Other studies have investigated alternative agent configurations: Xu et al. (2024) modeled learners as distinct "personas," enabling agents to tailor knowledge-tracing to heterogeneous learning profiles, while Scarlatos et al. (2025) leveraged dialogic exchanges to interrogate and refine estimates of students' conceptual understanding through interactive conversation.

c. Error correction and detection (ECD)

Error detection and correction constituted essential functions of intelligent tutoring systems, aiding learners with immediate, context-sensitive feedback that refines their conceptual grasp. In the past, research largely remained at the level of conceptual frameworks for computer-assisted language learning (CALL) or the development and preliminary evaluation of prototype AI applications. It was only in 2004 that empirical investigations began to appear, as NLP and ML techniques were employed to assess automatic error correction accuracy, writing-assistance feasibility, vocabulary acquisition efficacy—a period often termed "the age of MALL" (Bax, 2011). From 2010 onward, studies diversified in thematic focus, methodological design, and applied AI algorithms, inaugurating what has been called "the age of I-CALL" (Bax, 2011). Over the past decade, research has spanned the proposal and re-examination of conceptual frameworks for AI deployment, rigorous evaluation of system-level learning gains, examination of learner attitudes toward AI tools, and both quantitative and qualitative inquiries conducted in classroom-based and corpus-based settings. Wang (2024) found that AI feedback not only yielded greater gains in fluency and accuracy but also produced the largest reduction in writing anxiety. These results underscore the promise of AI-driven corrective feedback for enhancing writing performance and emotional well-being in language-learning contexts. Jiang (2024) presented an AI-assisted deep learning system for automated English composition correction that combines a front-end learning unit and a back-end support unit to preprocess student essays, extract features via inter-segment distance partitioning, apply dynamic time normalization for error correction and grading, and deliver feedback through an interactive module. However, traditional AI struggled with linguistic variability, domain shifts, and unseen constructions because they lack the broad contextual knowledge and flexible pattern-recognition capabilities of LLMs.

LLM agents detect and correct student errors across domains—such as writing, programming, and mathematical reasoning—by adjusting feedback to individual proficiency levels (Li et al., 2024c; Ye et al., 2022). These systems maintain internal state representations and use adaptive inference to track misconceptions over time (Bouzenia et al., 2024; Liu, Yin, et al., 2024; Park et al., 2024; Wang et al., 2025). Recent multimodal advances enable direct analysis of student drafts: Xu, Zhang, et al., 2025) converted handwritten and digital work into text for LLM analysis and feedback, while the CoT Rerailer framework (Wan et al., 2024) identified error "derailments" and issues targeted "rerailment" guidance. Zhang, Li, et al., 2025) introduced the MathCCS benchmark and a sequence-error analysis method employing multiagent collaboration, and ErrorRadar (Yan et al., 2024a) established the first multimodal dataset for evaluating error-detection agents.

3.2 Administrative automation

3.2.1 Scheduling & resource management

Efficient time management supports student achievement and successful course delivery (Sutjarittham et al., 2019). Manual timetable construction, however, relies on subjective decisions and often introduces conflicts and resource inefficiencies (Zepeda & Mayers, 2006). To mitigate these issues, recent studies integrate AI algorithms with empirical findings on time management and student learning patterns, generating timetables that respect room capacities, faculty preferences, and student course requests while maximizing overall satisfaction (Torelli et al., 2017). The timetable-generation task is NP-Hard, meaning its computational complexity grows rapidly with problem size and renders exact optimization impractical (Premananda & Muklason, 2021). AI-driven, constraint-aware schedulers tackle this challenge by heuristically searching the vast solution space, producing "smart" schedules that allocate resources efficiently and equitably. Study in (Bargavi & Mahesh, 2023) argues that effective class-schedule management demands holistic network virtualization reinforced by pervasive network intelligence. The proposed architecture—melding network slicing, digital-twin models, and AI-driven analytics—supports both service-centric and learner-centric connectivity, thereby enabling fine-grained resource allocation and continuously adaptive scheduling.

A notable application is dynamic timetable optimization using LLM-agent simulations. These agents combine natural language understanding with structured optimization processes to analyze scheduling constraints, detect conflicts, and recommend optimal allocations across teachers, classrooms, subjects, and student groups. Unlike traditional scheduling software, LLM-based systems can interpret complex natural language instructions/prompts (Zheng et al., 2024a), and assist in adapting schedules in real time to disruptions like teacher absences or room unavailability. Moreover, these systems can incorporate in-context learning (ICL) to dynamically adjust decision-making based on prior examples and user-specified constraints (Liu, Zhang, et al., 2024). ICL enables LLMs to infer scheduling logic from a few prompt examples, enabling fast adaptation without retraining. In the NATURAL PLAN benchmark, LLM models such as GPT-4 and Gemini 1.5 Pro showed improved calendar scheduling performance using long-context ICL, despite challenges from increased task complexity (Zheng et al., 2024a).

Another impactful application is automated lesson plan generation, where LLM tools assist educators in creating curriculum-aligned instructional content. Platforms such as MagicSchool AI (MagicSchool), Eduaide.Ai (Eduaide.Ai) and Diffit (Diffit.) allow teachers to input high-level learning objectives to receive tailored teaching materials. Fan et al. (2024) introduced LessonPlanner, a system that enables novice teachers to input basic course information—such as course name, lesson topic, and target student level—and receive a structured lesson plan organized according to Gagne's Nine Events of Instruction (Gagné, 1985). These generated plans typically include teaching goals, a structured lesson outline, instructional events and activities, customizable teaching materials, assessment and practice items, differentiation strategies, and optional slide creation guidance. Some systems also support batch generation of weekly or unit-based plans, significantly reducing teacher preparation time while maintaining alignment with pedagogical frameworks and educational standards (Zheng et al., 2024c).

3.2.2 Attendance & engagement analytics

AI-enabled student-monitoring system employing facial-recognition technology automates attendance collection, delivers visual analytics on attendance patterns to administrators, and issues real-time notifications to students and parents when irregularities arise (Alyammahi & Alrabaee, 2023; Alzaabi et al., 2023; Narkhede et al., 2023). By enhancing accountability and minimizing manual record-keeping, the platform frees instructional time and has been linked to higher academic performance. Study (Alyammahi & Alrabaee, 2023) presented an integrated classroom-management framework in which facial-recognition attendance supplies visual analytics and sends alerts to students and parents, interactive whiteboards enable richer bidirectional dialogue, an NLP-based Program Learning Outcomes (PLO) and Course Learning Outcomes (CLO) mapper ensures curricular alignment, and YOLOv5-powered attention monitoring quantifies learner focus in real time, refining teaching strategies and boosting student outcomes.

Integrating LLM agents into engagement analytics marks a significant evolution from traditional participation tracking toward context-aware and behaviorally informed monitoring of student involvement. Rather than relying solely on visual presence, LLM-based systems leverage multimodal data—such as speech transcripts, chat interactions, and LMS activity logs—to infer engagement with greater adaptability. For example, TeacherEye (Abdelhadi et al., 2025) explored how LLMs can support real-time engagement detection by converting classroom audio into text using speech recognition (e.g., Whisper API) and analyzing it through GPT-4 to detect signs of disengagement or suspicious behaviors, such as seeking unauthorized assistance. In addition to audio analysis, TeacherEye's visual module monitors facial expressions and behaviors—such as eye closure or gaze aversion—and aggregates this data through a Vision-Language Model (VLM) to recommend nonintrusive teacher interventions, like issuing general reminders, thereby enhancing classroom awareness and responsiveness. Similarly, Hao et al. (2025) developed an LLM-based collaborative learning platform that analyzes dialogue patterns to encode students' cognitive, behavioral, and emotional engagement features. Using hierarchical clustering and epistemic network analysis, their system classifies learners into profiles such as Active Questioners, Responsive Navigators, and Silent Listeners, revealing distinct patterns of participation, self-regulation, and emotional expression across group activities.

3.3 Collaborative learning enhancement

3.3.1 Group work facilitation

The development of teamwork competencies within educational context is garnering significant academic interest, driven by established pedagogical benefits and the prevalence of team-based structures in professional environments for managing complex projects. Despite the recognized importance of collaboration skills, a notable gap exists in systematic support for the team formation process itself. Further complicating this challenge is the inherent complexity of team dynamics, influenced by numerous factors, and the computationally intensive nature of optimizing team compositions. In the rapidly growing field of AI-enhanced collaborative learning, learner engagement and motivation are considered as key to success. Research is actively exploring AI-driven solutions to these challenges. For instance, Alberola et al. (2016) developed an AI tool leveraging coalition structure generation, Bayesian learning, and Belbin's role theory to optimize educational team formation. This system innovates by incorporating peer feedback over self-assessment, managing role uncertainty, and iteratively refining predictions based on interactions, demonstrably improving team dynamics and satisfaction in experimental settings. Concurrently, Yang et al. (2023) investigated co-orchestration technology to manage transitions between individual and collaborative learning, identifying tensions between student autonomy and teacher control that inform design principles for human-AI collaboration in personalized classrooms. Furthermore, Tastanova et al. (2023) highlighted AI's capacity to monitor learners' emotional and physiological responses, enabling tailored interventions to enhance engagement through a nuanced understanding derived from multimodal data, ultimately fostering more adaptive learning environments. Multimodal engagement strategies complement AI-supported collaborative learning because both deliver differentiated, attention-responsive experiences that strengthen participation and academic performance. This is supported by study (Zheng & Zheng, 2021), which confirmed that well-structured, AI-informed collaborative would generate emotional and social engagement, while Zheng et al. (2024b) illustrated that a feed-forward feedback model further enhances learners' motivation and active involvement in online team activities.

LLM will be substantial solutions which overcome common barriers in group brainstorming such as stagnation, expertise gaps, and disorganized dialogue. Collaborative Canvas (He et al., 2024) integrated LLM support into a shared digital workspace, enabling participants to generate AI-assisted sticky notes, cluster ideas, and maintain momentum during brainstorming sessions, thus facilitating more structured and inclusive group collaboration. LLMs also contribute to maintaining group cohesion by supporting conflict resolution within team-based activities. Aggrawal et al. (Aggrawal & Magana, 2024) explored ChatGPT 3.5 in simulating conflict scenarios where student teachers practiced resolving common teamwork challenges like missed deadlines, communication failures, and uneven participation.

Another important role of LLMs in group facilitation lies in automating peer feedback processes. Instead of relying solely on manual peer evaluations, LLMs assess review helpfulness and generate structured feedback based on rubric criteria. Liu, Cui, et al., 2024 introduced the RaFI (Reasoning after Feature Identification) framework, where LLMs were prompted to extract key features from peer feedback and reason about their alignment with evaluation rubrics, improving consistency and accuracy without additional supervised training. Similarly, Sajadi et al. (2024) developed an LLM-based peer review system to automatically summarize and refine peer and self-evaluation comments into coherent, actionable feedback reports, grounded in established educational frameworks.

3.3.2 Teacher-AI co-teaching

The rapid expansion of distance learning emphasizes the necessity for virtual laboratories and online collaborative platforms. de Novais et al. (2017) and Chang et al. (2023) showed that AI-enabled virtual labs and collaboration tools can replicate hands-on experiences, delivering real-time feedback and preserving instructional quality outside the traditional classroom. Their success, however, depends on AI's ability to provide individualized, continuous feedback loops (Ouyang et al., 2023). As Zheng and Zheng (2021) noted, careful task design is essential to align these AI-supported collaborations with innovative pedagogical aims and to maximize learning outcomes. Ultimately, the effectiveness of these initiatives rests on integrating e-learning theory with authentic practice while accounting for learner identity and contextual factors (Charbonneau-Gowdy & Galdames, 2024).

Rather than acting as replacements, LLMs serve as intelligent collaborators that reduce teachers' routine workloads, support real-time content scaffolding, and enable differentiated, student-centered learning experiences. One major function of LLMs in this domain is to act as co-facilitators in collaborative learning environments, offering real-time feedback, scaffolding instructional designs, and dynamically supporting student teachers during complex learning tasks. For instance, Li, Zhan, et al., 2025 demonstrated how integrating ChatGPT into STEM teacher training improved student teachers' autonomy, competence, and cognitive management by providing personalized feedback and suggestions during project-based learning activities.

Another role of LLMs is simulating classroom dynamics to support the development of instructional management skills. Zheng et al. (2025) illustrated how LLM-generated simulated students engaged in naturalistic dialogues, posed spontaneous questions, and exhibited diverse learner behaviors such as disengagement or confusion. Through interacting with these simulated students, preservice teachers practiced adjusting lesson pacing, managing diverse student needs, and fostering inclusive group engagement—skills essential for real-world classroom facilitation. In this sense, LLMs act not merely as informational assistants but as interactive learning agents that co-create dynamic teaching environments.

3.4 Accessibility & inclusion

3.4.1 Language learning

LLM agents are reshaping language instruction by providing dynamic, data-driven support for the core competencies that underlie effective communication and successful L2 acquisition (Ye et al., 2025). Through context-aware dialogue, real-time feedback, and adaptive task generation, these agents enhance learners' engagement with multiple skills, thereby accelerating both fluency and accuracy. The sections that follow survey recent innovations within each competency domain, illustrating how LLM agents—when thoughtfully integrated into pedagogical workflows—promote deeper language mastery for native-language students as well as second-language learners (Huang et al., 2023b; Ye et al., 2023).

a. Translation

In today's increasingly globalized classrooms, linguistic diversity presents both opportunities and challenges for educators and students. Language barriers can limit academic comprehension, hinder social participation, and exacerbate inequities, particularly for newly arrived immigrant students and English Language Learners (ELLs). LLM-based translation systems offer a promising solution by enabling real-time, context-sensitive multilingual support that bridges communication gaps and fosters more inclusive learning environments. Rather than relying solely on static translation tools, modern LLMs are capable of dynamically interpreting, contextualizing, and delivering translations that adapt to academic discourse and student needs, fundamentally reshaping multilingual education.

Recent work shows that LLM agents are reshaping both simultaneous (Guo et al., 2024c; Koshkin et al., 2024) and literary translation (Cheng et al., 2024; Škobo and Petričević, 2023) through tool integration and cooperative agent orchestration. In simultaneous interpretation, Agent-SiMT (Guo et al., 2024a) coupled a policy agent that decides when to read or speak with a generative translation agent, dynamically balancing these actions to achieve state-of-the-art performance. For literary translation, TransAgents (Wu et al., 2024) emulated the workflow of professional translation teams: multiple agents engage in collaborative reasoning to resolve cultural nuances, stylistic subtleties, and idiomatic expressions, thereby improving output quality in linguistically and culturally rich texts. Similarly, Patel & Chalageri (2023) introduced GlotBot, a mobile application built on EleutherAI's Pythia model, which translanguages live teacher speech in secondary mathematics classrooms. Collectively, these advances highlight how specialized tools and multiagent collaboration can elevate both translation practice and pedagogy (Zheng et al., 2024d).

b. Storytelling

Storytelling applications deployed LLM agents to create immersive, interactive language-learning experiences (Simon & Muise, 2022). STARie (Li & Xu, 2023) exemplified this trend: the peer-like embodied agent combines speech synthesis, facial animation, and other multimodal tools to scaffold children's storytelling, thereby nurturing narrative creativity and oral-communication skills (Beredo & Ong, 2021; Cassell, 2022). StoryAgent (Li et al., 2024) deployed top-down plot drafting with bottom-up asset generation, transforming brief prompts into coherent, multimodal narratives that automate complex storytelling workflows and democratize content creation (Liem et al., 2023). Expanding both expressiveness and consistency, the multimodal framework LLaMS (Zhang et al., 2025c) integrates a Story-Adapter module to illustrate lengthy image sequences, yielding human-level stories with rich visual accompaniment. Together, these systems demonstrate how tool-enhanced, multiagent LLM architectures can simultaneously advance the cognitive and creative facets of language education for children.

Despite their promise, these systems also face limitations that must be carefully considered. Vijayalakshmi et al. (Vijayalakshmi et al., 2025) noted that their dual-pipeline system exhibited high memory demands and performance degradation in noisy environments or when handling extreme dialectal variations. Similarly, Patel & Chalageri (2023) acknowledged that GlotBot translations are not always perfect, requiring teachers to actively monitor and adjust outputs to maintain instructional quality. Both studies emphasize that LLM-based tools should be viewed as assistive technologies—not replacements for human educators—and that human oversight remains critical to ensuring contextual appropriateness, cultural sensitivity, and pedagogical effectiveness.

3.4.2 Professional adaptation

Agents for professional adaptation exploit the capabilities of LLMs to offer scalable, adaptive, and context-aware learning experiences depended on to domain-specific needs. This subsection only focuses on two specific domains: medical and computer science professional.

a. Medical professional

LLM agent deployment in healthcare facilitates personalized, interactive, and scalable systems, evidenced by agents (Shusterman et al., 2025). For medical professional, Abd-Alrazaq et al. (2023) noted LLMs' potential for personalized curricula, adaptive plans, and assessments, alongside critical challenges like bias, misinformation, and privacy. Systems like MEDCO (Wei et al., 2025) simulated real-world training via multiagent collaboration (virtual patients, experts), boosting interdisciplinary learning. Abbasian et al. (2023) describe openCHA as an LLM framework integrating external resources for complex healthcare problem-solving using tool use and action planning (Ye et al., 2024). Nontraditionally, Steenstra et al. (2024) used LLMs for gamified adolescent health education, creating engaging, validated content that improved knowledge retention. Li et al. (2024a) developed Agent Hospital, where autonomous LLM agents improved significantly in medical reasoning and on benchmarks like MedQA (Jin et al., 2021) through simulated practice. These investigations reveal LLM agents' broad capabilities—reasoning, collaboration, tool use, adaptation—making them valuable for diverse medical education and clinical needs.

b. Computer science professional

LLM-based agents are reshaping computer science (CS) professional by delivering personalized support for coding, debugging, and foundational CS concepts (Lee & Song, 2024; Ma et al., 2024). Frameworks such as CodeAgent (Zhang, Li, et al., 2024) extended repo-level code generation with external utilities like WebSearch and DocSearch, while recent agentic systems—including ToolCoder (Ding et al., 2025), SWE-agent (Yang et al., 2024b), AgentCoder (Huang, Zhang, et al., 2023), and MapCoder (Islam et al., 2024)—demonstrated marked gains in students' coding efficiency. Beyond generation, Iris (Bassner et al., 2024) functions as a context-aware virtual tutor within the Artemis learning platform, offering tailored, conversational guidance. Complementing these tools, Chen et al. (2024b) advocated a Learning-by-Teaching approach that exploits LLM agents' capacity for dialogic interaction and demonstration-based learning.

3.4.3 Special needs adaptation

AI and LLMs are increasingly being explored for adapting educational experiences to better support students with special needs, such as dyslexia, ADHD, and other learning disabilities (Voultsiou & Moussiades, 2025). Broadly, LLMs offer flexible, accessible tools that can scaffold learning, reduce anxiety, and foster autonomy—critical elements for learners often marginalized by traditional education. Botchu et al. (2024) demonstrated how ChatGPT can empower students with dyslexia by assisting with proofreading, generating ideas, and providing a safe, anxiety-reducing practice environment, ultimately promoting educational inclusion and boosting self-esteem. Similarly, Zhao et al.'s survey (Zhao et al., 2025) revealed how LLMs help neurodiverse students by clarifying complex instructions, simplifying readings, overcoming writer's block, and enhancing drafts, all of which directly address executive function challenges common among students with ADHD and dyslexia.

Beyond basic task assistance, LLMs actively reshape how learners engage with content, giving them more agency over their learning processes. Tamdjidi (2023) emphasized ChatGPT's role as an on-demand reading companion for ADHD students, offering immediate summarization, clarification, and concept-checking functions. This immediate feedback loop reduces cognitive overload and helps maintain attention—both frequent hurdles for students with attentional disorders. Likewise, Zhao's findings underscore the LLMs' value in breaking down complex material into manageable pieces, fostering confidence and independent navigation of academic tasks. Together, these studies suggest that LLMs are not merely compensatory tools but can also act as catalysts for deeper educational participation among special-needs students when integrated thoughtfully into classroom practice.

However, these benefits are tempered by significant and recurrent limitations. All three studies highlight concerns about the reliability of LLM outputs—namely hallucinations and inaccuracies—which can mislead vulnerable learners without strong critical-reading skills. Studies (Botchu et al., 2024; Zhao et al., 2025) emphasized that over-reliance on text-based interfaces might frustrate students whose disabilities make reading itself challenging, while cost barriers risk exacerbating educational inequalities. Furthermore, inexperienced users often struggle to craft effective prompts, diminishing the potential benefit unless prior training is provided (Tamdjidi, 2023).

4. Challenges and future directions

In this section, we discuss key challenges that must be addressed to ensure the effectiveness, reliability, and ethical deployment of LLM agents for education. We focus on three critical areas: privacy, bias, and fairness concerns; integration with existing educational ecosystems and LLM hallucination and overreliance. For each challenge, we outline potential research directions to enhance the robustness and applicability of LLM agents in learning environments.

4.1 Privacy, bias, and fairness considerations

Assistant agents promise to transform smart classrooms, yet they also introduced serious privacy and bias risks that demand immediate attention. Since these systems ingest large volumes of sensitive behavioral and academic data, studies report "low technological readiness" for GDPR-level safeguards and differential-privacy controls in current educational platforms (Yan et al., 2024b). Model-inversion and membership-inference attacks have already extracted personally identifiable information from deployed chatbots, underscoring the need for privacy-by-design architectures (Zhao & Song, 2024). Recent researches highlighted additional privacy and security threats, urging the deployment of stronger, context-specific data-protection mechanisms (Chen et al., 2025; Zhang et al., 2024b). Equally pressing is algorithmic bias: models trained on vast, uncurated datasets can amplify stereotypes and exacerbate educational inequities, making comprehensive bias-mitigation strategies indispensable for fair learning environments (Huang, 2023). Resolving these challenges is critical to achieving the inclusive, unbiased educational outcomes envisioned for AI-enhanced classrooms.

To overcome the above issues, a number of future directions can be explored in Table 25.2.

4.2 Integration of assistant agents into existing educational ecosystems

Assistant agents offer considerable promise for personalized instruction, curriculum design, and project-based learning (PBL), yet they also pose challenges involving structured deployment, equitable access, and data privacy. A first obstacle is the absence of standardized frameworks for integrating LLM agents into smart classrooms. Although the FOKE framework (Hu & Wang, 2024) combined foundation models, knowledge graphs, and prompt engineering to deliver interactive, explainable services, its scalability and validity across diverse learning contexts remain untested. Likewise, studies exploring LLMs as catalysts for creativity and collaboration in PBL report limited impact when educators lack clear implementation guidelines (Zha et al., 2025). Equitable access constitutes a second challenge in smart classroom. Initiatives such as AI-VERDE (Mithun et al., 2025) attempted to democratize use through a unified "LLM-as-a-platform" offering with access controls, privacy safeguards, and budget management; however, broader adoption still hinges on cost-effective, scalable deployment models that accommodate varying institutional resources. Privacy and transparency concerns compound these issues. Student interaction logs can reveal sensitive learning trajectories, yet privacy-preserving fine-tuning, differential-privacy

TABLE 25.2 Future directions with different approaches for privacy, bias, and fairness considerations.

Technical limitation	Proposed approach	Expected outcome
Privacy-preserving memory management	Implement context-aware episodic buffers that encrypt, age-out, or locally hash sensitive logs after pedagogical use.	Minimizes long-term retention of student data while preserving learning continuity.
Federated/on-device fine-tuning	Shift incremental updates to edge devices; aggregate gradients with secure multiparty computation.	Keeps raw data in-classroom; aligns with regional data-sovereignty laws.
Real-time bias auditing pipelines	Embed automated fairness meters that flag biased explanations or assessments before delivery.	Enables immediate remediation and teacher oversight.
Explainable-AI interfaces	Pair model outputs with chain-of-thought rationales, confidence scores, and curriculum-aligned citations.	Enhances pedagogical transparency and teacher trust.

noise injection, and secure aggregation are seldom implemented at scale. Moreover, black-box outputs without human-readable rationales erode stakeholder trust; recent reviews recommend transparent prompt engineering and explainable dashboards as emerging countermeasures.

Future work should therefore prioritize (i) standardized, adaptable frameworks that embed LLM agents into personalized learning, PBL, and assessment; (ii) extensions of models such as FOKE to include adaptive strategies, multimodal content, and real-time feedback; (iii) interactive AI tutors that facilitate collaboration, project tracking, and context-aware guidance; (iv) cloud-based or decentralized deployment to lower costs and expand reach; and (v) comprehensive AI literacy training and oversight tools for educators to ensure responsible, effective integration.

4.3 Hallucination and overreliance

"Hallucinations," wherein LLM fabricate information that sounds authoritative yet is factually incorrect, threaten the pedagogical reliability of AI assistants (da Silva et al., 2024). In classrooms, such fabrications—for example, invented historical events or scientific data—can create misconceptions that take a lot of instruction time to resolve. Empirical studies further show that habitual dependence on AI-generated answers erodes learners' analytical reasoning and metacognitive monitoring, with critical-thinking scores declining as reliance on dialogue systems rises (Milano et al., 2023). Consequently, unchecked hallucinations not only mislead students but also impede deep skill acquisition and durable understanding. Some directions can be explored to mitigate hallucinations in LLM agents for smart classroom shown in Table 25.3.

5. Conclusion

In this book chapter, we provided a systematic review of assistant agents in smart classrooms, tracing their evolution from early implementations to current advanced applications across various domains. We categorized these agents into four primary areas: pedagogical applications, administrative automation, collaborative learning enhancement, and accessibility and inclusion. Through this structured classification, we clarified the roles and capabilities of assistant agents in enhancing educational environments.

TABLE 25.3 Mitigation techniques and proposals for hallucination and overreliance.

Mitigation techniques	Open questions and actions
Robust self-verification	Combine reflection loops with explicit confidence annotation so students see uncertainty levels; evaluate on multidisciplinary school datasets.
Domain-grounded RAG	Build subject-specific knowledge graphs and citation checkers that block release of ungrounded claims.
Adaptive teacher dashboards	Deliver real-time "AI accuracy alerts" and editable answer drafts within LMS interfaces to streamline oversight.
Pedagogical alignment	Train models with instruction-tuning datasets labelled by experienced teachers, ensuring outputs follow evidence-based teaching practices.

In addition, we discussed key challenges and future research directions critical to advancing these technologies, emphasizing ethical considerations, bias mitigation, and effective real-world integration. While AI assistant agents, particularly LLMs, continue to grow in sophistication and potential, ensuring their effectiveness, reliability, and ethical deployment remains a significant challenge. Ultimately, we aim for this chapter to serve as a foundational resource for future research and innovation, driving the development of AI-driven smart classrooms integrated with digital twin technologies and advancing educational equity.

References

Abbasian, M., Azimi, I., Rahmani, A. M., & Jain, R. (2023). Conversational health agents: A personalized llm-powered agent framework. *arXiv preprint arXiv:2310.02374.*

Abd-Alrazaq, A., et al. (2023). Large language models in medical education: Opportunities, challenges, and future directions. *JMIR Medical Education, 9*(1), e48291.

Abdelhadi, Z., Naseif, M., Alhejali, W., & Elhayek, A. (2025). TeacherEye: An AI-Powered system for monitoring student engagement in online education. In, *22. 2025 22nd International Learning and Technology Conference (L&T)* (pp. 25–30). IEEE.

Abdelrahman, G., & Wang, Q. (2023). Learning data teaching strategies via knowledge tracing. *Knowledge-Based Systems, 269*, 110511.

Abu-Rasheed, H., Abdulsalam, M. H., Weber, C., & Fathi, M. (2024). Supporting Student Decisions on Learning Recommendations: An LLM-Based Chatbot with Knowledge Graph Contextualization for Conversational Explainability and Mentoring. In *CEUR Workshop Proceedings, 3667* (pp. 230–239) [Online]. Available: https://www.scopus.com/inward/record.uri?eid=2-s2.0-85191971646&partnerID=40&md5=73647c62287adf729ae82fbe4ee3287c.

Abu-Rasheed, H., Weber, C., & Fathi, M. (2024). Knowledge graphs as context sources for llm-based explanations of learning recommendations. In *2024 IEEE global engineering education conference (EDUCON)* (pp. 1–5). IEEE.

Aggrawal, S., & Magana, A. J. (2024). Teamwork conflict management training and conflict resolution practice via large language models. *Fut. Internet, 16*(5), 177.

Alberola, J. M., Del Val, E., Sanchez-Anguix, V., Palomares, A., & Teruel, M. D. (2016). An artificial intelligence tool for heterogeneous team formation in the classroom. *Knowledge-Based Systems, 101*, 1–14.

Alyammahi, W., & Alrabaee, S. (2023). SES: Smart education system. In *2023 international conference on computer and applications (ICCA)* (pp. 1–4). IEEE.

Alzaabi, M., Almeheiri, M., Alqubaisi, S., & Shuhaiber, A. (2023). AI-Teacher'Assistant system: A smart attendance and participation tracking system for students. In *2023 international conference on smart applications, communications and networking (SmartNets)* (pp. 1–6). IEEE.

Asmussen, C. B., & Møller, C. (2019). Smart literature review: A practical topic modelling approach to exploratory literature review. *Journal of Big Data, 6*(1), 1–18.

Asthana, P., & Hazela, B. (2019). Applications of machine learning in improving learning environment. In *Multimedia big data computing for IoT applications: Concepts, paradigms and solutions* (pp. 417–433). Springer.

Barbieri, C. A., Rodrigues, J., Dyson, N., & Jordan, N. C. (2020). Improving fraction understanding in sixth graders with mathematics difficulties: Effects of a number line approach combined with cognitive learning strategies. *Journal of Educational Psychology, 112* (3), 628.

Bargavi, S. M., & Mahesh, T. (2023). The dynamic window-based scheduling framework for complex wireless sensor networks. In *2023 international conference on distributed computing and electrical circuits and electronics (ICDCECE)* (pp. 1–6). IEEE.

Bassner, P., Frankford, E., & Krusche, S. (2024). Iris: An ai-driven virtual tutor for computer science education. *Proceedings of the 2024 on Innovation and Technology in Computer Science Education, 1*, 394–400.

Bax, S. (2011). Normalisation revisited: The effective use of technology in language education. *International Journal of Computer-Assisted Language Learning and Teaching, 1*(2), 1–15.

Beredo, J., & Ong, E. (2021). Beyond the scene: A comparative analysis of two storytelling-based conversational agents. In *Proceedings of the Asian CHI symposium* (pp. 189–195).

Bhowmik, S., et al. (2024). Evaluation of an LLM-powered student agent for teacher training. In *European conference on technology enhanced learning* (pp. 68–74). Springer.

Botchu, B., Iyengar, K. P., & Botchu, R. (2024). Can ChatGPT empower people with dyslexia? *Disability and Rehabilitation: Assistive Technology, 19*(5), 2131–2132.

Bouzenia, I., Devanbu, P., & Pradel, M. (2024). Repairagent: An autonomous, llm-based agent for program repair. *arXiv preprint arXiv:2403.17134.*

Cagliero, L., Farinetti, L., & Baralis, E. (2019). Recommending personalized summaries of teaching materials. *IEEE Access, 7*, 22729–22739.

Cai, D., Zhang, Y., & Dai, B. (2019). Learning path recommendation based on knowledge tracing model and reinforcement learning. In *2019 IEEE 5th international conference on computer and communications (ICCC)* (pp. 1881–1885). IEEE.

Cassell, J., 2022. Socially interactive agents as peers, in: The Handbook on Socially Interactive Agents: 20 years of Research on Embodied Conversational Agents, Intelligent Virtual Agents, and Social Robotics. Interactivity, Platforms, Application. 2. pp. 331-66.

Celik, I., Dindar, M., Muukkonen, H., & Järvelä, S. (2022). The promises and challenges of artificial intelligence for teachers: A systematic review of research. *TechTrends, 66*(4), 616–630.

Chanaa, A., & El Faddouli, N.-E. (2020). Predicting learners need for recommendation using dynamic graph-based knowledge tracing. In *Artificial intelligence in education: 21st international conference, AIED 2020, Ifrane, Morocco, July 6–10, 2020, proceedings, part II 21* (pp. 49–53). Springer.

Chang, I., et al. (2023). Technology in higher education: A collaborative innovation ecosystem model. In *Proceedings of the Latin American congress on automation and robotics* (pp. 1–19). Springer.

Charbonneau-Gowdy, P., & Galdames, C. (2024). Identity matters: A framework for bridging contemporary e-Learning theory with online practice. *Electronic Journal of E-Learning, 22*(7), 1–19.

Chen, Y., Ding, N., Zheng, H.-T., Liu, Z., Sun, M., & Zhou, B. (2024). Empowering private tutoring by chaining large language models. In

Proceedings of the 33rd ACM international conference on information and knowledge management (pp. 354–364).

Chen, A., Wei, Y., Le, H., & Zhang, Y. (2024). Learning-by-teaching with ChatGPT: The effect of teachable ChatGPT agent on programming education. *British Journal of Educational Technology, 57* (1), 163–184. https://doi.org/10.1111/bjet.70001

Chen, J., et al. (2025). SAFEERASER: Enhancing safety in multimodal large language models through multimodal machine unlearning. *Proceedings of the Annual Meeting of the Association for Computational Linguistics*, 14194–14224. https://doi.org/10.18653/v1/2025.findings-acl.731. https://www.scopus.com/inward/record.uri?eid=2-s2.0-105028555742&doi=10.18653%2fv1%2f2025.findings-acl.731&partnerID=40&md5=7dbe42e87efbc13b195fb883e882bb29

Cheng, S., et al. (2024). Towards achieving human parity on end-to-end simultaneous speech translation via LLM agent. *arXiv preprint arXiv:2407.21646.*

Chin, D. B., Dohmen, I. M., Cheng, B. H., Oppezzo, M. A., Chase, C. C., & Schwartz, D. L. (2010). Preparing students for future learning with teachable agents. *Educational Technology Research and Development, 58*, 649–669.

da Silva, W.d. A., Fonseca, L. C. C., Labidi, S., & Pacheco, J. C. L. (2024). Mitigation of hallucinations in language models in education: A new approach of comparative and cross-verification. In *2024 IEEE international conference on advanced learning technologies (ICALT)* (pp. 207–209). IEEE.

Dai, C.-P. (2021). Designing artificial intelligence (AI) in virtual humans for simulation-based training with graduate teaching assistants. In *Proceedings of the 15th international conference of the learning Sciences-ICLS 2021.*

de Novais, A. S., Silva, M. B., & Muniz, J. (2017). Strengths, limitations and challenges in the implementation of active learning in an undergraduate course of logistics technology. *International Journal of Engineering Education, 33*(3), 1060–1069.

Dessì, D., Osborne, F., Recupero, D. R., Buscaldi, D., & Motta, E. (2021). Generating knowledge graphs by employing natural language processing and machine learning techniques within the scholarly domain. *Future Generation Computer Systems, 116*, 253–264.

Diffit. https://web.diffit.me/. (Accessed April 25, 2025).

Ding, H., et al. (2025). ToolCoder: A systematic code-empowered tool learning framework for large language models. *arXiv preprint arXiv:2502.11404.*

Dolianiti, F. S., et al. (2019). *Sentiment analysis on educational datasets: A comparative evaluation of commercial tools*. Educational Journal of the University of Patras UNESCO Chair.

Du, J., et al. (2024). Llms assist nlp researchers: Critique paper (meta-) reviewing. *Proceedings of the 2024 conference on empirical methods in natural language processing*, 5081–5099.

Eduaide.Ai. AI created for teachers. https://www.eduaide.ai/. (Accessed April 25, 2025).

El-Kassas, W. S., Salama, C. R., Rafea, A. A., & Mohamed, H. K. (2021). Automatic text summarization: A comprehensive survey. *Expert Systems with Applications, 165*, 113679.

Erkan, G., & Radev, D. R. (2004). Lexrank: Graph-based lexical centrality as salience in text summarization. *Journal of Artificial Intelligence Research, 22*, 457–479.

Estévez-Ayres, I., Callejo, P., Hombrados-Herrera, M.Á., Alario-Hoyos, C., & Delgado Kloos, C. (2024). Evaluation of LLM tools for feedback generation in a course on concurrent programming. *International Journal of Artificial Intelligence in Education*, 1–17.

Fan, H., Chen, G., Wang, X., & Peng, Z. (2024). LessonPlanner: Assisting novice teachers to prepare pedagogy-driven lesson plans with large language models. In *Proceedings of the 37th annual ACM symposium on user interface software and technology* (pp. 1–20).

Fan, X., Luo, W., Menekse, M., Litman, D., & Wang, J. (2015). CourseMIRROR: Enhancing large classroom instructor-student interactions via Mobile interfaces and natural language processing. In *Proceedings of the 33rd annual ACM conference extended abstracts on human factors in computing systems* (pp. 1473–1478).

Feng, M., Heffernan, N., & Koedinger, K. (2009). Addressing the assessment challenge with an online system that tutors as it assesses. *User Modeling and User-Adapted Interaction, 19*, 243–266.

Foong, O.-M., Yong, S.-P., & Jaid, F.-A. (2015). Text summarization using latent semantic analysis model in Mobile android platform. In *2015 9th Asia modelling symposium (AMS)* (pp. 35–39). IEEE.

Foster, M. E. (2024). Evaluating the impact of supplemental computer-assisted math instruction in elementary school: A conceptual replication. *Journal of Research on Educational Effectiveness, 17*(1), 94–118.

Gagné, R. M., (1985). The conditions of learning and theory of instruction.

Gandedkar, N. H., Wong, M. T., & Darendeliler, M. A. (2021). Role of virtual reality (VR), augmented reality (AR) and artificial intelligence (AI) in tertiary education and research of orthodontics: An insight. *Seminars in Orthodontics, 27*(2), 69–77.

Gottipati, S., Shankararaman, V., & Ramesh, R. (2019). TopicSummary: A tool for analyzing class discussion forums using topic based summarizations. In *2019 IEEE frontiers in education conference (FIE)* (pp. 1–9). IEEE.

Graesser, A., & McNamara, D. (2010). Self-regulated learning in learning environments with pedagogical agents that interact in natural language. *Educational Psychologist, 45*(4), 234–244.

Grljević, O., Bošnjak, Z., & Kovačević, A. (2022). Opinion mining in higher education: A corpus-based approach. *Enterprise Information Systems, 16*(5), 1773542.

Guo, S., Latif, E., Zhou, Y., Huang, X., & Zhai, X. (2024). Using generative AI and multi-agents to provide automatic feedback. *arXiv.*

Guo, S. T., Zhang, S. L., Ma, Z. R., Zhang, M., & Feng, Y. (2025). Agent-SiMT: Agent-assisted simultaneous translation with large language models. In *IEEE Transactions on Audio Speech and Language Processing, 33* (pp. 2074–2083). https://doi.org/10.1109/taslpro.2025.3566220

Guo, S., Zhang, S., Ma, Z., Zhang, M., & Feng, Y. (2024). Sillm: Large language models for simultaneous machine translation. *arXiv preprint arXiv:2402.13036.*

Hao, Z., Jiang, J., Yu, J., Liu, Z., & Zhang, Y. (2025). Student engagement in collaborative learning with AI agents in an LLM-Empowered learning environment: A cluster analysis. *arXiv preprint arXiv:2503.01694.*

He, J., et al. (2024). AI and the future of collaborative work: Group ideation with an LLM in a virtual canvas. In *Proceedings of the 3rd annual meeting of the symposium on human-computer interaction for work* (pp. 1–14).

Hooshyar, D., Pedaste, M., Saks, K., Leijen, Ä., Bardone, E., & Wang, M. (2020). Open learner models in supporting self-regulated learning in higher education: A systematic literature review. *Computers in Education, 154*, 103878.

Hu, S., & Wang, X. (2024). Foke: A personalized and explainable education framework integrating foundation models, knowledge graphs, and prompt engineering. In *China national conference on big data and social computing* (pp. 399–411). Springer.

Hu, B., Zhu, J., Pei, Y., & Gu, X. (2025). Exploring the potential of LLM to enhance teaching plans through teaching simulation. *Npj Science of Learning, 10*(1), 7.

Huang, L. (2023). Ethics of artificial intelligence in education: Student privacy and data protection. *Science Insights Education Frontiers, 16* (2), 2577–2587.

Huang, Z., et al. (2019). Exploring multi-objective exercise recommendations in online education systems. In *Proceedings of the 28th ACM international conference on information and knowledge management* (pp. 1261–1270).

Huang, H., et al. (2023). A frustratingly easy plug-and-play detection-and-reasoning module for Chinese spelling check. *Findings of the Association for Computational Linguistics: EMNLP 2023*, 11514–11525. https://doi.org/10.18653/v1/2023.findings-emnlp.771 [Online]. Available: https://www.scopus.com/inward/record.uri?eid=2-s2.0-85183292626&doi=10.18653%2fv1%2f2023.findings-emnlp.771&partnerID=40&md5=3e7b51bd9bc4326948babaf447e2af3b.

Huang, D., Zhang, J. M., Luck, M., Bu, Q., Qing, Y., & Cui, H. (2023). Agentcoder: Multi-agent-based code generation with iterative testing and optimisation. *arXiv preprint arXiv:2312.13010.*

Hwang, G.-J., Xie, H., Wah, B. W., & Gašević, D. (2020). *Vision, challenges, roles and research issues of artificial intelligence in education* (Vol. 1, p. 100001). Elsevier.

Ibrahim, Z. M., Bader-El-Den, M., and Cocea, M., A data mining framework for analyzing students' feedback of assessment. in: 13th European conference on technology enhanced learning doctoral consortium, 2018: CEUR workshop proceedings. p. 13.

Islam, M. A., Ali, M. E., & Parvez, M. R. (2024). Mapcoder: Multi-agent code generation for competitive problem solving. *Proceedings of the Annual Meeting of the Association for Computational Linguistics, 1*, 4912–4944. https://doi.org/10.18653/v1/2024.acl-long.269. https://www.scopus.com/inward/record.uri?eid=2-s2.0-85204435298&doi=10.18653%2fv1%2f2024.acl-long.269&partnerID=40&md5=55c424404f6dcc78b4c440f30a73d05d

Jiang, X. (2024). Automatic correction method for English composition errors assisted by artificial intelligence. In *2024 international conference on interactive intelligent systems and techniques (IIST)* (pp. 264–268). IEEE.

Jin, D., Pan, E., Oufattole, N., Weng, W.-H., Fang, H., & Szolovits, P. (2021). What disease does this patient have? A large-scale open domain question answering dataset from medical exams. *Applied Sciences, 11*(14), 6421.

Jin, H., Yoo, M., Park, J., Lee, Y., Wang, X., & Kim, J. (2024). TeachTune: Reviewing pedagogical agents against diverse student profiles with simulated students. *Conference on Human Factors in Computing Systems - Proceedings.* https://doi.org/10.1145/3706598.3714054. https://www.scopus.com/inward/record.uri?eid=2-s2.0-105005757598&doi=10.1145%2f3706598.3714054&partnerID=40&md5=09a0d97a97e5d3d78bd068487456b423

Jinxin, S., Jiabao, Z., Yilei, W., Xingjiao, W., Jiawen, L., & Liang, H. (2023). Cgmi: Configurable general multi-agent interaction framework. *arXiv preprint arXiv:2308.12503.*

Johnson, W. L. (2003). Interaction tactics for socially intelligent pedagogical agents. In *Proceedings of the 8th international conference on intelligent user interfaces* (pp. 251–253).

Jusoh, S. (2018). A study on NLP applications and ambiguity problems. *Journal of Theoretical and Applied Information Technology, 96*(6).

Kastrati, Z., Dalipi, F., Imran, A. S., Pireva Nuci, K., & Wani, M. A. (2021). Sentiment analysis of students' feedback with NLP and deep learning: A systematic mapping study. *Applied Sciences, 11* (9), 3986.

Kay, J., Bartimote, K., Kitto, K., Kummerfeld, B., Liu, D., & Reimann, P. (2022). Enhancing learning by open learner model (OLM) driven data design. *Computers and Education: Artificial Intelligence, 3*, 100069.

Kim, Y., & Baylor, A. L. (2016). Research-based design of pedagogical agent roles: A review, progress, and recommendations. *International Journal of Artificial Intelligence in Education, 26*, 160–169.

Koshkin, R., Sudoh, K., & Nakamura, S. (2024). TRANSLLAMA: LLM-based simultaneous translation system. *EMNLP 2024 - 2024 Conference on Empirical Methods in Natural Language Processing, Findings of EMNLP 2024*, 461–476. https://doi.org/10.18653/v1/2024.findings-emnlp.27. https://www.scopus.com/inward/record.uri?eid=2-s2.0-85217620843&doi=10.18653%2fv1%2f2024.findings-emnlp.27&partnerID=40&md5=649ab39d29929a738092f26faed7e041

Lee, S., & Song, K.-S. (2024). Teachers' and students' perceptions of AI-generated concept explanations: Implications for integrating generative AI in computer science education. *Computers and Education: Artificial Intelligence, 7*, 100283.

Li, J., et al. (2024). Agent hospital: A simulacrum of hospital with evolvable medical agents. *arXiv preprint arXiv:2405.02957.*

Li, Q., et al. (2024). Learning structure and knowledge aware representation with large language models for concept recommendation. *arXiv preprint arXiv:2405.12442.*

Li, Y., et al. (2025). Rethinking the roles of large language models in Chinese grammatical error correction. In *Proceedings of the Annual Meeting of the Association for Computational Linguistics, 6* (pp. 553–567). https://doi.org/10.18653/v1/2025.acl-industry.39 [Online]. Available: https://www.scopus.com/inward/record.uri?eid=2-s2.0-105020383087&doi=10.18653%2fv1%2f2025.acl-industry.39&partnerID=40&md5=f4c328957089670ae9059cca86a85985.

Li, H., et al. (2025). Exploring llm-based student simulation for metacognitive cultivation. *arXiv e-prints, p. arXiv: 2502.11678.*

Li, D., Sohn, S. S., Zhang, S., Chang, C.-J., & Kapadia, M. (2024). From words to worlds: Transforming one-line prompt into immersive multi-modal digital stories with communicative LLM agent. *17th Conference on Motion Interaction and Games, Arlington, VA, Nov 21-23 2024.* https://doi.org/10.1145/3677388.3696321

Li, F., He, Y., & Xue, Q. (2021). Progress, challenges and countermeasures of adaptive learning. *Educational Technology and Society, 24* (3), 238–255.

Li, Z., & Xu, Y. (2023). Designing a realistic peer-like embodied conversational agent for supporting children's storytelling. *arXiv preprint arXiv:2304.09399.*

Li, T., Zhan, Z., Ji, Y., & Li, T. (2025). Exploring human and AI collaboration in inclusive STEM teacher training: A synergistic approach based on self-determination theory. *The Internet and Higher Education, 65*, 101003.

Liem, J., Kusnick, J., Beck, S., Windhager, F., & Mayr, E. (2023). A workflow approach to visualization-based storytelling with cultural heritage data. In *2023 IEEE 8th workshop on visualization for the digital humanities (VIS4DH)* (pp. 13–17). IEEE.

Lippert, A., Shubeck, K., Morgan, B., Hampton, A., & Graesser, A. (2020). Multiple agent designs in conversational intelligent tutoring systems. *Technology, Knowledge and Learning, 25*(3), 443–463.

Liu, C., Cui, J., Shang, R., Jia, Q., Rashid, P., & Gehringer, E. (2024). Generative ai for peer assessment helpfulness evaluation. In *Proceedings of the 17th international conference on educational data mining* (pp. 412–419).

Liu, D., et al. (2024). Personalized tutoring through conversational agents. In *Using educational robots to enhance learning: An analysis of 100 academic articles* (pp. 59–85). Springer.

Liu, Z., Yin, S. X., Lin, G., & Chen, N. F. (2024). Personality-aware student simulation for conversational intelligent tutoring systems. *EMNLP 2024 - 2024 Conference on Empirical Methods in Natural Language Processing, Proceedings of the Conference*, 626–642. https://doi.org/10.18653/v1/2024.emnlp-main.37 [Online]. Available: https://www.scopus.com/inward/record.uri?eid=2-s2.0-85217814700&doi=10.18653%2fv1%2f2024.emnlp-main.37&partnerID=40&md5=845713724e1b24dd4f8bd391e08ac9ff.

Liu, Y., Zhang, H., Li, Z., & Miao, Y. (2024). Optimizing the utilization of large language models via schedule optimization: An exploratory study. In *Proceedings of the 18th ACM/IEEE international symposium on empirical software engineering and measurement* (pp. 84–95).

Lou, R., et al. (2024). AAAR-1.0: Assessing ai's potential to assist research. *arXiv preprint arXiv:2410.22394*.

Luan, Y., He, L., Ostendorf, M., & Hajishirzi, H. (2018). Multi-task identification of entities, relations, and coreference for scientific knowledge graph construction. *Proceedings of the 2018 Conference on Empirical Methods in Natural Language Processing, EMNLP 2018*, 3219–3232. https://www.scopus.com/inward/record.uri?eid=2-s2.0-85081754181&partnerID=40&md5=709a9ea70a6f4e7a6d4b10acc623c89a.

M Vijayalakshmi, D. N. N. K., Sriram, I., SivaBalan, G., & Tharun, N. (2025). Dual-pipeline intelligent framework for RealTime contextual multilingual communication using LLM and neural TTS. *International Journal for Research Trends and Innovation, 10*(4 April 2025).

Ma, Q., Shen, H., Koedinger, K., & Wu, S. T. (2024). How to teach programming in the ai era? Using llms as a teachable agent for debugging. In *International conference on artificial intelligence in education* (pp. 265–279). Springer.

MagicSchool. AI for educators. https://www.magicschool.ai/. (Accessed April 25, 2025).

Matelsky, J. K., Parodi, F., Liu, T., Lange, R. D., & Kording, K. P. (2023). A large language model-assisted education tool to provide feedback on open-ended responses. *arXiv preprint arXiv:2308.02439*.

Matsuda, N., Weng, W., & Wall, N. (2020). The effect of metacognitive scaffolding for learning by teaching a teachable agent. *International Journal of Artificial Intelligence in Education, 30*(1), 1–37.

Maxfielf, C.. Arizona state university to use adaptive learning technology. https://www.eetimes.com/arizona-state-university-to-use-adaptive-learning-technology. (Accessed April 25th, 2025).

Mihai, S., et al. (2022). Digital twins: A survey on enabling technologies, challenges, trends and future prospects. *IEEE Communications Surveys and Tutorials, 24*(4), 2255–2291.

Mihalcea, R., & Tarau, P. (2004). Textrank: Bringing order into text. In *Proceedings of the 2004 conference on empirical methods in natural language processing* (pp. 404–411).

Milano, S., McGrane, J. A., & Leonelli, S. (2023). Large language models challenge the future of higher education. *Nature Machine Intelligence, 5*(4), 333–334.

Miller, D. (2019). Leveraging BERT for extractive text summarization on lectures. *arXiv preprint arXiv:1906.04165*.

Mirata, V., Hirt, F., Bergamin, P., & van der Westhuizen, C. (2020). Challenges and contexts in establishing adaptive learning in higher education: Findings from a Delphi study. *International Journal of Educational Technology in Higher Education, 17*, 1–25.

Mithun, P., Noriega-Atala, E., Merchant, N., & Skidmore, E. (2025). AI-VERDE: A gateway for Egalitarian access to large language model-based resources for educational institutions. *arXiv preprint arXiv:2502.09651*.

Moon, H., Lee, J., Eo, S., Park, C., Seo, J., & Lim, H.-S. (2024). Generative interpretation: Toward human-like evaluation for educational question-answer pair generation. In *Findings of the association for computational linguistics: Eacl 2024* (pp. 2185–2196).

Nair, I., Tan, J., Su, X., Gere, A., Wang, X., & Wang, L. (2024). Closing the loop: Learning to generate writing feedback via language model simulated student revisions. *EMNLP 2024 - 2024 Conference on Empirical Methods in Natural Language Processing, Proceedings of the Conferenc*, 16636–16657. https://www.scopus.com/inward/record.uri?eid=2-s2.0-85217737484&doi=10.18653%2fv1%2f2024.emnlp-main.928&partnerID=40&md5=920d02ba5f8b145671c1ea2214a4ae47.

Narkhede, N., Menon, A., Mathane, I., Nikam, S., & Dange, S. (2023). Facial recognition and machine learning-based student attendance monitoring system. In *2023 3rd international conference on intelligent technologies (CONIT)* (pp. 1–7). IEEE.

Nussbaumer, A., Hillemann, E.-C., Gütl, C., & Albert, D. (2015). A competence-based service for supporting self-regulated learning in virtual environments. *Journal of Learning Analytics, 2*(1), 101–133.

Ouyang, F., Wu, M., Zheng, L., Zhang, L., & Jiao, P. (2023). Integration of artificial intelligence performance prediction and learning analytics to improve student learning in online engineering course. *International Journal of Educational Technology in Higher Education, 20*(1), 4.

Pal Chowdhury, S., Zouhar, V., & Sachan, M. (2024). Autotutor meets large language models: A language model tutor with rich pedagogy and guardrails. In *Proceedings of the eleventh ACM conference on learning@ scale* (pp. 5–15).

Pardos, Z. A., Tang, M., Anastasopoulos, I., Sheel, S. K., & Zhang, E. (2023). Oatutor: An open-source adaptive tutoring system and curated content library for learning sciences research. In *Proceedings of the 2023 chi conference on human factors in computing systems* (pp. 1–17).

Park, M., Kim, S., Lee, S., Kwon, S., & Kim, K. (2024). Empowering personalized learning through a conversation-based tutoring system with student modeling. In *Extended abstracts of the CHI conference on human factors in computing systems* (pp. 1–10).

Patel, M., & Chalageri, S. (2023). GlotBot: Hybrid language translator for secondary level mathematics classrooms. In *Proceedings of the 22nd annual ACM interaction design and children conference* (pp. 439–443).

Perez, S., et al. (2017). Identifying productive inquiry in virtual labs using sequence mining. In *Artificial intelligence in education: 18th international conference, AIED 2017, wuhan, China, June 28–July 1, 2017, proceedings 18* (pp. 287–298). Springer.

Premananda, I. G. A., & Muklason, A. (2021). Complex university timetabling using iterative forward search algorithm and great deluge algorithm. *Khazanah Informatika: Jurnal Ilmu Komputer dan Informatika, 7*(2), 39–46.

Putra, A., Gram, D., Stefanou, C., & Santoro, D. (2022). The use of adaptive learning technology to enhance learning in clinical veterinary dermatology. *Journal of Veterinary Medical Education, 49*(1), 118–125.

Ramírez-Moreno, M. A., et al. (2021). Eeg-based tool for prediction of university students' cognitive performance in the classroom. *Brain Sciences, 11*(6), 698.

Sajadi, S., Huerta, M., Ryan, O., & Drinkwater, K. Harnessing Generative AI to Enhance Feedback Quality in Peer Evaluations within Project-Based Learning Contexts, *International Journal of Engineering Education,* Article vol. 40, no. 5, pp. 998-1012, 2024. [Online]. Available: https://www.scopus.com/inward/record.uri?eid=2-s2.0-85205941405&partnerID=40&md5=c9c19948fc434670c7a-a3971004c1f3a.

Škobo, M., & Petričević, V. (2023). Navigating the challenges and opportunities of literary translation in the age of AI: Striking a balance between human expertise and machine power. *Društvene i humanističke studije, 8*(2), 317–336.

Scarlatos, A., Baker, R. S., & Lan, A. (2025). Exploring knowledge tracing in tutor-student dialogues using llms. In *Proceedings of the 15th international learning analytics and knowledge conference* (pp. 249–259).

Seo, K., Tang, J., Roll, I., Fels, S., & Yoon, D. (2021). The impact of artificial intelligence on learner–instructor interaction in online learning. *International Journal of Educational Technology in Higher Education, 18*, 1–23.

Shahzad, T., Mazhar, T., Tariq, M. U., Ahmad, W., Ouahada, K., & Hamam, H. (2025). A comprehensive review of large language models: Issues and solutions in learning environments. *Discover Sustainability, 6*(1), 27.

Shi, D., Wang, T., Xing, H., & Xu, H. (2020). A learning path recommendation model based on a multidimensional knowledge graph framework for e-learning. *Knowledge-Based Systems, 195*, 105618.

Shusterman, R., Waters, A. C., O'Neill, S., Bangs, M., Luu, P., & Tucker, D. M. (2025). An active inference strategy for prompting reliable responses from large language models in medical practice. *Npj Digital Medicine, 8*(1), 119.

Simon, N., & Muise, C. (2022). TattleTale: Storytelling with planning and large language models. In *ICAPS workshop on scheduling and planning applications*.

Sindhu, I., Daudpota, S. M., Badar, K., Bakhtyar, M., Baber, J., & Nurunnabi, M. (2019). Aspect-based opinion mining on student's feedback for faculty teaching performance evaluation. *IEEE Access, 7*, 108729–108741.

Sonlu, S., Bendiksen, B., Durupinar, F., & Güdükbay, U. (2024). The effects of embodiment and personality expression on learning in llm-based educational agents. *arXiv preprint arXiv:2407.10993*.

Steenstra, I., Murali, P., Perkins, R. B., Joseph, N., Paasche-Orlow, M. K., & Bickmore, T. (2024). Engaging and entertaining adolescents in health education using LLM-generated fantasy narrative games and virtual agents. In *Extended abstracts of the CHI conference on human factors in computing systems* (pp. 1–8).

Sutjarittham, T., Gharakheili, H. H., Kanhere, S. S., & Sivaraman, V. (2019). Experiences with IoT and AI in a smart campus for optimizing classroom usage. *IEEE Internet of Things Journal, 6*(5), 7595–7607.

Sutoyo, E., Almaarif, A., & Yanto, I. T. R. (2020). Sentiment analysis of student evaluations of teaching using deep learning approach. In *The international conference on emerging applications and technologies for industry 4.0* (pp. 272–281). Springer.

Tabakcıoğlu, M., Çizmeci, H., & Ayberkin, D. (2016). Neurosky EEG biosensor using in education. *International Journal of Applied Mathematics Electronics and Computers*, (Special Issue-1), 76–78.

Tamdjidi, R. (2023). In *ChatGPT as an assistive technology to enhance reading comprehension for individuals with ADHD*.

Tastanova, M., Jilkibayeva, R., Arkhabayeva, Z., & Tripathi, S. (2023). Enhancing predictive analytics in education through collaborative filtering algorithms in knowledge management. In *2023 4th international conference on computation, automation and knowledge management (ICCAKM)* (pp. 1–6). IEEE.

Torelli, J. N., Lloyd, B. P., Diekman, C. A., & Wehby, J. H. (2017). Teaching stimulus control via class-wide multiple schedules of reinforcement in public elementary school classrooms. *Journal of Positive Behavior Interventions, 19*(1), 14–25.

Veletsianos, G., & Russell, G. S. (2013). What do learners and pedagogical agents discuss when given opportunities for open-ended dialogue? *Journal of Educational Computing Research, 48*(3), 381–401.

Voultsiou, E., & Moussiades, L. (2025). A systematic review of AI, VR, and LLM applications in special education: Opportunities, challenges, and future directions. *Education and Information Technologies*, 1–41.

Wan, G., Wu, Y., Chen, J., & Li, S. (2024). CoT rerailer: Enhancing the reliability of large language models in complex reasoning tasks through error detection and correction. *arXiv e-prints, arXiv-2408*.

Wang, D. (2024). Teacher-versus AI-generated (poe application) corrective feedback and language learners' writing anxiety, complexity, fluency, and accuracy. *International Review of Research in Open and Distributed Learning, 25*(3), 37–56.

Wang, F., et al. (2020). Neural cognitive diagnosis for intelligent education systems. *Proceedings of the AAAI Conference on Artificial Intelligence, 34*(04), 6153–6161.

Wang, T., et al. (2025). LLM-Powered multi-agent framework for goal-oriented learning in intelligent tutoring system. *WWW Companion 2025 - Companion Proceedings of the ACM Web Conference 2025*, 510–519. https://doi.org/10.1145/3701716.3715244. https://www.scopus.com/inward/record.uri?eid=2-s2.0-105009207804&doi=10.1145%2f3701716.3715244&partnerID=40&md5=5e083d3d062f6e16eb3e2903cc837880

WestEd. Carnegie math pathways graduates to larger home in WestEd. https://www.carnegiefoundation.org/newsroom/news-releases/carnegie-math-pathways-graduates-to-larger-home-in-wested/. (Accessed April 25, 2025).

Wei, H., Qiu, J., Yu, H., & Yuan, W. (2025). Medco: Medical education copilots based on a multi-agent framework. *Lecture Notes in Computer Science, 15630 LNCS*, 119–135. https://doi.org/10.1007/978-3-031-91813-1_8. https://www.scopus.com/inward/record.uri?eid=2-s2.0-105006878360&doi=10.1007%2f978-3-031-91813-1_8&partnerID=40&md5=f6924e34a2ffe50c9f8a4bf6a225d98c

Wu, M., Yuan, Y., Haffari, G., & Wang, L. (2024). (perhaps) beyond human translation: Harnessing multi-agent collaboration for translating ultra-long literary texts. *Transactions of the Association for Computational Linguistics, 13*, 901–922.

Xu, S., Wen, H.-N., Pan, H., Dominguez, D., Hu, D., & Zhang, X. (2025). Classroom simulacra: Building contextual student generative agents in online education for learning behavioral simulation. *Conference on Human Factors in Computing Systems - Proceedings*. https://doi.org/10.1145/3706598.3713773. https://www.scopus.com/inward/record.uri?eid=2-s2.0-105005756893&doi=10.1145%2f3706598.3713773&partnerID=40&md5=65b9a6074ef6fa6bd847e6f4506c8ec4

Xu, T., Zhang, Y., Chu, Z., Wang, S., & Wen, Q. (2025). Ai-driven virtual teacher for enhanced educational efficiency: Leveraging large pretrain models for autonomous error analysis and correction. *Proceedings of the AAAI Conference on Artificial Intelligence, 39* (28), 28801–28809.

Xu, S., Zhang, X., & Qin, L. (2024). Eduagent: Generative student agents in learning. *arXiv preprint arXiv:2404.07963.*

Yan, Y., et al. (2024). Errorradar: Benchmarking complex mathematical reasoning of multimodal large language models via error detection. *arXiv preprint arXiv:2410.04509*, 2024.

Yan, L., et al. (2024). Practical and ethical challenges of large language models in education: A systematic scoping review. *British Journal of Educational Technology, 55*(1), 90–112.

Yang, K. B., et al. (2023). Pair-up: Prototyping human-AI co-orchestration of dynamic transitions between individual and collaborative learning in the classroom. In *Proceedings of the 2023 CHI conference on human factors in computing systems* (pp. 1–17).

Yang, K., et al. (2024). Content knowledge identification with multi-agent large language models (LLMs). In *International conference on artificial intelligence in education* (pp. 284–292). Springer.

Yang, J., et al. (2024). Swe-agent: Agent-computer interfaces enable automated software engineering. *Advances in Neural Information Processing Systems, 37*, 50528–50652.

Ye, J., et al. (2024). ProductAgent: Benchmarking conversational product search agent with asking clarification questions. *Proceedings of the 2025 Conference on Empirical Methods in Natural Language Processing: Industry Track*, 383–398.

Ye, J., et al. (2025). Position: LLMs can be good tutors in foreign language education. *arXiv preprint arXiv:2502.05467.*

Ye, J., Li, Y., Li, Y., & Zheng, H.-T. (2023). Mixedit: Revisiting data augmentation and beyond for grammatical error correction. *arXiv preprint arXiv:2310.11671.*

Ye, J., Li, Y., Ma, S., Xie, R., Wu, W., & Zheng, H.-T. (2022). Focus is what you need for Chinese grammatical error correction. *arXiv preprint arXiv:2210.12692.*

Zaiane, O. R. (2002). Building a recommender agent for e-learning systems. In *International conference on computers in education, 2002. Proceedings* (pp. 55–59). IEEE.

Zepeda, S. J., & Mayers, R. S. (2006). An analysis of research on block scheduling. *Review of Educational Research, 76*(1), 137–170.

Zhang, M., Dilling, A. P., Gondelman, L., Lyngdorf, N. E. R., Lindsay, E. D., & Bjerva, J. (2025). SEFL: Harnessing large language model agents to improve educational feedback systems. *arXiv e-prints*, Article arXiv-2502.

Zhang, Z., et al. (2024). Psysafe: A comprehensive framework for psychological-based attack. Defense, and Evaluation of Multi-Agent System Safety. *Proceedings of the Annual Meeting of the Association for Computational Linguistics, 1*, 15202–15231. https://doi.org/10.18653/v1/2024.acl-long.812 [Online]. Available: https://www.scopus.com/inward/record.uri?eid=2-s2.0-85198636685&doi=10.18653%2fv1%2f2024.acl-long.812&partnerID=40&md5=eae09eb44654baad3f2dce483f6377c6.

Zhang, R., et al. (2025). Let storytelling tell vivid stories: A multi-modal-lagent-based unified storytelling framework. *Neurocomputing, 622*, 129316.

Zha, S., Qiao, Y., Hu, Q., Li, Z., Gong, J., & Xu, Y. (2025). Designing child-centric AI learning environments: Insights from LLM-Enhanced creative project-based learning. *International Journal of Human-Computer Studies, 204*, 15202–15231. https://doi.org/10.1016/j.ijhcs.2025.103602, 103602.

Zhang, K., Li, J., Li, G., Shi, X., & Jin, Z. (2024). CODEAGENT: Enhancing code generation with tool-integrated agent systems for real-world repo-level coding challenges. *Proceedings of the Annual Meeting of the Association for Computational Linguistics, 1*, 13643–13658. https://doi.org/10.18653/v1/2024.acl-long.737. arXiv preprint arXiv:2401.07339 https://www.scopus.com/inward/record.uri?eid=2-s2.0-85204422336&doi=10.18653%2fv1%2f2024.acl-long.737&partnerID=40&md5=0b267ccad36ea311fb8c8173fa1f00f0.

Zhang, Y.-F., Li, H., Song, D., Sun, L., Xu, T., & Wen, Q. (2025). From correctness to comprehension: AI agents for personalized error diagnosis in education. *arXiv preprint arXiv:2502.13789.*

Zhao, X., Cox, A., & Chen, X. (2025). The use of generative AI by students with disabilities in higher education. *The Internet and Higher Education, 66*, 101014.

Zhao, G., & Song, E. (2024). Privacy-preserving large language models: Mechanisms, applications, and future directions. *arXiv preprint arXiv:2412.06113.*

Zheng, H. S., et al. (2024). Natural plan: Benchmarking llms on natural language planning. *arXiv preprint arXiv:2406.04520.*

Zheng, Y., et al. (2024). Automatic lesson plan generation via large language models with self-critique prompting. In *International conference on artificial intelligence in education* (pp. 163–178). Springer.

Zheng, J., et al. (2024). Fine-tuning large language models for domain-specific machine translation. *arXiv preprint arXiv:2402.15061.*

Zheng, L., Fan, Y., Chen, B., Huang, Z., LeiGao, & Long, M. (2024). An AI-enabled feedback-feedforward approach to promoting online collaborative learning. *Education and Information Technologies, 29* (9), 11385–11406.

Zheng, L., Jiang, F., Gu, X., Li, Y., Wang, G., & Zhang, H. (2025). Teaching via LLM-Enhanced simulations: Authenticity and barriers to suspension of disbelief. *The Internet and Higher Education, 65*, 100990.

Zheng, L., & Zheng, L. (2021). The model of task design in computer-supported collaborative learning. *Data-Driven Design for Computer-Supported Collaborative Learning: Design Matters*, 19–30.

Zhuang, Y., Liu, Q., Huang, Z., Li, Z., Shen, S., & Ma, H. (2022). Fully adaptive framework: Neural computerized adaptive testing for online education. *Proceedings of the AAAI Conference on Artificial Intelligence, 36*(4), 4734–4742.

Zimmerman, B. J. (1990). Self-regulated learning and academic achievement: An overview. *Educational Psychologist, 25*(1), 3–17.

Chapter 26

Digital twin technology for student healthcare in smart schools

Zahra Rizvi[1] and Aida Ghazizadeh[2]
[1]*College of William & Mary, Williamsburg, VA, United States;* [2]*Old Dominion University, Norfolk, VA, United States*

1. Introduction

The past few decades have witnessed a profound and accelerating transformation of educational environments, driven by relentless technological innovation. This evolution has culminated in the rise of smart schools—dynamic learning ecosystems that seamlessly integrate digital solutions to enrich learning experiences, optimize school management processes, and, crucially, enhance the overall welfare of students (Lee & Lee, 2020, pp. 301–320). Fig. 26.1 shows the integration of Digital Twin (DT) technologies in a futuristic smart school setting, illustrating both classroom-level applications and individual student monitoring. Central to this technological renaissance in education is DT technology, a powerful and multifaceted concept initially conceived and refined within industrial sectors to achieve operational excellence and predict future outcomes with remarkable accuracy (Vikhman & Romm, 2021). Now strategically adapted for the unique context of education, digital twins hold immense promise for revolutionizing student healthcare by enabling continuous real-time monitoring of vital parameters, generating

FIGURE 26.1 A futuristic smart school campus powered by Digital Twin technology, highlighting interconnected systems. The red inset shows an advanced classroom environment with immersive learning, while the yellow inset emphasizes wearable IoT devices for real-time physiological monitoring of students.

Digital Twins in the Smart Classroom. https://doi.org/10.1016/B978-0-443-41632-3.00027-8

data-driven predictions of health trends and potential risks, and facilitating proactive and personalized health interventions (Rivera et al., 2019).

The provision of comprehensive and effective student healthcare within educational settings transcends the mere treatment of immediate ailments; it is a fundamental pillar for fostering long-term academic success, promoting holistic personal development, and cultivating a supportive and nurturing school environment. Educational institutions today face escalating demands to monitor student health with greater vigilance, effectively manage the increasing prevalence of chronic health conditions among young people, and respond swiftly and efficiently to a wide spectrum of health emergencies (Gereige et al., 2022). DT technology, underpinned by an intricate network of sophisticated IoT sensors, advanced AI-driven analytics engines, and robust cloud-based infrastructure, emerges as an innovative and potentially game-changing solution to address these multifaceted challenges (Hassija et al., 2024).

The objectives of this chapter are fourfold: to provide a clear overview of the fundamental principles of DT technology, to evaluate its potential impact on student healthcare in smart schools, to examine the broader educational implications of its adoption, and to identify possible challenges along with practical solutions for effective and ethical implementation. By connecting technological progress with the application of medical knowledge and the evolving needs of education, digital twins offer a structured and forward-looking approach to student health management within smart school environments.

2. Digital twin technology: A technological overview

2.1 Definition and fundamentals

At its core, a DT represents a dynamic and continuously evolving digital counterpart of a tangible physical system, entity, or complex process (Maropoulos, 2003). By leveraging a continuous stream of data meticulously collected from a network of sensors strategically embedded within the physical object or environment, digital twins create a highly accurate and up-to-date digital representation that mirrors real-time operational conditions, intricate processes, and even subtle environmental nuances. This sophisticated digital simulation empowers stakeholders with the unprecedented ability to comprehensively analyze the physical entity's current performance, accurately predict potential future states and behaviors under various conditions, and ultimately optimize operations and decision-making processes through the lens of informed, data-driven insights. Fig. 26.2 depicts the fundamental data flow and a simplified functional architecture of DT systems in smart schools and student healthcare contexts. Table 26.1 shows how the DT technology is built upon the convergence of several fundamental concepts and enabling technologies.

2.2 Key technological components

The successful implementation and effective operation of DT technology necessitate the seamless integration of several critical technological components, each contributing uniquely and synergistically to its overall functionality and value proposition (Fuller et al., 2020). Fig. 26.3 presents the fundamental technological building blocks that enable DT systems in smart schools. These components are vital because they form the foundation of a fully functional DT ecosystem. IoT sensors provide continuous streams of environmental and physiological data, while AI & ML algorithms extract patterns and predictions from this information. Cloud computing provides scalability and centralized access, whereas edge computing allows for low-latency, real-time responses close to the data source. Finally, data analytics and visualization transform raw data into decision-support

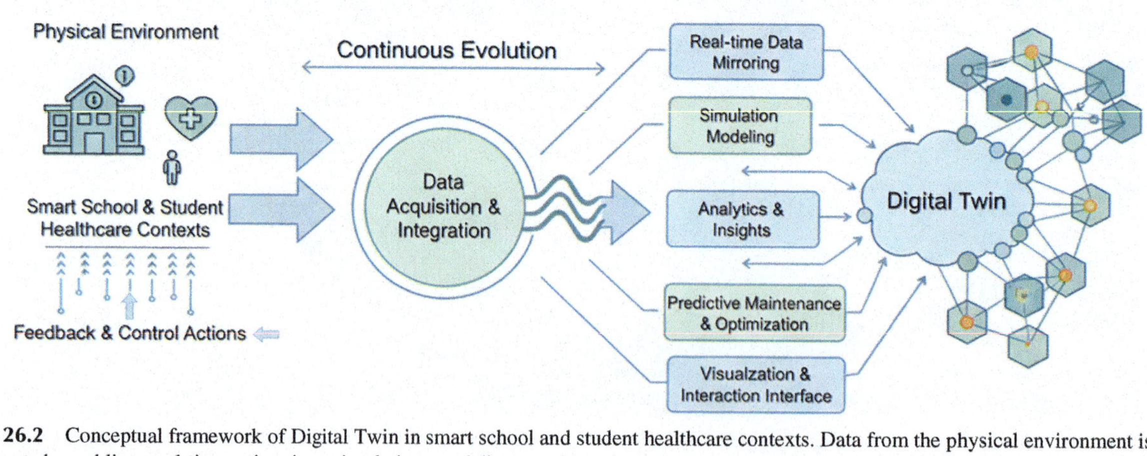

FIGURE 26.2 Conceptual framework of Digital Twin in smart school and student healthcare contexts. Data from the physical environment is acquired and integrated, enabling real-time mirroring, simulation modeling, analytics, predictive optimization, and visualization, with feedback loops for continuous evolution and control actions.

TABLE 26.1 Core components and enabling technologies of digital twin systems.

Component	Technical description	Core functions	Technological enablers
Real-time data integration	Continuous ingestion of high-frequency data streams from heterogeneous sensor networks to maintain a dynamically updated virtual replica of the physical system.	Ensures digital twin fidelity and synchronicity with real-world state.	IoT sensors, edge computing, time-series databases
Predictive analytics	Application of machine learning and statistical models on historical and streaming data to infer future states, detect anomalies, and recommend actions.	Enables forecasting, early warning systems, and proactive maintenance.	Big data frameworks, ML models, AI engines
Simulation and optimization	Execution of scenario-based simulations within a virtual environment to evaluate the impact of alternative decisions, enabling cost-free experimentation and optimization.	Assesses "what-if" situations, guides strategic planning.	Physics-based models, digital simulation platforms
Connectivity and interoperability	Secure, scalable integration across physical assets, digital models, and enterprise systems using standard communication protocols and data exchange formats.	Facilitates end-to-end system orchestration and feedback loops.	API gateways, middleware, OPC UA, MQTT, RESTful APIs
Visualization and interaction	Use of real-time dashboards, 3D models, and interactive UI/UX frameworks to represent and manipulate data and system behavior in an intuitive way.	Supports stakeholder decision-making and operational insights.	AR/VR, WebGL, dashboards, UX design, visualization APIs

FIGURE 26.3 Core technological components supporting Digital Twin in smart schools, comprising IoT sensors, AI & ML algorithms, cloud computing, edge computing, and data analytics & visualization. These elements work together to enable seamless data acquisition, intelligent processing, and actionable insights.

tools, enabling educators and healthcare providers to monitor student well-being, enhance learning outcomes, and optimize school operations. Table 26.2 presents the core technological components essential for the deployment and operational success of DT systems, particularly in the context of smart schools and student healthcare environments.

3. Benefits for schools

The adoption and effective implementation of DT technology offer a multitude of compelling benefits for smart schools, particularly in the critical area of enhancing student health and overall well-being (Villa & Chiaia, 2021). As detailed in Table 26.3, DT technology delivers a

TABLE 26.2 Components supporting DT in smart school and student healthcare contexts.

Component	Technical role	Primary functions	Key technologies
IoT sensors	Act as the data acquisition backbone by capturing environmental and physiological parameters in real time.	Continuous monitoring of air quality, temperature, humidity, heart rate, stress levels, etc.	Temperature/humidity sensors, wearables, CO_2 detectors
AI and ML algorithms	Analyze and learn from large volumes of real-time and historical data to detect patterns and forecast outcomes.	Health anomaly detection, outbreak prediction, personalized health recommendations	TensorFlow, scikit-learn, deep learning models
Cloud computing	Provides scalable infrastructure for data storage, high-performance computation, and system-wide accessibility.	Real-time data access, scalable storage, secure collaboration among stakeholders	AWS, Azure, Google cloud platform, Kubernetes
Edge computing	Enables low-latency processing by handling critical data near the source to ensure responsiveness and privacy.	Emergency alerts, on-site health analysis, latency-sensitive computation	Raspberry Pi, NVIDIA Jetson, Microservices frameworks
Data analytics & visualization	Converts raw sensor output into decision-support tools using visual dashboards and analytical platforms.	Trend analysis, KPI tracking, real-time reporting for health and environment management	Power BI, Tableau, Grafana, D3.js

TABLE 26.3 Benefits of DT technology for student health in smart schools.

Benefit category	Description	Impact on school environment	Enabling mechanisms
Real-time monitoring	Continuous tracking of both physiological and environmental parameters within school premises.	Early detection of health issues and environmental hazards; ensures rapid response.	IoT sensors, edge computing
Predictive capabilities	Analyzes data to forecast health trends and emerging risks using ML/AI models.	Enables proactive health interventions and outbreak prevention.	Predictive analytics, historical data modeling
Personalized healthcare	Tailors health strategies to individual student profiles and needs.	Supports chronic disease management and minimizes academic disruptions.	Wearables, EHR integration, ML personalization algorithms
Resource optimization	Informs efficient allocation of medical personnel, supplies, and facilities.	Reduces waste, ensures critical resources are available where and when needed.	Data analytics, operational dashboards
Improved communication	Facilitates stakeholder coordination via secure, integrated communication platforms.	Enhances coordination between students, parents, teachers, and healthcare providers.	Secure APIs, data sharing protocols, collaboration tools
Data-driven decision-making	Leverages objective data insights to shape health policy and planning.	Informs evidence-based strategies, improving school-wide health outcomes.	Data warehouses, decision support systems
Early intervention	Detects early signs of illness or stress, enabling timely action.	Decreases absenteeism, enhances preventive care, improves student safety.	Edge AI, rule-based alert systems
Environmental safety	Continuously assesses environmental factors to maintain optimal learning conditions.	Promotes comfort and health, boosting focus and academic performance.	Environmental sensors, real-time alerts

multifaceted set of benefits that enable smart schools to proactively manage student health and foster an optimal learning environment.

4. Smart schools and healthcare: Bridging technology and wellness

4.1 Concept of smart schools

Smart schools represent a paradigm shift in educational institutions, characterized by the pervasive and strategic integration of digital technology to create interactive, interconnected, and adaptive learning environments (Kwet & Prinsloo, 2020). Fig. 26.4 summarizes the transformative paradigm shift enabled by smart school ecosystems. These modern schools leverage a robust digital infrastructure, encompassing high-speed Internet connectivity, a diverse array of IoT-enabled devices, and seamlessly integrated software systems, to support more effective teaching methodologies and highly personalized learning experiences tailored to individual student needs. By thoughtfully incorporating advanced technologies across various facets of school operations, smart schools aspire to provide comprehensive solutions that enhance education management, foster greater student engagement, facilitate proactive health monitoring, and streamline administrative efficiency. The underlying principle is to harness the power of technology to create a more dynamic, responsive, and supportive educational ecosystem (Rafique et al., 2024).

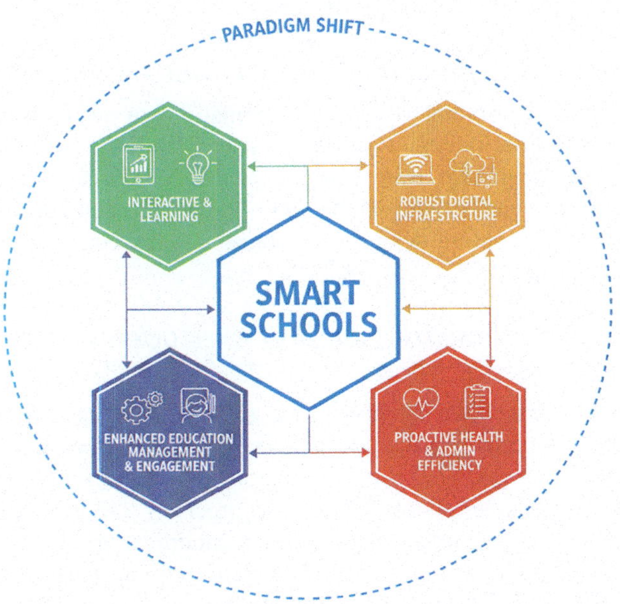

FIGURE 26.4 The paradigm shift toward smart schools, highlighting four core dimensions: interactive learning, robust digital infrastructure, proactive health and administrative efficiency, and enhanced education management and engagement.

4.2 Integration of sensor technology

Sensor technologies form the very bedrock upon which the functionality and effectiveness of smart schools, particularly in the realm of student healthcare, are built. These technologies enable continuous and automated monitoring of a wide range of critical parameters within the school environment and related to individual students. Environmental sensors, including sophisticated temperature, humidity, and carbon dioxide (CO_2) sensors, play a vital role in maintaining optimal learning conditions within classrooms and other school spaces (Zivelonghi & Giuseppi, 2024). These sensors ensure that the physical environment is conducive to concentration and well-being, minimizing discomfort and potential health risks associated with poor environmental conditions. Simultaneously, physiological sensors, often integrated into wearable devices or embedded in furniture or fixtures, continuously track students' vital health parameters, such as heart rate, body temperature, and activity levels. Furthermore, motion sensors and passive infrared (PIR) sensors can detect student activity patterns and classroom occupancy, providing valuable data for both security management and understanding student movement and engagement (Zhang et al., 2024). The wealth of data meticulously collected by this network of sensors empowers educators and administrators with the information necessary to make informed decisions regarding timely health interventions, efficient resource allocation, and the creation of a safer and more supportive school environment.

4.3 Case studies of successful implementations

The tangible value and transformative potential of DT technology in enhancing student healthcare within smart schools are increasingly being demonstrated through various successful implementations across the globe:

- CO_2 and air quality management: A compelling example comes from a European smart school project where classrooms were equipped with advanced CO_2 sensors (Zivelonghi & Giuseppi, 2024). These sensors continuously monitored the air quality in real-time, and the data was used to automatically adjust the school's ventilation systems. The result was a significant reduction in student absenteeism attributed to respiratory conditions, alongside reported improvements in student concentration levels and overall health and well-being. This demonstrates the direct impact of environmental monitoring on student health and academic performance.
- Wearable health devices: In several pioneering schools in the United States, the adoption of wearable health devices, such as smartwatches and fitness bands, has provided valuable insights into student health (Ometov

et al., 2021). These devices continuously monitor vital signs such as heart rate, track physical activity levels throughout the day, and even provide indicators of stress based on physiological data. This continuous health monitoring has enabled early identification of potential health concerns, leading to timely interventions and a reduction in health-related disruptions to learning. Moreover, the aggregated data has proven invaluable for informing and tailoring health education programs within the schools.

- Predictive health analytics: Educational institutions in Asia have been at the forefront of leveraging AI-driven digital twins to analyze historical and real-time health data to predict potential outbreaks of infectious illnesses, such as influenza or COVID-19 (Zhao et al., 2024). By identifying patterns and anomalies in health data, these predictive capabilities have significantly enhanced schools' preparedness and response times to public health threats, minimizing disruptions to the educational process and effectively protecting the health and safety of the student population.
- Personalized chronic disease management: In some specialized educational settings catering to students with chronic conditions, digital twins are being used to create highly personalized management plans (Rivera et al., 2019). By integrating continuous physiological monitoring with individual health records and medication schedules, these systems provide real-time feedback and alerts, helping students manage their conditions more effectively and reducing the risk of acute episodes.
- Mental health early warning systems: Pilot programs are exploring the use of digital twins to analyze behavioral data, such as changes in activity patterns, communication frequency, and even subtle linguistic cues, to identify early warning signs of mental health challenges such as stress, anxiety, or depression (Spitzer et al., 2023). Early detection allows for timely intervention and the provision of appropriate support services.

4.4 Benefits of sensor integration in smart school ecosystems

The strategic integration of sensor technologies into the digital twin framework of smart schools goes far beyond mere data acquisition. It forms the foundation for a responsive, predictive, and student-centric educational environment. When implemented effectively, sensor systems generate a range of benefits that span individual health, collective well-being, and operational efficiency. Following is an outline of the key domains where such benefits are most profound.

4.4.1 Enhanced health monitoring

Automated and continuous data collection from wearable devices and environmental sensors ensures real-time awareness of student health. This persistent visibility enables school health personnel to identify subtle physiological changes or environmental anomalies early—such as elevated body temperature or poor air quality—allowing for immediate intervention and reducing the risk of escalation (Kumar et al., 2025).

4.4.2 Improved learning environments

By maintaining optimal classroom conditions—such as temperature, CO_2 levels, humidity, and noise—sensor systems help mitigate factors that negatively impact cognitive performance. Adaptive environmental controls, informed by sensor feedback, contribute to a healthier, more focused, and more comfortable atmosphere for both students and staff (Zivelonghi & Giuseppi, 2024).

4.4.3 Proactive health management

Sensor data becomes exponentially more valuable when paired with predictive analytics. Patterns in student vitals or attendance can forecast illness trends, enabling schools to shift from reactive care to a preventive model. This transformation supports early interventions, reduces absenteeism, and fosters a culture of wellness.

4.4.4 Efficient resource utilization

Real-time insights into health incidents and environmental conditions help administrators allocate medical staff, supplies, and infrastructure more effectively. Instead of blanket policies, resources can be dynamically targeted to where they are most needed, enhancing both responsiveness and cost-efficiency (Rivera et al., 2019).

4.4.5 Early detection of infectious outbreaks

Aggregated biometric and symptomatic data allows early detection of infectious disease clusters. With the ability to detect deviations from baseline health conditions, school systems can trigger alerts and implement containment strategies before widespread transmission occurs (Zhao et al., 2024).

4.4.6 Personalized support for students with special needs

Sensor-based data streams offer granular insights into the daily experiences of students managing chronic conditions or requiring special accommodations (Rivera et al., 2019). This enables schools to deliver precisely tailored support—ranging from environmental modifications to real-time health interventions—thereby improving inclusivity and equity (Ometov et al., 2021).

4.4.7 Increased parental engagement

By providing parents with secure access to selected health metrics, schools foster a transparent and collaborative relationship with families. Real-time updates and trend visualizations help parents remain informed and participate meaningfully in decisions related to their child's care and wellness (Farooq et al., 2024).

4.4.8 Evidence-based policy development

Sensor-generated data, once anonymized and aggregated, serves as a robust empirical basis for school health policy development. Whether evaluating the effectiveness of air filtration systems or justifying investments in health personnel, decision-makers can craft data-driven policies that reflect real conditions and outcomes (Zivelonghi & Giuseppi, 2024).

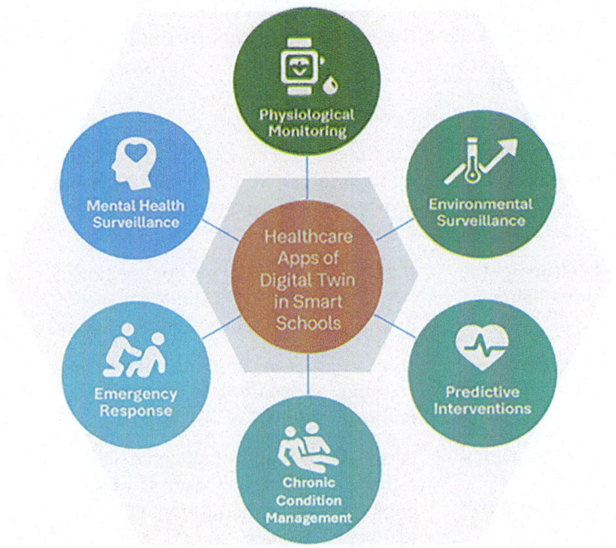

FIGURE 26.5 Key healthcare applications of Digital Twin technology in smart schools, including physiological monitoring, environmental surveillance, predictive interventions, chronic condition management, emergency response, and mental health surveillance.

5. Healthcare applications of digital twin in schools

The convergence of DT systems with real-time sensor networks, AI-driven analytics, and cloud-edge computing frameworks has introduced a paradigm shift in school-based healthcare ecosystems (Sheraz et al., 2024). Rather than serving as passive data repositories, modern DT architectures function as intelligent, adaptive agents capable of interpreting multidimensional health data and environmental parameters in real time. These systems operate not merely as monitoring tools but as proactive health management frameworks, facilitating continuous physiological surveillance, environmental regulation, risk forecasting, and individualized care delivery. Wearable-integrated digital twins, in particular, empower school health teams to detect early biomarkers of physical or psychological distress, preempt the onset of chronic or communicable diseases, and dynamically adjust healthcare protocols based on evolving biometric trends. Moreover, DTs support intelligent triaging and emergency response coordination, thus reinforcing resilience against acute health events while sustaining long-term student well-being. Fig. 26.5 outlines the primary healthcare applications enabled by DT systems in smart schools. Table 26.4 outlines the core healthcare applications of digital twin systems in smart school environments, categorizing their functional contributions across key healthcare domains.

6. Technological infrastructure for digital twins in smart schools

The successful deployment and effective operation of digital twin technology within smart schools rely on a robust and interconnected technological infrastructure comprising several key components:

- Wireless sensor networks: Wireless sensor networks (WSNs) form the fundamental technological infrastructure that underpins the deployment of digital twins in smart school environments (Kalyanaraman et al., 2024). These networks consist of a large number of interconnected, low-power sensor nodes strategically distributed throughout the school premises. These sensors continuously and autonomously gather critical data pertaining to both the physical environment and, through wearable integration, individual student biometrics. Environmental data collected includes parameters such as air quality (particulate matter, volatile organic compounds, CO_2), temperature, humidity levels, ambient noise levels, and lighting conditions. Student biometric data can encompass vital signs (heart rate, temperature), physical activity levels, and even indicators of sleep patterns or stress. These sensors communicate wirelessly using various protocols (e.g., Bluetooth Low Energy, Zigbee, Wi-Fi), transmitting the collected data to centralized servers or cloud-based platforms. This continuous data flow enables real-time analytics, immediate anomaly detection, and timely interventions when necessary. The scalability and energy efficiency of WSNs are crucial for their widespread adoption in school settings.
- Wearable technologies: Wearable technologies, including smartwatches, fitness trackers, smart clothing, and RFID-enabled badges equipped with sensors, play an increasingly vital role in student healthcare monitoring within DT frameworks (Ometov et al., 2021). These devices provide a continuous and personalized stream of health-related data specific to each student. The data collected can include a wide range of physiological metrics, such as heart rate

TABLE 26.4 Healthcare applications of digital twin technology in smart school systems.

Application domain	Functionality	Technological enablers	Impact on school healthcare
Physiological monitoring	Continuous, noninvasive tracking of vital signs (e.g., HR, BP, temp, respiration) using wearable sensors embedded in daily student gear.	IoT wearables, edge AI, real-time data Pipelines	Early illness detection, timely alerts, reduced health emergencies.
Environmental surveillance	Monitoring of indoor air quality, noise, humidity, CO_2 levels, and lighting to maintain optimal classroom conditions.	Ambient sensors, smart HVAC systems, WSNs	Improved respiratory health, cognitive performance, reduced fatigue.
Predictive interventions	Forecasting of health trends and early warnings using AI/ML to identify outbreak potential or risk factors.	Predictive analytics, historical data Mining, health modeling	Preventive care planning, targeted health campaigns, reduced absenteeism.
Chronic condition management	Personalized plans for students with asthma, diabetes, epilepsy, etc., integrating real-time monitoring with historical health profiles.	Digital health records, medication alert systems, decision engines	Better disease control, fewer class disruptions, improved quality of life.
Emergency response	Instantaneous detection of emergencies (e.g., cardiac episodes, seizures, falls) with real-time triaging and auto-notifications to medical staff.	Edge processing, motion sensors, alert protocols	Faster intervention, minimized health risk severity.
Mental health surveillance	Analyzing behavior and emotional indicators (via wearables and digital interaction) to identify early signs of anxiety, stress, or depression.	Ethical data Fusion, sentiment analysis, behavioral analytics	Timely counseling, reduced psychological distress, improved resilience.

variability (an indicator of stress), physical activity levels (steps taken, energy expenditure), sleep duration and quality, body temperature fluctuations, and even early detection of falls. Wearables empower schools to gain a more granular and continuous understanding of each student's health status, enabling the provision of more personalized interventions, tailored health recommendations, and early detection of potential health issues. Ensuring data privacy and security for information collected through wearable devices is paramount.

- Near-field communication: Near-field communication (NFC) technology, while not directly involved in continuous monitoring, enhances the functionality and security of various DT applications within schools (Ansari & Guo, 2024). NFC-enabled devices facilitate secure and contactless interactions, such as efficient and accurate attendance tracking, secure access to restricted areas such as classrooms or medical facilities, and contactless payment systems for school services (e.g., cafeteria, library). The secure and seamless exchange of information via NFC can also be used for quick and secure sharing of relevant health information with authorized personnel in emergency situations. This technology contributes to improved operational efficiency, enhanced security, and a more streamlined experience for students and staff within the school environment.
- Cloud computing: Cloud computing platforms form the robust and scalable backbone of DT technology. They provide the massive data storage capacity required to accommodate the vast volumes of data generated by the multitude of sensors and wearable devices deployed throughout the school (Al-Hawawreh & Hossain, 2024). Furthermore, cloud platforms offer the immense computational power necessary to process this data in real-time, run complex AI and ML algorithms for analytics and prediction, and support the seamless integration of data from diverse sources. The inherent flexibility and remote accessibility of cloud services enable educators, healthcare professionals, and administrators to access and analyze critical health information from anywhere with an Internet connection, facilitating timely and informed decision-making. Cloud providers also offer robust security measures and compliance certifications to ensure the confidentiality, integrity, and availability of sensitive student health data.
- Edge computing: Edge computing strategically complements cloud computing by bringing data processing and analysis closer to the data sources—i.e., at the "edge"

of the network, near the sensors and wearable devices (Liu et al., 2024). By analyzing critical data locally, at the network's edge, schools can significantly reduce data transmission delays (latency), which is crucial for enabling rapid responses in time-sensitive emergency situations. For example, if a wearable device detects a sudden drop in a student's heart rate or a fall, edge computing can process this information immediately and trigger local alerts without having to send the data to a distant cloud server for analysis. This near real-time processing can be lifesaving in critical scenarios. Additionally, edge computing can reduce the overall network load and enhance data privacy by minimizing the volume of sensitive data that needs to be transmitted to centralized cloud systems, as some initial processing and filtering can occur locally.

7. Medical and healthcare perspectives

The incorporation of DT systems into educational healthcare infrastructure represents a clinically significant advancement in proactive health surveillance and personalized care delivery. From a medical perspective, this integration demands a dual emphasis: maintaining clinical rigor in data interpretation and fostering collaborative healthcare networks beyond the school setting. To ensure reliability, DT platforms must adhere to rigorous standards for clinical validation, data calibration, and algorithmic accuracy, aligning their insights with established medical protocols. Equally essential is the creation of structured partnerships with healthcare providers, enabling real-time decision support, seamless referrals, and bidirectional data interoperability with systems such as electronic health records (EHRs). Moreover, DT-enabled physiological monitoring enhances emergency preparedness by furnishing school staff and medical personnel with critical real-time health intelligence during acute incidents. Concurrently, the ability to ethically monitor indicators of emotional distress supports a holistic model of care that prioritizes both physical and mental health. Table 26.5 outlines these key clinical and operational dimensions, emphasizing their technological underpinnings and institutional implications.

8. Challenges and solutions

The successful implementation of digital twin technology in smart schools is not without its challenges. Addressing these hurdles proactively is crucial for realizing the full potential of this transformative technology:

- Technical challenges: Implementing and maintaining the complex technical infrastructure required for digital twins in smart schools presents several significant challenges (Xames & Topcu, 2024). Data security and privacy remain paramount concerns, given the sensitive nature of the personal and health information collected from students. Robust cybersecurity measures, including encryption, access controls, and regular

TABLE 26.5 Medical and clinical considerations for DT implementation in smart school systems.

Clinical dimension	Core focus	Key mechanisms	Healthcare impact
Clinical validation & accuracy	Ensures that all sensor readings, algorithms, and outputs meet recognized medical standards.	Clinical trials, calibration protocols, algorithm verification	Enhances reliability and trustworthiness of school-based health decisions.
Partnership with healthcare providers	Establishes formal collaboration between schools and external medical institutions for integrated care.	EHR integration, referral networks, consent-based data exchange	Enables coordinated interventions and continuity of care across environments.
Emergency response integration	Enables rapid response to medical emergencies with real-time health data and access to medical history.	Edge monitoring, automated alerts, contextual patient profiles	Improves first response outcomes, accelerates time to treatment, reduces complications.
Preventive health Programming	Uses predictive models to identify trends and risk clusters, informing proactive wellness campaigns.	Risk modeling, environmental correlation analysis, AI forecasting	Supports early interventions and reduces preventable illnesses in student populations.
Mental health interventions	Identifies indicators of emotional distress and enables ethical, tailored psychological support.	Behavioral analytics, counselor alert systems, sentiment inference	Promotes emotional resilience, reduces stigma, and improves student mental health.
Ethical & Legal safeguards	Ensures privacy, consent, and transparency in medical and emotional data collection and usage.	GDPR/HIPAA compliance, role-based access control, anonymization	Protects student rights, builds public confidence, and aligns with legal frameworks.

security audits, are essential to protect this data from unauthorized access and cyber threats. Ensuring interoperability among a diverse range of devices (sensors, wearables), communication protocols, and software systems from different vendors can be a complex undertaking. Standardizing data formats and communication protocols is crucial for seamless data integration and system functionality. Furthermore, managing the immense volumes of data generated by continuous monitoring demands a robust and scalable IT infrastructure, necessitating substantial investment in reliable network connectivity, high-capacity cloud storage, and powerful edge computing resources for real-time processing. System maintenance, software updates, and technical support also represent ongoing technical challenges that need to be addressed through skilled IT personnel and proactive system management.

- Ethical and privacy concerns: The very nature of DT technology, involving the continuous collection, storage, and analysis of vast amounts of personal and health-related data from students, raises significant ethical and privacy concerns that must be addressed with utmost diligence (Saeed et al., 2024). Ensuring strict student data privacy and achieving full compliance with relevant data protection regulations, such as the General Data Protection Regulation (GDPR) in Europe or the Health Insurance Portability and Accountability Act (HIPAA) in the United States (depending on the context and data being collected), is absolutely critical (Tyagi & Mittal, 2025). Schools must establish transparent and clearly communicated data collection policies, outlining what data is collected, how it is used, who has access to it, and for how long it is retained. Implementing strict access controls and robust anonymization or pseudonymization techniques for data analysis are essential to protect individual student identities. Furthermore, obtaining informed consent from parents or guardians (and students themselves, where appropriate) regarding data collection and usage is a fundamental ethical requirement. Continuous education and training for all stakeholders—students, parents, educators, and administrators—on data privacy rights, responsible technology use, and the school's data security measures are crucial for fostering a culture of data privacy and building trust in the system.
- Financial constraints: The initial setup and ongoing maintenance of a comprehensive DT infrastructure in a school environment can be financially demanding. The costs associated with procuring high-quality and reliable sensors and wearable devices for a large student population, investing in sophisticated data analytics software and platforms, establishing secure cloud storage solutions, and providing for continuous system upgrades, maintenance, and technical support can represent a significant financial burden for many educational institutions (Abdillah & Wahyuilahi, 2024). Schools often face budgetary constraints and competing financial priorities that can make funding these advanced technologies challenging. Strategic financial planning, exploring phased implementation approaches, and actively seeking public–private partnerships, government grants, or philanthropic funding opportunities are potential strategies to overcome these financial barriers and make DT technology more accessible to schools. Demonstrating the long-term cost-effectiveness and the potential for improved student health outcomes and reduced healthcare-related disruptions can also help justify the initial investment.

8.1 Solutions

To unlock the transformative potential of DT systems in educational settings, institutions must proactively address the multifaceted challenges associated with integration, security, scalability, and stakeholder readiness. Success hinges on a systems-level strategy that combines technological foresight with institutional resilience. At the core of this strategy lies the imperative for interoperability—ensuring that heterogenous devices, platforms, and data formats can communicate seamlessly through adherence to open standards. Simultaneously, securing sensitive health and behavioral data requires the implementation of advanced cybersecurity frameworks that encompass both technological safeguards and governance policies. Financial sustainability can be achieved through strategic partnerships and targeted funding mechanisms that align institutional goals with public health and innovation priorities. Equally critical is the cultivation of digital competence among all stakeholders through comprehensive training programs, fostering both ethical awareness and operational fluency. Finally, a phased implementation approach—grounded in iterative evaluation and scalability—provides a pragmatic pathway to adoption, minimizing risk while maximizing adaptability. Table 26.6 delineates these interdependent solution domains and their operational implications.

9. Conclusion

Digital twin technology stands as a powerful and transformative convergence of cutting-edge technology, evolving educational paradigms, and the critical domain of student healthcare within the intelligent ecosystems of smart schools. By effectively and ethically integrating real-time monitoring capabilities, sophisticated predictive analytics, personalized healthcare interventions tailored to individual needs, and robust educational programs that

TABLE 26.6 Strategic solutions for implementing digital twin technology in smart school systems.

Solution domain	Core strategy	Operational mechanisms	Expected outcome
Standardization and interoperability	Adoption of open protocols and formats to unify system components and enable seamless integration across platforms.	Use of APIs, XML/JSON schemas, collaboration with standards bodies	Reduced integration cost, enhanced system compatibility and vendor neutrality.
Secure technologies and Practices	Implementation of advanced cybersecurity frameworks to protect sensitive health and behavioral data.	Encryption, MFA, penetration testing, role-based access, privacy policies	Strengthened data protection, compliance with legal and ethical standards.
Funding and partnerships	Leverage external collaborations and diversified funding streams to support technological deployment and sustainability.	Government and private grants, academic partnerships, cost–benefit analysis	Financial resilience, enhanced institutional buy-in, long-term project viability.
Education and training	Comprehensive stakeholder training to ensure ethical, informed, and confident use of DT technologies.	Cybersecurity literacy, ethical tech use education, age-appropriate user guides	Increased trust, user competence, responsible technology use across school communities.
Phased implementation and scalability	Gradual, data-informed deployment to identify challenges and scale solutions iteratively.	Pilot programs, feedback loops, modular and cloud-native architectures	Minimized deployment risk, scalable growth, evidence-based optimization.

foster digital and health literacy, digital twins hold the remarkable potential to significantly enhance student wellness, improve academic outcomes, and cultivate a safer and more supportive learning environment. While technical, ethical, and financial challenges undoubtedly exist, addressing these proactively through strategic planning, collaborative partnerships, and a commitment to data privacy and security is paramount for ensuring the sustainable and impactful implementation of this technology. As technological advancements continue at an unprecedented pace, DT systems promise an even greater potential to revolutionize the landscape of student healthcare and educational experiences, ultimately fostering healthier, more informed, and digitally competent future generations who are well-equipped to thrive in an increasingly interconnected and technology-driven world. The journey of integrating digital twins into smart schools is an ongoing evolution, one that requires careful consideration, ethical stewardship, and a steadfast commitment to the well-being and success of every student.

10. Acronyms

A comprehensive list of acronyms used throughout this chapter, along with their full forms and contextual relevance, is provided in Table 26.7.

TABLE 26.7 Acronyms.

Acronym	Full term	Context/Description
AI	Artificial intelligence	Used for predictive analytics and personalized health recommendations.
API	Application programming interface	Enables communication and data exchange between systems.
AR	Augmented reality	Potential future use in interactive health education experiences.
AWS	Amazon Web services	Example of a cloud computing platform for data storage and analytics.
CO_2	Carbon dioxide	Environmental factor monitored by sensors for air quality.
DT	Digital twin	A dynamic virtual replica of physical entities or systems.
EHR	Electronic health record	Used for integrating individual student health profiles into digital twin systems.

Continued

TABLE 26.7 Acronyms.—cont'd

Acronym	Full term	Context/Description
GDPR	General data protection regulation	European regulation for data privacy.
HIPAA	Health Insurance Portability and Accountability Act	US regulation governing medical data privacy.
HVAC	Heating, ventilation, and air-conditioning	School systems adjusted based on sensor feedback.
IoT	Internet of things	Network of physical devices collecting and exchanging data.
KPI	Key performance indicator	Metric tracked via analytics dashboards.
ML	Machine learning	Enables predictive modeling and anomaly detection.
MQTT	Message Queuing Telemetry Transport	Lightweight messaging protocol used in IoT systems.
NFC	Near-field communication	Used for secure contactless interactions in school operations.
OPC UA	Open platform communications unified architecture	Standard for secure and reliable data exchange.
PIR	Passive infrared	Sensor type used for motion detection in smart schools.
RFID	Radiofrequency identification	Used in smart badges for student monitoring.
STEM	Science, technology, engineering, and mathematics	Curricular area that integrates digital twin technologies.
UI/UX	User interface/user experience	Refers to the design of dashboards and interaction tools.
VR	Virtual reality	Explored as a tool for immersive health education.
WSN	Wireless sensor network	Core infrastructure for data acquisition in smart schools.

References

Abdillah, M. R. N., & Wahyuilahi, M. (2024). Digital twin technology in logistics: A narrative review of implementation, impact, and challenges. *Sinergi International Journal of Logistics, 2*(4), 225–238.

Al-Hawawreh, M., & Hossain, M. S. (2024). Digital twin-driven secured edge-private cloud industrial internet of things (IIoT) framework. *Journal of Network and Computer Applications, 226*, 103888.

Ansari, O., & Guo, H. (2024). Extending near field communication range for ultra-dense internet of things. *IEEE Journal of Radio Frequency Identification, 8*, 770–779. https://doi.org/10.1109/JRFID.2024.3453770

Farooq, M., Iqbal, I., & Ahmad, R. K. (2024). Parental involvement in the digital age: How technology is changing the parent-teacher dynamic. *Al-Mahdi Research Journal (MRJ), 5*(5), 302–310.

Fuller, A., Fan, Z., Day, C., & Barlow, C. (2020). Digital twin: Enabling technologies, challenges and open research. *IEEE Access, 8*, 108952–108971.

Gereige, R. S., Gross, T., Jastaniah, E., & Council on School Health and Committee on Pediatric Emergency Medicine. (2022). Individual medical emergencies occurring at school. *Pediatrics, 150*(1), Article e2022057987.

Hassija, V., Chamola, V., De, R., Das, S., Chakrabarti, A., Sangwoan, K. S., & Pandey, A. (2024). A survey on digital twins: Enabling technologies, use cases, application, open issues and more. *IEEE Journal of Selected Areas in Sensors, 1*(1), 84–107. https://doi.org/10.1109/JSAS.2024.3523856

Kalyanaraman, S., Ponnusamy, S., & Harish, R. K. (2024). Amplifying digital twins through the integration of wireless sensor networks: In-depth exploration. In *Digital twin technology and AI implementations in future-focused businesses* (pp. 70–82). IGI Global Scientific Publishing.

Kumar, V., Lee, D. J., & Park, S. S. (2025). Multifunctional composites-based sensors for health monitoring: Latest insights and prospects. *Materials Today Communications*, 112187.

Kwet, M., & Prinsloo, P. (2020). The 'smart'classroom: A new frontier in the age of the smart university. *Teaching in Higher Education, 25*(4), 510–526.

Lee, R. S., & Lee, R. S. (2020). *Smart education. Artificial intelligence in daily life*.

Liu, W., Hossain, M. A., & Ansari, N. (2024). Mobile edge computing for multi-services digital twin-enabled IoT heterogeneous networks. In *IEEE transactions on cognitive communications and networking*.

Maropoulos, P. G. (2003). Digital enterprise technology–defining perspectives and research priorities. *International Journal of Computer Integrated Manufacturing, 16*(7–8), 467–478.

Ometov, A., Shubina, V., Klus, L., Skibińska, J., Saafi, S., Pascacio, P., … Lohan, E. S. (2021). A survey on wearable technology: History, state-of-the-art and current challenges. *Computer Networks, 193*, 108074.

Rafique, H., Ul Islam, Z., & Shamim, A. (2024). Acceptance of e-learning technology by government school teachers: Application of extended technology acceptance model. *Interactive Learning Environments, 32*(6), 2970–2988.

Rivera, L. F., Jiménez, M., Angara, P., Villegas, N. M., Tamura, G., & Müller, H. A. (November 2019). Towards continuous monitoring in personalized healthcare through digital twins. In *Proceedings of the 29th annual international conference on computer science and software engineering* (pp. 329–335).

Saeed, M. M. A., Saeed, R. A., & Ahmed, Z. E. (2024). Data security and privacy in the age of AI and digital twins. In *Digital twin technology and AI implementations in future-focused businesses* (pp. 99–124). IGI Global Scientific Publishing.

Sheraz, M., Chuah, T. C., Lee, Y. L., Alam, M. M., Al-Habashna, A. A., & Han, Z. (2024). A comprehensive survey on revolutionizing connectivity through artificial intelligence-enabled digital twin network in 6G. *IEEE Access, 12*, 49184–49215.

Spitzer, M., Dattner, I., & Zilcha-Mano, S. (2023). Digital twins and the future of precision mental health. *Frontiers in Psychiatry, 14*, 1082598.

Tyagi, P., & Mittal, K. (2025). Digital twin technology in healthcare: Benefits and security considerations. *Artificial Intelligence and Cybersecurity in Healthcare*, 443–460.

Vikhman, V. V., & Romm, M. V. (2021). "Digital twins" in education: Prospects and reality. *Vysshee Obrazovanie v Rossii= Higher Education in Russia, 30*(2), 22–32.

Villa, V., & Chiaia, B. (2021). Digital twin for smart school buildings: State of the art, challenges, and opportunities. *Handbook of Research on Developing Smart Cities Based on Digital Twins*, 320–340.

Xames, M. D., & Topcu, T. G. (2024). A systematic literature review of digital twin research for healthcare systems: Research trends, gaps, and realization challenges. *IEEE Access, 12*, 4099–4126.

Zhang, X., Ding, Y., Huang, X., Li, W., Long, L., & Ding, S. (2024). Smart classrooms: How sensors and ai are shaping educational paradigms. *Sensors, 24*(17), 5487.

Zhao, A. P., Li, S., Cao, Z., Hu, P. J. H., Wang, J., Xiang, Y., ... Lu, X. (2024). AI for science: Predicting infectious diseases. *Journal of Safety Science and Resilience, 5*(2), 130–146. https://doi.org/10.1016/j.jnlssr.2024.02.002

Zivelonghi, A., & Giuseppi, A. (2024). Smart healthy schools: An IoT-enabled concept for multi-room dynamic air quality control. *Internet of Things and Cyber-Physical Systems, 4*, 24–31.

Chapter 27

Transformative impact of smart classroom/school/campus on the educational environment

Semiu Olawale Makinde[1] **and Wing Cheung Tang**[2, 3]
[1]Institute of Education, Al-Hikmah University, Ilorin, Nigeria; [2]Jesselton University College, Kuala Lumpur, Malaysia; [3]Institute of Vocational Education, Hong Kong Metropolitan University, China

1. Introduction

Digitalization has significantly changed the structure of traditional education, in tandem with the swift advancement of science and technology (Mahdi, 2023). Future educational innovations include smart classrooms, smart schools, and smart campuses. Smart classrooms employ the Internet of Things (IoT), artificial intelligence (AI), big data, and so forth to create an interactive and individualized learning environment, in which students can explore independently, learn collaboratively, and improve learning efficiency and interest. Interactive whiteboards, tablet PCs, and instant response systems in smart classrooms enable teachers and students to communicate and interact in real time, overcoming the limitations of space and time.

Smart schools take smart classrooms as the basis and further expand to digital management and services of the entire campus. Through the smart campus platform, schools can conduct intelligent operations in academic management, security monitoring, resource allocation, and so on to improve management efficiency and safety. At the same time, smart schools also emphasize data-driven decision-making, which helps schools to make personalized teaching plans based on students' learning behaviors and performance.

Smart campus (Valks et al., 2021) refers to the digitalization of the entire education system, such as teacher training, student administration, home-school communication, and integration of community resources. More advanced technologies like virtual reality (VR) and augmented reality (AR) will enable smart campuses to offer immersive learning and boost cross-regional and cross-cultural collaboration. In addition, the smart campus also emphasizes data privacy and security to ensure that student and teacher data is well-protected.

Smart classrooms, smart schools, and smart campuses together drive the innovation of digital teaching, making learning more flexible, interactive, and individualized, and offering infinite possibilities to the future learning environment. With the development of science and technology, smart education will be increasingly popular and become a solid base for developing talents with innovative strength and international competitiveness.

1.1 Definition and significance of smart learning environment

Rosmansyah et al. (2023) defined a smart learning environment as "a hybrid learning system that provides learners and other stakeholders with a joyful learning process while achieving learning outcomes as a result of the employed intelligent tools and techniques." It is an organization that incorporates state-of-the-art information technology, big data mining, artificial intelligence, and other technologies to provide students with a secure, interactive, and personalized study environment. In addition to classical class premises, such a learning space involves digital hardware and software equipment, networks, educational websites, intelligent operating systems, etc., and their purpose is to speed up the development of individual learning, collaborative learning, innovative ideas, etc. The smart learning environment emphasizes student-centeredness, providing individualized learning resources and support based on individual interests, learning habits, and differences in ability to improve learning effectiveness and efficiency.

Digital Twins in the Smart Classroom. https://doi.org/10.1016/B978-0-443-41632-3.00004-7

The smart learning environment is the direction of future education. With technology empowerment, it constructs a more flexible, diversified, and inclusive learning system and offers a robust foundation for future talent cultivation with innovative thinking and problem-solving capacities. Since learning environments continually change, the learning process itself is sure to undergo different levels of change (Cheung et al., 2021).

1.2 Transition from traditional classrooms to digital spaces

The education system has undergone a radical transformation, from the traditional model of learning through face-to-face interaction to a learning environment centered on digital space. This transformation is not only driven by technological advancement but also by transformations in social needs and diversification of learning processes.

The teacher-centered traditional classroom is textbook, blackboard, and oral lecture-based. The students are primarily taught in the classroom. The learning materials are relatively limited, and the learning mode is relatively passive. Although this model has certain organizational and controllable roles, it is also limited by space and time, and it is difficult to offer individualized support for different students, especially in remote or underdeveloped areas (Alzahrani and Alhalafawy, 2023).

With the rapid growth of information technology, cyberspace provides more flexible and varied platforms and resources for learning. With virtual courses, electronic teaching materials, and web-based interactive platforms, learners can design their own learning time and space, breaking away from the limitations of the physical classroom. Digital learning can also be blended with technology such as VR and AR (Al-Ansi et al., 2023) to enhance the learning process and render abstract concepts concrete, and improve learning performance.

1.3 Key components, advantages, challenges, and future trends of smart classrooms

A smart classroom is a teaching environment that combines high technology and emerging learning principles, and its functions are to improve the quality of education and learners' participation. The following is an overview of the key components, benefits, drawbacks, and opportunities of smart classrooms:

(a) Key components: Interactive tools (Conceicao, 2022) such as electronic whiteboards, projectors, and interactive displays to facilitate interaction between teachers and students. Learning management system (LMS) provides a platform to facilitate assignment management, the sharing of content, and assessment. Tablets, smartphones, and laptops are shared devices for students and teachers to utilize learning materials. Personalized instruction and effectiveness measurement by capturing and analyzing learning data.

(b) Advantages: Interactive and multimedia resources help to improve student understanding and retention. Students can learn at their own pace and on their own topics of interest. It is wise to utilize technology to maximize student engagement, increase interactivity, and promote collaborative learning. They provide personalized support and resources according to learning data.

(c) Challenges: Special technological skills need to be in place for both teachers and learners. The installation of equipment and technologies employed in smart classrooms requires a capital outlay. The collection of students' information and its utilization must be subject to compliance with existing legal requirements aimed at ensuring confidentiality. Differences at the school and locality levels of access to technology hardware can lead to learning disparities.

(d) Future trends: Adoption of AR and VR is a more authentic learning space, with an immersive experience. Smart support and suggestions based on students' learning behavior and needs. Interdisciplinary instruction emphasizes combined knowledge and skills, and supports incorporating science, technology, engineering, mathematics (STEM), and other disciplines. Smart classroom planning and administration will come to be more concerned with green issues and the efficient use of resources. In the near future, digital twin technology will be a bridge between the real and virtual classrooms. Classroom digital twin is also the mirror that reflects the real classroom into the virtual classroom.

2. Understanding smart classrooms

A smart classroom is an intelligent learning space that uses cutting-edge technology to support the learning process. It incorporates varied digital materials and interactive technologies (Heeter, 2023) to provide an efficient, adaptable, and interactive learning experience. But frequently, network connectivity disruptions can impinge directly on the smart classroom's efficacy and ruin the smart class experience of the teacher and the students (Ren, 2024).

2.1 Definition and main features

Smart classroom has been categorized by Saini and Goel (2019) as "a technology-assisted closed environment that enhances teaching and learning experience" (p.2). It is a highly smart, digitalized, and interactive class environment built through collaboration with advanced technologies, such as the most modern information technology, the

Internet, big data, and the IoT. It is oriented toward improving the efficiency of teaching, promoting customized learning (Pande et al., 2024), and achieving educational innovation. The key important features are: (a) information equipment integration, (b) data-driven management, (c) high interactivity, (d) resource sharing, (e) intelligent management, (f) personalized learning, and (g) environmental perception and adaptation.

2.2 Challenges and considerations

Although smart classrooms have transformed instruction, they still face numerous issues regarding technology, administration, and society in the very process of their promotion. Following is the close examination of some of the major issues and solutions:

(a) Infrastructure high cost: High-tech tools such as VR or smart boards can cost over $10,000, making them hard to adopt in rural schools. A phased rollout, prioritizing cost-effective tools and seeking government or corporate support, like Google's education programs, can help.
(b) System compatibility and maintenance: Frequent software updates and limited device compatibility challenge teachers. Choosing an open-standards platform ensures broader access, while setting up an in-school IT team helps with quick support and smoother adaptation.
(c) Network stability requirements: Real-time teaching needs fast Internet, often lacking in remote areas. Using offline modes with preloaded content and backup options like 5G or satellite networks (e.g., Starlink) can ensure continued access.
(d) Teacher technology anxiety and training shortage: Older teachers may hesitate to use new tech, with only 30% feeling confident. Gradual training and a peer mentoring system like "Teacher Technology Ambassadors" can build skills and reduce isolation.
(e) Difficulty in converting teaching models: Shifting from lectures to interactive teaching is challenging and adds to teachers' workload. Providing smart lesson templates and starting with simple tools like digital Q&A or polling can ease the transition.
(f) Distraction risk of the students: Multiscreen setups can distract students, reducing focus by up to 20%. Solutions include locking devices in "Classroom Mode," setting clear digital rules, and using tools like AI attention tracking to maintain engagement.
(g) Data privacy and information security risks: Student data, such as responses and facial cues, can be misused. To protect privacy, schools should follow laws like GDPR, choose ISO 27001-certified systems, limit data collection, and ensure parental consent
(h) It increases inequality: Many students from low-income homes lack devices or Internet, causing learning gaps for 37% globally. Solutions could include device loan programs, community digital centers, and low-tech tools such as SMS homework alerts.
(i) Health problems: Extended screen use can strain eyes, and wireless devices raise health concerns among parents. Promoting regular eye breaks and choosing certified, low blue-light devices can help ease discomfort and build trust.

3. Expanding to smart schools

Expanding to smart schools can highly transform the education sector, fostering innovation and enhancing the engagement of students. The principals should turn the schools into smart schools as soon as possible. By properly planning and executing technology-based initiatives, schools can create a more productive and inclusive learning environment. In fact, advanced technologies play an important role in pedagogical transformation for schools to achieve sustainable development (Makinde et al., 2024).

3.1 Concepts and infrastructure

(a) Essential concepts: Smart school is a future-ready learning space that infuses digital technology within the education infrastructure. It has the following three defining characteristics.
  (i) Data-informed decision making: Learning analytics for optimizing teaching styles.
  (ii) Smooth digital integration: Physical and virtual learning space integration is very high.
  (iii) Energy- and resource-saving smart management: Artificial intelligence of Things (AIoT) enables energy and resource optimization.
(b) Infrastructure: There is a multilayered technology architecture (Damodaram et al., 2022) in smart schools.
  (i) Perception layer: Environmental sensors are used for temperature, humidity, light, and air quality monitoring. Biometric devices are used for facial recognition, access control, or emotion recognition cameras. Interactive devices are used for touching a whiteboard, an AR marker, or a smart desk.
  (ii) Network layer: 5G private network and Wi-Fi 6E multiband coverage ensure advanced connectivity solutions that are tailored for differing needs. Edge computing nodes (Corneo et al., 2021) to reduce latency, improve performance, and maximize the effectiveness of data processing. Software Defined Network (SDN) Traffic Management to improve flexibility, efficacy, and ease of use (Chiang & Liu, 2025).

(iii) Platform layer: Shared data center for LMS, IoT, and administration system consolidation. Hybrid cloud architecture uses local storage of sensitive data and elastic scaling of the public cloud. Microservice application programming interface (Perrotta et al., 2021) is used for fast access to third-party educational apps.

(iv) Application layer: Intelligent teaching system for intelligent learning platforms. School management system for planning AI classes and power scheduling. The security monitoring center provides real-time notice of unusual behavior.

3.2 Key technologies of smart schools

The operation of smart schools relies on multi-level technology integration, which is divided into six basic technology areas below, and delineates their application scenarios and representative solutions in education:

(a) ***IoT (IoT) and sensing technology:*** Environmental monitoring (temperature, humidity, air quality, and noise), equipment status tracking (projector usage hours or book borrowing records), and personnel dynamics analysis (classroom flow heat map).

  (i) Low Power Wide Area Network (LPWAN): Remote water and electricity meter reading.

  (ii) Edge AI: Cameras monitor student behavior in real time, such as fall detection.

  (iii) RFID/NFC: Contactless attendance, smart textbook management.

(b) ***Data analysis and artificial intelligence:*** Prediction of learning effectiveness, resource allocation optimization, and abnormal event alert.

  (i) Learning analytics: Prediction of student dropout risk based on the XGBoost algorithm.

  (ii) Computer vision: Analyze class participation, such as hand raising, eye tracking.

  (iii) Natural language processing: Autogenerate teaching summaries (GPT-4 Education Edition).

(c) ***Extended reality (XR) educational technology:*** It combines features of VR, AR, and traditional environments to create interactive learning spaces.

  (i) VR: Meta Quest 3's simulated experiment (simulated chemical reaction).

  (ii) AR: Microsoft HoloLens 2's mechanical disassembly guide (superimposed 3D model entity).

  (iii) MR: Magic Leap 2's remote cocollaborative anatomy course.

(d) ***Blockchain and digital credentials:*** They offer a new way to issue, authenticate, and preserve educational and professional qualifications.

  (i) Microcredentials: Document detailed skills such as Python programming, Lv3.

  (ii) Anticounterfeiting transcripts: MIT has adopted the Blockcerts standard.

  (iii) Smart contract: Automatically award qualified scholarships.

(e) ***5G and network infrastructure:*** They offer faster speeds, reduced latency, and increased connectivity for smart school network planning.

  (i) Online learning: More authentic interaction between teachers and students.

  (ii) Cloud programming: Student code gets compiled in real time without prior notice.

(f) ***Robotic process automation (RPA)***: It can greatly increase efficiency, accuracy, and productivity in various learning processes.

  (i) Admissions process: Automatically pull the registration form from the email attachment. Fill out the database and send a confirmation letter to the sender. Synchronize the financial system to generate payment slips.

  (ii) Equipment repair: Chatbot identifies first the type of fault, then automatically sends work orders, and AI acceptance once repaired.

3.2.1 Future technology evolution

(i) Educational digital twin: Policy impact simulation on campus virtually, as if the impact of a new class schedule on student traffic.

(ii) Brain-computer interface (BCI): Use the electroencephalography (EEG) headband to track learning attention and dynamically adjust teaching content.

(iii) Generative AI teaching assistant: Answer student questions 24 h a day and 7 days a week, such as a ChatGPT education fine-tuning model.

3.3 Advantages of smart schools

Smart schools revolutionize the learning environment (Ullah et al., 2021) through the incorporation of technology. Their core advantages can be analyzed as follows:

(a) Personalized deployment of learning: Computer programs inspect students' patterns of responses and dynamically adjust the difficulty of the questions, similar to the Knewton platform (Hakkal & Ait Lahcen, 2021), reducing learning time by half. Biometric sensing, such as eye tracking, recognizes comprehension barriers in real-time.

(b) Revolution of educational content: AR textbooks make cell structures "three-dimensional and broken down," with intricate ideas simple to understand. Global high-level course sharing, such as MIT OpenCourseWare (Elias, 2022) fed directly into the classroom.

(c) Teacher empowerment: AI teaching assistants grade objective questions automatically, freeing up one-third of lesson preparation time (Gradescope test data). Heat map of teaching identifies the pattern of students' attention and adapts the rhythm accordingly.

(d) School automation: Robotic process automation (RPA) robots automate repetitive administrative work, such as the accuracy of conflict detection in the class schedule system is 99.7% (Lalitha et al., 2023). With electronic signature and blockchain storage of evidence, the procurement process is cut from 5 days to 2 hours.

(e) Best resource allocation: Pseudocode is applied to optimize the use of classrooms. When the sensor senses an unoccupied classroom and the timetable indicates no classes, it will automatically shut down air conditioning and lighting. It has been shown to save 35% of energy expenses.

(f) Risk early warning system: Aberrant student activity (extended absence or unanticipated grade deterioration) triggers the tutoring system. Structural building sensors project equipment aging 3 months before time.

(g) Immersive interaction: VR replicates past battles like BBC VR Normandy Landings, which is more effective than static textbook illustrations. Touch the whiteboard with a 3D function surface interaction to enhance team learning.

(h) Endless collaboration: The international project team collaborates to design solar cars in the Metaverse classroom, such as the ENGAGE XR platform. Rural schools connect via 5G with premier urban teachers in real time.

(i) Promoting equal rights in education: Visually impaired students learn geometry with tactile feedback diagrams (Zhao, 2023). Rural schools bypass geographical barriers with mobile smart school bus classrooms.

(j) Demonstration of sustainable development: Paperless operation reduces 12 tons of carbon dioxide annually (the same as 300 trees absorbing carbon). Solar power and an AI-powered grid raise campus energy self-sufficiency rate to 60%.

3.4 Implementation barriers

The formation of smart schools comprises technological, managerial, and cultural transformation, among others. In the promotion process itself, several significant obstacles typically arise, each requiring concrete solutions.

(a) Financial and cost barriers: Upgrading one smart classroom is estimated to be in the range of $20,000 to $50,000 (interactive whiteboard, IoT sensors, AI system) (Fernández-Batanero et al., 2021). For a university-scale infrastructure private 5G network or data center, the budget will run into millions of dollars. Hardware needs to be replaced every 3–5 years, and annual software license fees are 15%–30% of total costs. It can prioritize the construction of demonstration classrooms and gradually expand after verifying the benefits. It can use government funding or corporate CSR sponsorship in exchange for technology verification sites, such as the Microsoft Education Experimental School.

(b) Integrating technology complexity: The IoT devices from different vendors may not share a common protocol, which will cause issues with data disconnection from teaching platforms such as Google Classroom to school management systems like PowerSchool. Legacy analog systems, such as some aging broadcast systems, will be very difficult to digitize. It can make compulsory demands on newly acquired equipment to support European Development Finance Institutions (EDFI) data exchange standards. It can introduce an API gateway layer like Kong Gateway to bring interoperability between the heterogeneous systems.

(c) Resistance of organizational culture: 62% of educators discover that technology increases their workload. Elder teachers resist new tools. They may hurt mental health without proper training (Fernández-Batanero et al., 2021). Administrative staff can also resist tool change. It can establish a "Teacher Technology Committee" to allow users to participate in tool selection. It can establish a "tech sandbox" to allow teachers to try new methods in a safe arena. Insert application of technology as another item under the assessment of teachers, or come up with the "Digital Teaching Innovation Award" and provide incentives.

(d) Security and privacy risks: Student biometric data (such as facial recognition) is strictly controlled by the General Data Protection Regulation (GDPR) and California Consumer Privacy Act (CCPA). IoT devices become hacker launching pads, such as compromised surveillance cameras. Many parents oppose classroom monitoring of emotions. Sensitive data (such as faces) is only processed on edge devices, and the raw data is never transmitted (Figueroa et al., 2024). It might host a parent open day to demonstrate how data encryption functions.

(e) The digital divide widens: The low-income students lack home terminals and broadband. They may lack digital literacy among low-income students and widen the learning gap. It could provide a 1:1 device lending scheme and a free 4G hotspot. It could have low-tech interfaces, such as voice-controlled operating systems. It could preserve old paper options and phase out the lead digital transition.

4. Scaling up: Smart campuses in higher education

People's work habits, lifestyles, and learning methods have all changed significantly in tandem with technological advancements. Education reform and development have been driven by the slow transformation of the learning environment and the growing need for individualized and flexible learning. The smart campus emerged as the upscale version of a smart education system and has gained increasing global interest. A smart campus is an essential component of the smart city framework, as it provides residents with a smart learning environment that helps them become smart workers. The knowledge economy is also supported by the growth and popularity of smart campuses. Between 2018 and 2022, the worldwide smart education market is expected to expand at a compound yearly growth rate of 15.96% (Arrais et al., 2021). There is an urgent need to conduct active research and get a comprehensive grasp of the smart campus and its characteristics in such a rapidly evolving field.

4.1 Defining smart campuses

A smart campus is an educational environment enriched with advanced technologies to enhance learning while meeting stakeholder needs. In order to build an ecosystem that combines digital and physical spaces, smart campuses engage with stakeholders through information and communication technologies (ICT). In order to develop a productive, innovative, and sustainable environment, it implements responsive, intelligent, and enhanced services (Min-Allah & Alrashed, 2020; Villegas-Ch et al., 2020).

Universities' digital transformation through ICT, digital management, and online learning grew in 2020 due to the COVID-19 pandemic and other constraints (Ramírez-Montoya, 2020). According to Durán-Sánchez et al. (2018) and others, smart campuses provide technology while also attempting to enhance the liveability and quality of life of their stakeholders. There is currently disagreement around the definition, scope, and attributes of a smart campus, despite the concept being initially put forth in 2000 (Chiu et al., 2020; Min-Allah & Alrashed, 2020; Prandi et al., 2019).

Based on the aim of sustainability, Silva-da-Nóbrega et al. (2022) suggested a framework for a smart campus with aspects associated with the SDGs, namely economics, education, environment, living, management, mobility, security, and technology, which is a transversal one, i.e., present in all dimensions (Fig. 27.1).

This model suggests variables for each smart dimension in accordance with the SDGs and a smart city perspective (Silva-da-Nóbrega et al., 2022). The university implements or enhances the smartization process as it becomes more sustainable and meets the SDGs. It can serve as a monitor for both viewpoints, choosing which dimension the university's operations relate to.

4.1.1 *Digital ecosystems around the university*

DE is a collaborative concept referring to networks of interconnected organizations that collectively produce value by exploiting digital technology (Jacobides et al., 2018; Tan et al., 2020). DEs have been well-studied in management research and garnered attention in the newest DT study (Hanelt et al., 2021; Plekhanov et al., 2022; Tan et al., 2020).

Venkatraman (1994) established the conceptual foundations of DE research by arguing that IT and the external environment are becoming more and more important to businesses, boosting productivity both inside and outside of them, and creating new value propositions and business network connections. First, scholars have persisted in describing organizational theories through the use of ecosystem analogies (Bogers et al., 2019).

Second, without taking into account a specific value proposition, researchers referred to ecosystems as an association of organizations (for example, industrial ecosystems like the food or smart living ecosystem). Third, the literature on (platform) ecosystems has changed as a result of the growth of digital platforms (Hein et al., 2020). What a DE is and how it differs from other forms of interorganizational collaboration was clarified by the groundbreaking work of Jacobides et al. (2018), Kapoor (2018), and others. The modularity and complementarity of resources are now the main areas of DE study (Shipilov & Gawer, 2020). Member firms want to optimize their portion of value capture, even while they all share the objective of maximizing total value generation in a DE (Teece, 2018).

4.1.2 *Integration of smart infrastructure (energy management, AI-based campus services)*

Khan et al. (2025) highlight the role of AI and machine learning (ML) in optimizing various aspects of smart infrastructure, including predictive maintenance, energy optimization, fault detection, and smart traffic management. AI models such as deep learning, reinforcement learning, and neural networks improve infrastructure monitoring and power system efficiency (Ahmad et al., 2022). The use of IoT sensors and AI-based predictive modeling enhances real-time performance analysis, making infrastructure management more efficient. AI-controlled systems outperform traditional methods in structural monitoring, load forecasting, and grid management, resulting in lower costs, increased safety, and

DIMENSION	CODE	VARIABLE
Smart Economy	EC-01	Electronic Services
	EC-02	Entrepreneur and Innovation Support
	EC-03	Collaborative Economy
	EC-04	Local Development
	EC-05	Employability
Smart Education	ED-01	Smart Education Technology
	ED-02	Internet Access
	ED-03	Community Needs Awareness
	ED-04	Satisfaction
	ED-05	Results-Based Learning
Smart Environment	EN-01	SDGs
	EN-02	Biodiversity
	EN-03	Eco-Friendly Resources
	EN-04	Smart Buildings
	EN-05	Recycling
Smart Living	LI-01	Quality of Life
	LI-02	Health
	LI-03	Social Inclusion
	LI-04	Leisure
	LI-05	Extracurricular Activities
Smart Management	MA-01	Sustainable Management
	MA-02	Transparency
	MA-03	Participation
	MA-04	Process Efficiency
Smart Mobility	MO-01	Accessibility
	MO-02	Signaling
	MO-03	Sustainable Transport
	MO-04	Collaborative Economy
	MO-05	Eco-Friendly Transport
Smart Security	SE-01	Security
	SE-02	Biosecurity
	SE-03	Smart Technologies
	SE-04	Cybersecurity
	SE-05	Disaster Prevention
Smart Technology	TE-01	Internet Technology
	TE-02	Data Management
	TE-03	Information Systems
	TE-04	Internet Technologies

FIGURE 27.1 Smart campus framework (Silva-da-Nóbrega et al., 2022).

efficient energy use (Ahmad et al., 2022). Hence, this encourages effective maintenance of smart campus environments.

But, despite all the benefits of smart infrastructure, it does not go without challenges and risks. Part of the challenges are high computational demands, cyber threats, and data dependency (Cheung et al., 2021). Also, real-world implementation obstacles such as scalability, ethical concerns, and regulatory issues are not left out. According to Khan et al. (2025), hybrid AI models and security solutions, such as blockchain-based security and XAI explainability, could be the remedy.

4.2 Core technologies in smart campuses

Smart campuses integrate cutting-edge technologies to enhance learning, security, and operational efficiency. Here are some of the core technologies driving this transformation.

4.2.1 Big data and predictive analytics for student success

As institutions of higher education collect more and more data about their students, and as students' record databases have grown more complex and accessible, we are entering a new era of using data to improve student success, streamline processes, and more effectively utilize resources (Batool et al., 2023). Higher Education needs to record all academic-related data from various activities, such as students' data, registration data, assessment data, etc. Due to the advancement of information technology, social interactions have increasingly moved online. These online interactions can be traced and collected, and will form a huge amount of data and be considered as big data. Thus, data collection and analysis have become a challenging task in higher education sectors (Ziezo et al., 2021). Big data analytics refers to various techniques to analyze data, extract information, and gain insights from large-scale datasets with complex patterns because conventional data-processing views cannot be easily dealt with.

Data having a huge number of samples have higher statistical power, as overcomplicated data with a great dimensional feature space might lead to a greater rate of false discovery (Jamarani et al., 2024). Data capture, data storage, search, visualization, transfer, sharing, query, update, data analysis, information privacy, and data source are the most important challenges in big data (Jamarani et al., 2024). The recent application of the term big data seems to refer to using user behavior analytics, predictive analytics, or different advanced data analytics methods that obtain value from data, and rarely to a data set of a special size (Shah et al., 2021).

This approach will help in predicting a dropout student, a student who needs additional help, or even a student who needs a more challenging assignment. With the help of big data, it is possible for the teacher to formulate formative assessments aptly challenging or demanding according to each student's talent and learning ability.

4.2.2 Smart transportation and IoT-enabled facilities

The rapid pace of urbanization presents a complex and pressing global challenge that demands integrated and innovative solutions. Urban populations have grown significantly as more individuals migrate to cities in search of better opportunities. According to the United Nations projections, nearly 4.9 billion people are expected to live in urban areas by 2030. This surge brings with it a host of concerns, including increased pollution, traffic congestion, and strain on urban resources (Balamurugan et al., 2025).

In parallel with these demographic shifts, technological advancements, particularly in the IoT, have led to an explosion in the number of interconnected devices. These devices continuously gather and transmit data to be processed and analyzed by computing systems. With recent progress in deep learning, this data is now being used in increasingly sophisticated ways to drive automation and intelligent decision-making.

Modern education and e-learning systems cannot disregard the role of IoT (Makinde et al., 2025). Introduced the IoT model that connects actual physical things while researching how technology affects e-learning. They found that the adoption of IoT-enabled ML techniques can improve the e-learning system. In their study (Bayani et al., 2017), they conducted the theoretical and statistical analysis. Their findings demonstrate that the IoT's influence on the current educational system cannot be disregarded. Due to a variety of causes, digital learning and e-learning have emerged as a necessity. It gives you flexibility and chances to learn from anywhere at any time.

4.3 Benefits for universities and students

4.3.1 Sustainable and cost-effective campus operations

Many campuses are loosely grouped in a central area or consolidated into compact complexes. The composition of universities is changing, with more and more institutions consisting of numerous campuses located throughout a city or region, which are linked to form a decentralized cluster network, often referred to as a distributed campus. The isolation of university campuses from urban surroundings—and the resulting lack of diversity—once seen as an advantage, has become a disadvantage. Suburban campuses that are isolated and lack cultural and social diversity are disadvantaged compared to their urban counterparts. However, this disadvantage can be addressed by improving the quality of public transportation, enhancing the living spaces, offering community and commercial services, and making structural changes to the campus. The guidelines of sustainable campus design are an excellent and innovative model of sustainable development in urban areas. They are not limited to the physical appearance of a higher education institution but can also serve as a model for sustainable development in any urban setting.

On the other hand, encouraging users to adjust their habits contributes to sustainability by lowering overall energy demand. However, budget limitations often pose challenges for institutions aiming to implement and maintain sustainability initiatives. One effective solution is to allow these initiatives to generate their own funding. This can be achieved through energy-saving measures that lower operational costs (Amaral, 2023).

4.3.2 Enhanced research capabilities through virtual laboratory collaboration

A virtual laboratory is a sophisticated program that enables experiments not feasible in a real laboratory environment. These virtual laboratories allow students to learn remotely at any time. Additionally, as mentioned (Poo et al., 2023), this software reduces the costs of real laboratory resources and minimizes the risk of negative consequences associated with failed experiments or the improper use of materials, reagents, or instruments during training. In Fig. 27.2, each block reflects the main components of a virtual educational laboratory, such as the user interface, workspace, modules, and tools for experiments, display of results, as well as additional functions such as saving and settings. A virtual laboratory can be defined as a computer-generated environment that allows users to engage in realistic experiments. This technology has numerous applications, particularly in the training and education of professionals (Frøland et al., 2020).

As noted by Kennepohl (2023), an optimal approach to teaching laboratory procedures combines didactic methods, including theoretical instruction, the use of virtual technologies, and subsequent reinforcement of acquired competencies and skills in authentic laboratory settings. This combined approach allows students to become familiar with theoretical material, procedures, and the specifics of conducting experiments on complex equipment in advance (Romanenko et al., 2025).

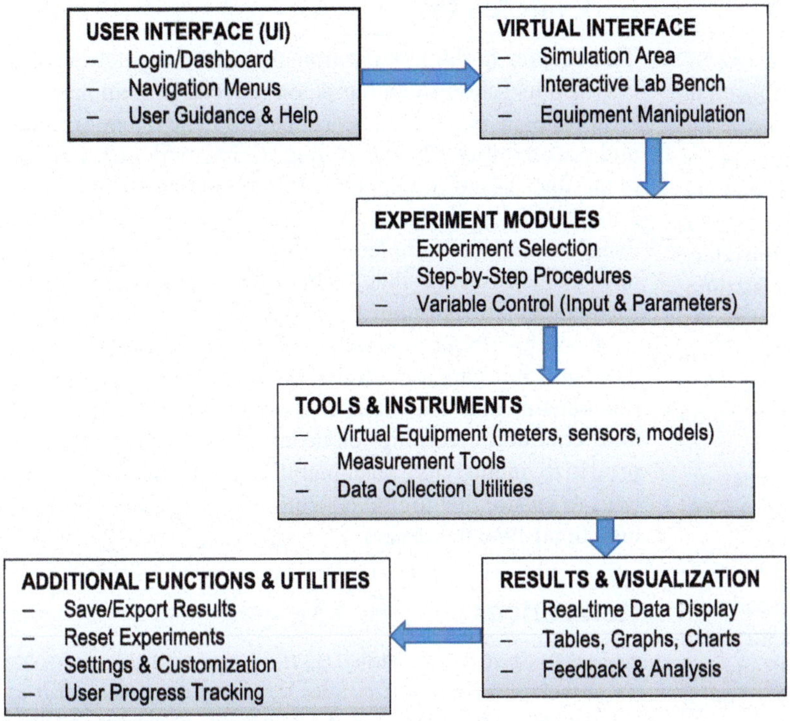

FIGURE 27.2 Components of a virtual laboratory system..

4.4 Challenges in deployment

One of the major technical challenges in developing smart campuses lies in ensuring that a wide variety of systems and technologies can work together effectively—a concept known as interoperability. This involves managing a complex ecosystem of devices, platforms, and applications, and requires deep knowledge of communication protocols and technical standards to ensure seamless integration (Hazra et al., 2021). Numerous scholars have raised concerns about the absence of universally adopted open standards, noting that interoperability is a recurring theme across multiple strands of the literature. In some cases, it is even met with skepticism; for example, Hazra et al. (2021) suggest that the belief in achieving full interoperability through a single universal standard may be overly optimistic. Despite these concerns, ongoing research continues to focus on developing standardized frameworks for IoT systems to help resolve these integration issues (Albouq et al., 2022).

University campuses occupy a unique position in terms of scale; they are large enough to generate meaningful insights, but still manageable in terms of implementation. Exploring innovative approaches to teaching and learning, along with the development of corresponding infrastructure, represents a rich area of academic inquiry. Nevertheless, only a limited number of studies addressing these pedagogical innovations appeared in the literature reviewed, suggesting that there is relatively little crossover between research on smart campuses and research focused on educational methods. Much of the existing scholarship expresses a strong interest in flexible, technology-enabled learning models—such as mobile, online, and ubiquitous learning that extend beyond the boundaries of traditional classrooms. While these approaches offer clear advantages, some studies also report drawbacks, such as increased dropout rates (Sitar-Tăut et al., 2024).

Smart governance allows individuals to take part in decision-making processes and engage with public services (Bastos et al., 2022). On university campuses, both faculty and students often express interest in accessing data from smart campus systems for learning and research. However, as noted by Chagnon-Lessard et al. (2021), obtaining such data through official university channels can be challenging. In some cases, administrative bodies may be reluctant to support these efforts, either viewing them as unproductive or raising concerns over issues such as data privacy and institutional competitiveness. The idea of open data is sometimes mistakenly seen as conflicting with privacy protections, when in fact, striking a balance between data accessibility and privacy is a key challenge that must be addressed in smart decision-making frameworks.

5. Future trends and innovations in smart learning environments

The Role of 5G and Edge Computing in Education: The combination of 5G technology and edge computing is expected to bring significant changes to the education sector. With faster internet speeds and lower delays, schools and universities can offer smoother online learning experiences, run virtual classes more efficiently, and introduce interactive tools such as AR and VR into the classroom. Edge computing adds to this by handling data processing closer to where it's generated, which eases pressure on cloud systems. This setup allows for quicker responses in digital classrooms, more effective personalized learning tools, and better data security by limiting how much information needs to travel long distances (Al-Ansi et al., 2023; Cheung et al., 2021; Mahdi, 2023; Jacob and Centofanti, 2024).

Blockchain for Secure Credentialing and Academic Records: Blockchain technology is transforming academic credentialing by providing tamper-proof, decentralized records. Traditional methods of storing and verifying academic credentials are prone to fraud and inefficiencies, but blockchain ensures secure, transparent, and immutable documentation of student achievements. Educational institutions can issue digital certificates stored on a blockchain, allowing students and employers to verify credentials instantly without intermediaries. This technology also supports lifelong learning records, enabling individuals to maintain a comprehensive portfolio of their skills and qualifications (Makinde et al., 2024; Ullah et al., 2021).

Ethical AI and Bias Mitigation in Smart Learning Environments: As AI becomes more integrated into education, ensuring ethical AI practices and mitigating bias is crucial. AI-driven learning platforms must be designed to promote fairness, transparency, and inclusivity. Bias in AI can stem from skewed datasets or unconscious developer prejudices, leading to disparities in educational outcomes. Strategies for bias mitigation include using diverse datasets, implementing algorithmic fairness checks, and fostering interdisciplinary collaboration to ensure AI systems align with ethical standards. Policymakers and educators must establish guidelines to regulate AI applications in education, ensuring they enhance learning without reinforcing inequalities (Ren, 2024; Sovacool et al., 2021; Valks et al., 2021). These innovations are shaping the future of education, making learning environments more connected, secure, and equitable.

Digital twin computing in Smart Learning Environments: Digital twin computing could not only reduce the disparity between real and virtual classrooms but also generate more varied new datasets.

6. Conclusion

The chapter highlights the transformative impact of smart classrooms/schools/campuses on education, emphasizing their role in enhancing learning efficiency, engagement, and accessibility. By integrating digital twin, AI, IoT, big data, and cloud computing, these environments create personalized and interactive learning experiences. However, successful implementation requires a balanced approach, ensuring that technology complements pedagogical principles rather than replacing them.

Looking ahead, the future of smart education ecosystems will be shaped by digital twins, 5G, edge computing, blockchain, and ethical AI, fostering secure, immersive, and equitable learning environments. Institutions must prioritize inclusivity, sustainability, and ethical AI practices to ensure that digital education remains accessible and beneficial for all learners.

References

Ahmad, T., Madonski, R., Zhang, D., Huang, C., & Mujeeb, A. (2022). Data-driven probabilistic machine learning in sustainable smart energy/smart energy systems: Key developments, challenges, and future research opportunities in the context of smart grid paradigm. *Renewable and Sustainable Energy Reviews, 160*, 112–128.

Al-Ansi, A. M., Jaboob, M., Garad, A., & Al-Ansi, A. (2023). Analyzing augmented reality (AR) and virtual reality (VR) recent development in education. *Social Sciences and Humanities Open, 8*(1), Article 100532.

Albouq, S. S., Abi Sen, A. A., Almashf, N., Yamin, M., Alshanqiti, A., & Bahbouh, N. M. (2022). A survey of interoperability challenges and solutions for dealing with them in the IoT environment. *IEEE Access, 10*, 36416–36428.

Alzahrani, F. K., & Alhalafawy, W. S. (2023). Gamification for learning sustainability in the blackboard system: Motivators and obstacles from faculty members' perspectives. *Sustainability, 15*(5), 4613.

Amaral, A. R. M. (2023). *Prospects for a methodological approach toward a sustainable campus-A case study.*

Arrais, C. A., Corcioli, G., & Medina, G. D. S. (2021). The role played by public universities in mitigating the coronavirus catastrophe in Brazil: Solidarity, research and support to local governments facing the health crisis. *Frontiers in Sociology, 6*, Article 610297.

Balamurugan, R., Selvakumar, M., Abid, Z., Azahad, S., & Nidhya, M. S. (2025). Industrial sensors in smart transportation system. In *Driving green transportation system through artificial intelligence and automation: Approaches, technologies and applications* (pp. 91–104). Cham: Springer Nature Switzerland.

Bastos, D., Fernández-Caballero, A., Pereira, A., & Rocha, N. P. (2022). Smart city applications to promote citizen participation in city management and governance: A systematic review. *Informatics, 9*(4), 89 (MDPI).

Batool, S., Rashid, J., Nisar, M. W., Kim, J., Kwon, H. Y., & Hussain, A. (2023). Educational data mining to predict students' academic performance: A survey study. *Education and Information Technologies, 28*(1), 905–971.

Bayani, M; Segura, A; Saenz, J; & Mora, B. (2017). Internet of things simulation tools: Proposing educational components.

Bogers, M., Sims, J., & West, J. (2019). What is an ecosystem? Incorporating 25 years of ecosystem research. *Academy of Management Proceedings, 1*, Article 11080. https://doi.org/10.5465/AMBPP.2019.11080abstract

Chagnon-Lessard, N., Gosselin, L., Barnabé, S., Bello-Ochende, T., Fendt, S., Goers, S., ... Zhang, P. (2021). Smart campuses: Extensive review of the last decade of research and current challenges. *IEEE Access, 9*, 124200–124234.

Cheung, S. K., Kwok, L. F., Phusavat, K., & Yang, H. H. (2021). Shaping the future learning environments with smart elements: Challenges and opportunities. *International Journal of Educational Technology in Higher Education, 18*, 1–9.

Chiang, H. S., & Liu, Q. Y. (2025). Traffic incident duration prediction based on deep learning methods. *Enterprise Information Systems*, Article 2448828.

Chiu, P. S., Chang, J. W., Lee, M. C., Chen, C. H., & Lee, D. S. (2020). Enabling intelligent environment by the design of emotionally aware virtual assistant: A case of smart campus. *IEEE Access, 8*, 62032–62041.

Conceicao, A. C. (2022). Dynamic and interactive tools to support teaching and learning. *Mathematical and Computational Applications, 27*(1), 1–19.

Corneo, L., Mohan, N., Zavodovski, A., Wong, W., Rohner, C., Gunningberg, P., & Kangasharju, J. (2021). (how much) can edge computing change network latency?. In *2021 IFIP networking conference* (pp. 1–9). IEEE.

Damodaram, A. K., Sreenivasa Chakravarthi, S., Venkateswara Reddy, L., & Reddy Madhavi, K. (2022). Multi-layered architecture to monitor and control the energy management in smart cities. In *Intelligent computing and applications: Proceedings of ICDIC 2020* (pp. 81–94). Singapore: Springer Nature Singapore.

Durán-Sánchez, A., Álvarez-García, J., Del Río-Rama, M. C., & Sarango-Lalangui, P. O. (2018). Analysis of the scientific literature published on smart learning. *Espacios, 39*, 14–27.

Elias, M. M. M. (2022). *Adapting OpenCourseWare based on the needs and preferences of disabled learners*. Doctoral dissertation. Universitats-und Landesbibliothek Bonn.

Fernández-Batanero, J. M., Román-Graván, P., Reyes-Rebollo, M. M., & Montenegro-Rueda, M. (2021). Impact of educational technology on teacher stress and anxiety: A literature review. *International Journal of Environmental Research and Public Health, 18*(2), 548.

Figueroa, E., Batista, E., Palau, R., Unciti, O., Ferre, M., & Ballesté, A. M. (2024). The use of artificial intelligence techniques in smart classrooms is in its infancy. *IEEE Access, 12*.

Frøland, T. H., Heldal, I., Sjøholt, G., & Ersvær, E. (2020). Games on mobiles via web or virtual reality technologies: How to support learning for biomedical laboratory science education. *Information, 11*(4), 195.

Hakkal, S., & Ait Lahcen, A. (2021). An overview of adaptive learning fee-based platforms. In , *1. In Proceedings of the 2nd International Conference on Big Data, Modelling and Machine Learning BML* (pp. 222–226).

Hanelt, A., Bohnsack, R., Marz, D., & Antunes, M. C. (2021). A systematic review of the literature on digital transformation: Insights and implications for strategy and organizational change. *Journal of Management Studies, 58*(5), 1159–1197. https://doi.org/10.1111/joms.12639

Hazra, A., Adhikari, M., Amgoth, T., & Srirama, S. N. (2021). A comprehensive survey on interoperability for IIoT: Taxonomy, standards, and future directions. *ACM Computing Surveys, 55*(1), 1–35.

Heeter, C. (2023). Implications of new interactive technologies for conceptualizing communication. In *Media use in the information age* (pp. 217–236). Routledge.

Hein, A., Schreieck, M., Riasanow, T., Setzke, D. S., Wiesche, M., Böhm, M., & Krcmar, H. (2020). Digital platform ecosys tems. *Electronic Markets, 30*(1), 87–98. https://doi.org/10.1007/s12525-019-00377-4

Jacobides, M. G., Cennamo, C., & Gawer, A. (2018). Towards a theory of ecosystems. *Strategic Management Journal, 39*(8), 2255–2276. https://doi.org/10.1002/smj.2904

Jacob, T., & Centofanti, S. (2024). Effectiveness of H5P in improving student learning outcomes in an online tertiary education setting. *Journal of Computing in Higher Education, 36*(2), 469–485.

Jamarani, A., Haddadi, S., Sarvizadeh, R., Haghi Kashani, M., Akbari, M., & Moradi, S. (2024). Big data and predictive analytics: A systematic review of applications. *Artificial Intelligence Review, 57*(7), 176.

Kapoor, R. (2018). Ecosystems: Broadening the locus of value creation. *Journal of Organization Design, 7*, 1–16. https://doi.org/10.1186/s41469-018-0035-4

Kennepohl, D. (Ed.). (2023). *Teaching science online: Practical guidance for effective instruction and lab work*. Taylor and Francis.

Khan, M. I., Ahmad, B., Shoaib, M., Anwar, S., & Ali, Z. (2025). Artificial intelligence and machine learning applications in smart infrastructure and electrical systems. *The Critical Review of Social Sciences Studies, 3*(1), 2005–2016. https://thecrsss.com/index.php/Journal/about.

Lalitha, V., Magesh, K., Selvanarayanan, A., & Youkesh, G. (2023). Automating timetable generation with conflict resolution algorithms in web-based systems for educational institutions. In *2023 intelligent computing and control for engineering and business systems (ICCEBS)* (pp. 1–4). IEEE.

Mahdi, M. F. (2023). The role of IT in transforming traditional education to digital learning. *International Journal of Computational & Electronic Aspects in Engineering, 4*(4), 134–147.

Makinde, S. O., Ajani, Y. A., & Abdulrahman, M. R. (2024). Smart learning as transformative impact of technology: A paradigm for accomplishing sustainable development goals (SDGs) in education. *Indonesian Journal of Educational Research and Technology, 4*(3), 213–224.

Makinde, S. O., Ibrahim, R. B., Ibrahim, T., & Ganiyu, R. S. (2025). Awareness of internet of things (IoT) applications for learning enrichment among undergraduate students. *ASEAN Journal of Educational Research and Technology, 4*(2), 161–170. https://ejournal.bumipublikasinusantara.id/index.php.

Min-Allah, N., & Alrashed, S. (2020). Smart campus—A sketch. *Sustainable Cities and Society, 59*, Article 102231.

Pande, L., Pandey, U., & Sengupta, S. (2024). *Big data to promote educational equity and customized learning: A theory-based approach.*

Perrotta, C., Gulson, K. N., Williamson, B., & Witzenberger, K. (2021). Automation, APIs and the distributed labour of platform pedagogies in google classroom. *Critical Studies in Education, 62*(1), 97–113.

Plekhanov, D., Franke, H., & Netland, T. H. (2022). Digital transformation: A review and research agenda. *European Management Journal, 1–24*, 821–844. https://doi.org/10.1016/j.emj.2022.09.007

Poo, M. C. P., Lau, Y. Y., & Chen, Q. (2023). Are virtual laboratories and remote laboratories enhancing the quality of sustainability education? *Education Sciences, 13*(11), 1110.

Prandi, C., Monti, L., Ceccarini, C., & Salomoni, P. (2019). Smart campus: Fostering the community awareness through an intelligent environment. *Mobile Networks and Applications, 25*, 945–952.

Ramírez-Montoya, M. S. (2020). Transformación Digital e Innovación Educativa En Latinoamérica En El Marco Del COVID-19. *Campus Virtuales, 9*, 123–139.

Ren, F. (2024). The construction and application of smart classrooms in China: A literature review based on 93 studies. *Science Insights Education Frontiers, 23*(1), 3705–3724.

Romanenko, T., Sergienko, V., Avramenko, O., Malezhyk, M., & Mosiienko, H. (2025). Impact of virtual laboratories on the development of research competence in students of technical specialties. *WSEAS Transactions on Computer Research, 13*, 65–76.

Rosmansyah, Y., Putro, B. L., Putri, A., Utomo, N. B., & Suhardi. (2023). A simple model of smart learning environment. *Interactive Learning Environments, 31*(9), 5831–5852.

Saini, M. K., & Goel, N. (2019). How smart are smart classrooms? A review of smart classroom technologies. *ACM Computing Surveys, 52*(6), 1–28.

Shah, S. I. H., Peristeras, V., & Magnisalis, I. (2021). Government big data ecosystem: Definitions, types of data, actors, and roles and the impact in public administrations. *ACM Journal of Data and Information Quality, 13*(2), 1–25.

Shipilov, A., & Gawer, A. (2020). Integrating research on interorganizational networks and ecosystems. *Academy of Management Annals, 14*, 92–121. https://doi.org/10.5465/annals.2018.0121

Silva-da-Nóbrega, P. I., Chim-Miki, A. F., & Castillo-Palacio, M. (2022). A smart campus framework: Challenges and opportunities for education based on the sustainable development goals. *Sustainability, 14*(9640). https://doi.org/10.3390/su14159640

Sitar-Tăut, D. A., Mican, D., & Moisescu, O. I. (2024). To be (online) or not to be? The antecedents of online study propensity and e-learning-dependent dropout intention in higher education. *Technological Forecasting and Social Change, 207*, Article 123566.

Sovacool, B. K., Hook, A., Sareen, S., & Geels, F. W. (2021). Global sustainability, innovation and governance dynamics of national smart electricity meter transitions. *Global Environmental Change, 68*, Article 102272.

Tan, F. T. C., Ondrus, J., Tan, B., & Oh, J. (2020). Digital transformation of business ecosystems: Evidence from the Korean pop industry. *Information Systems Journal, 30*(5), 866–898. https://doi.org/10.1111/isj.12285

Teece, D. J. (2018). Profiting from innovation in the digital economy: Enabling technologies, standards, and licensing models in the wireless world. *Research Policy, 47*(8), 1367–1387. https://doi.org/10.1016/j.respol.2017.01.015

Ullah, N., Mugahed Al-Rahmi, W., Alzahrani, A. I., Alfarraj, O., & Alblehai, F. M. (2021). Blockchain technology adoption in smart learning environments. *Sustainability, 13*(4), 1801.

Valks, B., Arkesteijn, M. H., Koutamanis, A., & den Heijer, A. C. (2021). Towards a smart campus: Supporting campus decisions with internet of things applications. *Building Research and Information, 49*(1), 1–20.

Venkatraman, N. (1994). IT-enabled business transformation: From automation to business scope redefinition. *Sloan Management Review, 35*, 73.

Villegas-Ch, W., Arias-Navarrete, A., & Palacios-Pacheco, X. (2020). Proposal of an architecture for the integration of a chatbot with artificial intelligence in a smart campus for the improvement of learning. *Sustainability, 12*, 1500.

Zhao, Y. (2023). *Designing TADA: Touch-and-audio-based diagram access for people with visual impairments*. Doctoral dissertation.

Ziezo, M. M., Osakwe, J. O., Ujakpa, M. M., & Iyawa, G. E. (2021). Challenges of implementing big data technology in higher institutions. *Journal of Information Systems and Informatics, 3*(3), 376–391.

Chapter 28

Role of artificial intelligence in enhancing school digital twin

Prabhakar Krishnan[1], Amit Kumar Tyagi[2] and Shabnam Kumari[3]
[1]Department of Management, St.Joseph's University, Bengaluru, Karnataka, India; [2]National Forensic Sciences University, Gandhinagar, Gujarat, India; [3]SRM Institute of Science and Technology, Chennai, Tamil Nadu, India

1. Introduction

The global education system is undergoing dramatic change, incited by rapid technological change and changing pedagogies. The digital transformation of education argues for the inclusion of digital tools, platforms, and teaching-learning methods that have expanded within the past few years mainly due to internet availability, mobile devices, and cloud computing services (Almarashdeh, 2016; Safitry et al., 2015). As learners, teachers, and administrators have transitioned from traditional physical-first education to hybrid and fully digital education, learning institutions have changed from using only old technology to utilizing new and emerging technology as it impacts the learning environment, teaching and learning processes, and the efficient utilization of resources while ensuring inclusive education (Teräs, 2022). In addition to this transition, global events such as the COVID-19 pandemic have quickened the pace of the digital transformation of education across the world (Crawford et al., 2020).

One of the most exciting implications of this transformation is the development of DT in education. DTs were originally developed in industry contexts manufacturing, aerospace, smart cities, etc., are virtual representations of physical entities that match real-time data and allow for simulation, prediction, and decision-making (Grieves & Vickers, 2016, pp. 85–113). In medical and automotive industries, DT has offered predictive maintenance, improved optimization of processes, and reduced risks based on data models (Tao et al., 2018). Transferring this idea into the educational realm also implies a new precedent, as educational institutions are now beginning to work toward developing "smart schools," whereby data from classrooms, student interactions, infrastructure, and administrative systems are modeled to construct a virtual space that is analyzable and able to be optimized (Kritzinger et al., 2018).

School DT is an advanced evolution of standard management systems in schools, integrating real-time data from all educational components—students, teachers, infrastructure, and learning systems—into a model that can simulate experiences, predict outcomes, and shape strategic operational decision making. Education DT, unlike traditional digital platforms, leverages a feedback loop: the physical space informs the digital space for action, which signals what the physical space does next in real-time data and predictive intelligence or scenarios (Jankovskis et al., 2024). These models can represent individual students, whole classes, or entire systems for the school as an organization.

AI is the main enabler of the intelligence and adaptability of DT. AI encompasses a range of computational technologies that imitate human cognitive capacities, including learning, reasoning, and decision-making. Within the educational context of DT, a few subfields of AI are particularly salient. More specifically, ML algorithms might discover learning patterns, help predict student performance, or suggest personalized education (Li et al., 2018). NLP permits students' questions to be interpreted, be used to analyze student feedback using sentiment analysis, and may result in intelligent tutoring systems (Liu et al., 2024). CV allows the analysis of visual data, including a student's engagement through facial expressions and classroom occupancy (Sholi et al., 2024). Collectively, these subfields empower static digital systems to become dynamic, intelligent models capable of recognizing and responding to the complexities of educational contexts.

The purpose of this research is to investigate AI's potential for enhancing School DT and how they will affect

Digital Twins in the Smart Classroom. https://doi.org/10.1016/B978-0-443-41632-3.00031-X

the educational ecosystem in which they operate because of the introduction of intelligent computational models. Uses of AI-enabled DT offer numerous benefits including customized pathways for learning; optimized use of resources; predictive academic intervention; and more informed administrative decisions (Chen et al., 2020; Selim et al., 2024). AI-enabled DT also offers the potential for equitable access to educational support, beyond geographic or social-economic boundaries (Ahuja, 2023). Fig. 28.1 illustrates the interrelationship among the various AI technologies machine learning (ML), natural language processing (NLP), and computer vision (CV) operating in the context of the DT framework of a smart school. The physical world actors (students, teachers, classrooms, buildings, infrastructure) are interfaced with the DT model through real-time data mining and/or streaming from IoT devices, sensors connected to school systems, and digital platforms. The AI technology interprets the information to enable predictive analytics, personalized learning, and adaptive management. The feedback loop represents the continuous and dynamic interaction between the physical layer and the digital layer, supporting the student/adult decision-making process, empowering enhanced educational outcomes.

This research analyzes the theoretical basis and implementation of AI in School DT. It explores the architectural structure, practical functions, and evaluative impact. The research takes place within K-12 education, but the frameworks and models presented are potentially scalable to higher (postsecondary) education and competence-based learning against skills recognition. Providing direction for the research, this study will aim to answer the following research questions:

- **RQ1:** How could AI enhance the capability and intelligence of School DT?
- **RQ2:** What are the architectural and technical considerations of introducing AI into the educational DT process?
- **RQ3:** How do AI-enabled School DT enhance teaching, learning, and administration?
- **RQ4:** What are the ethical, technical, and operational challenges of implementing such systems at scale?

These questions will undergird a multidisciplinary study at the intersection of AI and education through DT technologies.

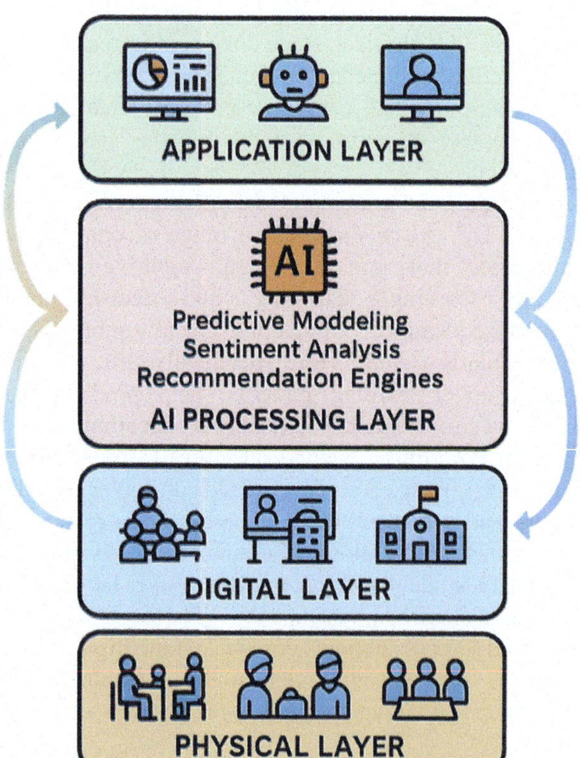

FIGURE 28.1 Scientific illustration of AI-enabled digital twin in education.

2. Related work

The emerging term DTs, initially referred to in terms of industrial and manufacturing contexts, have begun to be applied in different areas, such as healthcare, smart cities, and even in education. A DT is primarily a virtual reflection of a physical entity that takes into account the feedback from real-time data streams from sensors and networks. In education, DT can model different components, including school infrastructure to a particular student's behavior, providing potential opportunities for improved learning, planning, and workplace efficiency (Fuller et al., 2020). DT in education is distinguished by their use of continuous data transformation and feedback to dynamically imitate the states and processes of a physical entity. Their capacity for real-time data synchronization, analysis, predictive or prescriptive modeling, and simulation of strategic "what will happen if" processes allows their DTs to continually monitor and analyze performance. This synchronous relationships create an intelligent feedback ecosystem that supports automated and human decision-making. DT in education can also be categorized by type and purpose. Generally, there are three broad types of educational DT based on their focus on: facilities, student models, or administrative/organizational focus DTs. Facility-focused DTs focus on the physical environment of classroom, labs, HVAC systems, or on building security. Student-model focused DTs simulate or digitally replicate student learning trajectories, behavioral trends, and academic progress, enabling educators to adjust instructional design effectively. Administrative models enable institutional strategic planning, mid-course resource allocations, or to help educational leaders map overall strategic management in (Pathak & Upadhyay, 2024; Sinthamrongruk et al., 2024, pp. 387–390).

Establishing DT in educational contexts requires complex data architecture, IoT sensor integration, cloud computing, and increasingly AI algorithms. Although it is still in infancy compared with other industries, the increased digitalization of educational institutions because of the COVID-19 pandemic and the global enthusiasm for smart education, has created a surge of interest in DT-led systems that can adapt, optimize, and evolve with educational demands (Paolucci et al., 2024). DT in educational settings can take many different forms based on the intentions of the institution with regard to challenges and outcomes priorities. DT focused on facilities is typically established to monitor indicators, such as temperature, occupancy, lighting, and energy use. This can include routinely collecting sensor data coupled with predictive simulation to gain insight into efficient energy consumption, to ensure space usage is optimized, or to support policies for improved safety (Attaran & Celik, 2023). In contrast, student-model DT offers a more personalized and pedagogically relevant opportunity. These systems produce detailed digital representations of students through the integration of their learning behaviors, engagement levels, academic performance measures, and emotional states that have been measured through affective computing. These resulting virtual profiles are useful in providing personalized learning paths, detecting students who will struggle academically as early as possible and real-time instructional interventions. Prototype learning analytics dashboards and other forms of adaptive learning would fall under the student DT models (Luna, 2024; Wang et al., 2024).

As discussed, the administrative DT model has a more organizational focus and takes a high-level view of the institution (i.e., a policy simulation, resource allocation, or an operational model of a crisis) that combines historical institutional data, real-time operational data, and strategic plans to create awareness and flexibility for institution leaders to manage decision-making and governance. With current developments in cloud interoperability and federated data models, administrative DTs are able to provide compatibility and alignment among inconsistent data systems across departments so that the educational governance is digital and unified (Woolf, 2009). Despite the unequal distribution of these applications within global educational systems, pilot studies in countries such as South Korea, Finland, and the UAE show an increasing level of international interest in DT-related educational change. These systems show great potential not just in improving operational efficiencies but improving pedagogical innovation and equity (Thai-Nghe et al., 2010).

AI has become a decisive catalyst for DT, especially in Education. AI in Education (AIEd) spans a range of techniques and applications to replicate components of human cognition, assist decision-making, and improve learning (Algarni, 2016). AIEd uses machine learning, natural language processing, and neural networks to analyze data and determine patterns in large volumes of educational data, then create predictive inferences that inform personalized learning and administrative planning. One of the most visible areas where AI has shown transformative potential is personalized learning. The algorithms in AI use real-time engagement and historical responses to dynamically adjust to determine the content that should be least or most difficult, identify learning materials, and predict a student outcome. When this is coupled with DT, the algorithm produces a feedback loop that replicates and supports the personal learning pathway of a student. Furthermore, AI can identify cognitive and emotional engagement via multimodal inputs in ways that support real-time emotional and behavioral interventions (Matos et al., 2025).

Predictive analytics is another critical area in AIEd because it will allow institutions to be proactive in identifying risks associated with dropout, absenteeism, or a downward trend in performance. By drawing on a multitude of diverse data sources, for example, grades, attendance, SES, digital engagement, etc., AI systems will identify students that are at risk, and can then recommend interventions before a risk escalates. AIEd can also be used to help managers in education, for example, to reduce and streamline administrative actions, provide recommendations to optimize educational schedules or timetables, forecast enrolments, or assist in the design of curriculum. AIEd also provides students and administrative staff with chatbots or virtual assistants to facilitate communication and reduce the administrative load on human personnel (Kabudi et al., 2021). Many platforms and tools have been created and developed to realize AIEd into practice. Adaptive learning systems, like Carnegie Learning and Squirrel AI, are using machine learning to personalize instruction. Platforms like Knewton and DreamBox adapt the instructional materials in real time based on student learning data, while tools like Coursera and edX have recommendation systems that offer tailored suggestions about which courses or elements of courses to review. Learning analytics platforms, such as Civitas Learning and IBM Watson Education, provide insight and recommendation detail about student behavior and contexts, allowing instructors to design relative interventions (Gligorea et al., 2023).

Natural language processing tools, including Google's Dialogflow and IBM Watson Assistant, have a significant part to play in supporting educational systems. This includes automating assessment, providing real-time feedback, intelligent tutor support, and more. Academic integrity with AI-support systems is also ensured by AI applications in academic integrity such as plagiarism detectors, like Turnitin and Grammarly, which use natural language processing and semantic analysis. Furthermore, with the rise of generative AI systems, like GPT-4, more educational systems are looking to how to employ intelligent content generation, assessment creation, language translation in real time, and so forth, that continue to stretch the boundaries of personalized and inclusive education (Lin et al., 2023). These digital learning systems, when situated within the DT architecture, act as cognitive engines that allow the DT to learn from the data, make autonomous decisions, shift over time with the stated objectives of the institution and the needs of learners (Shi et al., 2022, pp. 284–289).

AI's use with DT has already produced significant benefits in sectors such as healthcare, manufacturing, and smart cities. In healthcare, the AI-enabled DT is used for personalized patient monitoring, modeling history of disease progression, and training for surgical activity. Commercial systems like Siemens Healthineers, and Philips HealthSuite have demonstrated the application of using AI real-time analytics to direct medical interventions from simulations of DT. In manufacturing companies, such as General Electric and Siemens are using AI DTnning to monitor equipment health, predictive maintenance, and streamline production workflow to provide cost-efficiency and operational resilience (Bruynseels et al., 2018; Philips, n.d.). Smart cities also use AI-enabled DT to simulate traffic patterns, monitor environmental conditions (i.e., air quality and temperature), and dynamically optimize other resource allocations. Urban planners use DT to support long-term development planning and respond to real-time crises while AI enables the continuous learning and adaptation of city twin models. The intersection of AI and DT in these examples showcases the immense value in combining real-time data, a simulative model, and intelligent analytics (Siemens Industrial IoT, n.d.).

It is a natural next step to bring these advances into the education realm. Education systems can also take advantage of common AI-enabled DTs for personalized learning paths, their learning environment, and overall organizational planning. Smart classrooms' ever-changing learning environment, for example, can access environmental DTs to create a learning environment that works well for all students, while AI algorithms process student behaviors and adjust teaching and learning approaches to fit current interactions in the classroom. Education systems can draw on successful models of health and smart city applications, and build context-aware, adaptable, and predictive DTs that result in better pedagogical and operational effectiveness (Ignatius et al., 2019). Although still a new domain within the academy, the landscape surrounding AI-integrated DTs and education is changing rapidly. In both China and Singapore, projects have already utilized real-time student data to produce personalized curriculum without human intervention. Pilot projects have even had students interact with AI-tutored DT who adapt live classroom pedagogies depending on real-time assessments of student understanding and emotional engagement. Furthermore, there are designs for virtual campuses and smart classrooms with embedded AI and DT architectures that allow ongoing assessments and adaptation of related academic and nonacademic metrics (Ağca, 2023). These uses also raised important issues about data privacy, algorithmic bias, and access, which point to a need for ethical frameworks. However, the post-COVID climate of the educational space presents a large-scale, resilient, and personalized need for education, which could act as motivator to incorporate AI-DT.

The issue is not with the technology, but rather the institutional readiness, teacher professional development, policy support, and physical infrastructure to facilitate its proper use. The journey from idea to solution is becoming much clearer with emerging evidence of studies, increasingly aligned international policies, and aligned technology (Kuzminska et al., 2023). While some initial progress has been made in adopting AI and DTs simultaneously in education, evidence of planned and purposeful integration has been limited. Many implementations captured today are piecemeal, often limited to either AI-driven analytics or IoT-based DT focused on products/facilities without understanding the dimensions of a DT model holistically. In addition, while there are theoretically a lot of documented trials with AI-enabled DT as applied to Health and Manufacturing, there are still limited documented trials of DT in Education precisely on the dimensions of modeling learning outcomes, teachers' interaction with students, and the longitudinal impacts of education on the individual learner (Zhang et al., 2024). Future research and development possibilities include the building of holistic AI-DT landscapes that combine facility, student, and administrative data within one responsive model and allow connected capabilities, from mental health monitoring to adaptable curriculum. There is also a growing demand for scalable systems that help small to medium-sized education institutions adopt technology without incurring excessive cost or technical capacity. The advancement of edge computing, 5G, and federated learning could also help democratize the systems (Ning et al., 2024). In addition, ethical design frameworks need to develop to accommodate the real-time data collection and real-time algorithmic decision-making challenges in education. Institutions of education owe it to their students to provide clear data transparency, to inform on the data being collected and how it is going to be used, informed consent from users being part of this transparency, and processes of accountability for any AI-based predictions—these factors underpin trust and equity in organizations in digitally led learning. Once the accountabilities are better handled for the proposed data and time, AI-developed DT functions in educational contexts can evolve from individual prototypes toward integrated, intelligent, adaptable, inclusive, and sustainable learning environments (Abdullahi & Lazarova-Molnar, 2024, pp. 612–617). Table 28.1 summarizes the key points of the discussed related works on DT and AI in Education, highlighting main themes, focus areas, applications respectively.

The literature reviewed presented a clear possibility for detecting disruption to learning throughout the education sector when combining DT with AI. The literature provides good examples of the use cases seen in property management, personalized learning, and administrative

TABLE 28.1 Summary of related works.

Aspect	Details	Key applications/Examples	References
DT Overview	DT are virtual replicas of physical entities that change dynamically in real-time through data integration. As such, they are useful for monitoring and assessing decision-making processes in education.	Facility data tracking, student conduct simulation, administration	(Fuller et al., 2020; Paolucci et al., 2024; Pathak & Upadhyay, 2024; Sinthamrongruk et al., 2024, pp. 387–390)
Types of school-based DTs	Types of virtual DT include: 1. Facility twins—focused on monitoring their environment, including temperature, occupancy, and energy; 2. Student-model twins—model student learning and behaviors; and 3. Administrative twins—focused on management of resources and policy simulation, among other tasks.	Energy conservation, individualized learning, institutional development	(Attaran & Celik, 2023; Luna, 2024; Thai-Nghe et al., 2010; Wang et al., 2024; Woolf, 2009)
AI in education (AIEd)	AI (e.g., ML and NLP) for personalized learning, predictive analytics, and administrative optimization.	Adaptable learning, prediction, and prevention of drop-outs, AI-enabled chatbots	(Algarni, 2016; Kabudi et al., 2021; Matos et al., 2025)
AI Tools/ Platforms in education	Examples of systems for evolving adaptive learning experiences, learning analytics systems, NLP-based tutoring and student assessments, plagiarism detection, and content generation.	Carnegie learning, Knewton, IBM Watson education, Turnitin	(Gligorea et al., 2023; Lin et al., 2023; Shi et al., 2022, pp. 284–289)
AI + DT Synergy	AI takes DT further by allowing for real-time decision-making and predictive modeling or a digital replica of the built environment that evolves autonomously to operate more efficiently; feasibility has been demonstrated in healthcare, manufacturing, and smart cities.	Siemens Healthineers (healthcare), GE (manufacturing), smart cities	(Bruynseels et al., 2018; Ignatius et al., 2019; Philips, n.d.; Siemens industrial IoT, n.d.)
Transition to education	There are emerging pilot projects and real deployments of AI and DT for personalized learning, smart campuses, and adaptive classrooms.	Virtual campuses, real-time adaptive classrooms	(Ağca, 2023; Kuzminska et al., 2023)
Research Gaps & opportunities	The current implementations of DT in education are fragmented, and need integrated and holistic AI-ED-twin ecosystems; scalable solutions need to be developed for small-medium-sized education institutions; ethical frameworks for monitoring, privacy issues, and fairness need to be considered too.	Cross-functional AI-DT models, edge computing, federated learning	(Abdullahi & Lazarova-Molnar, 2024, pp. 612–617; Ning et al., 2024; Zhang et al., 2024)

automation; however, these studies have mostly been small and taken mostly pilot studies to test. Coupling the two technologies provides an avenue for constructing integrated intelligent educational ecosystems that can change in real-time and contextualize decisions; this is a unique opportunity for transformation and innovation. Nevertheless, challenges due to scalability, interoperability, and ethical governance are all real areas of concern. Future research should consider how to develop or improve complete and privacy-first Internet-wide frameworks for building AI-integrated DT systems in education that hold notions of fairness, transparency, and ethical inclusivity.

3. Architecture and design of AI-enhanced school DTs

The architecture of AI-augmented school DTs combines intelligent data systems, real-time modeling, and adaptive learning infrastructures as aligned to support decisions, personalized student experiences, and maximize operational efficiency. These systems are complex and require a strong modular architecture design in integrating several loosely coupled design and technology domains, such as sensing technology, AI, and human-computer interaction interfaces.

3.1 Conceptual model

The design of a school-centric AI-DT is generally conceived as a multilayer model. There are four primary layers to consider: the Physical Layer, the Digital Layer, the AI Processing Layer, and the Application Layer. These layers work together to not only represent the educational places and processes, but to understand and react in real time. The bottom layer, the Physical Layer, comprises all of the physical items (the bricks and mortar) such as classrooms, laboratories, libraries, HVAC systems, biometric access points, and every other component that comprises a school. With that, there are IoT-based sensors and actuators that can collect operational and environmental data such as temperature, light levels, air quality, occupancy, and energy consumption. In addition, if we incorporate IoT-based wearables, RFID tags, and motion detectors, we can also collect data on student movement and engagement and attendance. Thus, we have a true depiction of the state of physical items, including the human and nonhuman actors.

Moving up, the Digital Layer, is the virtual counterpart of physical entities, consisting of DTs built from continuous data streams collected by the underlying physical system. These digital interfacings are foundational to the DT and are analogous to a real-time models of classrooms, students, and administrative hierarchy. The quality and versatility of digital representations increase with accuracy, granularity, and consistency of the data collected by each layer of input. Synchronized learning management systems (LMS), student information systems (SIS), and building management systems (BMS) offer structured digital input (or forms), while sensor data and unstructured social or behavioral data enhance the robustness of the virtual model.

Above this, is the AI Processing Layer that serves as the cognitive engine of the architecture. This is where a set of AI modules can be found (e.g., data preprocessing, feature extraction, predictive analytics, and inference generation). This layer enables the DT to do more than just mirror and model real-world systems, it allows the structure to learn. AI models that could be deployed are (for example) supervised classifiers to deep learning architectures, to predict student performance, identify emotional state through sentiment analysis, recommend adaptive learning content, and simulate administrative practices. There are many forms of AI to build custom-specific models depending on the needs of the educational institution. AI models are often trained with historical data but can be continuously improved based on incoming real-time streams. This allows for closed-loop learning and enhancements to predictive capabilities in stages over time.

Lastly, the Application Layer includes all user-facing interfaces and functional tools that allow each user to interact with the DT system. This could include dashboards for school administrators, analytics dashboards for teachers, mobile or web interfaces for students, and digital assistants or chatbots that allow real-time responses to questions. The application layer is crucial to ensure that the insights generated from the DT are accessible, meaningful, and usable by stakeholders in the educational ecosystem. Fig. 28.2 illustrates the conceptual model of proposed AI-DT framework.

3.2 Roles of sensors, data aggregators, and AI inference engines

Orchestrating a fully functional AI-driven DT for educational institutions builds on three broad types of components: sensors, data aggregators, AI inference engines. Sensors form the first physical contact with the physical world, the first contact point with physical model. Sensors gather multiple types of data attributes/temperatures (thermistors), presence (infrared, and ultrasonic devices), air quality(gas concentration sensors), and acoustic levels (microphones). In addition, in different applications such as wearables, sensors can monitor heart rates, facial expressions, and screen interaction times. As can be seen from the disparate data sources, the heterogeneity, and potentially the distributed nature of these data sources will require real-time preprocessing, aggregation. In this process, data aggregators assume an important role in reconciling different inputs, and downstream signal

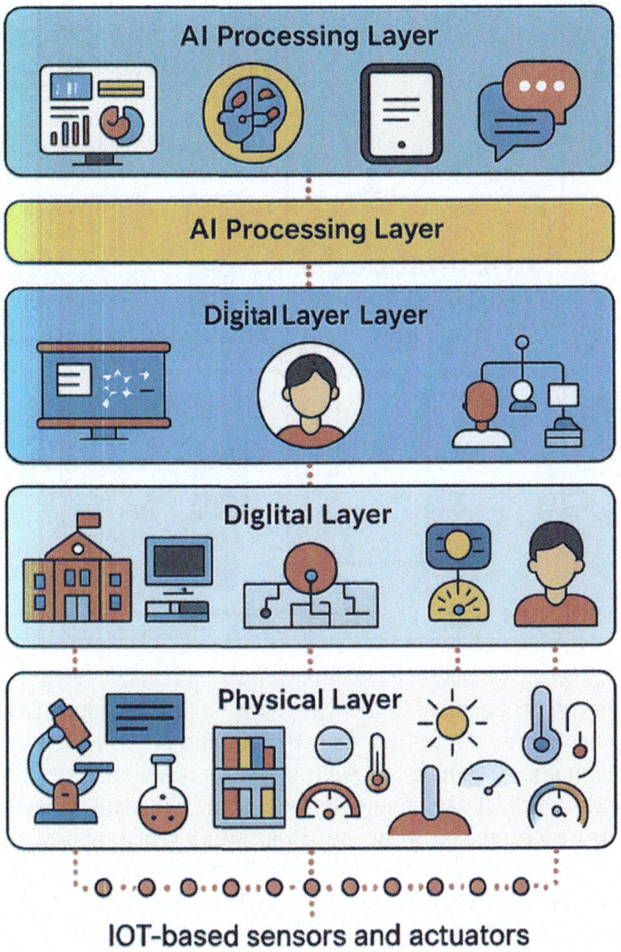

FIGURE 28.2 Conceptual model of AI-DT.

conditioning, noise filtering, and signal normalization. In their responsibility to convert unprocessed data into data structures, structural frameworks, representative formats as inputs to the AI processing layer, edge devices may also play a supporting role in preprocessing data through local aggregation and analytics monitoring network engagement rates—improving on network latencies or network load performance profiles. Adopting cloud aggregators approaches and hierarchical building blocks of large datasets provides structural framework for longitudinal-based analytics and interschool comparisons. AI inference engines will select and interpret training and composed datasets to generate actionable insights. Inference engines rely on decision trees, RNNs, CNNs, transformers, and other advanced analytic applications to draw inferences from the complex data and predict outcomes and provide recommendations. Inference engines complement overall intelligence in DTs. In addition to various referenced examples, inference engines are acquiring the capacity for automated communicative processes—such as sending Early Warning alerts.

3.3 System components

A holistic AI-enhanced school DT system contains several interconnected system components. Of course, the primary function is made possible by data collection mechanisms, which provide the input from various institutional sources. Embedded IoT devices or sensors located in physical spaces provide environmental information and usage; education systems (LMS, SIS, etc.); provide academic, demographic, and behavioral data. For example, various LMS systems (Moodle, Blackboard, Canvas, etc.) provide student-level logging of information related to student use of content, when they submitted work, how they actively participated in forums, and how they performed on quizzes. SIS provides student-level admission and or registration information, grade history, attendance, and behavior flags. AI functionality in the system is configured for particular types of analysis. For example, Predictive modeling engines predict the academic performance of students, flag students at risk, or predict trends for enrollment. Other functionality such as sentiment analysis uses natural language processing (NLP) to assess how students are feeling emotionally based on students' written feedback, classroom discussions, or social media interactions. Recommendation engines dynamically curate learning content, peer collaboration, and extracurricular opportunities based on each student's learning profile and preferences. The role of user interfaces is important to support the idea of turning insights provided by the AI and DTs into effectively used data. Dashboards for teachers demonstrate visual analytics supporting student progress, distributions in performance by class, and behavioral activity heatmaps. The virtual agents used in the learning platforms support personalized help for students with queries related to the subject, suggestions on revision topics, or help with administrative requests. Digital assistants provide parents with live updates on their child's status including attendance, performance trends, and scheduled meetings representing a more inclusive and transparent experience. Fig. 28.3 illustrates the discussed system components of AI-DT framework.

3.4 Data flow and feedback loops

A fundamental aspect of AI-enabled DTs is the incorporation of efficient data flow schemas and feedback loops for human and machine learning. Data flow can occur through streaming and batch modes in this space. Streaming processing is critical in cases where there is real-time or near-real-time action required, such as dynamic classroom management (e.g., responding to variations in students' behaviors or perceived behaviors), behavioral assessment, environmental condition monitoring, etc. The speed of processing allows for immediate

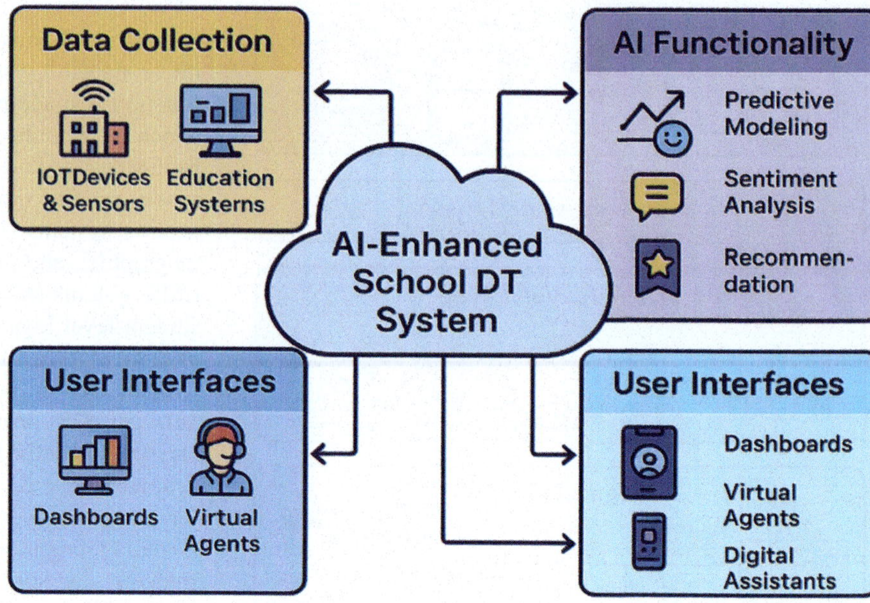

FIGURE 28.3 System components of AI-DT framework.

interventions, for example, managing an HVAC system based upon causal variables (e.g., occupancy counting and student activities), or alerting an educator to a student's disengagement (e.g., using facial recognition class activities). In contrast, batch processing accommodates longitudinal analysis, such as examining a cohort's performance pattern over a semester, investigating cohorts' differences, or providing training data for AI models on historical data. Batch jobs typically occur during nonpeak hours and supply cleaned models back into the inference engine for more predictive accuracy in streaming jobs.\n\in closed-loop feedback systems, the DT can learn from its initial outcomes and modify its actions based on its original recommendations. For instance, if a recommendations algorithm suggests a video resource, and students regularly quit watching it early on, the system could alter its recommendation process to prioritize interactive content, or to adjust the video into smaller components. These feedback loops enable the DT to continuously improve its recommendations in terms of personalization, efficiency, and accuracy of decisions based upon prior outcomes for a cohort or individual, thereby enabling self-evolution, self-optimization, and contextual adaptation.

3.5 Security and ethical considerations

While the potential for integrating AI and DT technology is tremendous, it also raises significant data security issues and questions about ethical governance. The educational setting is particularly notable for ethical concerns since it often involves data created by minors and other vulnerable populations. First, we consider student data. Privacy is an important consideration when we consider educational data. For example, the data that might be collected and maintained include biometric data, academic records, psychological data, and emotional state. The data collected must be protected from intrusion, interference, abuse, and breaches. Student data must be managed in compliance with data privacy regulations such as the General Data Protection Regulation (GDPR), the Family Educational Rights and Privacy Act (FERPA), and other applicable local cybersecurity laws and regulations. In addition to data protection legislation and regulations, the issues of data anonymization, data encryption, role-based access management, secure data transmission, and secure data storage are all components of an effective privacy-preserving DT architecture. Second, particular issues also arise regarding the ethical use of AI in the educational landscape since ethical use will immediately raise issues of transparency, fairness, and accountability. AI must be developed in ways that avoid biases that could disproportionately disadvantage students based on race, gender, socio-economic status, and disability. Furthermore, the opaque decisions during the analysis and inference steps in the AI process (the "black box" problem) must be addressed through explainable AI (XAI) model building, and allow educators and students to understand the bases for decisions about recommendations or predictions. Informed consent should also be treated as a best practice. Stakeholders (students, parents, and educators) need to be informed about the data that is collected, processed, and what it is used for. Also, students need a certain level of control over their digital profiles and a choice to opt-out of specific data, amend specific data, or remove specific parts of the data. It is important to create governance frameworks with ethical auditing, bias detection and remediation, and learner agency to ameliorate trust in animate learning environments.

4. Functional applications of AI in school DTs

The use of AI in School DT applications has led to a paradigm shift in education. From personalized learning to better use of school resources, AI-enabled DT is changing the way schools interact with data, how they teach, how they assess learning and student well-being. DT has given schools the ability to go beyond just replicating a physical process but into decision-making, simulations, predictions, and student support systems. As more educational institutions seek to use data to inform practice, the intentional integration of AI into DTs represents a unique opportunity to improve learning outcomes, reduce resource use, and provide timely and effective services for students.

4.1 Smart learning environments

AI-powered DT is intelligent learning ecosystems that can personalize instruction and adapt to learner profiles. They typically rely on a suite of machine learning algorithms contained within the learning management systems (LMS) of an organization to process data streams in real time, setting and adjusting question difficulty, instructional strategies, and recommending educational resources based on student achievement level and learning preferences. These adaptive learning systems typically rely on reinforcement learning models to update recommendation logic using data from student interactions with the site. Most adaptive learning systems also encourage mastery-appropriate learner movement through the educational content and concepts, allowing students to progress as they learn.

Another innovation enabled by these DT is the ability to digitally create engaging virtual laboratory environments and simulations. These platforms foster immersive learning for students, particularly in STEM education, allowing for real experiments, manipulating chemical components, visualizing biological systems, etc. These simulations, based on AI models, can enhance students' cognitive retention and conceptual understanding. The DT framework allows for the provision of collaboration-based virtual learning spaces, in which students from diverse geographical locations can engage in collaborative learning by participating in experiments and projects, sharing their experience and results, and engaging in discussions. AI agents support the experience by placing students in groups by skills and competencies, maintaining lively collaboration engagement and, removing students who are hampering participation when feasible.

4.2 Student performance monitoring

The introduction of machine learning student performance monitoring (DT) systems represents a major shift in monitoring student performance. The use of supervised machine learning leads to the development of predictive measures that will categorize students into risk categories and formalize the students who require intervention. The ability to monitor potential issues allows educators to switch from a reactive to a proactive strategy and to engage with a student before their academic or emotional due diligence becomes irreversible. An AI application can be used to identify patterns through clickstream data, interaction logs, discussion forums, and video metrics, plus variations in tone, sentiment, or conceptual understanding using natural language processing techniques on written submissions or discussion forums. Not only does this allow educators to differentiate instruction, modify pace or precision for at-risk skills, potentially impacting the learning outcome and satisfaction of learners; pupil behavioral analytics such as facial recognition, gesture analysis, and keystroke dynamics, are offering behavioral inferences such as levels of attention, confusion, frustration, or unfocused responses. Integrating behavioral analytics with punctual student academic performance at the micro learning and macro performance level can be a powerful way of achieving a complete picture of whether the learner is engaged, and can assist educators in developing appropriate motivational strategies, or pedagogical changes in a timely manner.

4.3 Infrastructure and facilities management

DT enhanced with AI has the potential to greatly enhance the efficiency and resiliency of physical school infrastructure. One use case is predictive maintenance, where AI models can detect anomalies and then predict failure modes in heating, projectors, computers, elevators, and other assets. Predictive maintenance reduces unplanned downtime, extends the life of an asset, and, ultimately creates a safe learning environment. One indirect benefit of predictive maintenance is cost effectiveness, as it prevents unnecessary replacements of gear and limits reactive repairs. An AI-based Computerized Maintenance Management System (CMMS) would usually exist within a DT. In addition, AI can optimize and allocate space and manage energy, combining real-time occupancy information, timetable scheduling, and environmental conditions. If it were possible to combine all class periods into one classroom, overall space could be used for other activities, such as a space to conduct extracurricular activities or a repair shop for maintenance. Similarly, potential energy savings could be realized if AI-based climate control systems could control lighting, ventilation, and heating based on occupancy and weather forecasts. For example, a reinforcement learning algorithm can be rewarding for optimal performance that minimizes footprints and compliance with human comfort. The model learns overtime what the best settings for heating and cooling the interior space, based on occupant location and number of occupants.

4.4 Administrative and decision support

Running an educational organization is an involved process, requiring administrators to schedule classes, write policy, staff, and budget. Artificial intelligent-powered DT can provide intelligent automation and decision support, which will help educational organizations do this level of planning. One application of this technology is timetable optimization, which allows AI algorithms to generate class timetables that minimize teacher conflicts, classroom conflicts, and student overloads. AI-powered DT can also help schools optimize the allocation of its resources, using preset AI models based on historically utilized utilization rates, student-teacher ratios, and projected demand. The AI model could produce data-driven budgets and staff hiring plans that properly align allocations with institutional strategic priorities. Furthermore, AI-based policymaking tools will provide school leaders with the capacity to project simulated outcomes, similar to DT, from proposed policies in a school context, such as a transition to block scheduling (with specified times) or a new grading system. To support leaders' right-of-from or managing workload constraints, DT solutions could be integrated with more conversational natural language processing interfaces on dashboards, allowing administrators to interact with or interrogate (while taking into account all relevant data/metrics) the AI analysis with immediacy and visibility. This would allow leaders to leverage AI-based data and reduce overweight on technical mediators between decision-makers (leaders) and data analysts.

4.5 Well being and mental health

AI in DT can greatly enhance student mental health and emotional well-being. Emotion AI, for example, employs face expression analysis, voice tone detection, keystroke pattern definition, and posture recognition to safely track emotional states in students. With all of the data that Emotion AI collects on students, educational staff can now detect stress, anxiety, or disengagement in students sooner than before. In the event that a student's emotions exceed certain thresholds, the DT will notify appropriate staff personnel like a counselor, or may even conduct automated check-ins with students on behalf of staff personnel. Through the use of NLP and chatbots, AI in DT can also provide mental health support and check-ins for students 24/7. AI can hold empathetic two-way conversations, monitor mood states, and provide potential strategies for counseling students. AI can also provide students in need of emotional wellness help with some nonjudgmental first steps to disclose their emotional distress, where they may have been hesitant to seek out those first steps openly. There are also AI sentiment analysis tools available that can analyze open-ended survey responses, means of discussions in online forums, or journal entries that reveal sentiment that has hidden patterns of students who may have had dissatisfaction, fear, or hope. Schools can aggregate and summarize trends of these sentiments to inform policy decisions, highlight dysfunctional cultural or structural conditions; reveal ways in which schools are inciting wellbeing-based initiatives. Of course, schools must responsibly and ethically deploy AI in schools, while protecting issues of consent, confidentiality, and transparency. It is important to emphasize that emotional AI does not offer emotional intelligence, or real human empathy. Emotional AI will supplement the capacity and capabilities of school-based staff and personnel to provide effective and holistic help and support to students in educational settings.

Table 28.2 summarizes these key functionalities of AI-DT along with their applications, benefits, and used technologies, respectively.

5. Case studies and practical implementations

While the concept of DT virtual representations of physical things is being introduced into education from a baseline of industrial and engineering applications, more and more schools and educational institutions are investigating or implementing artificial intelligence-enhanced DTs. These DT projects come in many different forms in a variety of scales some are smart school ecosystems, while others are limited pilot projects being developed in the environment of academic research. This section provides detailed examples of implementations and experimental prototypes of DT technologies in education; it presents real-world examples of how DT technologies are changing education in the present day; this section reflects on implementation challenges that may contribute to or detract from supporting future deployments.

5.1 Case study 1: AI-enhanced DT in a K–12 smart school

In terms of a demonstrable example of how to use an AI-powered School DT is a smart school project supported by the government in South Korea. Designed as a model for the next generation of digital learning environments, the K-12 school embraced an entire DT framework to advance

TABLE 28.2 Functional applications of AI in school digital twins—domains, technologies, and educational benefits.

Section	Key functionalities	AI technologies involved	Benefits
Smart learning environments	Personalization of instruction in real-time, adapting question difficulty, virtual labs, collaborative virtual environments	Machine learning (ML), reinforcement learning, AI agents	Tailored learning experiences, enhanced cognitive retention, global collaboration
Student performance monitoring	Student risk categorization and sentiment analysis, detecting attention/frustration	Supervised ML, natural language processing (NLP), computer vision, gesture & keystroke analysis	Proactive interventions, enhanced learning outcomes, full learner engagement profiling
Infrastructure & facilities management	Faults, predictive failures, space planning, energy management	AI-powered CMMS, reinforcement learning	Reduced downtime, cost savings, energy-efficient smart campuses
Administrative & decision support	Timetable generation, staff planning, AI-based policy simulation	Optimization algorithms, NLP dashboards	Improved decision-making, efficient resource use, responsive leadership tools
4.5 well-being and mental health	Emotion recognition, automated alerts, AI chatbot counseling	Emotion AI, NLP, sentiment analysis	Early issue detection, accessible emotional support, data-driven wellness insights

instructional and infrastructural function. At the project's center was the development of a multilayered DT system. The physical layer was made up of interrelated IoT devices that spanned across classrooms, corridors, labs, and administrative offices. This included temperature sensors, motion detectors, smartboards, student wearables, and facial recognition cameras. The digital layer created a real-time virtual rendering of the school environment capturing live student attendance, movement, equipment or learning space usage, and energy consumption.

The most dynamic aspect of the project was the inclusion of an AI processing layer, which had several distinct AI features. Specifically, predictive analytics modules provided early alerts for disengagement in students; machine learning models were used for performing behavioral assessments in the classrooms; and reinforcement learning systems optimized heating, lighting, and resource allocation. AI agents were able to monitor students' interactions with digital learning platforms by customizing the instructional content based on cognitive and emotional profiles determined by analyzing speech, facial expression, and performance results.

At the user level, administrators were able to access dynamic dashboards displaying relevant data on key indicators, including whether attendance was higher or lower than comparable cohorts, how well the HVAC system was working, and where students performed within grading distributions. The teachers were provided with AI-driven lesson plan recommendations, engagement prompts while teaching, and integration with an automated grading systems. Students interacted with intelligent digital assistants that facilitated tutoring, reminders, and wellness checks. The impact of the program was clearly measurable: attendance rates improved by 12%, energy usage decreased by 15%, and wellness and satisfaction improved for teachers about technology support. The school, on the contrary, compromised with the related concerns for student privacy, and the school created an ethical framework and a consent structure so the school could continue to use the technology in its second year. Fig. 28.4 illustrates in-depth insights in the case study.

5.2 Case study 2: Virtual campus twins for administrative simulations

Another interesting example came from a major European university that created a Virtual Campus Twin, again focused primarily on system planning and resource management. Although less focused on the K–12 teaching model, this example demonstrated how DTs may be used for large-scale academic administration and development through simulation and predicting outcomes based on data. The university created a detailed DT of their 40-acre campus, which included classrooms, laboratory spaces, libraries, dining locations, and dorms. The university collected operational data from their security system, student information system (SIS), timetable scheduling system, maintenance reports, and energy meters. Unlike previous platforms with static monitoring capabilities, the Twin adhered to the same principles as previous DTs but with the addition of AI-driven simulations.

In this instance, administrators were able to simulate the addition of hybrid learning on common space utilization and utilities use. The AI models tracked crowds, who

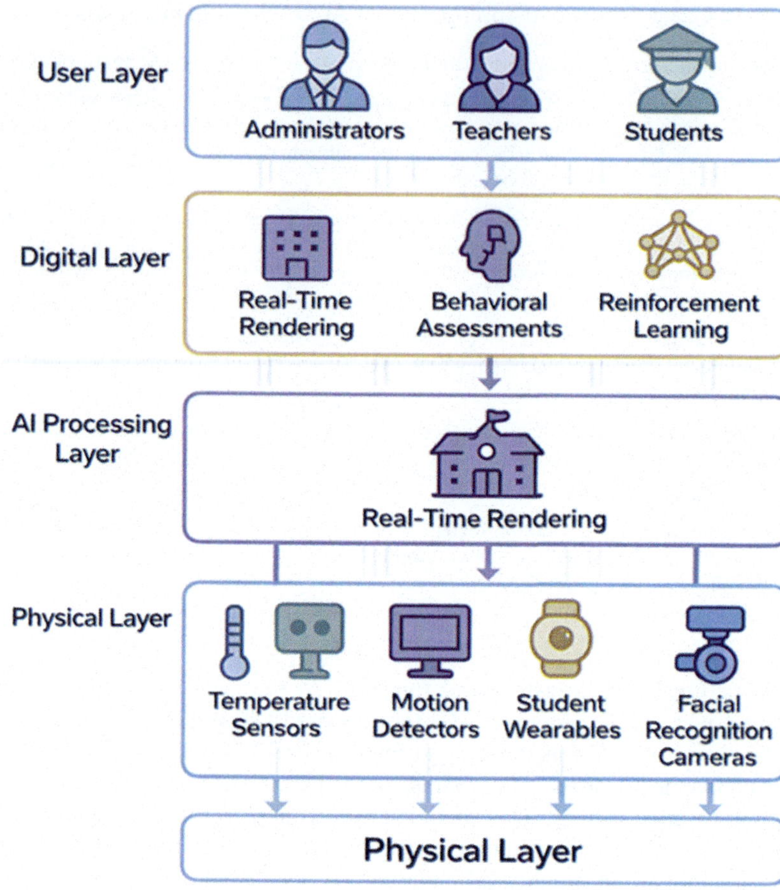

FIGURE 28.4 Illustration of case study 1.

would be on campus and when, monitored usage bottlenecks in both time and space, and provided the university best-fit classroom assignment choices, based on predicted flow of students into spaces. The university was also able to predict practical implications of budget cuts or department reorganizations in terms of faculty workload, student-instructor ratios, and classroom and space saturation. This DT was extremely useful in the response to the COVID-19 pandemic, allowing the modeling of multiple reopening possibilities. It was able to exhibit safe capacities, the movement of people to reduce likelihood of contact, or the relative effectiveness of staggered schedules in managing density, etc. This ability to test ideas and scenarios enabled informed and lightning-fast policy decisions, and to maintained the continuity of teaching and learning amidst an unprecedented environment.

Beyond just simulations, the university was also able to use the various AI modules of the DT to assess and verify elements of their long-term strategic planning. For example, the presumed return on investment (ROI) associated with constructing a brand-new engineering laboratory was driven by forecasts around future projected enrollments, depreciation of equipment, and future capacity in teaching spaces (all planned in the DT environment). This example highlights how DTs can go multidimensional, moving beyond classroom situations, to become complete, intelligent management systems across whole Academic ecosystems. Fig. 28.5 illustrates in-depth insights in the case study.

5.3 Pilot projects and prototypes

While the large-scale projects executed do illustrate the impact of an AI-assisted DT initiative, there numerous academic institutions and research organizations that are experimenting with AI user interfaces in their pilot projects and prototypes in lower-risk contexts. These pilot projects often provide a facility for testing new technologies, architectures, and analytics algorithms and pedagogical designs. One of the more notable projects was the "Smart Classroom Prototype" created by a consortium of European universities, supported by the Horizon 2020 research funding program. The project allowed selected classrooms to equipped with sensors, cameras, and audio recording to

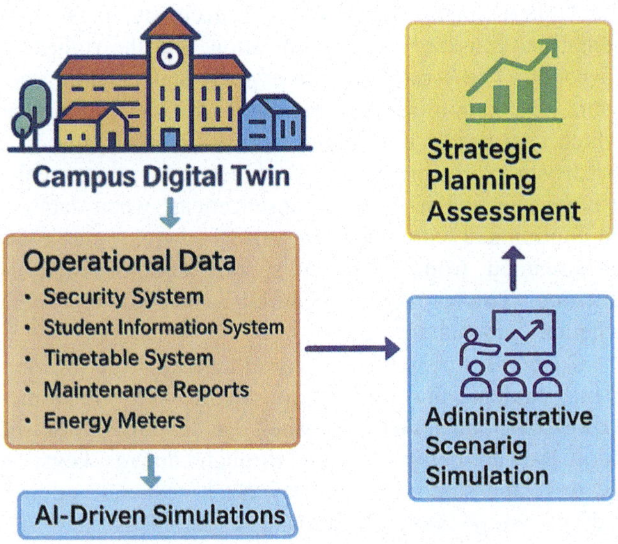

FIGURE 28.5 Illustration of case study 2.

collect multimodal data. That data were then used to train AI models for evaluating class engagement, determining the noise level of the class, and gauged how often the student had class interactions with their professors. Reports were created for awareness on various aspects of engagement using analytics dashboards and AI suggestions on how to improve teaching practices. The prototype also demonstrated that even small arrays of low-cost sensors can provide productive feedback loops about pedagogical choices using AI models trained on consequential behavior compared with all behavior. In the United States, a university-based project created a prototype DT with a focus on student mental health analytics. It used data from wellness surveys, activity logs, and sentiment analysis from student's public social media posts (with consent). The initial results indicated that the system was able to identify the at-risk students with 78% accuracy; however, the ethical and legal ramifications of using individuals' personal digital footprints led to ongoing discussions around what should be institutionally acceptable/implemented and the ethical lines that need to be drawn.

Another important line of prototypes tackles virtual simulation environments for the purposes of teacher training. Recent projects developed AI-enhanced DTs to run virtual classrooms with synthetic students that exhibited a range of different behaviors from engaged to disruptive. Preservice teachers interacted with and contributed to this environment to develop their own classroom management skills, pedagogical approaches, and emotional intelligence. The system used reinforcement learning to modify the students' behaviors based on the responses of the trainee, thereby simulating realistic and evolving classroom dynamics. These prototypes demonstrate that DTs can encompass and adapt to many different educational settings, levels, and contexts. In addition to serving as innovation incubators, they can also contribute to empirical outcomes related to educational applications of AI.

6. Challenges and limitations

Although AI-augmented Digital Twins DTs are a huge improvement in terms of innovation in education, we can expect many practical, pedagogical, ethical, and infrastructural challenges associated with their adoption and implementation. These challenges have the potential to undermine the effectiveness, sustainability, and ethical integrity of DT systems if not adequately addressed.

6.1 Technical challenges

A major challenge in developing and sustaining a viable DT system is the integration of data across multiple educational platforms and hardware devices. Schools typically use a variety of dissimilar systems, such as Learning Management Systems (LMS), Student Information Systems (SIS), biometric scanners, IoT sensors, and digital repositories of content, in a heterogeneous mix of capabilities. Developing a coherent Digital Twin out of these iterations requires integration that is standardized, temporally synced, and undifferentiated across various protocols. Unfortunately, many of these platforms are proprietary products that do not employ open APIs, and integration can be difficult and/or financially burdensome.

The second technical barrier relates to AI bias. Machine learning models that fuel an DT are only as good as the material they are trained on. If the training data are biased, either because of skewness or because of the reflection of an existing social or academic bias, AI models can maintain rather than correct inequities. For instance, a key feature of many DTs is predictive analytic tools to assess students' performance and identify intervention priorities. In this capacity, the predictive analytic tools are prone to reflect bias in historical data, which may lead to decision-making that is overwhelmingly biased against students from underrepresented backgrounds as high-risk students. These decisions lead to forced and poor interventions or labeling as being stigmatized. Adding to the problem, as with AI decision-making in general, there is sometimes limited accountability for decision-making due to the black-box nature of AI. This can also be an issue of explaining how specific decisions are made.

Interoperability is also a major hurdle. As DT ecosystems become increasingly larger and more complex, we must ensure interoperability between other tools, sensors, and AI modules. Interoperability is about technical specs too interoperability is always about how to connect pedagogical goals, administrative workflows, and data governance. Difficulty in interoperability, especially if the confusion is based on the absence of a standardized communication protocol, can cause inefficiency, create data silos, cause redundancy, and damage the real-time integrity and responsiveness of the Digital Twin.

6.2 Pedagogical issues

The use of AI and other automations for education is presented with a variety of pedagogical tensions. One of the more troubling concerns relates to the potential over-automation of teaching and learning; that is, smaller facets of education that get automated like automatic grading, behavioral engagement tracking, and personalizing recommendations of learning paths. While automation can positively relieve educators of mundane tasks, over-automation minimizes the profession to basic manager, with primary understanding being on data representation and data use.

Concern can also arise when the same automation disconnects students from more human experiences with educators. Certainly, AI-based platforms such as a tutoring system, chatbots, and feedback engines can do personalizing in education; it cannot duplicate empathy, sensitivity, and mentorship of an educator. Over time, students may even become familiar and attached to a digital agent as opposed to a teacher, weakening their sense of classroom community and interpersonal learning. A final consideration includes things like mental/emotional health, or cognitive readings of students can become general, and perhaps unlikely best position to be understood by AI. the restrictions could lead to generic intervention, or misguided programs. At an educational level, there is a danger of data-driven reductionism, in which students are only viewed as numbers engagement scores, feelings states, performance scorecards while omitting the qualitative dimensions of learning creativity, critical thinking, and social developmental growth. This can also contribute to a de facto constriction of the curriculum, and distort priorities of instruction, especially in testing-heavy situations.

6.3 Ethical and privacy concerns

Perhaps the most controversial consideration of using DT is a set of ethical and privacy issues related to data collection, processing, and data usage. Institutions of education collect and manage sensitive personal information—academic records, behavioral records, biometric data, and emotional information, all of which become more personal when they are continuously collected and processed in real time. There are serious considerations when we think about legal compliance with privacy rules (the General Data Protection Regulations (GDPR) in the EU or Family Educational Rights and Privacy Act (FERPA) in the United States) that emphasize accuracy, legal representation, consent, data minimization, and the right to be forgotten. These rules require explicit understanding consent, data minimization, and the right to be forgotten, which can almost impossible to fulfill in real time on an AI-driven platform. Basically, it becomes impractical to ask the question what does consent mean, when data may be continuously streamed via sensors in smart classrooms or streamed from video or speech through emotional tracking. This is very concerning when we considered whether students are minors.

Concerns surrounding a surveillance culture are prevalent. Students may feel anxious, stifle their freedom of expression, and lessen spontaneity in learning when they know they are being surveilled through cameras, keystroke logging, and sentiment analysis. Such academic environments complicate educational experiences and risk normalizing surveillance. The long-term consequences of this could cause permanent psychological and cultural damages. Moreover, parents, educators, and civil rights organizations are voicing strong concerns about who owns and controls the data, the storage mechanisms of the data, and the potential for secondary purposes for the data. The issues of bias and opacity in AI decision-making are other ethical concerns. In the absence of documentation or the ability to conduct an audit of the algorithms, students and parents may not be able to challenge automated decisions that are made about student academic risk or behavioral interventions, or even the sending of a mental health alert. Conducting ethical AI in education is dependent on the transparent design of models, diverse language datasets, processor accountability, and the oversight of the algorithm by various stakeholder groups and people.

Table 28.3 summarizes the notable technical, pedagogical, and ethical challenges related to joining and adopting AI-augmented Digital Twins in education with possible compromises to interoperability and integrity of instruction, potential harms to student well-being, and questions with overall governance of data.

7. Future directions and research opportunities

As AI-enhanced School DTs commence redefining educational landscapes, it is crucial to explore the boundaries of human imagination and examine the future innovations and possibilities. This exploration of future possibilities, innovations, and development trends spans not just wearables or digital technologies, but encompasses a range of possible future developments globally, including level of intelligence at the ecosystem level; governance structures; and interdisciplinary partnerships and collaborations. Knowing where that change might lead to today

TABLE 28.3 Challenges and limitations of AI-augmented digital twins in education.

Category	Challenges	Description
Technical challenges	Data integration	Issues integrating heterogenous platforms (LMS, SIS, IoT, etc.) with no open APIs.
	AI bias	Bias present in models due to historical or biased datasets creates unfair assessment and mitigation of student actions.
	Black-box decision-making	No accountability or transparency of the decisions made by AI.
	Interoperability	Incompatibility across tools, sensors, and platforms creates risk and inefficiencies improving the inverse of what we hope to achieve.
Pedagogical issues	Over-automation	Automation of current tasks, like grading, may mean we reduce educators to data monitors instead of drawing on their pedagogical purpose.
	Human Disconnection	Overreliance on AI may devalue teacher-student relationships and classroom community.
	Misguided personalization	Emotional or cognitive states may not be accurately assessed, leading to generic or inappropriate interventions.
	Data reductionism	Performance indicators at the expense of creativity, critical thinking, and overall learning.
Ethical and privacy concerns	Continuous surveillance	Our data is perpetually surveilled through cameras, sensors, and analytics in a way that could diminish student autonomy.
	Consent and legal compliance	Problems in obtaining consent (positive consent) in real-time; challenges in achieving GDPR, FERPA.
	Data ownership and control	Ambiguity on who owns student data and how it might be used, reused, or misused.
	Algorithmic opacity	Lack of algorithmic transparency can inhibit challenging and/or auditing decisions.
	Ethical AI design	Need diverse datasets, transparency, and accountability to maintain fairness and inclusivity to AI solutions.

allows us not only to rethink the design of next-generation DTs, but also to be clear about what ethical, pedagogical, and values-based traditions these innovations might invade.

7.1 Emerging technologies

The future of Artificial Intelligence-enabled Smart Data Transformations will be impacted significantly by Explainable AI (XAI), Edge AI, and Federated Learning. XAI will give further transparency and trust in AI-based decisions providing interpretable justifications for decisions when undertaken in a digital twin environment. Edge AI, on the contrary, will allow for quicker response times, reduced latency, and enhanced privacy by using localized intelligence on IoT devices and real-time data sources. Using localized intelligence will be foundational to the responsiveness and scalability of DTs. Federated Learning can support collaborative AI training and do so while preserving privacy. Rather than concentrating data in a central location, Federated Learning allows models to learn from data on separate sources while retaining data on the local device. Federated Learning enables institutions to share insights and collaborate as institutions using AI without compromising any individual student data. In addition, Federated Learning supports ethical AI development by enabling access to larger datasets for greater generalizability across schools and regions. In sum, those will all play major roles in the revolutionization of AI-enabled DTs.

7.2 Intelligent school ecosystems

The future of Smart Data Tools will be driven by the movement toward integrated multiagent systems, AI orchestration ecosystems, and experiential learning contexts. Multiagent systems will enable AI agents to communicate with others, allowing for interoperability and the collaboration between a variety of AI entities that may otherwise operate in isolation. In this scenario, AI entities will have an opportunity to operate in such a way that mirrors real-world school environments. With AI orchestration as an ecosystem layer, DTs can draw upon a wide variety of AI modules to create an intertwined digital ecosystem in schools, allocating resources on the fly, setting aspects of priority for learning tasks, and potentially resolving conflict. AI agents pertaining to learning will be able to work in tandem with AI agents monitoring well-being and mediation of competing priorities such as academic excellence versus inclusivity versus mental health. Orchestration ecosystems may be able to build "human-in-the-loop" models of AI-based recommendations to retain some level of human agency and/or accountability. Experiential learning contexts will only get more immersive, leveraging augmented/virtual reality, haptic feedback, and spatial computing. Digital Twins, a popular term these days, will allow content that involves history, science experiments that may not use live subjects, or simulations of complex social dynamics to be made available to learners in an engaging, personalized, and repeatable learning experience. AI will change how the Digital Twins are presented based on the learner's preferred speed of learning, types of learning address through lesson plans, and pedagogical goals. Immersive DTs can be transformative for training students, doing scenario planning, or testing policy.

7.3 Policy and governance models

As the applications of AI as an education system gets more involved, governance and policy responses need to ensure that ethical boundaries, regulatory oversight, and stakeholder trust are put in place. Future research should concentrate on developing strong defined models for accountability, inclusion, and transparency to AI as an education system. AI ethics committees in schools can be established that include educators, technologists, students, parents, and lawyers to set policies on data use, participate in managing the audit of AI models for bias or including legal considerations of new applications. Also, We need a regulatory framework to help to context relate to educational practices. The existing frameworks, such as GDPR, did not anticipate the complexity of dynamic analytics, or biometric monitoring, or AI-based assessment practices in schools. Future policy research needs to help to define consent for minors, data sovereignty in education, or certification for materials that are AI tools in classrooms. When governments are developing models that espouse strong principles of participatory design, students and teachers will co-create the DT system as active participants. Measuring students voice, having opt-in opportunities, flexible settings, and providing opportunities for continuous feedback will provide ownership. Having stakeholders be guided in alignment with human values and their educational mission will be a powerful tool.

7.4 Cross disciplinary research

Preparing for a new type of Smart Data Transfer is a complex challenge requiring a multilateral and interdisciplinary stance. The preparation must consider many contributors including education, computer science, psychology, sociology, and ethics. Informed teachers can provide a perspective on pedagogy and its underpinning theories (including learning styles), the dynamics of classrooms as socially defined spaces, as well as familiarity with the design and structure of curriculum. Psychologists and/or behavioral scientists might provide lots of insights for developing AI with emotional intelligence with a developed understanding of students' mood, motivation, and engagement in social interaction in classrooms. Ethicists and legal scholars can help write protocols and systems that respect rights as well as societal expectations. Collaborations like this could result in the development of AI literacy programs aimed at teachers and students where both groups have some understanding and capacity to reflect and consider the technologies being used including AI capabilities, limitations, and implications. Universities and research centers need to facilitate this collaborative activity by building interdisciplinary laboratories with access to sources of funding for education and collaboration on shared research. Engaging in global collaboration around the development of more robust technological systems could also help provide an educational solidarity across very diverse contexts as collaborative practice becomes more of a commonality than an exception.

8. Conclusion

The incorporation of Artificial Intelligence in School Digital Twins provides a revolutionary shift in how educational organizations operate, create connections, and advance. School Digital twins empowered with AI will bring issues to light about how to personalize education, manage resources, and inform policy across the school ecosystem with the ability to model real-time data, predict various outcomes based on patterns, and adapt learning practices in a continually evolving manner. Case studies show from the evidence presented, while the impediments are many, like interoperability of data, algorithmic bias, and the ethics of student privacy, the promise of creating learning spaces that are intelligent, thoughtful, equitable, and efficient is great. In the likeness of other disruptive technologies, School AI enabled Digital Twins will require more than simply resource investment, infrastructure, and policy. Schools will need to collaborate across disciplines to convene a design process with human-centered design practices, execute approaches embedding transparency with ethical and sustainable design, and inform their sustainability of process. As the education sector continues to grapple with digital transformation, the introduction of AI and School Digital Twins will create opportunities for us to rethink schooling from a process that is localized, static, and identity-driven to one that is fluid, co-created, and aware. Future research needs to look at developing integrated ethical and scalable AI-DT ecosystems providing learning that can span a lifetime, equity of access, and institutional agility, particularly when faced with a volatile environment.

References

Ağca, R. K. (2023). Using digital twins in education from an innovative perspective: Potential and application areas. *Education Mind*. https://doi.org/10.58583/pedapub.em2306

Abdullahi, S. M., & Lazarova-Molnar, S. (2024). *Toward a unified security framework for digital twin architectures*. https://doi.org/10.1109/csr61664.2024.10679442

Ahuja, V. (2023). Equity and access in digital education. In *Advances in educational technologies and instructional design book series* (pp. 45–59). https://doi.org/10.4018/979-8-3693-1826-3.ch005

Algarni, A. (2016). Data mining in education. *International Journal of Advanced Computer Science and Applications, 7*(6). https://doi.org/10.14569/ijacsa.2016.070659

Almarashdeh, I. (2016). Sharing instructors experience of learning management system: A technology perspective of user satisfaction in distance learning course. *Computers in Human Behavior, 63*, 249–255. https://doi.org/10.1016/j.chb.2016.05.013

Attaran, M., & Celik, B. G. (2023). Digital twin: Benefits, use cases, challenges, and opportunities. *Decision Analytics Journal, 6*, 100165. https://doi.org/10.1016/j.dajour.2023.100165

Bruynseels, K., De Sio, F. S., & Van Den Hoven, J. (2018). Digital twins in health care: Ethical implications of an emerging engineering paradigm. *Frontiers in Genetics, 9*. https://doi.org/10.3389/fgene.2018.00031

Chen, L., Chen, P., & Lin, Z. (2020). Artificial intelligence in education: A review. *IEEE Access, 8*, 75264–75278. https://doi.org/10.1109/access.2020.2988510

Crawford, J., et al. (Apr. 2020). COVID-19: 20 countries' higher education intra-period digital pedagogy responses. *Journal of Applied Learning & Teaching, 3*(1). https://doi.org/10.37074/jalt.2020.3.1.7

Fuller, A., Fan, Z., Day, C., & Barlow, C. (2020). Digital twin: Enabling technologies, challenges and open research. *IEEE Access, 8*, 108952–108971. https://doi.org/10.1109/access.2020.2998358

Gligorea, I., Cioca, M., Oancea, R., Gorski, A.-T., Gorski, H., & Tudorache, P. (2023). Adaptive learning using artificial intelligence in e-Learning: A literature review. *Education Sciences, 13*(12), 1216. https://doi.org/10.3390/educsci13121216

Grieves, M., & Vickers, J. (2016). *Digital twin: Mitigating unpredictable, undesirable emergent behavior in complex systems*. Springer eBooks. https://doi.org/10.1007/978-3-319-38756-7_4

Ignatius, M., Wong, N. H., Martin, M., & Chen, S. (2019). Virtual Singapore integration with energy simulation and canopy modelling for climate assessment. *IOP Conference Series: Earth and Environmental Science, 294*(1), 012018. https://doi.org/10.1088/1755-1315/294/1/012018

Jankovskis, D., Cirule, I., & Carbone, A. (2024). Digital twins and E-Learning: Navigating challenges and opportunities. In *Communications in computer and information science* (pp. 169–179). https://doi.org/10.1007/978-3-031-67351-1_12

Kabudi, T., Pappas, I., & Olsen, D. H. (2021). AI-enabled adaptive learning systems: A systematic mapping of the literature. *Computers and Education Artificial Intelligence, 2*, 100017. https://doi.org/10.1016/j.caeai.2021.100017

Kritzinger, W., Karner, M., Traar, G., Henjes, J., & Sihn, W. (2018). Digital twin in manufacturing: A categorical literature review and classification. *IFAC-PapersOnLine, 51*(11), 1016–1022. https://doi.org/10.1016/j.ifacol.2018.08.474

Kuzminska, O. H., Morze, N. V., & Osadchyi, V. V. (2023). Digitization of learning environment of higher education institutions: Conceptual foundations and practical cases. *Journal of Physics: Conference Series, 2611*(1), 012024. https://doi.org/10.1088/1742-6596/2611/1/012024

Li, H., Li, H., Zhang, S., Zhong, Z., & Cheng, J. (2018). Intelligent learning system based on personalized recommendation technology. *Neural Computing and Applications, 31*(9), 4455–4462. https://doi.org/10.1007/s00521-018-3510-5

Lin, C.-C., Huang, A. Y. Q., & Lu, O. H. T. (2023). Artificial intelligence in intelligent tutoring systems toward sustainable education: A systematic review. *Smart Learning Environments, 10*(1). https://doi.org/10.1186/s40561-023-00260-y

Liu, S., Guo, X., Hu, X., & Zhao, X. (2024). Advancing generative intelligent tutoring systems with GPT-4: Design, evaluation, and a modular framework for future learning platforms. *Electronics, 13*(24), 4876. https://doi.org/10.3390/electronics13244876

Luna, J. C. (2024). *AI in education: Benefits, challenges, and ethical considerations*. https://www.datacamp.com/blog/ai-in-education.

Matos, T., Santos, W., Zdravevski, E., Coelho, P. J., Pires, I. M., & Madeira, F. (2025). A systematic review of artificial intelligence applications in education: Emerging trends and challenges. *Decision Analytics Journal*, 100571. https://doi.org/10.1016/j.dajour.2025.100571

Ning, G., Luo, H., Yin, W., & Zhang, Y. (2024). Exploration of the application and practice of digital twin technology in teaching driven by smart city construction. *Sustainability, 16*(23), 10312. https://doi.org/10.3390/su162310312

Paolucci, C., Vancini, S., Bex, R. T., II, Cavanaugh, C., Salama, C., & De Araujo, Z. (2024). A review of learning analytics opportunities and challenges for K-12 education. *Heliyon, 10*(4), e25767. https://doi.org/10.1016/j.heliyon.2024.e25767

Pathak, N. D. R. K., & Upadhyay, N. D. P. (2024). Integration of digital twins technologies in education for experiential learning: Benefits and challenges. *Deleted Journal, 2*(03), 442–449. https://doi.org/10.47392/irjaeh.2024.0064

Philips. Philips healthcare. Philips. https://www.philips.co.in/healthcare.

Safitry, T. S., Mantoro, T., Ayu, M. A., Mayumi, I., Dewanti, R., & Azmeela, S. (2015). Teachers' perspectives and practices in applying technology to enhance learning in the classroom. *International Journal of Emerging Technologies in Learning (iJET), 10*(3), 10. https://doi.org/10.3991/ijet.v10i3.4356

Selim, A., Ali, I., Saracevic, M., & Ristevski, B. (2024). Application of the digital twin model in higher education. *Multimedia Tools and Applications*. https://doi.org/10.1007/s11042-024-20014-3

Shi, H., Li, C., Wang, Y., & Dong, C. (2022). *Design and implementation of intelligent education platform with film and television features centered on textbooks*. https://doi.org/10.1109/icis54925.2022.9882501

Sholi, N. R. T., et al. (2024). Application of computer vision and Mobile systems in education: A systematic review. *International Journal of Interactive Mobile Technologies (iJIM), 18*(01), 168–187. https://doi.org/10.3991/ijim.v18i01.46483

Siemens industrial IoT. siemens.com Global Website. https://www.siemens.com/global/en/products/automation/topic-areas/it-ot-convergence/siemens-iiot.html.

Sinthamrongruk, T., Narongtuwapan, K., Wongkum, P., Suangswang, P., Thongphun, K., & Akarajaka, T. (2024). *Smart facilities management using digital twin with 3D web-based technology*. https://doi.org/10.1109/ectidamtncon60518.2024.10480018

Tao, F., et al. (2018). Digital twin-driven product design framework. *International Journal of Production Research, 57*(12), 3935–3953. https://doi.org/10.1080/00207543.2018.1443229

Teräs, M. (2022). Education and technology: Key issues and debates. *International Review of Education, 68*(4), 635–636. https://doi.org/10.1007/s11159-022-09971-9

Thai-Nghe, N., Drumond, L., Krohn-Grimberghe, A., & Schmidt-Thieme, L. (2010). Recommender system for predicting student performance. *Procedia Computer Science, 1*(2), 2811–2819. https://doi.org/10.1016/j.procs.2010.08.006

Wang, S., Wang, F., Zhu, Z., Wang, J., Tran, T., & Du, Z. (2024). Artificial intelligence in education: A systematic literature review. *Expert Systems with Applications, 252*, 124167. https://doi.org/10.1016/j.eswa.2024.124167

Woolf, B. P. (2009). *Building intelligent interactive tutors: Student-centered strategies for revolutionizing E-learning*. Morgan Kaufmann.

Zhang, J., et al. (2024). The effectiveness of a digital twin learning system in assisting engineering education courses: A case of landscape architecture. *Applied Sciences, 14*(15), 6484. https://doi.org/10.3390/app14156484

Chapter 29

Multiple digital student attendance check through blockchain

S. Ramamoorthy[1] and R. Poorvadevi[2]
[1]*Computing Technologies, SRM Institute of Science and Technology, Kattankulathur, Tamil Nadu, India;* [2]*Department of Computational Intelligence, SRM Institute of Science and Technology, Kattankulathur, Tamil Nadu, India*

1. Introduction

Today, schools are quickly adopting digital tools to make administration easier and improve the learning environment. However, checking in students and keeping track of their attendance are still important tasks that have always been done by hand. Being present is not just a formality; it's also a very important part of judging academic success, enforcing rules, allocating funds, and following school rules. Traditional ways of keeping track of student attendance, like paper logs, biometric devices, or centralized websites, are being shown to be less effective, more likely to be hacked, and less clear. Paper-based systems take a long time and can be hacked or changed on purpose. Biometric attendance systems, on the other hand, often need expensive infrastructure, raise privacy concerns, and aren't always stable when used on a large scale. Even digital attendance systems that use centralized databases can have data stolen, managers change the data, or the system fails. Together, these problems show how badly we need a trustworthy, unchangeable, and verifiable attendance system that can work in both real and virtual classrooms.

The conversion to digital in education has made it possible for schools to run more efficiently and better. One change is the use of digital check-in systems. Students can mark their attendance with QR codes, mobile apps, GPS-based confirmation, or face recognition. The goal of these systems is to cut down on administrative work, allow monitoring in real time, and make attendance statistics more accurate. But even though digital check-ins fix some problems with the way things work now, they still use centralized systems and depend on trust in a single authority, usually the school or service provider. They are centralized, which makes them open to data theft, unauthorized changes, single points of failure, and slow synchronization in settings with multiple campuses. In mixed or online classrooms, where students arrive from all over and use different devices, it is much more difficult to ensure that attendance records are fair and accurate. Therefore, even if everything is now digital, the problems of trust, openness, and being able to be checked still exist, exposing a significant management system flaw in institutions.

One particularly robust and innovative approach to handling these issues is blockchain technology. Blockchain originated as the foundation for Bitcoin and other cryptocurrencies. Since then, it has evolved into a versatile framework for safe and distributed data management in various spheres, including healthcare, education, finance, and, more recently, supply chain. A blockchain is a shared digital record used for safe, unambiguous, nonchangeable tracking of occurrences. Every piece or block in the chain is linked to the one before it, creating an unchangeable record contingent on the agreement of the network. This cannot be modified characteristic ensures correct data; decentralization eliminates the necessity of a single authority. Blockchain is therefore ideal for applications that require a high degree of openness, security, and trust. Blockchain can fundamentally alter the way attendance is tracked by ensuring that every check-in is logged permanently, can be checked by everyone, and cannot be altered or erased in relation to student check-in systems. Common transgressions, such as faked records, proxy attendance, and accessing private data without authorization, can be stopped.

This work proposes a blockchain-based student check-in system. It will take advantage of the scalable and lightweight DLedger platform. Strong security elements abound on conventional blockchain systems such as

Digital Twins in the Smart Classroom. https://doi.org/10.1016/B978-0-443-41632-3.00006-0

Ethereum and Bitcoin, but they sometimes consume a lot of resources and require a lot of bandwidth and computational capability. For real-time applications in academic institutions, particularly those with restricted infrastructure, they are therefore not suitable. By substituting a Directed Acyclic Graph (DAG) structure for a straight chain, DLedger avoids these issues. Important for controlling regular check-ins at various sites is the DAG-based method's speed in authorizing transactions, decrease of latency, and scalability enhancement of the system. Specifically built for places with minimal computer capability, and the Internet of Things (IoT) is DLedger. This is fantastic on college campuses, where sensors, RFID systems, smart and mobile devices can all be used as nodes in a dispersed attendance network.

The project's blockchain-based check-in system is meant to interact with numerous digital attendance systems, including QR codes in the classroom, online login verification, entry sensors at the gate, and GPS-based verification throughout several departments and sites. Smart contracts construct and validate a new block with ID, timestamp, location, and course session every time a student checks in. They ensure that the check-in meets particular requirements, such as being in a registered course and showing up at the right place at the right time. After check-in, the transaction is entered into the DLedger network and stays permanently visible to all authorized users, including managers, educators, and students. Knowing this, everyone develops trust, as no one can change or delete attendance data using this distributed technique. Real-time dashboards and analytics tools combined will also enable automatic reporting for performance tracking, academic warnings, attendance-based grading, and institutional checks.

Apart from addressing waste and cheating issues, the suggested strategy tackles ideas of data security and empowering students. Unlike existing systems, which do not give access to their own records, a blockchain-based attendance ledger gives students access to their check-in history at any moment. It encourages people to be responsible while protecting data with restricted access and security. On the other side, institutions have less work to perform on administration, the possibility to observe trends in attendance, and peace of mind during internal or external reports. DLedger makes the system small and light, so it can be used in places with limited resources. Its scalability also lets it grow to bigger networks, such as interinstitutional platforms where students from different schools or universities can share and verify attendance records together.

In conclusion, adding blockchain technology to digital systems for checking in students is a huge step toward open, safe, and free school management. Blockchain fixes the problems with old and controlled digital systems, so that attendance records are correct, can't be changed, and can be checked in real time. DLedger is especially useful for academic deployments because it is efficient and scalable. This makes sure that the suggested system can work well in a variety of settings. As more and more school districts use digital tools for teaching and testing, safe attendance tracking becomes essential. This is needed not only for better management but also to keep the school community honest, fair, and trustworthy. For better responsibility and openness in managing students, this project not only suggests a new piece of technology, but also a change in how schools can use decentralization.

2. Literature survey

A student's attendance is always important in school because even one absence can have a big effect on their work and tasks related to discipline. Even though there are many ways to record students' attendance these days, it is still important to do it correctly (Geetha et al., 2023). Face recognition-based systems for keeping track of attendance are becoming more common in schools and universities. The automatic attendance management system has two problems. First, it needs an administrator to keep an eye on the students' attendance. Usually, it's hard to deal with a lot of kids at once. Another issue with the method is that records need to be kept safe from being faked. Blockchain technology is a form of independent management that can help keep private information safe. The purpose of this project is to create a web-based app with a notification system that will let teachers keep an eye on their mentees' attendance while they are sitting in their place. The work is split into two parts: first, using the CNN model to automatically find and analyze faces; and second, setting up warning systems and making logs to look over. The produced log is kept in the blockchain network.

Security cams are now used to keep an eye on schools and other educational facilities. Utilizing analytical functions like headcount, face detection, face recognition, and the black- and white-list filtering makes it possible to generate an accurate attendance record of all the students present in class. Making use of decentralized blockchain technology is a good way to store information safely. Building a blockchain by the university would be the best option because the school would be in charge of setting the rules for the operation (Bálint, 2021). A smart contract needs to be put into action in order to connect the blockchain to the NVR (Network Video Recorder) unit and get the system for writing attendance records up and running. When a smart contract is closed, it's important to make sure that the correct data was entered.

Blockchain is one of the newest technologies that is getting more and more attention from people in many areas, including the education sector. This is mainly because it has good qualities, like not being controlled by one group, being open and clear, being able to track transactions, being safe, and being dependable. Even though blockchain has some benefits, it still has some problems, and not many people are using it yet (Mohammad & Vargas, 2022). That's why this study's goal was to look at papers that have already been written about the problems with using blockchain in education. The study looked at scientific papers that came out between 2017 and 2022. Of the records that were looked over, 32 were read in full. Based on the technology-organization-environment (TOE) framework, 14 challenges were mentioned and put into groups in this review. This study also showed that, compared to technological barriers, organizational and environmental barriers didn't get as much attention in the literature.

The growing market for the IoT makes it more difficult for ledger systems to save and guard the massive volumes of consumer data that IoT devices acquire. Real, accessible, and correct records are absolutely crucial for clients as well as for corporate suppliers (Zhang et al., 2019). In this paper, we discuss DLedger, a distributed ledger system with both light and robust nature. Rather than one chain of blocks, DLedger creates the ledger over a directed acyclic graph (DAG). In this sense, its purposes can be fulfilled even in cases of a split in the network or occasionally limited connectivity. DLedger employs Proof-of-Authentication (PoA), which is simpler for IoT devices to utilize since it is less sophisticated than Proof-of-Work (PoW), which calls for a lot of processing capability. Named Data Networking (NDN), a data-centric network, underlines DLedger as well and facilitates data sharing between several IoT systems.

Two or more private individuals can bargain online using a smart contract (SC), which substitutes for the need of a third-party presence. It is a deal that runs itself and is written in computer code. The blockchain is what smart contracts use to work. This means that both the code and the deal are kept on a public database that is shared by many people and cannot be changed. There are many possible uses for smart contracts in the digital economy, such as in healthcare, banking services, management, and the IoT (Samanta et al., 2021). Ethereum and Hyperledger are two of the most popular open-source blockchain systems that are used across many industries. Big technical problems like security, privacy, correctness, and verifiability are some of the blockchain problems that haven't been fully fixed yet. In this paper, a full analysis of smart contracts is presented. Along with that, we show a case study of a University exam system with a mixed data format. With this deployment, we now have a full picture of the smart contract system. It has also been used to find and study the gaps in the current state of smart contracts.

The rise of blockchain technology in the past 10 years has led to the creation of many decentralized use cases in many areas, including education. This paper gives a unique bibliometric and qualitative analysis of the blockchain in education (Ocheja et al., 2022). It includes new information on how it has changed over time, new themes that keep coming up, and real-life examples of how it has been used and combined with other educational technologies. We try to figure out who the big players are in the space, what demographics are involved and adopting it, what the current hot topics are, where there are gray areas, and where new ideas could be developed. Based on what we've learned, the blockchain has been around for about 13 years, but it only became popular in schools 5 years ago. This study also shows that most of the work has been done on reporting and checking academic transcripts and certificates. Not much work has been done on reporting and linking more detailed academic records, such as learning behavior logs, learning contents, and evaluation data. This is a problem because the blockchain systems used for education right now don't take into account how different schools make and use academic data or how they can work together. Lastly, we talk about what our results mean for different groups of people, possible solutions, and parts of education blockchain research that can help make things better in the classroom.

Data are becoming more and more important to many businesses and the basis of whole systems. This is also true for intelligent transportation systems (ITS). Smart services for people are being made possible by the ability of smart vehicles and devices to collect and share data about their surroundings (Zichichi et al., 2020). We describe a system design that uses distributed ledgers and other related technologies to help the growth of ITS in this paper. These make it possible for people to create, store, and share data that is created by sensors on their phones or cars while they are being used. We plan to use Distributed Ledger Technologies (DLTs) to build a framework that will make data immutable, traceable, and verifiable. IOTA is a potential DLT for the IoT. It is used with Decentralized File Storage (DFS) to store and encrypt data (and the metadata that goes with it) that comes from cars or users' phones. Then, Ethereum is used as the base for smart contracts that manage the sharing of data by setting up access controls. When distributed key management methods and Zero Knowledge Proof are used, privacy is guaranteed. In order to find out if DLT and DFS technologies are ready to support complex services like those in ITS, we show the results of an experiment that used real traces. It is clear from the results that the plan is workable, but more work needs to be done on making DLT infrastructures more responsive.

Face recognition is being used more and more for authentication in almost all areas. Today, in the age of information, identification is very important. People want faster and safer ways to prove who they are. Face recognition has become more important, as image processing tools like OpenCV have become more common. With blockchain, data could be saved in blocks that are spread out across the network (Dasi et al, 2023). Blockchain is an incredibly safe way to store data and keep it safe from hackers. This new technology is very radical and could change every part of society. This study aims to use open-source computer vision (OpenCV) to create a model for detecting faces that will be used in a blockchain-protected Attendance Monitoring System. Not only will it simplify the process of taking attendance, but it will also make the system completely safe. Using OpenCV, this system will take a live video feed from a camera, find the faces of students, and record their attendance along with the time they came in. The data will be stored in a way that makes it available to everyone over the blockchain network. However, the data can't be changed.

The present work offers a thorough, systematic review of blockchain-based solutions applied in the educational field. Use cases were grouped by the authors into four categories: student credential verification, attendance tracking, transcript sharing, and administrative automation (Alammary et al., 2019). The paper underlines how blockchain provides tamper-proof, verifiable documents, therefore building confidence among institutions, students, and outside parties. The study also notes various limits, such as the need for standardization, regulatory uncertainty, and infrastructure lacking. These disclosures show how important and versatile blockchain is for decentralizing authority and enabling data immutability, thereby improving attendance systems.

The developers place in EduCTX, a blockchain-based solution designed to manage academic credits, student performance records, and attendance (Turkanovic et al., 2018). Although mostly concerned with credit transfer, the system also adds blockchain entries to verify student participation and completion. Emphasizing a framework upon which digital check-in systems can be built, the platform demonstrates how distributed ledger technology may be tailored to suit academic activities outside credentialing. EduCTX demonstrates how distributed systems could protect student privacy and data integrity while improving interoperability among educational institutions.

3. Proposed method

The suggested method consists of a blockchain-based digital check-in mechanism. This is aimed at making school checking-in student processes more dependable, straightforward, and efficient. Typical attendance-tracking methods, whether manual registers or centralized systems, are vulnerable to hacking and data loss. Our approach employs blockchain technology to provide a distributed, immutable, verifiable system for managing attendance that can be accessible through role-based interfaces, therefore addressing these challenges. Three core user modules, Admin, Faculty, and Student, formulate the architecture. Every one of these modules contributes to ensuring the safety and cooperation in the attendance area. The heart of the system is a multilayered identity and access control mechanism. The process begins on a login screen, informing users of actions depending on their jobs. Administrators who log in will find a dashboard allowing them to add faculty members. This makes sure only authorized teachers can access the system, so maintaining its safety. Teachers are granted access to a staff dashboard following successful registration, so they may register students and monitor attendance. Every user is checked out and added in a scheduled manner using this hierarchical registration system.

Staff members can register students and provide each one a unique ID comprising course details, an enrollment number, and biometric or RFID-based tags when they log in to their dashboard. These figures ensure that every check-in event relates to a particular student. Through the interface, the system lets teachers record attendance, therefore simplifying digital check-in. Using QR codes, facial recognition, or IoT sensors at the classroom door will help the teacher either manually or automatically validate the presence of a student. The attendance-taking aspect of this approach is among its most crucial elements. Every attendance event is seen as a transaction with the student's identity, the course ID, and confirmation from the teacher, all included. These occurrences are immediately entered straight into a blockchain record. Common in large institutions, a lot of people need to attend a lot of events, thus using a Directed Acyclic Graph (DAG)-based blockchain architecture like DLedger speeds up and widens the processing of transactions, making it ideal for such cases. Once a record is entered into the blockchain, anyone with authorization can check it, and it cannot be altered. Through logging into their Student Dashboard, students can examine their own attendance records in real time. This increases system openness and fosters faith in it.

The distributed character of the blockchain layer is a great advantage, since it guarantees that attendance records are kept on several nodes instead of one central server. This guards the system against potential data leaks, inadvertent deletions, or internal corporate threat generation. Every node maintains a copy of the ledger, all of it or only a portion of it. Consensus systems like Practical Byzantine Fault Tolerance or Proof of Authority verify every fresh transaction for legitimacy. The system makes sure it is highly available and resilient by getting rid of single points of failure. The suggested system also allows for auditability and traceability. Anytime, schools, accrediting bodies, or government agencies can check the attendance logs instead of just looking at their own records. Every transaction on the blockchain is time-stamped and cryptographically linked to previous transactions. This makes it impossible for fake attendance logs to exist. All data entries can be checked by anyone in the authorized network, which also protects faculty members from possible disputes or claims of bias.

The system makes operations more efficient by automating many routine chores. It also makes security and accountability better. Without any help from a person, alerts can be sent out for low attendance, reports can be sent automatically to parents or guardians, and managers can see real-time dashboards. This makes sure that attendance rules are followed, and lessens the workload of the teaching and administrative staff. The system architecture is made to be scalable and modular from a technical point of view. It can be put into use on either a private or a consortium blockchain, depending on what the institution needs. Also, the front-end platforms for Admin, Faculty, and Students are made with responsive web technologies or mobile apps, so they can be used on any device. It is the back end's job to verify transactions, carry out smart contracts, and talk to the blockchain network.

Finally, this suggested method creates a modern, reliable, and useful way to keep track of student attendance using blockchain technology. The system uses safe digital interfaces, decentralized ledgers, and smart contracts to make sure that everything is honest from start to finish. This cuts down on administrative work and gets students more involved. The method can be used not only in schools, but also in professional training centers, online certification programs, and any other situation where participants need to be reliably verified. It is a forward-looking way to run an educational institution because it fits with new digital governance standards.

4. DLedger-based student check data management

The employment of technology in various administrative and academic procedures is generating a major revolution in the educational field in the present digital era. Among them, one of the processes for controlling student attendance has changed significantly. Two classic techniques that are not only time-consuming but also prone to errors and manipulations, are roll calls and hand attendance records. Blockchain technologies seem to be a good solution, as companies aim for security, accuracy, and openness. In terms of student check-in systems, DLedger, a lightweight, distributed ledger architecture, provides a sensible and quick solution free of artificial intelligence (AI). DLedger runs in systems with limited processing capacity. DLedger makes use of a hash chain and a gossip protocol, unlike traditional blockchains employing energy-intensive consensus mechanisms, such as proof-of-work and high-performance nodes. This lets it operate without malfunction on common hardware, including departmental servers or cell phones. Its distributed architecture causes data to be disseminated among many nodes rather than gathered in one central location.

Students who log into a mobile app or scan a classroom QR code start the DLedger-based student check-in procedure. Important information included in the check-in request this action generates is course data, timestamp, and student ID. The system checks the student's identity using present credentials, either a password or a digital signature. After validation, the data becomes a block with crypto-security. This block then creates an unbroken and impermeable chain of records by connecting to the previous ledger entry. Every node linked also gets the new block, therefore guaranteeing network synchronization and consistency. One of the main benefits of DLedger is not depend on AI to run. Many modern technologies track attendance automatically, using AI for facial recognition or behavioral analytics. These ideas, fresh as they are, have flaws, including false positives, high energy use, privacy issues, and infrastructure needs. Moreover, usually requirements of AI systems usually require constant upgrades and training sets, which can be costly for educational institutions. On the other hand, the straightforward, dependable method provided by DLedger solves these problems while also permitting exact and open attendance control (Fig. 29.1).

The immutability of DLedger raises dependability even more. Data once recorded cannot be removed or modified without being observed. This guarantees that, when needed, every attendance record is accurate and can be used as either academic or legal proof. Real-time access to these figures also allows teachers to create reports or identify patterns in student participation. Other nodes guarantee business continuity and thereby avoid data loss in the distributed character of the system by maintaining the full ledger even in the case of a node failure or compromise. Moreover, capable of offline actions is DLedger. Local check-ins, timed with the main network when a link becomes available, are made possible by the technology. This is helpful in far-off or infrastructure-limited locations where internet access could be intermittent. Institutions in rural locations or those with inadequate internet access will especially gain from this capability. Imagine a university where, for instance, every department maintains a distinct DLedger node. Students note their classroom attendance using cell phones. The departmental ledger is instantly entered with the information. Every student's attendance record for every course is shown on a combined report created at the conclusion of the week.

FIGURE 29.1 Architecture diagram.

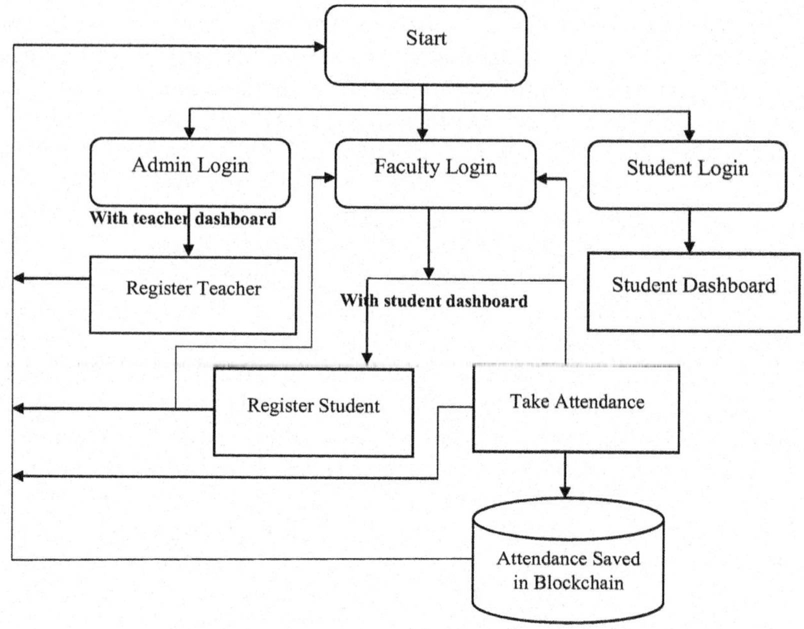

Though none of this is automated by AI, the system remains rapid, clear, and accurate.

All things considered, student check-in systems built on DLedger offer a quite good, safe, and reasonably priced replacement for traditional and AI-driven attendance systems. It's distributed, and tamper-proof architecture, offline operations support, and the ability to run without costly hardware help modern educational environments immensely. Academic data management may become more reliable, trustworthy, and sustainable over time if universities keep embracing digital transformation by means of blockchain-based platforms.

5. Digital student check-in process steps

5.1 Pre-check-in: Registration and onboarding

Before the blockchain-based digital check-in system is put into action, all users have to follow a set onboarding process. This first phase ensures that a flawless and safe attendance recording system is built upon. The registration procedure starts with enrolling all participants, including teachers, administrative staff, and the student body, into the system. Every student has a unique identity that acts as their digital fingerprint inside the blockchain, designed to maintain data integrity and traceability. This digital ID could be based on their present institutional enrollment count or developed as a new, distinct code. The identification is reinforced, depending on the institution's choice, by further security measures like mobile authentication or biometric verification. For instance, students could confirm their records. Nowadays, the administrative personnel are absolutely vital. Their main duty is to enter authenticated student data into the blockchain database, such that every student's identity exactly connects to their academic and personal background. System administrators also assign faculty members to teach courses and students to fit those courses. This ordered mapping is absolutely vital since it later enables very accurate tracking of course-by-course attendance.

Students also have to use a web-based dashboard or a mobile app created especially for the university to access the check-in system. Students link their mobile number or biometric data, if pertinent, to fully activate their digital identities after logging in. Their attendance each time they show up for class can then be recorded and kept permanently on the blockchain. Every student enters a permanent, verified identity into the blockchain network by the time they complete the registration and onboarding process. This guarantees the security and dependability of every attendance record to be created. On a distributed ledger, the data is tamper-proof and resistant to unwanted changes. This stage ensures validity and fairness in all upcoming attendance monitoring initiatives, therefore providing a firm basis for an accurate digital check-in system.

5.2 Class schedule and QR code generation

Priorities are user registration and onboarding; secondly, a crucial stage of the digital student check-in process is planning and scheduling academic sessions. This process starts once the academic calendar and subject-specific schedules have been finalized and system entry. The blockchain-based technology creates a dynamic Quick Response (QR) code or a time-bound session ID for every class session using this timetable, therefore serving as a unique digital identity. Since they are created especially for every class and include important metadata guaranteeing their authenticity and applicability, these scripts are neither general-purpose nor stationary. Several key factors affect the completely automatic development of the QR code or session ID by the system: the course code for that class, the lecturer's identity, the date and hour of the lecture, and the classroom or virtual conference venue where the session would take place. These elements provide the session with its unique framework and ensure that the code is tightly connected to a certain teaching moment. Furthermore, the QR code is constructed using safe encryption methods that prevent copying, reuse, or changing; it is thereby cryptographically protected. This function is absolutely important, since it ensures that the system cannot be influenced by fictitious scanning or proxy attendance.

Using their own safe login credentials, the faculty member scheduled for a lecture gains system access and starts the class when called for. Once the login process ends, the technology checks the instructor and lets them activate the QR code specific to their lesson. This QR code then appears on the classroom projector or is sent via the virtual platform, depending on whether the class is in-person or online. The code's time limit, which typically matches the real class timing, adds even more control and helps to avoid check-ins outside of the allocated window. Verify the integrity of the process; the system performs several important background tasks. Every created QR code has a compilation of hidden data, including the course title, subject name, faculty ID, exact creation time, and session-specific identifiers. All of this metadata is hashed and encrypted prior to being stored on the blockchain, therefore rendering it transparent and unbreakable. Once kept, this information is entered into the permanent digital ledger and allows for any necessary future audits and validation. Thanks to this clever, contemporary approach that replaces the outdated, error-prone methods of recording attendance, only students present during the session window can register their attendance. The solution provides a strong base for exact, unbreakable attendance tracking by linking the code with a specific class activity and protecting it with encryption and time limits.

5.3 Check-in by the student (live attendance)

Starting a class session sets the most crucial event of the blockchain-based attendance system: the digital check-in process of the pupils. By now, blockchain is essential for maintaining the accuracy, security, and integrity of attendance records. Unlike conventional systems, whereby attendance may be manually logged or vulnerable to manipulation, our digital solution ensures that every check-in is independently vetted, recorded, and permanently saved inside the blockchain network. Students should be expected to scan a QR code to confirm their classroom attendance. Scanning QR codes is the most widely used technique. Each lecture starts with the faculty creating and displaying a special QR code on the screen or projector in the classroom. As was previously said, this code is time-bound and unique to that session. Students utilize a specific program that the college provides on their mobile devices to scan the QR code. When the student scans the QR code, the system instantly displays their distinct digital identity, which was generated during the enrollment process. This identification is safely checked against the current class session's metadata, which includes the course, faculty, date, and time. When all the conditions are met, the student is recorded as present, and the attendance record is stored on the blockchain for future reference. For a seamless and frictionless experience, the entire procedure takes only a few seconds and occurs in real time.

Biometric authentication provides an extra or different way for students to check in at universities with more sophisticated technologies. Devices for facial recognition and fingerprint recognition may be installed in classrooms. Students only need to touch the scanner with their finger when they get to class, or they can let the system take pictures of their faces. The encrypted biometric information that was saved in the system throughout the onboarding process is then compared with the biometric input. The system verifies the student's identity if the input precisely matches the data that has been stored. Like the QR code system, the attendance is then instantaneously posted to the blockchain after being recorded. The principles of time stamping and verification are shared by both approaches. These are cryptographically protected entries in a distributed ledger that cannot be changed or removed, not only digital affirmations of existence. Students cannot check in before or after the planned session, since a timestamp guarantees that the attendance is recorded precisely at the moment it happens. Furthermore, it is nearly impossible for one student to check in on behalf of another, since each check-in is linked to a distinct and validated digital identity. This successfully resolves widespread malpractice problems that have long afflicted manual systems, such as proxy attendance. Ultimately, the check-in phase is a good example of the three primary benefits of blockchain technology: transparency, decentralization, and security. The academic monitoring system becomes more dependable and efficient, as each attendance record becomes an impermeable block in the distributed ledger. In addition to faculty members having less work to do, schools may rest easy knowing that their attendance records are accurate and traceable at all times.

5.4 Data validation and blockchain logging

- All related parameters (proper class, time, location, ID match, etc.) are checked by the system.
- The blockchain network is notified of the creation of a digital transaction.
- After being verified, each attendance record is put in the ledger as a separate block.

The block contains:

- Student ID (hashed for privacy)
- Session ID and subject
- Date and time
- Method of check-in (QR or biometric)
- Device/IP/location data (if applicable)

This information cannot be removed or changed after it has been added, guaranteeing the accuracy and traceability of the attendance record.

Faculty's Role: They are informed in real time about the number of students who have checked in and those who are still pending.

5.5 Post-check-in reporting and analysis

The blockchain-based system moves into the postsession phase, when recorded attendance data is gathered, processed, and displayed, after a class session ends and all check-ins have been finished. The capacity of the suggested digital check-in system to automatically create attendance dashboards for all parties involved, students, instructors, and administrators, is among its most advantageous features. This ensures accountability, transparency, and simple access to real-time data. Because of the role-based access feature included in these dashboards, each user is given visibility based on their rights and duties within the institutional ecosystem. The dashboard serves as a personal attendance record for pupils. They can instantly monitor their updated attendance status by logging in to their portal or mobile app following each session. Broken down by subject, the dashboard displays the student's attendance in classes missed, attended, or arrived late. This self-monitoring feature helps students to keep track of their academic status over the semester, and empowers them to take control of their attendance record. Even for students who often miss class, the system can deliver customized alerts, encouraging them to take corrective action before their attendance falls below the required standards. Administrators, in charge of institutional monitoring and compliance, have the most complete view. On their dashboard are comprehensive attendance statistics for every department, course, and student group. For performance reviews, university applications, and internal audits, among other uses, the system may automatically create monthly and semester-by-semester attendance summaries. Because the data is kept on a blockchain ledger, managers could rely on its immutability and integrity. Unlike conventional systems, whereby data may be either deliberately or unintentionally modified, blockchain guarantees that a record is tamper-proof once it is entered. This gives companies even more confidence in the accuracy and impartiality of their reporting.

The distributed character of blockchain is crucial in this environment. Although all relevant stakeholders have access to their own dashboards and data, none of the users, not even the system administrator, can change or delete attendance information. All of which were typical flaws in hand-operated or semiautomated systems, which removes the possibility of partial automation, human mistakes, or data fabrication. Blockchain allows access, thereby democratizing data, and also protects it. Apart from the normal attendance recording and display, the system incorporates sophisticated features, such as performance statistics and absentee notices. Students and, if necessary, their guardians could get automatic emails or SMS notifications if they surpass a designated absence level. In the meantime, academic authorities and educators should examine test results and attendance records to see whether low attendance lines up with poor academic performance. Encouragement of proactive intervention supports this data-driven strategy to increase classroom participation and academic success. Eventually, the postsession dashboard system transforms unprocessed attendance data into perceptive knowledge and helpful intelligence. Combining modern analytics with blockchain's advantages of transparency, decentralization, and immutability creates a robust, creative, and student-oriented academic administrative tool.

6. Implementation work

6.1 Administration activities and roles

- Establishing new members and making sure the system is run correctly falls to the administrative officer.
- To get to the Teacher Dashboard, administrators must log in through a private portal.
  This is where they can:
  - Add faculty members to the list and give them specific rights.
  - Check user activity and data flow.
  - Verify data consistency among every blockchain node.

6.2 Faculty roles and purposes

- Faculty members must authenticate using the Faculty Login page.

- Once logged in, students can view the Student Dashboard, where they can:
  - Register students and link them to suitable sessions or courses.
  - Track trends in attendance status.
  - Start the attendance check-in process in front of the class using:
    - Scan QR codes or
    - Biometric confirmation, that is, fingerprint or facial recognition.
    - Before being published to the blockchain, the attendance records are hashed and verified.

6.3 Student involvement and role

- Students log into the system via the Student Login portal.
- They are sent to the Student Dashboard once authorized, where they may:
  - View personal attendance history.
  - Real-time confirm check-in status.
  - Ask questions should disparities exist across groups.

Students are supposed to use their unique ID to check in at assigned times.

6.4 Integration of blockchain

- Each present event (the check-in time) is hashed, given a time stamp, and then added to the blockchain.
- This process makes sure that:
  - **Data consistency**: The audience can't be changed or destroyed after it has been saved.
  - **Openness**: Teachers and students can look up old records at any time.
  - **Protection**: Given that blockchain is distributed, changes that aren't supposed to be made can't happen.

6.5 A summary of workflow

The main flow is shown below:

- Administrators sign up teachers;
- Teachers sign up students.
- Students log in and check in.
- Faculty checks attendance.

Data that is put on the blockchain cannot be changed. Dashboards show changes that happen in real time.

6.6 Advantages of this method

- Gets rid of attendance fraud and notes made by proxy.
- Builds trust and responsibility between institutions and students.
- Makes sure that records can be kept for a long time and can be checked.
- Works perfectly with school enterprise resource planning (ERP) or learning management system (LMS).

Table 29.1 presents the data samples for this proposed method. A number that is unique to each student. It helps make sure that each student's record is unique. In this case, "STU001" stands for student Saran R., the whole name of the student who is coming to class. The course code for which the attendance is being tracked, for example, BCA401 for a BCA fourth-semester topic. Name the faculty member who is leading the class. "FAC101" could stand for Dr. S. Kumar, for instance. The date that the class met and the check-in took place. This helps put attendance information in order of when they happened. The exact time the student came in. This can help find entries that are on time or late. Let you know if the student was there or not during the lesson. This is usually taken care of immediately when a QR code is scanned or a biometric login is used. A one-of-a-kind transaction number is made when the attendance is added to the blockchain. This ID proves that the record is saved permanently and can be checked. For instance, 0xa93e2bfa ... 5d7c is a shorter form of a real blockchain hash.

7. Results

Through the blockchain-based multiple digital student check-in system in place, accuracy, openness, and data integrity are much better than with old-fashioned methods of managing attendance. Using smart contracts and decentralized ledger technology together, the system made

TABLE 29.1 Data samples.

Student ID	Name	Course Code	Faculty ID	Date	Time (AM)	Status	Blockchain Text ID
STU001	Saran R	BCA401	FAC101	19-06-2025	9:00	Present	0xa93e2bfa ... 5d7c
STU002	Ramya S	BSC301	FAC102	19-06-2025	9:02	Present	0xbc41e7d2 ... 3a8f
STU003	Manoj G	BCA401	FAC101	19-06-2025	9:05	Absent	0xcd68a93b ... f1b2
STU004	Sai G P	BCA401	FAC101	19-06-2025	9:07	Present	0xef34b1a7 ... 9c3e

sure that each attendance record could not be changed, was time-stamped, and could be tracked using a unique blockchain transaction ID. During the testing part, many students and teachers from different classes used the system to see how well it worked. All attendance information was recorded in real time, and each check-in took an average of 1.5 s to process. This shows that the system is very fast and works efficiently. Role-based access rules were properly put in place to make sure that only authorized users, such as teachers, could mark attendance, and that students could only see their own records. The blockchain's consensus system and smart contract validations also automatically turned down any attempts to access or alter the data without authorization, therefore demonstrating the robust security layer. By means of specific transaction hashes and block identities, the audit trail of the system lets managers and teachers follow individuals entirely, therefore enabling access to attendance statistics. Data recorded was safe, consistent, and accurate. Throughout many test sessions, no one saw any data duplication or loss. All things considered, the suggested solution not only provides a consistent and unchangeable method to track student attendance but also improves institutional openness, responsibility, and confidence in operations. These results suggest that blockchain could be a transformative tool for electronically monitoring school attendance.

8. Conclusion

In the field of digital record-keeping and the running of educational institutions, blockchain technology's inclusion in student registration systems marks a major advance. This chapter discussed the benefits of a distributed check-in system using blockchain technology to maintain attendance records safe, public, and immutable, together with the details on how to build such a system. Unlike conventional approaches that depend on centralized databases or manual entries, the proposed solution logs every check-in as a transaction unchangeable with distributed ledger technology. This approach stays away from administrative mistakes, unauthorized access, and data tampering events. It also increases the dependability of the operations carried out at the institution.

In spite of its flaws, the execution results reveal that the system is rather successful in terms of tracking attendance in real time, reacting fast, and being totally auditable. For administrative workers, instructional staff, and students, especially, the adoption of role-based access restriction is quite beneficial. Smart contracts and encryption offer protection of data privacy and security. Every single event attended has a unique Blockchain Transaction ID that, should a third party be able to confirm. This helps everyone to be more responsible and reduces the conflicts about attendance. This distributed strategy also helps to maintain data confidentiality and facilitates interactions with different departments of the university. It is used for the recording of fees, tests, and certifications, among other purposes.

Finally, for contemporary universities always striving to increase the openness and effectiveness of their operations, the blockchain-based digital student check-in system offers a fresh idea. As they become more digital, emerging technologies like blockchain will not only simplify daily operations but also allow educational institutions to manage data in a way that is both safer and more transparently viewable. In order to make these kinds of systems more valuable and ubiquitous worldwide, research may combine biometric verification, analytics driven by AI, and blockchain networks spanning many institutions going forward.

References

Alammary, A., Alhazmi, S., Almasri, M., & Gillani, S. (2019). Blockchain-based applications in education: A systematic review. *Applied Sciences, 9*(12), 2400. https://doi.org/10.3390/app9122400

Bálint, K. (September 2021). The connection of a blockchain with the students' attendance register based on security cameras. In *2021, IEEE 19th International Symposium on Intelligent Systems and informatics (SISY)* (pp. 67–70). IEEE.

Dasi, S. R., Gujar, E. S., Ansari, A. S. M. I., & Patil, Y. (April 2023). Real-time attendance monitoring system using machine learning and blockchain. In *2023 11th international conference on emerging trends in engineering and technology—signal and information processing (ICETET-SIP)* (pp. 1–6). IEEE.

Geetha, S. K., Kanakaprabha, S., Devipriya, A., Brindha, D., & Vetrimani, E. (January 2023). Implementation of a blockchain based attendance tracking system. In *2023 international conference on computer communication and informatics (ICCCI)* (pp. 1–5). IEEE.

Mohammad, A., & Vargas, S. (2022). Challenges of using blockchain in the education sector: A literature review. *Applied Sciences, 12*(13), 6380. https://doi.org/10.3390/app12136380

Ocheja, P., Agbo, F. J., Oyelere, S. S., Flanagan, B., & Ogata, H. (2022). Blockchain in education: A systematic review and practical case studies. *IEEE Access, 10*, 99525–99540.

Samanta, A. K., Sarkar, B. B., & Chaki, N. (2021). A blockchain-based smart contract towards developing secured university examination system. *Journal of Data, Information and Management, 3*, 237–249.

Turkanovic, M., Holbl, M., Kosic, K., Hericko, M., & Kamisalic, A. (2018). EduCTX: A blockchain-based higher education credit platform. *IEEE Access, 6*, 5112–5127.

Zhang, Z., Vasavada, V., Ma, X., & Zhang, L. (2019). DLedger: An IoT-friendly private distributed ledger system based on DAG. *arXiv*. doi: CoRR abs/1902.09031, 1–16.

Zichichi, M., Ferretti, S., & D'Angelo, G. (2020). A framework based on distributed ledger technologies for data management and services in intelligent transportation systems. *IEEE Access, 8*, 100384–100402.

Index

Note: Page numbers followed by "f" indicate figures and "t" indicate tables.

A

ABR streaming. *See* Adaptive-bitrate (ABR) streaming
Academic institutions, 169–170, 178, 391–400
Access and control, 298
  role-based access control (RBAC), 234
Accessibility and inclusion, 326, 336–338
  language learning, 336–337
  professional adaptation, 337
  special needs adaptation, 337–338
Acoustic sensor, 41
Adaptation and alignment with students' needs, 283, 285–286
Adaptive learning (AL), 328t, 332–333
  affective factors, 332–333
  contexts, 97
  ecosystem, 244
  EduAgent, 328t, 332
  OATutor, 328t, 332–333
Adaptive video streaming, 207–214, 216
  challenges, 207–208
  framework, 208–209, 208f, 211–212
Adaptive-bitrate (ABR) streaming, 207
Administration activities, 398
  faculty members, adding, 398
  user activity and data flow, checking, 398
Administrative automation, 326, 334–335
  attendance and engagement analytics, 334–335
  scheduling and resource management, 334
Administrative node, 125–126, 125t, 126f
Adoption and trust, 307t
Advantages of digital twins, 87–88
  innovation in classroom and educational models, 87, 90t
  personalized learning support, 87, 90t
  practical expertise, 87, 90t
Agent Hospital, 328t, 337
Agent-SiMT, 328t, 336–337
*See*AI and big data integration, 298
  behavioral analysis, 298
  maintenance forecasting, 298
  resource optimization, 298
AI and machine learning integration, 70
AI-driven chatbots and virtual tutors, 19
AI-enabled digital twin (AI-DT), 373–374, 377t, 378–389, 379f–380f, 383t, 384f–385f, 387t
  architecture and design, 378–381
  challenges and limitations, 385–386
  functional applications, 381–382
AI-powered learning, 14, 307–308, 308f
AI-powered tutoring and assessment tools, 16
*See*Air quality monitoring, 305
Ammonia (NH_3) sensor, 41
Analytics, 305–306
Anthropometry, 221t–222t, 227
  human body modeling, 151
  individual figure morphology analysis, 151
API. *See* Application programming interface (API)
Application layer, 62t, 64t, 69, 70t, 297–298, 304f, 374f, 378
  challenges and limitations, 69, 70t
Application platform, 306t
Application programming interface (API), 349t–350t
Applications of school digital twin, 164–165, 165t
  emergency monitoring and evacuation systems, 165, 165t
  real-time resource optimization, 165, 165t
  smart classroom management, 165, 165t
Architectural design, 378–381
  conceptual model, 378, 379f
  data flow and feedback loops, 379–380
  system components, 379, 380f
Artificial intelligence (AI), 1–5, 8–10, 13–19, 25, 27–28, 30–32, 73–78, 82f, 84t, 89t–90t, 95, 100, 102, 105, 105t, 109–110, 111t, 112, 115–117, 116t, 142–146, 143t–144t, 147t, 148–149, 149t, 169, 183–184, 187–188, 195–197, 201, 233, 239–245, 253, 255–256, 262, 264–266, 268–269, 271–272, 274, 280–283, 291–292, 297–299, 301–304, 303f, 307–309, 308f, 311–312, 314–315, 319, 321, 325–327, 329–340, 348–349, 349t–350t, 354t–355t, 361, 363–367, 370, 373–389, 374f, 377t, 379f–380f, 383t, 384f–385f, 387t
  AI and IoT integration, 14
  AI bias, 385–386, 387t
  AI Curation System, 199
  AI inference engines, 378–379
  AI literacy, 299
  AI models, 292, 298–299
  AI teaching assistant, 326
  AI-assisted lesson planning, 113
  AI-based dynamic learning paths, 31
  AI-based early intervention for learning disabilities, 31
  AI-based educational data mining, 74, 78
  AI-based smart class management and performance assessment, 73, 75
  AI-based smart educational agents, 76
  AI-driven assessment, 265, 272, 274
  AI-driven NPCs, 263, 266
  AI-driven optimization, 244
  AI-driven platforms, 240–241, 244–245, 253, 255–256
  AI-driven predictive analytics, 321
  AI-enabled student-monitoring system, 328t, 334–335
  AI-powered career and skill development programs, 31–32
  AI-powered recommendation algorithms, 195–196
  AI-supported collaboration tools, 328t
  algorithms, 95, 348–349, 350t, 352, 355, 355t
  data analysis, 364
  deep learning, 3–4
  digital twins, 111t, 116t
  Edge AI sensors, 88–90, 90t
  ethical AI, 370
  ethical considerations, 264, 272
  integration with DTs, 233
  integration with predictive analytics, 100
  machine learning, 3, 5, 9, 242, 253–254, 366–368
  machine learning (ML), 95
  personalization, 105t, 111t, 116t
  powered digital twins, 240–241, 244–245, 253, 255–256
  predictive modeling, 373–375, 377t, 378–379, 381–383, 383t, 385
  role in digital twins, 314
Asset and resource tracking, 142–143, 147–148, 147t, 151–154
Assistant agents, 326–329, 331–334, 336–340
  application review, 326–338
  challenges and future directions, 338–339
  methodology, 326

Attacks, 170, 176–178. *See also specific attacks*
application layer, 170
IoT-based, 170
malware, 170
Attendance monitoring, 66–68, 70–71
Attendance tracking, 391–400
digital check-in systems, 391–392, 394–398, 400
fraud, 351
traditional systems, 391, 394–400
Augmented reality (AR), 1–3, 10, 23–24, 27, 29–32, 37–39, 46–49, 51–52, 263–267, 272, 320–321, 349t, 361–362, 364, 370
applications in language education, 24
for students with special needs, 27
immersive learning, 244
in English language instruction, 264
integration with digital twins, 320
mobile devices, 264, 272
virtual worlds, 264
Authentication, 171–172, 174–178, 393–397
biometric, 396–397
facial recognition, 391–395
login and, 172, 174–176
QR codes, 391–392, 394–397, 399
robust, 176
Automated performance assessment, 73, 75
Avatar modeling, 221f
digital human models, 151
parametric modeling, 227

B

Bandwidth efficiency, 162
Benefits of school digital twin, 159, 159t
predictive maintenance, 159, 159t
resource efficiency, 159, 159t
smart learning, 159, 159t
Benefits of virtual smart environments, 26–28
cost-effectiveness, 27
effective school and campus management, 27–28
increased learning efficiency and safety, 26–27
BERT, 113
BERT-based rubrics, 113
BERT-based semantic analysis, 113
BERT-LSTM model, 113
Bias and fairness, 298
Big data, 195–197, 197f–198f, 361–363, 367–368, 370
big data center, 197f, 205
big data's dual role, 196
challenges, 367–368
predictive analytics, 367–368
Big educational data, 73–78
acquisition and mapping algorithms, 73, 77
mining and analysis, 73, 76–77
real-time analytics, 73–74, 77
Bilinear mapping, 171–172
BIM. *See* Building Information Modeling (BIM)
Biogenic amines sensor, 41
Biometric and access sensors, 292, 293f, 296–297
Biometric and behavioral sensors, 45–46
optical imaging sensors, 42
thermal imaging sensors, 42–43
Biometric scan. *See* Authentication; biometric
Blended learning, 84t, 89t
Blockcerts, 133–134, 135t
Blockchain, 1–3, 10, 73, 75–78, 101–102, 105–116, 105t, 106f, 106t, 107f–108f, 108t, 110f, 111t, 112f–113f, 114t, 116f, 116t, 119–120, 122–127, 122f, 124f, 125t, 126f, 171–172, 176–177, 195–199, 299
applications in education, 106t, 110–114, 111t, 116t
blockchain-anchored audit trails, 113
blockchain-anchored checksums, 199
blockchain-anchored competency passports, 198
challenges and limitations, 114–115
consortium, 176–177
cybersecurity, 3, 10
data block structure, 122–123, 124f, 127
data integrity, 3, 10
data security enhancement, 101
future of, 115–116
network, 122f, 125t, 126, 126f, 172, 174f
permission-grant (PGB), 176–177
security layer, 122, 122f, 127
student records management, 3, 10
transparency, 101
Blockchain and digital credentials, 364–365, 370
microcredentials, 364
smart contract, 364
Blockchain in education, 129–130, 132–139
benefits, 132–133
challenges, 135–137
future directions, 137–138
Blockchain technology, 30, 129–139, 264–266, 320–321, 391–400
characteristics, 130–131
decentralized data storage, 244
DLedger, 391–396
ethical implications, 264
immutability, 391, 393–400
privacy and security, 244
smart contracts, 392–393, 395, 399–400
types, 131
user security, 266
versus traditional databases, 131
Blockchain-powered educational platforms, 112–113, 113f, 114t
Blockcerts, 112–113, 113f, 115
ODEM (On-Demand Education Marketplace), 112–113, 113f, 115
Body–clothing system, 219–221, 225–226
interaction overview, 225
multidisciplinary foundations, 225
Building Information Modeling (BIM), 141–150, 143t, 147t, 149t, 153
Building Management System (BMS), 244

C

Campus data center, 172, 174f, 174t, 176–177
Campus management, 305, 307f
building maintenance, 214
energy management, 305
space utilization, 214
Campus sensor information center, 172–177, 174f, 174t
Carbon dioxide (CO_2), 350t, 351–352, 354t
Challenges for school digital twin, 165–166, 166f, 166t
data privacy and cybersecurity, 165–166, 166t
energy-efficient processing, 165–166, 166t
heterogeneity and interoperability, 165–166, 166t
Challenges of digital twins, 307t
computational challenges, 307t
data privacy and security, 307t
data standardization and integration, 307t
Check-in, 119–127, 122f, 125t, 126f
automated systems, 120–121
details, 123
scheme, 121–127
Chronic condition management, 353f, 354t
personalized, 374
Classroom 4.0, 186–187
Classroom digital twins, 19
Classroom scenario simulation (CSS), 328t, 329–330
CGMI framework, 328t, 329–330
Classroom Simulacra system, 328t, 329–330
Open Learner Model (OLM), 328t, 329
Clinical validation & accuracy, 355t
Cloud computing, 1, 9, 73–77, 86, 89t, 95, 99, 160–161, 163t, 169, 171, 302–304, 307, 307t, 348–349, 350t, 353–355
advantages, 160
fog–edge–cloud computing technologies, 74–75
key features, 160
limitations, 161
Cloud computing (CC), 23–24, 26, 30
Cloud platforms, 143–145, 147t, 148, 149f, 149t
Cloud storage layer, 122, 122f, 127
Collaboration, 320–321
across geographies, 244
parental involvement, 244
Collaborative engagement phase, 247t
cross-cultural interaction, 164
mentorship integration, 164
team-based activities, 164
Collaborative learning enhancement, 326, 335–336

group work facilitation, 335–336
teacher-AI co-teaching, 336
Communication & Transcoding Resource Allocation (CTRA), 216
Communication infrastructure, 94
Communication protocols, 62–63, 64t, 66–67, 70
data communication protocols, 62–63, 64t
selection criteria, 62, 63t
Computer science professional, 328t, 337
AgentCoder, 328t, 337
CodeAgent, 328t, 337
ToolCoder, 328t, 337
Computer vision (CV), 373–374, 383t
Computer vision-based performance assessment and classroom management, 73, 75
Computer-aided design (CAD), 219, 227
integration with 2D pattern drafting, 151
parametric design, 151
Computers and the internet, 14
Computing techniques comparison, 163, 163t
Connectivity and interoperability, 349t
standardization, 374
Consensus aggregation, 125
Consensus confirmation layer, 124
Consensus mechanisms, 119–120, 122–127, 125t, 138–139
Proof of Stake (PoS), 64
Proof of Work (PoW), 138
Consensus systems, 394–395, 399–400
Practical Byzantine Fault Tolerance, 394–395
Proof-of-Authentication (PoA), 393–395
Proof-of-Work (PoW), 393, 395
Consent and data privacy, 98–99, 101–102
data governance principles, 99, 101–102
informed consent, 99, 101–102
privacy-by-design models, 99, 101–102
Consent management, 187–190
collection of consent, 187–188
informed consent, 188–189
storage of consent, 187–188
Consortium blockchains, 119–120
Continuous assessment, 283–286
formative assessment, 284, 286
Cost efficiency, 106–107, 108f, 110, 111t, 112–115
administrative overhead, 106–107, 109–112, 111t, 114
operational optimization, 111t
Cost reduction, 6–8, 7f, 10–11
Cost-efficiency and sustainability, 17–18
Credential verification, 105–110, 105t–106t, 107f–108f, 108t, 111t, 112–116, 113f, 114t, 116t
blockchain-based, 105–110, 105t–106t, 108t, 111t, 112–114, 114t, 116t
fraud prevention, 105–114, 106t, 108f, 108t, 111t, 116t
MIT Digital Diplomas Initiative, 112–113, 113f
Credly, 114t
Cronbach's Alpha, 248, 250t
Cryptographic security, 105–107, 112, 115, 116f
CTRA. *See* Communication & Transcoding Resource Allocation (CTRA)
Cultural heritage preservation, 219, 226, 228–230, 233
digital archives, 226
historical costume reconstruction, 224t
museum exhibitions, 219, 226, 228–229, 231
Curriculum, 1, 3, 9–10
Customization, 226
custom garment design, 228
mass customization, 226
Cyber layer, 159, 159t
data analytics engines, 159
data processing and visualization, 159
simulation tools, 159

D

Dashboards, 392, 394–396, 398–399
Data access layer, 304, 304f
Data access platform, 306t
data management module, 306t
data processing module, 306t
related data modules, 306t
Data acquisition, 313–314
Data analysis, 246–250
ANOVA, 250, 250t
descriptive statistics, 246, 248t
t-test, 246, 250, 250t
Data analytics and visualization, 348–349, 350t
data-driven decision-making, 350t
real-time data integration, 349t
Data anonymization, 184
Data center architecture, 198–199, 198f
application center, 198f, 199
data center, 198f, 199
unified portal, 198–199
Data consistency, 399–400
Data flow and processing mechanism, 164
Data governance, 199
compliance, 197
quality assurance, 113
semantic harmonization, 113
Data integration, 312–313, 314t, 316–317, 319
Data interaction mechanism, 125–127
Data layer, 297
Data management, 1–3, 9–11, 391–392, 394–396, 398, 400
data analysis, 2–5, 9–11
data privacy and security, 1–3, 9–11
Data privacy, 170, 172, 176, 183–184, 186–192, 361, 363, 365, 369
ethical implications, 189–190
student data privacy, 184, 186–187, 189–192
vulnerability, 183–184
Data privacy and security, 311–313, 315, 317, 319–321
compliance with regulations, 311, 313, 315, 317
ethical considerations, 313, 317
Data privacy and security concerns, 13–15, 17–19
Data processing layer, 85f, 86
cloud processing, 86
edge computing, 82f, 86
Data protection, 262, 264, 274
ethical practices, 264, 272
privacy concerns, 264, 272, 274
Data retention policies, 188–190
Data security, 105–110, 105t, 108t, 111t, 112–116
privacy challenges, 108t, 111t, 114–115
risks, 108t, 115
Data security and privacy, 88, 89t–90t
Data submission, 125
Data-driven decision making, 98
Data-driven decision-making, 301–302, 302f, 316, 321
data-driven strategic planning, 241
long-term planning, 241
Decentralization, 129–132, 131t, 134, 136, 138, 391–395, 397–400
Decentralized autonomous learning systems (DALSs), 116t
Decentralized Identifiers (DIDs), 131–133
Decentralized identity (DID), 111t, 112, 114–115
Decision support platform, 199–201
individual-level analytics, 200
personalized learning engine, 197
predictive policy simulation, 201
Deep Deterministic Policy Gradient (DDPG), 212–214
Deep learning, 73, 75–78, 82f, 84t, 89t
deep collaborative neural networks, 76
deep reinforcement learning, 75–78
Deep reinforcement learning (DRL), 208, 212–214, 216
Digital accessibility, 138
Digital competence, 187, 189
Digital credentials, 129–134, 136–139
student empowerment, 132
verification, 132–133
Digital divide, 29, 32, 317
Digital ecosystems (DE), 366
Digital labs, 84t, 89t
Digital Layer, 374f, 378, 379f, 384f
Digital literacy, 76–77, 262, 274
Digital models (DMs), 219–220, 220t
functionality, 220t
scalability, 220t
vs. digital twins, 219–220, 220t
Digital sustainability, 188
Digital transformation, 93, 101, 242
Digital transformation (DX), 301, 309
Digital transformation in education, 279–282
cultural shift, 279
stakeholder engagement, 279
Digital twin (DT), 73–78, 362, 364, 370
classroom digital twin, 362

Digital twin (DT) (*Continued*)
classroom monitoring and analysis, 73, 76–77
digital twin computing, 370
education and training technology integration, 76
educational digital twin, 364
in education, 157
instructional design, 73–74
integration in smart schools, 158f
layers, 159t
Digital twin technology, 1–11
applications in education, 2, 5–8, 7f
fundamental principles, 5
predictive maintenance, 5–7, 10–11
Directed Acyclic Graph (DAG), 391–394
Distance learning, 224, 224t
access to virtual workshops, 224t
Distributed computing, 158
Distributed File Storage (DFS), 393
Distributed Ledger Technologies (DLTs), 393
DRL. *See* Deep reinforcement learning (DRL)
DT-assisted communication and computing resource allocation (DCTRA) algorithm, 213–214, 216

E

Early intervention, 350t
Ebbinghaus forgetting curve, 211
Edge AI, 387
Edge computing, 4–5, 9, 59t, 61–62, 62t, 65–68, 70–71, 95, 99, 144–145, 148, 151–152, 162–163, 163t, 320, 349t–350t, 353–355, 354t–355t, 363–365, 370
advantages, 162
key features, 162
limitations, 162
Education 4.0, 186
Education 5.0, 262, 272, 274
Education 6.0, 262, 272, 274
Education 7.0, 241–244, 262, 272, 274
Education informatization 2.0, 14
Education system, 373–389, 377t, 383t, 387t
digital transformation, 373, 376–377, 388–389
smart education, 375
Educational metaverse practices, 73–74
Educational outcomes integration, 147–149
Educational technology, 1, 3, 7–8, 11, 13–15, 17–19, 183–184, 186, 188–189
early integration, 13–14
rapid growth, 14
EduCTX, 394
Effective interaction and engagement, 283–286
collaborative learning, 284
interactive tools, 283–284, 286
Efficiency and scalability, 125
Efficient resource utilization, 301–302
Electronic health records (EHRs), 350t, 355, 355t
Electronic nose (e-Nose), 41–42, 49–50
Elliptic curve cryptograph (ECC), 172
Emergency evacuation planning, 153–154
Emergency management and safety, 141–143, 143t, 145–149, 147t, 153–154
Emergency response, 353f, 354t
integration, 355t
Emergency systems, 97–98, 101
Emotion recognition systems, 84t, 88–90, 89t–90t
Encryption, 184, 189
Energy consumption, 115
proof-of-work systems, 115
sustainability issues, 115
Energy efficiency and optimization, 141–143, 143t, 145–148, 147t, 150–151, 153–154
Energy management, 66–68, 70–71, 303t, 305, 307f
Energy-efficient processing, 165–166, 166t
Engagement and interaction, 246–248, 249t
Engineering education, 84t, 89t
English as a Foreign Language (EFL), 264, 268, 271–274, 273f
language acquisition, 271–274
teaching methods, 267–268
writing skills, 269, 271–274
Enhanced learning experiences, 301–302
Enterprise resource planning (ERP), 351
Environmental sensors, 93, 95, 97, 100–101, 292, 293f–294f, 297
Environmental surveillance, 353f, 354t
environmental safety, 350t
improved learning environments, 352
Ergonomics, 226, 228, 231
kinematics, 151
wearability analysis, 231
Error correction and detection (ECD), 328t, 333–334
CoT Rerailer, 328t, 334
ErrorRadar, 328t, 334
MathCCS, 328t, 334
Ethereum, 391–393
Ethical and legal safeguards, 355t
data privacy, 355–357, 355t
Ethical and privacy concerns, 380–381, 386, 387t
algorithmic opacity, 386, 387t
consent and legal compliance, 380–381, 383, 386, 387t
data ownership and control, 386, 387t
Ethical and transparent practices, 317
Ethical considerations, 98–102
Ethical implications, 165–166, 166t
Ethical oversight committee, 298
Ethics, 261–262, 264–265, 267–274
classroom discussions, 272
ethical dilemmas, 272, 274
policies and guidelines, 264, 272, 274
European Blockchain Services Infrastructure (EBSI), 134, 135t, 136–137
European Framework of Digital Competences for Citizens (DIGCOMP), 187, 189–190
European Smart Campus Initiative, 110
Evaluation metrics system, 201–202
curricular assessment, 201–202
institutional benchmarking, 201
student competency profiling, 202
Explainable AI (XAI), 380–381, 387
Extended reality (XR), 73–78, 364
XR educational environments, 73, 77
XR training and education tools, 77–78

F

Face recognition technology, 26, 30
Facial recognition with LBPH algorithm, 294, 295f, 299
accuracy, 294
false acceptance rate (FAR), 294
false rejection rate (FRR), 294
Faculty roles, 398–399
attendance check-in process, starting, 399
students, registering, 399
Family Educational Rights and Privacy Act (FERPA), 183–184, 187, 189–190, 311, 313, 315, 317
Fashion design education, 219, 221–224, 226, 228–231, 233
changes in core subjects, 222t
transformation by DTs, 219, 221–222
Federated learning, 166t, 197, 199–201, 203. *See also* Research opportunities; Security and privacy
Federated Learning, 376–377, 387
Feedback comment generation (FCG), 328t, 330–331
CourseMIRROR, 328t, 330
FreeText, 328t, 331
PROF, 328t, 331
Finalization, 125
5G and network infrastructure, 363–365, 370
cloud programming, 364
online learning, 364
Software Defined Network (SDN), 363
5G networks, 320–321
instantaneous updates, 241
Flexible access to content, 283, 285–287
independent learning, 285–287
multimodal content, 285–287
Fog computing, 161–162, 163t
advantages, 161
key features, 161
limitations, 162
For Students/Parents, 205
access to campus news, 113
streamlined administrative services, 113
targeted resource recommendations, 113
For Teachers, 205
AI-assisted lesson planning, 113
automated grading analytics, 113
mobile office tools, 113
Functional applications, 381–382, 383t

administrative and decision support, 382, 383t
infrastructure and facilities management, 381–382, 383t
personalized learning, 373–376, 377t, 381, 386–389
Funding and partnerships, 374
financial constraints, 355–356
Future innovations, 30–32
AI-based dynamic learning paths, 31
gamification-based personalized learning, 31
IoT-based real-time school safety and security system, 30–31
Future trends, 70, 166f, 166t
5G/6G connectivity, 166, 166t
AI-driven twins, 166, 166t
enhanced connectivity, 70
metaverse in education, 166, 166t
standardization efforts, 70
sustainable and energy-efficient solutions, 70

G

Gamification, 31, 111t, 113–114, 116t, 265–267, 272–274
incentives, 111t, 113–114
smart contracts, 113–114
GDPR compliance, 197, 199–200, 203
GDPR. *See* General Data Protection Regulation (GDPR)
General Data Protection Regulation (GDPR), 183–184, 187–190, 311, 313, 315, 317, 355t
compliance, 183–184, 187–190
impact on digital technologies, 187
GlotBot, 328t, 336–337
GRI Sustainability Guidelines (GRI), 191
GRI 418: Customer Privacy 2016, 191

H

Hallucination and overreliance, 338–339, 339t
mitigation techniques, 339t
Haptic sensors, 47–48
kinesthetic sensors, 48
tactile sensors, 48
Hardware and software virtualization, 25, 30
Hardware connectivity and maintenance, 88–90
Hardware technologies, 15–16
HCN sensor. *See* Hydrogen cyanide (HCN) sensor
Health Insurance Portability and Accountability Act (HIPAA), 355t
Health monitoring sensors, 292–293, 293f
Heating, ventilation, and air-conditioning (HVAC), 354t
Heterogeneity and interoperability, 165–166, 166t
High initial costs, 315–316, 317t
Higher education, 240, 242–246, 250–256
Higher education institutions (HEIs), 93–94, 97–101
Higher-order cognitive skills, 280
Highsimb simulator, 177
HIPAA. *See* Health Insurance Portability and Accountability Act (HIPAA)
Historical costume reconstruction, 219, 221t, 224t, 226–228, 230–233
cultural heritage preservation, 219, 226, 228–230, 233
digital replicas, 224t, 229
virtual environment, 223–224, 228, 231
History of digital twin models, 302–304, 303f
information mirroring model, 302–304
mixed reality, cognitive and AI, 302–304
Holberton School, 134, 135t
highlights, 64
Homomorphic Encryption, 137
Human activity recognition (HAR), 169–171
Human body digital twins, 219–220, 226–229
parametric modeling, 227
practical applications, 227–228
Human-machine collaboration, 261–262, 264, 268–272
ethical issues, 261–262, 264, 269–272
guidance for students, 272
teaching and learning process, 271–272
Humidity sensor, 41
HVAC systems, 144t, 145–148, 146f, 149f, 150–151, 151f, 153–154
HVAC. *See* Heating, ventilation, and air-conditioning (HVAC)
Hydrogen cyanide (HCN) sensor, 41
Hydrogen sensor, 41
Hyperledger Fabric, 105, 106f, 115

I

Identity verification layer, 121–122, 122f, 127
Immersive learning environments, 261–262, 265–266, 268, 271–274
Immersive learning experiences, 1–2, 6, 10, 73, 77
Immutability, 129–133, 131t, 135, 138–139
Impacts of smart technologies, 32
economic, 32
social, 32
sustainability, 32
Improved safety and security, 301–302
IMU sensors. *See* Inertial Measurement Unit (IMU) sensors
Indonesia, 239–242, 244–246, 261–262, 267–268, 274
digital infrastructure, 262, 274
Metaverse adoption, 261–262
resource constraints, 262, 274
Indoor environmental quality, 141–142, 144–145, 147–148, 147t, 150–151, 153–154
Industry 4.0, 185
Inertial Measurement Unit (IMU) sensors, 46–47
MEMS accelerometers, 46
MEMS gyroscopes, 46–47
MEMS magnetometers, 47
Information and Communication Technologies (ICTs), 186–187, 279
Informed consent, 298
Infrastructure, 1–2, 9–11
Infrastructure and facilities management, 381–382, 383t
Infrastructure design and cost, 88, 90t
Innovative attendance systems, 96
Instructional support platform, 202–203
core functional modules, 203
design principles, 203
empirical validation, 203
Intelligent resource repository, 199
AI Curation System, 199
federated resource sharing, 199
knowledge-mapped assets, 199
Intelligent school ecosystems, 387–388
Interaction layer, 159, 159t
control, prediction, and optimization, 159
dashboards, 159
predictive models, 159
Interaction sensors, 293, 293f
Interactive learning experiences, 66–68, 70–71
Interactive whiteboards, 14–16
Interdisciplinary collaboration, 219, 226, 228, 231
for teachers, 223t
in research, 226, 228, 231
Internet of Things (IoT), 1–5, 7, 9–10, 13, 15–19, 23–26, 30–32, 37–49, 51–52, 57–59, 57f–58f, 59t, 61–63, 61t–65t, 65–71, 68t–70t, 73–78, 81, 84t, 89t, 94, 105, 109–110, 110f, 112–114, 141–152, 143t–144t, 147t, 149t, 154, 169–171, 178, 183–184, 187–188, 195–197, 199–202, 264–266, 280–281, 283, 291–299, 301–304, 303f, 307–308, 308f, 311, 313, 319, 348–349, 349t–350t, 354t, 361–364, 368, 370, 373–375, 378–379, 379f–380f, 382–383, 385, 391–394
applications, 65–66, 65t
architecture, 61–62, 62t, 65–66
blockchain integration, 109–110, 110f
characteristics, 58–59, 58f
energy efficiency, 25–26, 30, 32
environmental monitoring, 364
integration methodologies, 293
IoT educational tools and technologies, 73, 77
IoT environment, 81
IoT for monitoring student engagement, 16–17
IoT sensor networks, 77
IoT sensors, 197, 199–200
IoT service development, 81
IoT-based sensors and actuators, 378, 379f
IoT-based smart classrooms, 75–76, 78
IoT-enabled classroom devices, 16
IoT-enabled facilities, 368

Internet of Things (IoT) (*Continued*)
occupancy monitoring, 25
sensing technology, 364
sensor typologies, 292–293
sensors, 373–379, 382–386
smart attendance and environmental sensors, 16–17
smart classrooms, 109
Interoperability, 105–110, 111t, 113, 115–116, 129–139, 130t, 365, 369
cross-chain protocols, 115, 116t
key needs, 64
standardization challenges, 115
Inventory management, 305–306
IoT. *See* Internet of Things (IoT)
IoT applications, 65–66, 65t
agriculture, 65t
healthcare, 65t
smart cities and transport, 65t
IoT architectures, 61–62, 62t, 65–66, 70–71
3-tier architecture, 61–63, 62t, 68–69
4-tier architecture, 61–62
5-tier architecture, 61–62
IoT in schools, 157, 159–162, 165–167
IoT networking technology, 62, 63t
IoT. *See* Internet of things (IoT)
IOTA, 393

J

Joint Resource Allocation (JRAT), 216
JRAT. *See* Joint Resource Allocation (JRAT)

K

Key performance indicators (KPIs), 148, 153–154
Kirkpatrick's four-level training evaluation model, 245–246
Knowledge graphs, 195–197
Knowledge tracing (KT), 325, 328t, 333
Knowledge Augmented Data Teaching (KADT), 328t, 333
multiagent knowledge-tracing architectures, 333
Neural Computerized Adaptive Testing (NCAT), 328t, 333

L

Large Language Models (LLMs), 325–326, 329–340
in-context learning (ICL), 325–326, 334
retrieval-augmented generation (RAG), 325–326, 339t
Transformer architectures, 325–326
Layered architecture, 163–164, 164f
advantages, 164
cloud layer, 102
device layer, 102
Learning analytics, 38–39, 42–43, 45–46, 51–52
Learning and curriculum enhancement, 305, 307f
simulated environments for learning, 214
virtual classrooms and remote learning, 214
Learning efficiency, 26–27
Learning environments, 239–241, 243–248, 250–256
immersive, 245–248, 250–256
personalized, 240–243, 253–256
technology-enhanced, 239–241, 245, 250–255
Learning experience, 246–248, 249t
Learning Machine, 133–134
Learning management systems,
Learning management systems (LMS), 14, 16, 19, 97, 100, 185–187, 189, 279, 282, 286
Learning resource recommendation (LRR), 328t, 331–332
NeuroSky EEG biosensors, 328t, 331
Legal and regulatory compliance, 135
challenges, 64
FERPA, 135
General Data Protection Regulation (GDPR), 134–135
LiDAR sensors, 47
Lifelong learning, 105–110, 106t, 111t, 113, 115, 281
microcredentials, 105–110, 106t, 111t, 113
portable skills verification, 107–108
Light sensor, 41
Limitations, 161–162
cloud computing, 161
edge computing, 162
fog computing, 162
LMS. *See* Learning management system (LMS)
Logistics and resource management, 305–306, 307f
inventory management, 214
smart scheduling, 214
transportation and mobility, 306

M

Machine learning, 74–78, 280
machine and deep learning algorithms, 77
machine and reinforcement learning algorithms, 78
machine learning-based predictive maintenance automation, 77
Machine learning (ML), 144t, 145–146, 148, 149t, 312, 314–315, 319, 321, 329, 332–334, 349t–350t, 354t, 373–375, 377t, 378, 381, 383, 383t, 385
algorithms, 348–349, 350t, 352, 354t
role in digital twins, 314
Malta Ministry of Education, 134, 135t, 139
considerations, 64
Management support platform, 201–202
evaluation metrics system, 201–202
framework design, 201
precision profiling engine, 202
Markov Decision Process (MDP), 212–213
components, 212–213
Massachusetts Institute of Technology (MIT) Media Lab, 134, 135t, 139
innovations introduced, 64
limitations encountered, 64
MDP. *See* Markov Decision Process (MDP)
Medical professional, 328t, 337
MEDCO, 328t, 337
openCHA, 328t, 337
Mental health surveillance, 353f, 354t
early warning systems, 374
interventions, 355t
Message Queuing Telemetry Transport (MQTT), 293–294, 294f–295f, 349t
Metaverse, 261–267, 269–274
adoption in education, 261–262, 267
ecosystem, 263–265
ethical concerns, 261–262, 264, 274
Metaverse and extended reality (XR) learning environments, 19
Metaverse exhibition, 233
Metaverse technology, 27
Methane (CH4) sensor, 41
Methodological approach, 282–283
data collection and analysis, 283
method design, 282
participants, 282
Microcredentials, 105–110, 106t, 111t, 113, 114t
Edubadges project, 114t
NZ micro-credential system, 114t
Open badges, 114t
Microservice-based open ecosystem, 195–197
critical microservices, 204
microservice orchestration, 199
strategic microservice deployment, 113
Microsoft Azure Digital Twins, 149t, 150
Mixed reality (MR), 263–266
ML. *See* Machine learning (ML)
Mobile apps, 391–392, 395–396, 398
MQTT. *See* Message Queuing Telemetry Transport (MQTT)

N

Named Data Networking (NDN), 393
Nanyang Technological University (NTU), 318–319, 318t, 321
NASA Apollo program, 184
National Natural Science Foundation of China project EduRSS, 110–112
National standards for IoT devices, 299
Natural language processing (NLP), 326, 327f, 329–331, 333–335, 373–376, 377t, 379, 381–382, 383t
Near-field communication (NFC), 353–355
Net generation, 280
Network controller, 208–209, 208f
Network layer, 62–63, 62t, 64t, 69t, 85f, 86, 297
Bluetooth Low Energy (BLE), 85f, 86
challenges and limitations, 69, 69t
LoRaWAN, 85f, 86
Wi-Fi, 85f, 86

Network model, 172–175, 174f
Networked virtual interactive learning environments, 73, 76
NFC. *See* Near-field communication (NFC)
NH3 sensor. *See* Ammonia (NH3) sensor
Nitric oxide (NO) sensor, 41
NO sensor. *See* Nitric oxide (NO) sensor
Node-RED, 149t, 151–152
Nonfungible tokens (NFTs), 109–110, 111t, 115–116, 116f, 116t

O

O2 sensor. *See* Oxygen (O2) sensor
Occupancy and motion sensors, 43–45
  microwave radar sensors, 44–45
  passive infrared (PIR) sensors, 43–44
  ultrasonic sensors, 44
Online learning, 1, 6, 11, 263, 266–267, 272–274
  blended learning, 1, 6
  distance learning, 1, 6
  virtual classrooms, 2, 6
OPC UA. *See* Open Platform Communications Unified Architecture (OPC UA)
Open Platform Communications Unified Architecture (OPC UA), 349t
OpenCerts, 134, 135t, 137
OpenCV, 393–394
Openness, 391–396, 398–400
Oxygen (O2) sensor, 41

P

Paired samples test, 269, 270t
Parametric modeling, 226–227, 228f
  digital avatar generation, 227
  human body, 227
Parent and community engagement, 303t, 305–306, 307f
Passive infrared (PIR), 351
Pattern drafting, 221f, 222t, 225
  2D designs into 3D structures, 151
  garment patterns, 151
Pedagogical applications, 326–334
  classroom interaction, 327–332
  personalized tutoring, 332–334
Pedagogical digital twin technologies, 73–74
Pedagogical issues, 385–386, 387t, 388
  data reductionism, 386, 387t
  human disconnection, 386, 387t
  over-automation, 386, 387t
Pedagogical shifts, 17
Perception layer, 62t, 68t
  challenges and limitations, 68, 68t
Performance evaluation, 171, 177–178
  computational overheads, 171–172
  transaction processing speeds, 176–177
  transaction verification duration, 177, 177f
Personalized healthcare, 350t
Personalized learning, 3–5, 7–8, 10, 23–24, 27, 31, 74, 76–78, 93, 96–97, 100–101, 301, 307–309, 311–319, 321, 361–364, 370
  ecosystems, 100
  paths, 93
  personalized recommendations, 241
  personalized teaching and learning simulation, 73–74
Personalized learning engine. *See* Decision support platform
Personalized QoE (PQoE), 207–216, 208f, 215f
  DT-assisted model, 211–212
  factors, 207, 210–212, 214–215
  model formulation, 211
PF. *See* Proportional Fair (PF)
Phased implementation and scalability, 374
Physical device layer, 121–122, 122f, 127
Physical layer, 159, 159t, 297
  IoT devices and sensors, 159
Physical Layer, 374f, 378, 379f, 384f
Physiological monitoring, 353f, 354t
  enhanced health monitoring, 352
Pilot programs, 316, 318–319
Pilot test smart classrooms, 299
PIR sensors. *See* Occupancy and motion sensors
PIR. *See* Passive infrared (PIR)
Policy and governance models, 388
Post-test, 244–246, 248t, 250, 250t, 251f
Postprocessing, 221f
PQoE. *See* Personalized QoE (PQoE)
Practical Byzantine Fault Tolerance (PBFT), 124, 126
Precision Teaching Support Platform (PTSP). *See* Instructional support platform
Predictive analytics, 93–94, 96, 98, 100, 102, 349t–350t, 354t
  integration with artificial intelligence, 100
  predictive capabilities, 350t
  predictive interventions, 354t
  student identification, 93, 98
Predictive and preventive maintenance, 141–148, 143t–144t, 147t, 150–152, 154
Preexperimental design, 268, 268t
Preparation phase, 247t
  customizing learning goals, 164
  designing digital twin environment, 164
  technical setup, 164
Pretest, 244–246, 248t, 250, 250t, 251f
Preventive health programming, 355t
PRISMA diagram, 326, 327f
Privacy, 129–130, 134–135, 137, 139
  considerations, 64
  data confidentiality, 135
  zero-knowledge proofs (ZKPs), 64
Privacy, bias, and fairness considerations, 338, 339t
  explainable-AI interfaces, 339t
  federated/on-device fine-tuning, 339t
  real-time bias auditing pipelines, 339t
Private blockchains, 123–125
Privileged insider attacks, 172, 176
Professional development of educators, 28
Projectors and overhead transparencies, 13–14
Proof of Teaching (POTH), 122–125
  core design, 124
  technical advantages, 124–125
Proportional Fair (PF), 216
Protection, 351
Psychological and social impacts, 299
Public blockchains, 119–120
Purdue University, 318t, 319, 321

Q

Q-Q plots, 269, 273f
QR scan. *See* Authentication; QR codes

R

Radiofrequency identification (RFID), 45
Real world environment, 85–86, 85f
Real-time audit layer, 124
Real-time data acquisition and knowledge transfer algorithms, 74, 78
Real-time data analytics, 162
Real-time data and performance analysis, 13, 17
Real-time data integration, 349t
Real-time monitoring, 1–8, 10–11
Real-time simulation, 219–220, 225, 231
  fabric behavior, 220, 228, 231
  garment modifications, 220
Real-time validation, 125
Real-world transfer phase, 247t
  community engagement, 164
  ongoing development, 164
  practical applications, 164
Rebuffer time, 210–215
Record ownership, 130t, 132, 136, 139
  data control, 130t, 134
Reflection and assessment phase, 247t
  peer and instructor feedback, 164
  self-reflection sessions, 164
  skill assessment, 164
Registration phase, 172–174, 175f, 176–177
Regulatory compliance, 111t, 114–115, 114t
  GDPR, 111t, 114–115
  legal challenges, 115
Remote and hybrid learning, 312, 316
Remote education, 84t, 89t
Rendering, 221f
Research opportunities, 165–166, 166f, 166t. *See also* Federated learning; TinyML
  semantic middleware, 166f, 166t
Research questions, 374
  RQ1, 374
  RQ2, 374
  RQ3, 374
Resource management, 98, 101, 315–316
  cost control, 99
  predictive maintenance, 98
Resource optimization, 6–7, 7f, 10–11, 350t
Reward, 213–214

RFID sensors. *See* Radio frequency identification (RFID) sensors
RFID tags, 145–148, 147t, 149f, 151–153, 152f
RFID tap. *See* Authentication; RFID tap
RFID. *See* Radiofrequency identification (RFID)
Risk mitigation, 6, 7f, 8
Rivest–Shamir–Adleman (RSA), 172
Robotic process automation (RPA), 364–365
Role-aligned verification, 124–125
Round Robin (RR), 216
RPA. *See* Robotic process automation (RPA)
RR. *See* Round Robin (RR)
Rule-based systems, 325, 332

S

Safety, 6–8, 7f
Saudi Arabia, 282–283
SCADA systems, 145, 148
Scalability, 114–115, 170, 172, 176–177, 317–318, 317t
  constraints, 114
  solutions, 114–115
Scalability and cost management, 165–166, 166t
School digital twin, 157–167
  architecture, 160f, 164f
  benefits, 159t
  challenges, 166f, 166t
School facilities management, 141–154, 142f
  conventional practices, 141–143, 150–151
  issues, 141–144, 151
  smart technologies, 141–148, 151–152, 154
Secure technologies and practices, 374
  cybersecurity, 374
Security, 129–132, 130t–131t, 138–139, 169–172, 176–178, 391–400
  analysis, 171–172, 176–178
  solutions, 170
  threats, 170, 176
Security and privacy, 66–68, 70–71, 164
  blockchain-based authentication, 166t. *See also* Federated learning; Zero-trust security models
Security Village system, 294, 295f, 299
  Firebase, 294
  MQTT protocol, 294
  usability score, 234
Self-healing networks, 88–90, 90t
Self-managing classrooms, 241
Sensor network, 73–78, 84–90, 84t, 85f, 89t, 311–313, 315–316
  sensor network data, 74–75
  sensor network-based smart schools, 73, 75–78
Sensor nodes, 172, 174–177, 174f, 174t
Sensor technologies, 351–356
Sensors, 57–62, 58f, 60f, 61t, 65–71, 68t
  characteristics, 61, 61t
  classification, 60–61, 60f
  functioning, 59, 60f
Service delay model, 209–210
Sharjah, 282
Simulating educational scenarios, 1–2, 6–7, 7f
Simulation and optimization, 349t
Simulation parameters, 214, 215t
Simulation phase, 247t
  interactive feedback mechanisms, 164
  personalized interaction, 164
  role-playing scenarios, 164
Simulations and web labs, 84t, 89t
Singapore, 134, 135t, 137, 139
  advantages of government-led models, 64
Situational interactive teaching systems, 267–268
SLEs. *See* Smart learning environments (SLEs)
Smart Blended Learning system, 295–296, 296f, 299
  MobileNet-SSD, 296, 296f
  object detection, 295–296, 296f
  teacher tracking, 296
Smart building operation and maintenance, 74
Smart campus, 23–32, 84t, 89t, 94, 96, 100–101, 195–198, 280–281, 283, 361, 366–370
  benefits, 368
  challenges, 367, 369
  definition, 281
  definition and scope, 23–24
  management, 27–28
  smart transportation, 368
  sustainability, 23–24, 26, 28–29, 32
Smart Campus Framework, 296–297, 297f
  Syam-OK LMS, 296–297
  Universitas Negeri Makassar, 291–292, 296–297
Smart classroom, 1–4, 7–8, 10, 13–19, 23–32, 37–48, 51–52, 57, 65–68, 70–71, 73, 75–78, 109–110, 109f, 112–114, 116t, 119–127, 122f, 125t, 186–187, 207–209, 208f, 211–212, 216, 280–283, 306–309, 311–316, 318–321, 325–326, 329–330, 335, 338–340, 361–363, 370
  advantages, 67, 362
  assistant agents, 326–338
  challenges, 362–363
  challenges to adoption, 67
  components, 109, 109f
  definition, 281–282
  digital twins, 109, 116t
  ecosystem, 3–4
  features, 2, 2f
  functional tools, 120
  integration, 66–68
  intellectual configuration, 326
  IoT-enabled, 109
  key components, 362
  literature surveys, 326
  proposed scheme, 121–127
Smart contracts, 105–116, 105t–106t, 111t, 112f–113f, 116t, 131–133, 138
  automated processes, 105–114, 106t, 111t, 116t
  automation, 131
  decentralized learning marketplaces, 138
  financial transactions, 109–110, 111t
  gamification, 113–114
Smart devices, 125–126, 125t, 126f
Smart education, 37–52, 84–87, 105, 105t–106t, 109–116, 116t, 166–167, 239–242, 255–256
  blockchain-enabled, 105, 105t, 106f, 106t
  digital classroom components, 109f
  digital twins for
  IoT integration, 110f
  sensor networks for, 84–87
Smart learning environment, 361–362, 370
  definition and significance, 361–362
  future trends and innovations, 370
Smart learning environment (SLE), 98, 101
Smart learning environments. *See* Smart classrooms
Smart learning environments (SLEs), 279–287
  characteristics, 280
  key objectives, 280–281
  models, 280–281, 283
Smart projectors and AR/VR devices, 15–16
Smart school, 280–281, 283
  definition, 281
Smart school applications, 4, 11
Smart schools, 1–4, 6–11, 23–32, 291–299, 347–357, 361, 363–365, 370
  advantages, 364–365
  benefits, 2, 2f, 4, 6
  challenges, 1–2, 9–10
  concept, 351
  definition and scope, 23–24
  healthcare integration, 351–353
  implementation barriers, 365
  key technologies, 364
  management, 27–28
  paradigm shift, 351, 351f
Smart sensor networks, 1–4, 10
Smart Sensor Networks (SSNs), 93–102
  administrative SSNs, 94
  applications in student activities, 95–96
  challenges in higher education, 98–100
Smart/digital campus, 169–172
Social interaction, 239–256
  collaboration, 239–241, 244–248, 250–256
  communication, 239–241, 243–246, 250, 252–255
  teamwork, 245–248, 250, 252–255
Software and digital platforms, 16
Software programs, 220, 221f, 232–233
  avatar modeling, 221f
  pattern drafting, 221f
  virtual try-on, 221f

Space and occupancy management, 141–148, 143t, 147t, 153–154
SPM sensor. *See* Suspended particulate matter (SPM) sensor
Standardization efforts, 62t, 69t, 70
State, 212–213
  state transition probability, 212–213
Statistical models, 325, 333
Smart Technology-Enhanced Learning Environment, 240–241, 244–256
  engagement, 245–246
  knowledge acquisition, 245–246
  performance improvement, 245–246
Stewardship theory, 189–192
Storytelling, 328t, 337
  LLaMS, 328t, 337
  STARie, 328t, 337
  StoryAgent, 328t, 337
Student and teacher performance monitoring, 305–306, 307f
  analytics, 214
  faculty development, 214
Student check-ins. *See* Attendance tracking
Student engagement, 93, 95, 97–98, 100–101
Student engagement and participation, 13–14, 16–17, 19
Student healthcare, 348–357
  personalized, 350t, 353, 354t, 355–357
  proactive management, 352–353, 355
  well-being, 347–353, 356–357
Student involvement, 399
  attendance history, viewing, 399
  check-in status, confirming, 399
Student performance monitoring, 381, 383t
Student records management, 129–130, 130t–131t, 132–133, 135–139
  administrative processes, 132–133
  cross-border academic mobility, 133
Student sensor networks, 170–174, 176, 178
Student terminal, 125, 125t, 126f
Student welfare and well-being, 93–96, 101
  European institutions' wearable tech, 96
Student-centered privacy model, 184, 189–190
Support for modern technologies, 283–286
Suspended particulate matter (SPM) sensor, 41
Sustainability, 23–26, 28–29, 32, 183–192, 226, 229
  accounting and accountability, 190–192
  environmental impact, 183–184, 186–190
  goals, 183–190
  material optimization, 151
  reducing waste, 220, 226, 229
Sustainability and cost savings, 301–302
Sustainability monitoring and compliance, 141–144, 146–148, 147t, 153–154
Sustainable and energy-efficient solutions, 70
Sustainable development, 365–366, 368, 370
Swarm intelligence, 50
System initialization phase, 172
System integration, 88, 90t

T

Tamper-resistant audit trail, 125
Teacher node, 125–126, 125t, 126f
Teacher readiness and training, 18
Teacher training, 10–11
Technical and academic support, 283–286
  academic advisors, 284–285
  technical support, 284–285
Technical challenges, 18, 385–386, 387t
  black-box decision-making, 385, 387t
  data integration, 385, 387t
  interoperability, 385–386, 387t
Technical complexity, 311–312, 315–317, 317t, 319, 321
Technical restrictions, 99–100
  energy consumption, 99
  sensor accuracy, 99
  system interoperability, 99
Technologies enabling virtual smart environments, 29–30
  Artificial intelligence (AI), 30
  Augmented reality (AR), 29–30
  Digital Twins, 30
Technology audit and logging, 298
Technology-organization-environment (TOE) framework, 392–393
Temperature sensor, 41
Textile materials digital twins, 219–220, 229–230
  applications, 229–230
  methodologies and simulation, 229
  physical simulation, 151
Texture painting, 221f
The improvement in quantitative performance was great, 150
Timestamp, 174t, 176–177
TinyML, 166f, 166t. *See also* Research opportunities
Toxic gases sensor, 41
TPACK theory, 241
Tracking student progress, 1–2, 6, 7f, 8
Training and education, 84t, 89t
Translation, 328t, 336–337
  TransAgents, 328t, 336–337
Transparency, 129–133, 130t–131t, 135–136, 138–139
Transparency and accountability, 298
Transportation and mobility, 303t, 305–306, 307f
Trust model, 131t, 132–133, 138–139
Twenty-first-century skills, 240–243

U

Ubiquitous application platform, 203–204
  core functional modules, 204
  empirical validation, 204
  implementation strategy, 204
Ubiquitous Educational Service Platform (UESP). *See* Ubiquitous application platform
Ubiquitous learning environments, 195, 198
  ubiquitous learning spaces, 198
Unified portal, 198–199
  multichannel adaptive interface, 198–199
United Arab Emirates, 282–283
United Nations Sustainable Development Goals (UN SDGs), 185–186, 190
  SDG 12 (Responsible Consumption and Production), 185, 189–190
  SDG 4 (Quality Education), 185–186
  SDG 9 (Industry, Innovation, and Infrastructure), 185
Unity3D, 149t, 153
Universitas Negeri Makassar (UNM), 291–292, 296–298
University 4.0, 185–186
University of Melbourne, 133, 135t, 139
  challenges, 64
  key benefits observed, 64
University of Southern Denmark (SDU), 318t, 319, 321
Usability and accessibility, 246–248, 249t

V

Validation and calibration, 151
  of digital twins, 151
Video quality variation, 210–215
Video streaming, 207–212, 216
Virtual and augmented reality (VR/AR), 74–75, 239–244
Virtual educational systems, 73–74
Virtual laboratory, 368, 369f
  components, 368, 369f
  enhanced research capabilities, 368
  remote learning, 368
Virtual libraries, 84t, 89t
Virtual prototyping, 219–220, 222, 224t, 226–227, 229, 231
  cost savings, 224t
  design iteration, 227, 231
  experimentation freedom, 224t
Virtual reality (VR), 1–3, 5–6, 10, 23–24, 27, 29–32, 37–39, 46–49, 51–52, 73–78, 228, 233, 261–268, 271–272, 274, 320–321, 349t, 361–364, 370
  applications in English learning, 267, 271–272
  applications in language education, 24
  campus tours, 24, 27
  hands-on experience, 241
  immersive educational environments, 233
  integration with digital twins, 320
  Metaverse integration, 261, 265–267
  teaching collaborative ethics, 272
  virtual garment fitting, 228
  VR and digital twin technologies, 77
  VR teaching technologies, 73–74
  VR-based operational training, 74
Virtual simulation teaching systems, 73–75, 77–78
Virtual try-on, 221f, 227–228
  3D objects, 151
  computer simulations, 151
  software programs, 228

Virtual worlds (VWs), 263–267, 272
educational applications, 263–264, 266–267
immersive experience, 263–265
user interaction, 263, 265
Virtualization, 24–25, 27
VR. *See* Virtual reality (VR)

W

W3C Verifiable Credentials, 131–133, 136–137
Wearability analysis, 231
functional aspects, 231
historical and modern garments, 231
Wearable devices, 93, 95–97, 99, 101
Wearable technologies, 347f, 350t, 351, 353–355, 354t
Well-being and mental health, 382, 383t, 384–385, 387–388
WhatsApp automation, 152, 152f
Wireless Sensor Networks (WSNs), 94, 353–355, 354t
WSNs. *See* Wireless sensor networks (WSNs)

X

XR. *See* Extended reality (XR) educational technology

Z

Zero-knowledge proofs (ZKPs), 109–110, 111t, 114–115, 137, 171–172, 176
Zero-trust security models, 166f, 166t. *See also* Security and privacy

Printed and bound by CPI Group (UK) Ltd, Croydon, CR0 4YY
21/09/2026
02220062-0003